THE CHEMICAL
TECHNOLOGY OF WOOD

Carl G. Schwalbe

THE CHEMICAL
TECHNOLOGY OF WOOD

Hermann F. J. Wenzl

TECHNICAL AND SCIENTIFIC CONSULTANT
LUCERNE, SWITZERLAND

Translated from the German by

Friedrich E. Brauns

PROFESSOR EMERITUS
THE INSTITUTE OF PAPER CHEMISTRY
APPLETON, WISCONSIN

and

Dorothy A. Brauns

1970

ACADEMIC PRESS   New York and London

ACADEMIC PRESS, INC.
111 Fifth Avenue, New York, New York 10003

United Kingdom Edition published by
ACADEMIC PRESS, INC. (LONDON) LTD.
Berkeley Square House, London W1X 6BA

LIBRARY OF CONGRESS CATALOG CARD NUMBER: 73-108155

PRINTED IN THE UNITED STATES OF AMERICA

Contents

III. The Chemistry of Wood

IV. The Acid Hydrolysis of Wood

V. Further Destructive Processing of Wood

VI. The Preparation of Wood for the Production of Pulp

VII. The Sulfite Pulp Cooking Process

VIII. The Production of Pulp by Alkaline Reagents

Foreword

We have endeavored to translate "The Chemical Technology of Wood" as accurately as possible, and for this reason, we have consulted the original literature whenever possible. Also, to ensure a correct translation, a copy of the manuscript was submitted to the author for his approval. In this respect, Dr. Wenzl has been most helpful and cooperative.

We feel that this book contains a great deal of interesting information, and therefore, we hope that this English version will be widely distributed.

Friedrich E. Brauns
Dorothy A. Brauns

Preface

The ever increasing volume of literature in the field of the chemistry of wood and its chemical-technological utilization has made it more and more difficult for even the expert who is actively engaged in this field to keep informed as to the latest developments. It is true that a great many excellent monographs dealing with the different areas of the chemistry and technology of wood have been published, and reference is made to these throughout this book, but an up-to-date presentation covering the whole field of the chemical technology of wood has, until now, been missing.

During the past ten years, I have published, in various German and American trade journals, progress reports which give a summarized review covering the more recent scientific results and technological improvements made over a period of ten to fifteen years in the field of the chemical-technological utilization of wood. Surprisingly, these reports have received an extraordinarily favorable reception and have given rise to the demand that these monographs should be collected and published as a book.

It is apparent that such an undertaking necessitated a complete revision and amplification by inclusion of extensive new material. Moreover, it was my desire to correlate the various fields of interest in the chemical technology of wood and in this way to make these relationships more easily comprehensible to the reader. From this point of view, the book should serve primarily as an introduction to this special field, by means of which the range of interested readers will encompass not only chemists and engineers, but also biologists, foresters, and economists. Even the versatile expert will find here and there something of interest, but, above all, the book should assist him in his search for additional trade literature. With regard to the existing monographs, it was the special goal of the author to confine the presentation to the most important topics; in other words, to keep the total book as brief as possible.

It therefore follows that, primarily, the latest literature has been dealt with, while older publications have also been referred to only in so far as seemed necessary for the understanding of the development and the present status of research and techniques. So far as certain machines and apparatuses are referred to in the presentation of the technical processes and the mechanical installations necessary for their operation, they must be regarded only as examples and in no way as representing an evaluation of specific machines or products.

In my endeavor to reach the above-mentioned goal, I have been supported by numerous professional colleagues with advice and by provision of illustrative material. My special thanks are due to Professors W. A. Coté and F. Stemsrud, to Dr. H. Bucher, and to the Cellulosefabrik Attisholz for their ready provision of microscopic and ultramicroscopic photographs. I also thank the Technical Association of the Pulp and Paper Industry and the Lockwood Trade Journal, Inc., for permission to reproduce some graphic charts published in their journals. The Food and Agricultural Organization of the United Nations has generously provided me with extensive statistical material.

For the exemplary translation of the original German manuscript and for numerous suggestions and advice I thank Dr. and Mrs. F. E. Brauns. My thanks and appreciation are due to Academic Press for the careful preparation of the book.

This book is dedicated to the memory of my teacher and friend, Carl Gustav Schwalbe, who passed away in 1938, and who introduced me, almost fifty years ago, to the field of wood chemistry and technology. Schwalbe, in 1906, established the Chair of Cellulose Chemistry at the Technical University in Darmstadt (Germany) and later headed the Institute of Wood Research at the Forestry School in Eberswalde until his retirement in 1933. Schwalbe's publications have become a major part of the scientific literature of the first four decades of this century. The first edition of his book, "Chemie der Cellulose," was published in 1911. Of the planned second edition, consisting of four sections in two volumes, only the first section was published, in 1938. The entire collected literature for the second edition, which Schwalbe had left to me for further exploration after his death, was lost in the turmoil of the Second World War. Thus, Schwalbe's work, unfortunately, remained unfinished. May this modest book, which originated as a completely different concept, be regarded as the acknowledgment of a debt of gratitude.

Hermann F. J. Wenzl

THE CHEMICAL
TECHNOLOGY OF WOOD

I

Forests as the Source of Raw Materials

INTRODUCTION

About one-third of the world's land surface is covered by forests. These therefore constitute a principal factor in the use of land and fulfill a series of functions which are of great importance for mankind. The production of wood is only one, although a very significant one, of these functions. Moreover, forests provide protection against erosion, and the significance of this protection becomes especially apparent in areas that have been subjected to indiscriminate denuding. This presentation will deal only with the productive aspects of forests.

It is extremely difficult to collect reliable data on the size and distribution of forests, and it is to the credit of the Food and Agriculture Organization of the United Nations (FAO) that it has carried out authoritative pioneer work in this field. Accordingly, this chapter is based primarily on the publications of the FAO.

All too little is known about the extent and the quantitative yield of the forests. Over wide areas of the world, forests have not yet been surveyed and many data refer only to the size of the forest areas and the condition they are in. Only the forests of the temperate zones have been studied to an extent that permits reliable data to be given with regard to the aggregate volume of the growing stock and its total annual volume increment. The utilization of wood in different countries also differs widely so that any statistics must be regarded as being only tentative. In any case, data can be compiled from the statistics to give an indication of the distribution of the wood on the earth and the possibility of its utilization.

1. THE WORLD'S FORESTS

Differences in soil conditions, climate, location, and history are reflected in the diversity of the various types of forests. FAO reports distinguish between six different types and these will be described briefly in the following pages with regard to their basic properties and their principal types of wood (see Table I-1).

TABLE I-1
THE WORLD'S FORESTS (FAO, 1)

Forest type	Principal features	Principal species
Cool coniferous forests	Occur only in the Northern Hemisphere, forming a broad forest belt circling the globe. Trees are fairly uniform in size. The wood has relatively homogeneous, long fibers. Exploitation is heavy, the primitive aspects of this type have been preserved over large areas due to lack of accessibility and sparce population.	Principal commercial wood species are white spruce, black spruce, balsam fir in North America; Norway spruce and Scots pine in Europe and northwestern U.S.S.R., and larch in northeastern U.S.S.R.
Temperate mixed forests	Occur in the middle latitudes of the Northern Hemisphere. The number of species is large and comprises a large amount of subtypes ranging from predominantly coniferous to pure broad-leafed types. The softwoods of this type are an important source of industrial wood. The predominantly coniferous areas are the northern parts of mixed forests in the U.S.S.R., the West Coast of North America and the mountainous regions of Europe, Mexico and the Himalayas.	These forests supply most of the world's beech, oak, and birch; they contain a large amount of conifers, especially the forests of the U.S.S.R. The Douglas fir and hemlock forests are on the West Coast of North America.
Warm temperate moist forests	Occur in the warm temperate zones of both hemispheres. They contain a large number of hardwood species but are being heavily exploited for their content of softwoods. Natural forests are being replaced by plantations with native or exotic softwood species.	Pines are the most important softwoods. The better known hardwoods include oaks, and eucalyptus in Australia.
Equatorial rain forests	Occur in the tropical regions and are predominantly broad-leafed. The types of species within a few hectares are innumerable. The heterogeneity gives rise to many problems for the industry. Little or no exploitation has taken place in Latin America, but more in Africa and	Commercially well-known species include mahogany, cedar and green-heart (Latin America); okoume, obeche, sipa, limba and mahoganies (Africa), and species of the dipterocarp

TABLE I-1 (continued)

Forest type	Principal features	Principal species
	Asia. The forests of these regions supply a large part of the quality woods, and large-sized hardwoods, but they steadily decrease due to clearing for agriculture.	family (Asia).
Tropical moist deciduous forests	Occur in areas that have longer dry seasons. Large areas have been cleared for permanent agriculture.	Only the forests of Asia contain commercially well-known species such as sal and teak.
Dry forests	Occur in all parts of the world with severe dry seasons. Trees are usually short and malformed and include a large number of species. Dry forests have been extended as the result of grazing and the destruction of original forests, particularly in the Mediterranean region. In Africa there has been a century-long advance of the dry forests, at the expense of the rain forests, due largely to fire, grazing and shifting cultivation.	No commercial timber is produced except poles and fuelwood for local use.

In Europe, the Soviet Union, North America, and Japan—the regions where wood consumption is highest—the forests are predominantly coniferous. Especially rich in wood are Canada, Northern Europe, the Siberian provinces of the U.S.S.R., and Northern Japan. These areas which are so rich in wood are situated, in general, close to heavily populated and industrially advanced regions and thus provide the main source of the raw material for wood-using industries. Moreover, in these areas, forestry management is well developed. Continental Europe, the European regions of the U.S.S.R., the United States of America, and Japan have at least a quarter of their land areas under forests. Although these forests are largely mixed in character, they contain a significant proportion of coniferous trees. Coniferous forests have a number of advantages. They contain a limited number of wood species and these are relatively uniform in character as regards the properties and size of the logs, thus facilitating their harvesting and handling. The deciduous woods of these forests also include a relatively small number of species which increases their usefulness.

The relative uniformity of the climate of the temperate zones limits the volume of the growing stock and the rate of its growth. As shown in Table I-2, European forests contain an average of about 75 cubic meters of wood per hectare and show a net annual increment of about 2.3 cubic meters per hectare. The considerably higher levels of growing stock in the U.S.S.R. and North America indicate the sizeable areas of old-growth

TABLE I-2

SELECTED FEATURES OF THE FORESTS OF THE NORTH TEMPERATE ZONE (FAO, 1)

| | | Growing stock[b] | | Average net annual growth[b] | |
Region	Forest area[a]	Total	Per ha	Total	Per ha
Europe					
Northern Europe	52.0	3,472	67	117	2.3
EEC	25.6	1,965	77	75	2.9
Great Britain	1.9	116	61	4	2.1
Central Europe	13.2	1,627	123	35	2.7
Southern Europe	32.6	1,218	37	37	1.1
Eastern Europe	26.5	3,006	113	76	2.9
Total	151.8[c]	11,404[d]	75[d]	344[d]	2.3[d]
U.S.S.R.					
Southwest	69.6	6,176	89	—	—
Northwest	72.6	7,641	105	—	—
Western Siberia	105.0	12,747	121	—	—
Eastern Siberia	458.7	51,391	112	—	—
Total	705.9[e]	77,955[f]	110[f]	874[f]	1.2[f]
United States					
North	69.5	3,865	56	137	2.0
South	81.4	3,797	47	212	2.6
Rocky Mountains	26.5	2,796	106	26	1.0
Pacific Coast	28.5	7,324	257	86	3.0
Total	205.9[g]	17,782[h]	86[h]	461[h]	2.2[h]
Canada					
Atlantic	19.0	1,037	55	19	—
Central	99.7	5,772	58	90	—
Prairie	73.4	2,807	38	37	—
British Columbia	58.8	10,357	193	65	—
Total	245.9[i]	19,973[h]	81[h]	211[h,j]	—
Japan	23.4[k]	1,981[f]	81[f]	—	—

   [a] To which data on growing stock and growth refer; million hectares.
   [b] Million cubic meters.
   [c] Total forest land 1960.
   [d] Total volume without bark.
   [e] Stocked forest land in centrally managed forests, 1961. There is an additional 204.1 million hectares of forest land in the U.S.S.R. of which 84% is unstocked.
   [f] Total volume with bark.
   [g] Commercial forest land 1962.
   [h] Total volume without bark less deduction for rot and other defects affecting use of lumber. Excludes branches and tops less than 10 cm.
   [i] Productive forest land, 1962, excluding protection reserves in which cutting of industrial wood is prohibited.
   [j] Total allowable annual cut from presently economically accessible areas, assuming mixed pulpwood and sawtimber rotations.
   [k] Total forest land excluding protection reserves in which cutting of industrial wood is prohibited as reported in the FAO world forest inventory, 1963.

forests which are included in the table. Such areas of old-growth forests are situated mainly in the north and east of the U.S.S.R. and in the west of North America. They represent the greatest reserves of coniferous woods in the world.

In contrast to these temperate zone mixed forests, tropical forests—with the exception of those at higher altitudes—consist almost entirely of deciduous woods. This is especially true for the equatorial rain-forests. The economic value of such forests is very limited because they are quite inaccessible and are located in relatively sparsely populated regions of underdeveloped countries. The tropical rain-forests, with their rich wood content, are concentrated in the Amazon basin of Latin America, in west Central Africa, in Southeast Asia, and, especially, in insular Southeast Asia (Malaysia and Indonesia).

Tropical rain-forests contain a wide variety of species, ranging from extremely hard, heavy, and slow-growing trees to very soft, light, and fast-growing types. Because these forests have not as yet been managed scientifically, very little can be said about their size and content. Estimates put the figures at 850 million hectares with about 125 billion cubic meters of wood. The volumes of growing stock in tropical rain-forests tend to be very high, but they include only a few species that are suitable for commercial use. In Latin America the amount of logwood in old-growth forests is estimated to be around 200 to 300 cubic meters per hectare, but this includes an unusually varied mixture of different species. Similar data are valid for old-growth forests in West Africa, but only about 80 cubic meters per hectare occur in dimensions suitable for commercial utilization. Although the moist tropical forests are more or less limited in size, in the tropical and subtropical zones there are also wide areas of so-called dry forests (see Table I-1) which contain very low volumes of wood per hectare. In comparing forest areas, one must take into consideration their accessibility and their yields. Thus it is found that some very extensive wooded areas give only low yields of wood.

The uneven distribution of wooded areas even in the densely populated regions of the world is shown in Table I-3, which is taken, in a somewhat shortened form, from the FAO report. Apart from the almost useless dry forests, there are large areas of the world that contain no forests at all. This is true in the case of most of North Africa, with its desert areas, and also of the Near East, much of South Asia, and the mainland of China. These areas contain almost one-third of the world's population but their consumption of wood is extremely low.

Because of the facts outlined above, the cultivation and preservation of forests are of vital importance. Whereas in past decades this meant primarily the conservation and reforestation of stands of timber and replenishing them with species of the same kind, while taking into consid-

TABLE 1-3
Selected Features of the World's Forests, as Reported for 1963 (FAO, 1,2)

Region	Total forest area[a] 1000 ha (1)	Forest as % of total land area % (2)	Coverage of col. 3-5[b] %	Composition Coniferous % (3)	Composition Broadleafed % (4)	Composition Mixed % (5)	Coverage of col. 6[b] %	Productive forest land % (6)	Coverage of col. 7[b] %	Area under management plant 1000 ha (7)
Europe										
Northern Europe	54,571	47.3	100	71	11	18	100	42,438	100	35,161
EEC	25,737	22.0	100	40	58	2	100	24,845	100	10,707
Great Britain	1,914	6.1	100	52	47	1	100	1,422	100	1,190
Central Europe	13,048	34.2	100	29	55	16	100	11,319	100	5,870
Southern Europe	46,149	30.6	100	46	50	4	100	24,496	100	18,058
Eastern Europe	27,243	26.8	30	35	55	10	96	24,614	86	22,024
Total	168,662	30.4	89	53	37	10	99	129,134	98	93,010
U.S.S.R.	910,009	40.6	100	76	24	—	100	710,844	100	299,965
North America										
Canada	443,109	44.4	100	61[c]	13[c]	26[c]	100	227,926[d]	100	418,259[d]
United States	307,100	32.8	100	47[f]	53[f]	—	100	198,020	100	84,378
Total	750,209	38.8	100	55	32	13	100	425,946	100	502,637
Latin America										
Mexico	39,700	20.1	—	—	—	—	—	—	—	—
Central America	29,466	57.2	49	24	76	—	77	18,644	49	513
Caribbean	77,126	31.3	42	18	82	—	42	2,030	42	400
Northern South America	195,340	67.7	60	—	100	—	60	79,762	60	1,391
Southwest South America	154,686	50.3	70	1	72	27	70	71,042	70	6,287
Southeast South America	92,599	28.0	76	1	92	7	76	60,000	76	2,510
Brazil	352,100	41.6	100	1	99	—	100	93,000	100	140
Total	871,053	42.6	76	1	94	5	77	324,478	76	11,241

Africa									
Western	495,691	39.5	56	—	100	—	68	271,452	56	4,223
Eastern	241,919	26.3	81	—	99	1	81	100,447	81	6,387
Northern	9,207	1.6	58	35	65	—	58	4,219	58	429
Southern	15,904	6.0	26	36	58	6	26	1,041	26	322
Total	762,721	25.4	63	1	98	1	71	377,159	63	11,361
Near East									
Mediterranean Basin	942	2.9	29	61	13	26	52	269	29	154
Southwest Asia	6,420	22.9	23	—	99	1	23	900	23	3
Arabian Peninsula	1,480	0.6	—	—	—	—	—	—	—	—
Total	8,842	1.6	20	10	86	4	23	1,169	20	157
Far East									
Continental Southeast Asia	127,535	65.1	49	1	99	—	49	49,949	49	1,187
Insular Southeast Asia	191,297	66.8	17	3	97	—	35	35,975	36	3,895
South Asia	81,937	18.3	88	5	90	5	98	66,625	88	32,288
Japan	25,053	68.1	100	37	52	11	100	22,273	100	9,770
East Asia (less Japan)	17,289	66.5	54	41	40	19	54	6,194	54	2,422
Total	433,111	44.7	45	7	72	21	56	181,016	54	49,562
Pacific	218,180[g]	27.2	4	36	14	50	100	41,078	99	7,270
China (Mainland)	96,380	9.9	—	—	—	—	—	—	—	—
World total	14,229,167	32.2	75	35	60	5	83	2,190,824	81	975,203

[a] For countries where 1963 data were not yet available, data for total forest area as given in FAO World Forest Inventory 1958 have been used.

[b] Total forest area of countries reporting expressed as a percentage of total subregional or regional forest area.

[c] Refers to classified productive forest land not reserved for protection purposes.

[d] Excludes some classified productive forest land.

[e] All publicly owned forests.

[f] Refers to commercial forest land only.

[g] Includes large areas of sparsely forested land used for grazing.

eration local conditions, forest management today strives for an improvement and a quantitative increase in the yield of wood. By suitable cultivation of the soil, the application of fertilizers, and by irrigation, not only are higher yields obtained but the production is accelerated. Thus, under the right conditions, in Latin America, pole-sized eucalyptus in a yield of 20 to 30 cubic meters per hectare per year and, in tropical Africa, in a yield of 15 to 25 cubic meters per hectare per year were obtained. From more rapidly growing pines, sawlog-sized wood was obtained in an amount of 12 to 17 cubic meters per hectare per year. Polewood and fuelwood can be produced in this way in less than ten years, pulpwood in about ten years, and logs in a size acceptable for lumber within fifteen or more years.

Increased efforts to improve the yield and quality of woods are being made primarily in countries that are situated in the north temperate zone. Here it has become apparent that the tendency is to replace deciduous woods, to an ever increasing degree, by coniferous woods. Thus, for example, in the Federal Republic of Germany the percentage of coniferous woods had increased from 30% at the beginning of the century to about 70% in 1960. Similar efforts have been made in the United Kingdom, where new coniferous forests have been planted. In the more southerly and warmer parts of Europe, poplar trees are preferentially planted and today in Italy these supply about 40% of the demand for wood for industrial uses. In the warmer regions of the United States, new plantations of fast-growing pine species have been established. In the United States and in Western Europe it has been possible, because of unusually great increases in crop yields, to retire areas hitherto used for agriculture and thus to make them available for forestry development and hence to increase the production of wood. In Japan, over-mature and coppice forests have been replaced by the cultivation of high-yielding coniferous species, thus increasing the forest area from 30% to 40%. Special efforts are needed for the wood-poor regions of the Near East and the Chinese mainland. Similar efforts are being made in New Zealand, the southern part of Africa, and in Chile, where the major part of the industrial wood demand is already covered by new plantations. These areas were not devoid of natural forests, but the existing wood species produced low yields and qualities that were unsuitable for industrial utilization. Fast-growing pine species are usually planted, but eucalyptus is also gaining greater importance. In Latin America about 900,000 hectares and, in Africa, about 600,000 hectares have been planted with eucalyptus.

Of the tropical species of valuable woods, teak is of predominant importance. There are about a million hectares of teak plantations, chiefly in Indonesia and Burma, but also in some other tropical countries. Whereas the fast-growing pine species, eucalyptus and poplars are beginning to

cover the demand for fuelwood, poles and pulpwood, teakwood serves primarily as high-grade lumber.

Man-made forests are the most important sources of industrial wood supply. Their contribution to the yield of all forests is proportionally very much greater than their share of the world's forest areas. The high yield per acre makes local concentration possible so that the costs for harvesting and transportation are lowered. The uniformity of dimensions and quality that is obtained in man-made forests also contributes to the lowering of costs. Intensified control assures a greater degree of the qualities desired by the consumer. However, there are limitations. Large areas of a single species are especially susceptible to attacks by insects and diseases. The farther one departs from the natural growing conditions, the more a species is prone to adverse effects, not only from biological sources but also from climatic and other environmental influences. Intensive forestry management therefore requires good soil, high-quality stock, and careful handling prior to, during, and after planting.

2. THE DEMAND FOR WOOD AND WOOD PRODUCTS

2.1 Distribution of the World's Consumption of Wood and Wood Products

Wood is supplied as a raw material and as an industrial material for the most diversified uses. In addition, large amounts of wood are used for the production of heat, with complete destruction of the substance resulting. It is therefore of interest, with regard to the future supply of wood for industrial uses, to study the various types of application of wood and the extent of the requirements for each, and to compare the development of the demand in the period from 1950 through 1963. Such a comparison is contained in the above-mentioned FAO report (1). Besides this, the yearly statistics of the same organization (2,3) provide further details.

Table I-4 contains a summarized compilation of the consumption of roundwood and fuelwood during the period in question and the division of this consumption for the various wood products. Table I-5 shows the distribution of the estimated average total wood consumption for the years 1960 to 1962. It should be mentioned that the data given for the years 1951, 1956, and 1961 are those for the average yearly wood consumption for the three-year periods of 1950 to 1952, 1955 to 1957, and 1960 to 1962. While the roundwood consumption in Table I-4 is given in cubic meters, the figures for wood products such as paper and boards are given in metric tons. This also applies to the pulp products given in Table I-5. Here the figures for the products are recalculated into the equivalent volume units for roundwood. The corresponding conversion factors are given in Table

TABLE I-4

CHANGE IN RECORDED WORLD USE OF WOOD AND WOOD PRODUCTS, 1950 TO 1963

Use	Million units	1950–1952	1955–1957	1960–1962	1963	Change index 1951–61 (1951 = 100)
Roundwood						
Sawlogs[a] and veneer logs	Cubic meters	493.4	582.2	648.3	656.7	131
Pulpwood[b] and pitprops	Cubic meters	185.6	233.5	255.6	257.1	138
Other industrial wood	Cubic meters	129.2	124.9	116.3	118.3	90
Total industrial wood	Cubic meters	808.2	940.6	1020.2	1032.1	126
Fuelwood	Cubic meters	865.6	876.1	876.5	886.5	101
Total	Cubic meters	1673.8	1816.7	1896.7	1918.6	113
Wood products						
Sawnwood[c]	Cubic meters	266.1	309.6	341.0	354.2	128
Paper and paperboard	Metric tons	44.3	59.4	77.3	84.0	174
Plywood	Cubic meters	6.8	11.3	16.8	20.1	247
Fiberboard	Metric tons	2.2	3.3	4.5	5.3	210
Particleboard	Metric tons	0.04	0.57	2.3	3.6	5900
Roundwood[d]	Cubic meters	129.2	124.9	116.3	118.3	90

[a] Includes logs for sleepers.
[b] Includes roundwood used for the manufacture of particleboard and fiberboard.
[c] Includes sleepers.
[d] Excludes pitprops.

TABLE I-5

THE DISTRIBUTION OF ESTIMATED TOTAL WOOD USE, 1960 TO 1962 (FAO, 1)

Region	Sawn-wood[a]	Panel products[a,b]	Pulp products[c,d]	Round-wood[a]	Total industrial wood[a,e]	Fuel-wood[a]
Europe	78.31	8.41	22.87	36.60	246.7	107.9
U.S.S.R.	99.72	2.24	3.47	67.3	249.7	100.9
North America	94.37	16.25	37.37	18.6	308.7	46.1
Latin America	12.39	0.52	2.66	8.5	38.7	192.4
Africa	4.05	0.37	0.90	13.3	23.5	182.7
Pacific Asia	57.33	2.75	10.20	44.1	176.0	457.6
World total	346.20	30.50	77.50	188.0	1043.0	1088.0

[a] Million cubic meters.
[b] Excludes veneer.
[c] Million metric tons.
[d] Includes some nonwood fiber products; excludes dissolving pulp.
[e] The product quantities have been converted into equivalent volumes of roundwood using the standard factors in reference 3. No allowance has been made at this stage for the variation in transformation ratio from region to region, or for variation within an industry or over time; nor has the volume of wood raw material supplied in the form of wood residues been subtracted. The present estimates of roundwood are intended to do no more than give a preliminary indication of the broad orders of magnitude of the wood raw material corresponding to the estimates of consumption of wood products presented here. Excludes wood raw material equivalents of veneer and dissolving pulp consumption.

TABLE I-6

ROUNDWOOD EQUIVALENTS (FAO, 3)

Product	Unit	Roundwood equivalents	
		Cubic meters	Cubic feet
Charcoal	1 metric ton	6.0	212
Coniferous sawnwood	1 cubic meter	1.67	59
Broad-leaf sawnwood	1 cubic meter	1.82	64
Sleepers	1 cubic meter	1.82	64
Plywood and blockboard	1 cubic meter	2.3	81
Veneer sheets	1 cubic meter	1.9	67
Particleboard	1 metric ton	2.0	71
Fiberboard	1 metric ton	2.0	71
Wood pulp			
Mechanical	1 metric ton	2.5	88
Chemical			
Sulfite	1 metric ton	4.9	173
Sulfate	1 metric ton	4.8	169
Dissolving	1 metric ton	5.5	194
Semichemical	1 metric ton	3.3	117
Newsprint	1 metric ton	2.8	99
Printing and writing paper	1 metric ton	3.5	124
Other paper	1 metric ton	3.25	115
Paperboard	1 metric ton	1.6	56

I-6. As is shown in Table I-4, the consumption of roundwood has not increased within this 11-year period, but the consumption of sawlogs has increased by about 30%. The increase in wood consumption for pulp and paperboard is considerably greater and amounts to around 75%, while that for fiberboard has doubled and that for particleboard, which is a new product in terms of 1950, is approximately the same as that for fiberboard and is still increasing.

It is apparent that there is a great variation in the development of the consumption of wood between the different groups. This is true not only with regard to the products but also with regard to the distribution of the total wood consumption, as is shown in Table I-5. Whereas in Africa and Latin America nearly 90% of the total wood consumption is in the form of nonprocessed wood, in North America this amounts to only about 20% while the remaining 80% is used for sawlogs, plywood, paper, etc. From this it follows that there are wide regional differences, not only in wood consumption but also especially in the increasing demand for wood, or, in other words, marked differences in the rate of development of the increasing demand. In the underdeveloped countries, where the per capita income is low, much wood is consumed as fuel; in the highly developed countries the consumption of wood increases with increasing per capita income be-

cause of increasing demand for finished wood products. The relationship between consumption and income, however, is only an indirect one because most wood products are production goods and not consumer goods.

2.2 Roundwood and Fuelwood

By "roundwood" is meant wood that is used as such, without further processing. It is supplied for different, generally simple, uses and usually has a fairly low unit value. So far as its use is concerned, only rough estimates are available and the figures are not very reliable. In comparison with the total world consumption of wood, that of unprocessed roundwood seems to be declining. It is much more uniformly distributed throughout the world, however, than is the consumption of processed wood. North America and Europe consume about one-third of the total roundwood, Africa (with the exception of South Africa), Asia (with the exception of Japan) and Latin America consume about the same amount. Roundwood is used chiefly for the construction of simple buildings, with the main consumption being in rural districts. With increasing living standards, roundwood is replaced by more valuable building materials.

Roundwood is also used for scaffoldings and framework and also for power-line poles, including telegraph and telephone poles, but these require certain standards with regard to length and quality. Furthermore, mines and coal-pits also use roundwood as pitprops. Table I-7 shows the development of the consumption of roundwood for the period of 1950

TABLE I-7

CHANGE IN RECORDED CONSUMPTION OF ROUNDWOOD, 1950 TO 1963, AND
ESTIMATED TOTAL CONSUMPTION, 1960-62 (FAO, 1)

Region	Recorded consumption[a,b]				Change 1951-61 (1951=100)	Estimated total consumption 1960-62[b,d]		
	1950-52[c]	1955-57	1960-62	1963		Pitprops	Other	Total
Europe	25.12	21.31	22.45	22.88	89	14.12	22.44	36.56
U.S.S.R.	58.18	59.89	45.64	51.10	79	21.71	45.64	67.35
North America	18.03	16.96	16.98	12.01	94	1.60	16.98	18.58
Latin America	1.54	2.77	2.62	2.66	170	0.89	7.61	8.50
Africa	7.63	7.57	8.36	8.22	110	1.86	11.44	13.30
Pacific Asia	18.68	16.46	20.11	21.30	108	10.40	33.73	44.13
World total	129.20	125.00	116.20	118.20	90	50.60	137.80	188.40

[a] Million cubic meters.
[b] Excluding pitprops.
[c] Production.
[d] Estimated total consumption in 1960-62 differs from that recorded by an allowance for unrecorded consumption.

TABLE I-8

<small>Change in Recorded Consumption of Fuelwood,[a] 1950 to 1963, and Estimated Total Consumption,[b] 1960-62 (FAO, 1)</small>

Region	Recorded consumption[c]				Change 1951–61 (1951 = 100)	Estimated total consumption[c,e] 1960-62
	1950-52[d]	1955-57	1960-62	1963		
Europe	117.6	110.4	107.9	103.4	92	108.0
U.S.S.R.	108.0	121.9	100.9	96.7	93	101.0
North America	66.7	60.1	46.1	37.1	69	46.0
Latin America	174.2	175.3	191.0	210.7	110	192.0
Africa	148.0	156.7	172.0	180.4	116	183.0
Pacific Asia	251.1	252.1	259.9	259.0	104	458.0
World Total	866.0	877.0	878.0	888.0	101	1088.0

[a] Including wood for charcoal.

[b] Figures for 1960–62 are estimated average total annual consumption.

[c] Million cubic meters.

[d] Production.

[e] Estimated total consumption in 1960–62 differs from that recorded by an allowance for unrecorded consumption.

through 1962. From this table it can be seen that consumption has decreased in the European countries, in the Soviet Union, and in North America, while it has increased in the Latin American countries, in Africa, and in Pacific Asia. The decrease in consumption noticed in connection with the first group of countries is due primarily to the modernization of building methods and improvements in mining techniques. It is estimated that the world's consumption of roundwood as such will be about the same in 1975 as it was in 1960 to 1962—i.e., about 200 million cubic meters annually.

The reservations expressed above with regard to the consumption of roundwood apply even more forcefully to the data on the consumption of fuelwood. Table I-8 shows clearly that the consumption of wood as fuel is subject to extreme variations in different parts of the world. As already observed in connection with the consumption of roundwood, the consumption of fuelwood is decreasing in the highly developed continents as well as in the Soviet Union, while it is stationary or is increasing slightly in the less developed countries. The consumption of wood for fuel depends primarily on whether or not other fuels, such as coal, oil, natural gas, or hydroelectric power, are available in sufficient quantity and at reasonable prices. Because of its weight and limited heat value, wood cannot be transported economically over wide distances. The increasing urbanization in the developing countries brings with it a falling off in the demand for fuelwood and its replacement by more compact fuels with greater heat values. Table I-7 also contains data on the consumption of wood for the production of charcoal, but only for the period of 1950 to 1952. More information on this subject may be found in Chapter V.

With regard to the fact that over a billion cubic meters of wood are burned each year, the question arises as to whether or not greater quantities of such wood could be used advantageously for industrial purposes, e.g., for the production of pulp and paper. This question will be further discussed in the following chapters of this book. But it should be noted here that the cost of collecting and transporting such wood in most cases prevents such an undertaking from being economically feasible.

2.3 Sawnwood and Wood-Based Panels

About two-thirds of all processed roundwood is used as lumber. In 1963, the world's consumption of sawnwood amounted to 345 million cubic meters. The data given in Table I-9 represent only the consumption

TABLE I-9

GROWTH IN RECORDED CONSUMPTION OF SAWNWOOD, 1950 TO 1963, AND
ESTIMATED TOTAL CONSUMPTION, 1960-62 (FAO, 1)

Region	Recorded consumption[a]				Change		Estimated total consumption[b] 1960-62
	1950-52	1955-57	1960-62	1963	1951-56 (1951=100)	1956-61 (1956=100)	
Europe	61.48	70.76	78.31	79.42	115	111	78.31
U.S.S.R.	55.57	75.80	99.72	98.62	136	132	99.72
North America	103.53	101.84	93.86	100.21	98	92	94.37
Latin America	12.14	13.06	12.32	11.79	108	94	12.39
Africa	3.06	3.41	3.70	3.64	112	109	4.05
Pacific Asia	30.34	44.69	53.13	60.47	147	119	57.33
World total	266.12	309.56	341.04	354.15	116	110	346.20

[a] Million cubic meters.

[b] Estimated total consumption in 1960-62 differs from that recorded by an allowance for unrecorded consumption.

that has been officially recorded. A considerable amount has not been reported, however, because it was produced in small local sawmills and in hand-operated plants. The figures given in column 8 for the average total consumption for the years 1960 to 1962 have therefore been corrected correspondingly (1). Of the total given for the year 1963 (354 million cubic meters), 265 million cubic meters or 77% was obtained from coniferous woods and about 80 million cubic meters or 23% from deciduous woods. The consumption according to species varies markedly in the various continents; in Europe, the Soviet Union, and North America considerably more coniferous than deciduous wood is used, while in Latin America, Africa, and Pacific Asia the two species are used in about equal amounts, with coniferous woods predominating in some cases. About two-thirds to four-fifths of the total sawnwood consumption is used for building material, furniture, and packing material, while smaller, but still appreciable amounts are used for pitprops and railway sleepers. Here, too, the degree of development of the various continents influences the consumption to some extent. Local conditions also affect the consumption. Thus, for example, the amount of sawnwood used for the construction of a new house in the United States averages about 20.5 cubic meters, in northwestern Europe, about 6.8 cubic meters, and, in South Asia, only about one cubic meter. The rate of increase in the total world consumption of sawnwood during the last ten years has decreased to some extent. This is shown in column 5 of Table I-8.

Coal mining in the United States and Europe is declining, hardly any new railway lines are being constructed, and the sleepers of existing railways are being replaced by sleepers of other materials. In the field of home construction, the building of multifamily dwellings and the use of new and improved construction methods have further reduced the demand for sawnwood. None of these processes is designed to eliminate entirely the use of sawnwood but each has as its goal the improved utilization of this valuable natural product.

It is also possible to substitute, in part, "wood-based panels" for sawnwood. Such panels differ from sawnwood with regard to wood consumption, as will be shown in the following section.

Three different types of wood-based panels are produced, depending upon the method of production: plywood, fiberboard, and particleboard. Blockboard and veneer are usually included with plywood. As shown in Table I-10, the consumption of these wood-based panels has increased enormously within the last twelve years. This applies especially to the production of fiberboard and, even more, to the production of particleboards, which have been fabricated industrially and on a large scale only since the Second World War.

Fiberboards are produced in a compressed form as hardboards or in a voluminous heat-retaining form as insulation boards. Both types are used in the building industry and also in the manufacture of furniture. This is also true to an even greater extent of particleboards. The world consumption of such wood-based panels increased from 12.5 million cubic meters in 1951 to 30.2 million cubic meters in 1961 and reached 73.1 million cubic meters in 1963. The various wood-based panels are obtainable in different units of measurement—according to area, volume, and weight. For this reason, a comparison between them is difficult and the extraordinary increase in consumption of particleboards in volume units seems to be

TABLE I-10

GROWTH IN RECORDED CONSUMPTION OF PLYWOOD, FIBERBOARD AND
PARTICLEBOARD, 1950 TO 1963 (FAO, 1)

	Consumption				Change	
	1950–52	1955–57	1960–62	1963	1951–61[a]	1956–61[b]
Plywood[c]						
Europe	1.52	2.10	3.02	3.53	199	144
U.S.S.R.	0.72	1.04	1.32	1.45	185	127
North America	3.99	6.88	10.16	12.31	255	148
Latin America	0.13	0.22	0.33	0.34	257	150
Africa	0.05	0.074	0.12	0.12	239	158
Pacific Asia	0.40	0.94	1.88	2.39	468	201
World total	6.81	11.26	16.84	20.14	247	150
Fiberboard[d]						
Europe	0.68	1.17	1.74	2.05	256	149
U.S.S.R.	0.024	0.071	0.27	0.35	1120	379
North America	1.30	1.70	1.98	2.28	153	116
Latin America	0.034	0.058	0.12	0.14	346	204
Africa	0.019	0.065	0.054	0.067	282	83
Pacific Asia	0.12	0.20	0.38	0.42	333	188
World total	2.17	3.27	4.54	5.31	210	139
Particleboard[d]						
Europe	0.027	0.37	1.57	2.49	5800	430
U.S.S.R.	—	—	0.17	0.28	—	—
North America	0.012	0.17	0.42	0.62	3500	250
Latin America	—	0.007	0.025	0.039	—	364
Africa	—	0.012	0.007	0.002	—	53
Pacific Asia	—	0.013	0.096	0.13	—	731
World total	0.039	0.57	2.29	3.56	5900	404

[a] 1951=100.
[b] 1956=100.
[c] Million cubic meters.
[d] Million metric tons.

somewhat overestimated because of the difference in the density and thickness of the boards.

The production of particleboards, moreover, is limited to only a few countries, mainly to the region of the European Economic Community (EEC) with 68%, the United States of America with 16%, and the U.S.S.R. with 7%. With regard to the production of plywood, the United States produced 45%, Canada, Japan, the U.S.S.R., the United Kingdom, and the EEC countries together about 30%, Africa (without South Africa), Asia (without Japan), and Latin America together only about 4%. The respective data on the consumption of fiberboard are 40%, 33%, and 5%.

These wood-based panels serve primarily as a substitute for sawnwood. The increased consumption of these products is based on the following factors: there must already have been an extensive consumption of sawn-wood; the difference in price between the wood-based panels and sawnwood must be significant and the probable increase in the cost of each must be apparent; the extent of the forest resources available and their relationship to the site of the industry concerned are also important.

A highly significant factor in the manufacture of fiberboard and particleboard is the relatively inexpensive production process. The products possess a high degree of uniformity with regard to quality, and relatively inferior grades of wood can be used, especially small pieces from the wood-processing industries which cannot be used for other purposes and would mostly be burned.

Table I-11 gives a survey of the supply of raw material for such fiberboard and particleboard plants from roundwood, industrial waste wood, and other fibrous raw materials in different regions. It is apparent that the supply of raw materials from the different sources varies with the region: European factories fill 59% of their needs for the manufacture of fiberboards from wood wastes, whereas North American mills get only about 14% from this source. The reason for this is the fact that the supply

TABLE I-11

SUPPLY OF RAW MATERIAL FOR THE MANUFACTURE OF FIBERBOARD AND
PARTICLEBOARD IN 1960 (FAO, 1)

	Roundwood		Industrial wood residues		Other fibrous materials	
	1000 m³	Percent of total[a]	1000 m³	Percent of total[a]	1000 tons	Percent of total[a]
Fiberboard						
Europe	1740	40	2580	59	20	1
North America[b]	2740	53	650	14	720	33
Asia	160	65	40	14	30	21
Other regions[c]	620	69	240	27	20	4
Total	5260	48	3510	35	790	17
Particleboard						
Europe	1480	49	1120	37	250	14
North America	200	26	580	74	—	—
Asia	90	67	50	33	—	—
Other regions[c]	70	80	20	20	—	—
Total	1840	45	1770	44	250	11

[a] Percentage of total raw material supply on basis by weight.
[b] U.S. data converted from tons of fiber material.
[c] Excluding U.S.S.R.

of roundwood is much more favorable in North America. In the case of particleboard the conditions are reversed: North America uses wood waste to the extent of 74% while Europe uses only 37%. This difference is due to the fact that sawmills and plywood mills in North America produce wood wastes that are more suitable for particleboards than the wood wastes from the sawmills in Europe, which deal with smaller logs. Table I-11 also shows that, in North America, fibrous materials other than wood, e.g., bagasse from the southeastern states, are used for the production of wood-based panels.

In summary, it may be said that the fiberboard and particleboard industries have succeeded, through the technological development of their production processes, by their ability to use low-grade types of wood and waste wood from the wood-processing industries, through the wide uniformity of their products, and through the extreme flexibility of their processes, in successfully replacing sawnwood for many purposes. On the other hand, the plywood industry, in spite of the higher standards required for its raw material, has not only held its own through the extension of the range of its products with regard to size and quality but has even enlarged its market.

2.4 Pulp, Paper, and Paperboard

Almost the total pulp production of the world is used for the manufacture of paper and paperboard, with only a small amount being used for the production of synthetic fibers on a cellulose basis. The proportion of dissolving pulp in the United States in 1961 was about 4%, in the Soviet Union about 5%, in Europe 8.5% and in Japan 10%. Dissolving pulp is also used in the manufacture of films, plastics, explosives, solvents, varnishes, and other chemical products. Its main application is in the production of rayon. In recent times, rayon has been partly replaced by other synthetic fibers; thus the consumption of dissolving pulp is unlikely to be very important in the immediate future.

The world's consumption of paper and paperboard increased from about 44 million tons in 1951 to 84.5 million tons in 1963. As is shown in Table I-12 the annual increase for the period from 1951 to 1961 was 5.7%. The consumption of paper and paperboard provides a certain criterion for the degree of civilization of the region that is statistically recorded; for this reason, the consumption per thousand consumers is also given in Table I-12 and from this can be seen the extremely high rate of consumption of paper and paperboard in North America. In 1963, this rate was more than three times that of Europe, while the total consumption of paper and paperboard in North America was only about 55% of that of Europe.

TABLE I-12

GROWTH IN RECORDED CONSUMPTION OF PAPER AND PAPERBOARD,
1950 TO 1963 (FAO, 1)

	Recorded consumption				Change		Average annual rate of growth[c] 1951–1961
	1950–1952	1955–1957	1960–1962	1963	1951–1956[a]	1956–1961[b]	
Total consumption[d]							
Europe	10.99	16.24	22.87	25.36	148	141	7.6
U.S.S.R.	1.63	2.64	3.47	3.87	162	131	7.9
North America	27.28	32.73	37.37	39.37	120	114	3.2
Latin America	1.44	1.95	2.63	2.77	136	135	6.2
Africa	0.38	0.58	0.82	0.89	152	142	8.0
Pacific Asia	2.62	5.28	10.17	11.79	202	193	14.5
World total	44.35	59.42	77.33	84.05	134	130	5.7
Consumption per 1000 capita[e]							
Europe	26	38	51	55	143	134	6.7
U.S.S.R.	9	13	16	17	150	120	6.1
North America	161	177	185	189	110	105	1.4
Latin America	8.6	10.3	12.2	12.2	119	119	3.5
Africa	1.8	2.5	2.9	3.2	138	118	5.0
Pacific Asia	1.9	3.5	6.1	6.8	183	177	12.4
World total	18	22	26	27	123	119	3.9

[a] 1951 = 100.
[b] 1956 = 100.
[c] Percent.
[d] Million metric tons.
[e] Metric tons.

Table I-13 shows the further distribution of paper and paperboard consumption for newsprint, for writing and printing papers, and for paper and paperboard products used in industry. From both tables it can be seen that the enormous rate of increase in consumption during the years immediately after the Second World War has fallen off. In this connection, the world consumption of newsprint increased from 9.09 million tons in 1951 to 14.5 million tons in 1961, corresponding to an annual rate of increase during this decade of 4.8%. At the same time, the demand for writing and printing papers increased from 7.97 to 14.08 million tons, an average annual increase of 5.9%. The world's consumption of "cultural" papers increased, in the same decade, by about 65%, whereas that of paper and paperboard products used in industry increased by 80%.

Paper and paperboard are used for a great variety of purposes, but two main consumer groups can be distinguished: printing and writing, and packaging. A main component of the first group is newsprint, which con-

TABLE I-13

GROWTH IN RECORDED CONSUMPTION OF PAPER AND PAPERBOARD BY
CATEGORY, 1950 TO 1963 (FAO, 1)

	Recorded consumption[a]				Change	
	1950–52	1955–57	1960–62	1963	1951–61[b]	1956–61[c]
Newsprint						
Europe	2.03	3.15	4.16	4.29	205	132
U.S.S.R.	0.22	0.32	0.43	0.52	198	134
North America	5.73	6.55	7.12	6.53	124	109
Latin America	0.42	0.55	0.73	0.70	175	133
Africa	0.08	0.12	0.17	0.17	204	140
Pacific Asia	0.61	1.23	1.89	2.10	309	153
World total	9.09	11.92	14.50	14.31	159	122
Other printing and writing paper						
Europe	2.37	3.39	4.79	5.46	202	141
U.S.S.R.	0.34	0.61	0.73	0.80	211	120
North America	4.40	5.33	6.36	7.14	144	119
Latin America	0.25	0.37	0.41	0.39	164	112
Africa	0.06	0.12	0.18	0.19	316	157
Pacific Asia	0.55	1.01	1.61	1.89	294	159
World total	7.97	10.82	14.08	15.87	177	130
Industrial paper and paperboard						
Europe	6.60	9.69	13.92	15.60	211	144
U.S.S.R.	1.07	1.72	3.32	2.55	217	135
North America	17.14	20.86	23.89	25.70	139	115
Latin America	0.77	1.03	1.48	1.68	194	144
Africa	0.24	0.34	0.47	0.53	195	137
Pacific Asia	1.46	3.03	6.67	7.81	456	220
World total	27.27	36.67	48.75	53.88	179	133

[a] Million metric tons.
[b] 1951 = 100.
[c] 1956 = 100.

sumes about half of the printing and writing paper. In 1962, the packaging industry in the United States used 47.5% of the total paper and paperboard production. For Europe in the same year this figure was 48%. So far as can be predicted, a further increase in the consumption of paper and paperboards, especially by the packaging industry, will occur.

The starting material for the production of paper and paperboard is wood, from which groundwood is produced by mechanical means, pulp by chemical means, and semichemical wood pulp by mechanical-chemical means. Table I-14 shows the production of such fibrous materials. The total production of wood pulp in 1962 amounted to 62.3 million metric tons, of which 70% was chemical wood pulp and 30% mechanical wood

TABLE I-14

RECORDED PRODUCTION OF WOOD PULP, BY CATEGORY OF PULP AND BY
REGIONS, 1950 TO 1963 (FAO, 1)

	Production[a]				Annual rate of growth (%)	
	1950–52	1955–57	1960–62	1963	1951–61	1956–61
Total woodpulp						
Europe	10,186	13,720	17,895	19,756	5.8	5.4
U.S.S.R.	1,750	2,615	3,457	3,907	7.0	5.7
North America	22,560	29,025	34,776	37,977	4.4	3.7
Latin America	237	334	705	874	11.5	16.1
Africa	23	67	156	228	21.1	18.5
Pacific Asia	1,325	2,872	5,319	6,326	14.9	13.1
World total	36,081	48,633	62,308	69,057	5.6	5.1
Chemical and semichemical woodpulp						
Europe	6,376	8,758	11,687	13,220	6.2	6.0
U.S.S.R.	1,256	1,851	2,431	2,757	6.8	5.5
North America	14,790	20,010	25,127	28,109	5.4	4.6
Latin America	102	170	444	528	15.8	19.9
Africa	10	61	137	197	29.9	17.6
Pacific Asia	697	1,809	3,786	4,669	18.4	15.9
World total	23,231	32,668	43,612	49,480	6.5	6.0
Mechanical woodpulp						
Europe	3,810	4,962	6,208	6,525	5.0	4.6
U.S.S.R.	494	764	1,026	1,150	7.6	6.0
North America	7,770	9,015	9,649	9,868	2.2	1.4
Latin America	135	155	261	346	6.9	11.1
Africa	13	6	19	31	3.8	27.9
Pacific Asia	628	1,063	1,533	1,657	9.3	7.6
World total	12,850	15,965	18,696	19,577	3.8	3.2

[a] 1000 metric tons.

pulp. The figures for chemical pulp also include those for semichemical pulp. More than half of the wood fiber production (34.8 million tons) is carried out in the United States and Canada. Finland, Sweden, and Japan together produce about 20% of the world's supply. The various types of wood fiber products show different production rate increases, not only regionally but also according to type.

The total world production of wood pulp increased during the decade from 1951 to 1961 by an average annual rate of 5.6%. Japan had the largest growth rate, 15% annually, while North America, with only 4.4%, still had the greatest absolute increase. The U.S.S.R. had an average annual increase of about 7%.

The above statement, that wood is the principal raw material for the

TABLE I-15

REGIONAL PULP PRODUCTION, BY TYPE OF FIBROUS RAW MATERIAL,
1960–62 AVERAGE (FAO, 1)

Region	Production[a]		Percentage of total pulp		
	Total pulp	Wood pulp	Wood pulp		Nonwood pulp
			Coniferous	Broad-leafed	
Europe	19,150	17,895	84	10	6
U.S.S.R.	—	3,457	99	1	—
North America	35,207	34,776	83	16	1
Latin America	1,035	705	57[b]	11[b]	32[b]
Africa	217	156	16	57	27
Pacific Asia[c]	1,195	689	58		42
Japan	3,954	3,948	50[d]	50[d]	—
China (Mainland)	2,200	683	(20)	(11)	69

[a] 1000 metric tons.
[b] Based on 1963 figures.
[c] Less Japan and China Mainland.
[d] Based on 1962 figures.

TABLE I-16

UTILIZATION OF WOOD RESIDUES FOR PULP PRODUCTION[a] IN SELECTED REGIONS
AND COUNTRIES, 1957–1963, AND ESTIMATES FOR 1965 AND 1966 (FAO, 1)

	1957	1960	1962	1963	1965[b]	1966[b]
Europe (less U.S.S.R.)						
Wood requirements	62,494	67,041	75,829	76,463	90,241	95,939
Residues utilized	5,323	8,328	9,629	9,933	11,839	12,347
Percentage of requirements	8.5	12.5	12.7	13.0	13.1	12.9
Finland						
Wood requirements	12,245	14,300	17,400	18,730	21,960	23,000
Residues utilized	822	2,400	2,510	2,150	2,800	2,800
Percentage of requirements	6.7	16.7	14.4	11.5	12.7	12.2
Sweden						
Wood requirements	21,300	22,600	25,500	23,845	28,100	30,600
Residues utilized	1,350	1,800	2,100	2,400	3,350	3,350
Percentage of requirements	6.3	8.0	8.2	10.0	11.9	10.9
U.S.S.R.						
Wood requirements	12,000	14,600	15,395	16,800	20,220	22,100
Residues utilized	320	1,800	2,207	1,837	2,600	2,700
Percentage of requirements	2.7	12.3	14.3	10.9	12.9	12.2
Canada						
Wood requirements	34,656	37,980	41,280	42,815	51,100	52,000
Residues utilized	2,182	4,050	6,200	6,923	8,400	8,700
Percentage of requirements	6.3	10.7	15.0	16.2	16.4	16.7

[a] 1000 cubic meters.
[b] Estimates.

production of paper and paperboard, must be supplemented by the statement that, in 1963, about 77% of this production was from wood fibers, 5% from other fibrous materials such as straw, bagasse, esparto, etc., and 18% from waste paper. There are also great differences between the various fibrous materials, not only with regard to the types of paper and paperboards produced, but also with regard to region—a discussion of which would lead too far afield. Within the scope of this study, those data are of primary interest which are related to the consumption of wood for wood-fiber products. Table I-15 contains some data that are of interest in this respect, while Table I-16 deals with the utilization of wood wastes and the demand for wood for the production of wood fiber for the years 1957, 1960, 1962 to 1963, and 1965 to 1966. In this connection, it is important to note that world-wide scarcity, especially of pulpwood, has led to the use of considerable amounts of wood residues for the manufacture of fiber-based panels. Thus, for instance, the consumption of waste wood has increased in the United States from 600 thousand cubic meters (corresponding to 2.2% of the total wood consumption for wood-based panels) in 1940 to more than 19 million cubic meters (corresponding to over 20%) in 1962, with a further increase to about 25% predicted by 1970. In a study carried out by the Forest Products Laboratory in Madison, Wisconsin (4) it is stated:

"While residues not used have only decreased about 8%, the amounts used for pulp and paper and other fiber have increased 640%. This increase is greater than for any other component of the forest products industry. Plywood, veneer, and hardboard, with the best growth during the same period, about doubled in production. Most of this growth in pulp-chip use appeared to be at the expense of residues used as fuel, which showed about two-thirds reduction. There appeared to be about a 14% reduction in amounts of residue produced that could only be attributed to improved manufacture.

"The improvement in residue utilization during the 10-year period (1952–1962) resulted in the present practice, of which the salient features are the following:

(1) More residues are now being used than wasted.

(2) The conversion of coarse residues to pulp chips has advanced so that, at present, 22% of the Nation's requirements are satisfied from that source.

(3) Of the total residues not being utilized about 60% are classified as fine residues, the least usable of those being produced.

(4) Of the coarse residues not being used (40% of the total unused residue) about 75% are of the softwood variety originating mainly in the Rocky Mountain area, where pulping facilities have not

been developed as extensively as elsewhere, and along the Pacific Coast. The South has made tremendous gains in amounts of softwood coarse residue utilization. In 1962, this area converted to fiber for pulp five and one-half times as much material as remained as unused residue of that kind. Only about 15% of the unused residue in that area is classified as softwood-coarse.

(5) Of the residue not being used, about 11% is classed as hardwood-coarse. Nearly all originates in the South and North. Mixed species, the fact that much is dry rather than green, and the fact that only a small percentage of existing pulp mills convert hardwood have been deterrents to utilization."

It is the author's opinion that this representation of the problem of the utilization of waste wood can lead to the integrated cooperation of the wood-processing industries in other countries where regional conditions are propitious. Even in Japan the amount of waste wood used in the manufacture of wood-based panels had already, in 1961, surpassed 28% of the total wood consumption. Table I-16 also shows the rapid increase in the utilization of waste wood in the different production areas. Whether this can be further increased depends primarily on whether there is a further extension of the wood-processing industries from which the waste wood originates. The many difficulties involved in the collection and utilization of waste woods from the forests have already been mentioned.

A further spectacular improvement in the supply of wood for the preparation of wood-based panels is provided by the increased use of deciduous instead of coniferous woods. The superiority of coniferous woods, which have longer and more uniform fibers, could be counteracted by an improvement in the alkaline pulping processes (kraft pulp and semichemical pulp) for many intended uses. For the production of dissolving pulps, too, deciduous woods are used in many cases. Table I-17 shows the trend toward the increased utilization of deciduous woods for some specific areas.

2.5   Wood for Silvichemicals

Not even approximately reliable data can be given on the utilization of wood for the production of chemicals. There are several reasons for this. The direct chemical processing of wood for the production of chemical intermediates, for example by hydrolysis (Chapter IV) or by pyrolysis, hydrogenation, or oxidation (Chapter V), so far as these operations are carried out at present on a large scale, is done, in most cases, with wood wastes of industrial or of forest origin. The statistics on such waste woods are recorded in other connections or are unavailable. However, it is possible to calculate the wood consumption from the chemical products ob-

TABLE I-17

TRENDS IN THE USE OF BROAD-LEAFED SPECIES IN PULP PRODUCTION IN
SELECTED COUNTRIES AND AREAS (FAO, 1)

Region	Pulpwood production[a]		Percent of total broad-leafed pulpwood
	Total	Broad-leafed	
U.S.S.R.			
1963	16.52	0.28	2
1965[b]	20.73	0.96	5
1970[b]	56.26	8.56	15
United States[c]			
1950	45.5	6.38	14
1960	88.0	17.82	20
1962	94.2	19.58	21
1970[b]	124.3	29.70	24
1980[b]	169.4	46.20	27
Western Europe[d]			
1950	27.5	1.2	4
1959	47.5	4.5	9
1960	52.3	5.3	10
1975[b]	90.0	20.6	23
Scandinavian countries			
1950	22.7	0.6	2.6
1959	36.4	1.6	4.4
1960	40.1	2.0	5.0
1975[b,e]	57.6	8.6	14.9
Other Western European countries			
1950	4.8	0.6	12.5
1959	11.1	2.9	26.1
1960	12.2	3.3	27.0
1975[b]	32.4	12.0	37.0
Japan[f]			
1956	8.6	1.3	15.1
1960	12.3	3.7	30.1
1961	14.2	4.6	32.4

[a] Million cubic meters.
[b] Estimates.
[c] Based on conversion factor 2.2 m³/cord.
[d] Roundwood removals.
[e] Estimate has proven overly conservative.
[f] Pulpwood consumption.

tained and this has been done, as far as possible, in the discussions of the
relevant processes in the respective chapters. Here, it is necessary only to
refer to the data given in these chapters.

TABLE I-18

ANNUAL VALUE OF SILVICHEMICALS PRODUCED IN THE
UNITED STATES IN 1963 (BRATT, 5)

Product	Annual production[a]	Annual value[b]
Naval stores except tall oil rosin and sulfate turpentine		
Gum rosin (lb)	239	26.3
Steam-distilled rosin (lb)	571	62.8
Gum turpentine (gal)	7.03	2.5
Steam-distilled turpentine (gal)	7.87	2.4
Pine oil (gal)	9.57	10.5
Rosin oil (gal)	1.96	1.3
Dipentene (gal)	1.27	0.6
Subtotal		106.4
Sulfate mill products		
Crude tall oil, total (lb)	989	
Crude tall oil, used as such (lb)	71	2.6
Distilled tall oil[c] (lb)	82	7.0
Tall oil rosin[d] (lb)	275	27.5
Tall oil fatty acids (lb)	239	21.5
"Heads" fraction, pitch, etc. (lb)	200	4.0
Crude sulfate turpentine (gal)	23.5	4.7
Dimethyl sulfide[e] (lb)	10	—
Dimethyl sulfoxide[e] (lb)	5	1.7
Sulfate lignin products (lb)	20	2.0
Subtotal		71.0
Sulfite mill products		
Lignosulfonates (lb)	415	20.8
Concentrated spent liquor (gal)	?	—
Ethyl alcohol, 190 proof (gal)	3.5	1.8
Vanillin (lb)	1.4	4.2
Torula yeast (lb)	8.0	0.8
Acetic acid, glacial[f] (lb)	8.0	0.8
Formic acid, 90%[f] (lb)	0.6	0.1
Subtotal		28.5
Wood pyrolysis products		
Acetic acid, glacial (lb)	17.7	1.8
Methanol (lb)	8.2	0.4
Charcoal, incl. briquettes (lb)	750	30.0
Subtotal		32.2
Miscellaneous products		
Arabinogalactan (Stractan) (lb)	?	
Hemicellulose extract (gal)	1.5	0.1
Levulinic acid[e] (lb)	5	?
Bark products (lb)	100	4.0
Subtotal		4.1
Grand total		242.2

[a] Millions.
[b] Million dollars, in car load or tank car lots or wholesale, f.o.b. mill.
[c] Includes acid-refined tall oil.
[d] Data pertain to crop year.
[e] Plant capacity.
[f] From NSSC spent liquors.

Other examples of chemical products from wood are the by-products of the sulfite and kraft pulping industries. Data on these are to be found in Chapters VII and VIII.

Finally, the products obtained by the extraction of wood must be mentioned. Some relevant data are given in Chapter III. Table I-18 gives a survey of the approximate production and importance of such silvichemicals. It shows that, in the United States, the annual value of these chemicals was estimated to be 240 million dollars in 1963, representing an increase of 12% over 1961. The sales revenue from the by-products of the kraft pulping industry had increased, since 1961, by 18% and totaled 71 million dollars. This increase can be attributed primarily to the increased production and price of fatty acids from tall oil and of sulfate turpentine. The by-products of the sulfite pulp industry also showed an increase in sales revenue of over 100% since 1961, reaching 28.5 million dollars in 1963. This increase is primarily the result of increased production and enhanced values of lignosulfonates, ethyl alcohol, and vanillin. Increased consumption of charcoal caused an increase of 20% in sales since 1961. Products derived from bark also showed improved sales revenues. On the other hand, plants for the hydrolysis of wood have been abandoned in all countries except the Soviet Union. Only in Japan is there a mill that processes the prehydrolyzate from the production of dissolving pulps into furfural, while a sulfite pulp mill produces nucleic acid from torula yeast (5).

3. ANTICIPATED DEVELOPMENT OF THE MOST IMPORTANT WOOD-CONSUMING INDUSTRIES AND FUTURE CHANGES IN THE WORLD'S WOOD BALANCE

The future development of the principal wood-consuming industries depends, on the one hand, on the demand for the products that these industries manufacture and, on the other hand, on economic conditions, on the demand for capital, the situation with regard to the labor market, and the supply of necessary raw materials which these industries use. With regard to future supplies, some information has already been given in the previous sections, So far as the raw material, wood, is concerned, this has also been indicated above.

There still remains to be examined, by what means the wood-consuming industries can, in the future, meet the increased demands made on them. One of the main factors in establishing an economically sound price level for all wood products is the buying price of the wood. In this regard, the industry is confronted with steadily increasing prices. To counteract the increasing cost of wood, further steady development of the techno-

logical processes must be sought, with the goal of producing from one volume unit of wood either a larger amount or a qualitatively improved product. Moreover, rational use of each wood species, including its waste, must be considered, and this, in turn, presupposes that the special suitability of the species for the different intermediate and final products has been investigated and established. A further prerequisite is the improvement of the forestry economy so that the costs of the cultivation, management, and use of the forests can be reduced, because these are the prime factors in establishing the price of the wood.

In this connection it must not be overlooked that, if these precepts are strictly followed, advantages can be realized from the mutual relationship between the various wood-processing industries, and these advantages cannot be won by a single plant or even by a group of plants producing the same products. Thus, for example, the plywood and veneer industries can advantageously process woods of large diameters and of highest quality, whereas woods of smaller diameters but of good quality are suitable for sawnwood. The pulping, fiberboard, and particleboard industries, on the other hand, can process woods of smaller diameters and also of medium quality. If these industries are combined into groups, significant advantages are obtained by forestry administrations with respect to the use and delivery of the wood. In addition, the pulping, fiberboard, and particleboard industries are in a position to process wood wastes from sawmills, veneer plants, and even from furniture factories. The more such industries are grouped together in neighboring plants, the greater are the reductions in the costs not only of the raw material but also of its transport and delivery. It cannot be denied that the requirements with regard to quantity and quality of woods overlap in the different industries or that their wants and requirements are in competition with each other.

As indicated above, these considerations stimulate a far-reaching horizontal integration of the wood-consuming industries, the initiative for which may be taken on a private ownership basis or, as in eastern European countries, by authoritative government orders. Such a horizontal integration is of special significance for the pulping and wood-based-panel industries which are increasingly dependent upon the utilization of wood wastes from other wood-processing industries. It must not be overlooked that fiberboards and particleboards are today, in many cases, taking the place of sawnwood. It is estimated that the world consumption of such wood wastes for the production of pulp and wood-panel products will, by 1975, reach the remarkable figure of 100 million cubic meters. This takes into consideration only the use of coarse residues. The industrial utilization of the huge quantities of sawdust and bark still requires a satisfactory solution. The figures below illustrate the vast amounts involved.

In Europe in 1960, 26 million cubic meters of coarse wood residues

were obtained and, of these, 8.5 million cubic meters were processed industrially. In the same year, 26 million cubic meters of fine residues were produced, but only one million cubic meters were processed. In 1962, 76 million cubic meters of wood waste were produced in the United States; of this, the pulp industry consumed 20 million cubic meters, the wood-panel and other industries used 21 million cubic meters, while over 36 million cubic meters remained unused. It is to be hoped that technological developments will lead to the utilization of this hitherto unused raw material.

In addition to the horizontal type of integration mentioned above, a vertical integration is also indicated in an arrangement whereby the hitherto preferentially processed varieties of wood are shifted vertically. Thus, for example, wood, which had hitherto been used as fuel or as roundwood, becomes, to an increasing extent, wood for industry, while more and more woods which have hitherto, because of their dimensions, been used by sawmills are now consumed within the industrial group of pulp and wood-panel mills. In this way the production from sawmills is reduced in favor of other wood-consuming industries because the latter return a greater profit per volume unit of the raw material. Within the pulp industry, a trend not to stop with the production of pulp as an intermediate product but to go further with the production of paper and paperboard is apparent.

These tendencies are, understandably, observed primarily in areas and countries having intensified forest management. In underdeveloped countries and in regions which are located in or near tropical forest zones, the importance of sawnwood will continue for a long time, especially if more emphasis is placed on the quality of the products than is the case at present.

Predictions and speculation about the world's balance of industrial wood for 1975 must be made with reservations, in particular with regard to the above-mentioned perspectives of the technological and commercial development of the wood-processing industries. In spite of this, the figures given in Table I-19 are of interest because they show that some regions with extensive wood demands show marked deficiencies in wood, whereas other zones have a large surplus of wood. From the point of view of an integrated worldwide forestry management, it should be possible to fill the requirements of the areas that have a deficiency of wood. Such an undertaking, however, encounters serious obstacles. In the first place, wood is not a raw material that can be transported economically over great distances. This was indicated already in the discussion of national and regional wood economies and their wood-related industries. Secondly, both horizontal and vertical integration mean that the industrial groups cooperating must strive to produce and market the highest-priced wood products. Wood supplies of any particular region will therefore favor only the

TABLE I-19

EXPECTED CHANGE IN THE WORLD'S INDUSTRIAL WOOD BALANCE,
1961 TO 1975[a] (FAO, 1)

Region	Consumption[b]		Production[c]		Surplus (+) or deficit (−)	
	1961	1975	1961	1975	1961	1975
Europe						
Northern Europe	33.3	39	85.2 (90.4)	99 (108)	+57.1	+69
EEC	103.4	160	54.1 (59.2)	64 (70)	−44.2	−90
Great Britain	40.2	60	3.0 (3.2)	5 (5)	−37.0	−55
Central Europe	16.9	24	21.0 (22.5)	29 (32)	+5.6	+8
Southern Europe	11.7	21	9.3 (9.4)	17 (18)	−2.3	−3
Eastern Europe	53.9	72	51.7 (53.6)	58 (64)	−0.3	−8
Total	259.4	376	224.3 (238.3)	272 (297)	−21.1	−79
U.S.S.R.	243.2	305	256.8 (259.0)	—[d] —	+15.8	—
North America						
Canada	32.5	44	89.0 —	—[e] —	+56.5	—
United States[f]	288.0	376	250.3 —	326 —	−37.7	−50
Total	320.5[g]	420[g]	339.3 —	— —	+18.8	—
Latin America	39.9	76	38.5 —	— —	−1.4	—
Africa	25.0	36	25.6 —	— —	+0.6	—
Near East	9.5	14	6.4 —	— —	−3.1	—
Far East						
Continental Southeast Asia	8.8	14	10.1 —	— —	+1.3	—
Insular Southeast Asia	15.0	25	22.0 —	— —	+7.0	—
South Asia	11.6	20	10.1 —	— —	−1.5	—
Japan	63.0	112	48.4 (52.6)	72 (82)	−10.5	−30
East Asia	6.1	9	3.1 —	— —	−3.0	—
Total	104.5	180	93.7 —	— —	−6.6	—
Pacific	18.1	26	15.4 (16.1)	21 (22)	−2.0	−4
China (Mainland)	34.0	62	34.0 —	— —	—	—
World total	1054.1	1495	1034.0 (1055.1)	— —	—	—

[a] Million cubic meters of roundwood and wood raw material equivalent.

[b] Consumption of processed wood products and other industrial roundwood expressed in equivalent volumes of wood raw material.

[c] Removal of industrial wood and, where possible with total domestic wood supply, roundwood removal plus usable wood residues in parentheses.

[d] The annual allowable cut in 1963 was reported to be almost 605 million cubic meters.

[e] Allowable annual cut from presently economically accessible areas is 210 million cubic meters; it could be raised to about 315 million cubic meters.

[f] Wood balance based on data from timber trends in United States.

[g] Excludes roundwood equivalent of wood residues consumed.

area that is its immediate neighbor. The sole exception is provided by those wood species which are used only for special purposes because of their unique properties. The wood-processing industries, especially those in highly industrialized countries, are facing great changes, the extent of which cannot at present be accurately predicted, There can be no question, however, that technological development is deeply involved (5,6) and this, insofar as it refers to chemical-technological processes, is discussed in the following chapters.

REFERENCES

1. Food and Agricultural Organization of the United Nations (FAO), "Wood world's trends and prospects." Basic Study No. 16, Rome, 1967.
2. FAO, "World's forest inventory 1963," Rome, 1963.
3. FAO, "Yearbook of forest products statistic 1965," Rome, 1965.
4. Lewis, W. C., Wood residue utilization. *Forest Prod. J.* **15** (8), 303 (1965).
5. Bratt, L. C., Trends in production of silvichemicals in the United States and abroad. *Tappi* **48** (7), 46A (1965).
6. FAO, "European timber trends and prospects. A new appraisal 1950-1975," New York, 1964.

II

Anatomy and Physical Properties of Wood

THE EVOLUTION OF FORESTS

Our present-day forests that originate from the Gymnosperms, to which the conifers belong, date back 200–300 million years. The first appearance of deciduous woods, the Angiosperms, may be assumed to have been more than 100 million years ago. In the Carboniferous period, seed ferns were the first predominant group of the Gymnosperms. These ferns already showed a cambium and a secondary growth in thickness. They contributed to the formation of mineral coal and died out toward the end of the Paleozoic era. Bennettitales were found widely distributed in the Jurassic period and faded out during the Muschelkalk epoch. Cycadales were extensively found in the Mesozoic era; some species are still to be found today in tropical and subtropical forests. They exhibit cambiums and secondary growths in thickness. Their wood consists of tracheids with bordered pits. While the Cycadophytae are short and unbranched, the Coniferophytae form long and strongly branched stems. Thick layers of wood provide the stem with strength, while pith and bark form minor components. The main group of Coniferophytae is the Cordaitales which formed the first large forests during the Carboniferous period. Trees of up to 30 meters in height are no rarity and tree trunks a meter in diameter have been found. The change from humid to arid growth periods caused the formation of weak growth zones in the tree, which are not identical with the annual rings. In the primary wood, steplike tracheids are found; in the secondary wood, only tracheids with bordered pits. These precursors of present-day conifers, which were of great importance in the formation of mineral coal deposits, died out toward the end of the Paleozoic era, but were succeeded by the Coniferales.

During the Permian period, the Coniferales appeared and, unlike the above-mentioned species, formed large forests in the temperate zones. These conifers are primarily responsible for the formation of the tertiary lignite deposits and even today are the most important species for the production of lumber. Next appeared the *Araucaria* species. In the Triassic and, especially, in the Jurassic periods, the coniferous species such as *Cypresse* and *Abies* predominated; in the Tertiary period, woods with rosin ducts were found.

In the Jurassic period and the upper Muschelkalk epoch, the Angiosperms, to which the deciduous species of dicotyledons belong, became widespread. They, also, participated in the formation of lignite deposits.

The climate during the Carboniferous and lower Permian periods in which the ferns flourished must have been uniformly mild and rainy. This then changed to a drier and cooler climate which suited the Gymnosperms better and caused them to predominate. In the Muschelkalk epoch, the Angiosperms prospered, as a direct consequence of the change in the climate. A subtropical rainy climate gradually gave way to a cooler, more moderate climate which, in the Paleocene epoch, was not unlike our present temperate-zone climate. The Glacial epoch then greatly decreased the tree flora. In spite of the vast changes that took place over long periods of time, the structure of the conifers, which contain primarily tracheids with bordered pits, remained unchanged. A pronounced formation of annual rings, with different formations of early- and latewood tracheids, can be found already in the Tertiary period. In lignite deposits woods can be found that scarcely differ from the woods of present-day trees of the same species. Spruces, firs, and pines can be identified.

The cells of deciduous woods exhibit certain functions, such as the rapid transportation of juices and the storage of nutrients. In spite of this, the conifers are not only older but they are also superior to the deciduous woods in their powers of reproduction. Deciduous and coniferous woods must have differed from each other at a very early stage of their development, as is shown not only by the difference in their cell formation but also by the dissimilarity of their formation of tension and compression woods. This will be discussed further in later sections (18,131,139).

1. THE FORMATION OF WOOD CELLS

Trees originate from their buds. These consist of a tissue that is very readily capable of division and has almost equilateral thin-walled cells, rich in plasma; it is called the "meristem." All subsequent types of cells originate from the meristem. In the center of this cell system the pith is formed, while the cells on the outside extend and divide. In this way

vascular vessels are formed in a circular arrangement around the axis of the branch and serve as conductors of the sap. The inner parts of these vessels or conduits soon die and lignify; the outer parts are formed of thin, living cells that consist of plasma and nuclei. The woody portion is also called the xylem and the outer, the bast or sieve-tubes, is called the phloem. Between the xylem and the phloem lies the true growth-layer, the cambium, which is capable of continuously forming new cells. The growth of the cells compresses the sap-conduction cells and changes their form. A small strip of the basic tissue remains, however, and connects the pith with the bark. With the steady growth of the cambium, new secondary medullary rays are constantly formed, in addition to the primary rays, and are uniformly embedded in the stems and branches of the tree.

Through the compression of the conductive vessels and the division of the medullary rays, a closed ring, the cambium ring, is slowly formed. This ring produces, toward the outside, bast cells and, toward the inside, wood cells and in this way causes the growth in thickness of the stem or branch. The cells formed toward the inside are not only more numerous but are also more thick-walled so that the cambium layer always moves, during the growth process, more and more toward the periphery of the cross-section. The bast layers correspond with the annual rings of the wood. In the bast portion the younger layers are inside and in the wood portion the opposite is true. The increasing diameter causes an expansion and cracking of the bast layer and the bark. The cracked layers dry out and die, forming the outer bark and in this way providing protection for the cambium and preventing the wood from drying out or being damaged.

In the temperate zones, the cambium annually resumes its cell-forming activities in uniform rhythm. As a result, the wood cells formed during the spring are thin-walled while those formed during the late summer are thick-walled and have a small lumen. Thus zones are formed which can be clearly distinguished and which show the annual growth in the form of annual rings while, on the other hand, the width of the rings permits conconclusions to be drawn with regard to environmental factors affecting the growth. Coniferous woods show especially pronounced annual rings whereas only those deciduous woods that form wide lumen vessels in the spring exhibit readily detectable ring formation.

The cambium is one of the most long-lived of vegetable tissues. In *Sequoia* trees it can reach the age of a thousand years or more. Even when it ceases to produce new cell-division it does not necessarily die. Cell-division can be stopped for several years and then resumed. Thus it is possible for the cambium to become inactive in the lower part of the stem while remaining active in the upper part. External factors can cause a strongly one-sided growth and, in this way, the activity of the cambium can vary in the radial direction, too. To achieve full activity, the cam-

bium needs adequate nourishment and when this is lacking its activity is hampered.

A criterion for many wood species is the difference between the inner part of the stem, the heartwood, and the outer layers, the sapwood; both are usually clearly distinguishable and they often differ markedly from each other. The reason for this is that only the outer layers of the older stem are required to conduct the sap and the inner layers no longer participate in this activity. Only the outer layers or annual rings are in contact with the leaves which evaporate the water and with the root hairs which take up the water. Because the inner layers have lost their significance for the vital processes of the tree, the cells of the water-conducting tissues are slowly closed and are deprived of their functional activity. The heart of the tree can develop a high degree of immunity from damage from fungi or insects by the incorporation of various nuclear substances.

The process of the conversion to heartwood has been studied extensively. In general, this process can be characterized by the fact that it consists chiefly of chemical and enzymatic processes. Of special significance for the conversion to heartwood is the disappearance of the starch from the parts undergoing the transition. The sugars, such as fructose, glucose, and saccharose, found in the sapwood also decrease toward the heartwood zone in wood felled during the winter. With progressive drying, a decomposition of the starch obviously takes place and glucose is converted by stages, near the heartwood zone, into aromatic compounds. The formation of heartwood can also lead to the increased embedding of numerous other chemical compounds such as tannins, rosins, phenols, alkaloids, and salts. The conversion to heartwood therefore represents not only a structural change in the inner portion of the stem but also a chemical conversion of certain wood components. As will be shown later, both changes, under certain conditions, may cause complications during a later chemical processing of the wood (9,31,50,55). For the utilization of the wood as lumber, the amount of heartwood and its proportion in the cross-section are of primary interest. There are different methods for differentiating between sapwood and heartwood and for revealing the transition zones in various woods. These methods can only be referred to here (104,105).

2. THE MICROSTRUCTURE OF WOOD FIBERS

For a study of the structure of wood, various types of sections can be made, as shown in Fig. II-1. A cut, vertical to the fiber direction, can be made, and this is called a transverse or cross-section. If a cut is made parallel to the fiber axis and vertical to the radial plane, i.e., parallel to the annual rings, a tangential section is obtained. The radial surface is

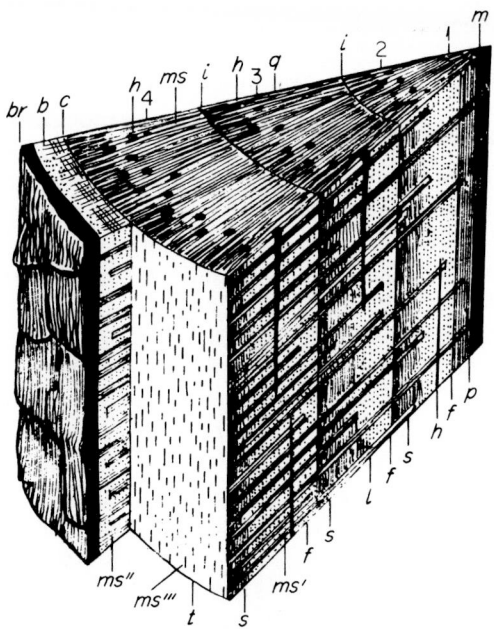

Figure II-1. Section of a four-year-old trunk of *Pinus silvestris*, showing a transverse section (top), a radial longitudinal section (right), and, where the bark has been removed, the tangential longitudinal section (front). Legend: *i*, boundaries or annual rings; 1,2,3,4, the four successive annual rings; *m*, pith; *f*, springwood; *s*, summerwood; *c*, cambium; *b*, inner bark of the bast zone; *br*, outer bark; *h*, resin canals; *ms*, transverse aspect of medullary ray; *ms'*, radial section of medullary ray, appearing as a band; *ms''*, the same, seen in the inner bark; *ms'''*, tangential section of medullary ray. Magn. 6 : 1.

shown when the stem is cut at a right-angle to the cross-section and parallel to the radius of the stem. Thus one can distinguish between a cross-section, radial section, and tangential section. In this way the structural elements of the stem become visible, the pith in the center with the annual growth of wood cells surrounding it and containing the predominant portion of the vessels of the stem. Toward the outside, the cambium, the bast, and then the bark follow.

2.1 The Coniferous Woods

By far the greater part of the wood of conifers consists of tracheids. These are elongated, tubelike structures which usually have a square cross-section. Their ends are closed. They are arranged parallel to the longitudinal direction of the stem. The length and breadth of the tracheids are of paramount importance for the technical utilization of the wood fibers. They are subjected to certain fluctuations within the length of the stem and therefore also within the annual rings. Thus, for instance, the average fiber length in black spruce (*Picea mariana*) increases until the tree attains

a certain height and then decreases again. With other spruce or pine species, this phenomenon is less pronounced.

Table II–1 shows the average length and breadth of the fibers of sprucewood which has grown slowly and rapidly. It can be seen that the fibers from fast-growing wood are short and wide and that the ratio between length and width is 45 to 60, whereas that of slowly grown wood may increase to 90 (49). Greater differences are observed with regard to the fiber length of pinewood, whereas the fiber breadth remains fairly constant. But with pine, also, the slowly grown wood contains longer and thinner fibers than the rapidly grown. The wood fibers of spruce are usually longer than those of pine. Because the two types of fibers are fairly similar

TABLE II-1

AVERAGE LENGTH AND BREADTH OF SPRUCE FIBERS IN RELATION
TO THE NUMBER OF ANNUAL RINGS PER CENTIMETER
AND THE DENSITY OF THE DRY WOOD (49)

Sample of spruce	Number of annual rings per cm	Density of dry wood	Average length (L) of fibers (mm)	Average breadth (B) of fibers (mm)	Ratio $L : B$
Fast-growing	1.7	0.28	2.22	0.050	44
Fast-growing	1.4	—	2.49	0.50	50
Fast-growing	2.5	0.32	3.04	0.048	63
Fast-growing	2.0	0.31	2.64	0.046	57
Slow-growing	18.5	0.40	3.78	0.041	92
Slow-growing	17.4	0.42	3.45	0.40	86
Slow-growing	23.2	0.40	3.35	0.039	86
Slow-growing	17.3	0.40	3.19	0.040	80

TABLE II-2

AVERAGE LENGTH AND BREADTH OF PINE FIBERS IN RELATION
TO THE NUMBER OF ANNUAL RINGS PER CENTIMETER
AND THE DENSITY OF THE DRY WOOD (49)

Sample of pine	Number of annual rings per cm	Density of dry wood	Average length (L) of fibers (mm)	Average breadth (B) of fibers (mm)	Ratio $L : B$
Fast-growing	2.8	0.36	2.23	0.044	51
Fast-growing	2.8	0.40	2.47	0.041	61
Fast-growing	2.8	0.36	1.81	0.044	41
Slow-growing	12.0	0.45	2.48	0.038	66
Slow-growing	14.0	0.40	2.40	0.036	66
Slow-growing	29.0	0.44	2.68	0.036	77
Slow-growing	21.0	0.39	3.22	0.037	86
Slow-growing	18.0	0.39	2.85	0.036	79

as regards width, the ratio of length to breadth is greater for spruce than for pine. Table II-2 shows the data for the average length and width of pinewood fibers.

The average fiber lengths for a number of conifers are given in Table II-3. From this it can be seen that the fiber lengths vary between 3.5 and 5.0 mm, depending upon the species and their location. In some exceptional cases, even greater lengths are reached (75). The various fiber lengths, as well as the differences in the thickness of the wall in spring- and summerwood fibers, have an effect upon the technical processing of the fibers for the manufacture of paper. Earlywood and latewood fibers can be clearly differentiated by fluorescent-microscopic investigations after staining with certain dyestuffs (59). The lengths of the fibers are closely related to the habitat of the tree. In colder climates, they become wider. With rapidly grown wood the proportion of summerwood is less than in slowly grown wood. The earlywood fibers show a pronounced lumen and are thin-walled; latewood fibers develop thicker walls and have a smaller lumen. The transition of the springwood tracheids to the summerwood tracheids forms the border of the annual rings. Figure II-2 shows a cross-section of sprucewood and clearly illustrates the change from early- to latewood.

In the early stages of their development, the tracheids serve primarily

TABLE II-3

APPROXIMATE AVERAGE LENGTH OF FIBERS OF VARIOUS CONIFEROUS
SPECIES IN MATURE TREES (75)

Species	Average fiber length (mm)
White spruce (*Picea glauca*)	3.5
Black spruce (*Picea mariana*)	3.5
Sitka spruce (*Picea sitchensis*)	5.5
Engelmann spruce (*Picea engelmanii*)	3.5
Balsam fir (*Abies balsamea*)	3.5
Grand fir (*Abies grandis*)	5.0
White fir (*Abies concolor*)	3.5
Douglas fir (*Pseudotsuga taxifolia*)	5.0
Hemlock (*Tsuga canadensis*)	3.5
Western hemlock (*Tsuga heterophylla*)	4.0
Western larch (*Larix occidentalis*)	5.0
Jack pine (*Pinus banksiana*)	3.5
Red pine (*Pinus resinosa*)	3.5
Southern yellow pine (Loblolly, Longleaf and Shortleaf)	4.6
White pine (*Pinus strobus*)	3.5
Lodgepole pine (*Pinus contorta* var. *latifolia*)	3.5
Ponderosa pine (*Pinus ponderosa*)	3.6
Bald cypress (*Taxodium distichum*)	6.0
Redwood (*Sequoia sempervirens*)	7.0

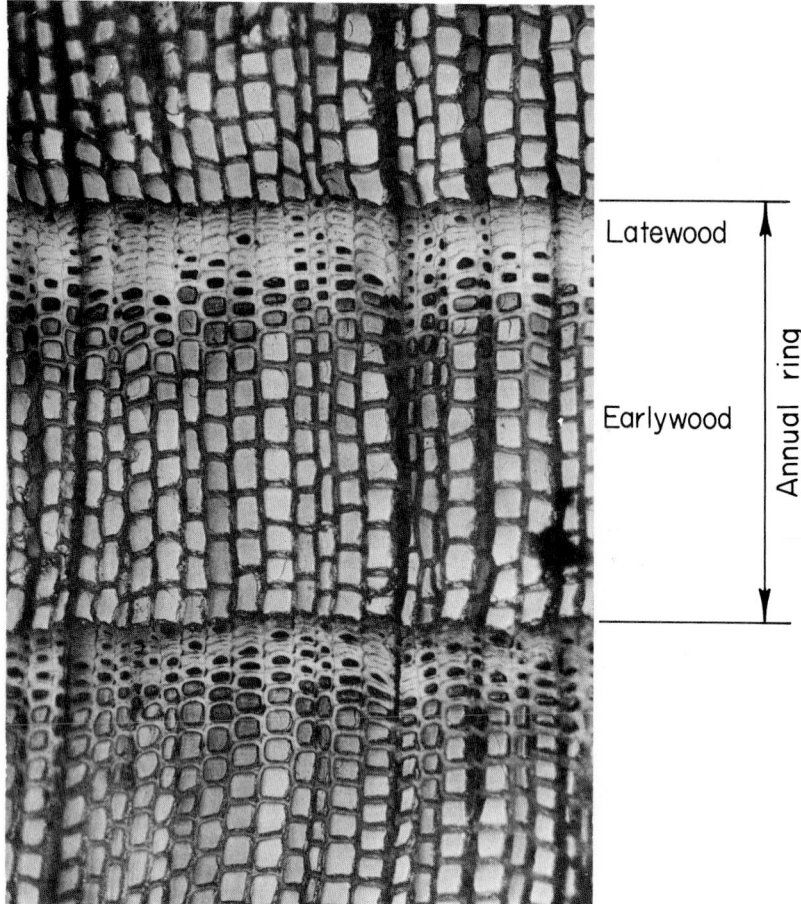

Figure II-2. Cross-section of sprucewood. Magn. 190 : 1. (Courtesy of Cellulosefabrik Attisholz.)

for the conductance of water. The passage of the water between the individual tracheids takes place through small porelike openings, called pits, and their submicroscopic structure has attracted the interest of many investigators. The structure of the pits will be disccussed together with the structure of the cell wall.

Parenchymatous cells are also found in the radial direction, i.e., transverse to the wood fibers or tracheids, and are called medullary rays. Single- and double-row cell layers can be seen. These medullary rays and the tracheids are connected by pits. In the same way, there is a parenchymatous tissue, arranged in the direction of the fibers, between the tracheids.

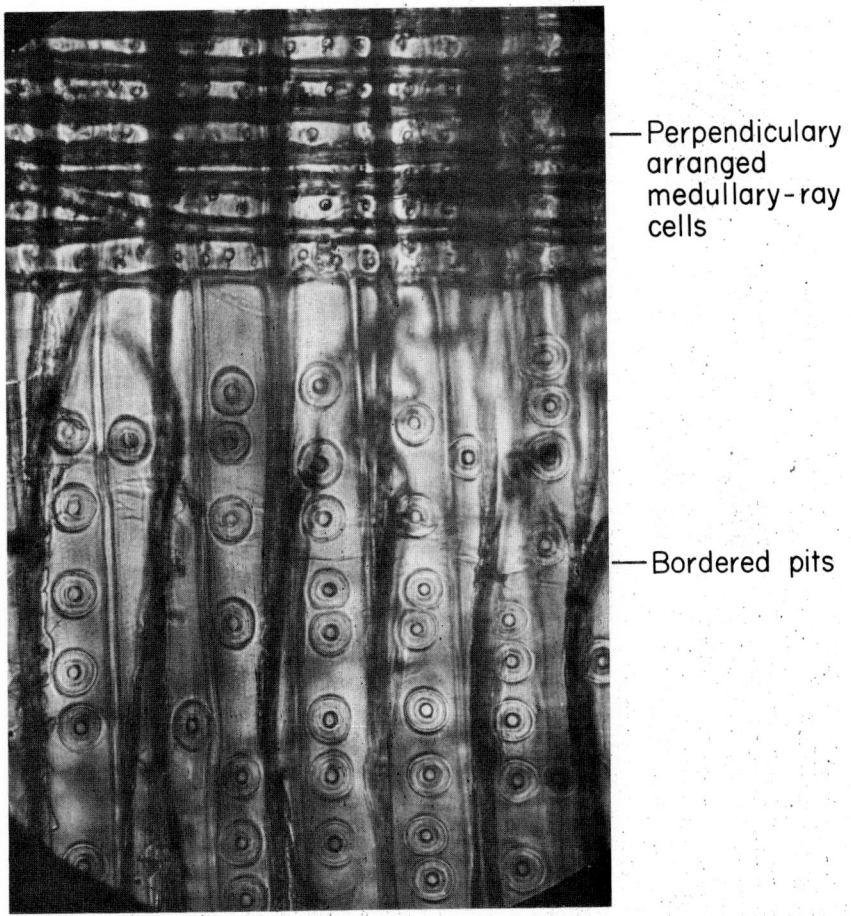

—Perpendiculary
arranged
medullary-ray
cells

—Bordered pits

Figure II-3. Radial cross-section of sprucewood. Magn. 360 : 1. (Courtesy of Cellulosefabrik Attisholz.)

This tissue is called the longitudinal parenchyma. Living plasma is found in the parenchyma cells. When these are pushed back, rosin ducts, which may run vertically or radially, are formed. The wood has three functions affecting the vital processes of the tree: it provides tissues that serve to conduct the sap from the roots to the crown, other tissues that lend strength to the trunk, and still a third type to provide storage for nutrients. The growth in length and thickness begins before the tree forms new foliage in a sufficient amount to produce the required nutrients. As a consequence, the latter must be taken from stored reserves. The cells that store the nutrients in the sapwood are called wood parenchyma. They are a com-

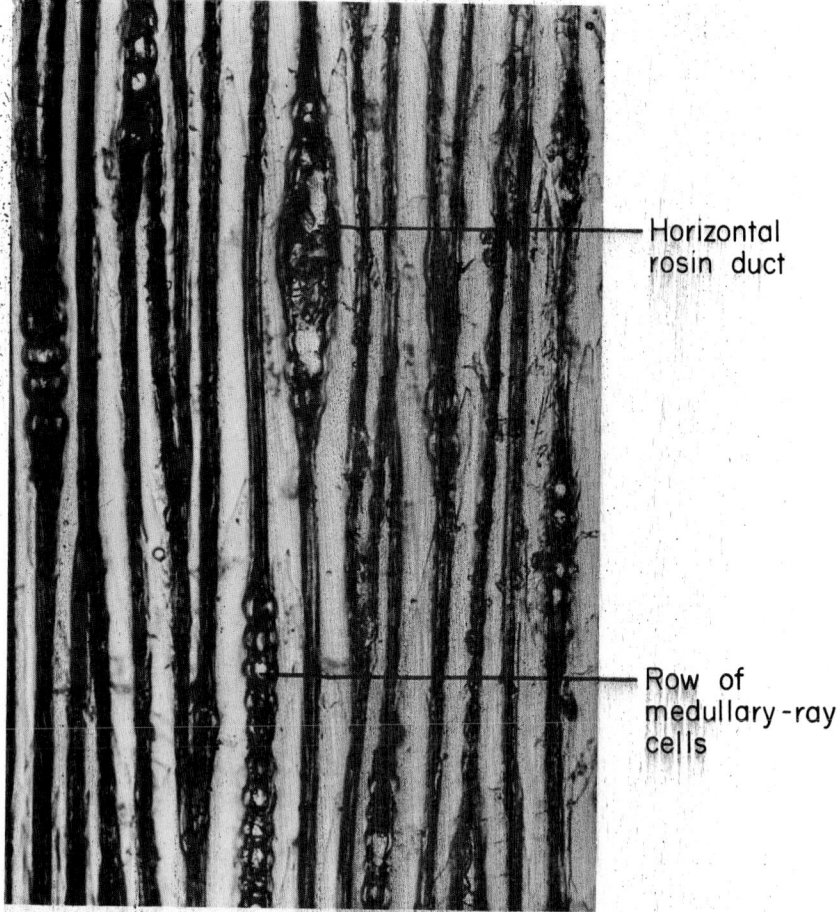

Horizontal
rosin duct

Row of
medullary-ray
cells

Figure II-4. Tangential section of sprucewood. Magn. 360 : 1. (Courtesy of Cellulosefabrik Attisholz.)

ponent of the sapwood and form a living parenchymatous system which is connected through the cambium with the parenchyma of the living inner bark (59).

The tracheidal and parenchymatous medullary rays of *Pinus silvestris* show marked differences as regards their length, width, and thickness. There are also differences between the outermost and innermost tracheid cell rows of the medullary rays with regard to their height. There is an interrelationship between the three cell dimensions, especially the tracheidal medullary rays. These differences in the dimensions of the cells are of importance economically and technically in the sulfite pulping process. By

means of fiber fractionation, not only the rosin-duct particles but also the parenchymatous cells, which contain rosin and therefore are incompletely delignified, are removed. The tracheidal medullary-ray cells are too short to be retained in the fractionation process. The average volume of the medullary-ray cells amounts to 3.9% in the woods under investigation. Less than 10% of this volume deals with medullary-ray cells having rosin ducts. In all, the medullary-ray cell volume consists of 45% tracheidal and 53% parenchymatous cells. The remainder consists of radial rosin ducts. The amount, by weight, of the medullary-ray and other parenchymatous cells in pulp from pinewood was estimated by fractionation experiments to be 8% (13).

Picea abies has only one type of parenchymatous medullary-ray cells. These cells are thick-walled and their relative amount of pits is 5% of the cell-wall surface. For *Pinus silvestris*, this amount is 50%. The pits are small and their dimensions, together with their small number, may explain the slow chemical deresinification of pulp from sprucewood. The medullary-ray cells do not vary greatly in their shape. Their relative occurrence in *Picea abies* as compared with the medullary-ray parenchymatous cells is about 1 : 2.7, in contrast to 2.8 : 1 for *Pinus silvestris*. Vertical parenchyma cells are similar to the medullary-ray parenchyma cells and occur on the vertical rosin ducts in a few perpendicular rows without forming a closed ring around the ducts. The parenchymatous medullary-ray cells of *Pinus sivestris* and *Picea abies* are considerably longer than the tracheidal cells and the scattering of the distribution of the cell lengths is almost twice as great as that of the medullary-ray tracheids (93,94).

The tissue which is arranged in the longitudinal direction of the stem and which forms the main part of the wood consists of prosenchyma. Unlike the parenchyma, the prosenchyma serves primarily to conduct the sap and to give strength to the stem. In coniferous woods, these prosenchyma are composed chiefly of fibrous cells, the tracheids, in which the radially arranged medullary-ray cells are embedded. Figure II–2 shows a cross-section of sprucewood, while Fig. II–3 shows a radial cross-section. They clearly indicate the arrangement of the tracheids with their bordered pits. In the upper part of Fig. II–3, the perpendicularly arranged medullary-ray cells are plainly visible. Figure II–4 shows a tangential section of sprucewood. It clearly shows the row of medullary-ray cells and the presence of a horizontal rosin duct formed by the separation of such cells (10).

2.2 Deciduous Woods

A cross-section of the stem or branch of a deciduous tree shows a large number of round hollow spaces or pores. These pores are the cross-sections of the vessels, which have a much greater diameter than the fiber cells

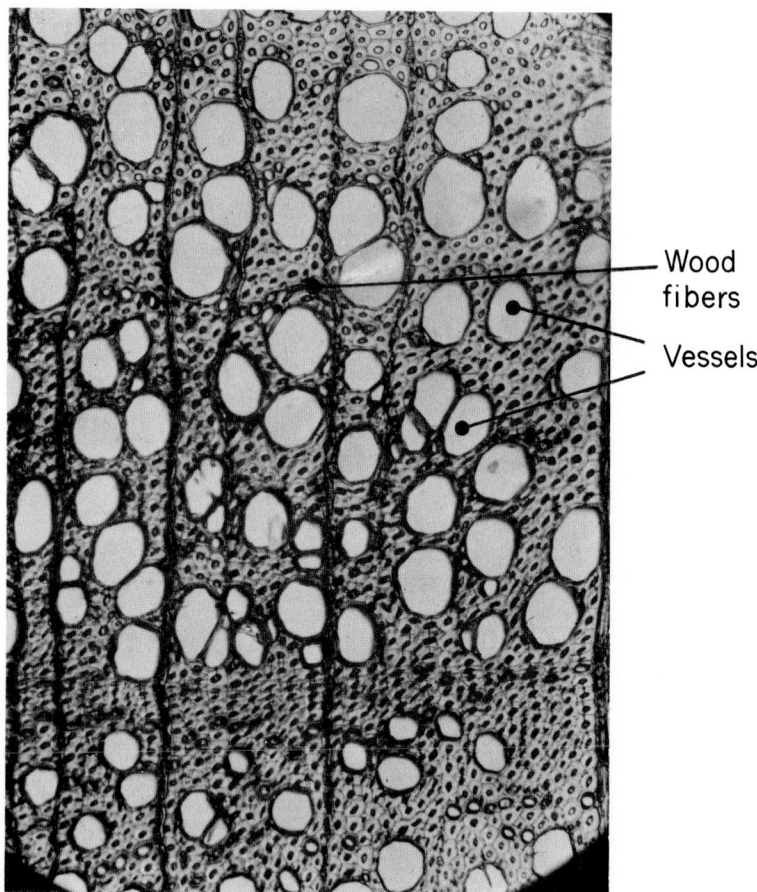

Wood fibers

Vessels

Figure II-5. Cross-section of beechwood. Magn. 300 : 1. (Courtesy of Cellulosefabrik Attisholz.)

themselves. The vessels form a long, continuous tube system which serves as a water conduit; the tubes have a wide lumen and are composed of single pipe segments. These tube systems run parallel to the tree axis, in a vertical direction. The cross-section, the form, and the arrangement of the vessels often make it possible to identify the wood species. Two different types of arrangement of these vessels can be distinguished. Some woods develop vessels that do not differ in the diameter of their early- and latewoods but which, in the earlywood, form a more or less closed circular zone. These woods are therefore said to be ring-porous. Other woods, on the contrary, during the annual growth period form vessels of almost uniform dimensions which are distributed uniformly throughout the cross-

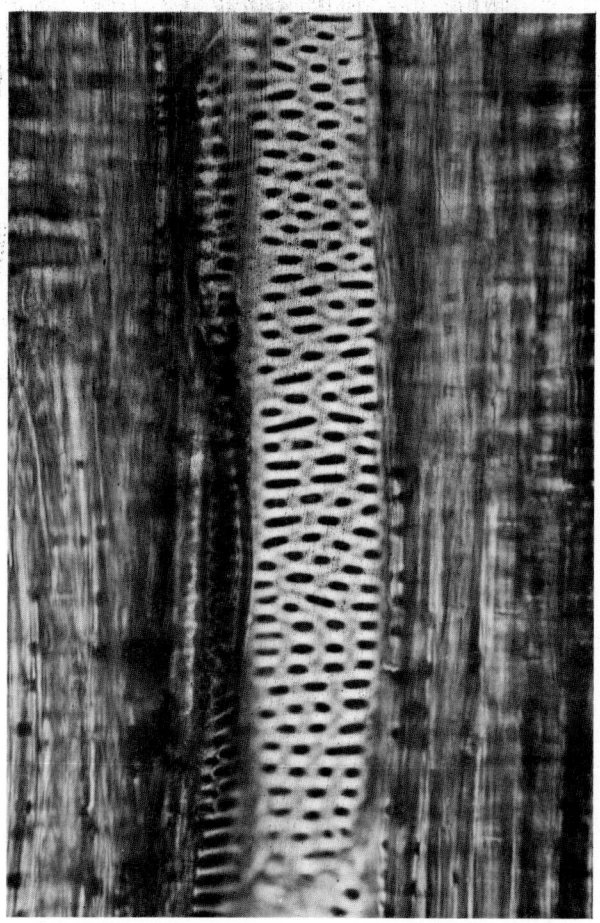

Figure II-6. Longitudinal section of beechwood. Magn. 230 : 1. (Courtesy of Cellulosefabrik Attisholz.)

section. Such woods are said to be diffuse-porous. Birch, maple, and poplar are such diffuse-porous woods in which the annual rings can often scarcely be distinguished.

In the sapwood of both types of deciduous woods the vessels are open. When the tree becomes older, the openings of the vessels in the heartwood of many deciduous wood species become closed through the growth of small cells, known as tyloses. Some deciduous woods close the openings with gumlike substances.

The wood fibers, also called sclerenchyma or libriform fibers, are the chief source of strength for the trunk of deciduous trees. Their fibers are

considerably shorter and thinner-walled than those of conifers. Their cross-section is usually circular or oval in shape and seasonal differences in the thickness of their cell walls are not found, as they are in coniferous woods. Figure II–5 shows a cross-section of beechwood. The vessels form wide pores between the fibers. In the radial section, the wood fibers surrounding the vessels can be seen. Figure II–6 shows the numerous pits in the walls of the vessels. The parenchymatous medullary-ray cells are much shorter than the prosenchymatous wood cells; they mostly have square ends. Some wood species, such as, for instance, poplar, have ribbonlike medullary rays that consist of a single row of cells and are therefore difficult to recognize. Other woods, such as North American oaks, have medullary rays which are 35 or more cells in width.

In addition to the sclerenchyma fibers and the vessels, there are intermediate forms present which are characterized as tracheids, fiber tracheids, or vessel tracheids. However, on the basis of the volume of beechwood, for example, they amount to only about 0.4% to 1.2%. The slitlike pores of the pits usually form an angle of about 45° with the longitudinal axis of the tracheids. In birchwood, these tracheids are about 1.0 to 1.4 mm in length.

The anatomical structure of deciduous woods consists, therefore, of a sequence of concentrically arranged layers of prosenchyma, which are chiefly vertical in position and into which radially arranged groups of parenchymatous medullary rays are often embedded. With the exception of the oaks, these parenchymatous tissues amount to between 5% and 20%,

TABLE II-4

APPROXIMATE AVERAGE LENGTH OF FIBERS OF BROAD-LEAF
SPECIES IN MATURE TREES (75)

Species	Average fiber length (mm)
White ash (*Fraxinus americana*)	1.20
Aspen (*Populus tremuloides*)	1.25
Aspen (*Populus grandidentata*)	1.25
Cottonwood (*Populus deltoides*)	1.30
Basswood (*Tilia americana*)	1.20
Paper birch (*Betula papyrifera*)	1.20
Yellow birch (*Betula lutea*)	1.50
Beech (*Fagus grandifolia*)	1.20
Yellow poplar (*Liriodendron tulipifera*)	1.80
Black gum (*Nyassa sylvatica*)	1.70
Red gum (*Liquidamber styraciflua*)	1.60
Elm (*Ulmus americana*)	1.50
Maple (*Acer saccharum*)	1.00
Sycamore (*Platanus occidentalis*)	1.70

or an average of about 14%, of the total wood volume. In the case of the oaks, this proportion increases to 20% to 30% (7,9,10,46,57,58).

In the technical utilization of the fibers, the length of the sclerenchyma fibers plays a decisive role. Table II–4 shows the average fiber lengths of a number of deciduous woods. The vessel cells, depending upon the extent which to they are present, have an important effect upon the development of the physical properties of a paper sheet made from pulp from deciduous woods. The vessel cells undergo beating much more readily than the sclerenchyma fibers, i.e., they are disintegrated into fragments and fines much more quickly. Unbeaten vessel cells, or those that have not been mechanically treated, usually possess strength properties lower than those of the fibers. With the vessel cells the strength properties rapidly pass through a maximum and then abruptly decrease on mechanical treatment, but, with the fibers, the strength of the paper sheet increases steadily when treated in the same way. In conclusion, it can therefore be said that the vessel cells of a pulp from deciduous wood contribute to the strength properties of a paper sheet so long as they are present in the same proportion as they occur in the wood. This will be discussed in greater detail in Section 6.

3. THE FINE STRUCTURE OF THE CELL WALL

3.1 The Formation of the Cell Wall and the Terminology of the Cell-Wall Layers

All cell membranes are formed by the appearance of cross-walls which divide the original cell into two. The synthesis of the new cell wall occurs through the plasma of the mother cell. The new cell wall, formed by plasma condensation, assumes the form of a convex lens and finally forms a semisolid layer, which is called a cell plate. By staining, it can be shown that this cell-wall plate already contains acid cell-wall substances, such as pectins. The new cell wall, which is still not connected with the outer cell walls, achieves this connection through further growth and combines with them, thus terminating the final division of the cell. Under the polarization microscope it can be seen at an early stage that the new cell wall consists of three layers, the middle layer of which is called the middle lamella; the other two are the primary layers or primary lamellas. Their double refraction is caused by small amounts of cellulose (38).

The newly formed wall represents, in general, the shortest connection between the opposite cell walls. There are also cell divisions, however, that originate from the insertion of a new wall in the longitudinal direction of the cell. The layers attached to the new cell walls are strongly hydrated

TABLE II-5

TERMINOLOGY OF CELL-WALL LAYERS (38,84)

Old terminology[a]	Layers in Fig. II-7	Layers in Fig. II-8	Ontogenetic (68)	Morphological (13,84)	New terminology (37)	
Middle lamella	O	M	Middle lamella	Middle lamella	Middle lamella	Middle layer
	I	P	Primary wall	Primary wall	Primary wall	Middle layer
Primary lamella	II₁	S₁	Secondary wall	Transition layer	Secondary wall	Outer layer
Secondary lamella	II₂	S₂	Secondary wall	Secondary wall	Secondary wall	Central layer
Tertiary lamella	II₃	S₃	Secondary wall	Tertiary wall	Secondary wall	Inner layer
	III	—	—	—	Tertiary or terminating lamella	—

[a] Dippel (1898), Ritter (1928) and Van Iterson (1927).

and contain only a little solid material of which about one-third is cellulose. The merismatic cell is now subjected to great changes in shape and size. While a meristem cell has a diameter averaging from 15 to 20 microns, fibrous cells can have a length of 500 microns and, with conifers and bast cells, this may even reach several millimeters, as can be seen in Tables II–3 and II–4. The question of how such an enlargement of the cell wall is possible has given rise to many extensive studies. Since no reduction in the thickness of the cell wall can be found, in spite of the expected severe stretching it undergoes during the growth of its surface, it must be assumed that new cell-wall material is simultaneously added to the cell-wall structure. This can occur anywhere on the cell wall, primarily at the ends of apically growing cells (42,106).

When the growth of the surface of the primary wall is completed, it is strengthened by the deposition of a thickening layer. This is called the secondary wall and is distinguished by a fibrillar structure; it may contain 90% or more of cellulose. When the secondary wall is swollen, or when wood tracheids are subjected to an enzymatic degradation, there remains a fine inner skin which, for a long time, was considered to be dried-out plasma residues but which has been shown to be a separate wall layer. It is called the tertiary wall and is not identical with the inner layer of the secondary wall. Since, in the older literature, the term tertiary lamella is used for this inner layer, the terminology of the cell-wall layers is given in Table II-5, based on the cross-section of a wood-fiber cell shown in Fig. II-7 and on the model of a coniferous tracheid shown in Fig. II-8 (13,37,38,68,84,87).

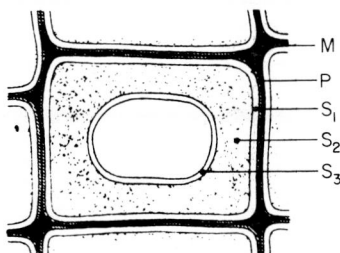

Figure II-7. Cross-section of a wood-fiber cell (68). M = middle lamella; P = primary wall; S_1 = outer layer of the secondary wall; S_2 = secondary wall; S_3 = inner layer of the secondary wall (also called tertiary wall).

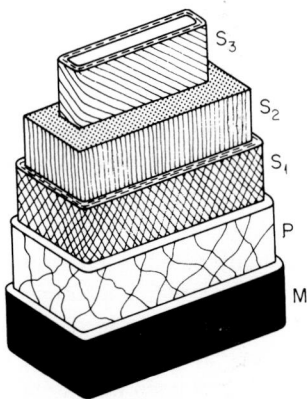

Figure II-8. Model of a conifer tracheid (87). M = middle lamella; P = primary wall; S_1 = transition layer (outer layer of the secondary wall); S_2 = secondary wall (central layer of the secondary wall); S_3 = tertiary wall (inner layer of the secondary wall).

3.2 The Middle Lamella and Primary Wall

Electron optical investigations with ultrathin sections of tracheids clearly show the middle layer, the layers of the secondary wall, and the tertiary wall of untreated wood. After partial delignification, the primary wall also becomes visible. Whereas the middle lamella appears to be homogeneous under the electron microscope, the primary wall shows a system of interwoven submicroscopic cellulose fibrils. This interweaving of the microfibrils prevents their formation on the surface of the middle layer; instead, a whole surface layer of cytoplasm, considerably surpassing in thickness that of the microfibrils, must take part in the formation of the fibrils. Since only one-third of the primary wall consists of cellulose, based on the dry weight, only about 2.5% cellulose is present in the fresh primary wall. These cellulose-containing microfibrils are loosely embedded in the

basic substance of the cell wall and therefore are not combined with each other at their points of contact, as might be concluded from the interweaving shown in the electron microscopic picture. During the growing of the primary wall, the structural substance itself, not its network, must become plasticized (38,90).

After a delignification for several hours with sodium chlorite, clearly visible changes in the middle layer are shown. It becomes wider and the primary wall partially separates from its cell wall. Thus connective microfibrils and contact points between neighboring primary walls are found. Through hydrolysis of the polysaccharide portion with sulfuric phosphoric acid and with hydrochloric acid, lignin frameworks showing the single wall layers beside each other can be obtained. The structure of the lignin framework confirms the long-known fact that the lignin is present in its most compact form in the middle layer, while its concentration is considerably less in the cell wall but increases toward the outer layers and those adjacent to the lumen (61,63,90).

Microanalyses of the middle wall of Douglas fir, for example, gave a lignin content of 72% and a pentosan content of 14%. The residual portion consisted of hemicelluloses of the glucuronic acid type (3). Through measurements of the ultraviolet absorption spectrum of the native lignin in sprucewood, it was found that the lignin content of the middle layer amounts to 60% to 90%. This great variation not only shows the inaccuracy of microoptical methods for such quantitative determinations but also shows the wide range in the chemical composition of the middle layer. No cellulose seems to be present in this layer. The lignin is obviously embedded isotropically, but a slight degree of orientation cannot be completely ruled out. Ultrathin sections through a middle layer that had been treated enzymatically and impregnated with osmium gave no indication of any orientation. If it is assumed that no structural changes occur as a result of the enzymatic treatment, the structure of the middle layer must be considered to be amorphous. Most of the investigations of the structure of the middle layer have been carried out with coniferous woods. It may be assumed, however, that there are no basic differences in the general structure of this middle layer in coniferous and deciduous woods (2,72).

The primary wall is the outermost layer of the single cell. There are basic differences between the primary and secondary walls as regards their formation and growth. The primary wall shows a pronounced growth in its surface and only a minor growth in thickness, whereas the secondary wall shows a growth in thickness but not in surface area. These different patterns of growth affect the microfibrils in the two layers. From observations made under the polarization microscope, it can be concluded that the fibrils in the primary wall are transversally orientated; however, they are in no way arranged parallel but exhibit instead the structure of a network.

Electron optical investigations show that the fibrils in this network are arranged more in a parallel direction to the fiber axis in the outer layer of the primary wall, while those in the layers toward the secondary wall are oriented more transversally (100).

The primary wall shows only a slight capacity to swell; even solvents for cellulose cause only a slow dissolution of the cell wall. There are also differences in the enzymatic degradation of the primary and secondary walls. The former shows a considerably greater resistance, and this justifies the assumption that the cellulose in it is not identical with that in the secondary wall. It is possible that the microfibrils of the primary wall contain some hemicellulose units. It is also possible that the course of the growth of the microfibrils plays a role here, since the microfibrils of the primary wall have probably grown under the effect of stretching forces (84).

3.3 The Secondary Wall

As mentioned above, a thickening layer is deposited on the primary wall after its surface growth has been completed, and this layer is called the secondary wall. It consists of three layers which, according to the terminology given above, are termed the outer layer (S_1), the middle layer (S_2), and the inner layer (S_3). The morphological analysis considers the outer layer, S_1, and the inner layer, S_3, not as constituents of the secondary wall but as morphological entities. Accordingly, these layers are termed the transition lamella (S_1) and the tertiary wall (S_3), as shown in Table II–5. The growth in thickness of the secondary wall is appreciable and the wall thickness can reach 1 to 10 microns. Its cellulose content may amount to 90% or more. The result is that the microfibrils are no longer arranged in a loose distribution, as in the primary wall, but are relatively densely packed in a parallel arrangement and are in contact with each other. The parallel texture, formed by the parallel arrangement of the submicroscopic fibrils, gives special mechanical and optical properties to the secondary cell wall. Stretched-out fiber cells can be dissected by lengthwise cission of the fibrils into microscopic fibrils. The principal orientation of the microfibrils in the outer layer has been found to be flat and spiral.

Recent investigations have shown that a fibril arrangement exists in which the layers run in opposite directions. The angle of this spiral-like arrangement varies. The slope is characterized by its angle, i.e., by the angle formed by the tangent of the helical line and the direction of the screw axis. With regard to the direction of the rotation of the spiral, statements that are often confusing have been made. In the more recent literature, it has been agreed that the terms S-screw motion (left-screwed) and Z-screw motion (right-screwed) should be used. Figure II–9 clearly illustrates the meaning of this terminology (34,35,84).

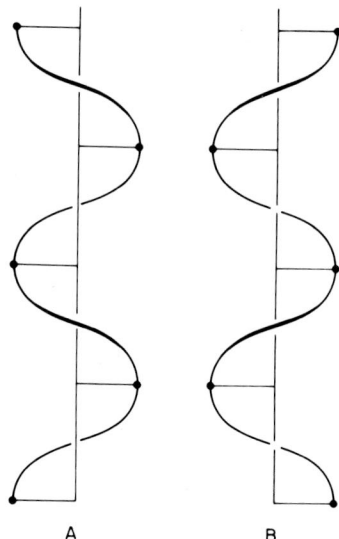

Figure II-9. A = S-screw motion (left-screwed); B = Z-screw motion (right-screwed).

In the tracheids of pinewood, two systems of parallel fibrils can be found in the S_1 layer; these systems run spirally at a wide angle to the screw axis, in opposite directions, and symmetrically to the axial direction at an angle of about 60° (35). It is doubtful, however, that these two systems are actually only in the S_1 layer, i.e., in the outer layer of the secondary wall. From an investigation of immature xylem elements, it is very difficult to decide whether, in truth, only a single cell layer is present— in fact, it is difficult to say, with certainty, whether only the S_1 or both the S_1 and S_2 layers are being seen. It may be supposed that only one single system of screw-shaped fibrils, arranged in a parallel direction, are present in the outer layer (84,134).

The outer layer of the secondary wall is more highly lignified and, in this respect, more closely resembles the primary wall. The microfibrils of the outer layer are also more resistant to attack by fungi than those of the middle layer (S_2), which represents the principal constituent of the cell wall. In the middle layer the cellulose fibrils are tightly packed in a parallel, spiral arrangement. The angle of the slope in the tracheids of some deciduous woods amounts to only a few degrees; in coniferous woods, as for example in the tracheids of pinewood, variations between 0° and 45° have been observed. These variations are the result of a number of internal and external influences which are difficult to identify. The variations in the slope angle are overlapped by orderly variations in the size of the slope angle within and between the annual rings. The variations therefore play

a more important role in the latewood than in the earlywood. Within an annual ring, the slope angle decreases from the beginning to the end of the latewood in a radial direction. In successive annual rings, the slope angle decreases continuously from the heart toward the bark until a state is reached at which the angle remains constant or is subjected to only minor changes. This constant state varies with the rate of growth of the tree. A similar observation can be made with regard to the stem, from the crown to the base, when the increment sheath of a certain calendar year is tested in the center of consecutive annual height increments. The slope angles of tracheids in compression wood are always as large as or larger than those in normal summerwood at the corresponding position within the stem. They, too, follow the general trend, but the decrease in the size of the slope angle is not as pronounced as that in normal summerwood (53).

From these investigations, a direct relationship between the thickness of the cell wall and the slope angle has been found. If it is assumed that the middle layers of the secondary wall actually show a uniform slope angle of the fibrils, then the close relationship between the wall thickness and the slope angle must be traced to a third factor. Since the variations in the thickness are greatest in the middle layer, it may be assumed that a thin cell wall has a relatively thin S_2 layer with a uniformly wide slope angle, and a thick cell wall has a thick S_2 layer with a uniformly narrow slope angle. As compared with the S_2 layer, the primary wall and the S_1 and S_3 layers are relatively thin. Each of these layers shows a different structure. The primary wall resembles a network, the outer (S_1) and the inner (S_2) layers of the secondary wall show spiral structures in the longitudinal direction of the parallel-arranged fibrils. The thinner the middle layer, the greater the influence of the other layers. Thus far, however, it cannot be proven with certainty whether the relationship of the wall layers shifts with cells of varying wall thickness (54).

3.4 The Tertiary Lamella (Tertiary Wall)

As can be seen from Table II-5, different concepts exist with regard to the designation of the S_3 layer shown in Fig. II-8. This layer can be set free by the swelling and dissolution of the secondary wall. The swelling agent has a macerating effect on this layer and makes its structure visible. The structural orientation can be found from the slope of the helixes. The slope angles are, for example, 30° and 65° in *Picea excelsa*, 65° in *Picea canadensis*, 70° in *Pseudotsuga taxifolia*, 80° in *Abies pectinata*, 80° in *Abies balsamea*, 84° in *Pinus silvestris*, and 89° in *Larix europea*. In sprucewood, the S_3 layer shows a fibrillar structure and a slope angle similar to those of the secondary wall; in other wood species, the slope is flatter than in the secondary wall. The structure of the S_3 layer corresponds to an S-screw

motion, whereas the secondary wall shows a Z-screw motion (11,12,13). From this it may be concluded that the S_3 layer, in fact, possesses specific properties. The S_3 layer differs from the secondary wall also in its increased resistance toward enzymatic degradation. Because xylan is an essential component of this layer and the hemicelluloses are more resistant to alkaline decomposition than to an acid cook, greater resistivity is found in the S_3 layer in the kraft pulping process and a lower resistivity in the sulfite pulping process (84,86,103).

The S_3 layer in *Pinus silvestris* and *Picea abies* is usually 700 to 800 A. thick; in the deciduous woods, however, it is considerably thinner. It is composed of microfibrils of cellulosic material and an amorphous matrix. The thickness of the microfibrils is between 120 and 180 A, and they usually form flat-screwed systems which cross each other at slope angles of between 20° and 30°. There are also layers, however, the microfibrils of which form angles of up to 20° with the longitudinal axis of the cell. In the deciduous woods the angles vary; in *Betula verrucosa* they range between 60° and 80°, in *Fagus silvatica* between 50° and 70°, and in *Populus nigra* between 70° and 80° (78).

There is not necessarily any reason to depart from the ontogenetic terminology of the cell-wall layers, as given in Table II–5. According to this, the S_3 layer is a constituent of the secondary wall. Residues of cytoplasm can accumulate after the cell has died and form an incrusting membrane on the inner surface. With the exception of reaction wood (see Section 4), the three-layered system of the secondary wall always observed can be traced to the heterogeneous, screwlike orientation of the cellulose microfibrils in the three layers, and there is therefore no reason to speak of a tertiary wall, a tertiary lamella, or a transition lamella (135). Consequently, a differentiation should be made between the possible presence of an inner layer of the secondary wall and the closing inner skin of the cell walls, which, according to its historic development, merits a special name and for which the term "closing lamella" (Abschluss lamella) or "formative skin" has been suggested (38).

The inner surface of this closing lamella is often covered with submicroscopic warts. Such warts are to be found not only in the tracheids of softwoods, especially in pine and spruce tracheids, but also in hardwoods, as in birch and beech. These wart layers are also found in the pit canals, pit coronas, and in the pit membranes, as is shown, for example, in Fig. II–10. The number of warts per area unit differs. Their diameter varies between 0.01 and 1 micron, and is usually between 0.1 and 0.2 micron. The warts are also found in reaction wood. In some species the warts seem to be on the S_3 layer but are sometimes covered by a membrane. They are fairly resistant toward hydrolytic decomposition and toward oxidation reagents (29,78,121).

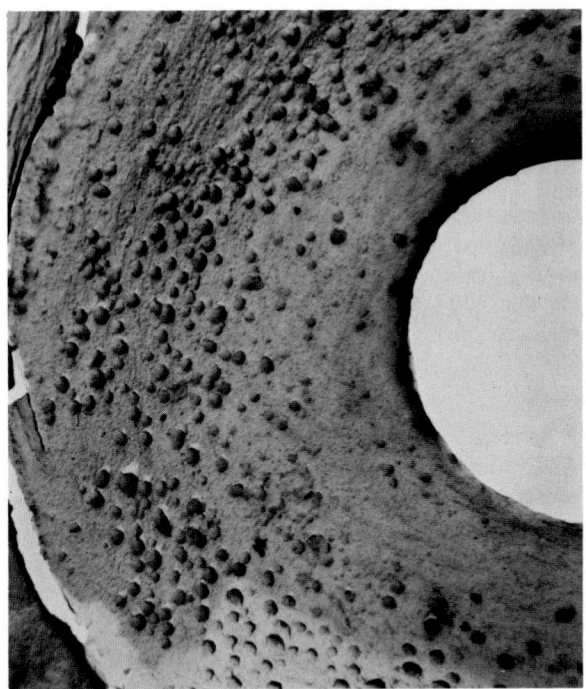

Figure II-10. Cutting of a pine pit with wart structure. Carbon film replica. Magn. 12000 : 1 (121).

On the inner surface of the lignified cell walls are found fibrous structures which are composed of irregularly distributed wider and narrower sections. It is assumed that these represent the beginning of the formation of cellulose microfibrils, or that they are bundles of such microfibrils formed through the polymerization of simple sugars; the latter are present in the protoplasm in a very dilute state. The wider structures would then have to be considered to be accumulations of hemicelluloses or low-molecular crystalline cellulose. The presence of undissolved fragments of fibrils does not seem to be required for the formation of such structures. The parallel orientation of these structures is explained as resulting from a parallel flow of protoplasm in the living cell (92).

3.5 The Pits

While the primary wall in the cell plate forms a homogeneous network of microfibrils, its texture soon begins to change. At the corners of the cells, microfibrils, running parallel to each other, are formed into reinforcing strips. Simultaneously, the formation of pits is begun. The

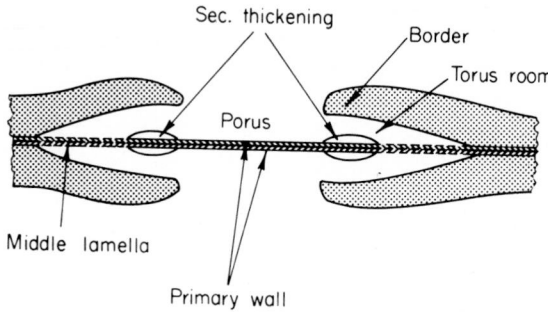

Figure II-11. Cross-section of the torus in the sapwood of spruce, fir and pine (121).

originally homogeneous texture is then obviously broken by increased local growths of plasma and the microfibrils already formed are pushed apart and are supplemented by fibers that run circularly. The cytoplasm of the growing cell tries to come into contact with that of the neighboring cells. Normally, the pit membrane of a single pit is formed from such a primary pit compartment. The microfibrils which run from the surrounding fibril structure into the pit compartment are combined to coarser fibril strands and form pores in the primary wall. Through these small canals, which are called "plasmodesma," a connection is formed between the cytoplasm and the contents of the neighboring cell (38,42).

The variations in the simple pits involve the longitudinal walls of the extended cells. On the transverse walls, no openings are visible. The pit membrane in the large window pits of the medullary-ray parenchyma of pinewood shows the regular network of the primary wall with a scattered texture. The thickening of the cell wall, i.e., the formation of the second-ary wall, does not occur in the bordered pits and, as a result, pit pores (simple pits) are formed at both sides of the middle lamella. In this way, a connecting pit canal is formed between neighboring cells, and this canal is interrupted only by the pit membrane. The structure of this membrane is an irregular network of microfibrils embedded in an amorphous material. The membrane between a longitudinal tracheid and the ray parenchyma consists of an intercellular layer and two primary walls. The membrane is not perforated and has no torus. In contrast to the bordered pits, which will be described later, the formation of a border occurs only on the longi-tudinal side of the pits, which may therefore be described as half-bordered pits. Incrusting material is found between the microfibril structure of the membrane and in the cell wall of the ray parenchyma of the sapwood of four different pine species. On the other hand, the membranes of pits in the heartwood are heavily incrusted and these incrustants cannot be re-moved even with an acid sodium chlorite solution (22,39,128).

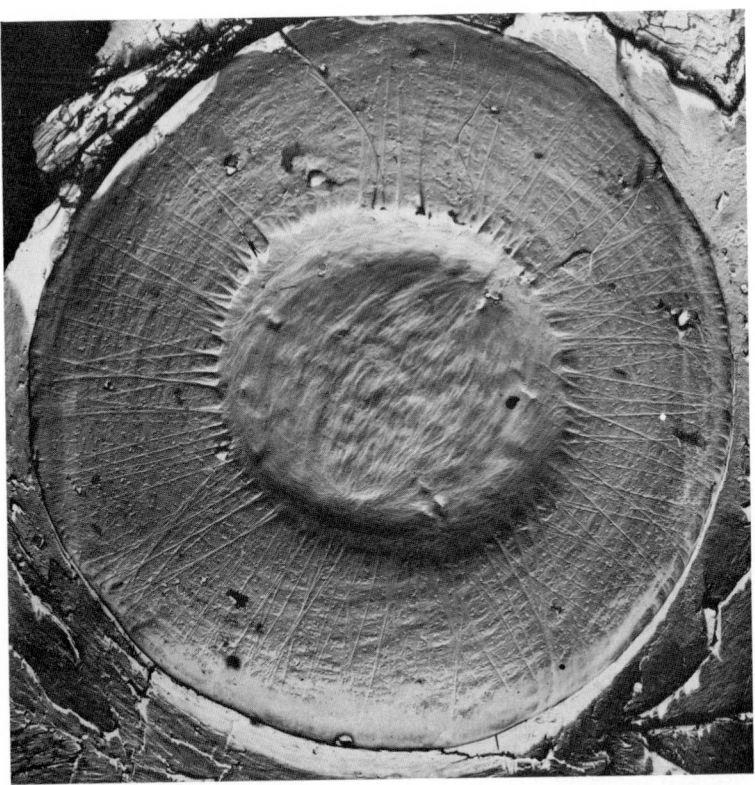

Figure II-12. Electron micrograph of a typical coniferous bordered pit showing the torus, supporting membrane and back dome. Carbon film replica, chromium pre-shadowed. Magn. 4920 : 1 (27).

These simple pits are in contrast to the so-called bordered pits, which are found especially in the tracheids of the conifers and, to a lesser degree, in the tracheidal elements of hardwoods. A schematic cross-section of such a bordered pit is shown in Fig. II–11. The secondary wall is greatly thickened on both sides of the pit membrane and surrounds the opening of the pits, the porus, like a dome. The thickened parts surround the dome-shaped pit chamber which is divided into two halves by a centrally arranged pit membrane from the middle lamella and the primary walls. The primary walls are thickened in the middle of the pit to form the torus. The main function of the pits is the regulation of the flow of the water, with the pits acting primarily as a sort of valve by means of which they can close the porus through pressure from one side.

Since the electron microscope has made possible a far-reaching revelation of the submicroscopic structure, a great many investigations have been

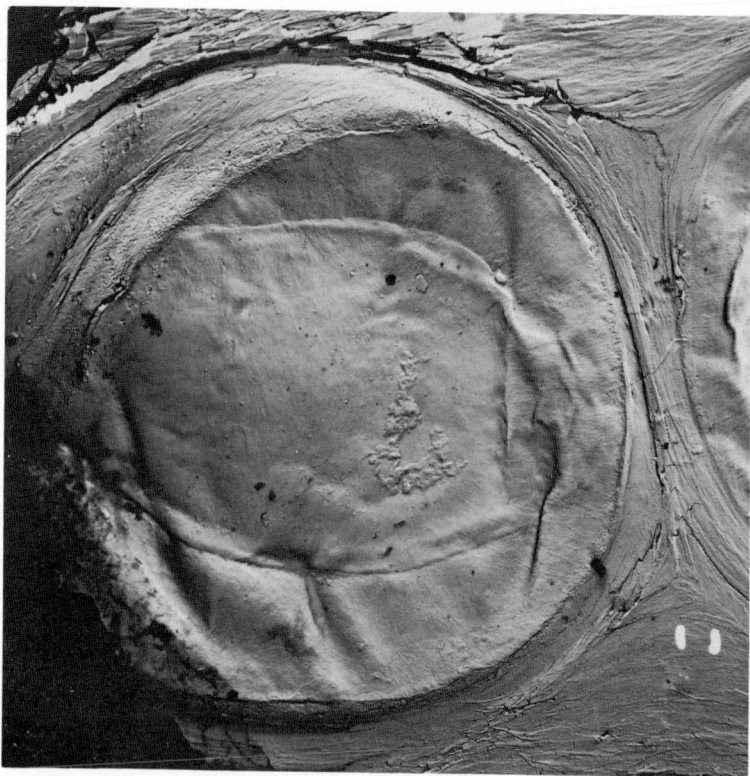

Figure II-13. Electron micrograph of the pit membrane surface in a half-bordered pit pair in soft pine. Microfibrillar structure and an impression of the overhanging border can be seen. Carbon film replica, chromium preshadowed. Magn. 4200 : 1 (27).

carried out into the fine structure of the membranes of the bordered pits. Figure II–12 shows a typical bordered pit of conifers, with the torus in the middle and the adjacent membrane. The thickened border, which forms the dome and surrounds the porus, is clearly visible. Figure II–13, on the other hand, shows the surface of the membrane of a half-bordered pit of pinewood; the microfibrillar structure and the matrix of the projecting thickened border are clearly seen. As described above, the microfibrils of the bordered pits are combined into strands by plasma. The torus shows circularly running microfibrils and is connected by strands with the surrounding primary wall (see Fig. II–14) (26,27,39).

The permeable border surrounding the torus is called the "margo." It is radially striped and its permeability is of importance not only for the exchange of the sap but also for the technical processes of impregnation and pulping so far as these processes do not take place by diffusion through

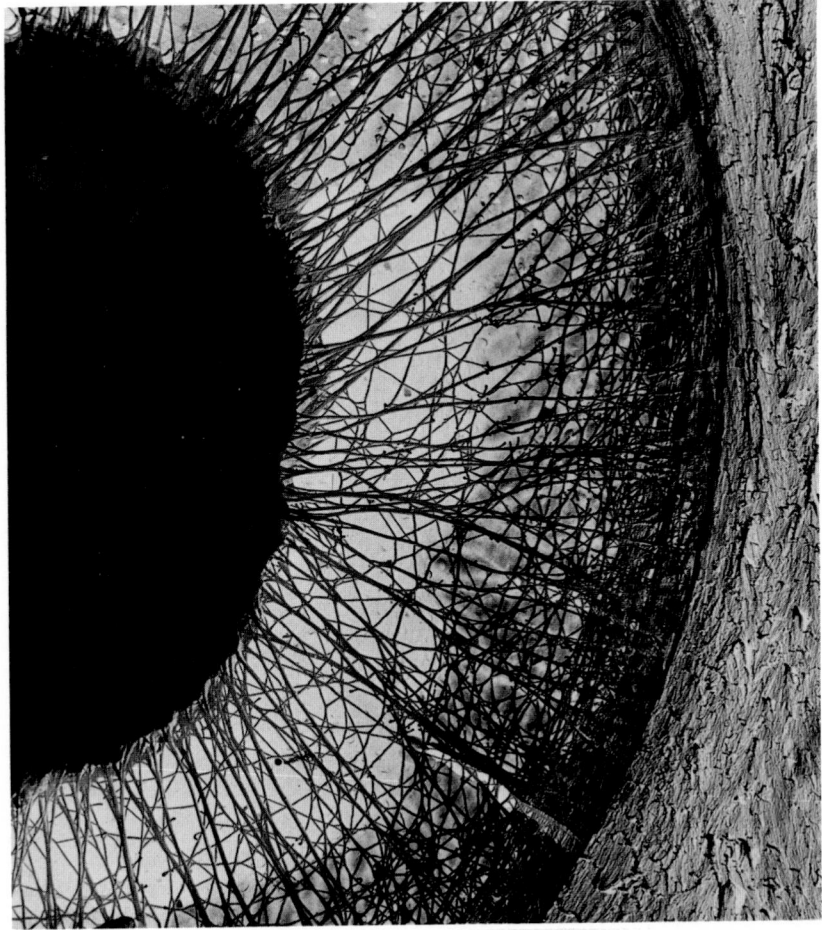

Figure II-14. Loblolly pine bordered pit membrane. Solvent-exchange dried from pentane. Magn. 10625 : 1 (26).

the cell walls. The porosity of the margo is appreciable, since even titanium dioxide particles, 150 millimicrons in diameter, can pass through (80). From the dimensions given for the pores between the radially running strands, it may be concluded that it is not, strictly speaking, a membrane but rather a porous network that is being dealt with here (79). It must be taken into consideration, however, that as regards size, these pores are within the submicroscopic range, whereas the vessels in the deciduous woods attain microscopic, and sometimes even macroscopic, dimensions; this explains the tenfold increase in velocity of the rise of the water in deciduous woods as compared with coniferous woods.

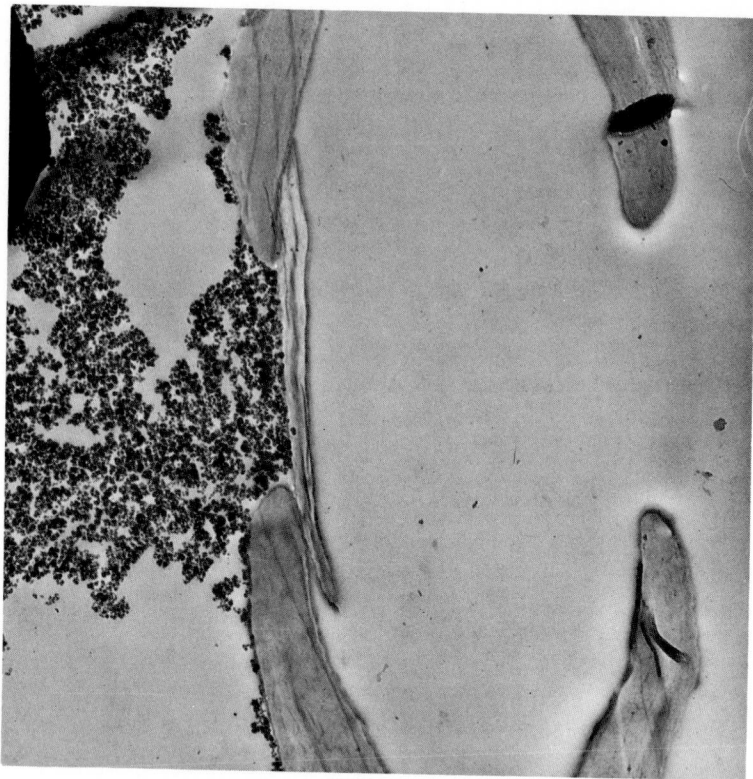

Figure II-15. Electron micrograph of an ultrathin tangential section of a closed pit of *Pinus strobus*. Magn. 2260 : 1 (27).

Bordered pits, circular in shape, are found chiefly in tracheids with a wide lumen; in those with a narrow lumen, the pits acquire an elongated shape and the pori are slit-shaped. With regard to their functional importance, it has already been mentioned that the opening of the pit can be closed by placing the torus over one of the pit openings. This closing of the pit tori is above all a reversible process. During the slow conversion to heartwood in the older layers of the stem, the bordered pits are definitely closed and the tracheids are relieved of their functions. The bordered pits of the latewood are always open, whereas those in the earlywood in the heart are closed; in the sapwood of the living tree they are open. As the wood dries, the pits of the earlywood (sapwood) also close. During the impregnation of air-dry wood with oily wood preservatives, their penetration is limited almost entirely to the latewood, with very slight penetration into the earlywood and none through the medullary rays (76).

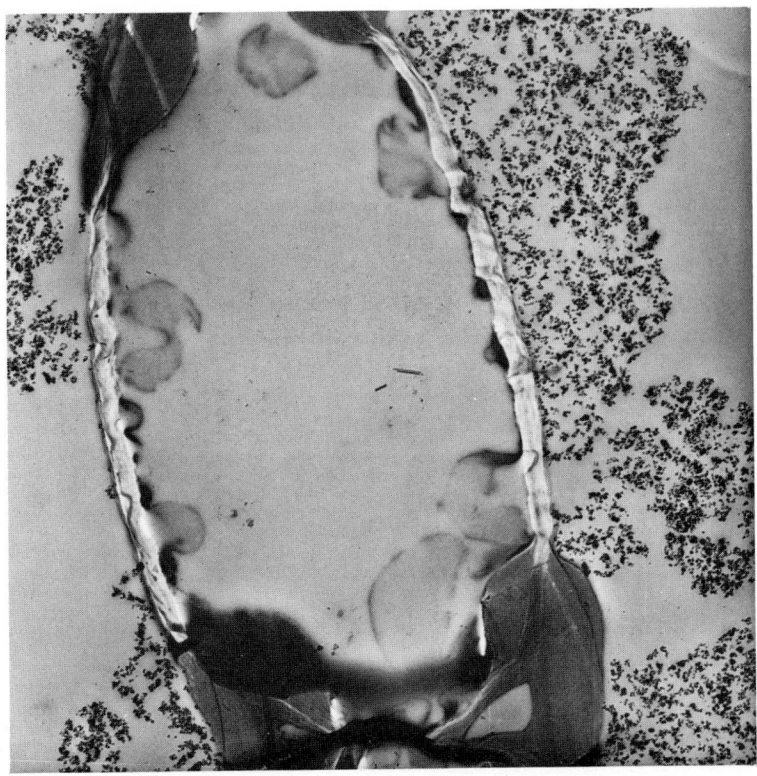

Figure II-16. Electron micrograph of an ultrathin tangential section showing two windowlike pit membranes of a ray parenchyma cell. Magn. 5400 : 1 (27).

The structure of the membrane in the bordered pits is not the same in all wood species. Thus, for example, Douglas fir (*Pseudotsuga taxifolia*) has a structure typical for most conifers, whereas the membrane of hemlock (*Tsuga canadensis*) has torus extensions other than the microfibrillar strands that run radially from the torus. In Western red cedar (*Thuja plicata*), the membrane does not possess a torus but consists exclusively of numerous, very densely packed strands. The distances between these strands, which to some extent represent the pores, amount to about one-tenth of those in Douglas fir. For this reason, the closing action of the bordered pits in the heartwood occurs differently. In Douglas fir, the torus is deposited over the pit aperture by aspiration and closes it hermetically; in hemlock, both aspiration and incrustation cause the closure of the pit. The wartlike projections at the borders of the pits can hinder a complete closure but the small openings that might have existed are closed by incrustation. In Western red cedar, the extensive incrustation on the pit membrane renders

it impermeable. Extractives from the heartwood may also contribute to the cementing of the membrane (71).

During the aspiration of the torus, a compression of the fibrillar network of the membrane must be assumed. In this way, a thickened pad of microfibrils is formed at the edge and they assume a more pronounced circular orientation. This padded edge effects the sealing (60,64).

In order to make visible the nature and characteristics of the bordered and half-bordered pits, they have been impregnated with various finely dispersed suspensions. The use of titanium dioxide and India ink for this purpose has already been mentioned. Figure II–15 shows an electron microscopic picture of an ultrathin tangential section of a bordered pit of *Pinus strobus*. Small blocks of this wood were treated with India ink in the fiber direction, with the application of a slight vacuum at one end of the block. From this impregnated woodblock, ultrathin tangential cuts were made. As Fig. II–15 shows, the torus is firmly pressed against the pit opening in such a way that the particles that had penetrated the neighboring cell could penetrate neither the pit dome nor the adjacent empty cell.

In contrast, Fig. II–16 shows an electronmicrograph of an ultrathin tangential section through two windowlike pit membranes of a ray parenchyma cell. The lumen of the ray cell does not contain any India ink particles, while the adjacent longitudinal tracheids show a clear accumulation of the ink (27).

When the tertiary lamella dies after the growth of the pit is complete, in many cases a characteristic layer of warts is formed, especially on the inner side of the padded edge. In pinewood species, the number of warts in each area unit and their size have been found to be a specific phenomenon so that it may be possible to apply the submicroscopic structure of the warts for the identification of those wood species (40,41,77,79).

The wart structure is formed in the concluding stages of the cell differentiation when lignification is complete or nearly so. The warts can be separated from the other cell layers and consist, for instance in *Pinus strobus*, of a membrane surrounding the lumen, with numerous pockets extending into the lumen. In these pockets there are small round particles, 0.1 to 0.5 microns in diameter. The structure of the warts differs from that of the secondary wall and the tertiary lamella (23,135).

4. TISSUES WITH SPECIAL FUNCTIONS; REACTION WOOD

When trees or branches are forced to grow in an abnormal position because of external influences and growth conditions, they try to compensate for the compression and tensile stress thus caused by eccentric growth

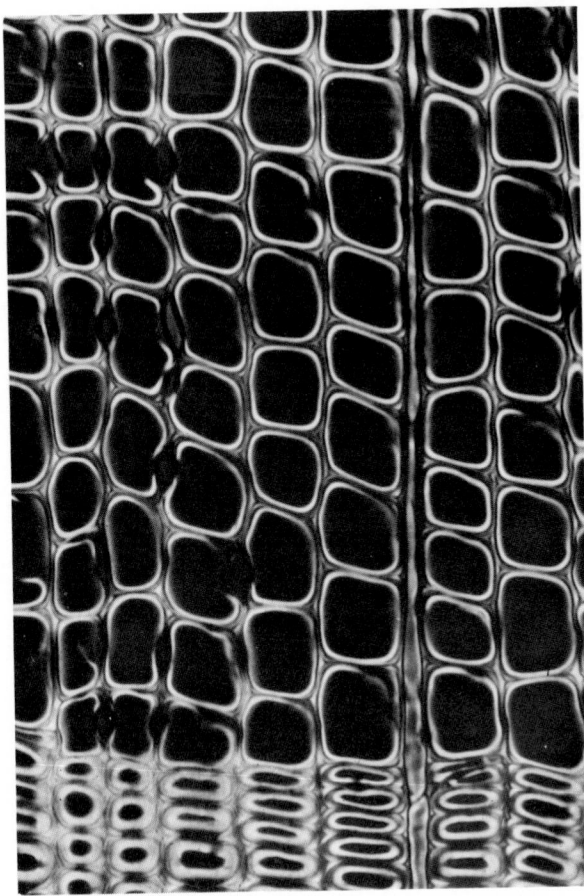

Figure II-17. Transverse section of compression wood of *Picea rubens*. The early spring-wood cells lack the rounded outline and thick cell wall typical of compression wood tracheids. Note the flat shape of summerwood tracheids. Light micrograph, phase contrast. Magn. 440 : 1 (24).

in thickness. With coniferous woods, the underside is thickened; with deciduous woods, on the other hand, the upperside is affected. In the tree trunk the wood with broadened annual rings is called compression wood in the case of coniferous woods, and tension wood in the case of deciduous woods. Compression wood is usually found in trees grown in an inclined position or in branches on the side of greater growth. It is interesting to note that compression wood can also be found on the upper side of branches of coniferous woods when they are forced by external conditions to grow upward at an abnormal angle. The formation of tension wood in deciduous trees occurs in the same way. Reaction wood, therefore, has the

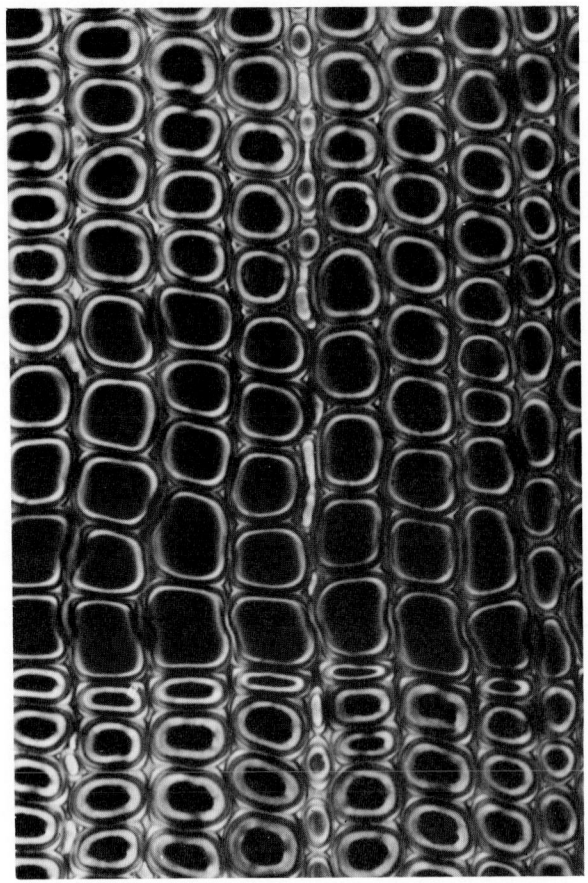

Figure II-18. Transverse section of compression wood of *Picea abies*. The first four or five rows of cells consist of typical compression wood tracheids. Magn. 440 : 1 (24).

function of returning to their normal position the parts of the tree which have been forced from their natural state of growth. Because this growth in thickness is a reaction against abnormal behavior brought about by external influences, the term "reaction wood" has been introduced for this type of wood (56).

In thin layers, the compression wood of coniferous trees appears to be opaque as compared with normal wood. The differences between early-wood and latewood are blotted out because their cells show uniformly thickened walls. The increased lignification causes a darker, reddish coloration of the compression wood so that it is sometimes designated "red wood." The tracheids of compression wood in conifers show rounded-off

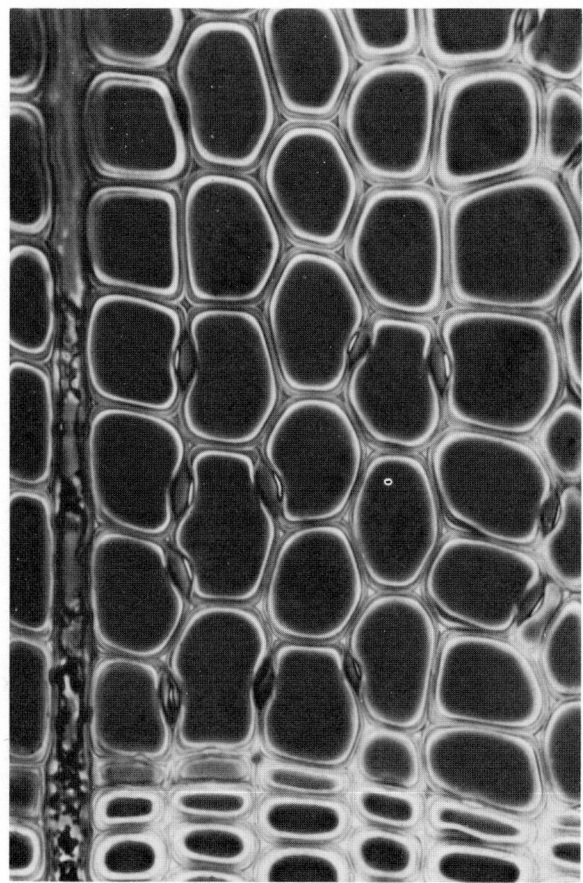

Figure II-19. Transverse section of compression wood of *Larix laricina*. The spring-wood tracheids are irregular in shape and thin-walled. Magn. 440 : 1 (24).

shapes in the cross-section, whereas those of normal wood have a more or less square-shaped form. Intercellular spaces are formed in the tissue of the xylem. The cell walls are greatly thickened, and the secondary wall has spirally arranged fissures. No tertiary lamella, as a seal against the lumen, can be observed (83,136).

The fissures in the secondary wall result from an irregular deposit of cellulose lamellas during the growth in thickness of the cell wall. The primary wall is wider than in normal wood and is usually more lignified. The spiral structure in the secondary wall is steeper than in normal wood and often attains an angle of 45°; this, however, is dependent upon the length of the fiber. The fibers in the compression wood are generally shorter than in the normal wood (16).

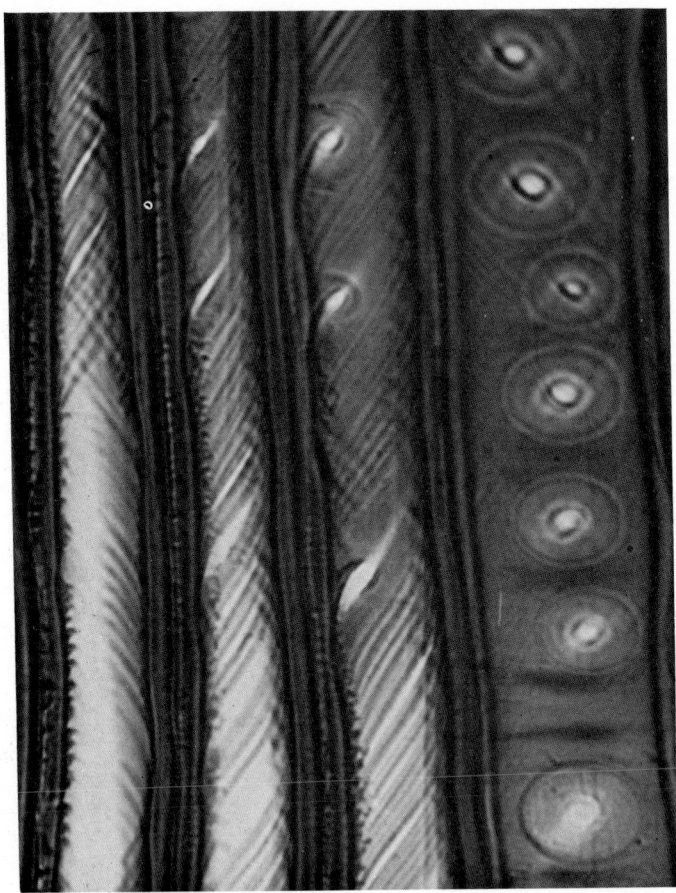

Figure II-20. Radial section of compression wood of *Larix laricina*. Helical striations are present in summerwood but not in the first-formed springwood tracheids. Magn. 1300 : 1 (24).

Figure II-17 shows a transverse section of compression wood of *Picea rubens*. In the lower part of the light-microscopic picture, the flattened and thick-walled cells of the compression wood, which are formed toward the very end of the growth season, are clearly visible. The cells of the early-wood have a square-shaped cross-section which is characteristic for the tracheids of normal wood. Figure II-18 shows a similar cross-section of *Picea abies*. In this case, only the first four or five rows of cells have a square-shaped cross-section, while the succeeding cells have a round-shaped one with the walls becoming thicker and the cavities between the cells be-coming wider. A transverse section of *Larix laricina*, illustrated in Fig. II-19, shows the more irregular outline of the tracheids of the earlywood, with

all of them being thin-walled. In a radial section, helical striations are shown in the tracheids of the latewood, but these are absent from the tracheids of the earlywood. This is clearly shown in Fig. II–20 (24).

Because of its harmful mechanical properties, compression wood is an undesirable component of coniferous woods. Its specific gravity is 20% to 40% greater than that of normal wood because of the greatly increased thickness of its cell walls. Based on its specific weight, all its strength properties, especially its tensile strength, are considerably lower than those of normal wood. Particularly unfavorable is its shrinkage in the longitudinal direction; this is normally 0.1% to 0.3% but in compression wood it may reach 4% or more. This increased shrinkage in the longitudinal direction has been thought to be related to the increased slope angle of the microfibrils in the secondary wall. More recent investigations have shown that there exists no linear relationship between the angle of the microfibrils and the shrinkage in either the longitudinal or the tangential direction. With an angle of 25° for the microfibrils, minimum values are found for the longitudinal shrinkage, which can even be negative. With greater angles, the shrinkage increases rapidly. If one visualizes the cell walls as being an amorphous matrix in which the crystalline microfibrils are embedded and are arranged at an angle to the cell axis, a swelling or shrinkage of the matrix can be assumed to occur through the absorption or emission of water. Such swelling would take place isotropically if the embedded microfibrils did not cause a deformation of the matrix. The anisotropic shrinkage of the wood must be attributed primarily to the microfibrils and less to the total anatomy of the wood (4,48).

That the chemical composition of reaction wood differs considerably from that of normal wood has already been mentioned. Thus, of considerable significance is the greater lignification, which is already apparent through its coloration. In the compression wood of *Picea excelsa*, a large quantity of 10% galactan has been found by hydrolysis experiments (122). Whereas the noncellulosic polysaccharides of normal wood contain little or no galactose, those of reaction wood have a relatively high galactose

TABLE II-6

LIGNIN CONTENT OF REACTION WOOD AND NORMAL WOOD (86)

Wood species	Lignin content in percent	
	Normal wood	Reaction wood
Picea abies	28.0	38.8
Pinus silvestris	29.0	35.5
Fagus silvatica	18.8	12.4
Populus tremula	19.2	13.8

content (110). Table II–6 shows the difference in lignin content of normal and reaction wood in coniferous and deciduous trees. It is apparent that the lignin content of the reaction wood is higher in conifers but considerably lower in deciduous woods than in the normal wood (85). It has not been definitely established whether or not there exist constitutional differences between the lignin of the compression wood and that of the normal wood. The nitrobenzene oxidation of lignin from compression wood from *Pinus radiata* gave good yields of *p*-hydroxybenzaldehyde, while that of normal wood and wood meal from compression wood gave none. The methoxyl-free aromatic rings which give aldehydes are therefore present in the lignin fraction of the compression wood. The infrared spectrum of the lignin from compression wood did not show the presence of carboxyl groups, whereas the presence of such groups has been shown in all isolated lignins. It is surmised that the methylation process is disturbed during the formation of compression wood lignin. The tension wood lignin, on the other hand, shows a normal behavior (5).

When compression wood from *Picea rubens* is delignified with chlorite solution, 80% of the total galactan can be extracted from the holocellulose. The structure of the galactan was elucidated by methylation and partial hydrolysis and it was found that it consists of at least 300 β-D-galactopyranose units which are combined by 1-4-linkages to a macromolecular structure. It is assumed that the polysaccharide in its isolated state was present in an amorphous form. Moreover, the compression wood contains water- and alkali-soluble galactoglucomannans, an acid arabinoxylan, and a product which, on hydrolysis, gives galactose, xylose, and glucose and corresponds in its composition to the amyloid (107).

From the holocellulose of the compression wood of *Abies balsamea*, galactan, galactoglucomannan, and glucomannan could be isolated. Considerable quantities of glucan, which is not identical with starch, have also been found in the compression wood. All the polysaccharides thus far isolated from normal woods are also present in compression wood, but usually in different amounts (28).

The composition of the primary-wall layer adjacent to the middle lamella is characterized by its high content of galactan, cellulose, and araban and is the same in normal and compression woods. The galactan with its β-D-1-4-linkages, which is characteristic of the compression wood, is present chiefly in the primary wall and the outer layer of the secondary wall, whereas the cellulose is mostly accumulated in the inner layer of the secondary wall (25).

Table II–7 gives the types and amounts of the sugars in the hydrolyzates of various conifers and the variations in them in normal and compression wood. From these data it may be concluded that compression wood almost always contains less cellulose than normal wood. Poly-

saccharides containing galactose are present to a much greater extent in compression wood than in normal wood. In the *Taxodiaceae* the xylan content is greater in normal wood than in compession wood, and in the *Pinaceae* the glucomannan content is greater in normal than in compression wood (85).

The tension wood of deciduous woods also has a higher specific weight than that of the normal wood. The porosity is reduced and this makes it appear more whitish than normal wood. This may be due to a lower lignin content. As in the compression wood of conifers, the shrinkage of the tension wood is greatly increased in the longitudinal direction. The crushing strength of tension wood is lower and the tensile strength greater than those of the normal wood. Tension wood is less suitable for mechanical processing. It is distinguished by its lignin-free cell-wall layer which is deposited on the middle layer of the secondary wall. This layer differs markedly from the secondary wall and the tertiary lamella by the arrangement of its fibril structure, which is almost parallel to the axis. There are also structures, however, in which this layer is superposed upon the outer layer of the secondary wall (S_1) or the tertiary lamella, i.e., the inner layer (S_3) of the secondary wall (17,133).

Drastic changes in the structure of the cell wall can be seen under the microscope. As compared with normal fibers, a thickening layer, which fills almost the whole cell lumen and which is free of lignin, is observed. The cells of tension wood, which are rich in cellulose, are not close to the vessels or the medullary rays. The difference in the cellulose content of the tension wood as compared with the normal wood of *Populus canadensis* is 8.27% for the former and 20.23% for the latter. For chemical utilization, therefore, the tension wood may be considered to be a more valuable starting material (62).

Table II–7 also shows the composition and the amount of the sugars in the hydrolyzates of normal and tension woods from various species of deciduous woods. Tension wood is shown always to contain more cellulose than normal wood does. This "excess cellulose" is present in the innermost, so-called gelatinous layer of the tension wood fibers. There is less xylan in the tension wood and in some species, but not in all, there are more polysaccharides containing galactose than in the normal wood (85).

The capacity of tension wood fibers to shrink is unusually great. When such wood is dried, cracks may be formed by the complete collapse of the wood fibers. The property of the tension wood to shrink in the fiber direction gives it the ability, as reaction wood, to compensate for the oblique growth of the stem by raising it. How this contraction occurs is not yet clear, but there is reason to assume that the cellulose present in the gelatinous layer is there first in a paracrystalline state and that the beginning of the crystallization process causes a contraction (132).

This hypothesis is disputed, however, by the fact that the texture of

TABLE II-7

AMOUNT OF SUGARS (%) IN THE HYDROLYZATES FROM NORMAL AND
REACTION WOOD OF CONIFERS AND ANGIOSPERMS (85)

	Galactose	Glucose	Mannose	Arabinose	Xylose
Angiosperms					
1. *Salicaceae*					
Populus Tremula L.					
normal	1.1	67.6	3.0	1.1	27.1
tension	2.0	73.3	0.8	1.1	22.8
2. *Betulaceae*					
Betula pendula Roth					
normal	1.3	63.8	2.8	0.8	31.3
tension	10.4	73.8	tr.	tr.	15.8
3. *Fagaceae*					
Fagus silvatica L.					
normal	1.8	57.4	4.9	1.0	35.1
tension	6.6	73.5	tr.	2.6	17.3
4. *Ulmaceae*					
Ulmus scabra Mill.					
normal	3.6	68.0	5.5	tr.	22.9
tension	5.0	74.7	3.7	tr.	16.6
5. *Aceraceae*					
Acer pseudoplatanus L.					
normal	4.7	68.6	3.5	tr.	23.2
tension	8.7	69.3	2.1	1.4	18.5
Conifers					
1. *Taxodiaceae*					
Sequoiadendron giganteum (Lindl.) Bucch.					
normal	4.1	64.2	8.0	4.3	19.5
compression	19.7	55.2	8.8	3.3	13.0
Taxodium distichum (L.) Rich.					
normal	6.0	65.8	13.0	2.2	13.0
compression	22.9	58.0	9.8	1.4	7.8
2. *Cupressaceae*					
Chamaecyparis lawsoniana (A. Murr.) Parl.					
normal	7.3	64.5	11.5	4.2	12.6
compression	26.2	51.9	8.5	2.9	10.7
3. *Pinaceae*					
Picea abies (L.) H. Karsten					
normal	2.1	70.7	17.2	1.3	8.7
compression	19.6	55.6	11.5	2.6	10.7
4. *Cedrus atlantica* mannetti					
normal	4.8	67.6	12.4	1.7	13.5
compression	18.7	60.0	10.0	1.2	10.0
5. *Pinus silvestris* L.					
normal	5.4	63.8	18.8	2.7	9.3
compression	20.6	53.9	11.7	2.9	10.9

the cellulose macromolecules that have been proven to be parallel to the cell axis is expected to lead to a transverse, but not to a longitudinal, contraction on crystallization. The longitudinal contraction can be achieved by isotropic cell-wall substances which, during the shrinkage, can contract not only in a transverse but also in a parallel direction to the fiber axis. For the swelling properties of the tracheids or their single layers, the accompanying substances of the skeleton components are decisive. With the tracheids of the compression wood of conifers, the high content of lignin and galactose-containing polysaccharides plays a certain role in its lesser degree of shrinkage. This may also be the case with the longitudinal contraction of the tension-wood fibers of deciduous woods and, in addition, the lower lignin content and the higher cellulose content may contribute to it. The lamella-like structure of the fiber wall is primarily responsible for the shrinkage, whereas the fibrillar texture determines the strength properties of the cell walls (38).

5. THE PHYSICAL PROPERTIES OF WOOD

5.1 The Specific Weight, Density, and the Weight of the Wood

The specific weight, i.e., the weight per volume of dry wood, plays a decisive role in the utilization of wood as raw material because it indicates how much solid wood substance is present in the unit of volume. It is apparent that, from a given volume of heavy wood, more cellulose is obtained than from the same volume of light wood. This is true for all types of chemical utilization of wood, even for the most primitive one, i.e., fuel. For the utilization of wood as an industrial raw material, too, its density is important. Because the strength properties, which are related to the specific weight and the density, are of only minor importance for the chemical utilization of wood, they will not be discussed here.

Wood is a multipurpose raw material. The variations in the specific gravity of each species are considerable. A comparison of different wood species, therefore, is possible only when the moisture content is taken into consideration—in other words, only the specific gravities of the absolutely dry woods can be compared, and the same holds true for the density. The density of the *wood substance* is about 1.53 and is almost unaffected by differences in the same species, by growth factors, and by chemical composition. On the other hand, the difference in the density of dry woods is extraordinarily great. The woods generally used for the production of pulp have densities of between 0.3 and 0.6—in other words, only 20% to 40% of the volume consists of wood substance while the remaining 60% to 80% is composed of cavities. These cavities can be filled with water and water

can also penetrate the wood substance itself. In this way, the weight and the volume of the wood increase. The volumetric shrinkage of wood on drying is between 10% and 20%. Usually, the volume of the wood is measured with the fresh, moist wood; the density, which is important in the chemical utilization, should therefore be defined as the absolutely dry weight in relation to the volume of the fresh wood. Table II-8 gives the density of some customarily used wood species.

The density expressed in grams per cubic centimeter is numerically identical with the specific gravity. The density expressed in pounds per

TABLE II-8

BASIC DENSITIES OF SOME COMMERCIAL PULPWOODS (9,75,129,131)

Wood species	Basic density: oven-dry weight / green volume
Balsam fir (*Abies balsamea*)	0.34
Western larch (*Larix occidentalis*)	0.48
Engelmann spruce (*Picea engelmannii*)	0.31
White spruce (*Picea glauca*)	0.37
Black spruce (*Picea mariana*)	0.40
Sitka spruce (*Picea sitchensis*)	0.37
Jack pine (*Pinus banksiana*)	0.39
Lodgepole pine (*Pinus contorta*)	0.38
Slash pine (*Pinus elliotii*)	0.56
Shortleaf pine (*Pinus echinata*)	0.46
Longleaf pine (*Pinus palustris*)	0.54
Ponderosa pine (*Pinus ponderosa*)	0.38
Scots pine (*Pinus silvestris*)	0.41
White pine (*Pinus strobus*)	0.34
Loblolly pine (*Pinus taeda*)	0.47
Douglas fir (*Pseudotsuga taxifolia*)	0.47
Redwood (*Sequoia sempervirens*)	0.42
Western red cedar (*Thuja plicata*)	0.35
Eastern hemlock (*Tsuga canadensis*)	0.38
Western hemlock (*Tsuga heterophylla*)	0.38
Sugar maple (*Acer saccharum*)	0.56
Red maple (*Acer rubrum*)	0.57
Paper birch (*Betula papyrifera*)	0.48
Silver birch (*Betula verrucosa*)	0.51
American beech (*Fagus grandifolia*)	0.56
European beech (*Fagus silvatica*)	0.58
White ash (*Fraxinus americana*)	0.55
Red gum (*Liquidambar styraciflua*)	0.44
Balsam poplar (*Populus balsamifera*)	0.35
European aspen (*Populus tremula*)	0.37
American aspen (*Populus tremuloides*)	0.35
Elm (*Ulmus americana*)	0.46

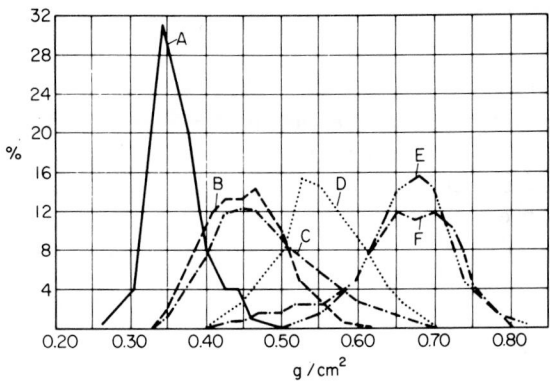

Figure II-21. Frequency distribution of specific gravity of various species of wood (130).

A = *Pinus strobus*, 844 samples, average 0.359 g/cm³
B = Spruce, 1962 samples, average 0.455 g/cm³
C = Pine, 2918 samples, average 0.489 g/cm³
D = Larch, 1910 samples, average 0.551 g/cm³
E = Beech, 1778 samples, average 0.671 g/cm³
F = Oak, 994 samples, average 0.647 g/cm³

cubic foot is obtained by multiplying the specific gravity by 62.4. No numerical data with regard to the specific gravity or density are complete unless they are based on the ratio of absolutely dry weight to moist volume, or absolutely dry weight to dry volume.

The fact that within a single species, even within the same tree, there are differences in the density makes a study of the causes of these variations of considerable interest. In the first place, each species has its own inherent capacity for adjusting its wood formation to external influences. In order to evaluate the fluctuations in density of a certain wood species, frequency curves must be set up; these must be made from a great many wood samples if reliable comparative values are to be obtained. Figure II–21 shows the frequency distribution of the density of a large number of different pulp-wood samples (130).

That certain growth properties are inherited may be assumed with certainty. Whether this can be used for the cultivation of desired tree qualities, such as, for instance, the production of a wood with increased density, requires further extensive research (115,123). Many environmental factors obviously have a lasting influence on the growth of trees and on certain properties that characterize the quality of the wood. Since these factors overlap it is difficult to distinguish the specific effect of an individual factor. As a whole, the effect of the various environmental conditions can be identified by the yearly growth (i.e., by the width of the annual rings), by the ratio of springwood to summerwood, or by the specific gravity. The relationship between the width of the rings and the density of the wood usually

permits unequivocal conclusions to be drawn as to its quality, but this relationship changes during the long life of the forest so that variations in environmental factors must be taken into consideration. Many investigations of the growth of the tree and its effect upon the properties of the wood have been made. It would transgress the limits of this monograph to discuss this subject in detail since excellent and comprehensive reports have already been published (124).

The control and regulation of the growing space within the tree stand provide means for the forester to influence the density of the wood. All wood species show the unmistakable effect of changes within the tree stand, whether it is increasingly dense or thinned out. With deciduous woods, thickening of the stand results in a decrease in the density of the wood, while thinning causes an increase in density. Wood with a uniformly high density can be produced by a fairly rapid rate of growth. With conifers, the conditions are different from those with deciduous woods. In this case, the density of the wood depends primarily on the quantitative proportion of spring- and summerwood within the single annual ring. In second-growth stands, the spacing and the development of the crown of the young tree have a paramount influence on the width of the springwood in the annual ring. This proportion of springwood is considerably less when the development of the crown is limited by a dense tree stand. In this case, the proportion of summerwood is greater and the wood becomes heavier because the wide rings of low density are missing, these being typical of widely spaced trees. The production of wood with a high degree of density and strength therefore requires a longer rotation (97).

The annual rings which are visible in a cross-section of a stem and which consist of spring- and summerwood become narrower toward the bark. With the increasing age of the tree the width of the ring decreases sharply, although the width of the summerwood remains almost constant, thus resulting in a continuous increase in the latewood portion, which can amount to over 30%. This increase causes an increase in the weight of the wood. Beside the age of the tree, the proportions forming the various parts of the trunk are of importance since the wood layers are formed at different periods of the tree's life. Thus, those parts of the stem which are in the crown are different from those below, but it must not be forgotten that the wood of the lower part of the stem was also once grown in the crown zone. These differences are especially pronounced with pines but are less noticeable with spruce- and beechwood (130).

Series of investigations with different species of pinewoods have shown that the density of the heartwood decreases up to a stem height of about ten feet and then remains almost constant to the crown. Within a single stem, the density of the heartwood can therefore be considered to be constant. On the other hand, the density of the outer wood decreases rapidly

with increasing height of the stem. The heartwood has a lower density than the sapwood of the same stem. The difference decreases with increasing height of the stem.

With larch, the proportion of summerwood exerts the greatest influence upon the density. This influence is closely related to the rate of growth and amounts to between 15% and 60%. The values for the density of earlywood and latewood of the larch species investigated lie between 0.3 and 0.7 grams per cubic centimeter (98).

Investigations of Norwegian red fir from a 20-year-old stand showed no relationships of significance between fiber lengths, density, proportion of summerwood, and width of annual rings, whereas such relationships could be shown to exist between the summerwood portion and the ring width of trees from a 44-year-old stand. No relationship between the fiber length and any of the other factors mentioned above could be established. Trees that flushed early showed wider rings, a lower percentage of summerwood, and therefore, also, a lower density of their wood than trees that flushed late. Up to a certain point it is possible to predict internal anatomical properties from external morphological characteristics (88).

There can be no doubt that the formation of the wood components is strongly affected by the location of the tree in the stand and, consequently, a comparison of trees of various locations can readily lead to false inferences. The fundamental changes in the factors that affect the quality of the wood are controlled by heredity, but they are also subject to the effect of the structure of the stand in which the tree grows and depend upon the geographical location and the growth conditions resulting from it. The direct influence of these environmental factors is exerted primarily on the growth of the tree crown and only indirectly on the growth of the xylem. The size and the distribution of the growth of the wood in the stem are, to a great extent, determined by those of the tree crown. The relationship between earlywood and latewood and the quality and structure of the wood within one annual ring depends, however, upon seasonal changes in the environmental factors that affect the growth of the tree crown and are subject to marked fluctuations (14,43,47,74,89,108,109).

It has been shown above that the specific gravity and density are of great importance in the processing of wood. Many variations in the physical and mechanical properties of certain wood species cannot be explained as resulting solely from variations in the specific gravity. The specific gravity is a measurement for the total proportion of wood substance per volume unit. The specific gravity does not indicate anything about the distribution of the wood substance itself in the wood structure, such as, for example, in the form of numerical data of the average diameter or the thickness of the cell wall. Such data, however, are of some importance with regard to the strength properties and other characteristics of pulps

and paper prepared from the wood. Recently, therefore, efforts have been made to develop testing methods that do not destroy the wood substance and which permit the rapid and reliable determination of the dimensions and thickness of the cell wall. The goal is to determine the specific gravity solely on the basis of the dimensions found for the cell wall. For this purpose, instruments have been developed that use reflected light or beta-rays for the determination of the differences in density within the single annual rings. These methods for measuring do not directly determine the cell dimensions but they show differences in the density and in the proportion of cell wall within an annual ring; thus the width of the annual ring and of the earlywood and latewood zones within the ring can be determined by measurement techniques from the density curves found. The average density of the ring can be calculated by integration of the surface area below the curve. Microphotometric methods and instruments have the advantage of providing an easier and more rapid measurement of the thickness of the cell wall and the diameter of the lumen, but this requires numerous microphotograms because of their great magnification and reduced field of observation, and the preparation of the microtomes entails further inaccuracies (14,15,44,69). The fully automatic measurement and recording of the light transmission through a cross-section of wood of 30 microns in thickness and over a region of several annual rings give diagrams from which the width of the annual rings, the proportion of earlywood and latewood, and the proportion of cell walls and lumen within a ring can be seen (45).

A fundamental improvement in this measuring technique is provided by a specially constructed instrument, the "dual-linear measuring micrometer," which permits a microscopic anatomic measurement directly on the surface of the wood sample. The wood sample needs no special pretreatment except for a superficial smoothing which can be achieved by means of a microtome or by hand with a razor blade. Because the microscopic observation is carried out by means of reflecting light, no thin cuts are required. By means of this micrometer, the thickness of the cell wall and the diameter of the lumen of a large number of cells can be determined and evaluated (112,113).

Thus far only the density of the wood has been discussed. Wood is a porous material and the total density values are composed of the density of the cell walls and the cavities. Investigations of the density of the cell-wall substance are very numerous and the methods used have been described in detail in publications already mentioned (112,113). The density of crystalline cellulose has been calculated from the known volume of the elementary cell and the known weight of the two cellobiose residues, i.e., from 1.592 and 1.583, respectively (51).

The values determined by the liquid displacement method depend

largely on the type of liquid used. The molecular weight and the affinity of the liquid for the cellulose influence the penetration. An attempt has therefore been made to determine the density of vegetable cell walls by planimetric measurements of the proportion of the volume of the cell wall in the total volume of the fiber sample and by division of this value by the weight of the sample. By this method the value for cotton was found to be 1.10 to 1.30 grams per cubic centimeter. To avoid confusing the terms "wood density" and "cell-wall density," the term "packing density" of the cell wall has been introduced, and this term will be used in the following sections (65).

The determination of the volume of the sample to be tested is carried out in the usual way according to the principle of water-displacement, then

TABLE II-9

SPECIFIC GRAVITY AND PACKING DENSITY OF THE CELL
WALLS OF DIFFERENT WOOD SPECIES

Observer	Reference	Wood species	Growth status	Specific gravity (g/cm³)	Cell-wall packing density (g/cm³)
Smith and Miller	108	Redwood	Young early	0.2575[a]	0.7315[a]
			Young late	0.6216	0.9220
			Old early	0.3004	0.8755
			Old late	0.6655	0.9739
Yiannos	115	Eastern spruce	—	0.5000	1.04
		Western hemlock	—	0.4160	1.04
		Silver fir	—	0.3960	1.10
		Douglas fir	—	0.5000	1.25
		Slash pine	—	0.6450	1.32
		Slash pine	Young early	0.3340	0.98
			Young late	0.7400	1.17
Smith	109	Douglas fir	Early	0.2690[b]	0.99[c]
			Late	0.7090	0.99
Jayme and Krause	113	Beech	—	0.5560–0.6610[d]	1.21–1.27[e]
		Poplar, 30% tension wood	—	0.4050–0.4720	1.19–1.20
		Birch	—	0.5600–0.6000	1.12–1.18
		Eucalyptus	—	0.5200–0.5220	1.02–1.06

[a] Calculated by the maximum-moisture method using 1.53.

[b] Determined by the maximum-moisture method after extraction with alcohol-benzene.

[c] Calculated from the specific gravity and average tracheid dimensions, assuming a rectangular lumen.

[d] Grams oven-dry per cubic centimeter air-dry.

[e] Grams oven-dry per cubic centimeter cell wall.

the sample, dried at 105°C, is weighed and the density is determined. The measurement of the percentage proportion of the cell walls in the total volume of the sample is made with the help of a special integration ocular which has cross hairs in the field of vision, consisting of two independently movable hairs which can be directed over the picture of the microscopic object by means of measuring screws. The specimens for such measurements are taken from two radial cuts made at right angles to each other. The values thus obtained for the packing density lie between 1.27 grams per cubic centimeter for beech and 0.71 grams per cubic centimeter for umbrella tree (*Musanga smithii*). Investigations with various deciduous woods have shown that the values for the packing density within the same species vary within narrow limits. With regard to the water absorption, it was found that, with an equal cell-wall proportion, the water take-up increases with decreasing packing density. This finding is of considerable technical importance (65).

The dual-linear measuring micrometer has also been used for the determination of the thickness and the proportion of the cell wall and of the diameter of the lumen. On the basis of a rectangular cross-section of the lumen and a density of the cell-wall substance of 1.53 grams per cubic centimeter, the estimated values for the packing density of the cell walls of redwood, as shown in Table II–9, were obtained. The data show that the cell walls in the latewood are denser than those in the earlywood and that, in old-growth wood, the cell walls are denser in both the earlywood and latewood than they are in the young-growth wood. This is in agreement with the finding that the cellulose in the secondary wall is packed more densely than in the primary wall, and that the ratio of the thickness of the secondary wall to that of the primary wall is greater in the summerwood than in the springwood (113).

The packing density of the cell walls of various woods was also determined on thin sections by planimetry. For this purpose, the density of the wood was first determined by the maximum-moisture method (114). Parts of a drill core from the wood to be tested are impregnated in vacuo with butyl acrylate containing some benzoyl peroxide. After polymerization at 50°C, microtome cuts of 10 microns in thickness are made and these are photographed at a 430-fold magnification with randomly selected focusings. The fibers are recorded and the area taken up by the cell-wall substance is calculated, disregarding the medullary rays and other nonfibrous portions. Column 6 of Table II–9 contains the values calculated for the packing density of the cell walls.

In summary, it may be said that the pycnometric determination of the density of the cell-wall substance is in good agreement with the theory; it does not, however, give any information as to the submicroscopic shape of the cell walls or, more especially, as to the presence of pores and cavities. The total volume of such cavities can be determined only by an accurate

estimation of the dimensions of the fibers and their weight. So far as present investigations of the packing density of the cell wall permit a preliminary appraisal to be made, the cell walls of the summerwood seem to be denser than those of the springwood. Furthermore, the abundant formation of heartwood in redwoods which are many hundreds of years old may be the reason for the increased packing density in their cell walls. The importance of the packing density of the cell walls of the dry fiber, however, should not be overestimated. The strength properties and the elasticity module of the fibers change with the size and distribution of the pores in the cell wall. It is known that the tensile strength changes markedly with the clamping length—for example, an enlargement of the clamping length can include more pores and therefore more weak places in the fiber. It is also known that the tension of the summerwood fibers is greater than that of the springwood fibers. The pore volume of the cell wall can also exert a lasting effect on the processing of the fibers to paper, as is also the case with regard to opacity and the absorption capacity of the fibers. These are, in part, still speculative concepts which must be confirmed by further experimental findings.

5.2 Moisture Content, Hygroscopicity, Sorption

The significance of the moisture content of wood has been pointed out in the preceding section. The moisture content, however, varies markedly within the tree and usually decreases from the bark toward the pith, i.e., it is highest in the sapwood and lowest in the heartwood. Even in the former, differences in the moisture can be observed; thus, for example, it can increase from the base to the crown of the tree. In other words, the increase or decrease in moisture content can vary with the height of the stem as well as from bark to the core. This variation in the conifers at different stem heights depends to a great extent upon the composition of the stem as regards sapwood and heartwood. The proportion of heartwood decreases in the upper parts of the stem and consequently they have a higher moisture content. The latter may also be influenced within the same species by environmental conditions and by seasonal changes which, however, are limited chiefly to the water-conducting layers of the stem and concern the heartwood only to a small exetent. With some deciduous woods, as for instance with birch, almost the whole stem participates as the water conduit (131).

The facts just discussed show that it is scarcely possible to obtain reliable data on the moisture content and the volume weight of freshly felled wood. Table II–10 gives some data of interest with regard to the relationship between the volume weight and the proportion of water by weight and volume of freshly felled wood, and the specific weight of the dry wood. It

is interesting to note, further, that the moisture creates a far-reaching equilibrium in the weight of the freshly felled wood (129).

Water is held in the wood substance by two different forces. The wood itself binds the water by adsorption. The adsorbed water is in an equilibrium with the atmospheric moisture. When the relative humidity of the atmosphere is 100%, the moisture content of the wood is about 30%, calculated on the basis of absolutely dry wood. This moisture content is also called "the fiber saturation point." In the hollow spaces of the wood the water is bound by capillary forces. At the fiber saturation point, these spaces are filled with air, but when more water is added to the wood, it penetrates the hollow spaces. When all the air in the wood is displaced, the maximum amount of water has been taken up. The amount of this take-up of water represents a measurement of the porosity of the wood. Adsorbed water is bound water. The structure of the cell wall indicates

TABLE II-10

RELATIONSHIP OF FRESH VOLUME WEIGHT, DRY VOLUME WEIGHT AND
MOISTURE CONTENT OF DIFFERENT WOOD SPECIES (129)

	Sap-wood	Heart-wood	Sap-wood	Heart-wood	Sap-wood	Heart-wood	Sap-wood	Heart-wood
	Pine		Spruce		Fir		Larch	
Fresh vol. weight, kg/m³	980	550	960	520	980	510	940	610
Dry vol. weight, kg/m³	420		390		370		470	
Moisture content								
by weight, kg	560	130	570	130	610	140	470	140
by volume, %	56	13	57	13	61	14	47	14
by weight, %	133	31	146	33	165	38	100	30
	Northern white pine		Douglas fir		Beech		Oak	
Fresh vol. weight, kg/m³	1020	580	910	540	1060	970	1000	
Dry vol. weight, kg/m³	320		420		560		570	
Moisture content								
by weight, kg	700	260	490	120	500	410	430	
by volume, %	70	26	49	12	50	41	43	
by weight, %	219	81	117	29	89	73	75	
	Birch		Alder		Ash		Maple	
Fresh vol. weight, kg/m³	950		930		860		970	
Dry vol. weight, kg/m³	510		430		570		540	
Moisture content								
by weight, kg	440		500		290		430	
by volume, %	44		50		29		43	
by weight, %	86		116		51		80	

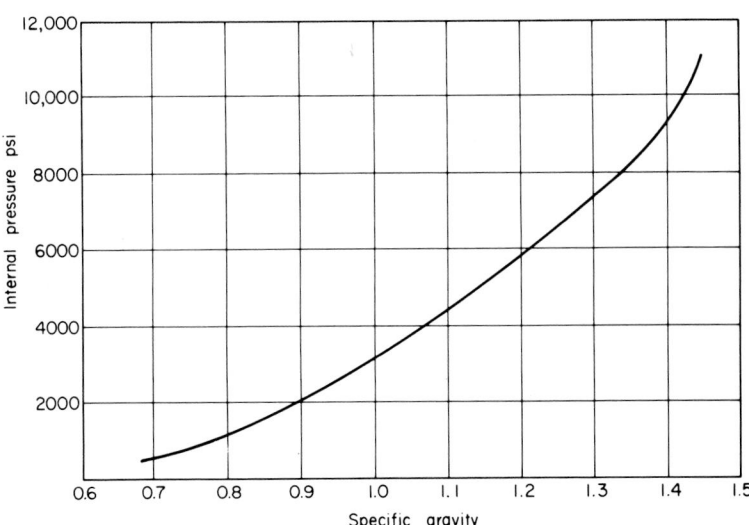

Figure II-22. Relationship between specific gravity of wood and the apparent swelling pressure of wood between 30% and 100% relative humidity (127).

that wood with thick cell walls contains more intermicellar cavities and can therefore take up and bind more water. The incorporation of water into the cell walls causes an expansion or swelling, whereas the drying out of the cell walls results in a shrinkage. Heavy wood with thick cell walls swells more than light wood with thin cell walls. In an analogous way, heavy wood also shrinks more. These processes play an important role in the processing of wood as a raw material. The apparent swelling pressure is always lower than the theoretically calculated or actual swelling pressure. Figure II-22 shows the relationships between the specific weight of the wood and the apparent swelling pressure at between 30% and 100% relative atmospheric humidity. The internal pressure increases rapidly with the specific weight and, at 1.44, reaches a value of 11,000 psi. This degree of magnitude should come very close to the actual swelling pressure (127). When the wood is subjected to a prior steaming, the swelling pressure is lowered markedly and thus an increase in the stability of the shape of timber is achieved (99).

Here it is necessary to discuss in some detail the sorption process. It has already been mentioned that an equilibrium is established between the moisture content of the atmosphere and that of wood. For a constant temperature, the respective equilibrium can be followed from the state of absolute dryness up to the state of complete saturation, i.e., up to the fiber saturation point. When moisture is taken up by wood, adsorption occurs, whereas loss of moisture to the surrounding drier air causes desorption.

The adsorption isotherms, typical for porous substances of colloidal character, show an *S*-shape, caused by the fact that the processes of adsorption, chemisorption, and capillary condensation overlap. In the range of low air humidity (about 0 to 20%), chiefly chemisorption occurs and this takes place monomolecularly and is accompained by a pronounced heat effect. In the next range of humidity (up to about 60%), pure adsorption takes place and leads to the formation of multimolecular adsorption layers with a considerably lower heat effect. With high humidity (up to 90% and over), capillary condensation is added to the multimolecular adsorption layers and the heat effect finally drops to zero. When moisture is removed from the wood, desorption occurs. With desorption, the hygroscopic equilibria are higher than with adsorption. This phenomenon, as applied to colloids, is termed "hysteresis" (70). Many explanations have been advanced to account for this phenomenon but they will not be discussed here because the problem has been dealt with in detail by experts in the field (116).

When the sorption isotherms of the wood are to be determined, it is necessary to establish the humidity equilibrium as quickly as possible. If the measurements are carried out in a vacuum, the transfer of the water vapor at the surface of the wood can take place unhindered and the wood sample can be reduced in size. Investigations in the absence of air with small samples of deciduous wood (*Auracaria klinkii*) have shown that the velocity of sorption decreases markedly in successive steps as the relative humidity increases. The diffusion of the water into the wood therefore may not be the determining factor for the sorption velocity. A moisture content at an equilibrium with an atmospheric relative humidity of up to 75% has been found to be completely independent of the shape of the wood sample and consequently it must be assumed that the sorption of the water vapor is not inhibited by the swelling phenomenon of the wood (20).

With regard to the differences in the sorption of water vapor by the individual chemical components of the wood, investigations in this field have shown that the sorption capacity of the wood is generally greater than that of the cellulose isolated from the wood. Higher and lower sorption capacities of the hemicelluloses and the lignins, respectively, have been confirmed (1,101,102). The water vapor sorption isotherms of a sample of *Eucalyptus regnans* were compared with those of its holocellulose, hemicellulose, and cellulose fractions as well as with those of the Klason and methanol lignins. From the results, the total sorption pertaining to the relative proportions of the main components of the wood were roughly calculated. The fractions of the total sorption calculated were 47% for cellulose, 37% for the hemicelluloses, and 16% for the lignin. The shapes of the isotherms of the wood fractions investigated were quite similar, but the size of the sorption hystereses differed appreciably. The latter were greater with the lignins than with the carbohydrate fractions. On the other

hand, the data of the integral heat of sorption per gram of adsorbed water in the range of 0 to 80% relative humidity were very similar for all fractions (21).

The integral and differential heats of adsorption must be separated basically from each other. The more easily determined integral heat of adsorption is obtained by allowing the degassed material being tested, in this case wood, to adsorb enough water vapor to establish an equilibrium, and measuring the amount of heat evolved. To determine the differential heat of adsorption, the adsorption process is separated into a theoretically infinite number of single steps and the heat evolved in each step is measured for one mole of adsorbed water. With increasing adsorption, the heat of adsorption generally decreases stepwise. If the heat of adsorption is expressed as the sum of the heat of evaporation and the heat of wetting, the heat of evaporation must be smaller than the heat of adsorption. The heat of wetting is then the integral heat of adsorption of the vapor at the saturation pressure, minus the heat of evaporation.

By measurements of the adsorption isotherms at a certain temperature with *Auracaria klinkii*, it could be shown that the dependence of the heat of adsorption on the relative water vapor pressure is the same for wood as those found by other investigators for cotton and other forms of pure cellulose. The integral heat of adsorption is dependent upon the temperature for all degrees of humidity above 0.12 gram of water vapor per gram of cotton, with the heat of adsorption formed decreasing with increasing temperature. Apparently, at low moisture contents, the heat of adsorption is independent of the temperature. The results of the changes in the heat of adsorption with the temperature show that the specific heat of the wood-water aggregate is greater than the temperature coefficient calculated from the specific heat of the pure components if the relationships which are concordant with those of simple mixtures are used (20,21,70).

Of special importance is the progress of the movement of the water when the wood is drying. Three different types of movement overlap: the movement of the liquid through the capillary system, the diffusion of water vapor, and the diffusion of bound water. For the investigation of the last process, the cavities of the capillary system were filled with an alloy with a melting point of 105°C. The low melting point was intended to prevent the formation of new capillaries through contraction during the solidification of the alloy. With radial sections, 0.1 to 0.2 inch in thickness, treated in this way, it was found that the diffusion coefficient for bound water on immersion in water is independent of the thickness and the specific weight of the wood sample and that only minor differences exist between the wood species tested. There were also only minor differences shown in these tests between sapwood and heartwood, while the differences in the deciduous woods were generally slightly greater. Thus, with increasing

temperature, the diffusion coefficient increases in the same way as does the vapor pressure of water. This is not unexpected since the water take-up depends on the frequency of the impact of highly energy-charged water molecules upon the surface of the wood, and the diffusion into the interior of the wood derives from the jump of a single water molecule from one site of adsorption to another with a greater adhesive force. Water therefore probably moves into the wood in this molecular jump fashion rather than in a continuous process. The activation energy of 12,000 cal per mole water corresponds in order of size to the formation of hydrogen bridges. Data for the adsorption of water vapor under various conditions of relative humidity show that the diffusion coefficient increases in the fiber direction in an exponential way with the average moisture content of the wood. The movement of the water is facilitated when the wood has a high moisture content. Each site of adsorption can bind, on an average, six moles of water (117).

At low moisture contents, the movement of the water in the wood is limited essentially to a water vapor movement through the capillaries filled with air. With increasing moisture content, the diffusion of bound water becomes of greater importance and the total diffusion also increases. The analogy between electrical conduction and diffusion, which was shown for the structure of wood filled with liquids, agrees with the conduction through, and the diffusion of bound water into the wood for the humidity region below the fiber saturation point. The specific electrical conductivity of wood in the radial direction is one-third to one-half of that in the fiber direction and is 17% to 25% greater than that in the tangential direction. This can possibly be traced back to the fact that the medullary rays play a role here. The swelling occurs without limitation in the diffusion direction but does so only in a right-angled direction to the latter (118,126).

Water vapor can diffuse unhindered through the fiber cavities and the pit chambers, but through the permanent pit membrane pores it is hindered. Bound water diffuses intermittently through the pit membranes and through the cell walls, continuously through the remainder of the wall. The term "intermittent" applies only to the part of the diffusion process in which the water diffuses through that portion of the cell walls which is turned toward the lumen. Continuous diffusion of bound water occurs in the remainder of the cell wall, lying in the direction of the diffusion. The diffusion coefficients calculated for an average moisture content of the wood of 20% show that the diffusion of bound water through the pit membranes within the pit system surpasses in its effect the diffusion of water vapor through the permanent membrane pores and through the pit chambers. The intermittent diffusion of bound water through the cell wall also surpasses the diffusion of water vapor through the pit system and the cell cavities that are connected with it. These findings are valid only for the heartwood of Sitka spruce with a certain degree of permeability of the

membrane pores; further investigations of wood with greater permeability and at higher or lower moisture contents are required to confirm the general applicability of the results (119).

5.3 Specific Heat

Of the physical properties of wood that are of importance for its chemical processing, the specific heat must be mentioned. In all processes that require a calculation of the heat demand or of the formation of heat in the course of a chemical reaction, the specific heat is important. With the more or less similar chemical composition prevailing in woods, the specific heat of the dry wood is nearly constant, varying between 0.317 for chestnut and 0.337 for long-leaf pine, with an average of 0.327 for 20 wood species. The specific gravity is almost without influence upon the specific heat but great variations occur in connection with variations in the temperature. The specific heat of wood can be calculated for each temperature from the equation:

$$H_{sp} = 0.266 + 0.00116t$$

where t is the temperature in degrees centigrade. The specific heat is 0.266 at 0°C and 0.382 at 100°C (32).

6. THE ANATOMY OF WOOD AND THE PRODUCTION OF FIBROUS MATERIAL

For many years, efforts of the forest management have been aimed at the cultivation of trees that are especially suitable for the pulp and paper industry. Sometimes, opinions have differed as to what constitutes "suitable." With regard to the various intended uses, it is established that wood can be processed, for example by purely mechanical means, to pulp. In this case, mechanical pulp or groundwood is obtained by degrading the wood by grinding it, without changing its chemical composition to any extent. Vast quantities of wood are converted in this way into groundwood, which is used primarily for the production of printing paper, especially newsprint. This process gives a high yield of fibrous material which is characterized by the mechanical properties of its highly lignified fibers and fiber bundles. It is chiefly coniferous woods that are used for this purpose, but in some countries where such woods are in short supply, deciduous woods are also used.

For papers of better quality, especially for medium fine and fine papers, or for papers for specialized use or requiring special strength properties, the wood must be decomposed to give a partially or completely

delignified pulp. In this case, a chemical wood pulp is obtained, consisting of a fibrous material produced by chemical means and by the separation of single fibers from the wood structure. These chemical processes, which entail a drastic change in the chemical structure of the wood, are described in detail in Chapters VII and VIII.

In the production of wood pulp, coniferous woods (gymnosperms) and deciduous woods (angiosperms) are both used. As has been shown above, the structures of these two wood species differ fundamentally from each other. In the next chapter, it will be shown that they also differ considerably in their chemical composition. The type of wood selected, therefore, is dependent upon the use for which the pulp is intended. The chemical composition of the wood and the cellulose content of its pulp are important for the production of artificial fibers based on cellulose. In the production of rayon, the cellulose is dissolved in a suitable solvent and is precipitated from this solution to give a completely new fiber. For the production of paper, the microscopic and submicroscopic structure of the fibers play decisive roles. Here, the fiber, which has been more or less chemically changed and is largely delignified, is subjected to a mechanical process that not only shortens it but also, under certain circumstances, fibrillates it. The term "fibrillate" indicates that the single components of the cell wall are separated from each other by the mechanical beating process.

Which morphological factors are now of importance? Some will be listed here, but, since this is a very complex problem, the list will not be complete (8). The following are especially important: the length of the fiber; the ratio of fiber length to fiber width; the diameter of the fibers, especially those of cylindrical shape; the thickness of the fibers, especially if they are ribbon-shaped; the ratio of fiber width to fiber thickness; the total cross-sectional area of the fiber; the net cross-sectional area, i.e., the total area minus the area of the lumen; the weight of the unit of length (known as the denier or titre in the textile industry); the shape of the fiber (round, flat, or ribbon-shaped); the thickness of the cell wall; the smoothness of the fiber surface; and the curling of the fiber.

Even a glance at this list shows that appreciable differencess may exist between the fibers of deciduous and coniferous woods. But there are differences not only between the fibers of these two species but also between those in the wood of the same species and even within the same tree, as, for example, between those in sapwood and heartwood, in earlywood and latewood, in normal and reaction wood, in those from the stem, crown, and branches—all of which have a permanent effect on the paper produced from them (6,19,52,114).

The basic problem facing the forest management is whether or not certain morphological properties can be developed by cultivation. Certain major structural properties are inherent in each species. Age and environ-

mental conditions can influence the cell's dimensions, but not its basic structure, which is genetically fixed. It has of late been shown repeatedly that anatomical properties of certain species, suitable for the production of paper, can be cultivated by hybridization (30).

It has been mentioned that the fibers isolated from the wood by chemical processes are subjected to a mechanical beating in the manufacture of paper. This is not the place for a detailed description of the beating process. The opinion prevailing during the first three decades of this century was that the cellulose in the fiber is hydrated by taking up water during the beating process and that the hydrate cellulose thus formed provides the strength necessary for the paper sheets made from the fibers because of its gelatinous character. This concept, however, may be considered to be out-dated because of results found in more recent research. The penetration of water into the submicroscopic regions of the cell wall causes a swelling of the fiber. This increase in volume is only one important effect, while the increased flexibility of the swollen fiber is of at least equal significance. At the beginning of the beating process, it is primarily the primary wall and the outer secondary wall that are affected, or even partially removed. In this way the swelling process is facilitated. The combined effect of the swelling and mechanical actions, in the latter of which the fibers are folded and compressed by the beaters, leads to a loosening of the bonds between the coaxial lamellas. The cell wall of the chemically disintegrated wood fiber begins to divide into separate concentric layers. This phenomenon does not occur with groundwood or with so-called high-yield pulp, which still contain considerable amounts of lignin. More water can penetrate into the cell walls or the lamellas and can be bound by the hemicelluloses and the amorphous portions of the cellulose. In this way, a plasticizing effect is also exerted on the crystalline portions of the cellulose and the coaxial lamellas can be shifted against each other.

The compressing action which is exerted on the fiber during the beating process is a reciprocal action not only between the beater and the fibers but also between the fibers themselves. This is shown by the fact that an increasing compression action can be observed when the consistency of the pulp is increased (33,34,82,95,96,111,120).

When a sheet of paper is formed from an aqueous pulp suspension of fibers treated mechanically in this way and the water is removed by drying, the fibers again come into close contact with each other. As the water is removed from the capillaries during the drying process, surface-tension forces exert an increased attraction between the fibers and the fibrils. When the distance between neighboring cellulose molecules becomes small enough, hydrogen bridges can be formed, and when these cellulose molecules belong to different fibers, the latter are bound to each other by forces similar to those occuring during crystallization. Hydrogen bridges, besides those be-

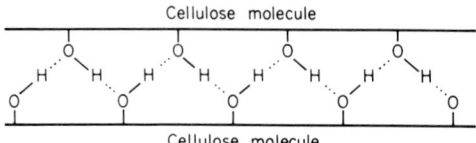

Figure II-23. Hydrogen bonding between two adjacent cellulose molecules (34).

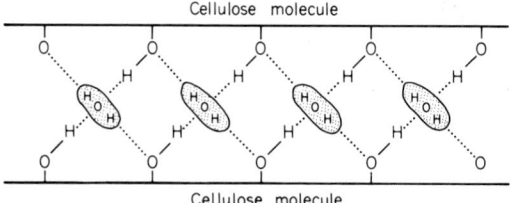

Figure II-24. Hydrogen bonding of two cellulose molecules through a monolayer of water molecules (34).

tween hydroxyl groups of neighboring cellulose molecules, can also be formed by means of a single or multiple layer of water molecules, as is shown schematically in Figs. II–23 and II–24. In this way, the marked decrease in strength on immersion of a paper sheet in water can be explained as resulting from an extensive loosening of the fiber-to-fiber bonds.

The behavior of the fibers on beating is therefore directly connected with their swelling properties. These properties, in turn, are influenced both by the morphological structure of the fibers, especially of the cell walls, and by their chemical composition. The amorphous polysaccharides, the amorphous regions within the cellulose molecules, and the hemicellulose have an important effect on the swelling. When the hydroxyl groups of the amorphous polysaccharides are replaced by other groups, e.g., by acetyl groups, the swelling capacity of the fibers is reduced and the possibility of fiber-to-fiber bonds is greatly diminished. The same holds true to a lesser degree when the fibers are deprived of substantial amounts of hemicellulose and amorphous cellulose by chemical reactions. The size of the bonded area, by which is meant that area which is in such close contact with the fibers that fiber-to-fiber bonds are possible, is also of some importance. In general, this area increases with increased beating. The purpose of the beating consists, therefore, in an enlargement of the binding surface by plasticizing the fibers so that, during the drying process, they can change their shape under the effect of surface-tension forces. This change in shape, alone, however, is not sufficient; it is more important that the surfaces of the fibers should come in close contact with each other so that a bond can be formed. This is probably promoted by a certain dispersion of the cellulose and other polysaccharides in the surface layers of the fibers (33,34,111,120).

REFERENCES

1. Algar, W. H., Giertz, H. W. and Gustafsson, A. M., *Svensk Papperstidn.* **54** (10), 335 (1951).
2. Asunmaa, S., *Svensk Papperstidn.* **58** (8), 308 (1955).
3. Bailey, A. J., *Ind. Eng. Chem. Anal. Ed.* **8**, 52 (1936).
4. Barber, N. F. and Meylan, B. A., *Holzforschung* **18** (5), 146 (1964).
5. Bland, D. E., *Holzforschung* **15** (4), 102 (1961).
6. Boutelje, J. B., *Svensk Papperstidn.* **69** (1), 1 (1966).
7. Braun, H. J., *Holzforschung* **21** (2), 3 (1967).
8. Britt, K. W., *Tappi* **48** (1), 7 (1965).
9. Brown, H. P., *in* L. E. Wise and E. C. Jahn, "Wood chemistry." Reinhold, New York, 1952.
10. Bucher, H., "Morphology and structure of wood fibers." Cellulosefabrik Attisholz, 1947.
11. Bucher, H., "Die Tertiärlamella von Holzfasern und ihre Erscheinungsformen bei Coniferen." Cellulosefabrik Attisholz, 1953.
12. Bucher, H., *Holzforschung* **11** (1), 14 (1957).
13. Bucher, H., *Holzforschung* **11** (4), 97 (1957).
14. Buijtenen, v. J. P., *Tappi* **47** (7), 401 (1964).
15. Cameron, J. F., Berry, P. F. and Phillips, E. W., *Holzforschung* **13** (3), 78 (1959).
16. Casperson, G., *Svensk Papperstidn.* **68** (16), 534 (1965).
17. Casperson, G., *Holzforschung* **21** (1), 1 (1967).
18. Chamberlain, G. J., "Gymnosperms, structure and evolution." Chicago, 1935.
19. Chidester, G. H. and Curran, C. E., *Forest Prod. Lab. Rept.* 1286, Nov. 1959.
20. Christensen, G. N. and Kelsey, K. E., *Holz Roh- u. Werkstoff* **17** (5), 178 (1959).
21. Christensen, G. N. and Kelsey, K. E., *Holz Roh- u. Werkstoff* **17** (5), 189 (1959).
22. Coté, W. A., *Forest Prod. J.* **8** (10), 296 (1958).
23. Coté, W. A. and Day, A. C., *Tappi* **45** (12), 906 (1962).
24. Coté, W. A., Day, A. C., Kutscha, N. P. and Timell, T. E., *Holzforschung* **21** (6), 180 (1967).
25. Coté, W. A., Kutscha, N. P., Simson, B. W. and Timell, T. E., *Tappi* **51** (1), 33 (1968).
26. Coté, W. A., Koran, Z. and Day, A. C., *Tappi* **47** (8), 477 (1964).
27. Coté, W. A. and Krahmer, R. L., *Tappi* **45** (2), 119 (1962).
28. Coté, W. A., Pickard, P. A. and Timell, T. E., *Tappi* **50** (7), 350 (1967).
29. Cronshaw, J., Davies, G. W. and Wardrop, A. B., *Holzforschung* **15** (3), 75 (1961).
30. Dadswell, H. E. and Wardrop, A. B., *Appita J.* **13** (5), 161 (1960).
31. Dietrichs, H. H., *Holzforschung* **18** (1/2), 14 (1964).
32. Dunlop, F., *US Dept. Agric. Bull.* **110** (1912).
33. Emerton, H. W., *Tappi* **40** (7), (1957).
34. Emerton, H. W., "The fundamentals of the beating process." Marshall Press, London, 1957.
35. Emerton, H. W. and Goldsmith, W., *Holzforschung* **10** (4), 108 (1956).
36. *Forest Prod. Lab.*, Lab. Rept. **1093** (1960).
37. Frey-Wyssling, A., *Protoplasma* **25**, 261 (1935).
38. Frey-Wyssling, A., "Die pflanzliche Zellwand." Springer, Berlin-Göttingen-Heidelberg, 1959.
39. Frey-Wyssling, A., Bosshard, H. H. and Mühlethaler, K., *Planta* **47**, 115 (1956).

40. Frey-Wyssling, A., Mühlethaler, K. and Bosshard, H. H., *Holz Roh- u. Werkstoff* **13** (7), 245 (1955).
41. Frey-Wyssling, A., Mühlethaler, K. and Bosshard, H. H., *Holz Roh- u. Werkstoff* **14** (5), 161 (1956).
42. Frey-Wyssling, A. and Stecher, H., *Experientia* **7,** 420 (1951).
43. Goddard, R. E. and Strickland, R. K., *Tappi* **45** (7), 606 (1962).
44. Green, H. V. and Worrall, J., *Pulp Paper Inst. Can.*, Tech. Rept **331** (1963).
45. Green, H. V. and Worrall, J., *Tappi* **47** (7), 419 (1964).
46. Hale, J. D., *Tappi* **42** (8), 670 (1959).
47. Hamilton, J. R. and Harris, J. B., *Tappi* **48** (6), 330 (1965).
48. Harris, J. M. and Meylan, B. A., *Holzforschung* **19** (5), 144 (1965).
49. Hägglund, E., *Svensk Papperstidn.* **37,** 133 (1934); **38,** 454 (1935).
50. Hägglund, E., "Chemistry of wood." Academic Press, New York, 1951.
51. Hermans, P. H., "Physics and chemistry of cellulose fibers." Elsevier, New York, 1949.
52. Hiett, L. A., Beers, W. L. and Zachariasen, K. A., *Tappi* **43** (2), 169 (1960).
53. Hiller, Ch. H., *US Forest Lab. Res.* Note FPL-034 (1964); *J. Forestry* **62** (4), 249 (1964).
54. Hiller, Ch. H., *Tappi* **47** (2), 125 (1964).
55. Hillis, W. E., Humphreys, F. R., Bamber, R. K. and Carle, A., *Holzforschung* **16** (4), 114 (1962).
56. Jaccard, P., see Frey-Wyssling, A., (37).
57. Jayme, G. and Azzola, F. K., *Holzforschung* **18** (1/2), 9 (1964).
58. Jayme, G. and Azzola, F. K., *Holzforschung* **19** (3), 69 (1965).
59. Jayme, G., Bauer, G., *Holzforschung* **11** (1), 16 (1957).
60. Jayme, G. and Fengel, D., *Holz Roh- u, Werkstoff* **17** (6), 226 (1959).
61. Jayme, G. and Fengel, D., *Holz Roh- u. Werkstoff* **19** (3), 50; (4), 97 (1961); *Das. Papier* **16** (10a), 519 (1962).
62. Jayme, G. and Harders-Steinhäuser, M., *Das Papier* **4** (7/8), 104 (1950).
63. Jayme, G. and Hunger, G., *Holz Roh- u. Werkstoff* **17** (2), 41 (1959).
64. Jayme, G., Hunger, G. and Fengel, D., *Holzforschung* **14** (4), 97 (1960).
65. Jayme, G. and Krause, Th., *Holz Roh- u. Werkstoff* **21** (1), 14 (1963).
66. Kelsey, K. E. and Clarke, L. N., *Nature* **176,** 83 (1955).
67. Kelsey, K. E. and Clarke, L. N., *Austral. J. Appl. Sci.* **7** (2), 160 (1956).
68. Kerr, T. and Bailey, J. W., *J. Arnold Arboretum* **15,** 327 (1934).
69. Keylwerth, R. and Kleuters, W., *Holz Roh- u. Werkstoff* **20** (5), 173 (1962).
70. Kollmann, F., *Holz Roh- u. Werkstoff* **17** (5), 165 (1959).
71. Krahmer, R. L. and Coté, W. A., *Tappi* **46** (1), 42 (1963).
72. Lange, P. W., *Svensk Papperstidn.* **48** (10), 241 (1945).
73. Lange, P. W., *Tappi* **42** (9), 786 (1959).
74. Larson, P. R., *Tappi* **45** (6), 443 (1962).
75. Lee, N. H. and Hale, J. D., "Pulp and paper manufacture" Vol. 1. McGraw-Hill, New York, 1950.
76. Liese, W., *Holz Roh- u. Werkstoff* **9** (10), 374 (1951).
77. Liese, W., *Holz Roh- u. Werkstoff* **14** (11), 417 (1956).
78. Liese, W., *J. Polymer Sci.* **C** (2), 213 (1963).
79. Liese, W. and Fahnenbrock, M., *Holz Roh- u. Werkstoff* **10** (5), 197 (1952).
80. Liese, W. and Johann, I., *Naturwissenschaften* **24,** 579 (1954).
81. Marton, R. and Agarwal, A. K., *Tappi* **48** (5), 264 (1965).
82. McIntosh, D. C., *Tappi* **50** (10), 482 (1967).

83. Meier, H., *Holzforschung* **11** (2), 41 (1957).
84. Meier, H., *Holz Roh- u. Werkstoff* **13** (9), 323 (1955); *Holzforschung* **11** (2), 41 (1957).
85. Meier, H., "On the chemistry of reaction wood." Symposium Intern. Grenoble, July 1964.
86. Meier, H. and Yllner, S. V., *Svensk Papperstidn.* **59** (11), 395 (1956).
87. Meier, H. and Wilkie, K. C. B., *Holzforschung* **13** (6), 177 (1959).
88. Mergen, F., Burley, J. and Yeatman, Ch., W,, *Tappi* **47** (8), 499 (1964).
89. Mitchell, H. L., *Tappi* **47** (5), 276 (1964).
90. Mühlethaler, K., Z. *Zellforschung* **39,** 299 (1953).
91. Necessany, V., *Svensk Papperstidn.* **60** (1), 10 (1957).
92. Necessany, V., *Svensk Papperstidn.* **62** (3), 73 (1959).
93. Nyrén, V. and Back, E., *Svensk Papperstidn.* **62** (17), 587; (19), 681 (1959).
94. Nyrén, V. and Back, E., *Svensk Papperstidn.* **63** (16), 501; (18), 619 (1960).
95. Page, D. H., *Pulp Paper Mag. Can.* **67** (1), T-2 (1966).
96. Page, D. H. and Grace De, J. H., *Tappi* **50** (10), 489 (1967).
97. Paul, B. H., *Forest Prod. Lab. Tech. Bull.* **1288** (July 1963).
98. Pearson, F. H. and Fielding, H. A., *Holzforschung* **15** (3), 82 (1961).
99. Perkitny, T., Lawniczak, M. and Marciniak, H., *Holz Roh- u. Werkstoff* **17** (2), 54 (1959).
100. Preston, R. D., "The molecular architecture of plant cell." London, 1952.
101. Runkel, R. O. H., *Holz Roh- u. Werkstoff* **12** (6), 226 (1954).
102. Runkel, R. O. H. and Lüthgens, M., *Holz Roh- u. Werkstoff* **14** (11), 424 (1956).
103. Saarnio, J. and Gustafsson, C., *Paperi ja Puu* **35,** 65 (1953).
104. Sandermann, W., Hausen, B. and Simatupang, M., *Das Papier* **21** (7), 349 (1967).
105. Sandermann, W. and Schmitz, G., *Holz Roh- u. Werkstoff* **23** (6), 221 (1965).
106. Schoch-Bodmer, H., *Ber. Schweiz. Botan. Gesellsch.* **55,** 313 (1945).
107. Schreuder, H. R., Coté, W. A. and Timell, T. E., *Svensk Papperstidn.* **69** (19), 641 (1966).
108. Schultze-Dewitz, G., *Holzforschung* **15** (3), 89 (1961).
109. Schütt, P., *Das Papier* **16** (11), 671 (1962).
110. Schwerin, G., *Holzforschung* **12** (2), 43 (1958).
111. Simmonds, F. A. and Chidester, G. H., *Forest Prod. Lab. Res.* Paper FPL 5 (May 1963).
112. Smith, D. M., *Forest Prod. J.* **15** (8) 325 (1965).
113. Smith, D. M. and Miller, R. G., *Tappi* **47** (10), 599 (1964).
114. Stairs, G. R., Marton, R., Brown, A. F., Rizzio, M. and Petrik, A., *Tappi* **49** (6), 296 (1966).
115. Squillace, A. E., Echols, R. M. and Dorman, K. W., *Tappi* **45** (7), 599 (1962).
116. Stamm. A. J., Surface Properties of Cellulosic Materials *in* Wise and Jahn, "Wood chemistry" 2nd ed., Vol. 2. Reinhold, New York, 1952.
117. Stamm, A. J., *Forest Prod. J.* **9** (1), 27 (1959).
118. Stamm, A. J., *Forest Prod. J.* **10** (10), 524 (1960).
119. Stamm, A. J., *Forest Prod. J.* **10** (12), 664 (1960).
120. Steenberg, B., *Svensk Papperstidn.* **66** (22), 933 (1963).
121. Stemsrud, F., *Holzforschung* **10** (3), 69 (1956).
122. Stockman, L. and Hägglund, E., *Svensk Papperstidn.* **51** (12), 269 (1948).
123. Stonecypher, R., Cech, F. C. and Zobel, B. J., *Tappi* **47** (7), 405 (1964).
124. Tappi Forest Biolog. Committee, *Tappi* Monograph series no. 24, New York, 1962.

125. Tappi Standard T 18 m-53 "Specific gravity of wood"; T12 m-59 "Extraction-free wood."
126. Tarkow, H. and Stamm, A. J., *Forest Prod. J.* **10** (5), 247; (6), 323 (1960).
127. Tarkow, H. and Turner, H. D., *Forest Prod. J.* **8** (7), 193 (1958).
128. Thomas, R. J. and Nicholas, D. D., *Tappi* **51** (2), 84 (1968).
129. Trendelenburg, R., "Das Holz als Werkstoff." Lehmann, München-Berlin, 1939.
130. Trendelenburg, R., *Papierfabrikant* **34** (43), 389; (44), 401; (45), 411 (1936).
131. Trendelenburg, R. and Mayer-Wegelin, H., "Das Holz als Rohstoff." Hanser, München, 1955.
132. Wardrop, A. B., *Austral. J. Botan.* **3**, 177 (1955); **4**, 152 (1956).
133. Wardrop, A. B. and Dadswell, H. E., *Austral. J. Sci. Res.* **B1**, 3 (1948).
134. Wardrop, A. B. and Dadswell, H. E., *Holzforschung* **11** (2), 33 (1957).
135. Wardrop, A. B., Liese, W. and Davies, G. W., *Holzforschung* **13** (4), 115 (1959).
136. Wergin, W. and Casperson, G., *Holzforschung* **15** (2), 44 (1961).
137. Wooten, T. E., Barefoot, A. C. and Nicholas, D. D., *Holzforschung* **21** (6), 167 (1967).
138. Zimmermann, W., "Die Phylogenie der Pflanzen." Jena, 1930.
139. Zobel, B., Webb, Ch. and Henson, F., *Tappi* **42** (5), 345 (1959).

III

The Chemistry of Wood

INTRODUCTION

Relating to the general topic, the chemistry of wood, a number of excellent book-length presentations, covering the various individual wood components, are available, and these will be referred to in this chapter under the heading Bibliography. To obtain an exhaustive insight into the different components of wood and their chemical and structural properties, it is suggested that such books should be consulted. Within the scope of this chapter, only those basic aspects of the chemistry of wood that are specifically related to the understanding of the chemical-technological utilization of wood will be dealt with.

As was shown in the preceding chapter, the analytical separation of the individual wood components meets with a good deal of difficulty, for several reasons. For one, specific agents that permit the selective isolation of only one component are rare, in spite of the elucidation of the constitution of the single cell-wall components; for another, the type and proportion of the various combinations of the cell-wall components are not sufficiently known. Thus it happens that one reagent will almost always react with several components. In spite of this, it has been possible to carry out a groupwise analysis and to sum up these group data. Consequently, "summative wood analysis" will be spoken of in the following section. This type of analysis gives a surprising degree of insight into the chemical structure of the various wood species and its importance for the study of the ecological conditions influencing the growth and decomposition, the resistance toward wood pests, and the suitability for the chemical-technology utilization of wood.

The chemical elementary analysis shows that, apart from minor amounts of nitrogen, wood is composed of about 50% carbon, 6% hydrogen, and 44% oxygen. Surprisingly, there is only a minor difference between the various wood species as regards their elementary composition. From this it might be concluded that wood is a uniform chemical substance. Over a hundred years ago, the successive treatment of wood with nitric acid, alkali, alcohol, and ether led to the isolation of a substance with the elementary formula $C_6H_{10}O_5$, identical with that of starch. This substance was called "cellulose," probably because it formed the main part of the vegetable cell wall (194). During the isolation of the cellulose from the wood, another part was dissolved and this was termed an "incrusting substance." Later, this substance, which is insoluble in concentrated acids was called "lignin." Simultaneously, it was found that the carbohydrate portion contained, in addition to cellulose, other polysaccharides and these were given the collective name, "hemicelluloses" (226,227).

The analytical methods thus far used, however, were unable to accomplish the isolation of the cell-wall components in an unchanged state. Based on systematic preliminary experiments (72,180) a method was developed in which the wood was treated with chlorine followed by an extraction with an alkaline agent such as sodium hydroxide or sodium sulfite, thus leading to the dissolution of the lignin while extensively protecting the cellulose and its associated carbohydrates (23). This method was further modified and improved by the use of alcoholic alkali solutions, pyridine-alcohol mixtures, or alcohol-monoethanolamine mixtures as extractives, thus increasing the selectivity of the method (206). The carbohydrate portion isolated in this way was designated "holocellulose." By adding to the latter the lignin, the extractives, and the ash, up to 99.6% of the wood substance could be accounted for analytically, at least so far as groups are concerned (18).

Instead of an alternating chlorine-alkali treatment, a similar method with chlorine dioxide and alkali, alkali sulfite, or pyridine was proposed (219,222,241). Although this method has lost its analytical importance because of the somewhat complicated preparation of chlorine dioxide and the fact that, during the lignin extraction, some carbohydrates are lost, the introduction of chlorine dioxide as a delignifying agent into the technical pulp bleaching plant is of historical significance. The salts of chloric acid, especially sodium chlorite, act similarly to chlorine dioxide and are used for the isolation of holocellulose since they have now been put on the market on a large scale by the chemical industry (135). In a modified form, this method is being used successfully in the summative wood analysis because a fractionation of the hemicelluloses can be achieved, especially by the extraction of the chlorite holocellulose with alkaline solutions of various concentrations. The remaining portion of "alkali-resistant" cellulose, also

called "alpha cellulose," corresponds fairly quantitatively to the amout of "pure cellulose" which is obtained by treatment of the chlorite holocellulose with 17.5% sodium hydroxide (274). Figure VIII-3 (see Chapter VIII) shows a chart of the components of the extractive-free cell wall (276).

1. SUMMATIVE WOOD ANALYSIS

The summative analysis of wood is based on the isolation and identification of certain groups of wood components and therefore does not deal with the determination of chemically uniform substances. These groups are chiefly cellulose, hemicelluloses, lignin, and extractives. A more exact definition of what is meant by these individual terms will be given below. To these group terms are often added other designations, e.g., "holocellulose," "alpha cellulose," or sometimes terms originating from the method used or the name of the discoverer of the process, such as "Cross and Bevan cellulose," which does not include any further chemical or physical definition but refers only to the method,

By "cellulose" is meant a linear polysaccharide built up from anhydroglucose units which are connected with each other by $1 \rightarrow 4\text{-}\beta\text{-}$ glucosidic linkages and possess an orderly structure. The "hemicelluloses" comprise all noncellulosic polysaccharides and related substances such as, for instance, the uronic acids and their derivatives. "Lignin" is a substance the structure of which has not yet been completely elucidated but which contains a basic skeleton of four or more substituted phenylpropane units per molecule. The "extractives" consist of a large number of organic compounds which can be extracted with organic solvents and also, partially, with water, and which do not belong to the hemicellulose group.

As mentioned above in the Introduction, the isolation of cellulose from the wood structure can be achieved by alternative treatments with chlorine and alkaline reagents; the cellulose thus obtained is called "Cross and Bevan cellulose" after the chemists who first described the method. Such cellulose still contains some noncellulosic components which the authors called "beta and gamma cellulose." These are now included in the "hemicelluloses."

When Cross and Bevan cellulose is treated with strongly alkaline solutions, e.g., with 17.5% sodium hydroxide, a residue which is resistant toward concentrated alkali remains; it is called "alpha cellulose" and consists chiefly of cellulose, although minor amounts of alkali-resistant hemicelluloses are still attached to it. The isolation of the cellulose is also achieved by other methods, especially the alternate treatment with acidified sodium chlorite solution and extraction with potassium hydroxide (274). The residue remaining after this treatment contains almost the total carbo-

hydrate portion of the wood and is called "holocellulose"; it represents a complex consisting of the cellulose and the hemicelluloses. The extraction of the holocellulose with strong alkali also gives the "alpha cellulose."

The methods described have the goal of delignifying the wood and of isolating the carbohydrate portion in a state as unchanged as possible. From the latter, the cellulose and the hemicelluloses can be separated by suitable procedures. Understandably, these methods are not perfect, because, in order to achieve a complete delignification, a certain decomposition of the hemicelluloses cannot be avoided because of their close association with the lignin. The same applies to the extraction of the hemicelluloses from the holocellulose. Here, too, certain low-molecular portions of the cellulose may possibly go into solution. The data given for the summative wood analysis must be accepted with reservations because of the testing methods used. An estimate of the reliability of these various methods may be found in the respective literature given in the Bibliography.

The term "hemicelluloses" has been further specified by the use of names such as polyuronides, mannan, galactan, xylan, and araban, although these are by no means homopolymer complexes. For the identification of a hemicellulose portion consisting chiefly of xylose and arabinose and called "pentosan," a hydrolysis with hydrochloric acid is used. Pentosans are converted by the addition of water into pentoses and the latter, by the splitting off of water, into furfural. The reaction, which takes place according to the following equation:

$$C_5H_8O_4 + H_2O \longrightarrow C_5H_{10}O_5 - 3H_2O \longrightarrow C_5H_4O_2$$

obviously is not a direct dehydration of the pentosan. This reaction will be discussed further in the technical wood hydrolysis (see Chapter IV, Section 5.6).

For the determination of xylan it is hydrolyzed to xylose, but its estimation is complicated and the results obtained are unsatisfactory. Biological methods, however, give more reliable data than the oxidation to xylonic acid and its crystallization by double salt formation.

Araban, per se, is not a wood component, but polysaccharides which give arabinose on hydrolysis are widely distributed. The arabans are easily hydrolyzable and therefore are not usually present in pulp. In beechwood 0.5%, and in sprucewood 0.7% arabinose has been found (85,272,276,278).

In coniferous woods, mannans are present in an amount of up to 10%, while in deciduous woods, only minor amounts are found. In this case also, no homopolymer mannan can be isolated from wood. The determination is carried out by hydrolysis of the polysaccharide to simple sugars and precipitation of the mannose with phenylhydrazine as its hydrazone (277).

Galactan is a regular component of the polysaccharides of the wood

and is often present in combination with anhydropentoses, for instance, as arabogalactan. With the exception of that in larchwood, the amounts of galactans in wood are small. Galactan also is readily hydrolyzed and therefore is present only in very small amounts in pulp. Its determination is carried out by hydrolysis to galactose and subsequent oxidation with strong nitric acid to mucic acid. Since dicarboxylic acids, such as saccharinic acid and oxalic acid, are formed simultaneously, the yield of mucic acid is not quantitative (51,142,199). Here, too, enzymatic methods have proven to be more reliable (271).

In evaluating the results of the summative wood analysis, it is important to take into consideration the analytic methods used. Moreover, it must be remembered that terms such as holocellulose, alpha cellulose, mannan, xylan, etc. do not represent uniform chemical substances but, as mentioned above, are products formed from the association of two or more wood polyoses. In spite of this limitation, the summative wood analysis gives a good insight into the chemical composition of woods. In addition, for many intended technical uses, a complete chemical analysis is not required. For the pulp industry, for example, the cellulose content of the wood is of prime interest with regard to the expected pulp yield. For the hydrolysis of wood, the amount of fermentable sugars is of interest, whereas for the production of furfural, the pentosan contents are important. Methoxyl groups play a role in the pyrolysis for the production of methanol, whereas the acetyl groups determine the yield of acetic acid. Extractives may be important for various commercial uses.

In Table III-1, the chemical compositions of ten North American woods are given with the various polysaccharides in the carbohydrate fractions being given quantitatively. In three hardwoods, beech, birch, and maple, the amount of glucan was found to be 45–47%, of xylan 16–17%, and of mannan 2–3%; galactan and araban were present only to the extent of 1% and 0.5%, respectively. Aspenwood contained more glucan (57%) and birchwood more xylan (25%) than all the other species. The comparable data for the softwoods (conifers) are 45–47% glucan, 8–13% mannan, 4–7% xylan, 1–2% galactan, and 0.5–1.5% araban. The data for the alpha cellulose have been corrected for the mannose, xylose, and noncellulosic glucose residues for an approximate determination of the "pure" cellulose. In this way, the cellulose content of aspenwood was reduced to 53%, that of the other hardwoods to 41%, and of the softwoods to 41–45%. All the woods mentioned in Table III-1 contained noncellulosic anhydroglucose units, the origin of which is unknown although those of the coniferous woods may derive from a glucomannan hemicellulose (247).

For the isolation of the alpha cellulose component, the chlorine-holocellulose method was used (18). The holocelluloses were extracted with 17.5% sodium hydroxide or with 24% potassium hydroxide. The

TABLE III-1

CHEMICAL COMPOSITION OF TEN SPECIES OF NORTH AMERICAN WOODS (247)

Wood species	α-Cellulose	α-Cellulose[b]	Lignin	Acetyl	Ash	Uronic anhydride	Galactan	Glucan	Mannan	Araban	Xylan
Trembling aspen (*Populus tremuloides*)	56.5	53.3	16.3	3.4	0.2	3.3	0.8	57.3	2.3	0.4	16.0
Beech (*Fagus grandifolia*)	45.2	42.1	22.1	3.9	0.4	4.8	1.2	47.5	2.1	0.5	17.5
White birch (*Betula papyrifera*)	44.5	41.0	18.9	4.4	0.2	4.6	0.6	44.7	1.5	0.5	24.6
Red maple (*Acer rubrum*)	44.8	44.1	24.0	3.8	0.2	3.5	0.6	46.6	3.5	0.5	17.3
Balsam fir (*Abies balsamea*)	47.7	44.8	29.4	1.5	0.2	3.4	1.0	46.8	12.4	0.5	4.8
Eastern white cedar (*Thuja occidentalis*)	48.9	45.4	30.7	1.1	0.2	4.2	1.5	45.2	8.3	1.3	7.5
Eastern hemlock (*Tsuga canadensis*)	45.2	42.4	32.5	1.7	0.2	3.3	1.2	45.3	11.2	0.6	4.0
Jack pine (*Pinus banksiana*)	45.0	41.6	28.6	1.2	0.2	3.9	1.4	45.6	10.8	1.4	7.1
White spruce (*Picea glauca*)	48.5	44.8	27.1	1.3	0.3	3.6	1.2	46.5	11.6	1.6	6.8
Tamarack (*Larix laricina*)	47.8	43.9	28.6	1.5	0.2	2.9	2.3	46.1	13.1	1.0	4.3

[a] Percent based on extractive-free wood.
[b] Corrected for nonglucan material.

former proved to be the more effective. For the determination of the monosaccharides, the wood was hydrolyzed with sulfuric acid into the corresponding sugar mixture (213) and the latter was chromatographed by means of two different solvent mixtures. All the hydrolyzates contained galactose, glucose, mannose, arabinose, xylose, and various uronic acids (247).

The alpha cellulose contents of the woods listed in Table III-1 were also determined by the nitration method which differs from the classical method. This method, which was originally designed for pulp (36,37, 232), has proved to be applicable also for the direct nitration of wood (233,244,245). In Table III-2 the data for the alpha cellulose determined according to the conventional method and according to the nitration method are compared. With the exception of the value for balsam fir, the alpha cellulose content of all the woods tested is somewhat higher than the value for the cellulose obtained by nitration. The difference is undoubtedly partly due to the fact that the alpha cellulose contains more nonglucan material than the cellulose regenerated from the nitrates. The latter was therefore carefully denitrated and both celluloses were subsequently hydrolyzed. The hydrolyzates were separated by paper chromatography and investigated quantitatively by a spectrophotometric method.

The composition of the two types of cellulose, i.e., the alpha cellulose and the cellulose obtained by nitration, is shown in Table III-3. It can be seen that the alpha cellulose contains minor amounts of araban while the cellulose obtained by nitration contains practically none. The sugars obtained from the cellulose nitrates from tamarack after denitration and

TABLE III-2

CHEMICAL COMPOSITION OF TEN SPECIES OF NORTH AMERICAN WOODS[a] (246)

Wood species	α-Cellulose	Lignin	Pentosan	Uronic anhydride	Cellulose in denitrated cellulose nitrate
Trembling aspen	56.6	16.3	18.9	3.28	52.1
Beech	45.8	22.1	21.0	4.76	39.1
White birch	44.5	18.9	25.5	4.63	42.3
Red maple	44.7	24.0	20.2	3.47	42.8
Balsam fir	47.7	29.4	7.7	3.40	48.5
Eastern white cedar	48.6	30.7	11.4	4.23	44.6
Eastern hemlock	45.2	32.5	7.0	3.27	41.3
Jack pine	45.0	28.6	11.4	3.93	41.3
White spruce	48.5	27.1	10.6	3.60	46.0
Tamarack	47.8	28.6	7.9	2.88	43.8

[a] Percentage based on extractive-free, oven-dry wood.

hydrolysis contain galactose because of the high galactan content of this wood. Alpha celluloses from hardwoods contain more pentosans than mannans, while, vice versa, the softwoods have more mannans than pentosans.

When the data for the alpha cellulose and the cellulose obtained by nitration are corrected for their nonglucan content, the values shown in Table III–4 for the cellulose content obtained by both methods are found. As the table shows, there is wide agreement between the cellulose content

TABLE III-3

Composition of α-Cellulose and Denitrated Cellulose Nitrates from Different North American Wood Species[a] (246)

Wood species	α-Cellulose			Denitrated cellulose nitrate		
	Glucan	Mannan	Xylan and araban	Glucan	Mannan	Xylan and araban
Aspen	94.9	1.7	3.8	98.6	1.0	0.4
Beech	91.8	1.1	7.1	98.2	1.3	0.5
Birch	92.2	1.4	6.4	98.4	1.3	0.3
Maple	91.8	1.2	7.0	99.0	0.8	0.2
Balsam fir	94.1	4.0	1.9	93.0	7.0	—
Cedar	92.8	3.9	3.3	93.7	6.3	—
Hemlock	94.8	3.8	1.4	94.0	6.0	—
Pine	92.4	5.0	2.6	96.6	3.4	—
Spruce	92.5	6.2	1.3	92.3	7.2	0.5
Tamarack	92.0	6.1	1.9	93.6	4.6	1.8[b]

[a] Percentage based on extractive-free, oven-dry wood.
[b] Galactan.

TABLE III-4

Comparison of the True Cellulose Contents of Woods Obtained from Their α-Cellulose and the Denitrated Cellulose Nitrate after Correction for Their Nonglucan Content (249)

Wood species	α-Cellulose[a]	Denitrated cellulose nitrate[a]
Aspen	53.3	51.4
Beech	42.1	38.4
Birch	41.0	41.6
Maple	41.4	42.4
Balsam fir	44.8	45.1
Hemlock	42.4	38.8
Pine	41.6	40.0
Spruce	44.8	41.0
Tamarack	43.9	42.5

[a] Percentage based on extractive-free, oven-dry wood.

TABLE III-5

SUMMATIVE ANALYSIS OF EIGHT SPECIES OF CANADIAN WOODS (47)

Component[a]	Balsam fir	White spruce	Black spruce	Jack pine	Eastern hemlock	White pine	White birch	Aspen poplar
Alcohol-benzene	4.25	3.06	2.55	6.54	2.18	9.77	4.88	3.82
Hot water extract[b]	0.39	0.69	0.75	1.83	1.69	1.16	0.69	0.82
Ash	0.40	0.22	0.21	0.19	0.28	0.18	0.29	0.38
Acetyl	1.52	1.08	1.14	1.08	1.32	1.15	4.94	3.41
Lignin	27.70	26.96	27.25	27.38	29.56	25.60	18.48	18.12
α-Cellulose	49.41	50.24	51.10	47.52	53.04	48.10	40.97	49.43
Hemicelluloses	15.41	16.39	15.18	16.18	12.51	14.13	27.25	21.18
Total	99.08	98.64	98.18	100.72	100.58	100.09	97.10	97.16

[a] All values in percent.
[b] Hot water extraction on residue from alcohol-benzene extraction.

TABLE III-6

CHEMICAL COMPOSITION OF THREE SCANDINAVIAN WOOD SPECIES (211)

Wood species[a]	Ash	Extractives[b]	Protein[c]	Lignin	Acetyl	Uronic anhydride	Glucan	Mannan	Araban	Xylan	Cellulose[c]	Hemicelluloses[c]	Lignin
Spruce (*Picea abies*)	0.4	1.8	1.3	28.6	1.4	2.5	44.3	10.3	0.5	7.6	43	27	29
Pine (*Pinus silvestris*)	0.4	5.3	1.2	27.8	1.6	2.4	44.8	7.6	0.6	7.2	44	26	29
Birch (*Betula verrucosa*)	0.3	3.1	2.5	19.5	4.8	5.9	37.5	0.5	0.5	24.6	40	39	21

[a] Percentage on unextracted dry wood basis.
[b] Alcohol-benzene extraction.
[c] Percent of gross composition.

obtained by nitration and the content of "corrected" alpha cellulose, but it must be kept in mind that the nitration of, for example, beechwood or hemlock may be scarcely quantitative. As mentioned above, the terms glucan, xylan, araban, and mannan must not be taken literally because they represent only the corresponding polysaccharides. Glucan may therefore be most nearly identical with cellulose (246).

Investigations of Canadian woods gave the values shown in Table III–5. The following woods were investigated: balsam fir (*Abies balsamea*), white spruce (*Picea glauca*), black spruce (*Picea mariana*), Jack pine (*Pinus banksiana*), white birch (*Betula papyrifera*), Eastern hemlock (*Tsuga canadensis*), white pine (*Pinus strobus*), and aspen (*Populus tremuloides*). The summative analysis covered the solubility in benzene-alcohol, the solubility of the residue of the benzene-alcohol extract in hot water, and the determination of the acetyl content, the lignin content, the alpha cellulose, and the hemicelluloses. The corresponding standard methods of TAPPI were applied for the analyses. A comparison of the results for coniferous and deciduous woods as given in Table III–5 again show that the deciduous woods have a higher pentosan, hemicellulose, and acetyl content, but a lower lignin content than the coniferous woods. With the exception of birch, which has an alpha cellulose content of about 41%, all the other woods have an average alpha cellulose content of around 50%, with minor deviations (47).

Table III–6 gives a comparison of the composition of three Scandinavian woods, but it should be mentioned that the results from the various woods have been obtained by different investigators. The difference between the composition of coniferous and deciduous woods is again shown by their lignin and hemicellulose contents (211).

2. CELLULOSE

2.1 Chemical Structure

Cellulose merits special consideration because it forms the main component of the cell wall and because of its special technical importance. At the beginning of this century, cellulose, as the component of the cell wall that is resistant toward alkalis and acids, was given the elementary formula $(C_6H_{10}O_5)_n$, in which a great many glucose units are combined into a chain, the ends of which are closed to form a ring. Not until the 1920s was the concept that cellulose is a linear macromolecule built up of anhydro-β-glucopyranose units confirmed. The methylation of cellulose provided clues as to its constitution through the formation of high yields of 2,3,6-trimethylglucose (54,259). The glucose residues in the cellulose molecule

therefore could be connected only at the 4- and/or 5-position in the form of α- or β-glucosides. Of special significance was the finding that the sugar units occurred normally in the form of pyranose rings, which made it appear probable that the units are combined by 1 → 4-oxygen bridges (11, 113,132).

The hydrolysis of octa-O-methylcellobiose to 2,3,6-trimethyl- and 2,3,4,6-tetramethylglucose and the oxidation of cellobiose to cellobionic acid, the methylation of the latter, and its subsequent hydrolysis to 2,3,4,6-tetramethylglucose and the γ-lactone of 2,3,5,6-tetramethylgluconic acid led to the complete elucidation of the constitution of cellobiose. The results of these two degradation reactions showed that the nonreduction moiety of the cellobiose is combined by a glucosidic oxygen bridge through the hydroxyl group in the 4 position of the reducing moiety (112,115).

The acetolysis of cellulose gave a cellotriose and a cellotetraose acetate (22,83). In the acetolytic degradation of cellulose, acetylated oligosaccharides were formed; they were separated by chromatographic analysis into the acetates of cellobiose, cellotriose, cellotetraose, cellopentaose, and cellohexaose and were identified as such in their crystalline forms (56). With these findings, all earlier constitutional formulas for the cellulose molecule became invalid, with the exception of the structural formula given in Fig. III–1. Finally, proof was presented that the glucose has a 6-membered ring structure and that the cellobiose is structurally a 4-glucopyranosylglucose and, consequently, the cellulose must be considered to be a linear macromolecule composed of numerous hexose units combined with each other by glucosidic main-valency linkages (110,112,114,122,238).

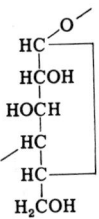

Figure III–1. Formula for cellulose (122).

Questions as to the number of glucose or cellobiose units in the cellulose molecule have resulted in an extensive study. At first, an attempt was made to establish the molecular size of the cellulose by the determination of the end groups on the assumption that not only aldehyde groups but also carboxyl groups are present. From the consumption of hypoiodite, the presence of aldehyde groups should be detectable and it should be possible to calculate the number present. Other proposals suggested the oxidation of the aldehyde groups to carboxyl groups and determination of the latter (20,130,220,221,265).

In this connection it is interesting to note that the acidity of cellulose is not due to the presence of carboxyl groups but results from the accumulation of hydroxyl groups. On the basis of the Donnan-equilibria theory, it could be shown that insoluble acids such as cellulose can be titrated only in the presence of a strong electrolyte. In the presence of 0.6 N sodium chloride solution the dissociation constant of native cellulose was found to be 8×10^{-8}, whereas the dissociation constants of α-D-galacturonic, glycolic, or lactic acid, for example, were in the order of magnitude of 10^{-4}. Cellulose is therefore 10^3 to 10^4 times weaker than the hydroxy acids mentioned, and native cellulose should not contain any carboxyl groups (279). If it is assumed that cellulose is built up of glucose molecules, then on its complete methylation and subsequent acid hydrolysis, two different types of methylglucose are to be expected. One of the two end-groups gives 2,3,6-trimethylglucose, the other 2,3,4,6-tetramethylglucose (132). On hydrolysis of trimethylcellulose, however, only 2,3,6-trimethylglucose and no tetramethylglucose were obtained. The cellulose chain, therefore, must consist of several hundred glucose units (81). The chain length of cellulose found in this way varies between 700 and 2000 glucose units (87,116).

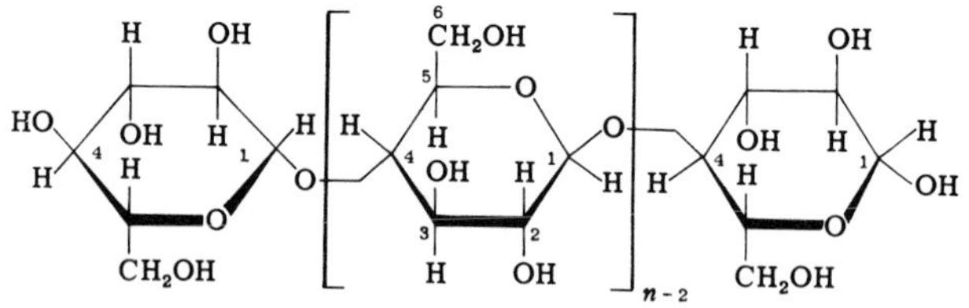

Figure III-2. Cellulose, Haworth formula (111).

In addition to the evidence provided by the isolation and identification of the oligosaccharides obtained on acetolysis of cellulose, its chainlike structure, formed from glucose units or cellobiose building stones, is confirmed by the development of the optical rotation from the bioses up to the cellulose, by investigations of the kinetics of the hydrolysis of cellulose with sulfuric acid, and also by studies of the preparation of polymer homolog series of cellulose derivatives (73,80,150,236,273). Moreover, the possibility of cross-linkages between the long, parallel-arranged cellulose macromolecules, thus forming a network such as has been shown with proteins, for example, plays an important role (91,175). Network formation, however, would render impossible the estimation of the average length of cellulose chains by end-group determination. That cross-linkages between neighboring chains are possible is shown by the fact that, when rayon fibers are treated with formaldehyde, a methylene content of as little as 1% to 2% is sufficient to change completely the properties of the fibers (121).

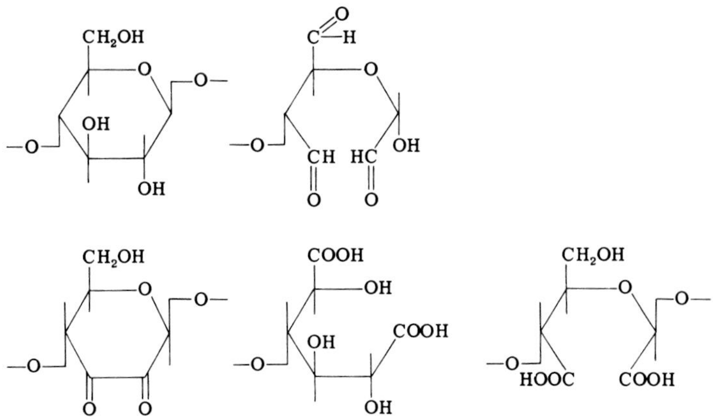

Figure III-3. Cellulose functional groups.

The oligosaccharide series from the cellobiose up to the cellopentaose very probably consist of anhydro-β-glucopyranose units which are combined with each other to form an open chain. Figure III-2 gives a good insight into the possible structure and the stereochemical relationships of the anhydroglucose building stones in these oligosaccharides. In this connection, it must be taken into consideration that the atoms in the anhydroglucose ring do not need to be arranged in a plane. These anyhydroglucose units are combined with each other by 1 → 4-glucosidic linkages and their end-groups differ from each other in the position of the carbon atoms in the ring to which the hydrogen and the hydroxyl groups are attached. The degree of polymerization (DP), i.e., the number of anhydroglucose units within the cellulose chain, is synonymous with *n*, which may be an even or an odd number. The stereochemical relationships are indicated by thick and thin lines in the three-dimensionally designed formula, with thick lines representing the foreground. The thick vertical linkages combining the substituted atoms lie above the ring plane, while the thin lines combine the atoms below the ring plane. In this way, the complete "transarrangement" of the ring substituents becomes apparent. The formula also shows that the subsequent rings are turned around 180°, but the puckered arrangement of the five carbon atoms in the ring is not discernable (111).

The functional groups in the cellulose molecule have an appreciable effect on its chemical and physical properties. The principal functional groups in pure cellulose are undoubtedly the hydroxyl groups. Cellulose, as a polyalcohol, undergoes oxidation and the groups formed in such a reaction are aldehyde, keto, and carboxyl groups. Aldehyde groups can be formed at the C-6 atoms of the cellulose chain and also at the C-2 and/or C-3 atoms when the pyranose ring has been split. Keto groups can be formed at the C-2 and/or C-3 atoms and carboxyl groups at the C-1 of the

reducing end group or the C-6 atoms and also at the C-2 and/or C-3 atoms after the pyranose ring has been opened. Figure III-3 shows these different possibilities (183).

For the determination of the reducing end groups, oxidative methods are used. These methods are purely empirical and have, primarily, only technical diagnostic importance, serving for purposes of comparison. Another group of methods attempt to obtain stoichiometric results. The oxidation with alkaline cupric sulfate solution belongs to the first group of methods and has gained importance in the industry under the term "copper number." The methods for the stoichiometric determination of reducing groups use, as oxidation agents, acidified sodium chlorite, alkaline sodium hypoiodite, and acidified potassium permanganate solutions. A detailed description of these methods can be dispensed with here, but may be found elsewhere (188).

The methods of oxidation and the reduction with borohydride are used for the determination of the carbonyl groups in hydrocellulose and in oxidized cellulose. The above-mentioned copper number method is also applicable here. The value of all these methods is limited and, in particular, there is no reliable method for differentiating between aldehyde and keto groups. In spite of their limitations, the technical importance of these methods should not be underestimated (97).

The presence of carboxyl groups gives the cellulose the character of a weakly acid ion exchanger. Almost all the methods for the determination of the acid groups are based on this property of ion exchange on the part of the cellulose. Alkalimetric methods are generally more suitable than methods that are based on the exchange of large ions, such as methylene blue; this is especially the case when no carbonyl groups, which disturb the test, are present. In the alkalimetric methods, the metal ions attached to the carboxyl groups are exchanged for hydrogen ions by treatment with acid. After being washed with deionized water, the cellulosic carboxyl groups can be neutralized with alkali. Of special interest are methods that permit the determination of the carboxyl groups in oxidized or partially hydrolyzed cellulose preparations. For this purpose, a modification of the above-mentioned methylene-blue method by the substitution of calcium acetate or zinc acetate, or, even better, the use of the alkalimetric method has been found to be suitable. In all these methods, the cellulose is first converted into its "free acid form."

There are other methods, however, that avoid the prior conversion into the free acid form. In these, the cellulose is conditioned by a sodium chloride solution adjusted to pH 7.5 to 8.0, the excess liquid is removed, and the adhering solution is determined by weighing. After elution with 0.01 N hydrochloric acid and washing, the combined eluate and wash-water are brought to the boiling point to expel any carbon dioxide formed, and the

free hydrochloric acid is titrated with barium hydroxide. In a modification of this method, the cellulose sample is treated with magnesium chloride solution, washed with deionized water, and the magnesium ions are eluted with 0.1 N hydrochloric acid and determined by complexometric titration (214).

2.2 The Submicroscopic Structure

Microscopic examination of the vegetable cell wall early led to the assumption that submicroscopic particles in the cell wall are responsible for its optical anisotropy. These particles are called micelles (90,185). Such concepts appeared in the literature as the micellar theory and were later confirmed as being characteristic for the crystalline structure of cellulose (6). In the early 1920s, the study of the submicroscopic structure of organic substances was given great impetus by the development of x-ray diagrams. Thus, at least a partially crystalline structure could be confirmed (120,190). Further investigations provided the basic data for the study of the crystal lattice in which rows of crystals are arranged chiefly parallel to the fiber axis (200,201). The diameter of the hexose molecule could be calculated to be 5.13 A and the interference lattice corresponded to the established pyranose formula. Cellulose fibers showed a periodic structure, with a periodical distance of 10.25 A along the fiber axis; this distance is nothing else than the length of a cell unit and exactly corresponds to the length of two glucose units. In this way, the concept (established on a purely chemical basis) that cellulose consists of cellobiose units was confirmed (234,235). Finally, it could be shown that the periodical repetition of the cellobiose units can be explained by the concept of a diagonal screwlike arrangment in which the two glucose groups are turned at an angle of 180° within a cellobiose unit.

The arrangement of the chains in the elementary cell of native cellulose is shown in Fig. III-4. The dimensions of this monoclinic cell are as follows: $a = 8.35$ A; $b = 10.3$ A (fiber period); $c = 7.9$ A; and $\beta = 84°$ (174,176). Thus the chains in the a-b plane are kept together, not only by polar forces between the OH groups, but also by hydrogen bridges. In the direction of the b axis the strongest forces are active; in the direction of the a axis the H bridges are weaker but still contribute to the strength of the intermolecular cohesion; in the direction of the c axis only the attractive forces of the OH dipoles and the permanent electrical moment of the C-O-C groups are active (176). The diameter of a glucose unit, according to the formula in Fig. III-2, is therefore calculated to be 5.13 A.

It must be mentioned that objections have recently been raised against the unit cell of native cellulose as shown in Fig. III-4. In particular, it has been said that possible intramolecular hydrogen bridges between the hydroxyl group at the carbon atom 3 of a glucose unit and the ring oxygen

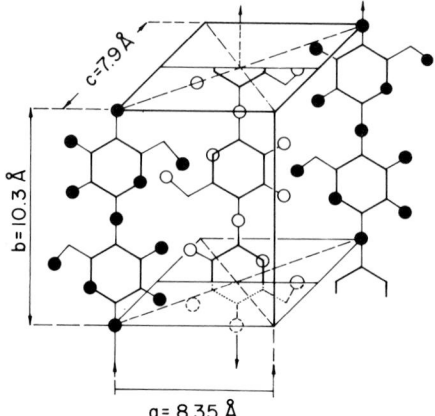

Figure III-4. Schematic representation of the unit cell of native cellulose (177).

of an adjacent glucose group cannot be explained by this formula. Intermolecular hydrogen bridges between the hydroxyl groups at the carbon atoms 6 of antiparallel chains and the bridge oxygen atom of adjacent parallel chains could also be possible (63,160).

Cellulose occurs in more than one crystalline modification. When native cellulose is converted into a derivative and the latter is reconverted into cellulose, the diagram of "regenerated cellulose" is obtained. Such regenerated cellulose is termed, not entirely correctly, "hydrate cellulose." Because celluloses which have this crystalline structure have the same elementary composition as native cellulose has, it is more correct to call them cellulose II than hydrate cellulose.

Figure III–5 shows the projection in the plane *a-c* of the elementary cell of native cellulose (cellulose I) and of mercerized cellulose (cellulose II). In cellulose II the molecules are displaced against each other along the *a* axis and are twisted 30° from the *a-b* plane. The dimensions of this monoclinic cell are: $a = 8.14$ A, $b = 10.3$ A, $c = 9.13$ A, and $\beta = 62.8°$ (7). According to Fig. III–5, the chain which lies in the middle of the cell runs in the opposite direction from the chains which are located in the corners. If one bears in mind that cellulose II is obtained by the regeneration of cellulose molecules from different solutions of cellulose and cellulose derivatives and that the single cellulose molecules have been completely in solution, it seems very probable that the cellulose chains are not all arranged in the same direction and that a certain random distribution of the chains in both directions occurs. This, however, leads to the assumption that also in the crystal lattice of cellulose I, the chains run in opposite directions (62,63).

The extremely high viscosity of cellulose solutions has given rise to

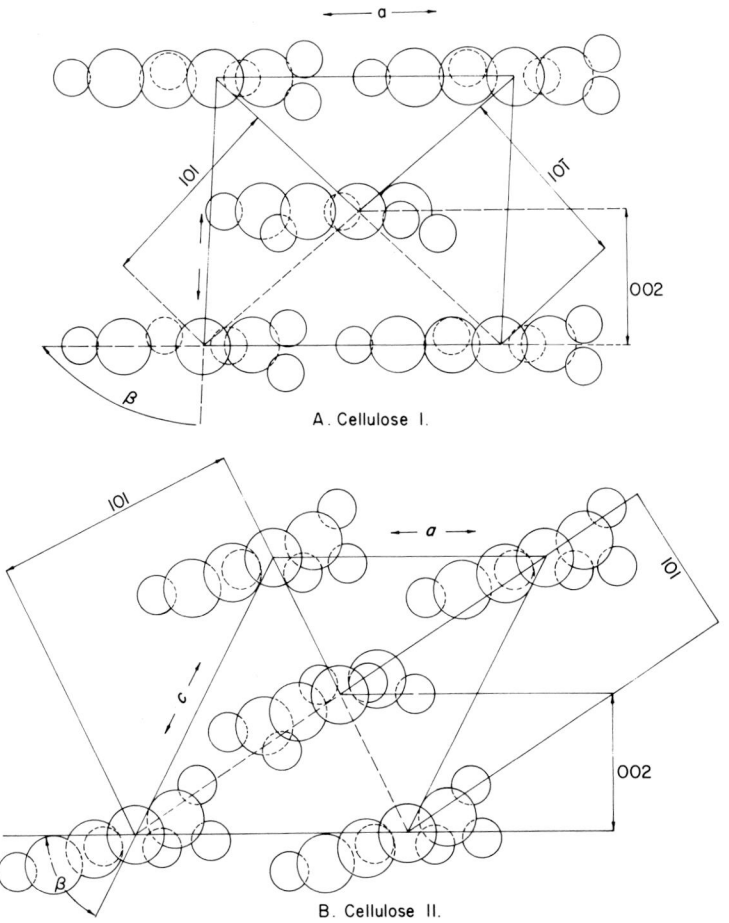

Figure III-5. View in the direction of the *b* axis of the unit cell of native cellulose (Cell I, A) and mercerized cellulose (Cell II, B).

	A.		B.	
a =	8.20 A		*a* =	8.14 A
b =	10.30 A		*b* =	10.30 A
c =	7.90 A		*c* =	9.13 A
β =	83.3°		*β* =	62.8°

the question of the state in which the cellulose and its derivatives are present in such solutions. The concept, long accepted, that main valence chains of 30 to 50 glucose units are combined into 40 to 60 chains and form a "micelle" (120,174), must be considered to be outdated by results obtained from a study of macromolecular chemistry. According to the latter, the high viscosity of cellulose solutions is caused exclusively by the extraordinary size of the molecules, because filamentary molecules are separated from each other in solution; the "micelles" dissolve into macromolecules which,

Figure III-6. The structure of bast fiber: above, transversal section; below, longitudinal section. The intermicellar spaces are shown in black; the white portions of the figure represent spaces occupied by cellulose chains (92).

in turn, are readily degraded, and celluloses of different degrees of polymerization can be formed. The theoretical and physical proofs for the behavior of macromolecules in solution cannot be discussed in detail here, but the relationships between viscosity, molecular size, and the degree of polymerization form the basis for the development of more recent concepts of the structure of high-polymer compounds and are therefore of great importance for the whole artificial fiber and plastic industries (236).

When the filamentary molecules are precipitated from a solution, they congregate again in bundles. In this way, amorphous or crystalline substances may be formed. While in the first case the molecules are arranged at random, in the second they are ordered in a latticelike form. It must be taken into consideration, however, that a macromolecular lattice is formed and this, in contrast to a normal molecular lattice, has no basic surface (240). In the cellulose, the filamentary molecules permeate the fibers in a parallel direction to the optical axis of the rod-shaped mixed bodies which are formed from the cellulose substance and the intermicellar system. The axis of these mixed bodies runs, in general, in a helical direction and only rarely at a slight angle, approximately parallel to the fiber axis. The chain molecules combine at certain intervals to form regions of ordered crystal lattices which are outlined in Fig. III-6 at the right side, by dashes. These crystal lattice regions correspond to the micelles (185) that have already been mentioned. These are not individual units but are combined by thread

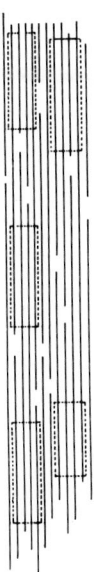

Figure III-7. Chain lattice of a fiber molecule with regions of a lattice order (240).

molecules which project beyond the ordered bundles or even traverse several ordered lattice regions and thus contribute to the structure of various crystalline regions. When the thread molecules consist of identical and repeating chain members, homogeneous lattice regions can be outlined, as shown in Fig. III-7. The regions of the crystal lattice which have ends of molecules, however, show inhomogeneities, but, in addition to these, flaws in the lattice occur with very long thread molecules in such a way that chains, the longitudinal periods of which are somewhat displaced against each other, do not fit into the crystal lattice because of their length and immobility or because they are hindered in their mutual association by hydrate water liberated during the crystallization. In this way, small flaws, running longitudinally, which are accessible to solvation agents and which are called in biological terms "intermicellar spaces," are formed. On the basis of such theories, starting from the macromolecular lattice, one arrives at the same scheme as that shown in Fig. III-6 and deduced from the phenomena of the rod double-refraction, the rod dichroism, the swelling on isotropy, etc. (92).

As regards the structure of gels, fibers, and similar products, one can therefore speak only of a micellar system and not of individual micells. This micellar system represents a dispersed system that consists of two components, i.e., the net structure and the hollow cavity system. One phase is represented by the cellulose substance, the other by gas or liquids in the cavities, or even by a solid material. In gels, when one component consists

of a network of single chain molecules that touch each other in some places, the chains within the cell structure can be arranged in a parallel order, forming crystalline regions, as is shown in Fig. III–7 (93).

From the width of x-ray interferences at half of maximum intensity, the thickness of the rods of ramie fibers has been determined photometrically to be 60 to 80 A and their length to be at least 600 A (117). Later measurements put the length at 1500 A (173). Electron optical investigations have shown that the diameter, sometimes given as up to 200 or 400 A, is explained by associations of individual microfibrils which, in turn, have a diameter averaging 100 A. The diameter of such microfibrils is obviously dependent upon the origin of the cellulose and increases in the following order: wood < cotton < bacterial cellulose < animal cellulose < ramie (202). The electron optically visible microfibrils are called micell-strings, thus showing a good relation with the dimensions found by x-ray studies (117).

The reactivity of cellulose is directly related to the ratio of crystalline to amorphous material. For the determination of this ratio, chemical and physical methods are available. For example, if the hydrogen atoms of the hydroxyl groups in the cellulose are exchanged by a reagent which cannot penetrate into the ordered regions of the fiber, then the ratio of reacted hydroxyl groups to unreacted groups gives a measurement of the proportion of reactive material present. Other methods utilize oxidizing or hydrolyzing reactions, the course of which can be determined exactly by their velocity. Such a reaction usually occurs rapidly at first, then slows down, and, finally, becomes constant. Assuming that the first phase affects the chains situated outside the crystalline areas, since they are more readily accessible, then the proportion of intramicellar material can be determined from the course of the reaction-time curve of the second part of the reaction by extrapolation. Such a conversion of the accessible hydroxyl groups has been carried out, for instance, with diazomethane. With dry cotton only up to 0.4%, with moist cotton up to 9.5%, and with mercerized cotton up to 23% "accessible" material was found. The reaction, therefore, obviously progresses only to the degree at which the capillaries and the pores of the amorphous region are opened by swelling (118,204,243).

Another method makes use of thallous ethylate in etheral solution, by means of which cellulose ethers with various molecular weights give different accessibility values. When these values are plotted against the molar volume of the ethers, a linear function is obtained from which the zero molar volume can be extrapolated. This number is defined as that portion of the cellulose fiber which is accessible to a liquid of a molar volume of zero, with no swelling properties toward cellulose and no penetrability into the crystalline regions. By this method, 27% of accessible material was found in mercerized linters (15,106).

The hydrolysis-oxidation method calculates the hydrolyzed cellulose

portion from the amount of glucose formed, with the latter being found from the amount of carbon dioxide formed on oxidation. From the hydrolysis-time curve, the degree of hydrolysis can be extrapolated to zero time by the use of some correction factors, and this zero time is used as a measurement of the amount of amorphous material. The strongest objections to these hydrolytic and oxidative methods are raised because of the fact that, during the hydrolysis, recrystallization may take place (32,189).

The accessibility of cellulose to water can also be determined by means of an isotopic exchange reaction between the hydroxyl groups and deuterium oxide. The reactive fraction of the cellulose can be determined from the difference in hydrogen content of the deuterium oxide before and after the reaction. Some of the hydroxyl groups of cellulose (up to 30%) which are hydrogen-bound in a way characteristic for a crystalline structure react with deuterium oxide as well as all the hydroxyl groups which are hydrogen-bound in an irregularly amorphous form. The difference in the hydrogen content of the heavy water before and after the reaction is determined by infrared spectroscopy (168).

The reaction of cellulose with liquid ammonia has also been used for the differentiation between the crystalline and amorphous regions of the cellulose. In this reaction, an ammonia cellulose is formed, capable of exchanging its hydroxyl hydrogen atoms for sodium. It is claimed that the method is in good agreement with the sorption method with heavy water (21,218,243).

The physical methods are based either on the evaluation of the diffused zone of x-ray diagrams by photometric means (15,170) or by colorimetric measurements of the sorption of water by cellulose (129,243). A modification of the method with heavy water is an exchange of the latter by tritium and subsequent measurement of the radiation. The hydroxyl groups of the crystalline cellulose are rendered accessible by their stepwise swelling by alkaline solutions of increasing concentration. This must be applied with caution because certain treatments with alkaline solutions may increase the crystallinity and also the packing density of the amorphous region.

The crystallinity of the wood cellulose that is present in the form of fibers cannot be measured directly by deuterium exchange and infrared absorption because its heterogeneous composition causes too great a diffusion of light. The tritium method, on the other hand, causes a partial exchange of hydroxyl groups. The total number of hydroxyl groups is calculated by taking into consideration the chemical structure of the wood components. The total average accessibility of the hydroxyl group is determined from the measurements of tritium exchange. The crystallinity of native cellulose is normally much lower (30% to 50%) than that of the regenerated cellulose. The accessibility of the different hydrogen bridges

varies, with the intramolecular bonds being more resistant than the intermolecular bonds. Analyses of wood, holocellulose, and purified wood cellulose gave accessibility values of 50% to 65% (203).

The use of interference heights in the x-ray-photometric method for the calculation of the empirical crystallinity leads to erroneous results. A planimetric method of the recorded diffraction diagrams has therefore been suggested (136). This method has been greatly simplified and improved by the use of a digital computer recorder (147). When pulverized fibers are used for the x-ray determination of the crystallinity, the preparation of the fiber sample can affect the result because an orientation of the pulverized fibers parallel to the surface necessarily occurs. To avoid this source of error, samples mixed with copper dust are prepared and are compressed into a pellet. From the latter, pieces are cut at varying angles to the direction of the pressure of the press and the diffractions of the resulting surfaces are measured. The integral intensity of the cellulose is compared with that of the copper. In comparison with 1.00 for bleached cotton linters, the crystallinity of purified flax fibers is calculated to be 0.95, of purified ramie 0.92, of spruce sulfite pulp 0.84, of polynosic fibers 0.64, of ordinary viscose rayon 0.51, and of high-tenacity rayon 0.56 (263).

With regard to the reactivity of cellulose, it can be said that it can react as a three-valent alcohol with one primary and two secondary hydroxyl groups per glucose unit. Disregarding reactions that cause a complete destruction of the chain structure of the molecule, the hydroxyl groups can be esterified by nitration, acetylation, or xanthation, and etherified by alkylation and benzylation, they can be replaced by other groups such as NO_2, NH_2, or halogen; the hydroxyl hydrogens can be exchanged by sodium and they can be oxidized to give aldehyde and carboxyl groups. Finally, addition compounds with acids, bases, or salts can be formed. Some of these reactions have already been mentioned and explained by examples, other reactions will be discussed in the following chapters in connection with various technological processes.

Consideration of the reaction mechanism on the basis of the simple formula of a three-valent alcohol explains only the type of reaction—it does not say anything about the way in which the reaction takes place with a high-polymer fiber-forming substance such as cellulose. The reactions with various inorganic and organic reagents are, to a large extent, controlled by the physical structure of the cellulose. Here, a differentiation must be made between purely surface-active reactions and reactions that are connected with swelling processes. In the first case, the capillaries, pores, and broken places remain inaccessible by the reagents (15,204). In reactions connected with swelling processes, macro- or microheterogeneous reactions may take place. When primary and secondary cell-wall components show different degrees of permeability for the reagent and also different degrees of swell-

ing, chiefly macroheterogeneous reactions occur, whereas a uniform swelling of the primary and secondary cell leads to microheterogeneous reactions. If the reagent can penetrate into the spaces between the microfibrils and into the amorphous regions, the reaction is strongly accelerated, but it still remains heterogeneous because the hydroxyl groups of the crystalline region cannot be reached. A homogeneous or uniform reaction can be achieved only when the reagent can penetrate both the amorphous and the crystalline regions. Even if all the hydroxyl groups of the cellulose are accessible to the reagent, a certain heterogeneity in the reaction product still remains because the substitution takes place at random along the chain molecule and three different hydroxyl groups are accessible, thus theoretically permitting eight possible substitutions.

The swelling therefore plays a decisive role in the course of the reaction. Purely surface reactions are very rare and can occur only when no swelling reagents are present. The acetylation in the presence of hydrophobic solvents and the first steps of the normal acetylation with acetic acid are macroheterogeneous reactions. It is possible that methylation and xanthation at the start of the reaction are also macroheterogeneous.

The hydrolysis of cellulose with hydrochloric and sulfuric acids, the oxidation with chromic acid and, to a limited degree, with periodic acid are microheterogeneous reactions. Most cellulose reactions take place microheterogeneously at the beginning. Homogeneous reactions, on the other hand, play a minor role, but they include the hydrolytic conversion of the primary to secondary cellulose acetate. Since it may be taken for granted that the cellulose is present in its solutions in the form of macromolecules, the reactions of dissolved cellulose take place homogeneously. The chain molecules, however, are not completely independent of each other except in very dilute solutions (236,243).

2.3 The Biosynthesis of Cellulose

Our knowledge of the biosynthesis of cellulose, as compared with that of numerous other polymers, is still at an early stage. It is known that nature converts certain basic building stones to a labile state and that organic catalysts, the enzymes, cause a transformation to a polymer. Only the important basic substances, such as sugars, fats, and amino acids, involved in the transformation to glycogen, starch, cellulose, and protein need be mentioned. The polymerization of the carbohydrates is basically a polycondensation, with the splitting off of water. When the reaction occurs in a water-free medium in the presence of inorganic catalysts, mixtures of polysaccharides, depending upon the reaction conditions and the catalyst used, are obtained, but these are either ill-defined or consist only of very low polymeric celluloses. Based on our knowledge gained in the synthesis

of the animal and vegetable polysaccharides, glycogen and amylose, experiments have been carried out to polymerize glucose by means of organic catalysts isolated from animal or vegetable organisms. The enzymatic polymerization was shown to consist of a series of reactions that depended upon three specific components. The monomer that is present in a labile form plays the role of a donor, while the primer represents a pattern of the system to be synthesized or a point of contact for the addition of further monomers. The enzyme is the catalyst which causes the splitting off of the monomers from the donor and the addition of the labile monomer to the primer without, itself, being a part of the reaction product. The action of the enzyme is entirely specific and occurs only with specific compounds.

An enzymatic synthesis of cellulose without the participation of living cells was carried out successfully first with cell-wall fragments that already contained cellulose, and inactivated bacteria of *Acetobacter xylinum*. The intact vegetable cell is surrounded by a thick cell wall which may be impermeable by intermediate products that can stimulate the synthesis of cellulose. Moreover, it is extremely difficult to determine the increased growth of cellulose in the presence of that already there in large amounts. Nor is it easy to isolate active cells from which the cell wall has been removed. These circumstances have led to a study of the synthesis of cellulose with active cellulose-forming bacteria. *Acetobacter xylinum* can convert 26% of the glucose in a sugar solution into cellulose. Of main interest now was the question, which enzyme system is capable of forming cellulose? On degradation of the cells by ultrasonic waves or other mechanical means, a pronounced reduction in the action of the cellulose-forming enzyme system was observed. By tracer technique the incorporation of glucose moieties from uridine diphosphoglucose (UDPG) into cellulose could be proven, whereas glucose itself or glucose-1-phosphate was not transformed into cellulose by the enzyme preparation. Uridine diphosphoglucose was therefore identified as a direct precursor of cellulose (55).

The enzymatic opening of the cell resulted in the almost complete preservation of the enzyme system capable of synthesizing cellulose (264). The preparation obtained by the enzyme lysozyme could be separated into the cell covering and the cell content; in this way it was found that, strangely enough, the cellulose-synthesizing capacity was situated in the cell covering and not in the cytoplasm. The cell covering, in turn, consists of the stiff bacterial cell wall and the cytoplasmic membrane. When the membrane is disintegrated with trypsin, the cell covering loses its capacity to synthesize cellulose, and thus it appears that the cytoplasmic membrane is the seat of the synthesis of the cellulose precursor (55).

How far these findings with bacterial cellulose can be applied to the formation of cellulose in the vegetable cell requires further investigation.

The cytoplasmic membrane in the vegetable cell is often firmly joined to the cell wall, but, on the other hand, it is surmised that the microfibril synthesis takes place chiefly in the outer regions of the cytoplasm (55). This would mean that the cellulose synthesis occurs preferentially at the inner surface of the cell wall. The cytoplasmic membrane in plants therefore could exercise the same function as *Acetobacter* does. By means of autoradiography it has been found that the synthesis of cellulose takes place over the whole cell wall (229); it is assumed, however, that the precursor of the cellulose molecule is formed within the cytoplasmic membrane and then transported through the cell wall to the site of the microfibril where the polymerizing enzyme is located. When the cell is destroyed by mechanical action, its synthesizing capacity is greatly reduced, and this leads to the assumption that an organized structure of the membrane is necessary for the synthesis of the cellulose precursor.

Among the cofactors that promote the formation of the cellulose precursor, phosphorylation plays an important role. In this, the uridine diphosphoglucose is undoubtedly an important intermediate. According to other investigations, a lipid-glucose compound participates in the transport of the glucose molecule to the site of the growing microfibril. The glucose is probably connected with the lipid by a glucosidic bond which is then used in the formation of the cellulose molecule. In this case, UDPG serves as the carrier of the glucose to the lipid and the energy of the UDPG is used for the formation of the lipid-glucose linkage (55).

With regard to the chain-length distribution in bacterial cellulose, its upper maximum, by and large, approaches the DP of vegetable celluloses. Recent estimates gave a DP of 6000 for bacterial cellulose, with a distribution curve having two maxima, at DP 2000 and DP 6000 (131). The distribution curve of the hairs of maturing cotton seeds shows two maxima, at DP 1500 and 5000, which correspond to those of a primary cellulose formed in a slowly occurring, widely distributed reaction. When the consumption amounts to only 2%, there is also formed a uniform secondary-wall cellulose with a DP maximum of 14,000. A comparison with equal distributions at 18% and 100% consumption proves the structurally orientated mechanism of the biosynthesis of secondary-wall cellulose. The average DP values of between 9000 and 12,000, measured at small consumptions, are therefore those of mixtures of primary- and secondary-wall celluloses. Two different mechanisms, parallel but separate, take place during the simultaneous synthesis of primary- and secondary-wall celluloses (171,172).

The rate of the formation of cellulose in the higher plants depends primarily on the availability of monomeric building stones. The reason for the rapid occurrence of the polymerization as compared with that of bacterial celluloses is primarily the presence of earlier-formed, ordered cell-wall cellulose which, as a primer and structural matrix, strongly promotes

the polymerization reaction. The longitudinal growth is thus limited to a great extent to the tips of the fibrils so that the further formation of fibril nuclei, by comparison, is greatly decreased. Whereas cellulose formed from single cells that are free in the substrate can follow its own rules of crystallization, that formed in cell tissues is spatially restricted. In this way, complicated structures are formed as a result of the direct influence of the living cells, and the cellulose thus formed is present in the single membranes in different states of order (55,94,178,179,228).

3. HEMICELLULOSES

3.1 Introduction

To distinguish them from cellulose, the readily hydrolyzable carbohydrates of the cell were originally called "hemicelluloses" (226). Numerous justifiable objections were raised to refute this collective term. Proposals were made to use the terms pentosans, hexosans, or wood polyoses instead of hemicelluloses (239). Thus the polysaccharides would be characterized according to their sugar components (141), whereas other investigators preferred a differentiation between cellulosans and polyuronides (192). For each of these suggestions there is a pro and con and when the term "hemicelluloses" is used with the necessary limitation on its meaning, as originally intended (226), the term is still appropriate, as is shown in the more recent literature.

The amount of hemicelluloses in wood has already been reported in connection with the summative wood analysis. It constitutes 20% to 35% of the weight of the wood. The hemicelluloses are associated with the cell wall, they are mostly soluble in alkali, and they are relatively easily degraded by an acid hydrolysis to simple sugars or sugar acids. The chromatographic analysis technique which was developed especially for sugar chemistry has proven to be extremely fruitful for the chemistry of the hemicelluloses. Three well-defined groups of hemicelluloses, i.e., the xylans, mannans, and galactans, are now recognized as single or collective components in all lignified plants, The xylans are present as arabinoxylans, glucuronoxylans, or arabinoglucuronoxylans; mannans are present in wood as glucomannans and galactomannans; the galactans, as such, are relatively rare but are often found in the form of arabinogalactans.

Absorption- and distribution-, as well as paper- and column-chromatography have gained great importance for the determination and identification of the sugars. Furthermore, gas-liquid distribution chromatography has been applied with great success (9,58,143,215). Fractionation by precipitation and extraction is based on the differences in solubility of

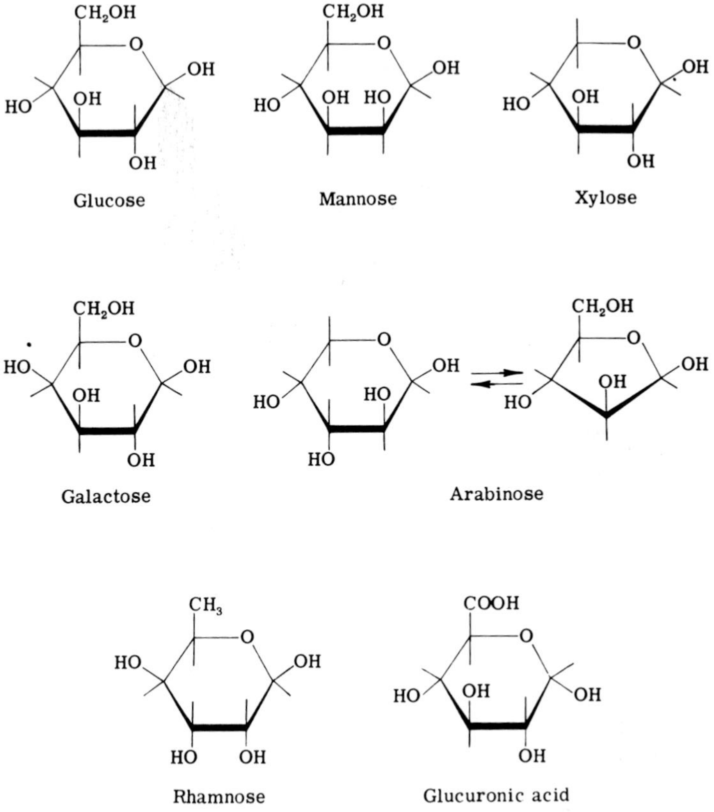

Figure III-8

the single components. With the hemicelluloses, differences in solubility occur because of differences in the average molecular weights, differences in the distribution of the molecular weight, and differences in the structure of the molecules, The configuration of the functional groups, the homogeneity of the composition, and the mutual relationships of the cell-wall components also exert an influence on the fractionation. The latter therefore depends not on only one of these solubility factors but on the combined effect of all of them, acting simultaneously in either a positive or a negative sense (19,148,187),

In addition to the above-mentioned methods, the electrophoretic separation of the sugars has been successfully carried out. For the elucidation of the constitution, x-ray investigations and infrared and magnetic nuclear resonance measurements have been used (137,196). The structural constitution of the hemicelluloses could largely be elucidated by these methods and the three groups were recognized individually. Within each separate

group there exists a series of polymers which differ with regard to their size, the relationship of their monomeric sugars participating in the structure, their branching, and in other respects. Figure III–8 shows the structural formulas of the monomeric sugars that play a role in the structure of the cellulose and the hemicelluloses. So far as cellulose is concerned, glucose is the prevailing sugar. In coniferous woods, mannose is an important monomer, while in deciduous woods, xylose predominates. Galactose is present in some deciduous woods and also occurs in larchwood. Arabinose is found in small amounts in some deciduous woods and, to a somewhat greater extent, in coniferous woods. Rhamnose occurs in small amounts in deciduous woods, while glucuronic acid is found in both wood species.

3.2 The Xylan Group

Analyses of holocelluloses isolated from two different species of coniferous and deciduous woods by means of buffered chlorite solution gave the results shown in Table III–7. The deciduous woods, according to these results, contain an average of 20% to 30% hemicelluloses, the principal component of which is a 4-*O*-methylglucuronoxylan with a high ratio of

TABLE III-7

ANALYTICAL DATA OF ACID CHLORITE HOLOCELLULOSES FROM
SOFTWOODS AND HARDWOODS (133)

	Western hemlock[a]	Southern pine[b]	Western red alder[c]	Gumwood[d]
Yield on oven-dry wood (%)[e]	65.3	62	68.5	71
Soluble lignin of holocellulose (%)	4.2	3.9	3.6	4.6
Yield of holocellulose	60.9	57.9	64.9	66.2
Galactose (%)	1.3	1.1	0.9	0.9
Mannose (%)	14.4	11.1	0.7	1.9
Arabinose (%)	0.7	0.7	0.1	0.3
Xylose (%)	2.9	6.8	19.2	16.6
4-*O*-Methyl-D-glucuronic acid[f]	present	present	present	present
Cellulose (%)[g]	41.7	44.3	45.0	46.5

[a] *Tsuga heterophylla.*

[b] An equal mixture of *Pinus elliotti* and *Pinus palustris.*

[c] *Alnus rubra.*

[d] A 3:2 mixture of red gums (*Liquidambar styraciflua*), Tupelo gum (*Nyssa aquatica*), and black gum (*Nyssa sylvatica*).

[e] Oven-dry yield on dry wood.

[f] 4-*O*-Methyl-D-glucuronic acid detected by paper chromatography; approximate content 10% of xylose for hardwoods and 15% of xylose for softwoods.

[g] Calculated from glucose by difference and corrected for glucose present in glucomannans and galactoglucomannans.

D-xylose to 4-*O*-methyl-D-glucuronic acid. Small amounts of mannose, galactose, and arabinose are also found. The greater part of the mannose seems to be combined with the glucomannan, which is accessible only with difficulty. The coniferous woods, on the other hand, contain 15% to 20% hemicelluloses, the main part of which consists of a glucomannan which is extracted only with difficulty. Smaller amounts of 4-*O*-methylglucuronoaraboxylan, which has a lower ratio of D-xylose to 4-*O*-methyl-D-glucuronic acid than that in deciduous woods, are present (103,104).

Investigations into the structure and properties of a methylglucuronoxylan isolated from birch and elmwood showed that an average of 197 β-(1 → 4)-D-xylopyranose units are present in the former, with about every eleventh unit having one side-chain of 4-*O*-methyl-D-glucuronic acid combined through an α-glucosidic linkage to the carbon atom 2 of the xylose moiety. The hemicellulose of the elm has a similar constitution but possesses an acidic side-chain at each seventh xylose unit. Its DP is 185. The isolation of 4-*O*-β-D-glucopyranosyl-D-galactopyranose makes the presence of a β-(1 → 4)-galactan seem probable. From birchwood, a water-soluble mixture of polysaccharides was obtained and, on hydrolysis, it gave galactose, glucose, mannose, and arabinose, and small amounts of uronic acid, xylose, and rhamnose (256).

A water-soluble arabino-4-*O*-methylglucuronoxylan could also be isolated from the bark of *Abies amabilis*. (1 → 2)-Terminally attached side-chains of 4-*O*-methyl-α-D-glucuronic acid and (1 → 3)-attached L-arabinofuranose units were joined to a chain of at least 124 (1 → 4)-β-D-xylose units (251). The enzymatic hydrolysis of 4-*O*-methylglucuronoxylan from birch gave a hydrolyzate containing 34% D-xylose and 26% neutral and 40% acid oligosaccharides. The neutral sugars consisted of a series of β-D-(1 → 4)-combined xylose oligomers, the amorphous acid oligosaccharides consisted of a series of uronic acids ranging from an aldotetra- to an aldoocta-uronic acid, while mono-, di-, and tri-uronic acids were not present (252).

The polymers generally known as xylan, i.e., 4-*O*-methylglucuronoxylan and 4-*O*-methylglucuronoarabinoxylan, have as their main structure a linear chain of monomeric xylose units that are connected through a β-D-(1 → 4)-bond, similar to the linkages of the glucose units in cellulose. In contrast to cellulose, the hydroxyl groups in xylan are substituted. The substituents are 4-*O*-methylglucuronic acid, arabinose, and acetyl groups. The glucuronic acid groups are primarily attached to the hydroxyl groups at the 2 position. It is possible that there are also pyranosidic bonds. Figure III–9A shows the structural formula for 4-*O*-methylglucuronoxylan and Fig. III–9B that for 4-*O*-methylglucuronoarabinoxylan (125,169).

The position of the acetyl groups in natural glucuronoxylan is not yet definitely established but they are probably attached to the hydroxyl

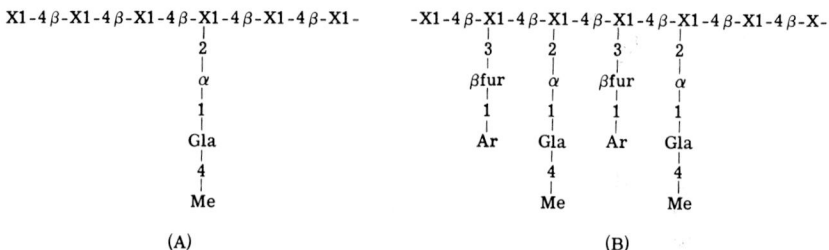

$$X1\text{-}4\beta\text{-}X1\text{-}4\beta\text{-}X1\text{-}4\beta\text{-}X1\text{-}4\beta\text{-}X1\text{-}4\beta\text{-}X1\text{-}$$

```
                                2
                                |
                                α
                                |
                                1
                                |
                               Gla
                                |
                                4
                                |
                                Me

               (A)
```

$$\text{-}X1\text{-}4\beta\text{-}X1\text{-}4\beta\text{-}X1\text{-}4\beta\text{-}X1\text{-}4\beta\text{-}X1\text{-}4\beta\text{-}X\text{-}$$

```
        3       2       3       2
        |       |       |       |
      βfur      α      βfur      α
        |       |       |       |
        1       1       1       1
        |       |       |       |
       Ar      Gla      Ar      Gla
                |                4
                4                |
                |                Me
                Me

                       (B)
```

Figure III-9. (A) Structure of 4-*O*-methylglucuronoxylan. X = monomer xylose; Gla = glucuronic acid monomer; Me = methyl. (B) Structure of 4-*O*-methylglucuronoarabino-xylan. Ar = arabinomonomer; fur = furanoidic bonding.

groups at the 2 and 3 positions (162). On treatment of 4-*O*-methylglu-curonoxylans with 5% sodium hydroxide at 170°C, the glucuronosyl groups are shown to be fairly resistant and the presence of traces of uronic acids could still be shown after 60 minutes of heating, while demethylation occurred rapidly. The xylan from birch contains one uronic acid unit as a side-chain per 27 xylose units (45). The more rapid disappearance of methoxyl groups as compared with the formation of glucuronic and ga-lacturonic acids is explained by the intermediate formation of an un-saturated carboxylic acid complex by the elimination of the 4-methoxyl group. Another suggested explanation attributes it to a combined mecha-nism of high temperature and ionized carboxyl. The electron-abstracting effect of the *trans*-methoxyl substituent may act as a reaction accelerator. From the 4-*O*-methylglucuronic acid function, a $\Delta^{4,5}$-unsaturated residue, which, on hydrolysis, would lead to a five-carbon deoxydialdehyde, would result. Methylation and hydrolysis of hemicelluloses containing xylose to which this artifact from the uronic acid group is attached would lead to the dimethyl ether of this compound. Neither the stability nor the chroma-tographic behavior of this five-cabon deoxydialdehyde or its methylated derivative is known (42).

A 4-*O*-methylglucuronoxylan isolated from American elm was con-verted into 4-*O*-methylglucoxylan by reduction with diborane. The loss of methoxyl, uronic acid, and anhydroxylose units from these two polymers on treatment with aqueous sodium hydroxide solution at various temperatures was determined and it was found that the presence of the carboxylic group renders the methoxyl group of the 4-*O*-methylglucuronoxylan, at tempera-tures of above 120°C, unstable toward alkali. Because no methylated uronic acid or acid-containing oligosaccharides could be detected in the hydrolyzate, it was concluded that the intermediate acidic reaction products were not true uronic acids. The loss of these groups on treatment with sodium hydroxide leads to the conclusion that they, too, were unstable toward alkali at higher temperatures. Most of the branches in 4-*O*-methyl-

glucuronoxylan are split off from the anhydroxylose backbone of the polymer as a result of the action of the alkali at higher temperatures. The reduction of the carboxylic acid group of the branch to a methylol group renders it stable toward hot alkali (207).

For the reactions of the hemicelluloses in the alkaline pulping process (see Chapter VIII) and during the chemical purification, it is important that the substituents of the glucuronic acid at the 2-hydroxyl group should not be very stable toward alkali at higher temperatures (104), but that, at lower temperatures, they should be capable of exerting a certain degree of protection against degradation. The acetyl groups are rapidly hydrolyzed in the alkaline pulping process, but they are relatively stable in the acid pulping process (see Chapter VII) (193).

That the main chain of xylan from deciduous woods has a linear structure has been widely doubted. However, it must be taken into consideration that the methylation for the determination of the end groups easily leads to a misinterpretation as regards the degree of branching. Combined gas- and paper-chromatographic investigations with completely methylated xylan from poplarwood (*Populus tremuloides*) indicated the presence of 3.0 2,3,4-tri-O-methyl-D-xylose and 2.0 2-O-methyl-D-xylose units per average macromolecule, indicating two branchings at the carbon atom 3. Mild extraction of xylan from wood in the presence of borohydride in order to lessen the degradation by alkali during the methylation and to obtain definite end groups gave, according to the methylation method, 1.6, and according to the periodate-formic acid procedure, 2.2 branchings (43).

For the indirect estimation of the relative length of the branches, the relationships between intrinsic viscosity and molecular weight were determined for a series of 4-O-methylglucuronoxylans and celluloses. It was found that xylan was very similar to linear cellulose and therefore must be considered to be an essentially linear macromolecule. The two or three branches per molecule are probably quite short. A survey of similar or related xylans from various angiosperms showed that most of these hemicelluloses have a main chain of sparsely branched xylan, the branching points of which are usually at the carbon atom 3 (280).

A largely unchanged but reduced arabino-4-O-methylglucuronoxylan was obtained from Norway spruce in a yield of 2.7% by extraction with potassium hydroxide in the presence of potassium tetraborate. After methylation, periodate oxidation, and partial hydrolysis of the polyaldehyde obtained by reduction with borohydride, the structure of the xylan was determined by gas-liquid-distribution chromatography in the form of trimethylsilyl derivatives (231,280). The polysaccharide contained one 4-O-methyl-α-D-glucuronic acid unit per 5.9, and one L-arabinofuranose unit per 7.4 xylose groups; each macromolecule contained an average of 128 xylose moieties and 1.8 branches. The branches were attached directly

to the carbon atom 2 or carbon atom 3 of the xylose unit. Gel filtration indicated a relatively low polydispersity, with values between 75 and 165. Long-chain fractions showed a higher proportion of arabinose and more branches. All xylans from coniferous woods seemed to have the same structure and were therefore sparsely branched and not linear (258,281).

In summary, it may be said that the xylans from deciduous and coniferous wood contain side-chains of 4-O-methyl-D-glucuronic acid and that those from conifers have more side-chains than those from deciduous woods. These side-chains are combined to the xylose units of the main chain by 1 → 2-linkages. Most of the wood xylans also contain 1-arabinofuranose groups which are connected with the xylose chain by 1 → 3-bonds. L-Arabinose is mostly found only in the xylans from coniferous woods.

3.3 The Mannan Group

On hydrolysis of the noncellulosic polysaccharide component of coniferous woods, D-mannose is also obtained. This fact has been known for a long time, but only in recent years have the methods for the isolation of mannose-containing polysaccharides and investigation of their structure been reported. It has been shown, in particular, that the mannose units occur in combination with glucose or galactose units. From the isolation and fractionation of mannose-containing polysaccharides from Western hemlock, white spruce, loblolly pine, and Norway spruce it was shown that these polysaccharides are not mixtures of glucans, mannans, and glucomannans but are true galactomannans (1,8,155,161).

The glucomannans are less soluble than the xylans in alkali. They are therefore isolated, first by an extraction of the plant material with dilute alkali, and then by an exhaustive extraction with strong alkali. For the isolation of glucomannans from deciduous woods, their holocellulose, for example, is first treated with potassium hydroxide in order to remove the greater part of the xylan. A subsequent extraction with sodium hydroxide containing borate then gives a crude glucomannan. On precipitation with barium hydroxide, a purified glucomannan is obtained. This method also gives good results with coniferous woods.

Glucomannans from the holocellulose of birch have a ratio of mannose to glucose of 1:1.1, while those from the holocellulose from other deciduous woods, such as maple, beech, poplar, and elm, have ratios of 1:1.5 to 1:2.0 (248). A galactomannan was isolated from Eastern hemlock in a yield of 4.8%; it contained glucose, galactose, and mannose units in a ratio of 1:1:3 and had a DP of 44. By means of electrophoresis and ultracentrifugation the homogeneity of the polymer was proven.

Galactoglucomannans could be isolated from fir (*Abies amabilis*), spruce (*Picea engelmannii*), pine (*Pinus strobus*), and cedar (*Thuja occiden-*

talis). All these polysaccharides resembled the galactoglucomannan from hemlock (*Tsuga canadensis*) but showed slight deviations in their sugar ratios. They can be considered to be homogeneous triheteropolymers. With few exceptions, all glucomannans isolated earlier had contained 2% to 4% galactose units. These units seemed to represent an integral component of these hemicelluloses and it is probable that the majority of coniferous woods contain 15% to 20% of a series of similar galactoglucomannans which differ only in their different proportions of galactose units and, probably, also in their molecular weight and degree of branching (249).

An alkali-soluble glucomannan from the bark of *Abies amabilis* gave, after methylation and partial hydrolysis, crystalline β-(1 → 4)-combined mannotetraose, mannotriose, mannobiose, mannosylglucose, and glucosylmannose. A minimum of 70 β-(1 → 4)-combined D-glucose and D-mannose moieties form a linear chain in which each tenth D-mannose and D-glucose unit probably has one (1 → 6)-combined D-galactopyranose group combined in the α modification (102). Wood and bark of coniferous woods seem to contain a series of closely related galactoglucomannans, the solubility properties of which depend primarily on the number of galactose side-chains (250,253,254,255).

A fundamental difference between the glucomannans from deciduous and coniferous woods is the absence of galactose units in the former. The results of investigations thus far reported lead to the general conclusion that the glucomannans have a linear structure so that no branchings at the carbon atoms 2 or 3 need to be assumed. The presence of short side-chains of galactose units in the glucomannans of coniferous woods may be considered to be certain. An interesting finding is the difference in the behavior of an acetylated and fractionated glucomannan from black spruce (*Picea mariana*) in osmotic and viscosimetric studies. According to these, a compact molecular configuration must be assumed for this glucomannan and this configuration would be in conflict with a linear polysaccharide structure. From the parallelism between the lignin content and the molecular weight and the hydrodynamic properties of glucomannan fractions isolated from a modified chlorite holocellulose, an association between the lignin and glucomannan molecules is deduced. A combination of the linear glucomannan chain with branched units of lignin would correspond to the deviation in the hydrodynamic behavior of these glucomannan fractions as observed. It is therefore assumed that glucomannan and lignin occur in the fiber in the form of a cross-linked matrix with a considerable number of lignin-carbohydrate bonds (165).

3.4 The Arabinogalactan Group

Arabinogalactan is a water-soluble polyose, found in a relatively large amount in all larch species. In its physical and chemical properties it shows

great similarity to natural gums and has therefore been of special interest for industrial production. During the course of the many investigations of this polyose, the most varied terms have been used; thus, the first extracts isolated from larch (*Larix occidentalis*) were termed ε-galactan. On hydrolysis with hydrochloric acid, a yield of furfural is obtained, corresponding to the presence of 10.5% pentosans although, in the isolated sugars, no pentose could be found. Later investigations confirmed the presence of arabinose in the original polysaccharide to the extent of 12% together with 84% to 85% galactose. A water-soluble polysaccharide, containing arabinose and galactose in a ratio of 1:6, was also found in tamarack (*Larix laricina*) and in European larch (*Larix decidua*). From a study of the esters and ethers of this water-soluble polysaccharide, it could be proven that the arabinogalactan in all the larch species investigated was identical (198, 223, 273, 275).

On methylation of arabinogalactan and methanolysis of the methylated product, the structure of this polyose could be defined as follows:

1. The arabinose units of the arabinogalactan are connected by a $1 \rightarrow 6$-bond to a di-linked galactose unit, with the formation of an arabofuranosidogalactan.

2. The galactose terminal units are connected to another galactose unit by a $1 \rightarrow 6$-bond and are, in turn, combined to a third galactose unit by $1 \rightarrow 6$- or $1 \rightarrow 3$-bond.

3. The $1 \rightarrow 6$-galactose linkages outnumber the $1 \rightarrow 3$-bonds.

4. The arabinogalactan has a branched chain structure that is terminated by groups of galactopyranose and arabofuranose (268).

Further investigations confirmed the combination of this water-soluble polyose of galactose and arabinose but led to the conclusion that the polyose is a mixture of the two components and not a uniform molecular structure (40, 127). By catalytic oxidation of the galactan from *Larix decidua* its uniform molecular structure could be confirmed (11). According to this result, the basic structure of the main chain of arabinogalactan from *Larix decidua*, as shown in Fig. III–10, may be considered to be proven. There still remains only the clarification of the position of the chains of the arabinose units. For these, the structure shown in Fig. III–11 has been proposed. When this structure is incorporated into that shown in Fig. III–10, under the assumption that the ratio of arabinose to galactose units is about 1:6, then the structural formula shown in Fig. III–12 is arrived at for the arabinogalactan (2, 105, 208).

Arabinogalactan from *Larix occidentalis* gives, on hydrolysis, methylation, and fractionation, two components that differ basically in their molecular weights. One fraction has a molecular weight of about 100,000, the other, 16,000. They have about the same composition but show different electrophoretic behavior. There is reason to assume that they are

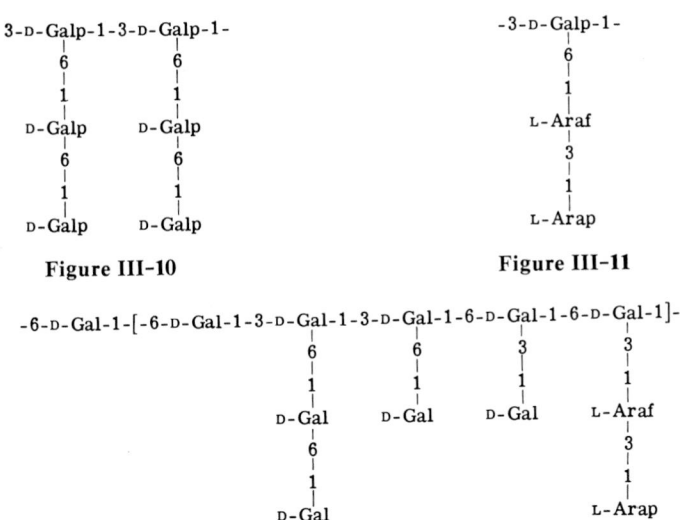

Figure III-10

Figure III-11

Figure III-12. Structure of arabinogalactan.
D-Gal = residue of D-galactopyranose
L-Araf = residue of L-arabinofuranose
L-Arap = residue of L-arabinopyranose

molecular fractions with different degrees of branching and that they are combined through galactopyranoside bonds of a weaker type (26). A similar conclusion was reached earlier on the basis of sedimentation and diffusion measurements (177).

The distribution of the galactan within the tree is not uniform. In the sapwood of tamarack or of Western American and European larch species there is little or no arabinogalactan present. From about the middle of the trunk, arabinogalactan and also ethanol-benzene extractives increase in quantity toward the border between the sapwood and heartwood. The relative amount of arabinogalactan increases with the height of the trunk in tamarack; in *Larix occidentalis* the reverse is true. Earlywood contains more galactan than normal wood does. The cell wall of tamarack contains almost no galactan, while most of it is deposited in the lumen of the tracheids. The galactan of the larch species seems therefore to be chiefly extracellular and shows a distribution in the tree identical with that of polyphenols and other extractives. The ratio of arabinose to galactose increases with increasing age of the tree from 1:5.4 for the outer heartwood of *Larix occidentalis* to 1:10.2 for the inner heartwood.

The average molecular weight of the polyose decreases linearly with the age of the tree from 70,000, close to the border of the splint-heartwood, to 46,500 near the heart. Since the acetic acid content increases toward the heart, it is assumed that the arabinogalactan of larch within the living

tree is subjected to an acid-catalyzed hydrolysis and that the hydrolytic decomposition is caused by the conversion of the acetyl groups into acetic acid (48,49,50).

Arabinogalactan from larchwoods can be used industrially as a gum, with a low viscosity and a high dispersion capacity. In the isolation of the polysaccharide from the wood, a crude galactan is obtained, containing about 80% arabinogalactan, 2% ash, and 4.7% tannin. The crude product is dark in color; its purification and decolorization entail a good deal of difficulty. The discoloration is chiefly caused by phenolic impurities, the removal of which is possible with magnesium oxide and with ion exchangers (60,61,69).

3.5 The Biogenesis of Hemicelluloses

Our present knowledge of the biosynthesis of hemicelluloses is still incomplete. The structural similarity between hexosans and pentosans has suggested an analogous reaction sequence for the biosynthesis of the hemicelluloses. Thus it may be assumed that the hexosans are oxidized to polyuronides and the latter are converted into pentosans by decarboxylation at the carbon atom 6. Since the monosaccharidic units in galactan, araban, and the polygalacturonic acid differ in their position at carbon atom 1, in the glucosidic linkages, and in the ring size, and since the structures of the pentosans and the combination of the various side-chains also differ completely, decarboxylation in the polymer state is impossible. The reaction therefore must occur in the monomeric state (138,139).

In the discussion of the biosynthesis of cellulose, the uridine diphosphoglucose (UDPG) was mentioned as a key component in the formation of cellulose precursors. It has now been shown that the conversion of a derivative of D-glucose to D-glucuronic acid and the decarboxylation of the latter can finally lead to a D-xylose derivative. This conversion takes place in a monomeric state (124,126,186,210). By the condensation of two triose fragments and under the influence of the enzyme, aldolase, D-fructose-1,6-diphosphate can be formed and can be cleaved by the action of the enzymes, *trans*-aldolase and *trans*-ketolase. Thiamine pyrophosphate serves as a cofactor in this reaction. From 2 moles D-fructose-6-phosphate and 1 mole D-glycerose-3-phosphate, 3 moles pentose-5-phosphate are formed (186). Another possibility for the formation of pentose is the cleavage of, for example, D-fructose-1,6-diphosphate by aldolase to dihydroxypropanone-phosphate and D-glycerose-3-phosphate. Because the enzyme is specific only for the ketose, the triosealdehyde can be replaced, for example, by a diose such as glycolaldehyde and, in this way, D-xylose-1,5-diphosphate is formed (128).

A more detailed presentation of other possibilities for the biosynthesis

of hemicelluloses may be omitted here, since the references given provide detailed information (138,139).

4. LIGNIN

No field of wood research has been the object of such extensive investigation during the past two decades as that relating to the chemistry and physics of lignin. The importance attached to this research is clearly indicated by the fact that numerous books on lignin have been published, giving an exhaustive survey of the work done. Some of the newer presentations on lignin are listed at the end of this chapter under the heading of general references. For a detailed study of the lignin problem, the reader is referred to these references. This section will be limited to a brief review of the present status of lignin chemistry. This can be done because the special importance of lignin in the technological processes is referred to in detail in the relevant chapters of this book.

4.1 Lignin Preparation

The lignin in plants is an almost insoluble substance. Its isolation from wood has engaged the attention of researchers for many years. For the quantitative determination of lignin in wood, the latter is treated with strong sulfuric acid, the cellulose and hemicelluloses are degraded by hydrolysis to low-molecular sugars, and the remaining dark brown residue is considered to be lignin. Such lignin is known in the literature as "Klason lignin." Highly concentrated hydrochloric acid can also be used for the isolation of lignin from wood; this is referred to in the literature as "Willstätter lignin." Both these processes will be discussed in detail in connection with the technical production of "wood sugars" (see Chapter IV). The sulfuric acid method is used as an analytical method for the quantitative determination of lignin in wood and other vegetable fibers although it has some basic sources of error (145,192,262,270).

It is understandable that the hydrolysis of the cellulose and hemicelluloses in wood with concentrated acids does not leave the lignin complex unchanged. The reaction capacity of the lignin isolated in this way is greatly reduced. Indeed, the lignin isolated with concentrated hydrochloric acid is lighter in color than the Klason lignin and is probably also less condensed, as is shown by its ready and complete solubility after sulfonation (101). Hydrofluoric acid, too, may be used for the hydrolysis of the carbohydrates and, hence, for the isolation of lignin, but this method is of only analytical importance (44,70,269),

Lignin can also be isolated from very finely divided wood, after prior

degumming with weakly alkaline solution, by alternate treatments with dilute sulfuric acid and ammoniacal copper oxide solution (89).

The degradation of the wood polyoses by oxidation with periodic acid and subsequent hydrolysis with boiling water of the dialdehyde formed give a periodate lignin (205).

All the methods mentioned thus far for the isolation of lignin preparations are based on the hydrolytic or oxidative degradation of the carbohydrate fraction of the wood to soluble products, leaving the lignin as an insoluble residue. To what extent the lignin itself has simultaneously undergone some changes is not entirely clear, but it depends to a large extent upon the method used for the isolation.

Further suggestions for the isolation of lignin preparations in a form as unchanged as possible are based on the concept that extractives which can combine with the lignin must be removed by a suitable solvent before the isolation of the lignin can be carried out. For the isolation of the lignin, 95% ethanol is used, and, after exhaustive extraction of the wood meal and distillation of the alcohol under reduced pressure, a resinous residue is obtained which can be dissolved in dioxane. By repeated dissolution in dioxane and precipitation into water and, finally, into ethyl ether, a light cream-colored precipitate with a constant methoxyl content is obtained in a yield of about 10% of the lignin in the wood, determined according to one of the above methods. This lignin preparation is known as "Brauns native lignin," after the originator of the process. The product has a methoxyl content of 14.8%, which corresponds exactly to the methoxyl content of the lignin prepared from the same sprucewood (28,29).

A unique method for the isolation of lignin is the enzymatic separation of the lignin from the carbohydrates. In the rotting of vegetable materials by fungi, two basically different types of fungi can be distinguished, one type, by its enzymatic action, affects especially the carbohydrates, leaving a brown residue, and is known as "brown rot," whereas the other type attacks the lignin, leaving a very light colored residue, and is known as "white rot." For the isolation of lignin, only the first type is used. When species of this type act on finely ground and sterilized wood meal for many months, a continuous decrease in the cellulose content and a corresponding increase in the lignin content take place.

When pinewood was subjected to the action of *Poria vailantii* for 15 months, up to 22.7% of the total lignin content could be isolated enzymatically. Comparison of the lignin preparation thus obtained with the above-mentioned native lignin justifies the conclusion that the native lignin is identical with the total lignin in the wood (225). On the other hand, investigations have shown that wood, from which the native lignin had been removed by exhaustive extraction, on subsequent enzymatic treatment gave a lignin, liberated by the action of the fungi used. This lignin there-

fore originated exclusively as the result of the action of the fungi on the carbohydrate fraction of the wood (149,225).

When wood meal is treated in a ball swing-mill in the presence of a nonswelling medium such as toluene, up to 50% or more of the lignin in the wood can be extracted with dioxane after the wood has been drastically disintegrated by the mechanical action. This product is called "milled wood lignin," or sometimes "Björkman lignin" after the originator of the method (24). Such a lignin is lighter in color than the acid lignins and has an average molecular weight of 11,000. It seems that the lignin obtained in this way is less condensed and less chemically changed than is the case with lignins isolated by other methods. It was shown that this lignin has the basic structure of a polymer phenylpropane derivative, that it has free phenolic hydroxyl groups, *p*-hydroxybenzyl alcohol units, and aromatic carbonyl groups combined to the propane chain (25). The absorption bands of the infrared spectrum support the homogeneity of this lignin.

Besides ethanol for the extraction of lignin, other solvents have been suggested, such as, for instance, extraction with alcohol or dioxane in the presence of small amounts of mineral acids. Concentrated aqueous solutions of sodium xylenesulfonate can dissolve considerable amounts of lignin at elevated temperatures and pressure. Such hydrotropic solutions dissolve lignin from deciduous woods especially readily, giving dark-colored products, while lignins from conifers remain largely undissolved (167,212,260).

Finally, the extraction of lignin by alkaline mediums must be mentioned. These, with or without the addition of sulfur components, form the basis for the industrial production of pulp according to the alkaline pulping processes, which will be described in detail in Chapter VIII. On the other hand, the extraction of lignin with aqueous monosulfite-bisulfite solution is discussed in Chapter VII, as the sulfite pulping process.

Whenever the term "lignin" is used, the type of isolated lignin referred to must always be specified because the history of its isolation may have a profound effect on the structure of the final product.

4.2 The Structure of Lignin

For the classical organic chemist, lignin is not an attractive subject. Because it is insoluble in most organic solvents, its isolation from the wood is accompanied by marked molecular changes. The preparation of simple and readily identifiable degradation products, as in the case of cellulose, is just as difficult and unsatisfactory as is the synthesis of ligninlike products from known model compounds. In spite of this, both processes, together with spectral analytical investigations, have, during the last two decades, contributed substantially to the widening of our understanding of the basic

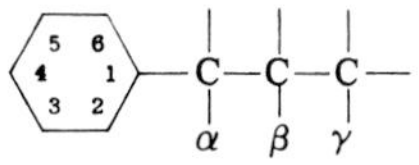

Figure III-13. Phenylpropane unit.

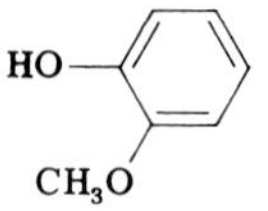

Guaiacyl 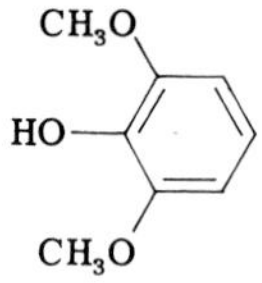   Syringyl

Figure III-14

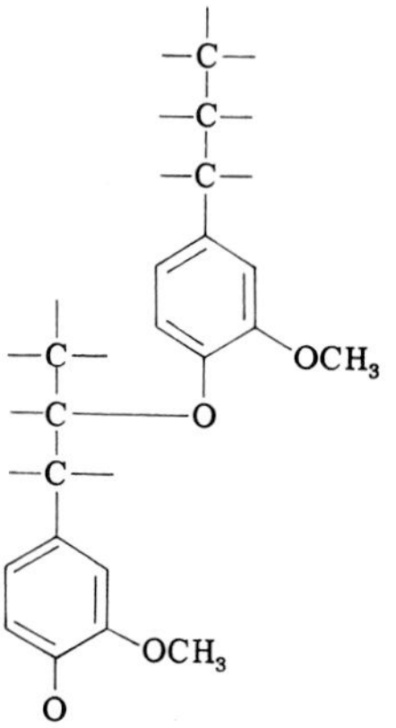

β-4' Ether bonding

(A)

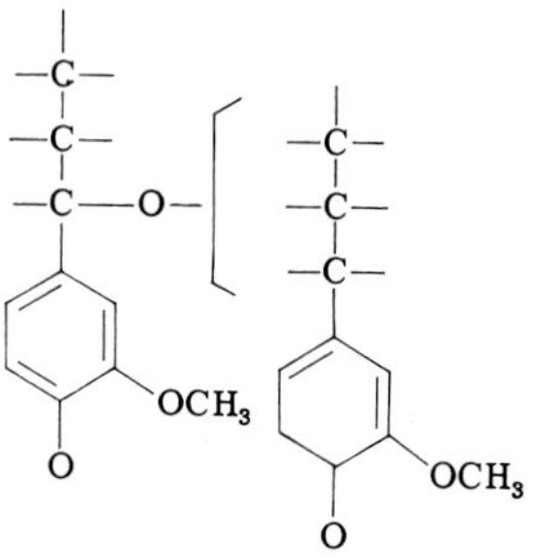

α-Alkylether

(B)

Figure III-15

structure of the lignin molecule. It must be remembered, however, that the term "lignin" refers to a mixture of substances that have similar chemical compositions but may have structural differences. The lignin molecules differ from isomers in that they may have the same molecular weight but different structures, and, from homologs, in that they may have different molecular weights but the same structure.

The aromatic structure of lignin has been proven by its characteristic ultraviolet absorption spectrum, which is very similar to that of some guaiacylpropane compounds (153). Morever, its aromatic structure could be deduced from the elementary composition of the wood and of the holocellulose prepared from it (205). But even much earlier, the aromatic structure of lignin and the identity of its chemical composition with that of a guaiacylpropane polymer was indicated (74,144).

From degradation products of the lignin from wood and other vegetable materials, a number of phenylpropane derivatives, containing the structure shown in Fig. III-13, could be isolated. It should be noted that coniferous wood lignins contain exclusively guaiacylpropyl units, whereas

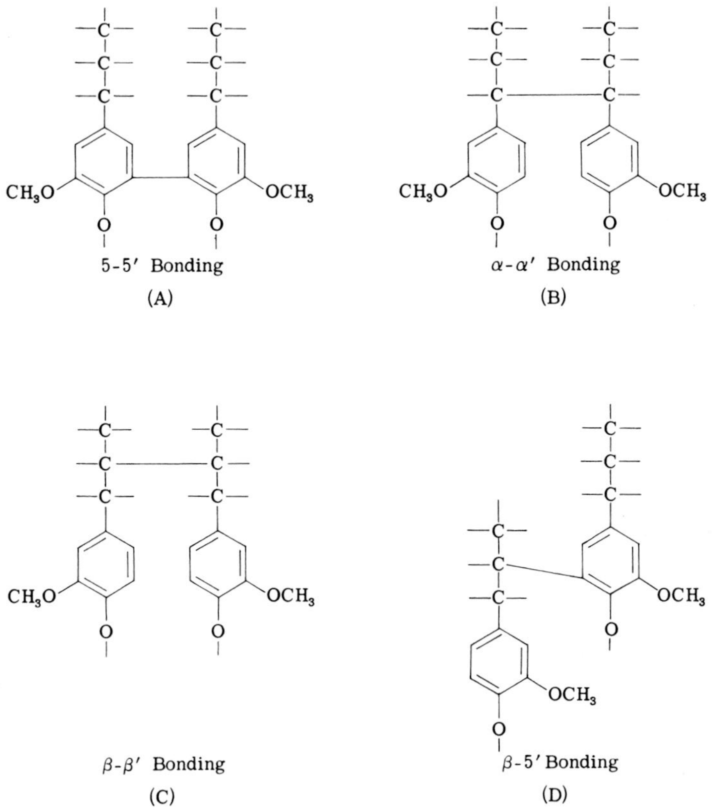
CH₃O
OCH₃
O
O
5-5′ Bonding
(A)
CH₃O
OCH₃
O
O
α-α′ Bonding
(B)
CH₃O
OCH₃
O
O
β-β′ Bonding
(C)
OCH₃
O
OCH₃
O
β-5′ Bonding
(D)

Figure III-16

deciduous wood lignins contain both guaiacyl- and syringylpropyl moieties. Figure III-14 shows the respective phenyl rings. On hydrogenolysis of organosolve lignins and even of wood meal from various and deciduous woods, a large number of cyclohexylpropane derivatives have been obtained (33,96,107,166,230,266).

The question still remains, in what way are the phenylpropane units combined in the macromolecule of lignin? This combination can occur either through ether linkages and/or by carbon-to-carbon linkages, as is shown in Figs. III-15 to III-17. Figure III-15 shows the various possibilities for the formation of ether linkages, while Fig. III-16 shows the various carbon-to-carbon linkages. Here it is of special technological significance that the carbon-to-carbon linkages are particularly resistant toward chemical attack. Although recent investigations into the formation of lignin in the living plant seem to support the hypothetical assumption of such connections between the phenylpropane units in the lignin molecule, experimental proof is still to be produced.

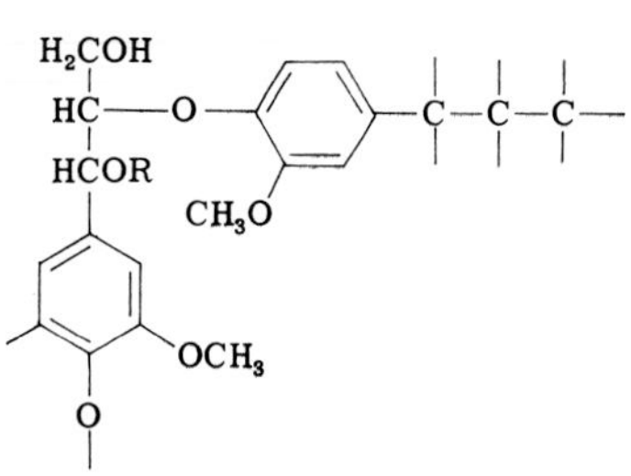

Guaiacylglycerolaryl
ether

(A)

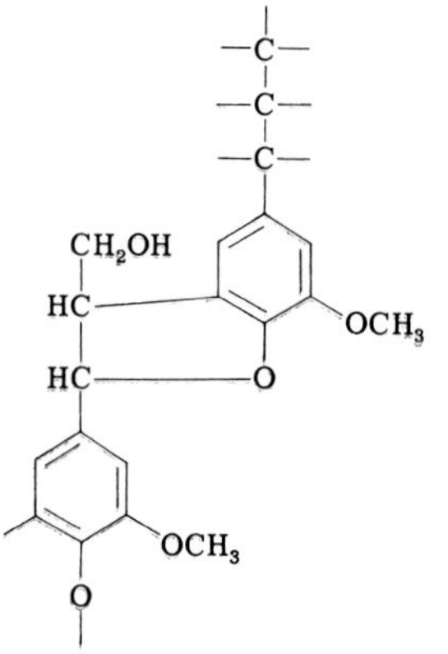

Benzofuran
structure

(B)

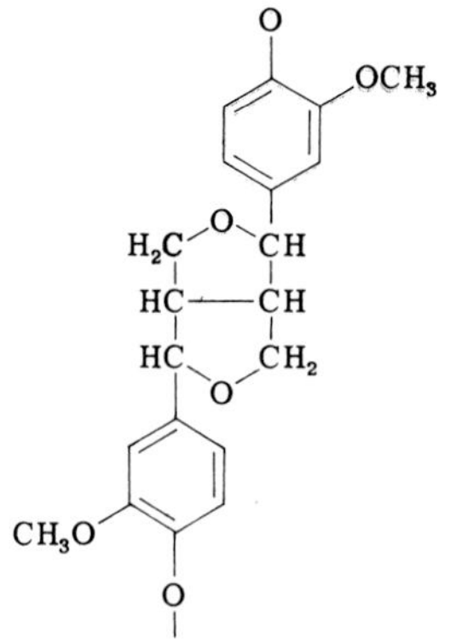

Pinoresinol
structure

(C)

Figure III-17

About 30% of the phenylpropane units contain a free phenolic hydroxyl group while the remaining 70% should therefore be connected by phenol ether linkages (4). In general, it may be assumed that the oxygen in the 4 position of one of the phenyl rings is combined with the carbon atom of the side-chain of another phenylpropane unit, thus forming an ether linkage. Many such phenol ether linkages probably belong to β-4'-ether linkages (Fig. III-15A), thus forming guaiacylglycerol-β-aryl ether structures (Fig. III-17A) (119,133,134). Carbon-to-carbon linkages may be

formed when two benzene rings are combined through their 5,5′ positions (Fig. III–16A). Such linkages may also be formed between the side-chains, thus giving an α,α′ linkage (Fig. III–16B) or a β,β′ bond (Fig. III–16C). There is also the possibility that the β carbon atom is connected to the 5′ carbon atom of a second phenylpropane unit, as is shown in Fig. III–16D. Furthermore, two phenylpropane radicals can combine in such a way that, for example, an ether and a carbon-to-carbon bond are formed, thus giving a benzofuran structure, as shown in Fig. III–17B, or two phenylpropane radicals can combine with the formation of a β,β′ carbon bond and two ether bonds between the γ,α′ and the α,γ′ carbon atom, thus giving a pinoresinol structure, as shown in Fig. III–17C (216). It must be emphasized, however, that some of these possibilities still require experimental proof. In recent years, studies of the biogenesis of lignin have lent support for such configurations (4).

There is a great temptation to give here a hypothetical structural formula for the lignin molecule, as has been done by competent experts in the field, but the reader is referred, instead, to the relevant publications where such structural formulas are presented with necessary reservations (4,30, 31,77,84).

4.3 Some Important Technical Lignin Reactions

One of the most important delignification processes is the treatment of wood with solutions of the salts of sulfurous acid. This process is discussed in detail in Chapter VII and only a brief presentation of the presumable reactions of the sulfonation will be given here. The heterogeneous composition of the lignin macromolecule, with its numerous reactive molecule groups, has led to speculation as to the course of the sulfonation reaction. The following theory relating to this reaction is based on results obtained by Swedish investigators (5,100,164). According to this, the reactive groups in sprucewood lignin are divided into three groups, X, Z, and B, depending upon the degree of their reactivity. Figure III–18 shows the structures of these various groups. Group X has a free phenolic hydroxyl group at the para position of the phenyl ring and an etherified hydroxyl group at the α carbon atom of the propyl side-chain. Group Z has an etherified phenolic hydroxyl group at the para position of the phenyl ring and a free hydroxyl group at the α carbon atom of the side-chain. Group B has etherified hydroxyl groups at both the para position of the phenyl ring and at the α carbon atom of the side-chain. On sulfonation in a neutral or weakly acid medium, only the X and Z groups are sulfonated at the α carbon atom, while the B groups remain unchanged (67,164). The activating action of the free phenolic hydroxyl group of the X group causes a rapid reaction, whereas the Z group, with its etherified phenolic hydroxyl

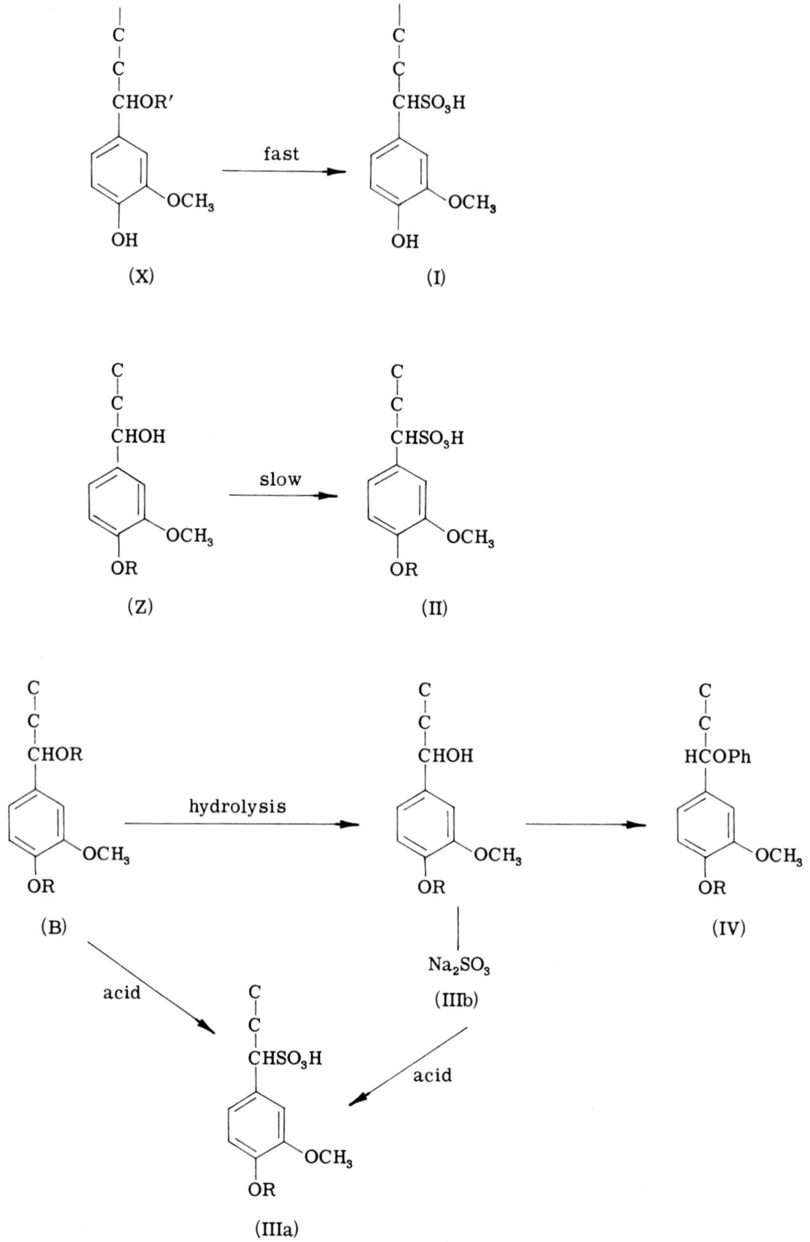

R = Alkyl; R' = H or alkyl; Ph = phenyl

Figure III-18

group, reacts more slowly. The α-sulfonic acid thus formed is called "solid lignosulfonic acid" and, since only a few linkages within the macromolecule and between the lignin and carbohydrates have been cleaved, it is still present in a macromolecular and, therefore, insoluble state (I and II in Fig. III–18).

The benzyl ether linkage in the B group is split by an acid hydrolysis and the lignin molecule is degraded into soluble products and goes into solution as "low-sulfonated lignosulfonic acid" (IIIa in Fig. III–18). Simultaneously with the hydrolysis, a reactive hydroxyl group at the α carbon atom (IIIb) is formed, giving a benzyl alcohol group (157,163). This group can be further sulfonated, not only by neutral but also by acid sulfite solution (68,157). Structures II and IIIa are identical, but this does not preclude the presence of structural differences at the β carbon atom of the side-chain.

In the technical sulfite pulping process the cooking acid has a pH value of 1.5 to 2.0 at the start of the cook so that the reactions shown in Fig. III–18 can take place practically simultaneously. The retarding action of polyphenols on the dissolution of lignosulfonic acid, to be discussed later, is explained as resulting from the addition of a phenolic complex to the hydrolysis product, IIIb, with the formation of a complex, IV. In the same way, the groups X and Z in an acid medium can react more rapidly with a phenolic compound than with bisulfite, thus forming insoluble lignin condensation products (67). Technological procedures for the prevention of such undesirable condensations will be discussed in Chapter VII, Section 3.1.

Another technologically important reaction is the alkaline hydrolysis, which is utilized in the alkaline pulping process. With isolated lignin, alkaline solutions cause the cleavage of alkyl-aryl ether linkages at the para position of the phenyl ring. In the sulfite process, such cleavages occur even at low temperatures. For the liberation of phenolic hydroxyl groups in lignin with alkali, on the other hand, higher temperatures, of 160°C or more, are required (86,156). Some carbon-to-carbon linkages in lignin are also sensitive toward alkaline hydrolysis. This is proven by the formation of vanillin and acetaldehyde from coniferyl aldehyde derivatives. These reaction products will be mentioned again in connection with the treatment of the sulfite spent liquor (Chapter VII, Section 9.5.2) and the destructive distillation of wood (Chapter V, Section 5). There are also indications that the linkages between the β and γ carbon atoms of the propyl side-chain are partially split.

In the hydrogenolysis of lignin from deciduous woods in a neutral medium in the presence of Raney nickel at 160° to 170°C, products are obtained from which syringyl- and guaiacylpropane derivatives can be isolated, whereas, in an alkaline medium, only monomeric phenylethane derivatives are obtained (33,197).

When wood is heated with alkaline solutions to temperatures of above 160°C, appreciable amounts of lignin go into solution. This reaction is utilized in the alkaline pulping process. It is interesting to note that the pure alkaline cook is suitable primarily for deciduous woods, while coniferous woods are delignified only with difficulty by pure sodium hydroxide solutions. The reason for this may be a condensation within the lignin molecule. New carbon-to-carbon linkages are probably formed between monomeric units (3). The addition of sulfur to the alkaline cooking liquor has facilitated the technical process (see Chapter VIII, Section 1). Sulfur-containing cooking liquor delignifies much more rapidly than alkali solutions without sulfur. It is assumed that hydrogen sulfide ions react, in part, with the groups that cause the condensation and thus prevent this reaction. This assumption was confirmed with model substances. Solid thiolignins with a high sulfur content are formed even at the beginning of a sulfur-containing cook by hydrogen sulfide ions or by thiosulfate ions at pH 7 and 100°C. During the further course of the reaction, these thiolignins are dissolved by the splitting off of sulfur and are converted into conventional sulfate lignin with a sulfur content of about 2% (65,66).

Another interesting reaction from the technical point of view is based on the analytical method for the determination of cellulose by alternating chlorination and extraction with alkali (97). This method has been utilized in a technical process for the production of pulp, especially from annual plants and agricultural residues. When applied to wood, however, this method requires a very extensive shredding of the material to be chlorinated. The process has the advantage of operating at low temperatures, thus yielding a cellulose that is only slightly damaged. Investigations with sprucewood lignosulfonic acid and with ground sprucewood meal have shown that the substitution by chlorine in the aromatic nucleus of the phenylpropane unit takes place chiefly at the 6 position. An exception occurs when the phenolic hydroxyl group in the para position is not etherified; in this case, a substitution may occur in the 5 position. It is unimportant whether etherified or nonetherified benzyl alcohol groups are present because chlorine splits the linkage between the benzene ring and the side-chain. Analogous to the fact that, on chlorination of veratryl alcohol as a model substance, substitution takes place at the 6 position of the benzene ring and formaldehyde is formed, the terminal carbon of the separated side-chain could be identified as an aldehyde group. Whether the phenolic ether linkages at the para carbon atom are split on chlorination remains in doubt. The methoxyl groups are split. As can be seen from Fig. III–19, not all phenylpropane units are split and this causes the formation of a number of molecule fragments, depending upon the type and number of chlorine treatments, and no single type of molecule predominates. The ionization-difference spectrum of the aqueous extract of chlorinated wood indicates that the aromatic nucleus cannot be highly chlorinated since it resembles

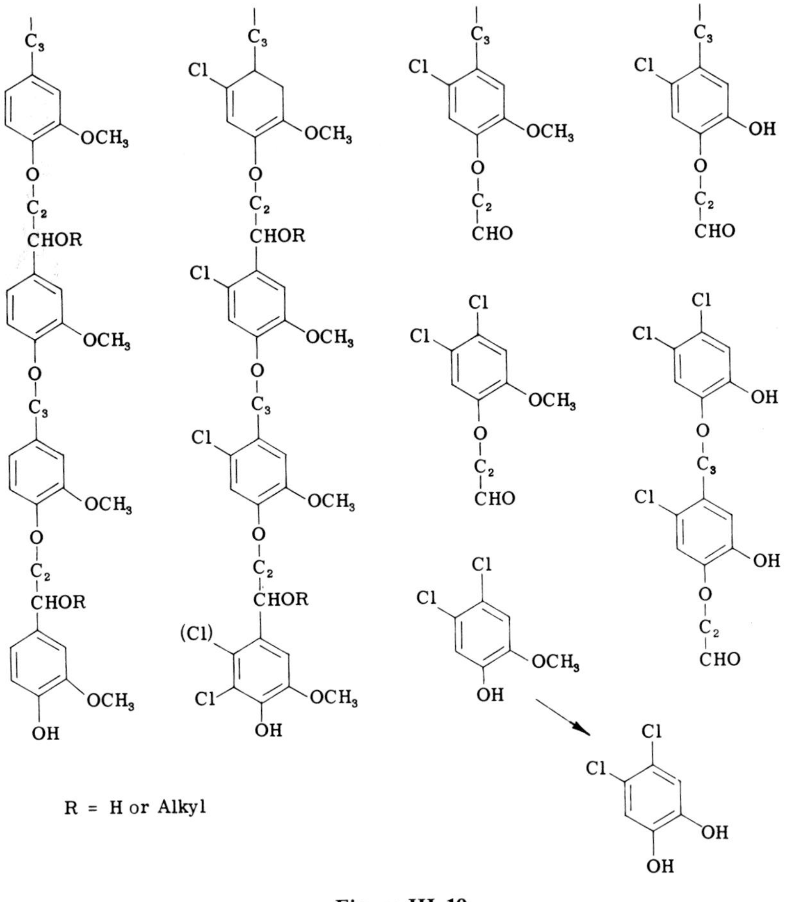

Figure III-19

the spectrum of 4,5-dichlorocatechol. This result is in good agreement with the analytical results of other investigators (53,181).

With regard to the dissolution with alkali of the chlorolignin compound formed, experiments with model substances, with the exception of certain catechols, have shown that they are only slightly dissolved on heating with 2% sodium hydroxide for one hour at 60°C. Chloro-*o*-benzoquinone and chloromuconic acid derivatives, which might be expected to be oxidation products during the chlorination sequences, showed a strong loss of chlorine in a similar treatment with alkali for two hours. This difference in behavior is primarily the result of the position of the chlorine substituent in relation to other functional groups. The greatest reactivity is shown by those compounds which have chlorine substituents in the side-chain. Even with weak bases, such as sodium bicarbonate and water, a

considerable degree of hydrolysis is observed. On extraction of chlorinated lignin, appreciable amounts of alkali are consumed through the hydrolytic elimination of chlorine atoms which need not necessarily be in the side-chains of the chlorolignin (157).

The nitration of lignin is quite similar to its chlorination. By alternate treatments with nitric acid and alkali, lignin can be extracted from the wood structure (195). This process forms the basis for methods for the determination of the carbohydrate or lignin portion in wood (98,151,152, 159). With a series of guaiacyl derivatives as model compounds it could be shown that an exchange by nitro groups is possible in the side-chain. On nitration of wood meal in ether, 2,4-dinitroguaiacol could be isolated. Aromatic methyl ethers also underwent partial demethylation on nitration. The nitration of lignin caused not only the loss of methoxyl groups but also the formation of methanol. Nitric acid was reduced to nitrogen oxides, ammonia, and hydrogen cyanide (82,99,209). The nitric acid process has also been utilized for the technical production of pulp.

4.4 The Biogenesis of Lignin

In recent years the biogenesis of lignin has become the subject of comprehensive book-length presentations which must be referred to here. These also contain extensive bibliographies relating to this subject (34,75, 76,78,191,216,224).

The concept which was developed during the second half of the last century—that sprucewood lignin is formed in nature from coniferin, a component of the cambial sap—has received strong support as the result of recent biochemical investigations (144,242). According to these, lignin is formed from coniferin by an enzymatic hydrolysis of the glucosidic linkage and the subsequent dehydrating polymerization of the liberated coniferyl alcohol. The presence of D-glucosidase, which is necessary for the hydrolysis of the D-coniferin, could be proved in the cambium cells which are in the process of lignification (88). Radioactive L-coniferin is not converted into lignin (79). An enzyme, laccase, is also present in the cambial sap and causes the dehydrating action on the liberated coniferyl alcohol; the latter is converted into insoluble lignin which is deposited in the middle lamella and in the cell walls. The treatment of coniferyl alcohol with the enzyme of mushroom extract or cambial sap in the presence of oxygen causes the formation of water-insoluble, ligninlike amorphous polymerizates which closely resemble spruce protolignin or "native lignin." The enzymatic action in the formation of such "dehydrogenation polymerizates" (DHP) and the biosynthesis of lignin are explained as resulting from the extraction of a hydrogen atom from the phenolic hydroxyl group of the coniferyl alcohol. It is probable that a recombining of different resonated

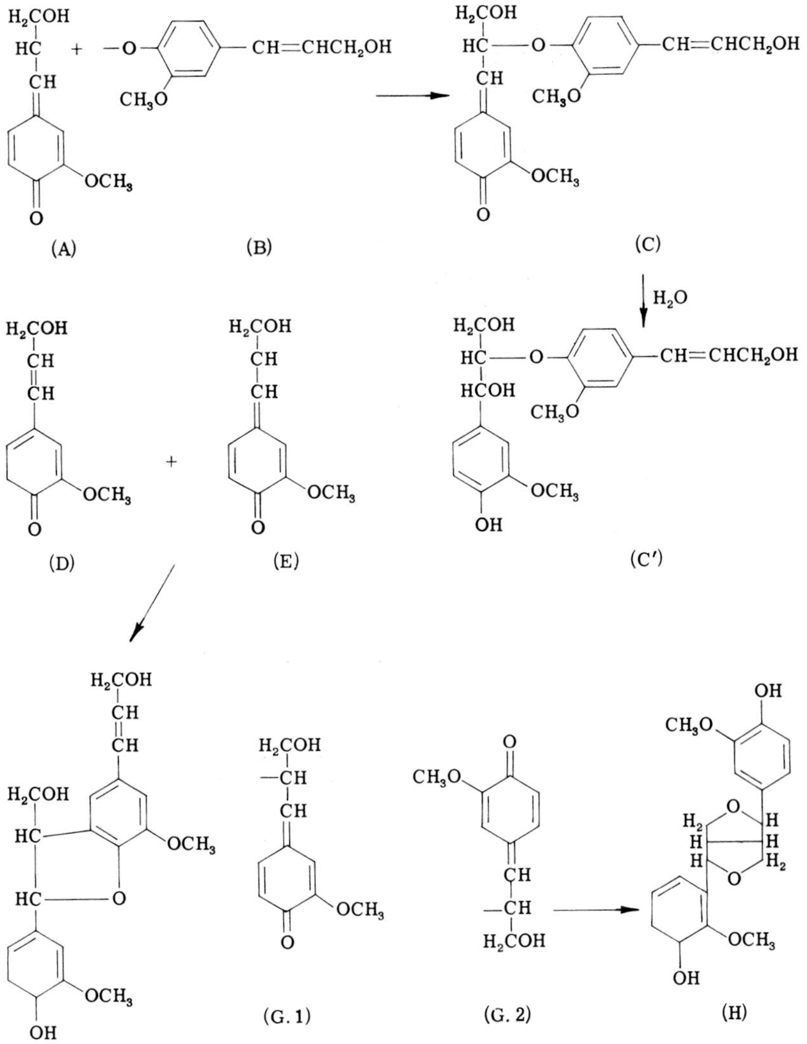

Figure III-20

structures of the coniferyl alcohol radical takes place. Thus, for example, the combination of two structural radicals, A and B, to C, as shown in Fig. III-20, is possible, and, after stabilization of the quinonemethide group, C, by the addition of water, guaiacylglycerol-β-coniferyl ether, C', is formed (see also Fig. III-17A).

In a similar way, dehydrodiconiferyl alcohol, F, may be formed from the corresponding resonance structures, D and E. Finally, pinoresinol, H, may be formed from two radicals, G (see also Fig. III-17C).

The stabilization of the quinonemethide intermediate may occur in the first case by an intramolecular addition of a phenolic hydroxyl group, and, in the second case, through the addition of a γ-hydroxyl group to both structures. Quinonemethides are very reactive compounds and readily undergo hydrolysis, alcoholysis, and dimerization. A reaction between a quinonemethide and a hydroxyl group which belongs to a carbohydrate molecule can therefore lead to a lignin-carbohydrate linkage. In the same way, two quinonemethides can form a carbon-to-carbon bond between two carbon atoms at the α position of the side-chain of two monomeric units. Thus the addition reactions of dimeric quinonemethide intermediate compounds can form one of the three lignin dimers, a lignin-carbohydrate bond, or a lignin primer, in the latter of which the third unit is combined by an α-alkyl ether or an α,α' carbon-to-carbon bond.

Successive dehydration processes, combination reactions, and addition of quinonemethide intermediates gradually lead to a high-polymeric end-product. If it is assumed that the above-described and other combination and addition reactions progress stepwise during the polymerization process, then the high-polymer lignin must be a randomly arranged molecule in which the phenylpropane units are combined with each other by the most varied types of linkages.

Although coniferyl alcohol definitely seems to be the precursor of lignin in coniferous woods, the biogenesis of lignin in deciduous woods has not yet been completely clarified. Sinapyl alcohol, a syringyl derivative, readily forms syringaresinol, a compound analogous to pinoresinol. Whether this compound corresponds to deciduous wood protolignin has not yet been established.

The role played by lignin in the general plant metabolism has been investigated by the introduction of radioactive compounds into living plants. Thus it was shown that shikimic acid, which plays a role in the biosynthesis of amino acids, also takes part in the formation of lignin in grasses, sugar cane, and maplewood (35,59). The formation of shikimic acid from D-glucose takes place through the condensation of a triose, formed by glycolysis, with a tetraose, formed via a pentose phosphate. In this process, p-hydroxyphenylpyruvic acid may play a role as an intermediate between the shikimic acid and the phenylpropane derivatives. Ferulic acid also has been found to be an effective lignin precursor (224).

5. EXTRANEOUS COMPONENTS OF WOOD

All woods contain a number of organic substances that can be extracted with organic solvents or, in some cases, even with water. Among these belong aliphatic, aromatic, and alicyclic compounds, hydrocarbons,

alcohols, ketones, and various types of acids, esters, and phenolic compounds. Furthermore, sterols, tannins, essential oils, resins, dyestuffs, lignans, proteins, waxes, and some alkaloids are found. Here, only a few technologically important groups of extractives, especially those that will be mentioned in following chapters in connection with various technical processes, will be discussed. In this field, too, the reader is referred to several book-length presentations (123).

5.1 Resins, Terpenes, Fats

The resin and fat contents of woods play quite an important role in technical delignification processes. It is therefore important to know the content of these substances in various woods. As is shown in Table III-8, coniferous woods contain considerable amounts of resin, with the pine species having an especially high resin content. Long-leaf pine (*Pinus palustris*) and slash pine (*Pinus elliottii*) are the main sources for the extraction of resin from their stumps for the production of oleoresin, gum turpentine, and gum rosin. From the spent black liquors of the kraft pulp industry, tall oil and turpentine are recovered (see Chapter VIII, Sections

TABLE III-8

ETHER-SOLUBLE EXTRACTIVES IN COMMON HARDWOODS AND SOFTWOODS (184)

Wood species[a]	Whole wood	Sapwood	Heartwood
Abies balsamea	1.0 –1.8	0.95	0.74-1.18
Abies concolor	0.23	0.20	0.25
Pseudotsuga menziesii	0.3 –2.6	0.4	1.0
Tsuga heterophylla	0.3 –1.3	0.2 –0.5	0.2 –1.0
Larix occidentalis	0.72-0.93	—	—
Picea glauca	0.4 –2.1	—	—
Picea rubra	—	0.6 –1.2	0.8 –1.5
Pinus banksiana	1.9 –4.3	1.6	5.2
Pinus palustris	2.1 –9.2	1.4 –2.7	3.6 –20.3
Pinus ponderosa	6.5 –9.5	3.2 –4.8	2.7 –9.9
Acer saccharum	0.22-0.89	0.23-0.26	0.25-0.33
Betula papyrifera	1.5 –3.52	0.79-2.97	2.19-3.89
Fagus grandifolia	0.3 –0.86	0.19-0.26	0.38-0.57
Nyssa sylvatica	0.27-0.40	0.44	0.54
Populus grandidenatata	0.86	1.02	1.03
Populus tremuloides	1.0 –2.7	—	—
Liriodendron tulipifera	—	0.13-0.27	0.43-0.58
Salix nigra	0.3	—	—
Ulmus americana	0.28	—	—

[a] All based on wood percent.

7.1 and 7.2). The term "resin" is a collective term, as will be shown in the subsequent discussion of the composition of this extractive.

As compared with coniferous woods, deciduous woods contain considerably less resin and quite varied amounts of it, depending upon the species. Exceptions are the birch and poplar species and these cause resin difficulties in the production of pulp. Table III–8 also shows the content of ether-extractable components of different deciduous woods. In summary, it may be stated that the resin content of coniferous woods is usually above 1%, based on the dry substance of the wood, with the pine species having up to 10% and more of extractable resin. On the other hand, deciduous woods generally have less than 1% of resin, with the exception of birch and poplarwoods which have 2% and more (184).

Wood resins can be separated into free resin acids, fatty acids, fatty acid esters, and unsaponifiable material. The basic difference between resins from deciduous and those from coniferous woods is the practically complete absence of resin acids in the former. Earlier findings of considerable amounts of such resin acids in birchwood could not be confirmed, but small amounts may be present in the resin of birch and poplarwoods (39). Apparently resin acids of the abietinic acid type, if present at all, are found only in traces in the resin of deciduous woods (182). On the other hand, resin acids are the main component in the resin extracts of coniferous woods.

Table III–9 shows the composition of the resin of some coniferous and deciduous woods. The amount of unsaponifiable material is apparently

TABLE III–9

COMPOSITION OF NONVOLATILE RESIN OF SOME SOFTWOODS AND HARDWOODS (184)

Wood species	Resin acids[a]	Fatty acids[a]	Fatty esters[a]	Unsaponifiables[a]
Pinus banksiana	33–39	8–11	37.5–46.5	12–13
Pinus elliottii	25–31	3–22	43–58	7–12
Pinus palustris	30–41	3–8	40–60	7–11
Picea mariana	31	—	42	21
Picea abies	24	8	40.5	20
Pseudotsuga menziesii	48	28	6	18
Abies lasiocarpa	3	37	31	32
Larix laricina	—	—	53	47
Betula papyrifera	2.5	9	52	32.5
Betula verrucosa	—	3–12	58–87	10–32
Populus tremuloides	—	35	50	14
Nyssa sylvatica	—	52	16	32
Liquidamber styraciflua	—	57	31	11

[a] All percent based on extract.

largely dependent upon the wood species. In some birch species, up to 30% of the extract was found to be unsaponifiable. Some poplar species and hazelnutwood contain only about 15%, and birch (*Betula verrucosa*) only a little more than 5% of unsaponifiable material. Sprucewood extracts contain about 20%, and pinewood extracts only 10% or less of unsaponifiable components; pinewood is therefore especially suitable for the production of tall oil.

The fatty acids in the extracts of coniferous and deciduous woods represent the main component and amount, in the former, to 40% to 60% and, in the latter, to 60% to 90% of the extracts. In freshly cut wood, these acids are present in an esterified form, but when the wood is stored the esters undergo hydrolysis. This phenomenon, which plays an important role in the technical utilization of wood, will be discussed later (see Chapter VI, Section 1).

As regards the distribution of resin within the wood, it is closely connected with the anatomical structure of the wood. In deciduous woods, the ray parenchyma cells are the chief depository of the resin. The short parenchymatic cells form the fines in the pulp and are therefore the carrier of the resin and are the cause of pitch troubles during the processing of the pulp. Sapwood, in general, does not contain any resin but it has been found in the heartwood (217,261).

Resin is also present in the ray parenchyma cells of coniferous woods and is found chiefly in resin ducts. Such resin ducts appear in the pine, spruce, and larch species and in Douglas fir, but in hemlock, white fir, and cypresses they occur chiefly as the result of injuries. The resin ducts contain oleoresin, a viscous liquid, consisting of a solution of a resin in a volatile solvent of a terpene type. The resin in the resin ducts consists chiefly of resin acids, whereas the parenchyma resin is a mixture of fatty acids and unsaponifiable material. It therefore appears that the parenchyma resin of coniferous and deciduous woods is the same. The resin in the resin ducts in the heartwood of spruce and fir contains the largest amount of free rosin acids and fatty acids as well as a part of the esters of the total resin. The ray parenchyma resin, on the other hand, consists chiefly of unsaponifiable material, together with some esters. The amount of free acids and esters in the resin ducts of sprucewood is approximately twice as great as that in the total resin, while the amount of unsaponifiable material is only about half of that in the total resin.

In the pulping of coniferous woods by the acid sulfite process, the resin in the resin ducts is freed, whereas the parenchyma resin chiefly remains in the cell walls. The distribution of these two types of resin is therefore important for the deresinification during, or after, the pulping process. The average amount of resin-duct resin in the nonvolatile total resin of sprucewood is about 45%, but great variations are found between

different trees (16,17,64). Differences in the resin content have been found in the various zones in the trunk of a Douglas fir tree. Thus, for example, the innermost heartwood gave 6%, mature heartwood 5%, and the sapwood 2% of ethanol-benzene extract. The earlywood of the whole trunk gave 2% of ether extract, while the latewood gave amounts increasing from 2% to 4% toward the center of the trunk. The latewood, which contains the largest amount of vertical resin ducts, gave almost the total yield of free resin acids. On the other hand, the ether extract of earlywood consisted chiefly of a mixture of fatty acids and unsaponifiable material (41).

With regard to the chemical structure of the oleoresin, it can be said that the volatile oil and the resin acids belong to the terpenoids, deriving from hydrocarbons known as terpenes. Most of these hydrocarbons are unsaturated and are built up of isoprene molecules (C_5H_8). They are acyclic and cylic in character. Depending upon the number of isoprene moieties in them, they are called monoterpenes $(C_{10}H_{16})$, sesquiterpenes $(C_{15}H_{24})$, and diterpenes $(C_{20}H_{32})$, etc. The volatile oil belongs chiefly to the mono- or sesquiterpenes, the resin acids to the diterpenes. Data on the composition of the gum-turpentines of various North American, including Canadian, woods are listed in Table VIII–13 (see Chapter VIII). Figure VIII–50 of Chapter VIII shows the constitutional formulas for some mono- and bicylic terpenes. For further details of the structure and the importance of terpenes, the reader is referred to the relevant technical literature (123,184).

The terpenes are unstable compounds which readily undergo condensation and rearrangement. This is true especially of the bicyclic terpenes. In the presence of an acid and a hydrogen acceptor, α-pinene is converted into *p*-cymene, a reaction that also takes place during the sulfite pulping process (see Chapter VII, Section 3.4).

The resin acids are the principal component of the nonvolatile parts of oleoresin and resin ducts of spruce and pine. They are diterpenic acids which may be divided into two groups, abietinic and pimaric acid types. Some of these resin acids will be discussed in connection with the tall oil of the kraft pulping process. The structure of the acids occurring in the natural resins is shown in Fig. III–21.

There is an appreciable difference between the resin acids which are extracted from pinewood chips and the resin in the pine oleoresin. The former are exposed to oxidation by the oxygen in the air because of their large surface. The amount of levopimaric acid is readily reduced, within a few hours, to about one-half; however, it is interesting to note that the pure acid is relatively stable toward oxidation, although this reaction is strongly accelerated by light and the presence of a sensitizer (158).

Pulpwood is often transported to the mill by being floated. The ether extracts from red pine (*Pinus resinosa*) and white pine (*Pinus strobus*) yield more resin from the heartwood of the former than from that of the latter.

Pimaric type

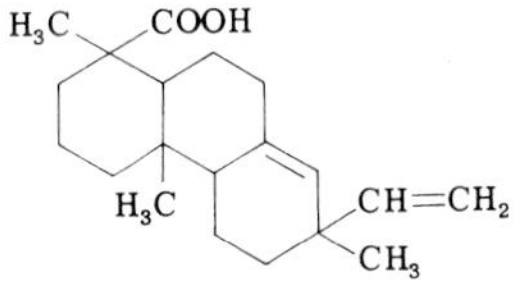

Dextropimaric acid

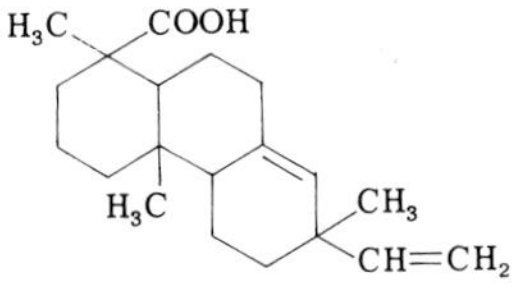

Isodextropimaric acid

Abietic type

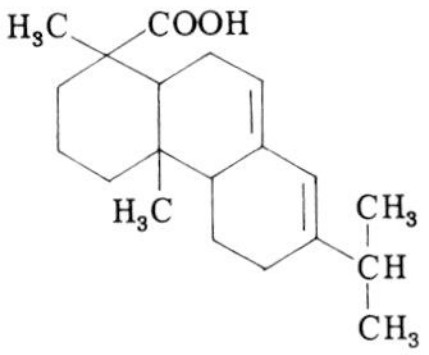

Abietic acid

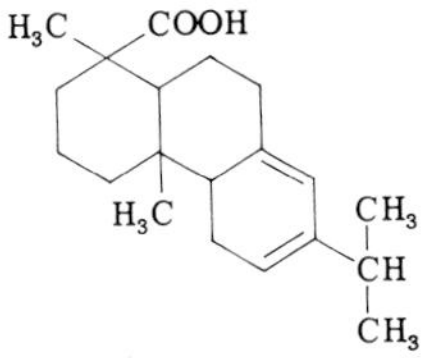

Levopimaric acid

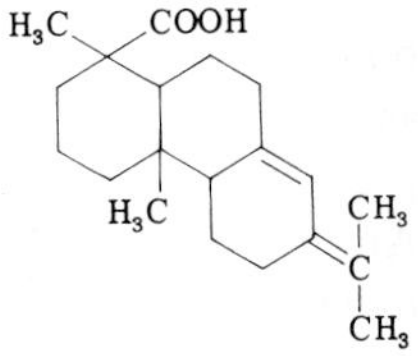

Neoabietic acid

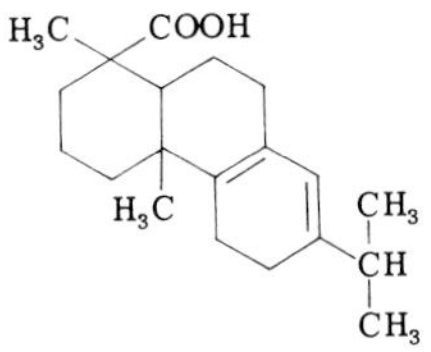

Palustric acid

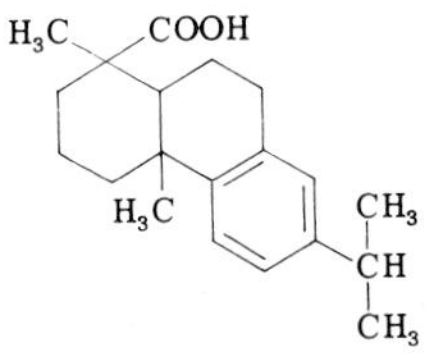

Dehydroabietic acid

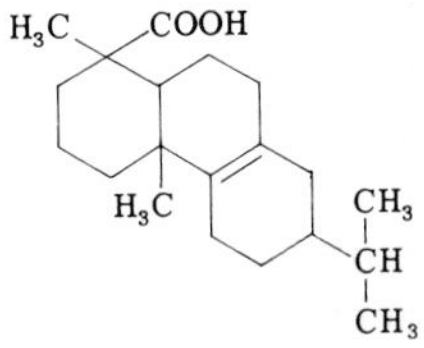

Dihydroabietic acid

Figure III-21. The resin acids.

The extracts from the sapwood of the two types of pine show no appreciable differences. The chemical composition of the extracts also shows no basic differences; they contain resin and fatty acids, fatty acid esters, and unsaponifiable material. However, there is a considerable difference between the heartwood and sapwood resins; the former contains more resin acids but less free and bound fatty acids than the latter. The bound acids are present as glycerides and sterol esters. The esters of the heartwood extracts are practically free of glycerol. Heartwood contains more unsaponifiable material than sapwood does. The extracts from the sapwood of water-stored wood contains more fatty acids and less fatty esters than those from air-stored wood. This effect, caused by hydrolysis, also results after storage in the air when it lasts for a prolonged period. The hydrolysis can also take place in the heartwood but requires a much longer time. The effect

of storage conditions on the quality and yield of pulps is reported in Chapter VI, Section 1 (158).

The main components of the resin from deciduous woods are fatty acids or neutral fatty-acid esters. When birchwood is stored, the fatty-acid esters are converted into the free acids and often a further decomposition occurs (182). In aspen and birchwoods, linolenic acid is the prevailing fatty acid, while linoleic acid and the corresponding saturated acids, such as palmitinic up to lignoceric acid, are also found. The fatty acids are present chiefly as triglycerides but some free acids and other alcohols in the esters have been found. About 4% of the fatty acids from birchwood are present as esters of sterols and 90% as triglycerides (38,39).

Chromatographic investigations of the methyl esters of the free and bound fatty acids from the ether extracts of various deciduous woods, such as birch, poplar, and basswood, showed that the unsaturated fatty acids prevailed in all fractions, with linoleic acid being the main component. Among the saturated acids, palmitic acid is the principal component, followed by stearic and other acids. Only aspen and birch (*Betula lutea*) contain true heartwood and sapwood fractions. While the amounts of saturated and unsaturated acids in the heartwood of aspen do not differ from those in the sapwood, the heartwood of the birch species is richer in saturated acids and poorer in unsaturated acids than the sapwood is. The chain lengths of the identified fatty acids vary between C_6 and C_{24} for the saturated, and between C_{12} and C_{18} for the unsaturated acids (46).

The resins and the fatty acids in wood are subject to chemical changes, especially when the wood is stored in the form of chips. The bound fatty acids, which amount, in fresh birchwood, to about 70% of the ether extracts, decrease rapidly during storage as the result of an enzymatic hydrolysis of the fatty-acid esters. After five to six months of storage, the amount of bound fatty acids is only about 10% of the "resin" content. The proportion of free unchanged fatty acid reaches a maximum after storage for three to four months and then decreases again. With sprucewood, the resin content also decreases as the result of enzymatic hydrolysis and of oxidation of the fatty substances, and these reactions take place much more rapidly during storage in the form of chips (12,13,14,52).

In the unsaponifiable neutral portions of the extractives, higher fatty alcohols and certain vegetable hormones, chiefly β-sitosterol and β-sitostanol, are found. These phytosterols are largely responsible for the difficulties encountered in the deresinification of pulps.

5.2 Phenols and Tannins

There is no real difference between phenols and tannins since both are phenolic in character. Tannins are divided into condensed and hydrolyz-

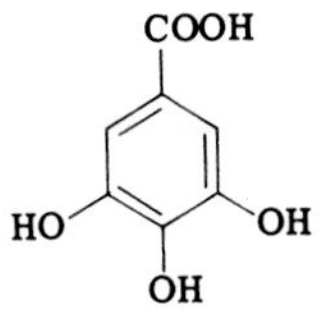

Gallic acid
(A)

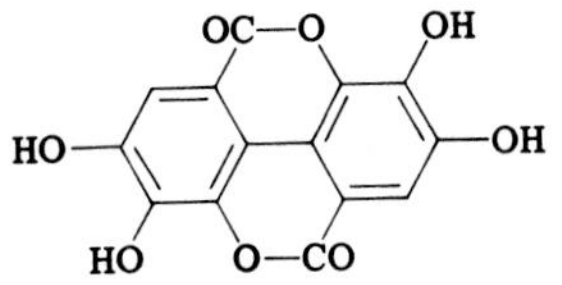

Ellagic acid
(B)

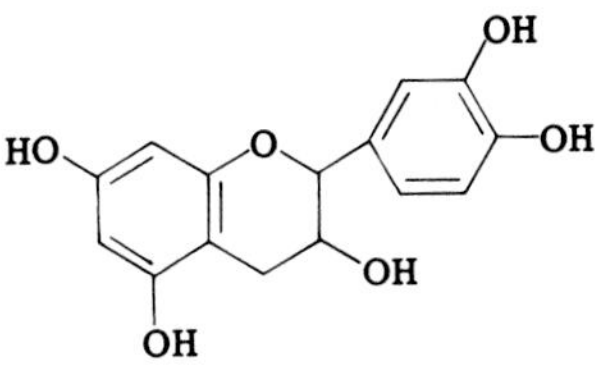

Catechin
(C)

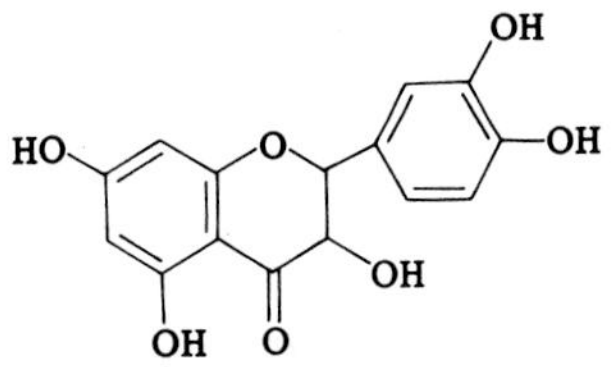

Dihydroquercetin
(D)

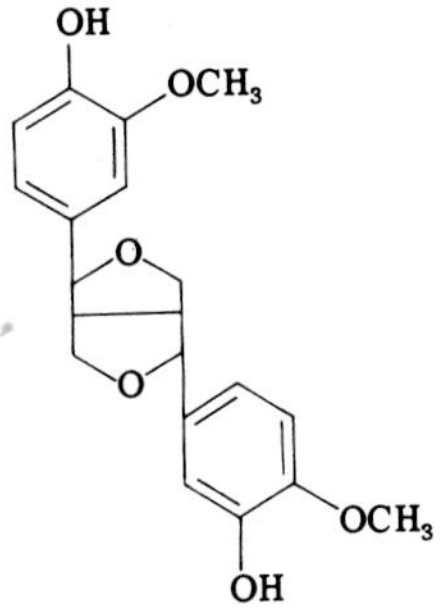

Pinoresinol
(E)

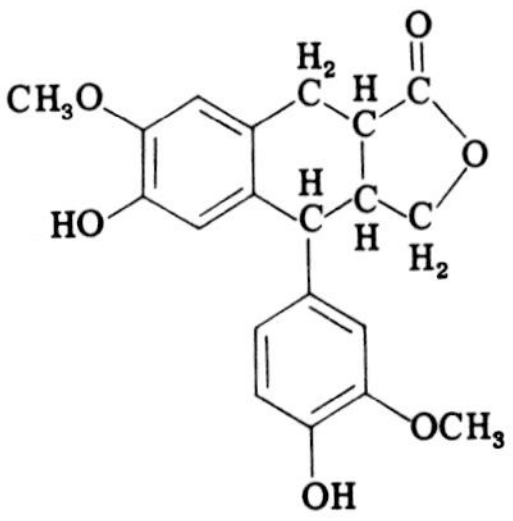

Conidendrin
(F)

Figure III-22

able tannins. They are primarily of technical importance for the tanning of animal hides; the hydrolyzable tannins are esters of a sugar, mostly glucose, with one or several polyphenolic carbocyclic acids. The ester linkages are cleaved by acids, alkalis, or enzymes. Depending upon whether gallic acid (Fig. III-22A) or ellagic acid (Fig. III-22B) is formed on hydrolysis, these tannins are termed gallo or ellagic tannins (109). The condensed tannins are built up of monomers of the catechin type (Fig. III-22C). They are found chiefly in the bark but sometimes also in the wood. A number of substances occurring in the bark and forming colored products are mentioned in Chapter VI, Section 2.2. They are known as flavanones (Fig. III-22D).

Another group of extratives are the lignans. They are formed by the combination of two phenylpropane groups through an α,α' carbon-to-carbon bond in the side-chains. Typical compounds of these lignans are the pinoresinol (Fig. III–22E) from spruce and pine, and conidendrin (Fig. III–22F) from spruce, fir, and hemlock (108).

In the pulping process, the phenolic compounds sometimes interfere with the course of the reaction. Thus, bisulfite can be reduced to thiosulfate and condensations can take place at the benzyl alcohol groups of the lignin. Pinosylvin, for example, is the principal cause of the difficulties occurring in the acid sulfite cooking of pine heartwood. Tannins and flavanones, in turn, can cause a discoloration of the pulp. Further details of reactions affecting the sulfite pulping process are discussed in Chapter VII, Section 3.4.

BIBLIOGRAPHY

F. E. Brauns, "The chemistry of lignin." Academic Press, New York, 1952.

F. E. Brauns, and D. A. Brauns, "The chemistry of lignin." Supplement Volume, Academic Press, New York, 1960.

B. L. Browning, "The chemistry of wood." Wiley (Interscience), New York, 1963.

E. Hägglund, "Chemistry of wood." Academic Press, New York, 1951.

P. H. Hermans, "Physics and chemistry of cellulose fibers." Elsevier, New York, 1949.

E. Heuser, "The chemistry of cellulose." Wiley, New York, 1944.

W. E. Hillis, "Wood extractives." Academic Press, New York, 1962.

IUPAC "Wood chemistry." Proceedings of the Wood Chemistry Symposium, Montreal, Canada 1961, Butterworths, London, 1962.

J. Marton, "Lignin structure and reactions." Symposium of the Division of Cellulose Wood and Fiber Chemistry, Atlantic City, 1965. *Am. Chem. Soc. Adv. Chem.* **59**, Washington, 1966.

N. I. Nikitin, "The chemistry of cellulose and wood." English Translation, Israel Program for Scientific Translations, Jerusalem, 1966.

A. G. Norman, "The biochemistry of cellulose, the polyuronides and lignin." Clarendon Press, Oxford, 1937.

E. Ott and H. M. Spurlin, "Cellulose and cellulose derivatives" 2nd ed., Vols. I, II, and III. Wiley (Interscience), New York, 1954.

W. J. Schubert, "Lignin biochemistry." Academic Press, New York, 1965.

R. L. Whistler, "Methods in carbohydrate chemistry." Vol. III, Cellulose, Academic Press, New York, 1963.

L. E. Wise and E. C. Jahn, "Wood chemistry" 2nd ed., Vols. I and II. Reinhold, New York, 1952.

REFERENCES

1. Adams, G. A., *Tappi* **40** (9), 721 (1957).
2. Adams, M. F. and Douglas C., *Tappi* **46** (9), 544 (1963).
3. Adkins, H., Frank, L. and Bloom, G. S., *J. Am. Chem. Soc.* **63**, 549 (1941).
4. Adler, E., *Das. Papier* **15** (10a), 604 (1961).

5. Adler, E. and Lindberg, B. O., *Svensk Papperstidn.* **55** (16), 563 (1952).
6. Ambronn, H., *Berichte Verhandl. Sächs. Akad. Wissenschaften* **63**, 249 (1911).
7. Andress, K. R., *Z. Phys. Chem.* **34**, 190 (1929).
8. Anthis, A., *Tappi* **39** (6), 401 (1956).
9. Applegarth, D. A. and Dutton, G. S., *Tappi* **48** (4), 204 (1965).
10. Aspinall, G. O., Hirst, E. R. and Ramstad, E., *J. Chem. Soc.* 593, Feb. 1958.
11. Aspinall, G. O. and Nicholson, A., *J. Chem. Soc.* 2503, June 1960.
12. Assarsson, A. and Akerlund, G., *Svensk Papperstidn.* **69** (16), 517 (1966).
13. Assarsson, A. and Croon, I., *Svensk Papperstidn.* **66** (21), 876 (1963).
14. Assarsson, A., Croon, I. and Donetzhuber, A., *Svensk Papperstidn.* **66** (22), 940 (1963).
15. Assof, A. G., Haas, R. H. and Purves, C. B., *J. Am. Chem. Soc.* **66,** 59 (1944).
16. Back, E., *Svensk Papperstidn.* **63** (19), 647 (1960).
17. Back, E. and Carlson, O. T., *Svensk Papperstidn.* **58** (11), 415 (1955).
18. Beckum, van, W. G. and Ritter, G. J., *Paper Trade J.* **104** (19), 49 (1937).
19. Beelik, A., Conca, R. J., Hamilton, J. K. and Partlov, E. V., *Tappi* **50** (2), 78 (1967).
20. Bergmann, M. and Machemer, H., *Chem. Ber.* **63**, 316, 2304 (1030).
21. Bernardy, G., *Angew. Chem.* **38**, 838, 1195 (1925).
22. Bertrand, G. and Benoist, S., *Compt. Rend.* **176**, 1583 (1923).
23. Bevan, E. J. and Cross, C. F., *J. Chem. Soc.* **38,** 666 (1880); "Cellulose." Longmans, Green & Co., London, 1895.
24. Bjorkman, A., *Nature* **174,** 1057 (1954).
25. Bjorkman, A., *Svensk Papperstidn.* **59** (13), 477 (1956).
26. Bouveng, H. O. and Lindberg, B., *Acta Chem. Scand.* **10,** 1515 (1956); **12,** 1977 (1958); **13,** 1877 (1959); **15,** 78 (1961).
27. Braddon, S. A. and Dence, C. W., *Tappi* **51** (6), 249 (1968).
28. Brauns, F. E., *Paper Trade J.* **108** (1), 12 (1939).
29. Brauns, F. E., *J. Am. Chem. Soc.* **61**, 2120 (1939).
30. Brauns, F. E., "Chemistry of lignin" pp. 621-669; see Bibliography.
31. Brauns, F. E. and Brauns, D. A., "Chemistry of lignin" Supplement Volume, pp. 616-629; see Bibliography.
32. Brenner, V. J., Frilette, J. and Mark, H., *J. Am. Chem. Soc.* **70**, 877 (1948).
33. Brewer, C. P., Cooke, L. M. and Hibbert, H., *J. Am. Chem. Soc.* **70**, 57 (1948).
34. Brown, S. A., "Chim. biochim., lignin, cellulose, hemicelluloses" *Proceedings Symposium Grenoble*, p. 247. Imprimeries Réunies de Chambéry 1964.
35. Brown, S. A. and Neish, A. C., *Nature* **175,** 688 (1955).
36. Bryde, O. , Ellefsen, O. and Smith, T. H., *Tappi* **36** (8), 252 (1953).
37. Bryde, O. and Ranby, B., *Svensk Papperstidn.* **50** (11B), 34 (1947).
38. Buchanan, M. A., Burson, S. L. and Springer, C. H., *Tappi* **44** (8), 576 (1961).
39. Buchanan, M. A., Sinnet, R. V. and Jappe J. A., *Tappi* **42** (7), 578 (1959).
40. Campbell, W. G., Hirst, E. L. and Jones, J. K., *J. Chem. Soc.* 774 (1948).
41. Campbell, J. R., Swan, E. P. and Wilson, J. W., *Pulp Paper Mag. Can.* **66** (4), T-248 (1965).
42. Casebier, R. L. and Hamilton, J. K., *Tappi* **50** (9), 441 (1967).
43. Chanda, S. H., Hirst, E. L., Jones, J. K. and Percival, E. G., *J. Chem. Soc.* 1289 (1950).
44. Clark, I. T., *Tappi* **45** (4), 310 (1962).
45. Clayton, D. W., *Svensk Papperstidn.* **66** (4), 115 (1963).
46. Clermont, L. P., *Pulp Paper Mag. Can.* **62** (12), T-511 (1961).
47. Clermont, L. P. and Schwartz, H., *Pulp Paper Mag. Can.* **52** (12), 103 (1951).

48. Coté, W. A. Day, A. C., Simson, B. W. and Timell, T. E., *Holzforschung* **20** (6), 178 (1966).
49. Coté, W. A., Simson, B. W. and Timell, T. E., *Holzforschung* **21** (3), 85 (1967).
50. Coté, W. A. and Timell, T. E., *Tappi* **50** (6), 285 (1967).
51. Creydt, R., *Chem. Ber.* **19**, 3115 (1886).
52. Croon, I., *Das Papier* **19** (10a), 711 (1965).
53. Dence, C. and Sarkanen, K., *Tappi* **43** (1), 87 (1960).
54. Denham, W. S. and Woodhouse, H., *J. Chem. Soc.* **103**, 1735 (1913); **105**, 2357 (1914); **111**, 244 (1917); **119**, 81 (1921).
55. Dennis, D. T. and Colvin, J. R., *Pulp Paper Mag. Can.* **65** (9), T-395 (1964).
56. Dickey, E. E. and Wolfrom, M. L., *J. Am. Chem. Soc.* **71**, 825 (1949).
57. Dolmetsch, H., *Das Papier* **17** (12), 710 (1963).
58. Dutton, G. S., "Chim. biochim., lignin, cellulose, hemicelluloses" *Proceedings Symposium Grenoble*, p. 279. Imprimeries Réunies de Chambéry 1964.
59. Eberhardt, G. and Schubert, W. J., *J. Am. Chem. Soc.* **78**, 2835 (1956).
60. Ekman, K. H., *Tappi* **44** (11), 762 (1961).
61. Ekman, K. H. and Douglas, C., *Tappi* **45** (6), 477 (1962).
62. Ellefsen, O. and Kringstad, K., *Faserforschung-Textiltechn.* **15** (12), 582 (1964).
63. Ellefsen, O., Kringstad, K. and Tönnesen, B. A., *Norsk Skogind.* **13** (11), 419 (1904).
64. Engström, P. and Back, E., *Svensk Papperstidn.* **62** (16), 545 (1959).
65. Enkvist, T., *Tappi* **37** (8), 350 (1954).
66. Enkvist, T., Moilanen, M. and Alfredson, B. *Svensk Papperstidn.* **52** (21), 517 (1949).
67. Erdtman, H., *Svensk Papperstidn.* **43**, 225 (1940); *Cellulosechemie* **18**, 83 (1940).
68. Erdtman, H., Lindgren, B. O. and Petterson, T., *Acta Chem. Scand.* **4**, 228 (1950).
69. Ettling, B. V. and Adams, M. F., *Tappi* **51** (3), 116 (1968).
70. Fredenbach, K. and Cadenbach, G., *Z. Angew. Chem.* **46**, 113 (1933).
71. Frei, E. and Preston, R. D., *Proc. Roy. Soc. London, Ser. B*, **154**, 70 (1961).
72. Fremy, E., *Compt. Rend.* **66**, 456 (1868).
73. Freudenberg, K., *Papierfabrikant* **35**, 247 (1937).
74. Freudenberg, K., *Das Papier* **1**, 209 (1947).
75. Freudenberg, K., IUPAC, pp. 9–20; see Bibliography.
76. Freudenberg, K., "Chim. biochim., lignin, cellulose, hemicelluloses" *Proceedings Symposium Grenoble*, p. 217. Imprimeries Réunies de Chambéry 1964.
77. Freudenberg, K., *Holzforschung* **18** (1/2), 3 (1964).
78. Freudenberg, K. *in* Marton "Lignin structure and reactions" pp. 1–21; see Bibliography.
79. Freudenberg, K. and Bittner, F., *Chem. Ber.* **86**, 155 (1953).
80. Freudenberg, K. and Blomqvist, G., *Chem. Ber.* **68**, 2070 (1936).
81. Freudenberg, K. and Braun, E., *Annalen* **460**, 288 (1928); **461**, 130 (1928).
82. Freudenberg, K. and Dürr, W., *Chem. Ber.* **63**, 2713 (1930).
83. Freudenberg, K., Friedrich, K. and Baumann, J., *Annalen* **494**, 41 (1932).
84. Freudenberg, K. and Harkin, J. M., *Holzforschung* **18** (16), 166 (1964).
85. Freudenberg, K., Molter, H. and Dietrich, G., *Chem. Ber.* **80**, 53 (1947).
86. Freudenberg, K. and Muller, H. F., *Chem. Ber.* **71**, 1281 (1938).
87. Freudenberg, K. and Plankenhorn, E., *Naturwissenschaften* **26**, 124 (1938).
88. Freudenberg, K., Reznik, H., Boesenberg, H. and Rasenak, D., *Chem. Ber.* **85**, 641 (1952).
89. Freudenberg, K., Zocher, H. and Dürr, W., *Chem. Ber.* **62**, 814 (1929).
90. Frey, A., "Die Micellar Theorie von Carl Nägeli." Akademische Verlagsgesellschaft, Leipzig, 1928.

91. Frey-Wyssling, A., "Submikroskopische Morphologie des Protoplasmas und seiner Derivate." Borntraeger, Berlin, 1938.
92. Frey-Wyssling, A., *Protoplasma* **25,** 261 (1936); **27,** 372, 533 (1937); *Papierfabrikant* **36,** 212 (1938).
93. Frey-Wyssling, A., "Submikroskopische Morphologie." Berlin, 1938.
94. Frey-Wyssling, A. and Mühlethaler, K., *Makromol. Chem.* **62,** 25 (1963).
95. Glaser, L. J., *Biol. Chem.* **232,** 627 (1958).
96. Goddard, H. P., McCarthy, J. L. and Hibbert, H., *J. Am. Chem. Soc.* **62,** 988 (1940).
97. Green, J. W., *in* Whistler "Methods of carbohydrate chemistry" pp. 49–54; see Bibliography.
98. Grushnikov, O. P. and Shorygina, N. N. *Izv. Akad. Nauk SSSR, Ser. Khim.* (8), 1774 (1967).
99. Gustafsson, C. and Andersen, L. *Paperi ja Puu* **37,** 1 (1955).
100. Hägglund, E., "Chemistry of wood" pp. 196–212; see Bibliography.
101. Hägglund, E. and Johnson T., *Biochem. Z.* **202,** 440 (1928).
102. Hamilton, J. K., Partlov, E. V. and Thompson, N. S., *J. Am. Chem. Soc.* **82,** 451 (1960).
103. Hamilton, J. K. and Thompson, N. S., *Pulp Paper Mag. Can.* **59** (11), 233 (1958).
104. Hamilton, J. K. and Thompson, N. S., *Tappi* **42** (9), 752 (1959).
105. Haq, S. and Adams, G. A., *Can. J. Chem.* **39,** 1563 (1961).
106. Harris, C. A. and Purves, C. B., *Paper Trade J.* **110** (6), 29 (1959).
107. Harris, E. E., D'Ianni, J. and Adkins, H., *J. Am. Chem. Soc.* **60,** 1467 (1938).
108. Hathway, D. E., The lignans *in* W. E. Hillis "Wood extractives" p. 159; see Bibliography.
109. Hathway, D. E., The condensed tannins *in* W. E. Hillis "Wood extractives" p. 191; see Bibliography.
110. Haworth, W. N., *Nature* **116,** 430 (1925).
111. Haworth, W. N., "The constitution of sugars." Arnold, London, 1929.
112. Haworth, W. N., Charlton W., and Peat, S., *J. Chem. Soc.* 89 (1926).
113. Haworth, W. N. and Leitch, C. G., *J. Chem. Soc.* **113,** 191 (1918).
114. Haworth, W. N., Long, C. W. and Plant, J. H., *J. Chem. Soc.* 2809 (1927).
115. Haworth, W. N. and Machemer, H., *J. Chem. Soc.* 2372 (1932).
116. Haworth, W. N., Montonna, R. E. and Peat, S., *J. Chem. Soc.* 1899 (1939).
117. Hengstenberg, J. and Mark, H., *Z. Kristallographie* **69,** 271 (1928).
118. Hermans, P. H., "Physics and chemistry of cellulose fibers;" see Bibliography.
119. Hernestam, S. and Adler, E., *Svensk Kemisk Tidskr.* **67,** 37 (1955).
120. Herzog, R. O. and Janke, W., *Chem. Ber.* **43,** 2162 (1920); *Z. Phys.* **3,** 196 (1920).
121. Heuser, E., *Paper Trade J.* **122** (3), 43 (1946); *Tappi* **35** (11), 481 (1952).
122. Hibbert, H., *Ind. Eng. Chem.* **13,** 256, 334 (1921).
123. Hillis, W. E., "Wood extractives;" see Bibliography.
124. Hirst, E. L., *J. Chem. Soc.* 522 (1949).
125. Hirst, E. L., IUPAC, p. 53; see Bibliography.
126. Hirst, E. L. and Jones, J. K. N., *J. Chem. Soc.* 496 (1938).
127. Hirst, E. L., Jones, J. K. N. and Campbell, W. G., *Nature* **147,** 25 (1941).
128. Hough, L. and Jones, J. K. N., *Adv. Carbohydrate Chem.* **11,** 185 (1956).
129. Howsmon, J. H., *Text. Res. J.* **19,** 152 (1949).
130. Husemann, E. and Weber, O. H., *J. Prakt. Chem.* **159,** 334 (1941).
131. Husemann, E. and Werner, R., *Makromol. Chem.* **59,** 43 (1963).
132. Irvine, J. C. and Hirst, E. L., *J. Chem. Soc.* **123,** 529 (1923).

133. Ishikawa, H., Schubert, W. J. and Nord, F. F., *Biochem. Z.* **338**, 153 (1963).
134. Ishikawa, H., Schubert, W. J. and Nord, F. F., *Arch. Biochem. Biophys.* **100**, 140 (1963).
135. Jayme, G., *Cellulosechemie* **20**, 43 (1942).
136. Jayme, G. and Knolle, H., *Das Papier* **18** (8), 249 (1964).
137. Jayme, G. and Tio, P. K., *Das Papier* **32** (6), 322 (1968).
138. Jones, J. K. N., IUPAC, p. 21; see Bibliography.
139. Jones, J. K. N., "Chim. biochim., lignin, cellulose, hemicelluloses" *Proceedings Symposium Grenoble.* Imprimeries Réunies de Chambéry 1964.
140. Jurd, L., The hydrolyzable tannins *in* W. E. Hillis, "Wood extractives" p. 229; see Bibliography.
141. Karrer, P., "Die polymeren Kohlenhydrate." Akademische Verlagsgesellschaft, Leipzig, 1925.
142. Kent, W. H. and Tollens, B., *Annalen* **227**, 221 (1885).
143. Kircher, H. W., *Tappi* **45** (2), 143 (1962).
144. Klason, P., *Svensk Kemisk Tidskr.* **9**, 133 (1897).
145. Klason, P., *Berichte Hauptversammlg. Zellcheming* 1908.
146. Klugsoyr, S., *Nature* **18**, 104 (1960).
147. Knolle, H. and Jayme, G., *Das Papier* **19** (3), 106 (1965).
148. Kringstad, K. and Ellefsen, O., *Das Papier* **18** (10a), 583 (1964).
149. Kudzin, S. F. and Nord, F. F., *J. Am. Chem. Soc.* **73**, 4619 (1951).
150. Kuhn, W., *Chem. Ber.* **63**, 1503 (1930); *Z. Phys. Chem.* A **159**, 368 (1932).
151. Kürschner, K., **41** (9), 168A (1958).
152. Kürschner, K. and Hoffer, A., *Technolog. Chem. Papier-, Zellstoff- Fabr.* **26**, 125 (1929).
153. Lange, P. W., *Svensk Papperstidn.* **47**, 262 (1944); **48**, 241 (1945).
154. Lawrence, R. V., *Tappi* **42** (10), 867 (1959).
155. Leech, J. K., *Tappi* **35** (6), 249 (1952).
156. Leopold, B., *Acta Chem. Scand.* **4**, 1523 (1950).
157. Leopold, B., *Acta Chem. Scand.* **6**, 57 (1952).
158. Levithin, M., *Pulp Paper Mag., Can.* **63** (3), T-169 (1962).
159. Lewis, H. F. and Laughlin, E. R., *Paper Trade J.* **93** (22), 43 (1931).
160. Liang, C. Y., and Marchessault, R. H., *J. Polym. Sci.* **37**, 385 (1959); **39**, 269 (1959).
161. Lindberg, B. and Meier, H., *Svensk Papperstidn.* **60** (21), 785 (1957).
162. Lindberg, B., "Chim. biochim., lignin, cellulose, hemicelluloses" *Proceedings Symposium Grenoble*, p. 275. Imprimeries Réunies de Chambéry 1964.
163. Lindgren, B. O., *Acta Chem. Scand.* **4**, 1365 (1950); **5**, 603, 616 (1951).
164. Lindgren, B. O., *Svensk Papperstidn.* **55** (3), 78 (1952).
165. Linnell, W. S. and Swenson, H. A., *Tappi* **49** (10), 444, (11), 491, 494 (1966).
166. McInnes, A. D., West, E., McCarthy, J. L. and Hibbert, H., *J. Am. Chem. Soc.* **62**, 2803 (1940).
167. McKee, R. H., *Ind. Eng. Chem.* **38**, 382 (1964).
168. Mann, J. and Marrinan, H. J., *Trans. Faraday Soc.* **52**, 481, 487, 492 (1956).
169. Marchessault, R. H., "Chim. biochim., lignin, cellulose, hemicellulose" *Proceedings Symposium Grenoble*, p. 287. Imprimeries Réunies de Chambéry 1964.
170. Mark, H., *J. Phys. Chem.* **44**, 764 (1940).
171. Marx-Figini, M., *Makromol. Chem.* **97**, 282 (1966); *Das Papier* **18** (10A), 546 (1964).
172. Marx-Figini, M. and Schulz, G. V., *Naturwissenschaften* **53**, 466 (1966).
173. Meyer, K. H., *Chem. Ber.* **70**, 266 (1937).

174. Meyer, K. H. and Mark, H., *Chem. Ber.* **61**, 593 (1928).
175. Meyer, K. H. and Mark, H., "Der Aufbau der hochpolymeren Naturstoffe." Akademische Verlagsgesellschaft, Leipzig, 1930.
176. Meyer, K. H. and Misch, L., *Helv. Chim. Acta* **20**, 232 (1937).
177. Mosimann, H. and Svedberg, T., *Kolloid Z.* **100**, 99 (1942).
178. Mühlethaler, K., *Beihefte Z. Schweiz. Forstverwaltg* **30**, 55 (1960).
179. Mühlethaler, K., *Das Papier* **17** (10A), 546 (1963).
180. Müller, M. H., *Hoffmann Berichte Entwicklg. Chem. Ind.* **III**, 27 (1877).
181. Müller, O. A., *Papierfabrikant* **32**, (29), 329; (30), 338; (31), 347; (32), 354 (1934).
182. Mutton, D. B., *Tappi* **41** (11), 632 (1958).
183. Mutton, D. B., *Pulp Paper Mag. Can.* **65** (2), T-41 (1964).
184. Mutton, D. B., *in* W. E. Hillis, "Wood extractives" p. 331; see Bibliography.
185. Nägeli, C. v. and Schwender, S., "Das Mikroskop." Engelmann, Leipzig, 1877.
186. Neish, A. C., cited by J. J. N. Jones, *Proc. Intern. Congr. Biochem. 4th Vienna* 1958.
187. Nelson, R., *Tappi* **43** (4), 313 (1960).
188. Nevell, T. P., *in* R. L. Whistler, "Methods in carbohydrate chemistry" pp. 43–48; see Bibliography.
189. Nickerson, R. F., *Ind. Eng. Chem., Anal. Ed.* **15**, 423 (1941).
190. Nishikawa, S. and Ono, S., *Proc. Math. Phys. Soc. Tokyo* **7**, 131 (1931).
191. Nord, F. F., in "Chim. biochim., lignin, cellulose, hemicelluloses" *Proceedings Symposium Grenoble*, p. 217. Imprimeries Réunies de Chambéry 1964.
192. Norman, A. G., "Biochemistry of cellulose, polyuronides, lignin;" see Bibliography.
193. Ohrn, O. and Croon, I., *Svensk Papperstidn.* **63** (18), 601 (1960).
194. Payen, A., *Compt. Rend.* **7**, 1052, 1125 (1838); **8**, 169 (1839); **9**, 149 (1839).
195. Payen, A., *Compt. Rend.* **29**, 493 (1849).
196. Peceny, R., *Svensk Papperstidn.* **70** (21), 719 (1967).
197. Pepper, J. M. and Hibbert, H., *J. Am. Chem. Soc.* **70**, 67 (1948).
198. Peterson, F. C., Maugham, M. and Wise, L. E., *Cellulosechemie* **15**, 109 (1934).
199. Pigman, W. W., Browning, B. L., McPherson, W. H. and Calkin, C. R., *J. Am. Chem. Soc.* **71**, 2200 (1949).
200. Polany, M., *Naturwissenschaften* **9**, 288 (1921).
201. Polany, M. and Weissenberger, K., *Z. Phys.* **9**, 123 (1922).
202. Ranby B. G., *Dissertation Uppsala*, 1952.
203. Ranby, B., *Das Papier* **18** (10A), 593 (1964).
204. Reeves, R. E. and Thompson, H. J., *Contr. Boyce Thompson Inst.* **11**, 55 (1939).
205. Ritchie, P. F. and Purves, C. B., *Pulp Paper Mag., Can.* **48** (12), 74 (1947).
206. Ritter, G. J. and Kurth, E. F., *Ind. Eng. Chem.* **25**, 1250 (1933); *J. Am. Chem. Soc.* **56**, 2720 (1934).
207. Ross, R. J. and Thompson, N. S., *Tappi* **48** (6), 376 (1965).
208. Roudier, A., *Rev. ATIP* **16** (5), 343 (1962).
209. Routala, O. and Sevon, J., *Cellulosechemie* **7**, 113 (1926).
210. Ruben, S., Hassid, W. Z., and Kamen, M. D., *J. Am. Chem. Soc.* **61**, 661 (1939).
211. Rydholm, S. A., "Pulping processes" p. 96. Wiley (Interscience), New York, 1965.
212. Rydholm, S. A., "Pulping processes" p. 175. Wiley (Interscience), New York, 1965.
213. Saeman, J. F., Moore, W. E., Mitchell, R. L. and Millett, M. A., *Tappi* **37** (8), 336 (1954).
214. Samuelson, O., *in* R. L. Whistler, "Methods in carbohydrate chemistry" pp. 31–38; see Bibliography.
215. Samuelson, O. and Wictorin, L., *Svensk Papperstidn.* **67** (14), 555 (1964).

216. Sarkanen, K. V., Wood lignins *in* B. L. Browning, "The chemistry of wood" pp. 248–3111; see Bibliography.
217. Sarten, P., *Das Papier* **8** (17/18), 376 (1954).
218. Scherrer, P. C. and Husey, R. E., *J. Am. Chem. Soc.* **53**, 2344 (1931).
219. Schmidt, E. and Graumann, E., *Chem. Ber.* **54**, 1860 (1921).
220. Schmidt, E., Meinel, K., Jandebeur, W. and Simson, W., *Cellulosechemie* **13**, 129 (1932).
221. Schmidt, E., Schnegg, R. and Hekker, W., *Naturwissenschaften* **21** 206 (1932).
222. Schmidt, E., Tang, Y. X. and Jandebeur, W., *Cellulosechemie* **21**, 201 (1931).
223. Schorger, A. W. and Smith, D. F., *Ind. Eng. Chem.* **8**, 494 (1919).
224. Schubert, W., "Lignin biochemistry" Chapter III; see Bibliography.
225. Schubert, W. J., *J. Am. Chem. Soc.* **72**, 977, 3835 (1950)
226. Schultze, E., *Chem. Ber.* **24**, 2277, 2286 (1891).
227. Schultze, E., *Z. Physiol. Chem.* **16**, 387 (1892); **19**, 38 (1891).
228. Schurz, J., *Das Papier* **17** (12), 1963.
229. Setterfield, G. and Bayley, S. T., *Can. J. Botany* **37**, 861 (1959).
230. Shorygin, N. N., Kefeli, T. Y. and Semechkina, A. F., *Dokl. Nauk. Akad. SSSR* (5), 689 (1949); *Gidrolizn. Promishl.* (2), 6 (1949).
231. Smith, B. and Carlson, O., *Acta Chem. Scand.* **17**, 455 (1963).
232. Smith, T. H., *Norsk Skogind.* **1** (5), 268 (1947).
233. Snyder, J. L. and Timell, T. E., *Svensk Papperstidn.* **58** (23), 851; (24) 889 (1955).
234. Sponsler, O. L., *J. Gen. Physiol.* **9**, 221 (1925); *Ind. Eng. Chem.* **20**, 1060 (1928); *Technol. Chem. Papier-Zellstoff-Fabr.* **28**, 20 (1931); *Cellulosechemie* **11**, 186 (1930).
235. Sponsler, O. L. and Doré, W. H., *Colloid Symposium Monography* **4**, 171 (1926); *J. Am. Chem. Soc.* **50**, 1950 (1928).
236. Staudinger, H., "Die hochmolekularen Verbindungen Kautschuk und Celloluse." Springer, Berlin, 1932; "Organische Kolloidchemie." Springer, Berlin, 1940.
237. Staudinger, H., *Chem. Ber.* **59**, 3019 (1926).
238. Staudinger, H., *Helv. Chim. Acta* **5**, 785 (1922).
239. Staudinger, H. and Reinicke, F., *Holz Roh-u. Werkstoff* **2**, 321 (1939).
240. Staudinger, H. and Signer, R., *Z. Kristallograph. Mineralogie, Petrograph.* A, **70**, 193 (1929).
241. Tang, Y. C. and Wang, H. L., *Cellulosechemie* **16**, 57 (1935); **17**, 21 (1936).
242. Tiemann, F. and Haarmann, W., *Chem. Ber.* **7**, 606 (1874); **8**, 509 (1875).
243. Timell, T., "Studies on cellulose reactions." Esselte AB, Stockholm, 1950.
244. Timell, T., *Pulp Paper Mag. Can.* **56** (7), 102 (1955).
245. Timell, T., *Svensk Papperstidn.* **58** (1), 1 (1956).
246. Timell, T., *Tappi* **40** (1), 30 (1957).
247. Timell, T., *Tappi* **40** (17), 568 (1957).
248. Timell, T., *Svensk Papperstidn.* **53** (15), 472 (1960).
249. Timell, T., *Tappi* **44** (2), 88 (1961).
250. Timell, T., *Svensk Papperstidn.* **64** (20), 744 (1961).
251. Timell, T., *Svensk Papperstidn.* **64** (20), 748 (1961).
252. Timell, T., *Svensk Papperstidn.* **65** (11), 435 (1962).
253. Timell, T., *Tappi* **45** (9), 734 (1962).
254. Timell, T., *Tappi* **45** (10), 799 (1962).
255. Timell, T., *Svensk Papperstidn.* **65** (21), 843 (1962).
256. Timell, T., Glaudemans, C. P. J. and Gillham, J. K., *Pulp Paper Mag. Can.* **50** (10), 243 (1958).

257. Timell, T. and Tyminski, A., *Tappi* **40** (7), 519 (1957).
258. Timell, T. and Zinbo, M., *Tappi* **50** (4), 195 (1967).
259. Tollens, B., "Kurzes Handbuch der Kohlenhydrate." Barth, Leipzig, 1914.
260. Traynard, P. and Eymery, A., *Holzforschung* **9** (6), 172; (1955); **10** (2), 43 (1956).
261. Treiber, E., Stenius, A. and Rehnström, J., *Svensk Papperstidn.* **61** (3), 55 (1958).
262. Ungar, E., *Cellulosechemie* **7** (5), 73 (1926).
263. Watanabe, S., Akahori, T. and Matsubara, R., *Hokkaido Daigaku Kogakubu Hokoku* **43**, 111 (1967).
264. Webb, T. and Colvin, J. R., *Can. J. Microbiol.* **8**, 841 (1962).
265. Weber, O. H., *J. Prakt. Chem.* **158**, 33 (1941).
266. West, E., McInnes, A. D. and Hibbert, H., *J. Am. Chem. Soc.*, **65**, 1187 (1943).
267. Whistler, R. L., *Tappi* **43** (2), 177 (1960).
268. White, E. V., *J. Am. Chem. Soc.* **63**, 2871 (1941); **64**, 302, 1507 (1942).
269. Wiechert, K., *Papierfabrikant* **37**, 325 (1939); *Cellulosechemie* **18**, 57 (1940).
270. Willstätter R. and Zechmeister, L., *Chem. Ber.* **46**, 2401 (1913).
271. Wise, L. E. and Appling, J. W., *Ind. Eng. Chem. Anal. Ed.* **16**, 28 (1944).
272. Wise, L. E. and Appling, J. W., *Ind. Eng. Chem. Anal. Ed.* **17**, 182 (1945).
273. Wise, L. E., Hamer, P. L. and Petersen, F. C., *Ind Eng. Chem.* **25**, 184 (1933).
274. Wise, L. E., Murphy, M., and D'Addieco, A. A., *Paper Trade J.* **122** (2), 35 (1964).
275. Wise, L. E. and Petersen, F. C., *Ind. Eng. Chem.* **25**, 184 (1933).
276. Wise, L. E. and Rattliff, E. K., *Anal. Chem.* **19**, 459 (1947).
277. Wise, L. E., Rattliff, E. K., and Browning, B. L., *Anal. Chem.* **20**, 825 (1948).
278. Wise, L. E., and Rittenhouse, R. C., *Tappi* **32** (9), 397 (1949).
279. Wyck, A. J. A. van der, and Studer, M., *Helv. Chim. Acta* **32**, 1698 (1949).
280. Zinbo, M. and Timell, T., *Svensk Papperstidn.* **68**, (19), 647 (1965).
281. Zinbo, M. and Timell, T., *Svensk Papperstidn.* **70**, (19), 597; (21), 695 (1967).

IV

The Acid Hydrolysis of Wood

INTRODUCTION: WOOD COMPOSITION AND HYDROLYSIS

The reactions of wood with water, salt solutions, acids, and bases have attracted the attention of numerous investigators for almost a hundred years because of their importance in the elucidation of the chemistry of the wood and its technological utilization. Many of the reactions to which wood has been subjected in technical and chemical processes involve a hydrolysis, either directly or indirectly. The heterogeneous composition of wood, consisting of cellulose, with primary, secondary, and tertiary hydroxyl groups, of lignin, with phenolic, alcoholic, and pseudo-acidic groups, of hemicelluloses, with lactone, acetal, ether, and ester groups, of hydrogen bridges of reactive groups, and of the partially crystalline and partially amorphous cellulose structure, affords innumerable reaction possibilities. Many microorganisms, such as fungi, enzymes, and bacteria, the oxygen of the air, light and heat may promote or intensify the hydrolysis of some wood constituents. Thus the study of the hydrolytic processes has long engaged and stimulated the attention of wood chemists (19,82,88,95,160, 207).

As will be shown in detail, cellulose is converted quantitatively into glucose upon hydrolysis with acids. Under these conditions, the β-glucosidic linkages of the cellulose chain molecule are split by the addition of water, thus forming fragments of shorter chain lengths but of unchanged basic structure. One of the newly formed end-groups of the chain members is a potential aldehyde group and possesses reducing power. Hemicelluloses, on the other hand, give a mixture of sugars and sugar derivatives. Table IV-1 shows the composition of various woods with regard to the polysac-

TABLE IV-1

COMPOSITION OF THE CARBOHYDRATE PORTION OF VARIOUS WOODS (70)

Wood species	Botanical name	Glucosan (%)	Galactan (%)	Mannan (%)	Araban (%)	Xylan (%)
Spruce	*Picea excelsa*	95.5	6.0	16.0	3.5	9.0
White pine	*Pinus silvestris*	65.0	6.0	12.5	3.5	13.0
Larch	*Larix sibirica*	63.0	17.5	7.5	3.0	9.0
Poplar	*Populus tremula*	64.5	1.5	3.0	1.0	30.0
Birch	*Betula verrucosa*	58.5	1.5	0.5	0.5	39.0
Linden	*Tilia cordata*	58.5	1.5	3.5	2.0	34.5
Maple	*Acer platanoides*	60.5	2.0	4.0	1.0	32.5
Beech	*Fagus silvatica*	65.0	4.0	1.5	1.5	28.0
Oak	*Quercus excelsior*	68.5	2.5	2.0	1.0	26.0
Ash	*Fraxinus excelsior*	60.0	3.0	2.5	2.5	32.0
Willow	*Salix alba*	74.0	3.0	2.5	1.0	26.0
Elm	*Ulmus scabra*	68.5	2.5	2.0	1.0	27.0
Alder	*Alnus incana*	67.0	3.5	1.5	1.0	27.0

charides formed in their hydrolysis with acids. The data show the proportion of the monosaccharides as determined by chromatography and calculated in percent, based on the total carbohydrate content of the wood. No other carbohydrates could be detected in these woods. The noticeable differences in galactan, mannan, and xylan contents in the coniferous and deciduous woods are clearly apparent. These differences in the composition of the carbohydrate portion have been recognized for a long time by the pulp industry and they play a special role in the production of pulps for chemical conversion (see Chapter VIII, Section 4.5).

To gain some insight into the course of the reaction and the mechanism of wood hydrolysis, it is necessary to study the hydrolysis of the single wood components and their hydrolysis products. This will be done in the next two sections, and the technical processes of wood hydrolysis will then be dealt with.

1. THE CHEMISTRY OF WOOD HYDROLYSIS

1.1 The Hydrolysis of the Cellulose

1.1.1 HYDROLYSIS WITH CONCENTRATED ACIDS

The degradation of cellulosic materials to sugars seems, at first, to be a hydrolytic cleavage of the glucosidic bonds, as is shown in a simplified form in Fig. IV-1. However, cellulose behaves fundamentally differently from other carbohydrates in the hydrolysis. The glucosidic bonds are

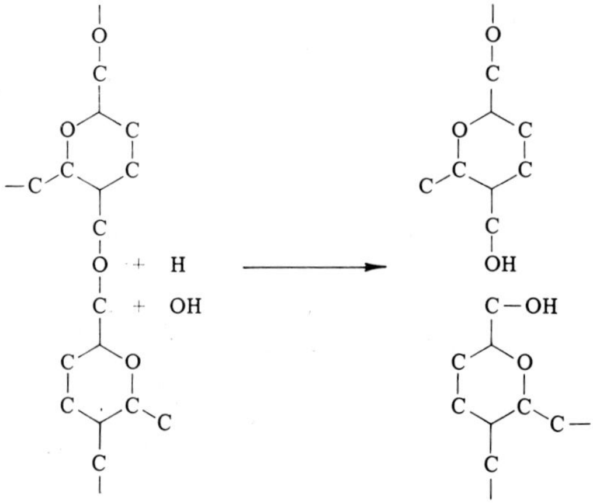

Figure IV-1

cleaved relatively easily, but the crystalline structure of the cellulose is far more resistant toward the heterogeneous hydrolysis by dilute acids than are similar, but noncrystalline, carbohydrates.

Over a hundred years ago it was found that, of the concentrated acids, highly concentrated hydrochloric acid is a very effective hydrolytic agent. If the hydrolysis is followed polarimetrically, it has been observed that, at the beginning, the solutions are optically inactive, but they become active after about an hour and, after 24 to 48 hours, they show the rotation value of glucose. The importance of the concentration of the hydrochloric acid for the course of the hydrolysis is shown by the use of supersaturated acid containing 42% to 45% hydrogen chloride, which has been found to be especially effective and has become the basis of the well-known lignin determination method and of a technical wood hydrolysis process (6,234,239).

Consecutive action of concentrated and dilute hydrochloric acids, in which the concentrated acid hydrolyzes the wood and is drawn off by evacuation, and the acid residue is then boiled with water, provides the first indication that such a combination can be applied for the technical hydrolysis of wood, with the extensive recovery of the concentrated acid being achieved, thus assuring its economical operation. In this way, a yield of 61% of glucose may be obtained and 76%, or almost the total carbohydrate content, of the wood may be recovered in the form of a 30% solution of simple sugars (78,79,237).

It seems, however, that a treatment with dilute acid prior to the main hydrolysis with concentrated acid is preferable to the reverse process. It is obvious that the hydrolytic degradation of cellulose is affected disadvantageously by the hydrolysis products of the hemicelluloses that are formed simultaneously. In the prehydrolysis with dilute acid, these hemi-

celluloses are hydrolyzed, whereas the cellulose itself is more resistant because of its crystalline structure, already referred to. It is interesting to note that the lignin, also, is affected in its methoxyl content, because only the lignin treated in a prehydrolysis shows the normal methoxyl content of an acid lignin obtained by hydrolysis with concentrated acid (172).

The hydrolysis of wood differs fundamentally from that of cellulose. In the hydrolysis of a sulfate pulp for 4 hours with a hydrochloric acid of up to 35% hydrogen chloride, only a minor degradation of the cellulose takes place; with a hydrochloric acid of 40% hydrogen chloride for 4 hours at 40°C, on the contrary, a quantitative hydrolysis of the cellulose occurs, and, at 60°C, the time required is only 10 minutes. In this case, the reaction velocity of the hydrolysis surpasses the rate of the formation of reversion products. Up to about 30°C, the hydrochloric acid degrades the cellulose to oligosaccharides and only at higher temperatures are the latter split to simple sugars. The highest yields of sugars are obtained with a hydrolysis period of 20 minutes at 60°C, 30 minutes at 50°C, or 60 minutes at 40°C. Longer periods cause an increased decomposition of the sugars. The hydrolysis follows a first-order reaction, with the velocity constant increasing sharply with the elevation of the temperature. In contrast to this, the most important factors influencing the hydrolysis of wood with 40% hydrochloric acid are the diffusion velocity of the acid into the wood structure and the migration of the dissolved carbohydrates from the wood structure into the surrounding acid solution. The diffusion velocity can be increased by reducing the particle size of the wood, by lowering the molecular size of the dissolved carbohydrates, by increasing the concentration gradient between the dissolved carbohydrates in the wood itself and the surrounding solution, and by decreasing the viscosity of the solution. Molecular size and viscosity can be lowered by raising the reaction temperature, whereas the concentration gradient can be controlled by a continuous countercurrent hydrolysis (131,132).

When lignocellulose which has been prepared by prehydrolysis of ground wood with 2% hydrochloric acid at 100°C is treated with a 38% to 41% hydrochloric acid for 16 hours at room temperature, the following amounts of lignocellulose are hydrolyzed: with 41% hydrochloric acid, 100%; with 40%, 72.6%; with 39%, 44.7%; and with 38%, 21.7%. The reaction velocity during the first 10 hours corresponds to that for wood but becomes considerably higher during the last 6 hours. With the exception of the 38% acid, wood is hydrolyzed more rapidly than lignocellulose, probably because, at this acid concentration, it is primarily the hemicelluloses that are hydrolyzed. The hydrolysis with 41% hydrochloric acid does not follow a unimolecular reaction equation because the reaction constant increases with the time. The hydrolysis with 40% hydrochloric acid, however, is unimolecular and takes place in three different phases. The 39% and 38%

hydrochloric acid hydrolyses show a decrease in the reaction constants with time. In the 41% hydrochloric acid hydrolysis, the amount of acid bound by the lignin in the lignocellulose is 16% at the start and drops to 6% to 9% toward the end of the reaction. This is in contrast to the hydrolysis of wood with the same acid concentration, where the amount of bound hydrochloric acid increases steadily. From this it may be concluded that, during the prehydrolysis, chemical changes occur in the lignin molecule and a possible cleavage of the polysaccharide-lignin complex takes place (170).

A variation in the hydrolysis with hydrochloric acid is the use of gaseous hydrogen chloride. Cellulose, such as cotton linters, is completely converted by dry hydrogen chloride at 44 atmospheres pressure and 20°C within 10 hours into water-soluble low-molecular glucoanhydrides (186).

The fact that the hydrolysis with 40% hydrochloric acid still requires a considerable amount of time has led to the suggestion that a still more concentrated acid should be used. With 44% to 48% hydrochloric acid concentrations, polysaccharides in vegetable materials can be hydrolyzed within 6 hours. Such highly concentrated acids, however, are stable only at temperatures of 8°C or lower and therefore cannot be used in diffusers; they can, however, be obtained at higher temperatures by saturating the moistened material with gaseous hydrogen chloride. The yield of polysaccharides from sprucewood chips, which at atmospheric pressure are in an equilibrium with hydrogen chloride gas, increases with the moisture content originally present in the wood. With a moisture content of 100% to 120%, calculated on the basis of the weight of the dry wood, the yield of polysaccharides reaches a constant value of 70% after 6 hours. The amount of hydrogen chloride is then 43.5% to 49.5% and therefore is higher than the hydrogen chloride concentration in the moisture of the wood at 20°C. This fact can be attributed to the reaction of the hydrogen chloride with the polysaccharides and the lignin. An undesirable heat formation, caused by the swelling and solution of the hydrogen chloride in the water, can be avoided by saturating the wood at $-12°$ to 8°C to a hydrogen chloride content of 45% to 49.5% and then increasing the temperature to 23° to 25°C. In this manner, yields of 64% to 65% polysaccharides are obtained within 6 hours, causing an acid consumption of 1.75 parts of hydrogen chloride per 1 part of absolutely dry wood. The acceleration of the hydrolysis with 44% to 45% hydrochloric acid is the result of the fact that the vapor tension of the hydrogen chloride over hydrochloric acid solutions which contain sugars is lower than that over pure hydrochloric acid (35).

Turning now to the hydrolysis with concentrated sulfuric acid, this, also, was observed and investigated at an early date. While the first investigations were more or less limited to ascertaining that cellulose and cellulose fiber-containing plants can be hydrolyzed to lower sugars, later experiments established that the reaction of cellulose with sulfuric acid

occurs in two stages, with the formation first of a sulfuric acid-cellulose addition compound, and then of glucose. This intermediate addition compound was called "amyloid" and was compared to "parchment cellulose," which is formed by the parchmentization of filter paper with sulfuric acid. Actually, these two "celluloses" differ in their degree of hydrolyzability and more nearly represent different stages of swelling (4,16,69,103,194,227).

In order to elucidate the configuration and constitution of cellulose and its accessory carbohydrates, the degradation velocity of cellulose in concentrated sulfuric acid was investigated. Iodometric determinations of the aldehyde groups served as the means for following the hydrolytic degradation; simultaneously the changes in the optical behavior of the solution of the hydrolysis products were used for comparison. The concept that, from the triose upward, each fragment with a terminal bond will react according to the dissociation constant, k_2, of the cellobiose, and with all other bonds according to the initial hydrolysis constant, k_n, of the cellulose, gave an acceptable basis for the calculation. However, this is only an approximation because, for an exact calculation of the cellulose degradation, the initial hydrolysis constant, k_3, of cellotriose also must be considered. With cellotetraose, the end bonds probably react with a velocity lying between k_3 and k_n and differing little from k_3. The constant for the middle bonds may be similar to k_n. By means of degradation reactions of the methylcellobiose, methylcellotriose, methylcellotetraose, and of methylcellulose, and by synthesis of the cleavage products of these reactions, it could be shown that these polysaccharides contain only one type of linkage, that of cellobiose. The molecular rotation changes from the biose upward, in close approximation to the rotation of a center glucose anhydride unit of the triose and of two center units for the tetraose, etc. The rotation of these center units is almost the same as the molecular rotation of the cellulose. All bonds of the polysaccharides are sterically identical (61,62).

The energy of activation of the hydrolysis was found to be, for cellobiose, 27,300; for cellotriose, 28,600; for cellotetraose, 28,900; and for cellulose, 29,800 cal per gram molecule. From this it can also be concluded that the hydrolytic reaction always attacks the same type of bond.

With regard to the technical application of the hydrolysis with concentrated sulfuric acid, the questions of a suitable concentration of the acid and of an adequate acid-to-wood ratio have received special attention. In a process in use in the U.S.S.R., 75% sulfuric acid and high ratios—1.04 for beechwood and 1.22 for pinewood per one part of sulfuric acid hydrate ($H_2SO_4 \cdot H_2O$)—are used. Without going into technical details, which will be discussed later, it is apparent that a uniform distribution of the highly concentrated acid in the material to be hydrolyzed is necessary. To achieve this, the material must be extensively reduced in size and thoroughly mixed with the acid. In another process, the material to be hydrolyzed is impreg-

nated with a dilute acid and the desired concentration is reached by subsequent drying of the material. During the drying, the acid is distributed in the microcapillaries and the concentrated acid is absorbed by the reactive groups of the cellulose and lignin. Because of the strong swelling caused by the acid, the rate of evaporation of the acid solution is greater than that of pure water. At a wood-to-acid ratio of 0.1 and a drying temperature of 90°C, only 19% of the cellulose is dissolved at a sulfuric acid concentration of 72%. Mechanical action on the acid-wood mixture increases the dissolved fraction to 44.6%. An increase in the temperature accelerates the hydrolysis, but also the decomposition of the sugars formed (156,174).

The swelling and the solubility of cellulose in concentrated sulfuric acid at 20°C increase until a maximum of 62% sulfuric acid is reached. The presence of hemicelluloses has no effect upon the solubility characteristics of the cellulose in such an acid. Lignin, on the other hand, exerts a lasting influence, obviously because it hinders the swelling and the solubility of the wood polysaccharides. In a temperature range of 20° to 40°C, the swelling and the solubility maxima lie in a concentration range of 62% to 63% sulfuric acid; if the temperature is raised or lowered, the swelling maximum drops (228,229).

The hydrolyzability and solubility of cellulose in concentrated sulfuric acid do not always run parallel. For example, filter paper from sulfite pulp, swollen in a sulfuric acid of 62% for 1 hour at 20°C, showed a solubility of 93% after a hydrolysis with 10% sulfuric acid, but only 55% of the cellulose was hydrolyzed. This difference indicates the probable formation of an intermediate compound, such as amyloid. In 55% to 62% sulfuric acid, the hydrolyzability was greater than the solubility. The presence of up to 16% water in the filter paper had no effect on the hydrolyzability, but the solubility increased with increasing moisture content. The presence of ethyl alcohol in the treatment with concentrated acid also increased the solubility and had no effect on the hydrolyzability. The acid-to-cellulose ratio had no appreciable effect on the solubility of the cellulose at a sulfuric acid concentration of up to 61%. With higher acid concentrations, the solubility decreased noticeably when the acid-to-cellulose ratio was decreased. The addition of glucose or xylose to the acid lowered the swelling of the cellulose. This makes it appear probable that the decreased solubility at lower acid-to-cellulose ratios is caused by the formation of increasing amounts of soluble products which may react or combine with the acid. Celluloses with dense macromolecular packing show S-shaped solubility curves without a pronounced maximum. On the other hand, less densely packed wood celluloses show a clear solubility maximum with 62% sulfuric acid. In the region of low swelling (H_2SO_4 concentration of below 60%), the solubility velocity increases with decreasing density, whereas in the region of high swelling (acid concentration of above 60%), the solubility rate increases

with increasing density. From this it may be assumed that, at lower acid concentrations, the penetration of the acid is inhibited by the swelling of the cellulose on the surface of the fibers (230). An effective concentration of the sulfuric acid in the hydrolysis of cellulose must therefore be above 60%; on the other hand, it cannot be above 75% in the cellulose-sulfuric acid-water system (185).

Because it was found in the hydrolysis of polysaccharides with concentrated hydrochloric acid that the acid concentration decreases steadily during the reaction with the polysaccharides and that the reaction comes to an end after the establishment of an equilibrium, it seemed reasonable to expect a similar reaction sequence in the hydrolysis with sulfuric acid. Samples of cotton hydrocellulose were impregnated with sulfuric acids of 65% to 90% concentrations and stored at room temperature until an equilibrium was established. The analysis of the hydrolyzates before and after inversion with 4% acid provided an unequivocal relationship between the concentration of the acid and that of the glucose in the hydrolyzate after the establishment of the equilibrium as long as the further hydrolysis of the difficultly hydrolyzable portion of the cellulose was inhibited. At 20° to 25°C, the calculated equilibrium concentration for 62.5% sulfuric acid and the composition of the hydrolyzate corresponded to a solution of $C_6H_{12}O_6 \cdot 1.37H_2SO_4$. Calculated on the elementary unit of the cellulose, the composition of this compound was the same as in the hydrolysis with hydrochloric acid. The ratio of the acid, calculated as monohydrate, for the complete hydrolysis of cellulose was 2.93 for 70%, and 1.03 for 90% sulfuric acid and, at room temperature, was considerably higher than the values calculated for hydrochloric acid. With higher temperatures, substantially more favorable values were obtained. At 50°C, the equilibrium concentration of the acid was calculated to be 53%, and the acid ratio to be 0.75 with 70% acid and 0.44 with 90% acid. The corresponding acid ratios for wood would be 0.44 and 0.26, i.e., the amount of acid can be sharply reduced even without mechanical action (37).

The kinetics of the course of the reaction of the hydrolysis of the cellulose can be elucidated by molecular weight determinations of solutions of cellulose in phosphoric acid. In this way it was found that the type of reaction taking place with native cellulose is different from that with hydrate cellulose. While the former is porous, the latter has a more compact structure. In addition, for the former, a definite mixture of phosphoric acid and water is used but, for the latter, a less concentrated impregnation acid and more concentrated dispersing acid are required. Native cellulose dissolves more slowly than hydrate cellulose. Very low molecular hydrate celluloses dissolve even in the impregnation acid. For every defined concentration of an acid, a definite dispersion region could be ascertained, and this became smaller with increasing molecular weight of the cellulose.

With the acid-water mixture, cellulose forms addition compounds of the following composition:

in phosphoric acid	$(C_6H_{10}O_5 \cdot 2H_2O \cdot H_3PO_4)_n$
in nitric acid	$(C_6H_{10}O_5 \cdot H_2O \cdot HNO_3)_n$
in sulfuric acid	$(C_6H_{10}O_5 \cdot 4H_2O \cdot H_2SO_4)_n$
in hydrochloric acid	$(C_6H_{10}O_5 \cdot 4H_2O \cdot HCl)_n$

By determining the decline in the viscosity of the cellulose solution during the hydrolysis, the decrease in molecular weight and, with it, the rate of the hydrolytic decomposition could be calculated (53). According to other authors, these experiments contain some sources of error because the degree of polymerization at the starting point of the reaction was not accurately determined, and the K_m constant for the calculation of the degree of polymerization of cellulose in phosphoric acid solution was not known exactly. Finally, the fact that the polymolecularity of the cellulose changes during the decomposition was not taken into account. The degree of polymerization, as determined viscosimetrically, in such cases can be twice as high as that found osmometrically or by end-group determination. The decomposition of the cellulose by hydrolysis follows a first-order reaction, in which the degradation constant with degrees of polymerization of 130 to 1500 is independent of the degree of polymerization within the limits of error. This proves that there cannot be any linkages present that are much more rapidly split than the β-glucosidic bonds (193).

From the fact that, during the hydrolysis of cellulose in phosphoric acid, a sharp drop in viscosity occurs at the beginning and later changes to a uniformly moderate decline (206), it was again concluded that semi-acetal-like linkages are split first and then glucosidic ones. The course of the reaction of the hydrolysis with concentrated acid becomes still more obscure as the result of the formation of condensation and reversion products of lower polymers which are water-soluble and show the properties of oligo-saccharides (101,238). Whereas with cotton and with isolated wood cellulose a clear solution is obtained in 84% phosphoric acid, with the cellulose present in wood this is not possible. The effectiveness of the phosphoric acid can be considerably increased by the addition of hydrochloric acid (219,231)

1.1.2 The Hydrolysis with Dilute Acids

While the hydrolysis with concentrated acids has long been the subject of exact investigations into the course and conditions of the reaction, the hydrolysis with dilute acids has been studied systematically only during the last three decades. This does not mean that dilute acids have not been suggested or used much earlier. Thus, wood was hydrolyzed with 3% to

5% sulfuric acid at 180°C; other experiments used dilute hydrochloric acid, and still others, sulfur dioxide, as a hydrolyzing agent. Some of these processes were tried in pilot plants, but the yield of sugars was low and the process could not be carried out economically (42,147,167,199,241).

Later investigations confirmed the findings as to the effect of the acid concentration, the acid-to-wood ratio, the reaction temperature, and the duration of the reaction on the yield of hydrolysis products, with deciduous and coniferous woods, under certain conditions, giving approximately equal yields of reducing sugars. Because of the high pentosan content of deciduous woods, the yield of sugars capable of being fermented to alcohol drops to about half that of coniferous woods (77,124,176).

Systematic investigations of the course of the hydrolysis of cellulose with dilute acids showed that the loss in fiber strength is parallel with the decrease in the viscosity in cuoxam solution. Through the determination of the decrease in viscosity of the dissolved cellulose, the hydrolytic action of water, salt solutions, and dilute acids could be determined numerically (15,208). It was shown that mercerized and regenerated cellulose fibers underwent hydrolysis more readily than native cellulose fibers. According to recent investigations, all cellulosic materials contain a certain proportion of "stable cellulose" or a "hydrocellulose fraction" which is cleaved in accordance with the law of a first-order reaction. This reaction gives a residue of an approximately constant molecular weight. From this it may be concluded that the degradation of this "stable cellulose" must lead to water-soluble products or to products that can readily be converted to a water-soluble form. These water-soluble products are then, in turn, rapidly converted into simple sugars, providing the following reaction scheme:

$$\text{native cellulose} \xrightarrow{\text{(I)}} \text{stable cellulose (hydrocellulose)} \xrightarrow{\text{(II)}} \text{soluble polysaccharide} \xrightarrow{\text{(III)}} \text{simple sugars}$$

The reaction, I, occurs very quickly and under relatively mild hydrolytic conditions; reaction II follows the law of a first-order reaction, takes place slowly, and is the controlling reaction in the cellulose hydrolysis; reaction III corresponds in its rapid course to the hydrolysis of oligosaccharides (93).

When wood cellulose, viscose fibers, and cotton were subjected to a hydrolysis under identical reaction conditions, it was found that the proportion of "stable cellulose" was, in each case, smaller than would correspond to the total cellulose portion of the material being hydrolyzed. In a specially prepared cellulose sample, the average degree of polymerization (DP) and its distribution were found by nitration and fractionated precipitation to be 913, with 36.5% of the cellulose having a DP of over 200. After a hydrolysis with 10% sulfuric acid for 6 hours, the distribution of the DP and the glucose content in the hydrolyzate were determined. It

was found that the reaction occurred very rapidly during the first 90 minutes and the DP dropped from 913 to 281. The reaction then slowed down and, at the end of the hydrolysis period, the DP reached a constant value of 170. The hydrolyzate contained 45% (based on the weight of the cellulose) of a fraction with a DP of up to 50, whereas the maximum of the DP was 400. From the reaction velocities it was established that cotton stalks contain 12.1% of an easily hydrolyzable fraction, while cotton linters have 2.4% and sulfite pulp 2.84% (149).

Since the amorphous fraction of the cellulose should be more readily accessible to hydrolysis than the crystalline portion, this variation in the behavior of different types of cellulose was considered to be a criterion for the proportion of crystalline and amorphous cellulose (45). Such conclusions must be drawn with some reservations because, in the first stage of the hydrolysis, recrystallization can occur (18). The percentage of increase in the crystallinity can be calculated from the data for the density of the crystalline cellulose, i.e., 1.59 grams per ml (specific volume 0.629) and 1.50 grams per ml (specific volume 0.667) (99,144). Thus, for example, on hydrolysis, viscose fiber showed a 10% loss in weight and, at the same time, a 16% increase in crystallinity. When the amorphous cellulose in hydrolyzed viscose stable fiber and ramie fiber was dissolved, the absorption capacity for water vapor decreased sharply in the former but remained unchanged in the latter. From this it was calculated that there was an increase of 39% to 49% in the crystallinity of the viscose fiber (54,100).

An investigation of the action of a heterogeneous hydrolysis with dilute sulfuric acid on the specific surface area of the cellulose showed that three different reactions take place. First, a degradation of accessible and especially reactive bonds occurs, accompanied by a strong increase in the specific surface regions and a decrease in the DP. This is followed by a recrystallization of free macromolecules, this time combined with a decrease in the surface areas. Finally, the reaction follows a stable and moderate course, the specific surface regions undergo only a minor increase, and the DP drops still further. These reactions occur only at the surface of the crystallites while their inner structure remains untouched. The microstructure of the cellulose is not changed during the heterogeneous hydrolysis with dilute acid. The heterogeneous hydrolysis therefore confirms the theory of the existence of more highly and less highly ordered regions in the cellulosic material (126).

The reaction velocity, K, of the hydrolysis with dilute sulfuric acid of the stable cellulose was found to be represented by the following equation:

$$K = HC_s^M e^{-\Delta H_a/RT}$$

in which H_a is the heat of activation, C_s is the acid concentration, M is the slope of the curve which is obtained when K is plotted against the logarithm

TABLE IV-2

FIRST-ORDER REACTION CONSTANT FOR THE HYDROLYSIS OF CELLULOSE WITH
AQUEOUS SOLUTIONS OF INORGANIC ACIDS (92)

Temper-ature (°C)	Acid concen-tration (mole)	First-order reaction constant, K			
		Sulfuric acid (min^{-1})	Phosphoric acid (min^{-1})	Sulfur dioxide (min^{-1})	Hydrochloric acid (min^{-1})
160	0.04	—	—	—	0.0013
	0.08	—	—	0.0019	0.0033
	0.16	—	—	0.0025	0.0078
	0.32	—	—	0.0031	0.0183
	0.88	—	—	0.0045	0.0672
170	0.04	0.00355	—	—	0.0043
	0.08	0.00886	—	0.0042	0.0105
	0.16	0.0222	—	0.0055	0.025
	0.32	—	—	0.007	0.058
	0.88	—	—	0.01	0.230
180	0.04	0.00995	0.00097	—	0.0137
	0.08	0.0258	0.00161	0.011	0.033
	0.16	0.0664	0.00291	0.0145	0.078
	0.32	—	0.00474	0.0185	0.18
	0.88	—	—	0.027	—
190	0.04	0.0299	0.00256	—	0.045
	0.08	0.0725	0.00395	—	0.105
	0.16	0.183	0.0067	—	0.25
	0.32	—	0.0117	—	—
195	0.04	—	0.0062	—	—
	0.08	—	0.0088	—	—
	0.16	—	0.0014	—	—
	0.32	—	0.025	—	—

of the acid concentration, T is the absolute temperature, R is the gas constant, e is the basis of the logarithm naturalis, and H is a constant. The reaction constants for the hydrolysis of cellulose from Douglas fir for various acids, various acid concentrations, and at different temperatures are given in Table IV-2 (92,183).

As can be seen from this table, the degree of hydrolysis increases with increasing acid concentration and temperature, with an increase of 10°C in temperature being more effective than a doubling of the acid concentration. The four acids tested differ substantially in their hydrolyzing action. When the concentration of the sulfuric acid is expressed in moles, the degree of hydrolysis for this acid is of the same magnitude as that of hydrochloric acid, indicating that the second hydrogen atom in the sulfuric acid does not take part in the hydrolysis. Sulfurous and phosphoric acids have low

TABLE IV-3

HYDROLYSIS OF DOUGLAS FIR STABLE CELLULOSE (183)

Temperature (°C)	Sulfuric acid concentration (%)	(mole)	First-order reaction constant K (min^{-1})	Half-life stable cellulose (min)
170	0.4	0.04	0.00355	195.0
	0.8	0.08	0.00886	78.2
	1.6	0.16	0.02220	31.2
180	0.4	0.04	0.0950	69.6
	0.8	0.08	0.2580	26.8
	1.6	0.16	0.06640	10.4
190	0.4	0.04	0.02990	23.2
	0.8	0.08	0.0725	9.56
	1.6	0.16	0.18300	3.78

dissociation constants and are only slightly catalytically active in the hydrolysis of cellulose.

Data for the hydrolysis of wood meal from Douglas fir, carried out with different concentrations of sulfuric acid and at various temperatures are shown in Table IV-3. After the hydrolysis is complete, the sugars formed are washed out and the remaining solid residue is dried in the air, saccharified, and the newly formed sugars are determined. The logarithm of the sugar values obtained are plotted against the reaction time for the various acid concentrations. Straight lines are obtained, indicating that the hydrolysis is a monomolecular reaction. The reaction constants and the half-life time for the stable cellulose are given in the same table.

As already mentioned, celluloses of various origin and those that have undergone special treatments show differing reactivities. Mercerized cotton, for example, contains 22.7% of readily hydrolyzable material, whereas pulp from pinewood has only 10% to 11% and flax fiber only 7.7%. These data are obtained in a hydrolysis with 0.8% sulfuric acid at 180°C. Thus the half-life time of the stable cellulose is 50 minutes for mercerized cellulose, 24 to 25 minutes for pulp from pinewood, and 72 minutes for flax fiber. Sulfite pulp from pinewood shows the same half-life time as the stable cellulose in the wood from which the pulp was obtained (183).

1.2 Hydrolysis of the Hemicelluloses

Compared with cellulose, hemicelluloses are much more readily hydrolyzed. The use of water, alone, at a higher temperature causes a conversion of the hemicelluloses into soluble products. When these are subjected to an after-hydrolysis, they can be converted by fermentation into acetone,

TABLE IV-4

HYDROLYSIS OF ASPENWOOD AND DOUGLAS FIR SAWDUST WITH SULFURIC ACID OF
VARIOUS CONCENTRATIONS AT VARIOUS REACTION TIMES (92)

Acid concen- tration	Time (hr)	Loss in weight[a] (L)		Reducing sugars by hydrolysis[a]		Reducing sugars after rehydrolysis[a] (Rr)		L/Rr	
		Aspen	Fir	Aspen	Fir	Aspen	Fir	Aspen	Fir
0.5	1	7.6	—	2.26	—	2.40	—	3.16	—
	2	9.1	—	2.88	—	3.10	—	3.16	—
	3	10.1	—	3.12	—	3.30	—	3.06	—
	4	11.5	—	3.48	—	3.70	—	3.10	—
1.0	1	—	9.45	—	2.36	—	5.75	—	1.64
	2	—	12.0	—	4.43	—	7.40	—	1.63
	4	—	14.75	—	7.38	—	10.75	—	1.37
2.0	1	11.6	11.6	3.98	4.59	4.25	9.00	2.73	1.28
	2	16.1	15.9	8.18	8.04	8.70	11.30	1.85	1.41
	3	18.3	—	9.78	—	10.20	—	1.79	—
	4	20.2	18.6	10.38	11.10	11.00	13.00	1.83	1.43
8.0	1	—	19.5	—	12.24	—	13.54	—	1.44
	2	—	20.7	—	14.12	—	14.76	—	1.40
	4	—	25.7	—	17.24	—	18.34	—	1.40
16.0	1	27.1	—	16.60	—	17.7	—	1.53	—
	2	30.6	—	18.60	—	19.6	—	1.56	—
	4	31.6	—	19.0	—	20.3	—	1.56	—

[a] All percent of weight of wood; rehydrolyzing by heating 0.5 hr at 120°C.

butanol, and lactic acid. This type of prehydrolysis has been utilized, especially in the production of kraft pulp for artificial fibers of cellulose derivatives. This will be discussed in Chapter VIII, Section 4.5. The cellulose itself remains completely unchanged by this hydrolytic treatment. There are at least two types of hemicelluloses that differ in their resistance toward hydrolysis. Thus, with a sulfuric acid concentration of 0.05%, a sharp decrease in Cross and Bevan cellulose occurs. With increasing acid concentration, the amount of sugars in the hydrolyzate increases more rapidly than would correspond to the amount of degraded cellulose. When the acid concentration increases to over 0.75%, the amount of degraded cellulose remains constant up to an acid concentration of 3% (148,235).

The mathematical formulation of the kinetics of the hydrolysis of the hemicellulose is facilitated by the fact that the two hemicellulose fractions have different reactivities but both involve first-order reactions. In this case the ash content of the wood plays a not unimportant role, because the actual concentration of the acid is lowered in the presence of salts (120). The salt effect has been the subject of various investigations with the goal

TABLE IV-5

DECOMPOSITION OF WOOD SUGARS IN 0.8% SULFURIC ACID AT 180°C (183)

Sugar	First-order reaction constant K (min^{-1})	Half-life (min)
D-Glucose	0.0242	28.6
D-Galactose	0.0263	26.4
D-Mannose	0.0358	19.4
D-Arabinose	0.0421	16.4
D-Xylose	0.0721	9.6

of obtaining a measurement for the acidity of the reaction agent. For this purpose, xylose solutions containing 10 grams per 100 ml, and mixtures of 0.05 to 0.2 mole sulfuric acid and 0 to 1.563 mole sodium 2,4-dimethylbenzenesulfonate, at temperatures of 120° to 150°C, were used. The degree of decomposition observed was found to have an exponential relationship to the ion strength and the order of magnitude expected for a reaction between an ion and an uncharged molecule. The primary salt effect, however, showed deviations dependent upon the concentration of the catalyst. Similar experiments, using 0.1 to 2 moles hydrochloric acid and 0 to 1.71 mole sodium chloride, indicated that the latter had no catalytic effect but caused a primary salt effect that was determined solely by the total stoichiometric strength of the ion. Since potassium and lithium chlorides showed the same action, the anion, alone, apparently influences the degree of decomposition of xylose solutions in aqueous hydrochloric acid and in the presence of alkali metal chlorides (203).

The hydrolysis of aspen and firwood with sulfuric acid of increasing concentrations and for different hydrolysis periods showed that the first stage, carried out under relatively mild reaction conditions, rapidly formed water-soluble products, causing the dissolution of material which is not made up of sugars and which cannot be converted into sugars by an after-hydrolysis. By extending the reaction time, the percentage portion of the dissolved material can be increased, but the ratio of the sugars formed to the loss in weight of wood substance remains more or less constant. An increase in the acid concentration also increases the ratio of sugars formed to the loss in weight of the wood and, especially at low acid concentrations and short reaction times, an after-hydrolysis causes a considerable increase in the yield of reducing sugars. The reaction conditions favorable to a saccharification of the hemicelluloses cause no saccharification, or only an incomplete one, of the water-soluble polysaccharides. Consequently, under these conditions, an after-hydrolysis is necessary to achieve an increase in the total sugar yield. Reaction conditions which cause a rapid hydrolysis of the cellulose lead to a complete decomposition of the sugars, causing a

more rapid degradation of the pentoses than of the hexoses. The experimental results given in Table IV-4 refer to the hydrolysis of aspen and fir sawdust at a reaction temperature of 95°C (92).

The reaction constants for the various sugars and the half-life time for the hydrolysis with 0.8% sulfuric acid at 180°C are given in Table IV-5. With regard to the reaction time and the acid concentration, maxima are obtained for the total sugar yield, although the yield of fermentable sugars may increase still further (125,197).

1.3 Lignin and Hydrolysis

The question of the nature of lignin as the lignifying component of plants has occupied the attention of numerous investigators for almost 150 years. Two methods have suggested themselves for the separation of lignin, in a state as unchanged as possible, from the accessory carbohydrates: (a) the hydrolysis of the latter with strong mineral acids, and (b) the extraction of the carbohydrates by means of suitable solvents. It is not the intention of this book to discuss in detail these analytical investigations, but it must be mentioned that, after various preparatory investigations by other authors, Peter Klason recommended, at the beginning of this century, the isolation of lignin by hydrolysis of the carbohydrate portion with 64% to 72% sulfuric acid. Lignin isolated in this manner is therefore known as "Klason lignin" (114). Later investigations, however, have shown that the concentration of the acid must be lower for deciduous than for coniferous woods (64). Lignin isolated in this way contains 3% to 6% "bound" sulfuric acid and certain parts of carbohydrates which can be removed in an after-hydrolysis with hot dilute hydrochloric acid (218).

Concentrated hydrochloric acid does not necessarily effect the total hydrolysis of the polysaccharides. On the other hand, supersaturated hydrochloric acid with 40% to 42% hydrogen chloride content was found to be suitable for the isolation of the lignin. This method gives hydrochloric acid lignin or "Willstätter lignin," after the discoverer of the method (234). In this connection, the question arises as to whether the lignins obtained by the hydrolytic degradation of the polysaccharides are identical with the natural lignin or whether they have undergone changes in the course of their isolation. As shown by the yields, such changes, if they have occurred, must consist in intermolecular rearrangements which are often termed "condensations," "polymerizations," or "humifications." A discoloration appears on the surface, causing the residual lignin to become increasingly darker. The assumption has also been made that there are different lignin fractions, one of which is insoluble, while the other is dissolved but separates again on heating of the hydrolyzate (164). Finally, the suggestion has been made that, at the beginning of the hydrolysis, a lignin-

carbohydrate complex is dissolved and, on continuous action of the hydrochloric acid, the carbohydrate component is hydrolyzed and dissolved while the lignin separates (80).

In the hydrolysis of wood with 72% sulfuric acid, water is taken up during the conversion of the polysaccharides into simple sugars and this increases the concentration of the acid. This change in the concentration was measured by conductivity determinations and thus the course of the reaction could be followed. After the hydrolysis curves for cellulose and xylan had been established, those for untreated beechwood, for beechwood extracted with 4% sodium hydroxide, and for beechwood prehydrolyzed with 0.7% hydrochloric acid were determined. It was found that the curve for the untreated wood showed a minimum which was attributed to the decrease in acid concentration caused by the splitting off of water from the native lignin. The curve for the prehydrolyzed wood did not show the pronounced minimum which was expected to be caused by the increase in lignin content as a result of the prehydrolysis, but it showed that the protolignin had undergone changes, which were attributed to an intramolecular splitting off of water. The prehydrolysis therefore could by no means be considered a mild treatment (150). These results were confirmed with unextracted beechwood, but hydrolysis curves of this type were not found for pure cellulose, holocellulose, or soluble native beech lignin. It is therefore doubtful that the course of the hydrolysis curves is the result of the splitting off of water from, or its addition to the lignin (46).

A study was made with dioxane lignin to determine whether or not condensations occur during the treatment with 1% sulfuric acid at temperatures of 100° to 210°C. For this purpose, chemical and spectrographic methods were used. It was shown that up to a temperature of 100°C, no substantial changes in the functional groups of the lignin occurred. In the temperature range of 130° to 150°C, noticeable changes took place. At temperatures of 180°C and above, a decrease occurred in hydroxyl, benzyl alcohol, and carbonyl groups, and in hydroxyl groups that are attached to the γ-carbon atom of the propyl side-chain. At the higher temperatures, the ultraviolet absorption spectra showed a gradual elimination of chromophoric groups, and the infrared spectra showed a loss in functional groups. A comparison of these results with those obtained with condensed lignins prepared by alkaline reactions showed that the formation of aromatic rings in the dioxane lignin can be assumed.

An attempt to condense lignin and lignin model substances with furfural gave negative results. From this it may be concluded that lignin which has been isolated from vegetable materials by means of hydrolysis has undergone chiefly an autocondensation in which the functional groups of the side-chain, the phenolic hydroxyl groups, and the reactive carbon atoms of the aromatic rings are involved (40).

2. DEGRADATION OF THE HYDROLYSIS PRODUCTS

When concentrated acids are used at room temperature and for normal periods of time for the hydrolysis, the conversion of the cellulose, the hexoses, and the pentoses into water-soluble oligosaccharides occurs relatively smoothly and without appreciable decomposition of the sugars formed. However, when the hydrolysis solution is heated or the reaction time is prolonged, darkening and, finally, a black coloration occur, due to the humification of certain sugars. It should be mentioned that these decomposition reactions were, some time ago, even considered to constitute the formation of lignin and, according to this theory, the lignin isolated from vegetable fibrous raw material is not a natural product but an artifact formed by humification of carbohydrates in the acid hydrolysis. However, this theory has now been completely refuted. Dilute acids at higher temperatures do not cause humification but do cause a decomposition of the sugars formed. Although there is available a great deal of informative literature on the hydrolysis of cellulose with concentrated acids and at low temperatures, it is only during the last two decades that detailed investigations have been reported on the hydrolysis of cellulose and its accessory components by dilute acids and at high temperatures.

To establish the relative degree of decomposition of various types of sugars, 5% sugar solutions in 0.8% sulfuric acid were heated at 180°C for

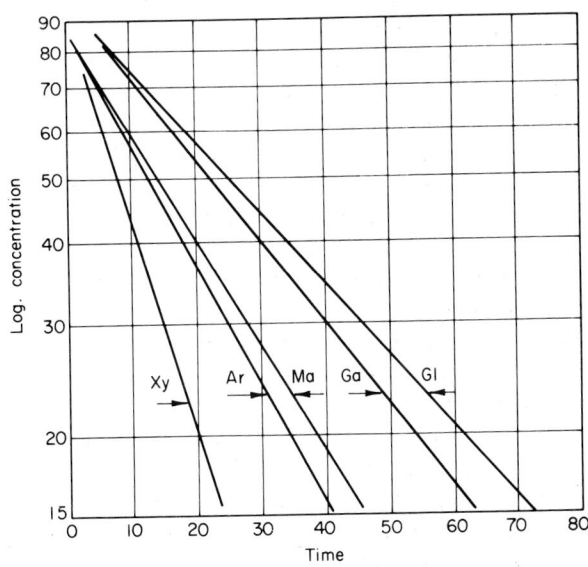

Figure IV-2. Decomposition of various sugars at 180°C by 0.8% sulfuric acid (183). Xy = xylose; Ar = arbinose; Ma = mannose; Ga = galactose; Gl = glucose.

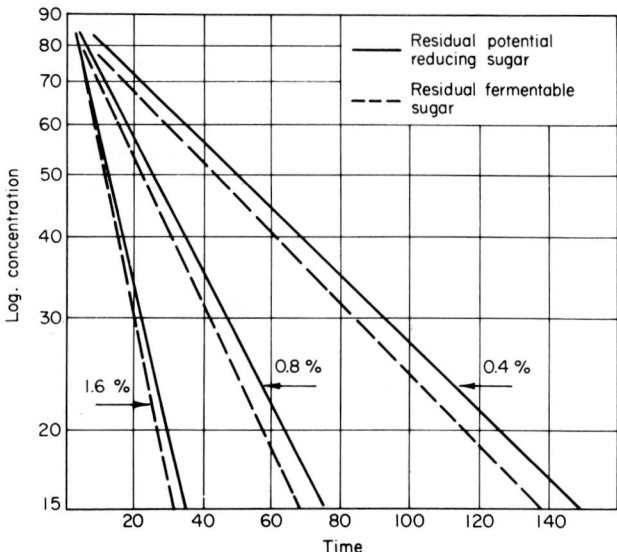

Figure IV-3. Decomposition of glucose by dilute sulfuric acid of various concentrations at 180°C (183).

various periods of time. The proportion of undecomposed sugar was calculated in percent of its initial concentration. The results are shown in Fig.
IV-2. When the logarithm of the concentration of the undecomposed sugar
is plotted as a function of the time, straight lines are obtained for all the
sugars investigated. This is a criterion for the fact that a first-order reaction
is involved. In such reactions, the degree of conversion is directly proportional to the concentration of the reacting substances. By multiplying the
slope of the line in Fig. IV-2 by −2.303, the reaction constant can be
calculated. In this way, the reaction constants shown in Table IV-5 were
determined. Figure IV-2 shows that, when the value zero is extrapolated for
the reaction time, none of the curves reaches the initial concentration. This
is probably due to the fact that, already during the heating up or the cooling
down period of the solution, a reaction with the acid has taken place. In
addition to this time factor, other factors are also decisive for the greater
initial decomposition of the sugars.

Glucose solutions are completely fermentable. On heating of the sugar
solution in the presence of acids, the amount of fermentable sugar in the
solution is, after a certain time, less than the amount of reducing sugar.
Figure IV-3 shows the decomposition of glucose in sulfuric acid at 180°C
and for various reaction times. The slope of the lines for the fermentable
as well as for the reducing sugars in the acid solutions, at the concentrations
and temperatures being investigated, shows the occurrence of first-order
reactions. Table IV-6 shows the reaction constants and half-life times for

TABLE IV-6

DECOMPOSITION OF GLUCOSE BY 0.4, 0.8 AND 1.6% SULFURIC ACID AT
TEMPERATURES OF 170°, 180° AND 190°C (183)

Temperature (°C)	Sulfuric acid concentration (%)	First-order reaction constant calculated on decreasing reduction power K (min^{-1}) (a)	Half-life (min) calculated like (a)	First-order reaction constant calculated on decreasing content of fermentable sugar K (min^{-1}) (b)	Half-life (min) calculated like (b)
170	0.4	0.00534	130.0	0.00569	121.9
	0.8	0.01057	65.6	0.0112	62.8
	1.6	0.0223	31.1	0.0234	29.6
180	0.4	0.0123	56.3	0.0129	53.8
	0.8	0.0242	28.6	0.0259	26.8
	1.6	0.0505	13.7	0:0531	13.0
190	0.4	0.0270	26.6	0.0292	23.8
	0.8	0.0535	13.0	0.0579	12.0
	1.6	0.109	6.4	0.119	5.8

the decomposition of glucose in sulfuric acid at various concentrations and temperatures.

When the logarithms of the reaction constants shown in Table IV-6 are plotted as a function of the logarithms of the acid concentration, a number of straight lines are obtained. The reaction rates increase by a constant multiple when the acid concentration is increased by a constant multiple. The formula relating acid concentration and the corresponding reaction constant at a certain temperature is:

$$\frac{\log k_2 - \log k_1}{\log C_2 - \log C_1} = M$$

in which M has the value 1.02 within the temperature range of 170° to 190°C. An increase of 100% in acid concentration corresponds to an increase in the rate of decomposition of the sugars of 103%. These relationships are shown in Fig. IV-4.

When the logarithms of the reaction constants, given in Table IV-6, are plotted against the reciprocal value of the absolute temperature, a series of straight lines is obtained, as can be seen in Fig. IV-4; this proves that the reaction follows the general law of Arrhenius, thus relating the reaction velocity to the temperature:

$$k = s^{-\Delta H_a / RT} \qquad \text{or} \qquad \log k = \frac{-\Delta H_a}{2.303 \, RT} + \text{constant}$$

in which s is a constant and ΔH_a is the activation energy.

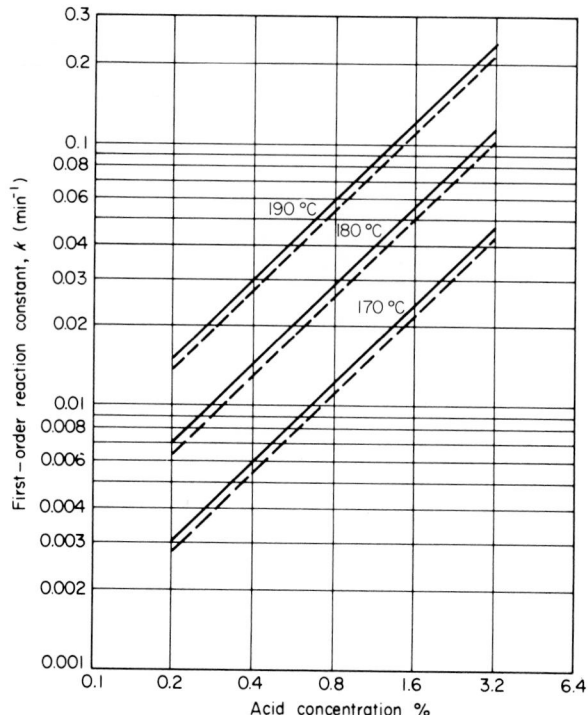

Figure IV-4. Decomposition of glucose with dilute sulfuric acid at various temperatures. Relationship between the first-order reaction constant, k, and the acid concentration (183).

Table IV-7 gives the values for the activation energy in relation to the acid concentration, as determined by the loss in reducing power and the loss in fermentable sugars. As an approximation, it may be said that, within the temperature range of 170° to 190°C, the rate of decomposition of glucose increases 125% with each increase of 10°C in temperature (183).

The hydrolysis of glucose is a first-order reaction; this holds for the use of hydrochloric, phosphoric, and sulfurous acids, the latter two of which are less ionized, thus causing a strong decrease in their catalytic action (92,93).

The saccharification of cellulose in wood therefore occurs in two consecutives steps, as follows:

$$\text{cellulose} \xrightarrow{k_1} \text{reducing sugar} \xrightarrow{k_2} \text{sugar decomposition products}$$

The maximum net sugar yield can be calculated from the equation:

$$(C_B)_{\max} = \left(\frac{k_2}{k_1}\right)^{\frac{k_2}{k_1-k_2}}$$

TABLE IV-7

ACTIVATION ENERGY OF THE DECOMPOSITION OF GLUCOSE IN SULFURIC
ACID SOLUTION OF VARIOUS CONCENTRATIONS (183)

Sulfuric acid concentration (%)	Activation energy in calories calculated on the	
	decreasing reduction power	decreasing content of fermentable sugar
0.4	33,100	33,200
0.8	32,900	33,300
1.6	32,100	32,100

TABLE IV-8

RELATIONSHIP $k_1 : k_2$ IN DEPENDENCE ON SULFURIC ACID
CONCENTRATION AND TEMPERATURE (183)

Temperature (°C)	Sulfuric acid concentration (%)	$k_1 : k_2$
170	0.4	0.62
	0.8	0.84
	1.6	1.00
180	0.4	0.81
	0.8	1.07
	1.6	1.31
190	0.4	1.11
	0.8	1.36
	1.6	1.68

The values for the ratio of $k_1 : k_2$ are given in Table IV-8. This shows that an increase in temperature of 10°C causes a more pronounced improvement in k_1 over k_2 than, for example, does the doubling of the acid concentration. It is therefore recommended that the hydrolysis be carried out at a higher temperature but the sugar formed should be removed rapidly from the reaction zone (183).

Figure IV-5 shows the course of the hydrolysis of Douglas firwood. In a one-step hydrolysis in 0.8% sulfuric acid at 180°C the maximum of fermentable sugars is obtained after 20 to 30 minutes heating. This result is in good agreement with the findings mentioned earlier and indicates that several successive short hydrolysis periods must lead to increased sugar yields (92,125,197).

Xylose and arabinose are the chief hydrolysis products of the pentosans in wood. The pentoses formed are usually converted immediately into furfural:

$$C_5H_8O_4 \xrightarrow[+H_2O]{k_1} C_5H_{10}O_5 \xrightarrow[-3H_2O]{k_2} C_5H_4O_2$$

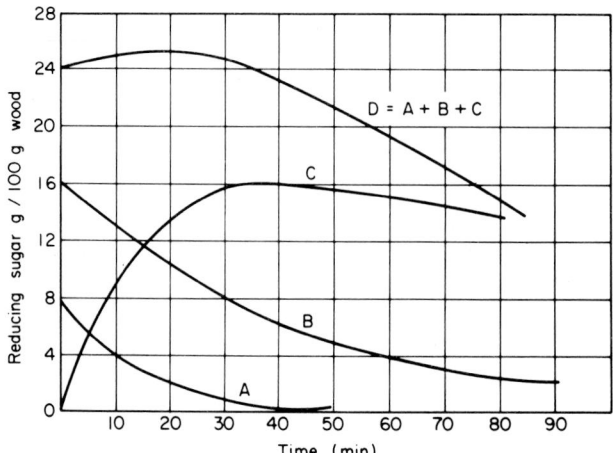

Figure IV-5. Yield of sugar in the hydrolyzate of Douglas firwood after hydrolysis with 0.8% sulfuric acid at 180°C and various reaction times (92). A = xylose; B = glucose ex hemicellulose; C = glucose ex cellulose; D = total sugar yield calculated.

Furfural, in turn, readily undergoes polymerization and resinification. By a one-stage hydrolysis process, about 2% to 3% furfural, based on the weight of the wood, can be obtained from deciduous woods (113). Multistage or continuous saccharification processes give better yields of furfural; if the furfural formed is rapidly removed from the reaction mixture by steam distillation, almost quantitative yields can be obtained. On the other hand, it has been found that the amount of furfural present in the solution has no appreciable effect upon the course of the hydrolysis of xylose. It may therefore be assumed that the conversion of xylose into furfural takes place via intermediate products which it has not yet been possible to isolate. The reaction velocities, k_1 and k_2, were determined for a temperature range of 160° to 280°C, with sulfuric acid concentrations of 0.8 N to 0.00625 N and xylose concentrations of between 200 and 3.125 grams per liter. In these experiments, the half-life time of xylose varied between one second and five hours. The yield of furfural increased with increasing concentration of sulfuric acid up to 0.1 N; above that concentration, the reaction velocity increased in proportion to the increase in acidity, whereas the maximum yield of furfural did not surpass the amount obtained with 0.1 N acid. For the complete range of experimental conditions, it may be said that an increase in the temperature also causes an increase in the yield of furfural. On the other hand, an increase in the concentration of the xylose causes a reduction in the yield of furfural, with the loss being greatest at low temperatures. From these experiments it may be concluded that good yields of furfural can be obtained from pentose or from pentosan-containing material by hydrolysis at higher temperatures and in dilute solutions without simultaneously removing the furfural formed (182).

Hydrolysis of hexoses by heating them in dilute mineral acids forms levulinic acid. Thus, on hydrolysis of glucose with 2.5% to 7.5% sulfuric acid at temperatures of between 145° and 162°C, 34% to 42% levulinic acid was obtained, while formic acid was split off simultaneously (56):

$$CH_2OH(CHOH)_4CHO \longrightarrow CH_3COCH_2CH_2CO_2H + HCO_2H$$

Hydroxymethylfurfural is obtained from hexose by hydrolysis, with water being split off simultaneously. The reaction occurs similarly to the formation of furfural from pentoses and possibly represents an intermediate step in the formation of levulinic and formic acids as just described (213).

Some of the methoxyl groups that are not associated with the lignin are hydrolyzed to methanol. The yield of methanol from coniferous woods amounts to about 0.5% and from deciduous woods, to about 1.5%, based on the weight of the wood. Acetyl groups in the carbohydrate portion of the wood are split off as acetic acid (7).

3. SUMMARY

The hydrolysis of cellulose with concentrated acids takes place via the formation of an addition compound. The hydrolytic decomposition of cellulose is a first-order reaction and the reaction constant is independent of the degrees of polymerization between 130 and 1500. From the yield of biose it may be concluded that its bond is constitutionally and configuratively the only one present in cellulose and all the bonds in the latter and in the polysaccharide are sterically uniform.

The hydrolysis with hot dilute acids proceeds via the formation of a stable cellulose or hydrocellulose to soluble polysaccharides and, further, to simple sugars. The real controlling reaction is the hydrolysis of the hydrocellulose to soluble polysaccharides. During the first step of the hydrolysis recrystallization can occur. The velocity of the hydrolysis increases with increasing temperature and concentration of the acid. The hydrolyses of the hydrocellulose and the sugars are also first-order reactions.

In comparison with cellulose, the hemicelluloses are more readily hydrolyzed. There are at least two types of hemicelluloses with different degrees of reactivity, but in both cases they are first-order reactions. The most favorable reaction conditions for a hydrolysis of the hemicelluloses result in no hydrolysis, or an incomplete one, of the water-soluble polysaccharides, but reaction conditions which lead to a rapid hydrolysis of the cellulose cause a complete decomposition of the sugars. In this case, the pentoses are more rapidly degraded than the hexoses. In order to synchronize the hydrolysis of the cellulose with the opposing reaction of the decomposition of the sugars, a short reaction period at a high temperature and

the rapid removal of the hydrolyzates from the reaction mixture are required for the technical hydrolysis of wood. A separation of the hydrolysis process into two steps, a prehydrolysis and the main hydrolysis, has the advantage of making it possible to adapt the reaction conditions to the reactivity of the various polysaccharides. A complete hydrolysis of the water-soluble polysaccharides, in this case, requires a posthydrolysis.

Lignin, by itself, is largely resistant toward acids of any concentration. It may be assumed, however, that lignin, under the conditions of the wood hydrolysis, undergoes autocondensation in which the functional groups of the side-chain, the phenolic hydroxyl groups, and the reactive hydrogen atoms of the aromatic rings are involved.

4. TECHNICAL PROCESSES FOR THE HYDROLYSIS OF WOOD

4.1 Introduction

Investigations, dating back for over a hundred years, on the hydrolysis of wood and its components have been undertaken for scientific reasons. They served primarily for the elucidation of the composition of the wood and the constitution of the single wood components. The resulting analytical processes such as, for example, the isolation of lignin by hydrolysis of the carbohydrates with acids, have given rise to the desire to use these hydrolytic processes technically and thus to produce low-molecular sugars. It was not originally intended, of course, to use wood that was suitable for the production of pulp for this purpose, but instead to use waste wood from forests and wood-processing industries or agricultural fibrous waste materials.

The adaptation of a process that has been developed experimentally in the laboratory for industrial use establishes a number of conditions which must be fulfilled to make the process economically feasible. These are:

1. a detailed chemical and technological study of the individual steps of the process;
2. investigations of the most suitable apparatus for the process, including, especially, the materials that are corrosion-resistant because of the involvment of strong mineral acids;
3. investigations of the application of the raw materials and the chemicals, the consumption of the raw materials, the yield of the final product, and the consumption and recovery of the chemicals;
4. investigations of the consumption of power and heat energy;
5. investigations of the market conditions for the raw materials used—

in this case, wood and waste wood—with regard to their availability, their cost, and their location;

6. investigations into the demand and market value of the final products, in this case, as compared with competing products from other sources;

7. finally, the location of the plant to be constructed and other local factors such as the labor force, water supply, disposal of waste water, availability of auxiliary power, transportation problems, and other factors which must be taken into consideration.

Although it may truly be said today that, with the exception of a few countries where the conditions are especially favorable for the fulfillment of the above requirements, the majority of wood saccharification plants have been closed down; the reason is probably that one or several of these prerequisites have not been met. To this must be added, as a major handicap, that all the wood saccharification plants were constructed in the period between the First and Second World Wars, partly because, at that time, there were material shortages, and partly in an attempt to make the country self-sufficient. For example, during the First World War, the question of the partial saccharification of wood and of pulp became important in some countries, primarily in order to produce additional fodder for farm livestock. With the end of the war, the economic basis for the production of such fodder became obsolete. Toward the end of the 1920s, a world-wide economic depression led to a resumption of these programs. The endeavor to be self-sufficient and reasons of economic defense led to the production of alcohol and, later, to the production of yeast for concentrated foodstuffs. Thus, in Germany, several pilot plants were erected, using first the process with concentrated hydrochloric acid and, later, the one with hot dilute sulfuric acid. Such plants, operating according to the sulfuric acid saccharification process, were built in Germany in Regensburg, Tornesch, Holzminden, and Dessau, in Switzerland in Ems, and also in Korea. Today none of the plants in Germany are still in operation. The Swiss plant, which produced alcohol with financial support from the government, has given up such production and changed to other products. During the First World War, in the United States too, pilot plants were built for the production of alcohol from wood wastes. Later, the method using dilute sulfuric acid was used there too, and underwent a series of modifications during the Second World War. But these plants also have ceased to operate. So far as is known, only in Russia and Japan are industrial wood sugar plants today in operation or under construction. The methods used in these plants will be discussed later.

For the development of a wood sugar industry, only those countries will be interested that have (*a*) a sufficient supply of raw materials—primarily waste wood that is unsuitable for the production of pulp, (*b*) no do-

mestic sugar industry or possibility of becoming self-sufficient in this respect, and (c) an extensive and steady demand for sugar. It is obvious that only a few countries will meet these requirements. The so-called "developing countries" must mostly be excluded because they do not have the necessary supply of wood or because local conditions for the construction of such chemical plants are not favorable.

The present economic situation makes the production of additional fodder supplies from wood hydrolyzates unattractive to those countries which fulfill the conditions required for developing large-scale plants. For countries that are suffering from a food deficiency, the technical and economic conditions for the construction of such plants are lacking. The production of alcohol as the principal product of the wood hydrolysis is no longer economical because it is produced much more cheaply synthetically by the cracking of ethylene and it is also obtained as a by-product from the sulfite pulp industry. The production of yeast has also become obsolete because of the cultivation of oil-bearing plants. The synthesis of butanol by the petrochemical industry has made its production from wood hydrolysis uneconomical. Whether the production of crystalline glucose from wood has a future cannot at present be predicted.

For these reasons, it is difficult to offer an accurate prognostication for the future of the hydrolysis of wood. It is certainly desirable to utilize in a profitable way the enormous amount of waste wood which today rots in the forests or is used as fuel, thus destroying the wood substance as such. It is therefore not only useful but even necessary to discuss the studies and the successes that have been made, especially with regard to the technical aspects of the wood saccharification. This will be done in the following sections.

4.2 Technical Processes with Hydrochloric Acid

4.2.1 THE RHEINAU-BERGIUS METHOD

The saccharification process known as the "Rheinau process" is based on the use of supersaturated, or fuming, hydrochloric acid. The hydrolyzing action of this acid was described in the German Patent 11,836, issued in 1880. This old patent obviously attracted little attention, but since it deals with the technical hydrolysis with supersaturated hydrochloric acid and the recovery of the latter, its essential claims may be briefly cited:

1. The conversion of cellulose into glucose by the action in the cold of gaseous hydrogen chloride on wood which has been impregnated with water or with hydrochloric acid of suitable strength.

2. The recovery of the used hydrochloric acid by distillation of the glucose solution in vacuum.

3. The use of vertical earthenware cylinders, arranged one above the other and connected with each other, which are cooled by being sprayed with water.

4. An evaporator, with troughs made of lavamonolytes and arranged in a shelf-like manner, in which the mass to be evaporated is exposed to the action of heat, and, in addition, an arrangement for the condensation of the vapors (47).

It is understandable that this process could not be successful for material-technical reasons and because its various steps had not been sufficiently investigated. In any case, the basic conditions for a technically usable wood saccharification process with concentrated hydrochloric acid were established (233). Later investigations of the actual course of the reactions of the hydrolysis, reported in the German Patent 273,800 of 1913 (162), and the finding that a cellulose-acid complex compound is formed in the hydrolysis with hydrochloric acid, as is the case with sulfuric acid, and that this compound is then hydrolyzed have provided the prerequisites for a mature technical process (10).

All these investigations and publications have contributed to the technical development of the hydrolysis process with concentrated hydrochloric acid, the crucial technical details of which are reported in the German Patent 391,969 of 1917 (65,78,79,81,83).

The hydrolysis according to this simplified hydrochloric acid process gave solutions containing only 3% to 4% sugars. The discovery that, after a certain time, the reaction reaches an equilibrium and comes to a stop and that the acid hydrolyzate is capable of continuing the hydrolysis if it comes into contact with fresh wood chips, has led to the development of a countercurrent system which assures a continuation of the hydrolysis and produces sugar solutions with up to 30% sugar content. For the large-scale technology of the process and the development of the apparatus required, which was extremely difficult at that time, Bergius must receive the credit. Figure IV–6 shows a flow-sheet of the process as it was carried out up to the end of the Second World War.

A description of the process, which has repeatedly been published in the literature, is unnecessary here. It should be mentioned, however, that the low dissolution velocity of the hydrochloric acid necessitated an enrichment of the hydrolyzate with sugar in the countercurrent process by means of a series of diffusers in order to obtain concentrated sugar solutions. This, in turn, made the separation of the hydrolyzate from the acid difficult. Such a separation was necessary not only for the production of acid-free sugars but also for economic reasons in the recovery of the acid. Since higher temperatures could not be applied because they caused the decomposition of the hydrolyzate, a vacuum had to be used. This idea was not new, but

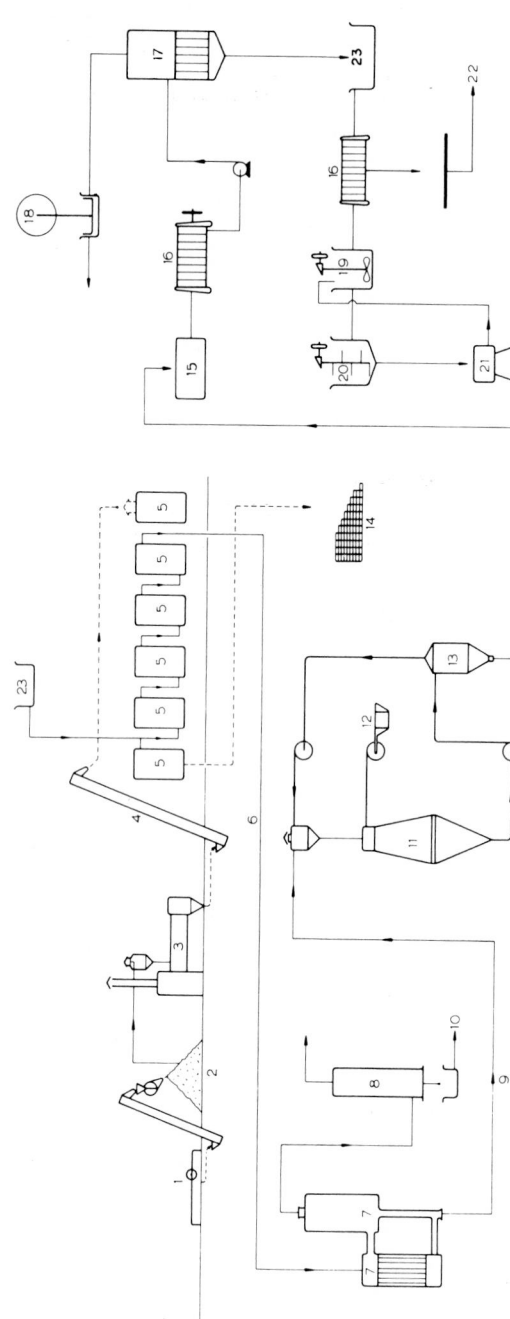

Figure IV-6. Flow-sheet of the Rheinau wood hydrolysis process. 1 wood chipper; 2 wood chip storage; 3 wood chip dryer; 4 elevator; 5 diffusion battery; 6 pipe line for wood sugar solution; 7 evaporator; 8 condensator; 9 pipe line for wood sugar syrup; 10 hydrochloric acid; 11 spray dryer; 12 air heater; 13 cyclone; 14 lignin storage; 15 inversion; 16 filter presses; 17 evaporator; 18 vacuum pump; 19 wood sugar dissolution; 20 crystallization; 21 centrifuge; 22 sugar solution to processing; 23 hydrochloric acid tank.

the difficulty lay in the transfer of the heat required for the evaporation. Because of the corrosive action of the acid, metals could not be used, and, because of their low heat conductivity, ceramic linings were also out of the question. Consequently, a mineral oil fraction—which remained liquid at the required temperatures, which did not react with hydrochloric acid, and did not give an emulsion with the hydrolyzate—was used as the means of heat transfer. This hot oil was brought into contact in a finely divided state with the hydrolyzate in an evacuated vessel and, in this way, the heat was directly transferred. The vapors, consisting of hydrochloric acid and steam, were removed by a vacuum pump and, after condensation, gave a hydrolyzate largely freed of hydrochloric acid but mixed with oil. The oil and the hydrolyzate were separated by decantation and the former was returned to the cycle. The separation could also be carried out in a special type of centrifuge (49).

One of the main problems in the Rheinau-Bergius process was undoubtedly the economical recovery of the hydrochloric acid. In spite of the use of the countercurrent process, approximately three parts of 41% hydrochloric acid were needed for one part of wood. The sugar concentration in the final hydrolyzate depended upon the number of diffusers and, while it increased uniformly, the hydrochloric acid concentration decreased sharply, especially in the last diffuser. This decrease was caused by the strong absorption of the acid by the lignin. Because the final sugar solution still contained some hydrochloric acid, it had to be evaporated in vacuo by spraying. The condensation of the vapors from the liquid being evaporated required the addition of water in the form of dilute hydrochloric acid. The wash-water used for the recovery of the acid absorbed by the lignin had to be removed by distillation. Since the system, $HCl-H_2O$, formed an azeotropic mixture, the boiling point of the water was raised by the addition of a saturated solution of calcium chloride. The steam, enriched with hydrogen chloride, could be dewatered by partial condensation and the water taken up by the calcium chloride could be removed by distillation. The evaporation was carried out in evaporators equipped with tantal-coated boiling tubes.

A reversion of the sugars parallels the saccharification. This reversion increases with increasing concentration of the sugar solution. If the sugar is to be used for fermentation or for the production of crystalline sugar, an inversion is necessary, because only then can crystallization be carried out after filtration and evacuation. The mother liquor from the crystallization can be used for the production of alcohol. The sugar obtained according to this process from coniferous woods, in a yield of 65% to 70% of the weight of the wood, contains 60% glucose, 17% to 21% mannose, 5% galactose, 13% to 16% xylose, and about 1% fructose. In addition to the sugars, about 33 kg lignin containing resins and humic substances, and 2 to

2.5 kg acetic acid are obtained from 100 kg wood. Fermentation of the carbohydrates gives 33 to 35 liters of 100% alcohol. From deciduous woods (beech) the yield amounts to 72 kg reducing sugar, consisting of about 67% glucose, 28% xylose, and 5% of other types, 25 kg lignin, and 6 kg acetic acid, based on 100 kg dry wood substance.

When the sugar solution is used for fermentation to alcohol, the hemicellulose fraction remains unused unless it is converted into fodder or nutritional yeast (133).

4.2.2 THE MODIFIED RHEINAU PROCESS

The high glucose content of the sugar solution obtained in the Rheinau process suggests the production of crystalline glucose. In addition to the difficulties encountered in the hydrolysis with concentrated hydrochloric acid, other complications arise due to the presence of decomposition products of the hemicelluloses. These not only cause losses in glucose but also have an unfavorable effect on its crystallization. As already outlined in the previous sections, a prehydrolysis was used prior to the main hydrolysis in order to remove the readily hydrolyzable hemicellulose fraction from the wood. This modification makes use of the findings that had been made some time before and results in the removal of the hemicellulose with hot, very dilute acid (172,191). The hydrolyzate from the prehydrolysis with 1% hydrochloric acid gives the chromatographic spectrum of the hemicelluloses, whereas, in the main hydrolysis after several hours, considerable amounts of cellobiose and 6-β-glucosidic gentiobiose are detected, the latter being formed by condensation of glucose. The process of the anhydrization and recondensation of simple sugars to oligosaccharides under the effect of strong acids had, for a long time, been known as "reversion." Already in 1880, Emil Fischer had found that glucose, under the action of concentrated hydrochloric acid, is converted into a disaccharide, which he called "isomaltose," but which was later identified as gentiobiose (12,59, 236). A trisaccharide, too, has been obtained by the action of hydrogen chloride on glucose. In this case, condensation probably leads only to a disaccharide and the anhydrization then forms the trisaccharide (60,187). The degree of reversion depends, apart from other factors, upon how much water is present in the system and how concentrated the glucose solution is. The greater the glucose concentration, the greater is the reversion. This shows the necessity for a posthydrolysis of the main hydrolyzate (9,191). The maximum yield of glucose is higher at low than at high temperatures, whereas the latter require shorter reaction times. From the necessary compromise between the reaction time and the glucose yield, it was found that the economical optimum for the posthydrolysis is reached at much lower concentrations of total sugars than had hitherto been assumed.

The complete removal of salts, originating partly from the wood and

partly from the neutralization of the remaining acid, is important for a trouble-free crystallization; this is accomplished by the use of ion exchangers and by a special kind of preliminary purification of the solutions (2,191). Because of the complex composition of the remaining acids, especially selective ion exchangers had to be found. Prehydrolysis, careful main hydrolysis, and optimum posthydrolysis lead to a sugar solution of 90% pure glucose. In a one-step crystallization, glucose of a purity of 99% is obtained, in 85% of the theoretical amount, as hydrate.

Another important criterion for the economical operation of the hydrolysis process is the recovery of the acids. In order to separate the acid absorbed by the lignin in a dilution as low as possible, it is necessary to synchronize the flow velocity of the wash-water with the rate of diffusion of the acid from the lignin into the water. Only in this way can the concentration of the acid be kept constant and the amount of wash-water be kept to a minimum. The last, very dilute part of the acid is suitable for the prehydrolysis. The principal difficulty, however, lies in the condensation of the hydrochloric acid vapors which are distilled off in vacuo from the sugar solution of the main hydrolysis. Of prime importance is the separation of the distillation into two successive steps and ensuring that the vapors contain sufficient steam so that they can be condensed with tap water at a vacuum of 50 mm. From the isotherms of the system, HCl-H$_2$O, it was found that the concentration must not surpass 30% hydrogen chloride, corresponding to a condensation temperature of 37°C. The vapors from the pre-evaporation then consist of 95% hydrogen chloride. The steam consumption per ton of crystalline glucose can be lowered by this simplification of the acid cycle from 27.3 to 8.8 tons per hour (177,191).

A flow-sheet of the modified operation of the Rheinau process is shown in Fig. IV-7. The prehydrolysis is carried out with 1% hydrochloric acid at 130°C, batchwise, in two vertical autoclaves. The sugar solution from the prehydrolysis is either evaporated to give sugar for fodder or fermented to alcohol. The prehydrolyzed wood is dehydrated in centrifuges and then dried. The main hydrolysis is carried out in a series of four hydrolysis towers in which the hydrochloric acid passes at a constant rate through the prehydrolyzed wood. The front of the advancing acid stream contains a highly concentrated sugar solution but it slowly changes to a zone of pure hydrochloric acid. This acid zone is followed by the wash zone, with a sudden drop in concentration. The liquid advances at the same speed as that at which the dry wood is fed in at the entrance. The charging of one tower takes about ten hours. The lignin in the last tower is removed at the top by water and air currents and is conducted onto an oscillating sieve. The hydrochloric acid in the sugar solution is continuously distilled off under vacuum in a series of circulating evaporators. In the first evaporator, pure hydrogen chloride is obtained at a vacuum of 100 mm, in the main

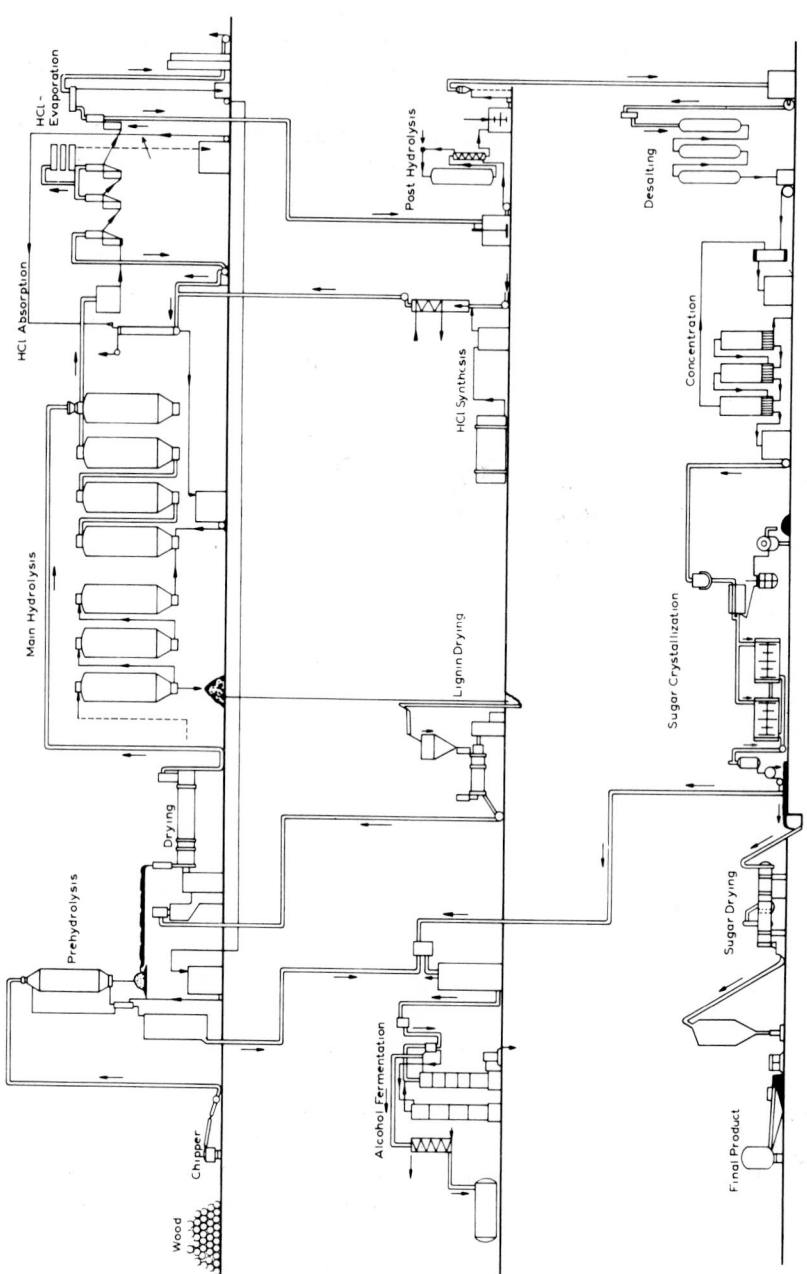

Figure IV-7. Flow diagram of the modified Rheinau process (191).

evaporator, a 30% hydrochloric acid at 50 mm. By the injection of steam, a 10% hydrochloric acid is recovered from the last evaporator. The 30% hydrochloric acid is brought up to a concentration of 41% by the hydrogen chloride from the first evaporator and by that produced by the combustion of hydrogen in a chlorine atmosphere. The heat formed by the absorption of the hydrochloric acid is recovered by a special absorption apparatus made from graphite. The continuous operation of the whole process makes possible an extensive utilization of the heat formed for the heating of the dilute hydrochloric acid used in the prehydrolysis and of the sugar solution in the posthydrolysis.

The glucose solution thus obtained is evaporated in several steps, using pressure and high temperatures to lower the viscosity and to improve the heat transfer. The last step uses thin-layer evaporators to shorten the evaporation time and thus to avoid the decomposition of the syrup. It is planned to fill the hydrochloric acid requirements by the combustion of hydrogen in a chlorine atmosphere.

In a later modification, the prehydrolysis, which was carried out with dilute acid under pressure at high temperatures, was transferred into the main hydrolysis tower and carried out in one operation at 20°C with a hydrochloric acid graduated in concentration. Apart from the simplification, a special advantage was that the structure of the wood remained intact, thus allowing the use of finely divided raw material, such as sawdust. The simultaneous use and recovery of the hydrochloric acid in the pre- and main hydrolyses resulted in a considerable reduction in the acid consumption. The sugars of the prehydrolysis were also obtained in a much purer and salt-free form and in high concentration (191).

A modified Rheinau process, known as the Udic-Rheinau process, differs in some essential details from the original. Figure IV–8 shows a flow-sheet of the Udic-Rheinau process (178). The dried wood chips are introduced into tower 3 and there subjected to a prehydrolysis with 32% hydrochloric acid. Then the wood is removed from tower 3 into tower 4 which is filled with 41% hydrochloric acid from the top. From there the acid passes through the column and is conducted from the bottom of column 4 to the bottom of column 3. In this way the prehydrolyzate in column 3 is replaced by the main hydrolyzate from column 4 and flows into the evaporator. As soon as the prehydrolyzate is displaced, the main hydrolyzate also flows from the top of column 3 into another evaporator. This improvement in the process avoids the drying of the prehydrolyzed material, reduces the heat consumption for the evaporation of the water, and increases the concentration of the prehydrolyzate.

The recovery of the acid is carried out in the same way as for the modified Rheinau process. The sugar solution, largely freed of hydrochloric acid, is diluted to 12%, subjected to a posthydrolysis, filtered, decolorized,

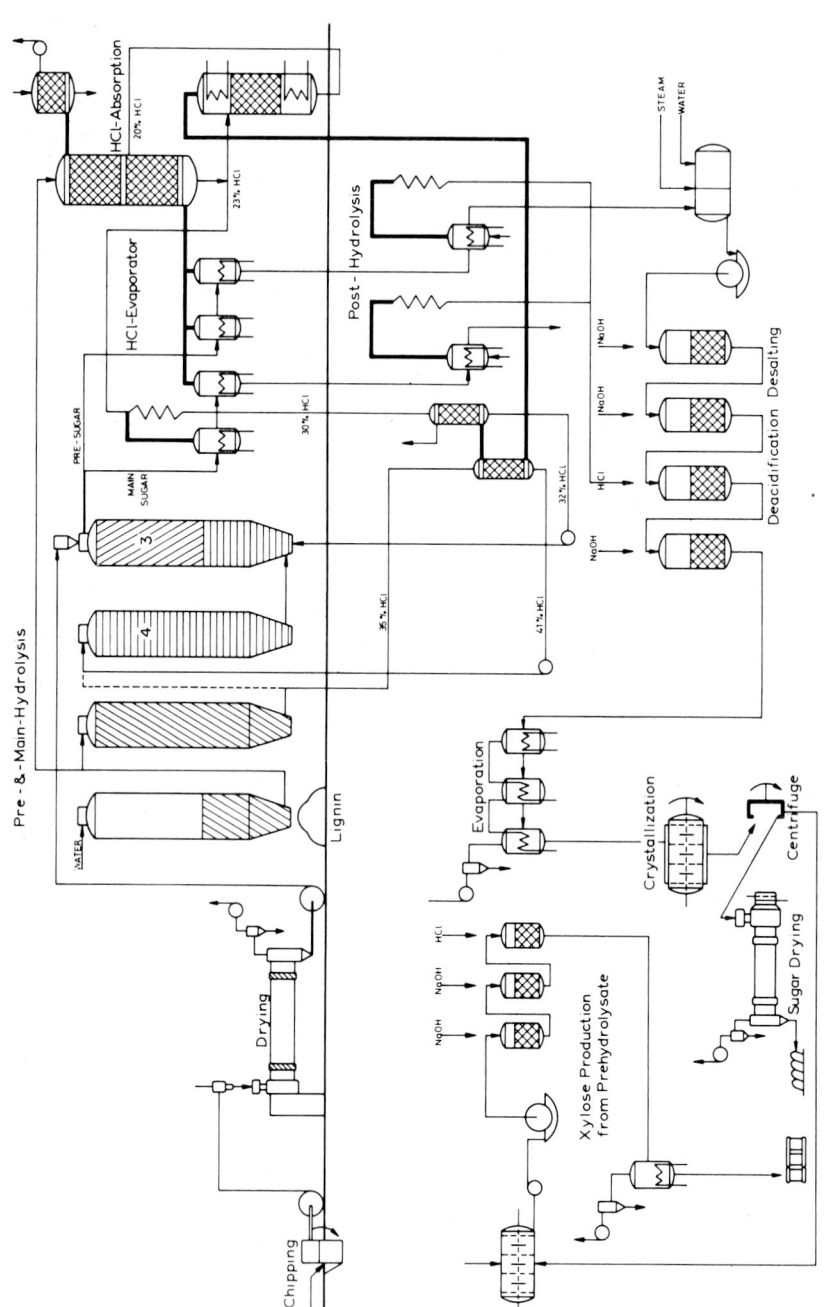

Figure IV-8. Flow-sheet of the Udic-Rheinau process (178).

and treated in a column charged with an ion exchanger; it is concentrated to 70% in a multiple evaporator and the glucose is then allowed to crystallize. From 100 kg dry wood, 22 kg crystalline glucose can thus be obtained in a single crystallization process. The prehydrolyzate, treated in a similar way, can be used for the production of crystalline xylose. The mother liquor from the latter contains xylose, mannose, and glucose and can be processed by hydrogenation to polyalcohols. The lignin can be used as a starting material for plastics. The material balance from 100 kg dry wood is then 22 kg crystalline glucose as monohydrate, 7 kg crystalline xylose, 25 kg polyalcohols, and 28 kg wood sugar lignin.

4.2.3 HYDROLYSIS WITH HYDROGEN CHLORIDE GAS

As has been shown, in the hydrolysis process with supersaturated hydrochloric acid, hydrogen chloride is used to increase the concentration of the acid. No sharp distinction can therefore be drawn between the process with concentrated hydrochloric acid and that with hydrogen chloride gas. The use of the latter was mentioned in German Patent 11,836 of 1880, to which reference has already been made (47). The basic objective in the use of hydrogen chloride was to shift the hydrolysis reaction into the interior of the wood particles and, simultaneously, to facilitate the recovery of the acid. The principle underlying the process is that the wood should be impregenated with a highly concentrated hydrochloric acid, while avoiding any excess, and then the concentration of the acid should be increased by passing hydrogen chloride into it. The acid can be regenerated relatively easily by warming the hydrolysis material.

Several methods which apply this principle, with major or minor deviations, have been proposed. The so-called Prodor process uses a reactor divided into twelve shelves into which the sawdust, mixed with concentrated hydrochloric acid, is injected at the top and is moved by stirring paddles toward the center, where it drops through an opening onto the next shelf and, from there, onto the outer edge of the third shelf, and so on. In a countercurrent fashion the continuously descending wood encounters a hydrogen chloride stream which is injected from the bottom. The upper part of the reactor is equipped with cooling equipment to absorb the heat of the reaction while, in the lower part, heat can be applied to accelerate the hydrolysis. In general, the reactor resembles the furnace used for the burning of pyrite ores (see Chapter VII). After an eight-hour hydrolysis, the material is transferred to a drying section, which is constructed similarly to the reactor, and the material is dried and the acid expelled by hot air. Because this process relies heavily on the structural material, which must be corrosion-resistant, it has not been used for large-scale operations (226).

A variation of the use of concentrated hydrochloric acid combined

with hydrogen chloride is the process of Hereng. Here, wood chips are impregnated with 30% hydrochloric acid in a mixer and then transferred by a conveyor into the reactor. Because the chips are not predried, the acid concentration in them is reduced to 20%. The reactor consists of a hexagonal upright vessel equipped with numerous inclined trays. The impregnated wood is moved from the top toward the bottom by means of a vibrator and thus the prehydrolysis is completed after 45 minutes. The acid, containing chiefly pentoses, is filtered off and part of it is returned to the mixer. In this way, the sugar concentration in the prehydrolyzate is increased to about 20% to 25%.

In a second step, the material, which has again been mixed with 30% hydrochloric acid, is passed through the reactor and hydrogen chloride gas is injected at the bottom in a countercurrent. In this way, the acid concentration is increased to 40% to 41% or more. The hydrolyzed material is conveyed from the bottom of the reactor to a drying chamber and is dried there with anhydrous hydrogen chloride gas which has previously been passed through a calcium chloride solution. The hydrolyzed material, consisting of lignin and sugar, is extracted with water and the sugar solution is subjected to a posthydrolysis. This process, too, has failed to progress beyond a pilot plant stage (51).

Fundamental investigations and experiences gained in pilot plant experiments on hydrogen chloride gas hydrolysis are reported in Japan and have resulted in the development of a process known under the name of "Noguchi-Chisso process." The bases of this are as follows:

1. The raw material is sawdust or waste wood reduced to sawdust size.

2. The wood is subjected to a prehydrolysis to remove the hemicelluloses; the residue then consists substantially of cellulose and lignin.

3. The prehydrolyzed wood is treated with small amounts of hydrochloric acid and cooled in order to permit the absorption of an adequate amount of hydrogen chloride gas.

4. The material thus obtained is heated to complete the hydrolysis of the cellulose.

5. Further heating leads to the recovery of hydrogen chloride gas and hydrochloric acid and these are returned, in part, to the reaction cycle.

6. The hydrolyzed wood is then posthydrolyzed in the presence of water and the remaining acid.

7. The dextrose solution is decolorized, deionized, concentrated, and crystallized.

8. The acetic acid is separated from the prehydrolyzate and the latter is treated in the same way as the dextrose solution, and crystalline xylose is obtained (138).

Problems incidental to this process were the cooling and heating of the wood during the various hydrolysis stages because its heat conductivity is relatively low. The decomposition of the sugars during the recovery of the hydrochloric acid at high temperatures caused difficulties that were augmented by the strong swelling of the wood by the acid, causing the material to become sticky and to cake. These difficulties could be overcome by fluidization, by flash recovery of the acid, and by the addition of antisticking agents.

The following optimum conditions were established for the various hydrolysis steps. For the prehydrolysis the acid-to-wood ratio should be 0.75. When the hydrolysis is carried out at ordinary pressure, an acid concentration of 5% and a reaction time of 3 hours at 100°C are found to be the best. The sugar yield then amounts to 22.8%, based on xylose. The duration of the hydrolysis can be shortened when the temperature is raised and pressure is applied. In this case, the acid-to-wood ratio should be 0.5, the acid concentration 2% when hydrochloric acid is used, or 3% to 5% when sulfuric acid is used, the duration of the hydrolysis, at 130°C, is 30 minutes, and the yield of sugar is 19.6% to 22.5%. For the main hydrolysis, when 60% to 70% concentrated hydrochloric acid is absorbed, the maximum degree of hydrolysis is 62.3% to 65.6%. At a lower absorption, the impregnation of the wood is incomplete. The particle size has a marked influence on the impregnation period. The absorption of the hydrogen chloride must be carried out at temperatures of below 10°C. The optimum temperature in the main hydrolysis is 40°C and, when the fluidized state of the material is maintained, the heat of absorption of the hydrogen chloride is constantly removed. To prevent the caking of the material, diatomite is added. The recovery of the hydrogen chloride is achieved by blowing hot air of 125° to 139°C through the material for a few seconds. The residual hydrochloric acid in the hydrolyzed material then amounts to about 6% to 7%, and the decomposition of the saccharides does not surpass 5%. Instead of hot air, hot hydrogen chloride may be used.

The residue from the main hydrolysis is almost black in color. The sugars formed are mostly in a polymer state. The depolymerization is carried out in a posthydrolysis with dilute acid. With 3% hydrochloric acid at 100°C and atmospheric pressure, the posthydrolysis requires 4 to 5 hours, with 6% to 7% hydrochloric acid, around one hour. Only 30 to 45 minutes are needed with one atmosphere pressure at 120° to 125°C and an acid concentration of 0.5% to 1%. Higher temperatures cause decomposition of the sugars and the formation of organic acids.

The particle size of the wood also has an important effect on the fluidization, with larger particles containing 50% to 55% water being used. Lignin may be substituted for the diatomite. This assures a continuous operation over a long period of time and gives 95% of the theoretical yield

of sugars. The relatively short reaction periods are characteristic of this process. For the impregnation with acid, these amount to between 30 and 60 minutes; for the hydrogen chloride absorption, to less than one minute; and for the main hydrolysis, to about 10 minutes. The recovery of the hydrogen chloride requires only a few seconds so that, for the whole process, only between 40 and 70 minutes are needed.

Beechwood chips, for example, were hydrolyzed with 1% hydrochloric acid for one hour at 130°C and then subjected, in a fluidized state, to a main hydrolysis with hydrogen chloride gas. The hydrolyzate was decolorized with active charcoal, treated with ion exchangers, concentrated in vacuo at 52° to 54°C, again decolorized, and gave a 71.2% sugar solution, consisting of 94.2% glucose, 1.3% xylose, and 4.5% polysaccharides. The maximum yield of crystalline glucose, after three crystallizations, amounted to 84.8%.

The advantages of this process are the possibility of using sawdust, the simplicity of the acid recovery, the reduced cost of the equipment due to the fluidized hydrolysis, the ease with which the lignin can be separated from the hydrolyzate, the formation of highly concentrated sugar solutions, the purity of the sugar obtained, and the elimination of the corrosion problem (111,130,161).

So far as can be learned from the available literature, the Russian Chalov process is very similar to the Japanese. In this case, also, the wood is impregnated with concentrated hydrochloric acid before the treatment with hydrogen chloride gas. The impregnation serves primarily for the reduction of the heat of absorption, which amounts to 58 kcal per kg at a hydrochloric acid concentration of 40% to 41%, regardless of the amount of impregnated acid. To this, the heat of the swelling of the wood, amounting to 34 kcal per kg, must be added. The total amount of heat formed is therefore 92 kcal per kg wood. This heat formation is controlled by saturating the impregnated material with deeply cooled hydrogen chloride gas. The hydrogen chloride is cycled and new gas is continuously added to keep the volume constant. A large contact surface between the wood and the gas is maintained, thus lowering the temperature of the reaction mixture by 20° to 25°C. Simultaneously, the decomposition of the sugars formed is reduced to 3.7% (21,32).

The material to be hydrolyzed is kept in a suspended state, which means that this process, too, utilizes the principle of fluidization, described above. The moisture content of the wood is 4% to 6%. The concentration of the impregnation acid amounts to about 43.5% and the ratio of wood to acid is 0.9 to 1.1. On saturation with hydrogen chloride, the wood should be kept at a temperature of below 35°C. The acid used for the impregnation is cooled to −6.4°C. The saturation of the wood with hydrogen chloride gas requires only one minute and, in this case, the final concentration of

the acid amounts to 46% to 48% hydrogen chloride. The duration of the hydrolysis is 6.5 hours and the total acid consumption is 91%, based on absolutely dry wood. The hydrolyzate contains 25.7% monosaccharides. When the main hydrolysis is preceded by a prehydrolysis, the acid consumption can be considerably reduced (33).

To study the effect of the moisture content of the wood on the course of the hydrolysis, pinewood meal containing 29.2%, 39.6%, and 48.8% moisture was saturated at 20°C with hydrogen chloride and heated in closed vessels for various periods of time to 40° and 60°C. During the heating, the partial pressure increased to several atmospheres. At 40°C, the hydrolysis was completed while the material was still being heated. The wood with the lowest moisture content required the longest hydrolysis period. The highest yield of reducing sugars was obtained from the wood with the highest moisture content. No noticeable decomposition of the sugars occurred. At 60°C, the hydrolysis progressed considerably more rapidly but there was an increase in the decomposition of the sugars when the reaction lasted more than 10 minutes. At 20° and 40°C the hydrolyzate contained the same amount of hydrochloric acid, corresponding to the formula $C_6H_{12}O_6 \cdot 5HCl$. When the saturation with hydrogen chloride gas was carried out at 0° to 15°C, the hydrolysis at 60°C was expected to yield the theoretical amount of sugars (121).

In the same way, absolutely dry pinewood meal was treated with hydrogen chloride gas for 6 hours at 0° to 70°C. The absorption curve obtained was an almost straight line and the absorption decreased by about 0.43% for every 10°C increase in temperature; at 0°C it amounted to 10.8% and, at 70°C, to 7.79%, based on the wood. At low temperatures no swelling occurred and, at above 30°C, the decomposition of the polysaccharides by dehydration, splitting off of water, and formation of furfural became apparent. At 20°C, the residual wood became light brown, at 50°C dark brown, and at 70°C almost black. The hydrogen chloride gas, together with the dehydration water, initiated the hydrolysis of the polysaccharides, of which a part remained unaffected, while the hydrolysis products underwent further decomposition. The decomposition of the polysaccharides after a 6-hour reaction period amounted, at 40°C, to about 5%, at 60°C to 15.3%, and at 70°C to 23.5%. In the praxis, the saturation with hydrogen chloride of the wood already impregnated with concentrated hydrochloric acid occurs so rapidly that no appreciable decomposition of the polysaccharides can occur, when an increase in the temperature is avoided (23).

The amount of hydrogen chloride absorbed is directly proportional to the moisture content of the wood and inversely proportional to the temperature. Hydrogen chloride concentrations of up to 60% to 65% can be obtained and, when the wood has a high moisture content, the hydrogen chloride concentration in it is always higher than that reached under the

same conditions in water. This difference in behavior can be explained by the complexity of the reactions, the effect of the porosity of the wood, swelling reactions, the formation of adducts, the hydrolysis, and the formation of linkages between the acid and the hydrolysis products. All these factors influence the partial pressure of the hydrochloric acid and their relative importance depends upon the temperature (34). Saturation data for hydrogen chloride gas in moist wood, with 20% to 180% moisture content, show that the hydrolysis with hydrochloric acid in a concentration range of 38% to 50% is possible. A hydrolysis at still higher acid concentrations and lower moisture contents is entirely possible. Pinewood meal, with 0% to 40% moisture content, was saturated at $-70°$ and $40°C$ with hydrogen chloride. The sorption isotherms showed an increase in the hydrogen chloride content with decreasing temperatures and increasing moisture content. Hydrogen chloride concentrations of over 80% could be obtained experimentally. As already mentioned, absolutely dry wood can be hydrolyzed with supersaturated hydrochloric acid. The required saturation temperature can be calculated for moisture contents of between 0% and 180%. At 0% water content the saturation temperature would be greater than or equal to $-54°C$ (135).

A study of the penetration of acid into wood pores provided data which could be used for the development of a continuous process of wood hydrolysis with gaseous hydrogen chloride or concentrated hydrochloric acid. Pinewood chips of uniform size and a constant temperature of $20°C$ were used. After hydrolysis with 38% to 41% hydrochloric acid at various wood-to-acid ratios, the loss in weight of the wood, the composition of the liquor within and outside the wood, and the volume of liquor retained in the wood pores were determined. After a hydrolysis of two hours, the amount of liquor retained in the wood ranged from 196% at 41% acid concentration to 125% at 3% acid concentration. These amounts increased with the reaction time and reached a maximum after 32 hours, when all the polysaccharides were dissolved, of 270% to 280% in the case of 41% acid. The solubilization of the polysaccharides began at the moment the acid was added and the amount of liquid retained was directly proportional to the amount of wood dissolved, until approximately 53% had gone into solution. Thus the amount of liquor retained could be determined by the degree of dissolution and was independent of the acid concentration. There was a straight-line relationship between the amount of sugars in the retained liquor and the percentage of dissolved wood. For a wood-to-acid ratio of 1:6 this relationship was 1:0.43. With decreasing amounts of acid, the proportionality factor increases and should be equal to 1 when all the liquor is retained in the wood. On the basis of these and earlier findings, the critical concentration at which all the liquid would be retained was calculated to be 46% to 47%. At higher acid concentrations, only a part of

the wood pores will be filled with the hydrolyzate. The absence of liquid outside the wood affords the best conditions for continuous hydrolysis with either concentrated acid or gaseous hydrogen chloride (134a).

For the recovery of the hydrochloric acid from the hydrolyzate, an adiabatic absorption column has been suggested. In this process, the hydrolyzate is continuously evaporated in vacuo, giving a secondary vapor of constant composition which, after condensation, gives a hydrochloric acid solution and nonabsorbed hydrogen chloride. The latter is passed into an absorption column rinsed with water or dilute acid. The liquid in the column serves for the absorption of the gas and simultaneously for the elimination of the heat of absorption (67). Another method is the absorption of the hydrogen chloride by an azeotropic hydrochloric acid solution from which the gas can be recovered at atmospheric pressure. By these methods, the acid is not diluted with water but is even partially concentrated, so that finally a 40% hydrochloric acid is obtained although the acid concentration is only 38.8% in the original hydrolyzate (30).

The importance of a knowledge of the boiling point equilibriums during the distillation of the hydrogen chloride from the concentrated hydrolyzate has already been mentioned (177). When the equilibrium curves of this system are projected in an *x-y* diagram, the curves shift toward the right and the azeotropic concentration increases. When the sugar concentration is reduced from 70.6% to 65.8%, the steam consumption increases threefold (68). It is therefore advisable to carry out the evaporation in two stages, in the first to evaporate the hydrolyzate continuously to a solid content of 30% to 50%, in the second, to 65%. The continuous evaporation can be controlled by constant checking of the boiling point in the evaporator (122). Another proposal recommends the extractive rectification of the hydrolyzate for the recovery of the hydrogen chloride. The proposed apparatus consists of a distillation column divided into two sections, with the hydrogen chloride solution being injected into the upper and a calcium chloride solution into the lower. Before being added, the calcium chloride solution is heated to a temperature close to the boiling point of the hydrolyzate. The hydrogen chloride is dried in the upper section in a condenser and the condensate is recycled. The liquid that leaves the lower section is a dilute calcium chloride solution which is concentrated again and returned to the cycle (217).

An interesting variation of the preceding process is a mechanical-chemical method. For this, disintegrated and dry wood is impregnated with hydrogen chloride and simultaneously subjected to grinding. In this way, sugar, in 90% yield, can be obtained from spruce- or pinewood meal at acid concentrations of up to 5%, based on the wood, and a grinding period of up to 40 minutes. When the wood mixture is diluted to a hydrochloric acid content of 0.2%, after the mechanical treatment and hy-

drolysis, and is posthydrolyzed for 10 to 80 minutes at 150°C, solutions containing 10% to 20% sugar are obtained (196). In a conventional quartz mill, the conversion of the polysaccharides into sugar at room temperature and the above acid concentration requires about 160 minutes. This period can be shortened by increasing the temperature to 60°C or by heating the material, after the grinding, to this temperature (28). Sulfuric acid may be used instead of hydrochloric acid. The velocity of the reaction increases linearly with increasing temperature during the grinding. The readily hydrolyzed part of the hydrolysis products decreases with increasing grinding temperature and becomes of minor significance at 100°C. From this it may be concluded that the readily hydrolyzable fraction of the polysaccharides, which retain their crystallizability, decreases with increasing temperature. It seems that grinding at a higher temperature and heating of the degradation products which have been obtained at relatively low temperatures have the same effect. The changes caused by the higher temperature are therefore more secondary in nature since they reduce the recrystallizability of the hydrolysis products. Thus the optimum conditions for the mechanical-chemical process are short grinding periods with a high-power input and avoidance of prolonged exposure to high temperatures (66). The time required for a complete hydrolysis depends upon the concentration of the acid and the acid-to-wood ratio. The results thus far obtained show that a continuous hydrolysis of wood can easily be carried out after a prior mechanical-chemical treatment (29).

4.3 Hydrolysis with Sulfuric Acid

4.3.1 PROCESSES USING CONCENTRATED SULFURIC ACID

Concentrated sulfuric acid very readily converts cellulose into glucose. This has been discussed in detail in Section 1.1.1, with special emphasis on

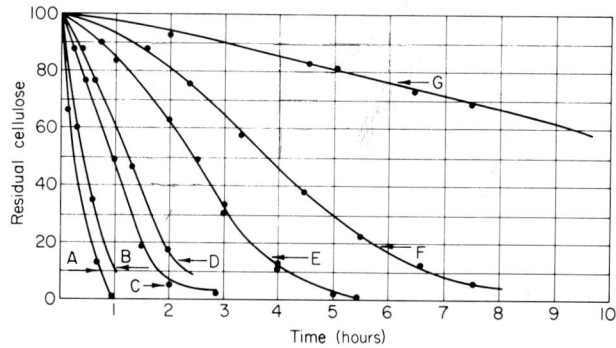

Figure IV-9. Hydrolysis of cellulose by concentrated sulfuric acid (116). A = 40°C, 62%; B = 0°C, 80%; C = 45°C, 51%; D = 15.5°C, 72%; E = 0.5°C, 75%; 20°C, 68%; F = 15.5°C, 68%; G = 0.5°C, 72%; 30°C, 51%.

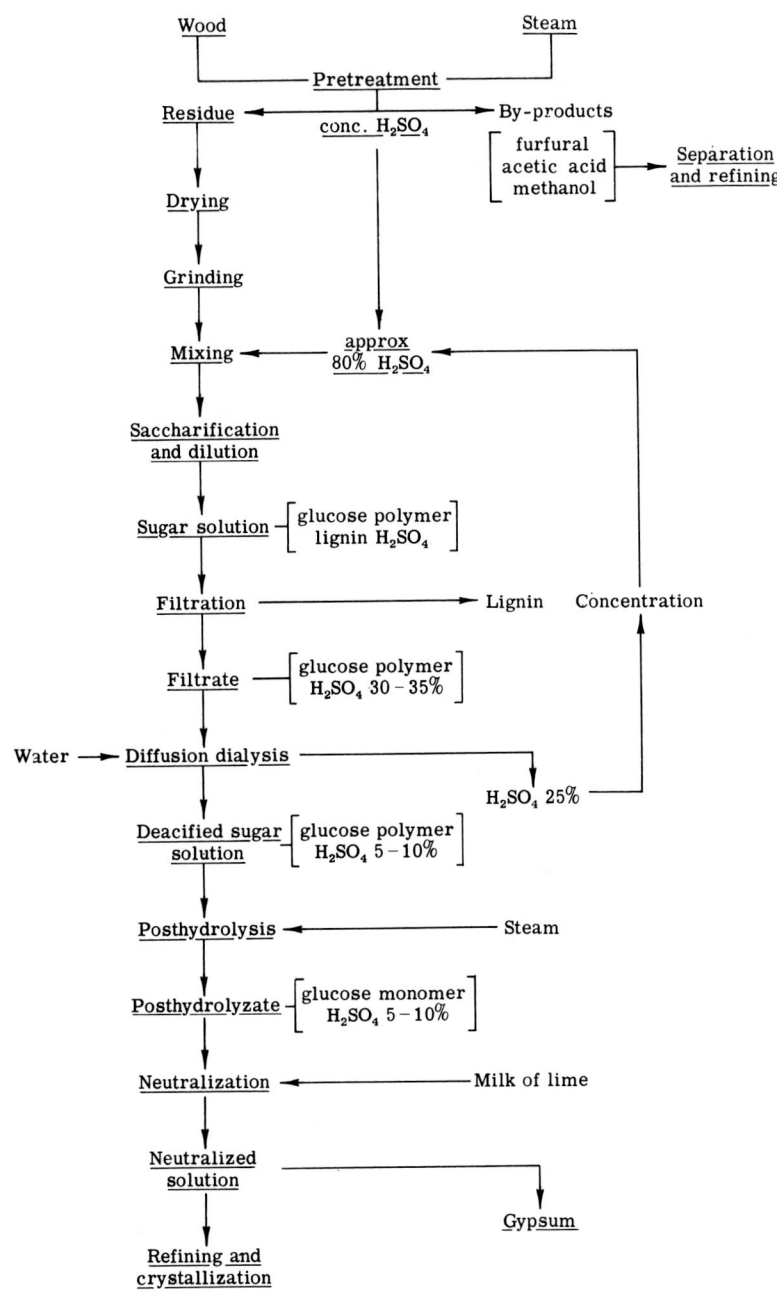

Figure IV-10. Flow-sheet of the Hokkaido wood hydrolysis process (160).

the formation of an intermediate addition compound. Figure IV-9 shows the effect of the acid concentration and the temperature on the velocity of the reaction. For the technical application of such a process, the acid consumption is the decisive factor in its economical operation. It has been found that the ratio of sulfuric acid to cellulose is of great importance with regard to the time involved in the reaction. At 30°C an 80% sulfuric acid, in a ratio of 100 : 1 with cotton cellulose, causes the conversion of the latter within one minute into a water-soluble polysaccharide; at a ratio of 1 : 1, about 35 hours are required (116).

For the application of this process on an economical technical basis, the goal must be to keep the ratio of acid to wood as low as possible, i.e., as close as possible to 1 : 1. The processes used before and during the Second World War applied relatively large amounts of concentrated acid and consequently were uneconomical unless justified by an emergency such as a war or a food crisis. This applies to the research carried out at the Northern Regional Research Laboratory in Peoria, Illinois, and in Bozen in Italy. In Japan, a process was developed in 1948 with the support of the government and is known as the "Hokkaido process." Figure IV-10 shows a flow-sheet of this process.

The wood, in the form of chips, is subjected to a prehydrolysis under conditions that depend upon whether furfural or xylose is to be the product recovered from the hydrolyzate. When the hydrolysis is carried out with steam at 180° to 185°C, the pentose formed from the pentosans is directly converted by dehydration into furfural, in a yield of 65 to 75 kg per ton dry wood. When the prehydrolysis is carried out at 140° to 150°C with 1.2% to 1.5% sulfuric acid, xylose is obtained. The prehydrolyzed wood is then pressed out and dried. The main hydrolysis is carried out with 80% sulfuric acid at room temperature. To obtain the desired low ratio of acid to wood, a new mixing process has been developed. In this, the dried and powdered prehydrolyzed wood is mixed with acid of the requisite concentration by spraying them together. Only about 30 seconds are required for the mixing, and the product is immediately filtered under pressure and washed. With a mixture in a ratio of 0.9, the yield of sugar amounts to over 90%, in a ratio of 1, to around 96%. The concentration of the sulfuric acid from the combined filtrate and wash-water is 30% to 40%. The objective must be to extract the sugar solution, which contains sulfuric acid, from the residual lignin in a concentration as high as possible. The sugar solution is then treated in a diffusion-dialyzer, which is equipped with a membrane capable of ion exchanging, and in this way 80% of the total sulfuric acid is regenerated. The acid is concentrated by evaporation to 80% strength. The sugar solution then contains 5% to 10% glucose polymer and 5% to 15% sulfuric acid. The loss of sugar during the dialysis amounts to 1.8% to 2% of the total sugars. The posthydrolysis of the acid sugar solution

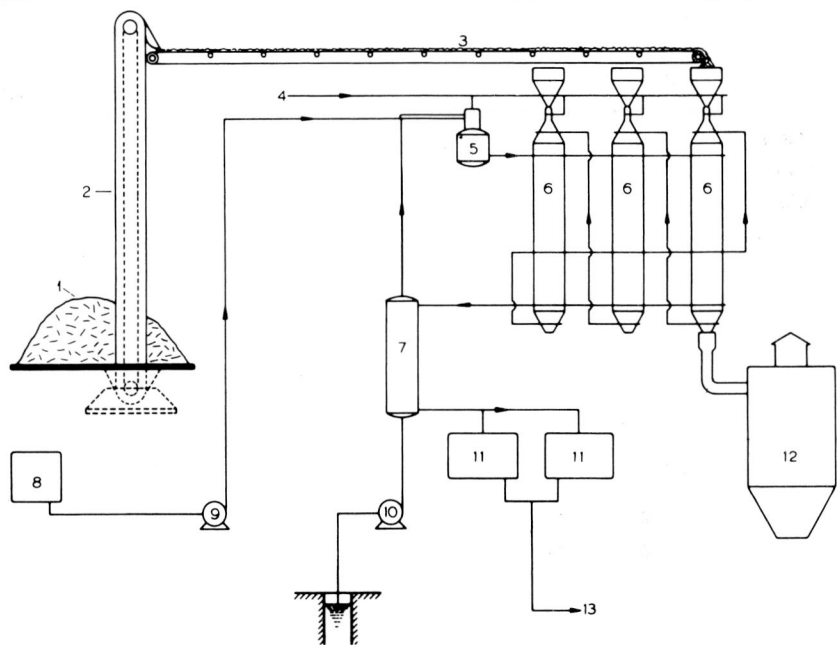

Figure IV-11. Flow-sheet of the Scholler-Tornesch wood hydrolysis process. 1 wood chip storage; 2 elevator; 3 conveyor belt; 4 steam inlet; 5 purging chamber; 6 percolators; 7 heat exchanger; 8 acid storage tank; 9 acid pump; 10 water pump; 11 storage tank for wood sugar solution; 12 lignin storage tank; 13 wood sugar solution to processing.

is carried out for about 100 minutes at 100°C, the solution is neutralized with milk of lime, and the calcium sulfate is filtered off. The sugar solution is then adjusted to a pH of 2.5 to remove the calcium ions and is concentrated to 50% to 60%. The sugar is separated as a double salt by the addition of the theoretical amount of sodium chloride. After a washing with cold water, the glucose is isolated in a crystalline form. With a total yield of 83% to 85%, 280 to 290 kg crystalline glucose is obtained from one ton of dry wood (115,160).

A modified process starts with moist prehydrolyzed wood which is impregnated with dilute sulfuric acid and is then dried with hot air. In this drying process, the sulfuric acid in the wood is brought up to the required concentration. The advantage of the process is the saving in energy through the omission of the mixing procedure.

4.3.2 PROCESSES WITH DILUTE SULFURIC ACID

4.3.2.1 *The Scholler-Tornesch Process*

Investigations carried out toward the end of the 1920s led to the establishment of a technical wood saccharification plant, named after its

developer and its location, the "Scholler-Tornesch" process (139,189). The process is based on the fact that the hydrolysis of cellulose and the decomposition of the sugars formed are two independent reactions and that, to obtain a high yield of glucose, certain reaction conditions must be met. The kinetics of the hydrolysis of cellulose, described in Section 1.1, require the rapid removal from the reaction zone of the sugars formed, as well as the maintenance of certain acid concentrations and temperatures. These conditions are, to a great extent, fulfilled in this process. Figure IV-11 shows a schematic representation of the reactions involved. In contrast to the hydrolyses that use concentrated acids, this process requires no prior drying of the wood. For the hydrolysis, three narrow perpendicular percolators are used, each having a capacity of between 25 and 50 cubic meters. The wood is charged into the percolators and packed by means of steam thrusts which are applied several times to obtain a uniformly full charge. Dilute sulfuric acid, 0.4%, is injected, at 170°C and 8 atmospheres pressure, into the charged percolator at the top and passes through the wood. The discharged solution, containing 4% sugar, is immediately cooled to avoid further decomposition. An appreciable improvement in the process is provided by the use of intermittent percolation. According to this, the sulfuric acid does not pass through the cellulose material in an uninterrupted flow, as described above, but is added in limited quantities at certain intervals by means of its own pressure or by the application of additional steam. After one such quantity of acid has left the percolator, the moist cellulosic material is surrounded by steam at a normal reaction temperature, thus continuing the hydrolyzing reaction with the adsorbed acid. During such intermissions, no sugar solution is removed from the percolator. A new quantum of acid, passing rapidly through the material, carries the sugar with it. This intermittent passage of the hydrolyzing liquor results in a higher concentration of sugars in the hydrolyzate at equal yields. After the hydrolysis is completed, the lignin present in the lower quarter of the percolator has a moisture content of about 50% and is expelled by a pressure of 8 atmospheres and separated from the steam in the cyclone.

The processing of the hydrolyzate is shown in Fig. IV-12. In large clearing vats the solution is neutralized with calcium carbonate. After filtration and cooling of the wort to the requisite temperature, the liquor is passed into large fermentation vats where the yeast and its nutrients are added. The ethyl alcohol is recovered by distillation in the conventional way. The average yields, based on dry coniferous wood, of reducing sugars amount to between 50% and 55% and of fermentable sugars to somewhat above 40%, giving an average alcohol yield of 24 liters from 100 kg wood. In addition, 30 kg lignin is obtained. Deciduous woods give about 35% fermentable sugars, 20% unfermentable pentoses, and about 20% to 25% lignin. When the sugars are used for the production of yeast, 25 to 30 kg

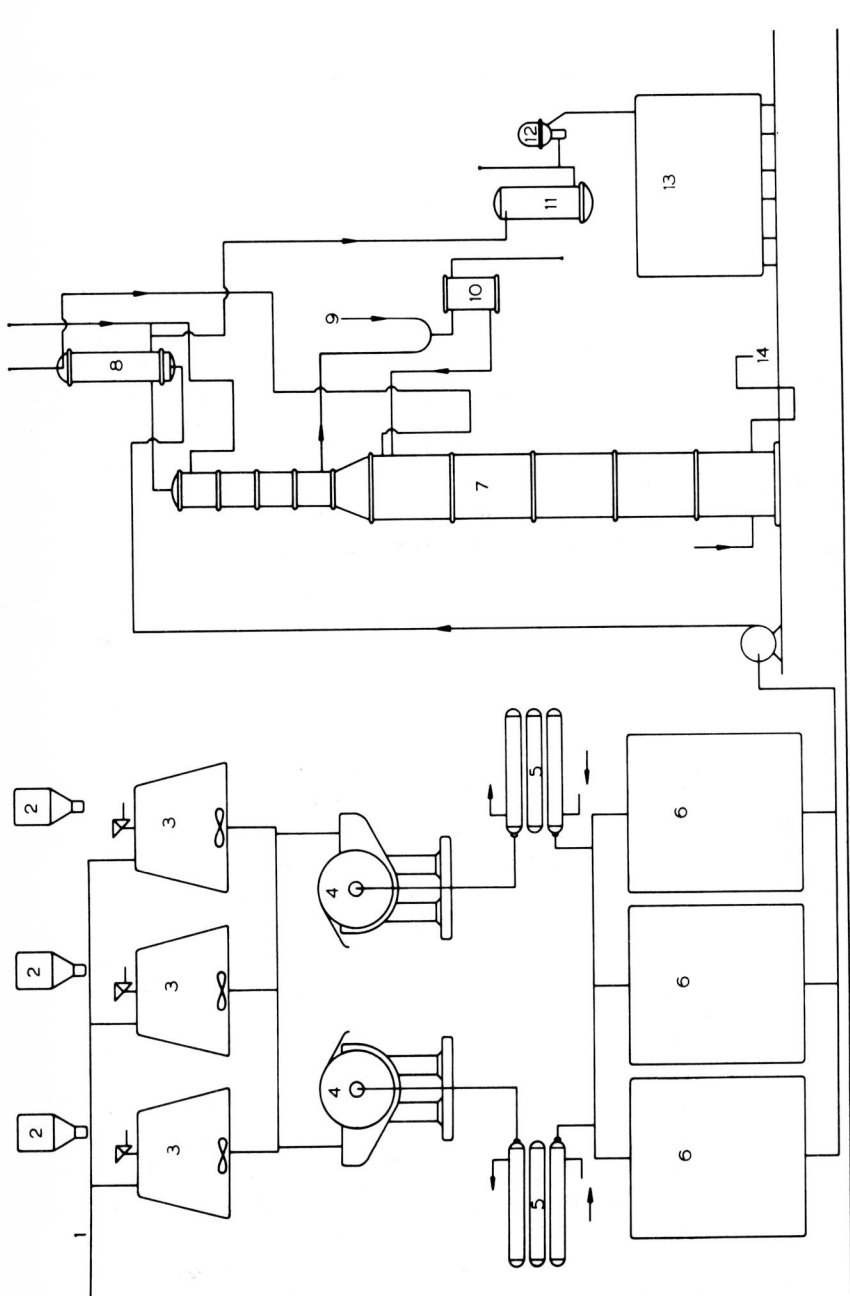

Figure IV-12. Processing of the wood hydrolyzate. 1 hydrolyzate from percolators; 2 neutralization medium; 3 clearing vats; 4 filters; 5 coolers; 6 fermentation vats; 7 distillation; 8 preheater-condensator; 9 water inlet; 10 separator; 11 cooler; 12 alcohol separator; 13 alcohol storage tank; 14 alcohol-free wort.

dry yeast, with about 50% protein content, is obtained from 100 kg dry wood (189).

In a much later patent, a prehydrolysis with 1% to 2.5% sulfuric acid at 140°C and a separate processing of the hydrolyzates from the two hydrolyses are suggested (179,184,190).

4.3.2.2 *The Madison Process*

The American patents for the Scholler-Tornesch process were assigned to the Dow Chemical Company and the Tennessee Eastman Company. The scarcity of ethyl alcohol during the Second World War made it necessary for the United States to resume experiments aimed at the production of alcohol from wood hydrolyzates. The American War Production Board initiated new investigations and assigned further research to the Forest Products Laboratory in Madison, Wisconsin. Experiments in the pilot plant of the Cliffs Dow Chemical Company with the Scholler-Tornesch process did not give satisfactory results. It was necessary, therefore, to develop a new process, now known as the Madison process. With the support of the government, a large wood saccharification plant was built in Springfield, Oregon, capable of processing 220 tons of waste wood daily, an amount which later could be raised to 300 tons per day. The plant consisted of five vertical hydrolyzers, 12 meters high and 2.5 meters in diameter, with a capacity of 15 tons of waste wood each. One hydrolyzer was operated for 6 months on an experimental basis, but before the whole plant could be put into operation, the war ended and, with it, the shortage of alcohol. So far as is known, the plant has never been used by private industry (89,90).

Unlike the Scholler-Tornesch process, the Madison process operated on a continuous basis. Instead of the 12 to 15 batches of acid intermittently used in the Scholler-Tornesch process, in which the acid was stationary during the hydrolysis, in the Madison process the acid flowed continuously, at a predetermined rate, through the wood chips. In this way, the reaction period was reduced from 15 to 18 hours to 3 to 3.5 hours. Although the residual lignin still contained 10% carbohydrates, a higher yield of sugars was obtained.

Figure IV–13 shows a flow-sheet of the Madison process. Monel metal proved to be the best material for the hydrolyzers. The huge hydrolyzers at the Springfield plant were made from steel sheets lined with silicon bronze. The discharging of the lignin from the hydrolyzers into cyclones is achieved by means of hydraulically operated quick-opening valves. The size of the wood chips is of special importance for the carrying out of the hydrolysis process in the short time required. The diffusion of the acid into the chips and the extraction of the sugar solution depend to a great extent upon the size of the chips since the rate of flow of the acid is decreased as the size of the chips is decreased. The formation of channels must be prevented

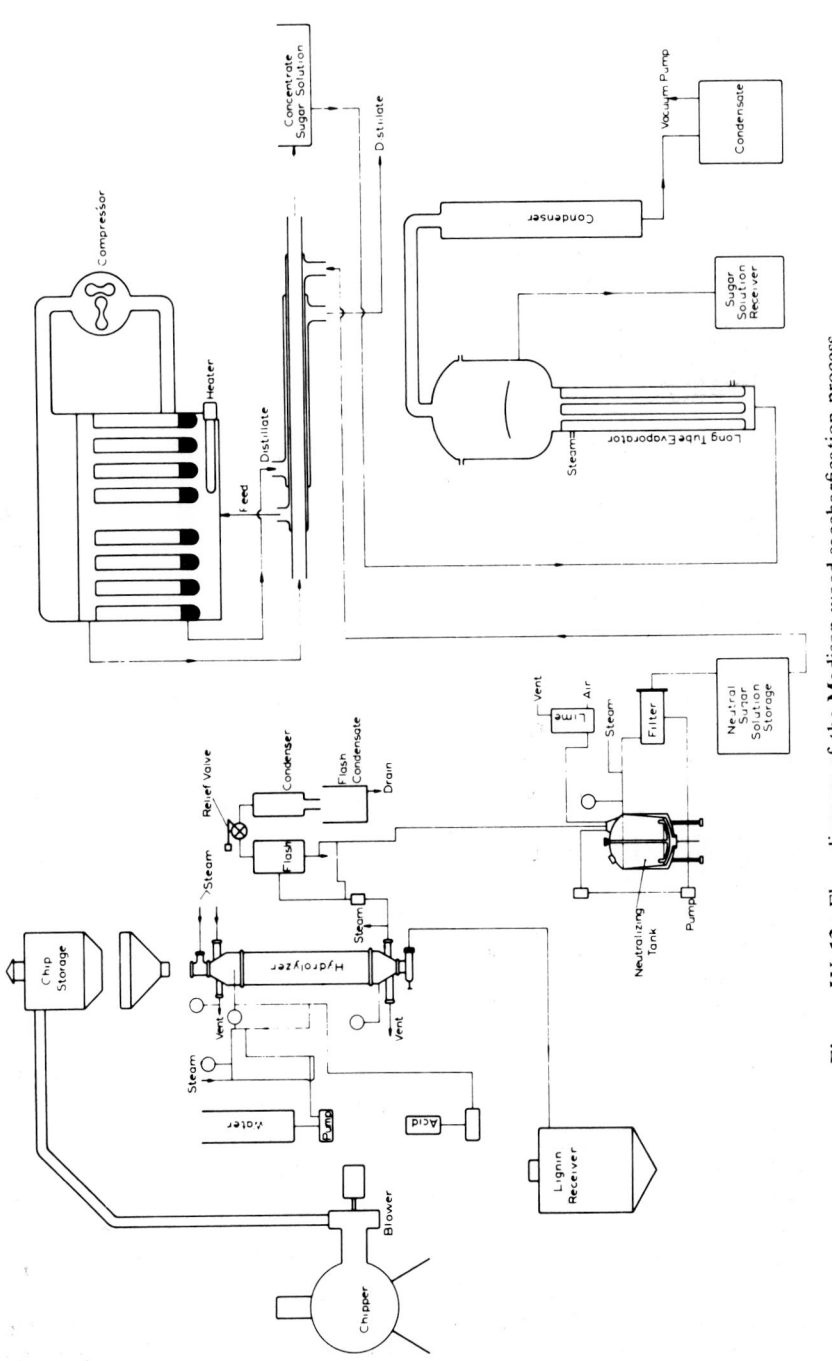

Figure IV-13. Flow diagram of the Madison wood saccharification process.

if a uniform hydrolysis is to be achieved. A satisfactory charge of chips is obtained in a way similar to that of the Scholler-Tornesch process, by alternate steam-pressure and relief. At the bottom of the hydrolyzer there is a vent to permit the exit of air released through the compression of the wood. The charge with coniferous waste wood is about 220 kg per cubic meter, with deciduous wood, up to 350 kg per cubic meter. Before the start of the hydrolysis, the wood is heated with steam to the desired hydrolysis temperature. The acid is injected into the preheated moist wood, under pressure, at a concentration necessary to give the concentration desired for the prehydrolysis. The acid strength needed depends upon the moisture content of the wood. The weight ratio of wood to acid is normally 1 : 3; by contrast, the wood-to-acid ratio in the Scholler-Tornesch process is 1 : 10 or 1 : 12 (189). The concentration of the sugar solutions in the Madison process is about 5%. The main hydrolysis is preceded by an impregnation period of about 30 minutes at about 150°C. The acid concentration then amounts to 0.5%. At 150°C the hydrolysis of the cellulose is relatively limited and the half-life time is 10 hours, while under these conditions the hemicelluloses are already almost completely hydrolyzed. The impregnation acid is then drawn off, giving a sugar solution with between 5% and 15% sugar, depending upon the hemicellulose content of the wood. Fresh acid is then added and the temperature is raised to 185°C at a rate of 0.5°C per minute. About 3 hours after the addition of the acid, the sugar content of the discharging solution drops to about 1% and, at this stage, the hydrolysis is interrupted. During the whole hydrolysis, the sugar solution is removed at the same rate as the acid is added and is passed into a flash chamber where volatile components, such as methanol, furfural, and steam, are removed. The heat from the flash chamber is recovered by heat exchangers and is used for heating the water for the preparation of the dilute acid. The progress of the hydrolysis is followed by determining the sugar content in the acid hydrolyzate. The reuse of the acid sugar solution for another hydrolysis, by passing it through a second hydrolyzer, causes an increase in the sugar content of the hydrolyzate. Calcium carbonate or milk of lime is used for the neutralization of the acid sugar solution. Since the solubility of calcium sulfate decreases with increasing temperature, it is necessary to keep the temperature during the neutralization higher than that to which the solution is later subjected during the evaporation; otherwise, calcium sulfate will be precipitated on the walls of the evaporator.

Basically, all wood species can be hydrolyzed, but they differ in the rate at which the liquor diffuses into them; resins, fats, and waxes may interfere. Stored wood wastes may lose up to 20% of their carbohydrates through the action of fungi and consequently give lower yields of sugar. The potential total yield of sugars from 20 different wood species was found to vary between 37% and 72%, and the yield of fermentable sugars, between

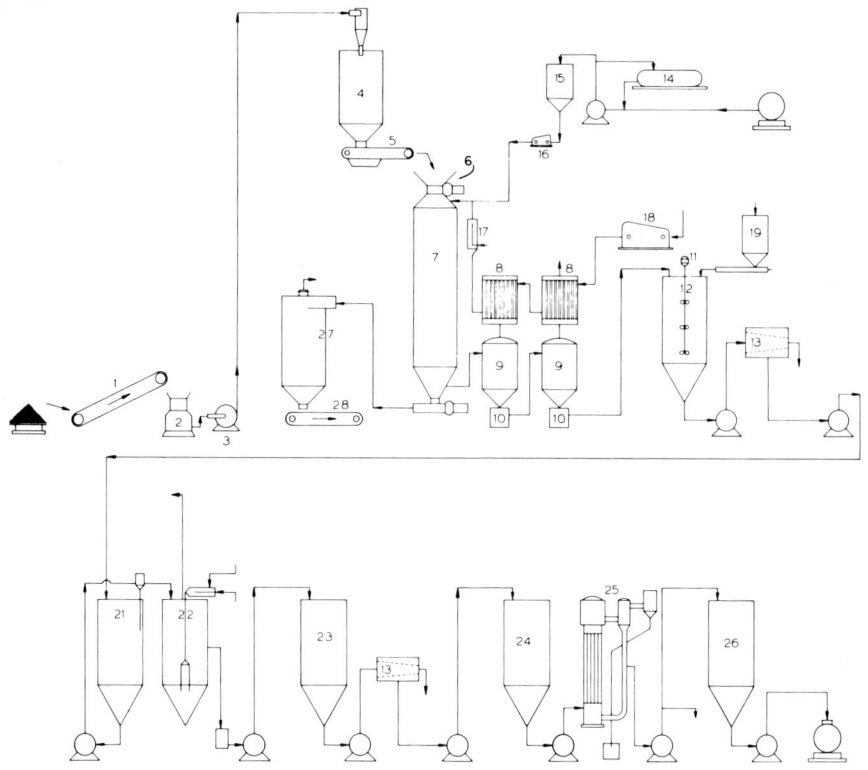

Figure IV-14. Flow-sheet of TVA wood hydrolysis plant. 1 cordwood conveyor; 2 wood hog; 3 chip blower; 4 chip storage bin; 5 chip conveyor; 6 feedoweight; 7 digester; 8 heat exchangers; 9 flash-tanks; 10 sludge separators; 11 agitator; 12 neutralizer; 13 centrifuge; 14 acid storage tank; 15 acid feed tank; 16 acid feed pump; 17 steam injection heater; 18 water feed pump; 19 limestone storage tank; 20 limestone feeder; 21 dilute solution tank; 22 submerged combustion evaporator; 23 25% solution storage tank; 24 clarified solution storage tank; 25 vacuum evaporator; 26 molasses storage; 27 lignin receiver; 28 lignin loader.

26% and 57%. The presence of bark increases the acid consumption and the duration of the extraction of the sugars from the residual lignin; it also affects the density of the charge in the hydrolyzer.

4.3.2.3 *The Tennessee Valley Authority Process (the TVA Process)*

A variation of the Madison process is that of the TVA. At Wilson Dam, the TVA built a pilot plant that was larger than that of the Forest Products Laboratory but smaller than the plant in Springfield. The goal was to simplify further the Madison process and to obtain more concentrated molasses that could be used as fodder. Figure IV-14 shows a flow-sheet of the TVA process. Here, too, a 0.5% to 0.6% sulfuric acid was used and

the density of the charge was 230 kg per cubic meter, or about the same as in the Madison pilot plant. Unlike the latter, the TVA plant used a pressure-actuated valve for the regulation of the acid flow. In this way, a hydraulic pressure of 14 to 16 atmospheres was maintained in the hydrolyzer, this being much higher than steam-pressure at 180°C. The packing of the charge in the hydrolyzer, the presteaming, the deaeration, and the addition of the acid were about the same as in the Madison process except that the hydrolyzate from a previous hydrolysis, containing less than 1% sugar, was added to the fresh sulfuric acid. After the hydrolyzer had been heated to the desired reaction temperature, it was kept at that heat for 30 minutes. After that, fresh acid was pumped in, with the temperature and pressure being kept constant. At this point, the sugar solution was continuously removed and at the same time the pressure was maintained by the addition of acid. The flow rate of the acid was then about 14 liters per minute per cubic meter hydrolyzer capacity. During the course of the hydrolysis, the temperature, which for the impregnation had been 140°C, was raised to 180°C at a rate of 3° to 4°C per minute. The total time for the hydrolysis was 2.5 to 3 hours. The hydrolyzate had a total sugar content of 6%.

The hydrolyzate was then neutralized and evaporated in two steps. The first step was carried out by the use of a submerged combustion burner in which a mixture of carbon monoxide and air was burned below the level of the liquor, thus assuring complete utilization of the heat. During the evaporation, calcium sulfate separated but did not form a precipitate or cause scaling. The evaporation was continued until a sugar concentration of about 25% was reached. The solution was then cooled and, in a second step, was further evaporated in vacuo until a final concentration of 45% to 50% sugars was obtained.

These two processes, the Madison and the TVA, are based on the same reaction kinetics. They differ in that, in the former, dilute sugar solutions were obtained which, without further evaporation, were fermented to alcohol, whereas, in the latter, highly concentrated sugar solutions (molasses) were obtained and could be used directly for fodder. So far as is known, however, neither process is at present being used (71,87).

4.3.2.4 *A Russian Modification of the Percolation Process*

According to Russian investigations, the hydrodynamic flow conditions have not been sufficiently taken into account in the percolation process with vertical percolators (119). For this reason they have combined the latter with a horizontal flow. This method has the apparent advantage that a practically unlimited flow of the hydrolyzate is assured. The yield of reducing sugars has thus been increased by 15% to 20% through the increased flow (32 cubic meters per hour) of the hydrolyzate as compared with that from the vertical percolation alone. No lignin remains in the percolator.

The process is much less dependent upon the particle size of the wood (52). In a further development, a third flow direction was introduced by passing the acid through the wood in an additional percolator, attached at a certain angle of inclination. By alternating the flow of the acid in horizontal, vertical, and inclined directions, the efficiency of hydrolysis plants can be markedly improved. This combined percolation reduces the production costs of alcohol through increased yields of sugars, and appears to have certain advantages over the alternating-flow percolation (123). A continuous system for the neutralization of the acid hydrolyzate, consisting of two groups of three neutralizers each, permits the throughput of 120 cubic meters hydrolyzate per hour. The whole neutralization process is controlled by level regulators, electromagnetic lime metering, pH measurements, and flow metering devices. The process requires a constant concentration of gypsum in the neutralized solution; this can be achieved by "direct crystallization." Ammonium sulfate or sulfuric acid is added to the lime in such an amount that 0.25 kg gypsum per cubic meter is formed (142).

4.3.3 MECHANICAL-CHEMICAL PROCESS

A series of Russian publications have appeared during the last ten years and they show that extensive work on the hydrolysis of wood, including its mechanical-chemical saccharification, have been carried out in that country. From this work has evolved the idea of subjecting very finely divided wood, preimpregnated with relatively small amounts of sulfuric acid, to a powerful mechanical treatment during the hydrolysis process. Thus, for example, finely ground coniferous wood, containing 56.5% polysaccharides, was impregnated with 1.5% to 5% sulfuric acid and dried. The longer the wood was dried and the higher the temperature, the greater was the proportion of sulfuric acid deposited in the cell walls. The wood was then passed through a vibratory mill and heated for 3 hours at 100°C with 4% sulfuric acid. From the nonimpregnated wood, 6.8% of readily hydrolyzable material was isolated; after drying at 105°C, this part increased to 15.9% with 49% sulfuric acid, and to 18.9% with 78% sulfuric acid present in the cell walls. Further heating to 60° to 70°C for one hour in the presence of 49.5% sulfuric acid increased the hydrolyzed cellulose fraction to 72.6%. With 78.7% sulfuric acid, the same yield was obtained in only 18 minutes. With a ratio of wood to acid of 1 : 0.03, only a partial hydrolytic degradation occurred, but with a ratio 1 : 0.1 it was almost complete. This led to the conclusion that the action of the sulfuric acid consists more in the uptake of free radicals formed during the hydrolysis than in the formation of hydrates along the cellulose chains. This is in agreement with the finding that the rate of hydrolysis at high and low ratios of wood to acid is similar, and that the hydrolysis occurs almost quantitatively at a wood-to-acid ratio of 0.05 : 0.045.

The effect of the acid concentration on the mechanical-chemical decomposition of the impregnated and dried wood in the vibratory mill increases markedly when the acid concentration in the cell walls reaches 60%. At this concentration, 95% of the polysaccharides are converted, within one hour of mechanical action, into the readily hydrolyzable form. At an acid concentration of 80%, this degree of hydrolysis is obtained in only 30 minutes grinding time. The range of 60% to 80% concentration can therefore be considered to be the optimum for the mechanical-chemical degradation of the polysaccharides. Drying alone, even for greatly prolonged periods of time, leads to a hydrolysis of only 16.4% of the polysaccharides. This clearly indicates the significance of the mechanical action, which considerably surpasses the hydrolytic cleavage in effectiveness.

The concentration of the sulfuric acid used for the impregnation can therefore be low, as little as 2%, so long as the concentration in the cell walls is increased, on drying, to 91% to 92%. This means, moreover, that the material from a prehydrolysis, which normally contains 2% sulfuric acid and can be used for the production of furfural, can be used for the hydrolysis of the cellulose in the mechanical-chemical process without the further addition of acid. To test this assumption, pinewood chips were prehydrolyzed with 4% sulfuric acid and, after separation from the hydrolyzate, the residual lignocellulose, containing about 60% polysaccharides, was washed and dried. It was then moistend with 2%, 4%, and 6% sulfuric acid and the concentration of the acid was increased, by drying, to 92%, 86%, and 83%, respectively. The material was then treated in a vibratory mill. The rate of the mechanical-chemical degradation increased only a little with increasing acid concentration and even after 30 minutes the differences in the rate were still very small. At an acid concentration of 92% and 86%, a quantitative hydrolysis of the polysaccharides was reached within 40 to 50 minutes. At the 83% concentration, the hydrolysis was accompanied by a partial decomposition of the polysaccharides. The mechanical-chemical hydrolysis followed a first-order reaction until 82% to 90% of the polysaccharides were dissolved. The degradation of the remaining 10% to 18% no longer followed a first-order reaction. The greater part of the degradation products consisted of soluble oligosaccharides with a DP of 4.

A further variation consists in the thermal post-treatment of the impregnated, dried (to 80% sulfuric acid), and mechanically treated wood meal for one hour at 80° to 130°C. The duration of the mechanical action and of the thermal treatment are in a reverse linear relationship. The thermal treatment seems, therefore, to cause a stabilization of the readily hydrolyzable fraction which loses its tendency toward recrystallization under the effect of water. The maximum yield at each stabilization temperature is independent of the acid consumption. The cleavage of the

glucosidic linkages can be attributed primarily to the mechanical treatment and to the fact that the catalytic action of the sulfuric acid manifests itself only with regard to the already decomposed polysaccharides and acts as a stabilizer for these degradation products. The readily hydrolyzed polysaccharide fraction increases rapidly during the thermal treatment and, simultaneously, undergoes a decomposition. The decomposition outweighs the thermal treatment after 20 minutes heating at 100°C, after 10 minutes at 110°C, and after 4 minutes at 120°C (22,24,25,26,27,31).

Experiments have shown that the effect of the capacity of the vibratory mill increases with an increasing charge of wood. At the start of the grinding, the specific surface of the wood particles increases rapidly and, during this period, the degradation of the micromorphological structure occurs through the opening of the channels and other cavities. Simultaneously, the main valence linkages of the polysaccharides are cleaved. In the second stage of the grinding, a clear reduction of the specific surface occurs and, at the end of this stage, the yield of the readily hydrolyzed fraction amounts to about 90%. At the same time, an almost complete degradation of the micro- and macromorphological and supermolecular structures is observed, with an extensive breaking of the molecular chains. This explains the high yield of a readily hydrolyzed fraction and the reduction of the specific surface by the compression of fragments. If the grinding is continued further, first a slight increase, and then again a decrease in specific surface occurs. It must be assumed that in the last stage of the decomposition, not only the supermolecular but also the molecular structure is decomposed (39).

If cellulose is subjected, in the presence of small amounts of sulfuric acid as a catalyst, to a grinding process, a solid hydrolysis product is obtained and this, after inversion, gives sugars in almost quantitative yield. When such a product is stored for a prolonged period with exclusion of moisture and the proportion of water in the system, cellulose-sulfuric acid-water, is determined, it is found that the proportion of free water increases during the first hour of grinding. This can be explained as resulting from the migration of bound water from the cellulose into the sulfuric acid. When the grinding is continued, the proportion of free water decreases again, and this can be explained by the participation of the water in the hydrolysis reaction. Even after prolonged storage of the powder-like hydrolysis product, no disintegration of the sugars formed can be observed and the proportion of free water remains unchanged. Apparently the sulfuric acid is firmly bound to the cellulose and cannot abstract water from the di- and oligosaccharides formed (157).

The "Riga hydrolysis process" begins with finely ground wood meal which is mixed by a nozzle sprayer with small amounts of concentrated sulfuric acid. Although the uniform distribution of the acid in the wood

meal causes considerable difficulty, it affects the hydrolysis and the yield of sugars. When the concentration of the acid is reduced, the degradation of the polysaccharides is retarded. With increased grinding of the material thus prepared, a 63% sulfuric acid gives a higher sugar yield than a 93% acid does. In a rotating hydrolyzer the decomposition of the cellulose during the hydrolysis of the wood meal can be carried out with 60% sulfuric acid. The yield of sugar increases with increasing duration and intensity of the mechanical treatment and amounts to 47%, based on the dry wood, when prehydrolyzed wood is used (110).

Birchwood meal was treated in a vibratory mill in the presence and absence of sulfuric acid. Under identical grinding conditions, the xylan molecules were the most readily degraded. The presence of 0.05 mole sulfuric acid monohydrate per weight unit of oven-dried wood increased the mechanical effect on the polysaccharides in the solid phase. Their degradation continued, even after the grinding, at room temperature, and, at the same time, an activation of the heterolytic destruction of the polysaccharide molecules occurred. When the temperature was increased to 80°C, a further acceleration took place. The proportion of water-soluble degradation products from cellulose and xylan depends, therefore, not only upon the effect of the acid but also on the intensity of the mechanical action. An almost quantitative conversion of the polysaccharides in birchwood into readily hydrolyzable carbohydrates by grinding in a vibratory mill in the presence of 0.5 to 1 mole sulfuric acid, based on the weight of the wood, is entirely possible (8).

It is not known to what extent these mechanical-chemical hydrolysis processes are being carried out on a technical scale in the U.S.S.R. In any case, the processes described above deserve close attention.

5. THE PROCESSING OF THE HYDROLYSIS PRODUCTS

5.1 Introduction

As already briefly mentioned, the construction of wood saccharification plants has taken place mostly in times of scarcity and under economic conditions peculiar to the times. The calculations that were based on the economy of the process used in plants in operation before and during the Second World War were therefore adapted to economic conditions then prevailing and were of little use during times of unimpeded trade. Moreover, market demand for the end product can change just as the supply and the cost of the wood can change from time to time. With regard to the final hydrolysis products, they must be able to compete with those produced by the petrochemical industry. The question therefore is justified:

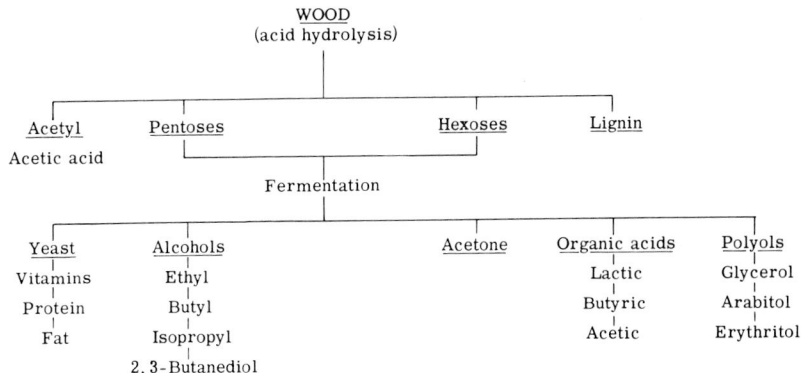

Figure IV-15. Fermentation products of acid wood hydrolysis.

whether, under present economic conditions, the hydrolysis of wood has any future. One is tempted to answer in the negative. As regards the enormous amounts of waste wood that are obtained annually, the problem of its profitable utilization still demands a prominent place in research.

In the United States alone, 75,000,000 tons of waste wood from forests and wood-processing industries are obtained annually. Of this, 60% is not used profitably, while 40% is used only as fuel. However, wood contains 40% to 55% cellulose, 15% to 25% hemicelluloses, 15% to 30% lignin, and 2% to 15% extractives. This waste wood, therefore, contains about 40,000,000 tons of carbohydrates. Considering that the world production of sugar amounts to over 40,000,000 tons, it becomes evident that a vast reserve of carbohydrates is contained in the waste woods of the whole world.

The hydrolysis of wood represents only one way in which a large-scale industry might utilize the waste wood. The difficulties involved in such an undertaking have already been indicated. Even if only the utilization of industrial wood wastes is taken into account, tremendous difficulties are involved in the collection and transportation of such wood, not only because it is bulky to move but also because it varies to a great extent in its composition and moisture content in accordance with its origin. From an economical point of view, such hydrolysis plants must be located close to a large wood-processing plant, such as a sawmill, etc., in order to avoid long-distance hauling. So far as the process is concerned, complicated methods with expensive regeneration plants for the acid used are out of the question. In every case, the market demand for the products obtained from the hydrolysis is the deciding factor. Considering the food shortages in many developing nations, the production of glucose or yeast, or both, appears promising. In such countries, however, the transportation problem

for the amounts of wood required is, at least for the present, impossible to overcome.

The production of alcohol, polyalcohols, ketones, and organic acids must compete with the chemical industries that can produce these products much more cheaply by synthetic processes. In countries where the chemical industry is poorly developed, the marketing of the hydrolysis products is difficult. Another problem, completely unsolved up to this time, is the utilization of the lignin.

In the following sections, the preparation of the different compounds obtained from the hydrolyzate will be discussed. Figure IV–15 shows schematically the isolation of the various fermentation products. From the hydrolyzates of the Udic-Rheinau and Scholler-Tornesch processes and their modifications, only ethyl alcohol and yeast were produced. Although the production of crystalline glucose was possible, so far as is known it was never carried out on a commercial scale. The plants existing or under construction in Japan are chiefly adapted for the production of glucose and xylose, with the possibility of preparing furfural also in mind. The main problem here, too, is the utilization of the lignin by-product. According to reliable sources, fourteen hydrolysis plants are now in operation in the Soviet Union. The methods used there have been discussed. The principal products are yeast and ethyl alcohol, but furfural is also produced in two of the plants (84,160).

5.2 The Production of Ethyl Alcohol

The fermentation of wood hydrolyzates to ethyl alcohol had become important as early as the First World War. The yields were poor, however, because the reactions taking place in the wood hydrolysis had not been sufficiently investigated and the hydrolyzates contained numerous fermentation inhibitors. A great deal of research has been done on these antifermentative substances, among which are formic acid, decomposition products of the carbohydrates, and furfural (105,112,136,140). A great many of these originate from the hydrolysis of the pentosans. Since the latter are considerably more readily and rapidly hydrolyzed than the hexosans, their removal is effected by means of a prehydrolysis. With high yeast concentrations (6% to 12% by volume) hydrolyzates can be fermented in a continuous process in 4 to 6 hours (134). The yeast is recovered from the fermented solution and added to a fresh wood sugar solution. In this way, the yeast becomes acclimated to the hydrolyzate. If the yeast is used in a series of fermentation vats, optimum alcohol yields can be obtained in a relatively short fermentation period (91). Pentoses are not fermented by *Torula utilis*. The arrangement of the plant for the fermentation of wood

sugar solutions has been described in connection with the Scholler process and a schematic diagram of it is shown in Fig. IV–12

5.3 Fermentation to Butanol and Butylene Glycol

In addition to ethyl alcohol, butanol has been produced by fermentation. *Clostridium felsineum* and *Clostridium butylicum* ferment both hexoses and pentoses. The yield of fermentation products varies with the wood species from which the sugars were obtained and amounts to between 29% and 35%, based on the dry wood; this corresponds to the utilization of 87% to 92% of the sugars present. Of special significance is the fact that the biochemical oxygen demand (BOD) of the residual liquid from the distillation liquor from this fermentation process is considerably lower than that from the fermentation to ethyl alcohol. The utilization of the pentoses by these yeasts makes this difference understandable. When spent sulfite liquor is fermented to butanol, the BOD of the final solution is about 35% lower than that from the ethyl alcohol fermentation. The distribution of the reaction products from the butanol fermentation of wood sugar hydrolyzates is given in Table IV–9.

TABLE IV-9

DISTRIBUTION OF PRODUCTS FROM BUTANOL-PRODUCING ORGANISMS (232)

Organism	Butyl alcohol (%)	Ethyl alcohol (%)	Acetone (%)	Isopropyl alcohol (%)
Clostridium felsineum	57	18	24	—
Clostridium butylicum	60	8	4	25

A pretreatment of the hydrolyzates seems to be advantageous in the butanol fermentation, too, because furfural strongly inhibits the fermentation. This pretreatment may be a steam distillation or the hydrolyzate may be treated with lime, adjusted to a pH of 10, and then filtered (165,232).

The production of 2,3-butylene glycol (2,3-butanediol) from wood hydrolyzates of both coniferous and deciduous woods is possible by fermentation with *Aerobacter aerogenes*. Here, too, a pretreatment of the hydrolyzate is necessary. Concentrated sugar solutions must be adjusted to a pH of 10 to 11 by the addition of lime, then filtered, and finally adjusted to pH 6. Solutions with up to 18% sugar content can be fermented to give a yield of 35% butylene glycol, based on the fermented sugar (168).

5.4 The Production of Glycerol

For the production of glycerol, four technical processes are available: (*a*) by the saponification of fats, (*b*) by synthesis, with propylene as starting

material, (c) by hydrolysis of carbohydrates and subsequent catalytic hydrolytic hydrogenation, and (d) by fermentation of sugars. The first two methods are of no interest here. Glycerol can be prepared from carbohydrates, such as sugar, especially cane sugar, starch, and wood flour by hydrolysis and subsequent catalytic hydrogenation. Such a process was developed by the former I. G. Farbenindustrie and was used during the Second World War. Starting with cane sugar, the course of the reaction is the formation of D-glucose and D-fructose by inversion and then, in the first step of the catalytic hydrogenation, their conversion into sorbitol and mannitol. In a second step, these form ethylene glycol, glycerol, erythritol, and, on further hydrogenation of the glycerol, 1,2- and 1,3-propylene glycol. No pure glycerol is therefore obtained, but a mixture consisting of about 40% glycerol, 25% to 30% propylene glycol, 5% to 10% ethylene glycol, and some nonhydrogenated hexite and water. The inversion of the cane sugar is carried out with 0.001% oxalic acid at 70° to 75°C. The hydrogenation takes place with a 40% to 50% sugar solution at pH 7.5 and 120°C at a hydrogen pressure of 300 atmospheres, with nickel as a catalyst. In a second step of the hydrogenation, the temperature is slowly raised to 210°C and the sugar solution then passes through a system of vertical hydrogenators, with the flow of the hydrogen being considerably more rapid than that of the sugar solution to avoid settling of the catalyst.

In the Degussa process, carbohydrate solutions are hydrogenated at 200 atmospheres hydrogen pressure at 150° to 170°C in the presence of 1% phosphoric acid, then the acid is neutralized, and the hydrogenation is continued at 200° to 230°C (50).

During the catalytic hydrogenation of xylose to xylitol in the presence of a nickel catalyst, inactivation of the latter occurs during the reaction. The catalyst must therefore be removed and reactivated, and this is time-consuming and troublesome. In a Russian plant, the catalyst is separated in a cyclone and regenerated in a vibratory mill by mechanically removing the inactivating film. Preliminary studies have shown that the final pH value of hydrogenation solutions of monoses in alkaline mediums can be influenced by the movement of the reaction mixture. Pure glucose and xylose solutions which had been purified by means of ion-exchange resins were hydrogenated under vigorous stirring at 200 atmospheres with nickel as a catalyst. In the experiments, calcium oxide, barium oxide, and sodium hydroxide were used as reaction promotors. The reaction-promoting properties of these additives may be attributed, not to the alkalinity, but to a specific action in which the cleavage of the carbon linkages at 3,4 positions of the xylitol is particularly facilitated. The rate of conversion of the xylitol increases with increasing temperatures (200° to 240°C) and the yield of reaction products, such as glycerol, ethylene glycol, 1,2-propylene glycol, and erythritol, also increases, whereas an increase in pressure from 200 to 250 atmospheres reduces the conversion velocity. If the pH of the sugar

solution is adjusted with calcium oxide to 9 to 10, and the hydrogenation is carried out at 125°C and 60 atmospheres hydrogen pressure in the presence of a nickel-kieselguhr catalyst, with vigorous stirring, the hydrogenation will be completed within 40 to 60 minutes. During the reaction, a small drop in the pH occurs and this can be traced to the decrease in the solubility of the calcium oxide during the heating. No acid intermediate products are formed, and the use of a suspended catalyst and the strong turbulence caused by the stirring prevent the occurrence of a Cannizzaro reaction. These findings have resulted in the development of a process which permits the conversion of vegetable polysaccharides into such polyalcohols as hexitol, pentitol, glycerol, ethylene glycol, and propylene glycol; of special interest is the fact that the hydrolysis and the hydrogenation can be carried out simultaneously in the presence of noble metals as catalysis (152,195,220, 221,222).

Galactose, the main component of the hydrolyzate of larchwood, and arabinose, xylose, and glucose were hydrogenated in the presence of Raney nickel to study the effect of the temperature, the pH, and the concentration of the catalyst. The yields of alcohols in the hydrogenation products were determined by the ebullioscopic, gravimetric, calorimetric, and chromatographic methods, with emphasis on the chromatographic separation of mixtures of sugars and alcohols. Hydrogenation of galactose at pH 8, 120° to 130°C, and 5% to 7% catalyst gave high yields of dulcitol (98% to 99%). In acid media the yields were considerably lower.

Hydrogenation of the water-soluble polysaccharides of larchwood without a preliminary hydrolysis, i.e., hydrolysis and hydrogenation occurred simultaneously, was carried out at an acidity just sufficient to hydrolyze the arabogalactan. To prevent the solubilization of the catalyst, small amounts of nickel sulfate or nickel phosphate were added. The yield of alcohols from arabogalactan of high purity (99.5%) at 145° to 150°C in the presence of 14% to 15% catalyst and 3% nickel sulfate was 87% to 89%, with a large percentage of the alcohols being dulcitol. Purification of arabogalactan prior to hydrogenation can be achieved with ion-exchange resins. The crude dulcitol obtained from the hydrogenation of pure arabogalactan contained 80% dulcitol, up to 17.3% arabitol, and 1% to 1.5% residual sugars (8a).

It was found, long ago, that minor amounts of glycerol are formed during the fermentation of sugar solutions to alcohol. The yield of glycerol can be considerably increased by the addition of sulfite or alkali. The course of the reaction of the fermentation to alcohol in the presence of sulfite is as follows:

$$C_6H_{12}O_6 + NaHSO_3 \longrightarrow CH_2OHCHOHCH_2OH$$
$$+ CH_3CHO \cdot NaHSO_3 + CO_2$$

According to this equation, the acetaldehyde formed as a by-product is fixed and a triose becomes the hydrogen acceptor, thus giving, theoretically, an equimolar amount of glycerol (153). These theoretical yields are not realized in the praxis because some normal alcoholic fermentation always takes place simultaneously.

During the First World War, large amounts of glycerol were produced from molasses. The fermentation was carried out in a neutral or weakly alkaline medium with baker's yeast. The maximum yield, after 2 to 3 days fermentation at 30° to 35°C, was 20%. Strong aeration increased the yield. The yeast formed could be reused after a regeneration fermentation in a weak acid medium (44,225).

A mixture of sodium mono- and bisulfite is less toxic than bisulfite alone. The bisulfite is added slowly during the reaction (43). The fermentation should be of shorter duration and the yield of glycerol should be higher than in an alkaline sulfite fermentation medium.

In recent years, especially in the United States, investigators have studied the question of the technical production of glycerol by fermentation. They have found that the reaction of sodium bisulfite and acetaldehyde is of fundamental importance for the course of such a process. The reaction between these two compounds, and the complex compound formed have been the subject of much discussion. Today it can be assumed that an α-hydroxysulfonic acid is formed. This acid is strongly ionized and has a strength equivalent to that of hydrochloric acid. The reaction scheme:

$$CH_3CHO + H_2SO_3 \rightleftharpoons CH_3\overset{\displaystyle OH}{\underset{\displaystyle H}{C}}-SO_3H$$

and

$$CH_3\overset{\displaystyle OH}{\underset{\displaystyle H}{C}}-SO_3H \rightleftharpoons CH_3\overset{\displaystyle OH}{\underset{\displaystyle H}{C}}-SO_3^- + H^+$$

shows that the number of complex ions in the solution depends on the concentration of the bisulfite. This concentration is determined by the pH of the solution and therefore the amount of free and bound sulfurous acid also depends upon the pH value. The addition of acid to a solution containing the complex compound shifts the equilibrium of the equation to the left and causes an increase in the concentration of sulfurous acid. In this case, sulfur dioxide and acetaldehyde create a high vapor pressure in the strongly acid solutions, corresponding to that which would be formed even without the formation of a complex compound. The addition of a base also shifts

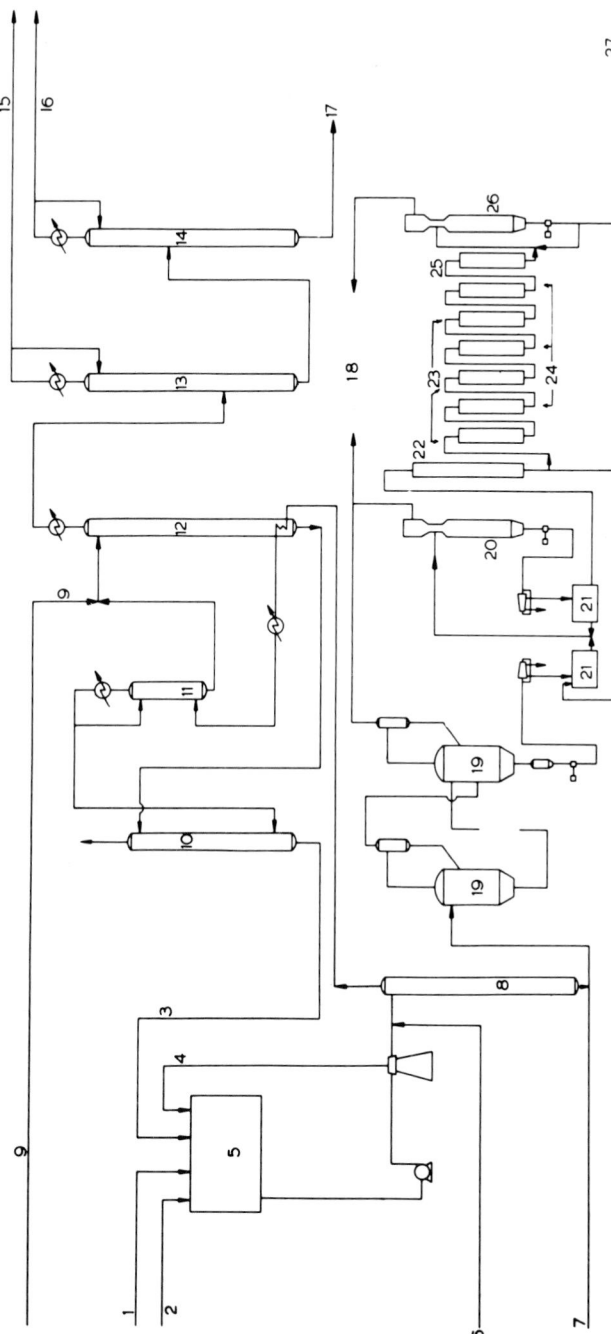

Figure IV-16. Flow-sheet of the fermentation glycerol process. 1 sugar solution inlet; 2 nutrients; 3 Na₂SO₃–Na₂CO₃ solution; 4 yeast recycle; 5 fermentator; 6 sulfuric acid; 7 caustic solution; 8 stripper; 9 soda ash and make up Na₂SO₃; 10 SO₂-absorber; 11 SO₂-desorption; 12 volatiles stripper; 13 acetaldehyde column; 14 alcohol column; 15 acetaldehyde exit; 16 ethyl alcohol exit; 17 waste; 18 condenser vacuum system; 19 multiple-effect evaporators; 20 finishing evaporator; 21 Na₂SO₄ tanks; 22 ion exclusion; 23 cation; 24 anion; 25 mixed bed; 26 finishing evaporation; 27 glycerol exit.

the equilibrium to the left:

$$H_2SO_3 \rightleftharpoons HSO_3^- + H^+$$

and the concentration of bisulfite ions is reduced. The sulfur is then chiefly present as sulfite ions and the partial pressure of the sulfur dioxide above the solution is low, while the acetaldehyde is not combined and exerts a strong partial pressure. Based on these data, the following process has been developed and is shown schematically in Fig. IV-16.

TABLE IV-10

FERMENTATION GLYCEROL PROCESS (94)

Reagents used	lb
Sugar	3910
Sulfuric acid	1070
Soda ash	1055
Sodium sulfite (make up)	105
Ion-exchange regenerants	
Sulfuric acid	38
Sodium hydroxide	106
Fermentation products	
Glycerol	1000
Ethyl alcohol	650
Acetaldehyde	420
Sodium sulfate (anhydrous)	1150

A wood sugar solution containing the necessary nutrients, yeast, sodium carbonate, and sodium bisulfite is passed continuously into the fermentation vat. The mash formed, containing glycerol, alcohol, and bound acetaldehyde, is continuously drawn off and freed of yeast which, after a mild acid wash, is returned to the cycle. The clarified mash is acidified with sulfuric acid to a pH of 1 and the volatile reaction products, such as alcohol, acetaldehyde, sulfur dioxide, and carbon dioxide, are distilled off. These are separated by fractional distillation in a regeneration aggregate after the adjustment of the pH with sodium hydroxide to separate the sulfur dioxide from the acetaldehyde. The fermented liquor thus obtained, containing about 5% glycerol, is neutralized and concentrated in a multiple-effect evaporator to about 60% solid content. The evaporation sometimes encounters difficulties because the solution has a high viscosity and contains certain residual sugars, sodium lactate, and sodium acetate. These fermentation products are of great importance and necessitate constant control of the fermentation. During the evaporation, sodium sulfate decahydrate separates. The concentrated solution passes columns of anion and cation exchangers, alternately, to remove impurities. The aqueous glycerol solution is then evaporated to give

TABLE IV-11

GLYCEROL FERMENTATION (94)

Input	lb	Output	lb
Sugar (20% solution)	3521	Glycerol	1000
Yeast (80% moisture)	390	Acetaldehyde	473
Nutrients			
Urea	14.6	Ethyl alcohol	735
Disodium phosphate	9.4	Sodium bisulfite (bound)	1117
Potassium chloride	3.1	Sodium bisulfite (free)	13
Magnesium sulfate	0.8	Residual sugar	32
Blackstrap molasses	48.0	Organic acids	36
Sodium bisulfite	1130	Inorganic salts	21
Sodium carbonate	281		

a highly valuable product. The consumption of raw materials and the yields of final products are listed in Table IV-10.

The yield of organic compounds, based on the sugar, is 25.6% glycerol, 16.6% ethyl alcohol, and 10.7% acetaldehyde. These figures represent a 90% recovery of the glycerol and 80% of the acetaldehyde and ethyl alcohol.

The technical data of the glycerol fermentation are shown in Table IV-11. The yield of glycerol, based on the sugar, is 55.51%, that of ethyl alcohol, 40.83% of the theoretically possible amounts. The aldehyde is produced in a ratio of 1 mole to 1 mole glycerol and is 55.16% of the theoretical amount (94).

Further experiments were carried out to find a yeast species that would produce glycerol as the sole high-boiling component so that the recovery of the glycerol would not be impeded by other volatile distillation products. For this purpose, a series of osmophilic yeasts were tested. As was to be expected, the tests of 22 cultures showed that the growth of the yeast and the degree of its utilization of the sugar differed widely. Some cultures fermented a 20% glucose solution in 72 hours while others did not do so even in 144 hours. Alcohol and glycerol were produced by all cultures but the latter, in most cases, only in traces. A culture of *Torulopsis magnoliae*, termed I_2B, was outstanding in that it gave good yields of glycerol and no other polyhydric alcohols. A fermentation temperature of 35°C was found to be the best for this culture. The fermentation had to be interrupted after the sugar content was exhausted, as otherwise a decomposition of the glycerol occurred with further fermentation. A yeast concentration of between 0.5% and 1% fermented all the glucose within 48 hours. High yeast concentrations promoted the formation of alcohol, whereas low concentrations produced only traces. The nitrogen content of the yeast increased with increasing nitrogen content of the mash, but even at the highest con-

TABLE IV-12

FERMENTATION OF VARIOUS SUBSTRATES BY *Torulopsis magnoliae*, I₂B (85)

Substrate	Fermen-tation time (hr)	Final pH	Yeast volume (%)	Concen-tration (g/100 ml)	Substrate fermentated (%)	Glycerol yield (%)
L-Arabinose	96	5.8	3.7	11.8	18.2	0.0
D-Galactose	96	4.8	3.7	10.3	18.3	0.0
D-Glucose	42	3.3	6.0	10.2	98.3	35.4
Glycerol	72	5.8	10.0	4.8	75.8	—
Lactose	72	7.3	0.3	5.9	0.0	0.0
Levulose	48	2.9	6.3	11.0	99.5	31.3
Maltose	96	8.1	0.3	3.8	0.0	0.0
D-Mannitol	96	5.6	5.5	8.0	24.1	0.0
D-Mannose	48	3.3	6.0	9.0	96.1	44.4
D-Sorbitol	96	4.8	5.5	10.3	31.7	0.0
Sucrose	48	3.0	5.5	11.2	99.2	34.5
D-Xylose	96	4.5	3.7	11.3	53.6	0.0

centration of yeast, its nitrogen content was only 6.2% and it decreased steadily with continued fermentation. The addition of urea caused primarily an acceleration of the fermentation. Without urea, considerable amounts of glucose remained unfermented even after 72 hours, but the addition of 0.1% resulted in a complete fermentation after 48 hours. The reason for this acceleration by the urea is not yet completely clear. The addition of relatively large amounts of phosphate had no effect on the formation of glycerol, but the formation of alcohol was increased. On the other hand, small amounts of phosphate increased the rate of the fermentation. Sufficient aeration was necessary for the formation of good glycerol yields. The amount of oxygen that had to be injected into the system also depended upon the glucose content of the sugar solution. To obtain high concentrations of glycerol, it was necessary to start with high glucose concentrations; this means that vigorous aeration was needed and this, in turn, caused strong foaming. With a rate of aeration of 42 mmoles per liter per hour and intermittent addition of the sugar solution, glycerol concentrations of up to 7.9% could be obtained.

Table IV-12 gives the amount of glycerol that was obtained with aerobic fermentation of different sugars with *Torulopsis magnoliae*, I₂B, and Fig. IV-17 shows the relationship between the yield of the products formed and the time of fermentation. After a minor correction for the formation of carbon dioxide during the fermentation has been made, a molar relation of carbon dioxide to glycerol of 3 : 1 is obtained. The formation of glycerol from glucose in the fermentation can be expressed by the following equation:

$$2C_6H_{12}O_6 + 5O_2 \longrightarrow 2CH_2OHCHOHCH_2OH + 6CO_2 + 4H_2O$$

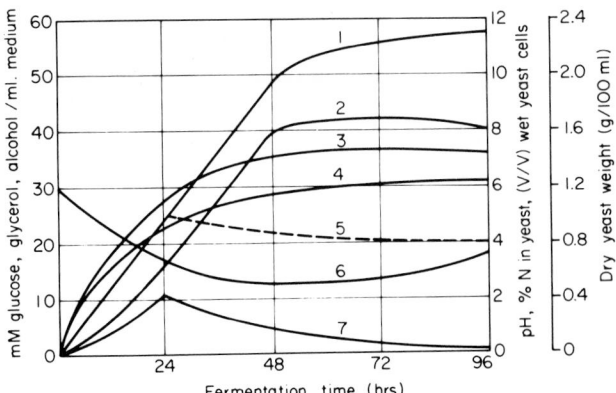

Figure IV-17. Relation of yield of products produced by *Torulopsis magnoliae*, I_2B, to time of fermentation (85). 1 = glucose used; 2 = glycerol yield; 3 = dry yeast weight; 4 = yeast volume; 5 = % N in yeast; 6 = pH; 7 = ethyl alcohol yield.

The ratio of carbon dioxide to oxygen in the equation is 1.20, compared to the respiratory quotient of 1.17 as determined by measurement. The efficiency of this fermentation is approximately 50%. If the growth of the yeast is limited by lack of nitrogen or by a short anaerobic interval, the amount of glycerol formed during the fermentation is also reduced. The decrease in the formation of glycerol results from a deficiency of phosphate and can be considered to be a type of suppression reaction of the normally weak aerobic system, with the consumption of the glycerol being the final result. The phase of the glycerol formation is repeatedly extended by the periodic addition of sugar. The presence of phosphate keeps the number of yeast cells constant during this phase. A small amount of phosphate is added during the prolonged fermentation in order to maintain an active culture. The final product is a largely pure glycerol in a reasonably high concentration (20).

5.5 Fodder and Nutrient Yeast from Wood Sugar Solutions

Yeast from wood hydrolyzates and spent sulfite liquor was of special importance before and during the Second World War because of its nutritive value. Although the plants that produced yeast from wood hydrolyzates were closed down soon after the end of the war, those that used spent sulfite liquor remained in operation. The yield of Torula yeast is about 20% to 25%, based on the dry wood. Of special significance is the fact that *Torula utilis* grows on both pentose and hexose sugars and, consequently, the hydrolyzates of coniferous and deciduous woods can both be used as nutrients. Spent sulfite liquors from coniferous woods contain about 25% pentoses and their hydrolyzates about 20%; the hydrolyzates from deciduous woods contain up to 85%, none of which can be fermented to alcohol.

In Germany there are about eleven plants for the production of fodder yeast in operation. Alcohol was prepared by sufite pulp mills long before the Second World War. This arose from the fact that most of these mills used coniferous woods for their pulping. It was only after there was an increased use of deciduous woods for the production of sulfite pulps that spent liquors with a high pentose content were obtained. After the rayon industry had discovered that they could use sulfite pulps from deciduous woods, their use for the production of sulfite pulp increased at a rapid rate. It must further be mentioned that the residual liquor from the distillation of the alcohol from fermented spent sulfite liquor contains considerable amounts of pentoses. In addition to the spent sulfite liquors, wood hydrolyzates were also available for the production of yeast. Of the numerous varieties of yeast tested, *Torula utilis* and *Monilia candida* were found to be especially suitable. For their large-scale production, however, many difficulties had to be overcome. At first, it was found that Torula could decompose xylose but caused little reproduction. Even after the permanent culture of Torula was established, its rate of growth in xylose solutions was much less than in glucose solutions. It was also found that the acetic acid in spent sulfite liquor and in its vinasse can contribute to an increased growth of the yeast.

By the term "nutrient" or "fodder" yeast is meant a yeast that has been dried to a stable form, fit for shipment. During the First World War a microbiological synthesis of protein from molasses and inorganic salts was carried out on an industrial scale with *Torula utilis*, with vigorous aeration. After the necessary plants had been completed, a shortage of molasses prevented their operation. In any case, the knowledge gained during the development of this process was of considerable value for later attempts to cultivate yeasts, rich in protein, in wood sugar solutions. Plants for the production of yeast from wood sugars were closely attached to wood saccharification plants. If large amounts of deciduous woods are hydrolyzed together with coniferous woods, it is advantageous to convert the pentoses to yeast after they have been isolated by a prehydrolysis process. The hexoses from the main hydrolysis then can be worked up to alcohol or to yeast. It is also possible first to recover glucose from this hydrolysis and then to produce yeast from the mother liquor. Spent sulfite liquors contain only 25% to 35% sugars, based on their solid content. About 80% to 85% of the reducing sugars can be processed to yeast. Because of their large proportion of hexoses, the spent sulfite liquors from coniferous woods are usually first fermented to alcohol and their vinasse then used for the production of yeast. Sulfite pulping of deciduous woods gives spent liquors that are rich in pentoses, and these form a good nutrient substrate for the production of yeast. Spent sulfite liquors are sterile and therefore can be used for the continuous production of yeast without interference from other organisms because these are more sensitive toward sulfur dioxide than

the yeast is. In the production of kraft pulp for the rayon industry, the wood is subjected to a prehydrolysis; the hydrolyzate, from pinewood, contains sugars consisting of about 30% of pentoses and 70% of hexoses. Such a hydrolyzate can be used to advantage for the production of yeast after nutrients have been added.

So far as nutrients are lacking in the hydrolyzate, they must be added to the fermentation vat before, or with, the liquor. Ammonia, ammonium sulfate, or ammonium phosphate is used as a nitrogen source; instead of the phosphate, aqueous extracts of superphosphates may be used. Potassium chloride or potassium sulfate supplements the potassium requirements. Magnesium and trace elements are usually provided in sufficient amounts by the wood; spent liquors from magnesium-base cooks may even have an excess of magnesium. The air used in the aeration of the mash should be as free as possible of foreign organisms. As already mentioned, it is mainly *Torulopsis utilis* and, more recently, certain *Candida* species that are used as microorganisms. For the industrial production of protein, mold fungi are utilized. In this case, a micelle layer is formed and can be readily separated although it shows a certain pH-sensitivity. Hydrolyzates often contain intracellular toxins and growth-inhibiting substances.

For the production of yeast from wood sugar solutions, the hot solution is adjusted to pH 4.5 to 5 by the addition of crude phosphate, limestone, or milk of lime and supplied with the required nutrients. The filtered solution is then passed into the fermentation vats. The hydrolyzate from the hydrochloric acid hydrolysis, after concentration and crystallization of the glucose, is diluted to 3% sugar content, adjusted to a pH of 4.5, and fermented in the same way as other hydrolyzates. The vigorous aeration causes strong foam formation which is counteracted by the addition of so-called fermentation fat, the consumption of which is of some magnitude. The industrial production of yeast is usually carried out in a batch process in which a certain amount of pitching yeast, cultivated by a special method, is added to the dilute mash and the mixture is aerated by compressed air. More hydrolyzate, containing nutrients, is added at the rate at which the sugars are consumed and the yeast multiplies. This addition is continued until the vat is filled and must be emptied to separate the yeast.

Later, a continuous process was applied and this has the advantage that the foam formation not only does not cause trouble but actually contributes to the aeration. In this case, it is unnecessary to add fermentation fat. For the aeration, a number of methods have been proposed, of which the "Waldhof aerator" (240) and the "I. G.-Scholler aeration system" (107) should be mentioned. The former consists of a fermenter which has in its center a cylinder suspended a little above the bottom of the vat; inside the cylinder there is a hollow axle, equipped at the lower end with propeller blades that suck the liquor through the cylinder from the surface to the bottom; the

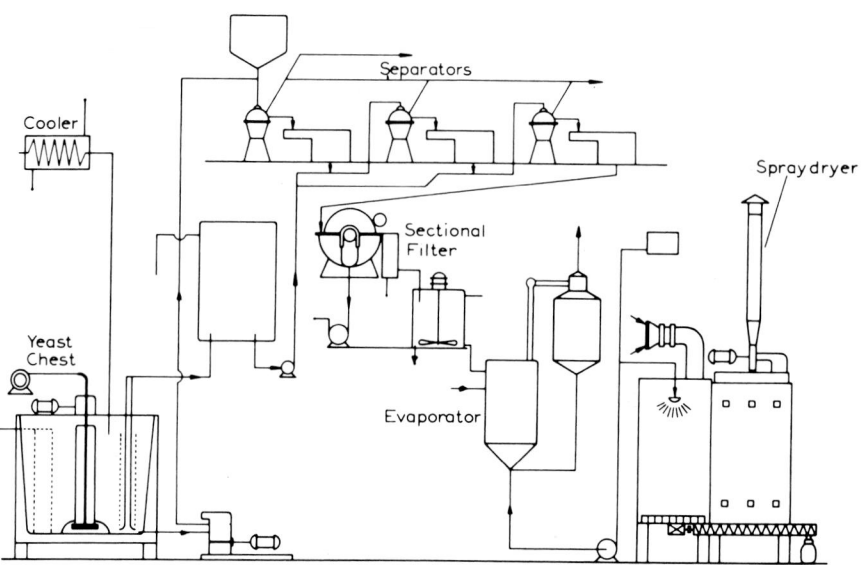

Figure IV-18. Flow-sheet of the Waldhof process for the production of yeast from beech sulfite spent liquor.

hollow axle has holes through which air is forced and thus is mixed with the liquor. At the same time, any foam formed is also drawn into the cylinder.

The I. G.-Scholler aeration uses a circulating system in which the liquor is drawn off at the bottom of the vat and is returned at the top. Nozzles are installed in the circulating system and the air which is forced through them promotes the circulation of the liquor; in this way, the mash is circulated on the principal of a Mammut pump and is sprayed over the surface of the foam layer, causing it to disintegrate.

The most favorable sugar concentration for the continuous fermentation process is between 2% and 3%. The time needed for the fermentation depends upon the concentration of the mash and varies between 3 and 6 hours. The temperature should be 36° to 38°C and must not surpass 40°C. The pH of the mash for wood sugar solutions is between 4.8 and 5.2, for spent sulfite liquor, between 4.8 and 6, and for sulfite liquor vinasse, between 4.8 and 5.0. Spent sufite liquors with a higher sulfur dioxide content require a higher pH.

The heat formation per kilogram of yeast formed increases with increasing pentose content and amounts to about 3500 kcal for wood sugar solutions containing 15% pentoses, to 3750 kcal for spruce spent sulfite liquors with 20% pentoses, and to 4600 kcal for beech spent sulfite liquors with 90% pentoses.

Figure IV–18 shows a flow-sheet of the Waldhof aerator system for

TABLE IV-13

AMINO ACID CONTENT OF DIFFERENT TYPES OF YEAST, IN
PERCENT ON DRY YEAST WEIGHT

Amino acid	Brewer's yeast	Bakers' yeast	Torula yeast
Alanine	4.80	5.20	5.78
Aminobutyric acid	0.03	0.02	—
Arginine	2.60	1.43	2.33
Aspartic acid	5.60	5.60	3.97
Asparagine	0.20	0.14	—
Cystine	0.74	0.61	0.54
Glutamic acid	10.60	8.50	6.74
Glycine	3.15	2.76	—
Histidine	1.76	1.39	1.19
Leucine	3.83	3.29	3.80
Isoleucine	2.76	2.45	3.70
Lysine	3.53	3.30	3.11
Methionine	1.05	1.12	0.60
Oxyproline	—	—	2.57
Phenylalanine	2.29	2.23	2.01
Proline	2.08	1.82	1.59
Threonine	3.26	2.49	2.50
Tryptophan	0.82	0.74	0.68
Tyrosine	2.09	2.33	2.77
Valine	3.23	3.21	3.30

the production of yeast from beechwood spent sulfite liquor. No significant changes in the apparatus are needed for the production of yeast from wood sugars.

The significance of the production of yeast from spent sulfite liquors and from the hydrolyzates of vegetable carbohydrates is the synthesis of protein, and—so far as pulp mills are concerned—the profitable utilization of oxygen-consuming components in the liquors and, with it, a considerable reduction in the pollution of the drainage canals into which the spent liquor is discharged.

Tables IV-13 and IV-14 show the importance of the yeast as a nutrient and give the amounts of amino acids and vitamins in various types of yeast.

In contrast to the formation of alcohol, the mechanism of the biological synthesis of protein is still largely unknown, but the importance of the role of the nucleic acids is known. For the synthesis of protein, the carbohydrates are probably broken down into acetaldehyde. So far as the pentoses are concerned, their assimilation is accomplished in such a way that the C-3-moiety serves for the cell synthesis, and the C-2-moiety is split off as carbon dioxide.

Under certain conditions, some yeasts can produce large quantities of

TABLE IV-14

VITAMIN CONTENT OF DIFFERENT TYPES OF YEAST, IN MILLIGRAMS
PERCENT OF DRY YEAST WEIGHT

Vitamin	Brewer's yeast	Bakers' yeast	Torula yeast
B_1, Thiamine	7.0–25.0	2.5–8.9	0.9–3.5
B_2, Riboflavin	1.7–5.6	2.5–8.5	3.9–7.2
B_6, Pyridoxin	2.3–10.0	1.6–5.6	3.3–9.4
Nicotinamide	30.6–63.0	29.0–53.7	45.3–70.7
Pantothenic acid	1.0–20.2	6.9–26.0	2.3–10.0
Biotin	0.11	0.06–0.18	0.1–0.23
Ergosterin	130–310	700–850	300–600

fat from sugars. Here, too, a breakdown to C-2-moieties and subsequent condensation are assumed. The efforts made thus far to carry out a technical microbiological production of fat have not met with success, chiefly because of the difficulty of cultivating good fat synthesizers in a continuous process (48,58,102,107,166,171,188,202,240).

5.6 Furfural

When strong acids react with pentoses, water is split off, with the formation of furfural:

$$CH_2OH-CHOH-CHOH-CHOH-C{\overset{H}{\underset{O}{\diagup}}} - 3H_2O \longrightarrow \quad (1)$$

Hexuronic acids can split off CO_2, with the formation of pentose:

$$HO_2C(CHOH)_4CHO - CO_2 \rightarrow CH_2OH(CHOH)_3C{\overset{H}{\underset{O}{\diagup}}} \quad (2)$$

This, in turn, can form furfural. Xylose can be converted almost quantitatively into furfural. Other pentoses give various yields of furfural as the result of the different degree of energy required for the splitting off of the water molecule, and drastic reaction conditions can cause some decomposition of the furfural. Furfural has been produced technically for a long time by the hydrolysis of corncobs. Furthermore, other plant materials and the majority of deciduous woods contain considerable amounts of pentosans which can be hydrolyzed to pentoses, and these can be dehydrated to furfural. Hydrolyzates of this type of wood therefore contain more or less large amounts of furfural, depending upon the pentosan content, and, in a two-stage hydrolysis, the pentoses are enriched during the prehydrolysis. The

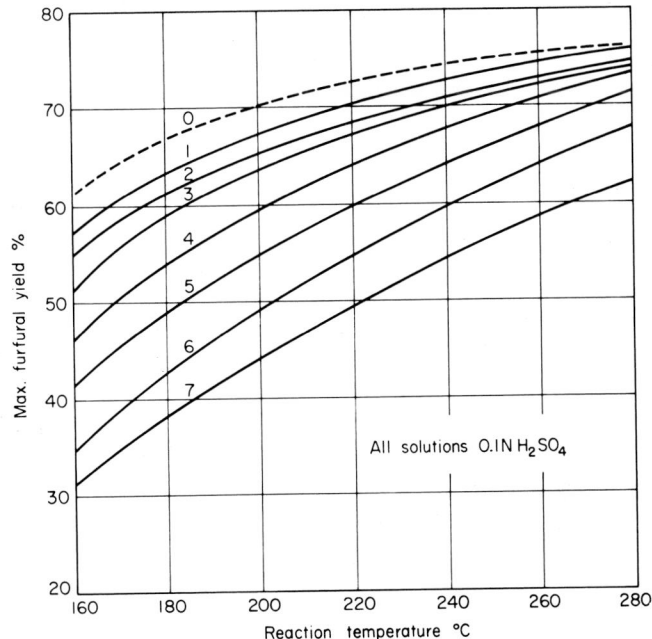

Figure IV-19. Maximum furfural yields at various xylose concentrations and reaction temperatures. Curves apply to all sulfuric acid concentrations greater than 0.1 *N* (181). 0 = 0 mg/cc; 1 = 3.125 mg/cc; 2 = 6.25 mg/cc; 3 = 12.5 mg/cc; 4 = 25 mg/cc; 5 = 50 mg/cc; 6 = 100 mg/cc; 7 = 200 mg/cc.

formation of furfural from pentosans thus occurs in two steps: in the first step, the pentosans are hydrolyzed to pentoses by the addition of water, and, in the second step, these are converted into furfural through the loss of water.

An investigation of the kinetics of the formation of furfural from xylose by the catalytic action of sulfuric acid of 0.1 or higher normality has shown that a close relationship exists between the maximum yield of furfural, the concentration of the xylose in the solution, and the reaction temperature. Acid concentrations below 0.1 *N* have an adverse effect on the reaction. The effects of the concentration of the xylose and the temperature of the reaction are shown in Fig. IV-19. The yield of furfural increases as the temperature is raised but decreases with increasing concentration of the xylose (181). To pursue this further, a concentrated hydrolyzate—containing 44% xylose, 12% glucose, 9% lignin, and 3.85% ash—from a hardboard mill was investigated. The results are shown in Figs. IV-20 and IV-21. All the concentrations in these graphs are based on the solid content of the solutions and this applies also to the concentration of the sulfuric acid. Figure IV-20 shows the relationship between the maximum yield of furfural and the acidity at various solid contents of the solution. Figure IV-21 shows the

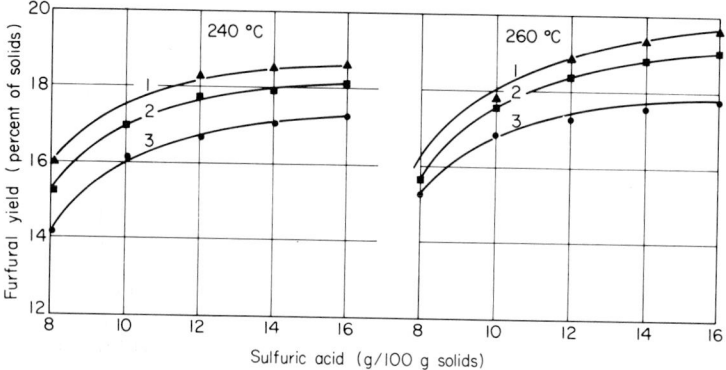

Figure IV-20. Maximum yield of furfural as a function of the acidity at various solids concentrations (98). 1 = 4 g solids per 100 ml; 2 = 9 g solids per 100 ml; 3 = 16 g solids per 100 ml.

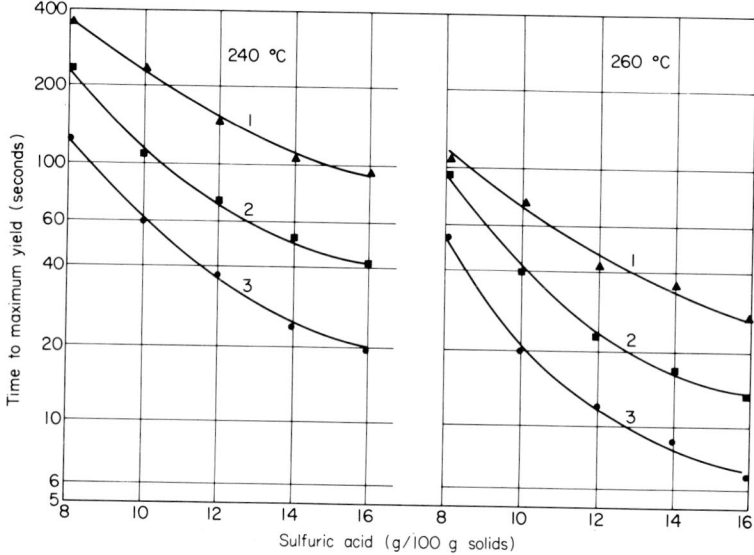

Figure IV-21. Time at which maximum yield of furfural is reached as a function of acidity at various solids concentration (98). 1 = 4 g solids per 100 ml; 2 = 8 g solids per 100 ml; 3 = 16 g solids per 100 ml.

time required to reach the maximum yield of furfural as a function of the acidity at various concentrations of the solids. A comparison of the two graphs indicates that the results of the changes in temperature and concentration is similar. With a decreasing acid-to-solid ratio, the yield decreases proportionally. This can be explained as resulting from the strong buffer action of the solids in the solution. Because of its low ionization constant, the acetic acid has no significant effect on the yield of furfural and the

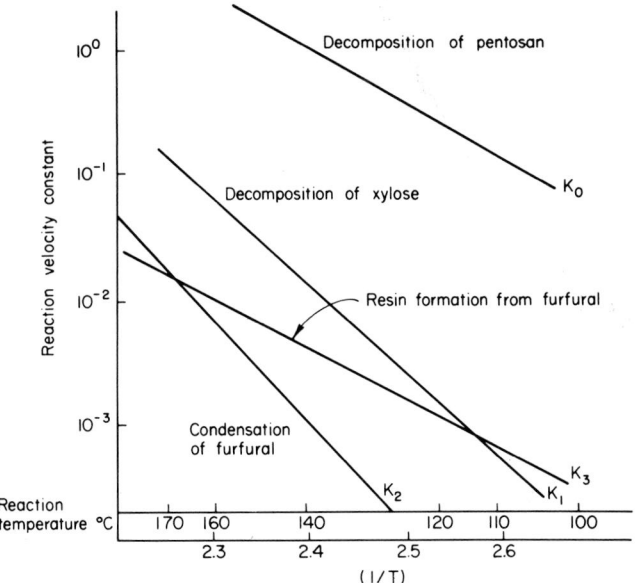

Figure IV-22. Relation between reaction velocity constants and temperature (192).

hydrogen ion concentration resulting from the acetic acid exerts no influence in the presence of sulfuric acid as a catalyst. On the other hand, glucose can have a detrimental effect on the furfural yield, especially at low temperatures. The results given in Fig. IV-22 show that the presence of glucose had only a minor effect at the concentrations and temperatures applied in this study. From technical pentose solutions, about 40% furfural is obtained, i.e., for each pound of furfural, 2.5 pounds of xylose is required (98).

The conversion of pentose into furfural involves not only the reaction scheme given in equation 1, but also side reactions which lead to condensation and to the formation of resins. The following reaction scheme can therefore be set up:

$$\text{pentose}\ \text{(xylose)} \xrightarrow{k_1} \text{intermediate compound} \xrightarrow{k'k_1} \text{furfural} \xrightarrow{k_3} \text{resin}$$
$$\big\downarrow{k_2}$$
$$\longrightarrow \text{condensation} \longleftarrow$$

The various reactions have different reaction constants. The following equations are obtained for the reaction velocities:

$$dX/dt = -k_1 X \tag{3}$$

$$dF/dt = k_1 X - k_2 X^n F^m - k_3 F \tag{4}$$

$$dR/dt = k_3 F \tag{5}$$

$$dC/dt = k_3 X^n F^m \tag{6}$$

In these equations, $X = $ xylose, $F = $ furfural, $R = $ resin, $C = $ condensation product; the constants k_1 and k_3 were found experimentally by measuring the decrease in xylose content caused by decomposition and from the decrease in the furfural content by resinification in the solution. These measurements were carried out with xylose in the presence of 1% hydrochloric acid at 140° to 160°C. From the constants, k_1 and k_3, and the yield of furfural, the values for n, m, and k_2 in equation 4 could be calculated. By applying the values found, the following equation can be established:

$$\frac{dF}{dt} = k_1X - k_2X^{7/6}F^{1/2} - k_3F \qquad (7)$$

Figure IV–22 shows the relationship between the temperature and the velocity constants of the various reactions. The increase in velocity, k_1, becomes greater than the increase for k_3 with increasing temperature. A higher temperature, therefore, will give a higher yield of furfural (192). From a technical point of view, the following conditions must be met in order to obtain a good yield of furfural: (a) The smallest possible amount of furfural should be left in the reaction mixture, i.e., the furfural should be removed as quickly as possible in order to prevent side reactions and condensations. (b) A high reaction velocity should be maintained and the reaction time should be as short as possible. (c) The heat energy should be kept to the minimum required.

Of reactions 3, 4, 5, and 6, which proceed simultaneously, reactions 5 and 6 should be suppressed and reaction 4 should be promoted. This can be accomplished as indicated in (a), by keeping the amount of furfural in the reaction system as low as possible. This amount can be represented by F/W, in which W represents a parameter. Equation 7 can then be presented as follows:

$$\frac{dF}{dt} = k_1X - k_2X^{7/6}\frac{F^{1/2}}{W} - k_3\frac{F}{W}$$

The amount of furfural in the reaction mixture that can react according to equations 5 and 6 can be expressed by F/W. When the amount of furfural removed from the reaction mixture is designated F_g, and the amount left in the mixture, F_r, the following equation is valid:

$$W = \frac{F_r + F_g}{F_r}$$

The yield of furfural increases with the increase in W (192).

Oat hulls contain about 32% to 35% pentosans and when heated with dilute mineral acids at 130° to 140°C the latter are hydrolyzed to pentoses which are converted, by the splitting off of water, into furfural. This process has been in operation for many years on a commercial basis and is known

as the Quaker Oats process. In this process, 2200 kg oat hulls are moistened in a digester with 660 kg of about 7% sulfuric acid and subjected to a steam distillation at a constant pressure of 4 atmospheres for 6 to 8 hours. The escaping steam, containing up to 10% furfural, is blown into a kind of cyclone and the condensate is distilled, to give 3 fractions. The yield of furfural is 10% to 11%, or only about 50% of the optimum yield of 22% to 23%. Although the necessary raw material is available in adequate supply, the disadvantages of the process are the relatively high steam consumption and the problem of finding an economical use for the 70% residue, since only 30% is converted into furfural (175).

The "Natta process" uses hydrochloric acid as the catalyst. The raw material used is moistened with dilute hydrochloric acid and is fed continuously into the top of a vertically arranged reactor. Superheated steam at a temperature of 190° to 200°C is injected at the bottom of the reactor, passes through the material in a countercurrent stream, and leaves the digester at the top, taking the furfural with it. The condensate contains only a small amount of hydrochloric acid and this is neutralized before the furfural is purified. The process operates continuously, uses considerably less steam and acid than the sulfuric acid process described above, and gives furfural yields of 70% to 80% of the theoretical amount. The aqueous hydrochloric acid is constantly enriched by the acid in the steam during the descent of the material in the digester because there exists in the ternary system, $HCl-H_2O$-furfural, a binary azeotropic mixture of $HCl-H_2O$ with a maximum boiling point. The process also permits the recovery of the acetic acid formed as a by-product. The strongly corrosive properties of the acid give rise to some difficulties (151).

The dehydration of the pentoses to furfural can also be achieved with steam under certain reaction conditions and without mineral acids. Thus it has been suggested that vegetable raw material should be passed continuously through a long reactor with steam of up to 250°C being forced through it in a countercurrent direction. First, acetic acid is formed by hydrolysis, and this hydrolyzes the pentosans to pentoses and converts the latter into furfural which is constantly removed from the reaction mixture with the steam (201). There can be no doubt that, under certain reaction conditions, furfural can be obtained in a high yield by this method.

According to investigations of the kinetics of the decomposition of xylose and the formation of furfural, with 14 different mineral acids and 6 organic acids at 0.2 N concentration, with a 4% xylose solution and a reaction temperature of 160°C, the highly dissociated acids, such as hydrochloric, hydriodic, hydrobromic, and perchloric acids, showed the greatest catalytic activity. Formic, acetic, propionic, and butyric acids were the least active as catalysts. Within the pH range of 0.7 to 1.2, a straight-line relationship was obtained between the reaction constants of a first-order

reaction and the hydrogen ion concentration. Acids that are only slightly soluble, such as vanadic, tungstic, and molybdic acids, greatly accelerate the decomposition of the xylose because of their oxidative properties. The highest yield of furfural, up to 83%, was obtained with hydroxymono-carboxylic acid while dicarboxylic acids gave somewhat smaller yields, e.g., trihydroxyglutaric acid gave 74% furfural and oxalic acid gave 70%. Of the mineral acids, hydriodic and phosphoric acids gave the highest yields, up to 76%, whereas hydrochloric and sulfuric acids gave 66% to 70%. The lowest yields, 18% to 21%, were obtained with the oxidizing acids (118).

Experiments have been carried out to find ways of preventing the re-sinification of furfural and thus to increase its yield. With the conventional batch hydrolysis only about 4 to 5 kg furfural per ton of dry wood can be obtained; with the intermittent steam-thrust method, the yields are considerably higher. Thus, for example, steam thrusts of only 10 to 20 minutes duration and 10 atmospheres pressure are recommended because, in this way, the removal of the furfural from the reaction zone is facilitated, the possibility of its resinification is reduced, and its yield is increased. If the production of furfural is desired in addition to that of yeast from the hydrolyzate, the steam thrusts must be curtailed since, otherwise, a too drastic dehydration of the pentoses will cause a considerable reduction in the yield of yeast. The steam-thrust technique reduces the costs of the rectification of the furfural because of its higher concentration in the steam condensate (108).

For a continuous process for the production of furfural, constant reaction conditions are necessary for trouble-free operation and high yields. Consequently, the reaction temperature should be kept at $170 \pm 2°C$ and the duration of the passage of the material should be 60 ± 4 minutes. The moisture content of the sawmill waste should be 20% to 25% and the vapors containing the furfural should be carefully controlled. The sawdust is charged by a feeder into the impregnation vessel and a measured amount of 3% to 10% sulfuric acid is added simultaneously. From there the material is passed in a measured amount by a screw conveyor into the hydro-lyzer. The latter is heated with saturated steam at 6 to 7 atmospheres, and superheated steam at 10 to 15 atmospheres is injected at the bottom of the reactor. The pressure in the hydrolyzer is kept at 7 atmospheres and is controlled by a pressure regulator which simultaneously keeps the inside temperature constant. The discharging is achieved by a control system consisting of a liquid-level regulator, a pump, a hydraulic Servo motor, and an automatically adjusted discharge valve (169).

The use of salts for the acceleration of the conversion of xylose into furfural on heating of pentosan-containing vegetable material has also been suggested. Aluminum and chromium chlorides and their sulfates are the most effective of the salts tested. The reaction constants for the xylose

degradation vary within a wide range. The reaction for most of the salts is a first-order reaction and the reaction constants of alkaline salts which readily undergo hydrolysis are time-dependent. The highest furfural yields—60% to 65% of the theoretical amount—are obtained with salts that have alkaline properties (lithium, sodium, barium, magnesium, calcium, cadmium, strontium, manganese, iron, cobalt, and nickel); the typical amphoteric elements (aluminum, tin, antimony, chromium, and lead) gave the lowest—32% to 37%. When salts are added to a reaction mixture containing 0.2 N sulfuric acid as a catalyst, the degradation of the xylose increases somewhat with the addition of sodium chloride, sodium bisulfate, or potassium bisulfate, whereas the addition of aluminum sulfate causes a marked degradation. Sodium chloride causes an increase in the yield of furfural as compared with that obtained when acid alone is used, whereas aluminum sulfate has no effect (117).

The method known as the "Agrifuran process" utilizes alkaline earth phosphates as acid catalysts for the production of furfural from vegetable materials. The use of these salts simultaneously causes the phosphatizing of the reaction vessels, thus preventing corrosion. The amounts used, however, are quite high. To decrease the consumption of the acid phosphates, the volatile organic acids formed during the hydrolysis are passed from one to another of a series of five autoclaves. In this way the amount of phosphate can be reduced to one-third (146,180).

In all the processes described above, the furfural is recovered by distillation with steam. A modified method is an extraction with solvents, in which the furfural is taken up as quickly as possible by a suitable solvent and thus is protected from further decomposition. Such a process can be carried out only in a two-step hydrolysis. A solvent for furfural with a suitable distribution coefficient toward aqueous solutions is not easy to find. It is therefore doubtful that such a process can be carried out economically. A vacuum-distilled furfural itself has been suggested as a solvent. Toluene may be suitable because it is inert toward furfural, it has a low boiling point and low density, and it is insoluble in water. The addition of sodium chloride and sodium sulfate as salting-out agents increases the extractive action. Dimethyl phthalate has a high boiling point, low solubility in water, and a high distribution coefficient for furfural for the extraction of aqueous solutions that also contain acetic acid and methyl alcohol as impurities (5,74,216).

Furfural solutions are readily susceptible to oxidation by air during storage or during the distillation, which may last up to 20 hours. It has therefore been suggested that the storage and distillation of such solutions should be accomplished in a nitrogen or carbon dioxide atmosphere, thus diminishing the loss of furfural (72,128,215).

With regard to the utilization of furfural, it may be stated that it is

an excellent solvent and is widely used in the industry. Furfural is a solvent for cellulose esters, for waxes, and for resins; it is a selective solvent for the purification of animal, vegetable, and mineral oils and for wood rosins (colophony). As an aldehyde, furfural is used for the production of artificial resins; with phenol it forms condensation products that are heat-resistant and show good dielectric properties. Similar reactions can be carried out between furfural and urea, aromatic amines, melamine, and ketones.

The hydrogenation of furfural gives furfuryl alcohol, tetra-hydrofurfuryl alcohol, 2-methylfuran, 1,4-pentanediol, and 1,2,5-pentanetriol, depending upon the pressure, temperature and catalyst used. In recent years, furfural has been an important raw material for the production of nylon. Of its oxidation products, maleic acid, $HO_2CCH:CHCO_2H$, is of special importance.

Hydroxymethylfurfural has recently attracted the attention of scientists because it, together with levulinic acid, is the most important of the many components formed by the decomposition of glucose when it is heated in the presence of mineral acids. The interest of industry in hydroxymethylfurfural is due to its polyfunctional character; as an alcohol and aromatic aldehyde it is capable of undergoing all the reactions expected of it, even including ring formation. In the presence of 0.1 N sulfuric acid at 240°C, glucose gives hydroxymethylfurfural, levulinic acid, formic acid, and humic substances. The yield of hydroxymethylfurfural passes through a maximum and reaches 40%, and more, of the theoretical amount. The prospects for its production on an industrial scale cannot as yet be predicted (97,159, 214).

Levulinic acid is the end-product of the hydrolysis of hexoses via hydroxymethylfurfural as an intermediate:

$$HOCH_2CHOHCHOHCHOHCHOHCHO \longrightarrow$$

$$\longrightarrow CH_3COCH_2CH_2CO_2H + HCO_2H$$

Levulinic acid is used chiefly for the formation of diphenolic acids for the production of artificial resins. In its relationship to other acids and ketones, it is capable of forming heterocyclic compounds (198).

5.7 Organic Acids Formed by Fermentation

Wood sugar mashes can be fermented to organic acids as well as to alcohols. It has been suggested that the sugars remaining in the vinasse could be fermented to lactic acid. *Lactobacillus pentosus* can convert 75% of these sugars into lactic acid. Hexoses are also fermented in this way.

The acid formed consists of 90% to 95% lactic acid and 5% to 10% acetic acid. Sugar solutions containing 7% to 10% sugars require 5 to 7 days for a complete fermentation. Lactic acid can also be obtained from spent sulfite liquors. The spent liquor must first be stripped by steam of sulfur dioxide, then adjusted to a pH of 8.5 with milk of lime, filtered, and again adjusted to pH 6. The total yield of acids is then 95% of the theoretical amount. In this way, 300 pounds of lactic acid and 80 pounds of acetic acid per ton of pulp may be obtained from spent liquor vinasse (7,134,145).

The production of butyric acid is of little economic interest. With a mixture of *Clostridium pastorianum* and *Bacterium amylobacter*, wood sugar solutions can be converted into butyric and acetic acids. At 40°C the fermentation requires about seven days. The calcium salts of the acids formed can be converted by pyrolysis into the corresponding ketones, and thus solvent mixtures are obtained which may be used for many purposes. Whereas the theoretical yields from hexoses amount to about 50% for butyric acid and 67% for acetic acid, the actual yields are only about 30% to 40% and 3.5% to 9%, respectively (13,86).

5.8 Crystalline Glucose

Crystalline glucose has been made for many years from starch hydrolyzates and is available in most countries. Its production from the saccharification of cellulose was, therefore, an obvious next step, in spite of the fact that numerous difficulties are involved. In the total hydrolysis of wood with acids, the hexoses, including the cellulose as its main component, are degraded to hexoses, chiefly glucose, and the hemicelluloses, consisting chiefly of pentosans, into pentoses. The main difficulty in the crystallization of the glucose is caused by the presence of other sugars and high-molecular compounds. The fact that the hydrolysis of wood can be carried out in two steps makes the production of crystalline glucose from wood hydrolyzates of greater technical interest. After the removal of the pentosans and other accessory components in the prehydrolysis, glucose alone is the principal product remaining in the hydrolyzate of the main hydrolysis and it can readily be separated from the residual lignin. It is of no importance what process is applied in the main hydrolysis; if it is carried out at low temperatures, high acid concentrations are necessary to hydrolyze the cellulose in a technically reasonable time, and the decomposition of the sugars formed is negligible. If the main hydrolysis is carried out with dilute acids, high temperatures are required and this may cause some decomposition of the sugars. The steps that can be taken to keep this decomposition to a minimum have already been discussed.

The main hydrolysis of cellulose does not lead exclusively to the formation of glucose because half the sugars in the solution are di- and

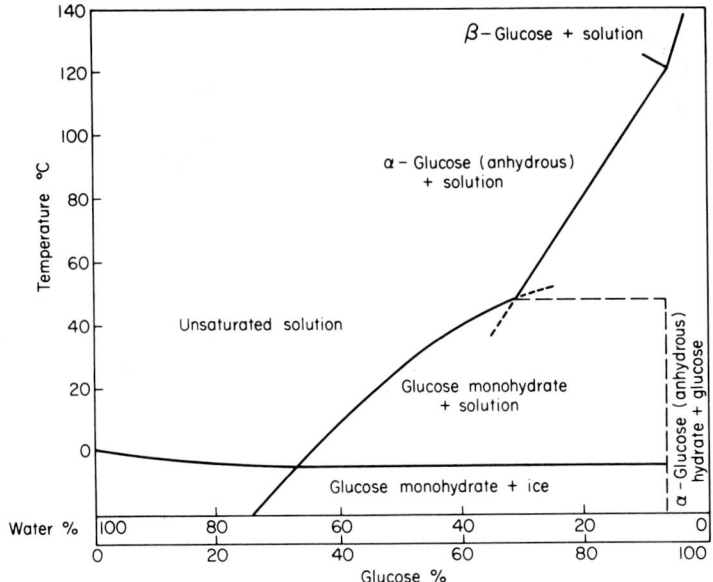

Figure IV-23. Phase diagram of glucose-water system (154).

trisaccharides which must be converted into the monosaccharide by a post-hydrolysis. If a supersaturated glucose solution is inoculated with a glucose crystal and the solution is stirred at below 50°C, the monohydrate of α-glucose begins to crystallize. At 50° to 115°C, chiefly anhydrous α-glucose, and, at above 115°C, chiefly anhydrous β-glucose crystallize. The conditions causing an equilibrium between the formation of crystals and the degree of supersaturation of the sugar solution are not clear, but, when the conditions are not right, crystals of different habits are formed and are difficult to separate from the mother liquor. Impurities in the solution reduce the crystallizability of the glucose to a considerable degree. Figure IV-23 shows a phase diagram of the glucose-water system. The removal of organic impurities, originating from the wood, from the glucose solution by special ion exchangers makes the purification process more complicated but results in 90% pure glucose. The purified sugar solution is as clear as water and contains 90% glucose, 4% disaccharides, 3% mannose, 2% xylose, 1% galactose, and 0.1% mineral salts (154,191).

Another possible way of obtaining crystalline glucose is via the formation of a double salt of glucose and sodium chloride, of the composition $(C_6H_{12}O_6)_2 \cdot NaCl \cdot H_2O$, which can also be obtained from crude sugar solutions. This does not mean that crystalline glucose can also be obtained via the double salt from the hydrolyzate of a one-step hydrolysis. The decomposition of the double salt is carried out by the addition of an adequate

amount of water, and the glucose thus obtained contains only 0.1% sodium chloride (55,75,104).

Whether or not the crystalline glucose from wood hydrolyzates can compete with that from starch hydrolyzates depends upon the local conditions relating to the raw materials, the cost of the chemicals and investment in the plant, and the market demand.

6. WOOD SUGAR LIGNINS

In the saccharification of wood, 30% of the wood is obtained as an acid lignin. The utilization of this lignin is still, today, the main problem of the wood hydrolysis industry. There has been no lack of research directed toward the utilization of the lignin as a raw material for the production of organic compounds. Although the lignin precipitated from the kraft black liquors has found some uses (see Chapter VIII, Sections 7.3 to 7.5), the acid lignins cause some difficulty because of their slowness in reacting and their lack of solubility. Kraft lignin is, of course, very soluble in alkali, in contrast to the acid lignins. This characteristic of the latter has greatly limited their utilization and many attempts have been made to increase their reactivity by activating them. Such an activation is usually carried out by heating the lignin, suspended in alkali, under pressure. Of great importance in this reaction are the following: the ratio of lignin to sodium hydroxide, which should be about 3 : 1; the reaction temperature, which is generally 160°C or higher; and the reaction time, which, in some cases, may amount to several hours. In this way, lignin solutions are obtained which can be processed economically only at their place of origin. Experiments to increase the lignin-to-sodium hydroxide ratio were only partially successful because the yield of soluble lignin was greatly reduced. A certain improvement was achieved when lignin and sodium hydroxide, at a high lignin-to-sodium hydroxide ratio, were treated in a vibratory mill in a way similar to that used in the wood hydrolysis process (129,211).

Another report claims that the optimum conditions are a ratio of lignin to sodium hydroxide of 1 : 2, a reaction time of 4 hours, and a temperature of 180°C (163). It has also been suggested that the activation of the lignin should be combined with a reaction with phenol, in which an excess of phenol is added to the alkaline solution. The yield of this phenol-treated lignin is 96% to 100%, based on dry acid lignin, and it contains 0.4 to 1.4 moles phenol per mole lignin. This type of lignin contains more carbon, less oxygen, fewer total hydroxyl groups, but more phenolic hydroxyl groups; its molecular weight varies between 250 and 410, as compared with 540 for the acid lignin (158). By taking advantage of the low vapor pressure of concentrated sodium hydroxide solutions, it should be possible to obtain

highly soluble lignin preparations directly from acid lignins without the application of pressure, without the addition of water, and with a relatively small amount of sodium hydroxide, so long as the moist lignin-sodium hydroxide mixture is heated rapidly to a sufficiently high temperature while avoiding a drying out of the mixture. Such a lignin should have a solubility of 97%. Some difficulties may arise from the heat transfer of the concentrated lignin-sodium hydroxide paste (127).

A hypothesis that explains the action of acids on lignin, including the activation of hydrolysis lignin, assumes that, when native lignin is subjected to the action of concentrated acids at normal temperatures or of dilute acids at an elevated temperature, it is converted into an active form in which it is capable of reacting with various compounds. Such lignin can undergo oxidation, reduction, sulfonation, phenolation, hydroxyalkylation in the presence of alkalis, and polymerization. According to present concepts of the lignin structure, the active form may be assumed to be a quinonemethide (see Chapter III, Section 4.4). Because of the high reactivity of quinonemethides in acid media, this form of lignin is a short-lived intermediate, its reaction products depending on the composition of the reaction medium. In the absence of substances with which activated lignin can react, it undergoes polymerization, yielding an inactive product of the hydrolysis lignin type. In the presence of an oxidizing agent, e.g., when wood is hydrolyzed with sulfuric acid in the presence of nitric acid or hydrogen peroxide, no polymerization or inactivation occurs and the lignin thus obtained is soluble in alkali and in many organic solvents. In the presence of an alcohol, e.g., methyl alcohol, the active lignin reacts in an acid medium to yield an ether with the methyl group in the α position of the side-chain. In the presence of phenols, the acid catalyzes phenolation of the lignin (154a).

The results of the pyrolysis of lignin correspond to those from the pyrolysis of wood, which is discussed in Chapter V. In the former, a tar is obtained, containing a series of phenols of which 60% are partially methylated. The isolation of specific phenols from this tar is difficult.

The structural properties of acid lignins make them especially suitable for the production of activated charcoal. For this purpose, the lignin, containing 45% to 50% water, is granulated in a screw press and carbonized at 450°C for one hour. A higher water content gives a product with larger pores. According to another suggestion, the lignin is mixed with birchwood tar, granulated, and then heated for 3 hours at 360°C. The product is then activated by treating it in a horizontal tube furnace with superheated steam at 850°C. Such an activated charcoal has strongly adsorptive powers (209, 210,223,224).

Experiments have also been carried out to modify acid lignins in such a way that they can be used as ion exchangers. Already during the Second

World War, attempts were made in this direction in Germany, but the products obtained did not satisfy the high specifications. On sulfonation with concentrated sulfuric acid (specific gravity 1.84) at a temperature of up to 220°C, satisfactory cation exchangers should be obtainable (73), while by chloromethylation and subsequent amination, anion exchangers should be obtained (212).

Many investigation have been carried out on the hydrogenolysis* of lignin and lignin-containing materials. Most of these experiments had as their goal, not the finding of a satisfactory means for the utilization of the acid lignins, but, primarily, the elucidation of the constitution of the lignin molecule. They are discussed in Chapter III. Moreover, reference may be made to the relevant books and recent publications (3,14,17,204). Already in 1925, the hydrogenolysis of hydrochloric acid lignin, at 250 atmospheres pressure in the presence of a nickel catalyst, was described. In this case, 15.5% coke, 50.2% aqueous acid reaction products, 19.7% tar with 43% phenolic components, and 19% gases and ash were obtained (57). Somewhat later, two Swedish patents were issued which described the hydrogenolysis of lignin and waste wood (137). On hydrogenolysis of sulfuric acid aspen lignin in aqueous supension with Raney nickel as a catalyst, at 225° to 250°C, the chief products were propylcyclohexane, propylcyclohexanol, and several other reaction products containing the propylcyclohexane moiety. When sprucewood lignin, in a suspension of decalin, was hydrogenated, cyclohexane, methyl-, ethyl-, and propylcyclohexane, cyclohexanone, cyclohexanol, *p*-ethylphenol, and toluene were obtained (41,96).

When hydrolysis lignin was heated in 90% ethyl alcohol in the presence of calcium hydroxide at 320°C, a crude tar was obtained in 87% yield. Hydrogenolysis of the tar in decalin at 430°C gave aromatic hydrocarbons, cyclic alcohols, and phenols (205). In alkali and in the presence of copper oxide as a catalyst, a hydrogenolysis at 250°C gave 45% to 60% phenols and 15% monocyclic compounds. Raney nickel as a catalyst, on the other hand, gave chiefly cyclohexanol (63).

A more detailed study has been made of the destructive hydrogenation (hydrogenolysis) of lignin and wood in alkaline mediums. This study showed that there are two characteristic temperature intervals, the first at 250° to 260°C, the second at 340° to 360°C. In the first case, a catalyst is required, but in the second, the reaction occurs even without a catalyst.

---

* In the early experiments on the reaction of lignin with hydrogen in the presence of a catalyst, with a high hydrogen pressure and at high temperatures, the term "hydrogenation" was always used although it was actually a hydrogenolysis that was involved. Hydrogenation means the addition of hydrogen to a compound without causing any change in its fundamental structure, whereas a hydrogenolysis involves the cracking of the product, with the complete degradation of its original structure and the addition of hydrogen to the fragmentation products. Since the treatment of wood and its components with hydrogen always involves the latter type of reaction, the term "hydrogenolysis" will be used in referring to it in this translation. F.E.B.

The hydrogenolysis of hydrochloric acid spruce lignin in the presence of a copper-nickel-kaolin catalyst gave 50% of an ether-soluble oil and 25% of a distillable oil. The oil contained chiefly phenols when it was produced with a catalyst of low activity, and chiefly cyclohexanols when a catalyst with high activity, such as Raney nickel, was used. In the destructive hydrogenation at temperatures of above 300°C, about 45% ether-soluble, and 25% distillable substances were obtained even without a catalyst (141).

A series of experiments on the destructive hydrogenation of spruce hydrochloric acid lignin in various solvents and in the presence of different catalysts, at an initial pressure of 95 to 100 atmospheres and a reaction temperature of 260° to 270°C (maximum pressure 230 atmospheres), showed that nickel oxide and dioxane gave the best results. In this case, one mole hydrogen was taken up by 40 grams lignin, the absorption of hydrogen stopped after 20 to 25 hours, and 46.7% ether-soluble products were obtained (76).

A stepwise hydrogenolysis, at 400 to 600 atmospheres and 350° to 600°C for 10 to 30 minutes, resulted, in the first step, in a partial cleavage of the lignin. The reaction mixture from the first step gave a gaseous fraction, a low-boiling fraction, and a residual liquid containing 30% phenols (173). Only the liquid was subjected to another or, if necessary, to several more hydrogenolysis steps, with each successive step being carried out at a higher temperature. A part of the liquid from the first step was used to stir the residue of the lignin into a paste for the next step. The hydrogenolysis of lignin in the presence of tar or phenol as a solvent, and with a catalyst based on ferrous sulfide, gave an aqueous layer of acetone and methanol and an oily layer, containing monophenols, catechols, and a heavy oil, as reaction products. At 350° to 370°C and 200 atmospheres pressure, 92.1% of the lignin could be converted into liquid products within 2 hours (155).

It is unnecessary here to discuss further the numerous suggestions that have been made for the utilization of lignin. These deal more with the elucidation of the structure of the lignin molecule than with the development of possible new industrial uses for lignin. Of importance are the different processes for the oxidation of lignin. One of the main products from this reaction is vanillin, which is produced on a technical scale primarily from spent sulfite liquor. This is discussed in detail in Chapter VII, Section 9.5. Whether vanillin, itself, can be used as a starting material for other organic syntheses which can utilize lignin from wood hydrolyses cannot, at present, be definitely decided.

7. REVIEW AND OUTLOOK FOR THE FUTURE

Wood consists chiefly of cellulose, hemicelluloses, and lignin. When wood is decomposed by an acid hydrolysis, cellulose and the other hexosans

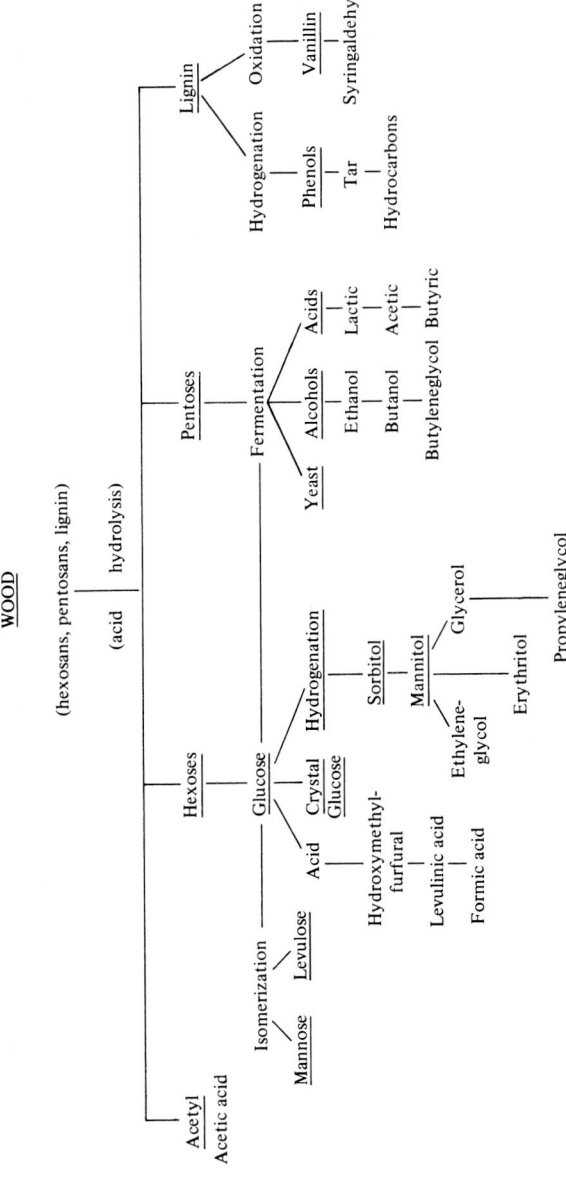

Figure IV-24. Wood hydrolysis by-products and their utilization.

are cleaved into hexoses, such as glucose, and the pentosans into pentose. The lignin, undoubtedly changed structurally, remains as insoluble acid lignin. Because the composition of the various woods differs, the yield of scission products will also vary. Deciduous woods contain more pentosans and therefore give higher yields of pentoses (xylose). The lignin content, too, varies with the different wood species. The various compositions and the morphological differences of the woods affect the wood hydrolysis. The pentosans are hydrolyzed more readily than the hexosans and these, in turn, more readily than the cellulose. If the pentoses and hexoses are to be obtained separately, the former must be dissolved by a mild prehydrolysis, while the latter, especially the cellulose, must be hydrolyzed in a main hydrolysis under more drastic reaction conditions (low acid concentration and high temperature, or high acid concentration and low temperature).

Figure IV–24 gives a simple schematic diagram of the main reaction products obtained in the hydrolysis of wood and the possibility of their being further processed. This diagram, of course, does not show whether the possible utilization has any promise for a large-scale development. Many of the possible uses have already been shown to be technically feasible. Whether the processes in question have any prospect of being of commercial value is, for many of them, at least doubtful.

These limitations do not mean that certain local conditions may not make the use of such processes economically practicable. For example, this may be the case when wood wastes are available in large amounts and at a reasonable price, and when the chemicals required for the hydrolysis can be obtained from the chemical industry, preferably as by-products. Proof that these statements are correct is provided by the growing wood hydrolysis industry in the Soviet Union and, perhaps on a smaller scale, in Japan.

Of special importance is the hydrolysis of wood when additional nutrients and more fodder are needed, or in periods of scarcity. Glucose, perhaps in crystalline form, yeast and fodder protein, and, possibly, furfural seem to be the most desirable reaction products, but even these can be obtained from other rapidly growing vegetable raw materials. One of the main problems of wood hydrolysis remains the utilization of the great amounts of lignin obtained, if its heat value as a fuel is discounted.

REFERENCES

1. Allgeier, R. J., Peterson, W. H. and Fred, E. B., *Ind. Eng. Chem.* **21**, 1039 (1929).
2. Apel, A. and Rossler, G., German Patent 845,780.
3. Arlt, H. G., Sonja, K. G. and Schuerch, C., *Tappi* **41** (2), 64 (1958).
4. Arnoud, J. E., *C. R. Acad. Sci.* **39**, 807 (1855).
5. Bagaev, A. N. and Melnik, N. A. *Gidrolizn i Lesokhim Prom.* **19** (2), 26–8 (1966).

6. Bechamp, A., *Ann. Chim.* **48**, 458 (1856); *C. R. Acad. Sci.* **42**, 1210 (1856), **51**, 255 (1860).
7. Beckum, van, W. G. and Ritter G. J., *Paper Trade J.* **108** (7), 27 (1939).
8. Beinarts, J. and Kalnins, A., *Latvijas PSR Zinatnu Akad. Vestis Khim. Ser.* **3**, 329–36 (1964).
8a. Belozerova, L. A., *Tr. Leningrad. Lesotekh Akad.* **105**, 89 (1966).
9. Bergius, F., *Chem. Z.* **57**, 475 (1933).
10. Bergius, F., *Zellstoffaser* **32**, 50 (1935); *Ind. Eng. Chem.* **29**, 247 (1937).
11. Bergius, F., Farber, E. and Jellinek, O., *Ergänz. Angew. Phys. Chem.* **1**, 199 (1931).
12. Berlin, J., *J. Am. Chem. Soc.* **48**, 1107 (1926).
13. Berthelot, C., "Combustibles et lubrificants de remplacement." Paris, 1943.
14. Bhattacharya, A. and Sondheimer, E., *Tappi* **42** (6), 446 (1959).
15. Birtwell, C., Clibbens, D. A. and Geake, A., *J. Textile Inst.* **17**, T140 (1926).
16. Braconnot, H., *Gilberts Ann. Phys.* **63**, 348 (1819); *Ann. Phys. Chim.* **12**, 172 (1820); *Dinglers Polytechn. J.* **312** (1820).
17. Brauns, F. E., "Chemistry of lignin." Academic Press, New York, 1952; Brauns, F. E. and Brauns, D. A. Supplemental Vol., 1960.
18. Brenner, F. C., Frilette, V. and Mark, H., *J. Am. Chem. Soc.* **70**, 877 (1948).
19. Burney, Mc. L. F., Heterogeneous hydrolysis *in* Ott and Spurlin "Cellulose" Vol. I. Wiley (Interscience), New York, 1954.
20. Button, D. K., Garver, J. C. and Hajny, G. J., *Appl. Microbiology* **14** (2), 292 (1966).
21. Chalov, N. V., USSR Patent 119,491, Apr. 1959.
22. Chalov, N. V., *Gidrolizn. i Lesokhim. Prom.* **15** (7), 4–7 (1962).
23. Chalov, N. V., Aleksandrova, O. A. and Leshuk, A. E., *Izv. Vyssikh Uchebn. Lesn. Zh.* **6** (2), 141–4 (1963).
24. Chalov, N. V. and Blinova, N. N., *Izv. Vysshikh Uchebn. Zavadenii, Lesn. Zh.* **7** (7), 164–69 (1964).
25. Chalov, N. V. and Blinova, N. N., *Izv. Vysshikh Uchebn. Zavadenii, Lesn. Zh.* **7** (4), 132–36 (1964).
26. Chalov, N. V. and Blinova, N. N., *Izv. Vysshikh Uchebn. Zavadenii, Lesn. Zh.* **9** (3), 140–44 (1966).
27. Chalov, N. V. and Goryachikh, E. F,. *Izv. Vysshikh Uchebn. Zavadenii, Lesn. Zh.* **6** (3), 137–44 (1963).
28. Chalov, N. V. Goryachikh, E. F. and Sharkov, V. I., *Sb. Tr. Vses. Nauchn.-Issled. Inst. Gidrolizn. i Sulfitno-Spirt. Prom.* **13**, 23–30 (1965).
29. Chalov, N. V., Goryachikh, E. F. and Sharkov, V. I., *Sb. Tr. Vses. Nauchn.-Issled. Inst. Gidrolizn. i Sulfitno Spirt. Prom.* **14**, 166–79 (1965).
30. Chalov, N. V. and Korotkov, N. V., *Sb. Tr. Vses. Nauchn.-Issled. Inst. Gidrolizn. i Sulfitno Spirt. Prom.* **10**, 61–78, 228 (1962).
31. Chalov, N. V. and Lappo-Danilevskii, V. K., *Izv. Vysshikh Uchebn. Zavadenii, Lesn. Zh.* **6** (1), 156–60 (1963).
32. Chalov, N. V. and Leshchuk, A. E., *Izv. Vysshikh Uchebn. Zavadenii, Lesn. Zh.* **5** (1), 155–62 (1962).
33. Chalov, N. V. and Leshchuk, A. E., *Izv. Vysshikh Uchebn. Zavadenii, Lesn. Zh.* **5** (3), 141–8 (1962).
34. Chalov, N. V. and Leshchuk, A. E., *Izv. Vysshikh Uchebn. Zavadenii, Lesn. Zh.* **9** (6), 139–43 (1966).
35. Chalov, N. V., Leshchuk, A. E. and Aleksandrova, O. A., *Zh. Priklad. Khim.* **33**, 2743–50 (1960).
36. Chalov, N. V., Leshchuk, A. E., Korotkov, N. V., Goryachikh, E. F., Aman, A.

K., Paasikivi, L. B. and Aleksandrova, O. A., *Zh. Priklad. Khim.* **34** (12), 2737–50 (1960).

37. Chalov, N. V., Leshchuk, A. E., Kozlova, L. V. and Volkova, T. M., *Gidrolizn i Lesokhim. Prom.* **19** (2), 2–6 (1966).
38. Chalov, N. V. and Paasikivi, L. B., *Izv. Vysshikh Uchebn Zavadenii, Lesn. Zh.* **7** (2), 137–43 (1964).
39. Chalov, N. V. and Shabalin, A. S., *Khim. Pererab. Drev.* **21**, 9–11 (1966).
40. Chudakov, M. I., Okun, M. G., Samsonova, A. P., Raskin, M. N. and Pilipchuk, Y. S., *Sb. Tr. Vses. Nauchn.-Issled. Inst. Gidrolizn. i Sulfitno Spirt Prom.* **14**, 238 (1965).
41. Clark, J. T., Hicks, J. R. and Harris, E. E., *Tappi* **34** (1), 6 (1951).
42. Classen, A., German Patents 118,540, 118,542 118,544 (1899), 130,980 (1901).
43. Cocking, A. T. and Lilly, C. H., US Patent 1,425,838 (1920); British Patent 164,034 (1919).
44. Connstein, W. and Ludecke, K., *Chem. Ber.* **52**, 1386 (1919).
45. Conrad, C. C. and Nelson, M. L., *Textile Res. J.* **18**, 165 (1948).
46. Correns, E. and Edelmann, K., *Faserforschg. Textiltechn.* **7**, 533 (1956).
47. Dauziville, E. S., German Patent 11,836 (1880).
48. David, B. D., *TIIC Rept.* May 1945.
49. Debro, A., *Holzhydrolyse AG. Durchschr.* Ser. II/3 1928.
50. Degussa, A. G., German Patent 892,590 (1941).
51. Desforges, J., *Chim. et Ind.* **67**, 753 (1952).
52. Efros, I. N., *Kompleksn. Ispolz. Drev. Petrozavodsk, Sb.* **1964**, 277–82.
53. Ekenstam, A., af *Svensk Kem. Tidskr.* **46**, 157 (1934); *Chem. Ber.* **69**, 549, 553 (1936).
54. Ellefsen, O., Gjonnes, J. and Norman, N., *Norsk. Skogind.* **13** (11), 411 (1959).
55. Emelyanov, I. Z., Lebedev, N. V. and Vakhrusheva, K. P., *Sb. Tr. Gos. Nauchn.-Issled Inst. Gidrolizn. i Sulfitno Spirt. Prom.* **11**, 66–72 (1963).
56. Erdman, E., *Chem. Ber.* **43**, 239 (1910).
57. Fierz-David, H. E. and Hennig, M., *Helv. Chim. Acta* **8**, 900 (1925).
58. Fink, H. and Lechner, R., *Biochem. Z.* **286**, 83 (1936); Fink, H. and Just, F., *Biochem. Z.* **300**, 84 (1938); **303**, 234 (1939); Fink, H. and Hock, A. Z., *Naturforschg.* **2b**, 187 (1939).
59. Fischer, E., *Chem. Ber.* **23**, 3687 (1890).
60. Frahm, H., *Chem. Ber.* **74**, 622 (1941).
61. Freudenberg, K. and Kuhn, W., *Chem. Ber.* **65**, 484 (1932); Freudenberg, K., Kuhn, W., Dürr, W., Bolz, F. and Steinbrunn, G., *Chem. Ber.* **63**, 1610 (1930); Freudenberg, K. and Kuhn, W., *Chem. Ber.* **63**, 1517 1528 (1930).
62. Freudenberg, K. and Blomqvist, G., *Sitzungsber. Heidelberg Akad. Wiss.* No. 7 (1936).
63. Freudenberg, K., Lautsch, W., Piazola, G. and Scheffer, A., *Chem. Ber.* **74B**, 171 (1941).
64. Freudenberg, K. and Plötz, Th., *Chem. Ber.* **73**, 754 (1940).
65. Goldschmidt, Th. A. G. and Hägglund, E., German Patent 391,969 (1917).
66. Gorokhov, G. I., *Sb. Tr. Vses. Nauchn.-Issled Inst. Gidrolizn. i Sulfitno Spirt. Prom.* **13**, 39–46 (1965).
67. Grinshtein, I. M., *Sb. Tr. Vses. Nauchn.-Issled. Inst. Gidrolizn. i Sulfitno Sprit. Prom.* **13**, 175–83 (1965).
68. Grinshtein, I. M. and Tyshetskaya, O. V., *Sb. Tr. Gos. Nauchn.-Issled Gidrolizn. i Sulfitno Spirt. Prom.* **11**, 23–30 (1963).

69. Guignet, L., *C. R. Acad. Sci.* **108**, 1258 (1889).
70. Gustavsson, Ch., Sundmann, J. and Saarinio, J., *Finn. Paper Timber J.* **31**, 467 (1949); **32**, 306 (1950); **33**, 115, 300 (1951).
71. Gilbert, N., Hobbs, I. A. and Levine, J. D., *Ind. Eng. Chem.* **44**, 1712 (1952).
72. Goldovskaya, I. F., Kulnevich, V. G. and Rubavtsova, M. N., *Gidrolizn. i Lesohim. Prom.* **19** (4), 15-17 (1966).
73. Gorfinkel, F. Z. and Kuzmina, O. N., *Izv. Tomskogo Politekhn. Inst.* **136**, 41-3 (1965).
74. Goryaev, M. I., Mirfaizov, K. M. Saraikina, M. G., Pugachev, M. G., Kuril'skaya, V. V. and Lysenkov, L. T., *Gidrolizn. Lesokhim. Prom.* **19** (7), 11-12 (1966).
75. Glazkova, E. L., *Sb. Tr. Gos. Nauchn.-Issled. Inst. Gidrolizn. i Sulfitno Spirt Prom.* **12**, 165-71 (1964).
76. Hachihama, Y., *J. Soc. Chem. Ind. Japan* **43**, 280 (1940); **44**, 773 (1943); **47**, 215 (1946).
77. Hägglund, E., *J. Prakt. Chem.* **91**, 358 (1915).
78. Hägglund, E., "Die Hydrolyse der Cellulose und des Holzes." 1915.
79. Hägglund, E., *Svensk Kemisk Tidskr.* **35**, 2 (1923).
80. Hägglund, E., *Cellulosechemie* **4**, 84 (1923).
81. Hägglund, E., *Papierfabrikant* **25**, 52 (1927).
82. Hägglund, E., "Chemistry of wood." Academic Press, New York, 1951.
83. Hägglund, E. Koch, F. and Lofman, N., German Patent 382,463 (1920).
84. Hajny, G. J., *Forest Prod. J.* **9** (5), 153 (1959).
85. Hajny, G. J., Hendershot, W. H. and Peterson, W. H., *Appl. Microbiology* **8** (1), 5 (1960).
86. Hajny, G. J. Ritter, G. J. and Gardner, C. H., *Ind. Eng. Chem.* **43**, 1384 (1951).
87. Hall, J. A., *FAO Rept.* 52/2/767 (1954).
88. Harris, E. E., Wood Hydrolysis *in* Wise and Jahn, "Wood chemistry" Vol. II. Reinhold, New York, 1952.
89. Harris, E. E., *Forest Prod. Lab. Rept.* R 1446 March 1944; R 1475, March 1945; R 1704 Dez. 1947.
90. Harris, E. E., Beglinger, E., Hajny G. J. and Sherrard, E. C., *Ind. Eng. Chem.* **37**, 12 (1945); Harris, E. E. and Beglinger, E., *Ind. Eng. Chem.* **38**, 890 (1946).
91. Harris, E. E., Hajny, G. J., Haman, M. L. and Rogers, S. C., *Ind. Eng. Chem.* **38**, 896 (1946).
92. Harris, E. E. and Kline, A. J., *J. Phys. Colloid. Chem.* **53**, 344 (1949).
93. Harris, E. E. and Lang, B. G., *J. Phys. Colloid Chem.* **51**, 1430 (1947).
94. Harris, J. F. and Hajny, G. J., *J. Biochem. Microbiol. Technol. Eng.* **2** (1), 9 (1960).
95. Harris, J. F., Saeman, J. F. and Locke, E. G. *in* Browning, "The chemistry of wood." Appleton, New York, 1963.
96. Harris, J. F., Saeman, J. F. and Sherrard, E. C., *Ind. Eng. Chem.* **32**, 440 (1940); US Patent 2,220,624 (1939).
97. Harris, J. F., Saeman, J. F. and Zoch, L. L., *Forest Prod. J.* **10** (2), 125 (1960).
98. Harris, J. F. and Smuk, J. M., *Forest Prod. J.* **11** (7), 303 (1961).
99. Hermans, P. H., "Contributions to the physics of cellulose fibers." Elsevier, New York, 1949.
100. Hermans, P. H. and Weidinger, A. W., *J. Polym. Sci.* **4**, 317 (1949).
101. Hiller, L. A. and Pacsu, E., *Textile Res. J.* **16**, 318 (1946); **17**, 405 (1947).
102. Holderby, J. M., *Fiat Rept.* No. 619, May 1946.
103. Hönig, M. and Schubert, A., *Monatsh. Chem.* **6**, 709 (1885); **7**, 455 (1886).

104. Hosaka, H. and Suzuki, H., "FAO technical panel on wood chemistry." Tokyo, 1960.
105. Hunter, M. J., Wright, G. F. and Hibbert, H., *Chem. Ber.* **71**, 734 (1938).
106. I. G. Farbenindustrie AG, German Patent 541,362 (1927).
107. I. G. Farbenindustrie AG, German Patents 697,835, 714,391 (1937).
108. Ivanova, M. A., Morosov, E. F. and Kholkin, V. I., *Khim. Pererabotka Drev. Sb.* **1**, 6-9 (1964).
109. Jerkeman, P., *Svensk Papperstidn.* **70** (18), 587 (1967).
110. Kalina, V., Vedernikov N. A. and Veips, A., *Latvijas PSR Zinatnu Akad. Vestis Khim. Ser.* **3**, 313-22 (1964).
111. Kasuma, J. and Ishii, T., *Kogyo Kagaku Zasshi* **69** (3), 469 (1966).
112. Katagiri, H., *Biochem. J.* **20**, 427 (1944).
113. Katzen, R., Aries, R. S. and Othmer, D. F., *Ind. Eng. Chem.* **37**, 442 (1945).
114. Klason, P., "Berichte Hauptversammlung Zellcheming." Elsner, Berlin, 1908.
115. Kobayashi, T., "FAO technical panel on wood chemistry." Tokyo, 1960.
116. Kobayashi, T. in Oshima, M., "Wood chemistry process engineering aspects." Noyes Development Corp., New York, 1965.
117. Konovalov, V. K. and Sharkov, V. I., *Izv. Vysshikh Uchebn. Zavadenii, Lesn. Zh.* **9** (2), 142-50 (1966).
118. Konovalov, V. K. and Sharkov, V. I., *Izv. Vysshikh Uchebn. Zavadenii, Lesn. Zh.* **9** (3), 145-50 (1966).
119. Korolkov, L. I., *Khim. Pererabotka Drev. Nauchn.-Tekhn. Sb.* **11**, 31-5 (1962).
120. Korolkov, I. I. and Likhonos, E. F., *Sb. Tr. Vses. Nauchn.-Issled. Inst. Gidrolizn. i Sulfitno Spirt Prom.* **13**, 53-62 (1965).
121. Korotkov, N. V. and Chalov, N. V., *Sb. Tr. Vses. Nauchn.-Issled. Inst. Gidrolizn. i Sulfitno Spirt Prom.* **14**, 180-91 (1965).
122. Korotkov. N. V., Glazkova, E. L., Chaloc, N. V. and Konovalova, L. Y., *Sb. Tr. Vses. Nauchn.-Issled Inst. Gidrolizn i Sulfitno Spirt Prom.* **13**, 203 (1965).
123. Kozlov, A. I. Krestan, E. Sh., Kutrueva, I. A. and Parmenova, I. V., *Sb. Tr. Vses. Nauchn.-Issled Inst. Gidrolizn i Sulfitno Sprit Prom.* **14**, 42-7 (1965).
124. Kressmann, F. W., *Ind. Eng. Chem.* **6**, 625 (1914); **7**, 920 (1915).
125. Kressmann, F. W., *US Dept. Agric. Bull.* No. 983 (1922).
126. Kudlacek L. and Ruzicka, J., *Vysokomolekul. Soedi* **6** (4), 587-93 (1964).
127. Kühl, K. E. and Bohm, R., *Holzforschung* **16** (2), 47 (1962).
128. Kulnevich, V. G., Kardailova, K. M. and Abranyants, S. V., *Gidrolizn i Lesokhim. Prom.* **18** (4), 7-9 (1965).
129. Kuidubov, J. F. and Rubina, S. J., *Gidrolizn Lesokhim. Prom.* **12** (5), 6 (1959).
130. Kusama, J., US Patent 3,067,065 (1962).
131. Lebedev, N. V. and Bannikova, A. A., *Sb. Tr. Gos. Nauchn.-Issled. Inst. Gidrolizn. i Sulfitno Spirt Prom.* **9**, 7-19 (1961).
132. Lebedev, N. V., Bannikova, A. A. and Paasikivi, L. B., *Sb. Tr. Gos. Nauchn.-Issled. Inst. Gidrolizn. i Sulfitno Spirt Prom.* **9**, 20-35 (1961).
133. Lehmann, F., *Holzhydrolyse AG. Druckschrift* Ser. II/4 (1928).
134. Leonard, R. R. and Hajny, G. J., *Ind. Eng. Chem.* **37**, 390 (1945).
134a. Leshchuk, A. E. and Chaloc, N. V., *Sb. Tr. Gos. Nauch-Issled, Ist. Gidroliz. Sulfitno-Spirt. Prom.* **15**, 156 (1966).
135. Leshchuk, A. E., Nagalyuk, E. A. and Chalov, N. V., *Gridolizn. Lesokhim. Prom.* **19** (7), 5-8 (1966).
136. Liang, H., *Z. Physiol. Chem.* **244**, 238 (1936).
137. Lindblad, A. R., Swedish Patent 70,795 (1930).

138. Locke, E. G. and Garnum E., *Forest Prod. J.* **11** (8), 380 (1961).
139. Luers, H. Z., *Angew. Chem.* **43**, 455 (1930); **45**, 369 (1932); *Holz Roh- u. Werkstoff* **1**, 35, 342 (1937).
140. Luers, H., Fries, G., Huttinger, W. Moncke, E. and Endres, C., *Z. Spiritusind.* **60**, 1 (1937).
141. Lautsch, W., *Cellulosechem* **19**, 69 (1941); Lautsch, W. and Piazolo, G., *Chem. Ber.* **76**, 486 (1943); Freudenberg, K. and Piazolo, G., US Patent 2,390,063 (1945).
142. Malyshev, D. A., *Gidrolizn. i Lesokhim. Prom.* **18** (5), 24-6 (1965).
143. Mamatov, Y., Vasyuina, N. A. and Balandin, A. A., *Gidrolizn. i Lesokhim. Prom.* **19** (2), 10-11 (1966).
144. Mark, H., "Physik und Chemie der Cellulose." Springer, Berlin, 1932.
145. Marten, E. A., Sherrard, E. C. Peterson, W. H. and Fred, E. B., *Ind. Eng. Chem.* **19**, 1162 (1927).
146. Marty, C., Swiss Patent 378,899 (1964).
147. Melsens, G. F., *Dinglers Polytechn. J.* **138**, 426 (1856).
148. Miller, R. N. and Swanson, W. H., *Ind. Eng. Chem.* **17**, 843 (1925).
149. Minina, V. S., Sushkevich, I. I. and Agamalova, V. G., *Khim. i Fiz.-Khim. Prirodn. Sintetich Polimerov. Akad. Nauch. Uz. SSR. Inst. Khim. Polimerov* **2**, 104 (1964).
150. Müller, O. and Dobberstein, H., *Makromol. Chem.* **8**, 156 (1952).
151. Natta, G., US Patent 2,689,250 (1954).
152. Nemanov, F., *Gidrolizn. i Lesokhim. Prom.* **19** (4), 23 (1966).
153. Neuberg, C., *Biochem. Z.* **78**, 264 (1916).
154. Newkirk, W. B., *Ind. Eng. Chem.* **28**, 760 (1936).
154a. Nikitin, V. M., *Izv. Vyssh. Ucheb. Zaved., Les. Zh.* **10** (3), 148 (1967).
155. Noguchi Research Foundation, US Patent 3,223,698 (1965).
156. Odincovs, P. and Murashchenko, N. F., *Khin. Pererabotka i Zashchita Drev. Akad. Nauch Latv. SSR. Inst. Khim. Drevesiny* 1964, 53-60.
157. Odincovs, P. and Murashchenko, N. F., *Latvijas PSR Zinatnu Akad. Vestis, Khim. Ser.* **1964** (3), 307-11.
158. Opolnova, G. V., Arkhipov, M. I. and Zotki, I. V., *Gidrolizn. i Lesokhim. Prom.* **19** (1), 13-15 (1966).
159. Orlov, V. I. and Sharkov, V. I., *Izv. Vysshikh Uchebn Zvadenii, Lesn. Zh.* **8** (2), 127-33 (1965).
160. Oshima, M., "Wood chemistry process engineering aspects." Noyes Development Corp., New York, 1965.
161. Oshima, M., Japan Patent 8670 (1956); US Patent 2,900,284 (1959).
162. Ost, H., *Chem. Ber.* **46**, 2995 (1913).
163. Osusky, A., *Drevar. Vyskum* **1965**, 4, 211-21.
164. Paloheimo, L., *Cellulosechem.* **9**, 35 (1928).
165. Partansky, A. B. and Heury, B. S., *J. Bacteriol.* **30**, 559 (1935).
166. Pavcek, P. L., *TIIC Rept.* May **1945**; June (**1945**).
167. Payen, A., *Dinglers Polytechn. J.* **185**, 308 (1867).
168. Perlman, D., *Ind. Eng. Chem.* **36**, 308 (1944).
169. Peskovoi, B. I., *Gidrolizn. i Lesokhim. Prom.* **19** (2), 6-10 (1966).
170. Petkevich, A. A., Ochneva, N. V., Korotkov, N. V., Revzina, E. D., Chalov, N. V., Leshchuk, A. E. and Goryachikh, E. F., *Sb. Tr. Gos. Nauchn.-Issled Inst. Gidrolizn i Sulfitno Spirt Prom.* **8**, 47-65 (1960).
171. Peukert, M. E., *Cellulosechemie* **21**, 32 (1943); German Patent 744,677 (1943).
172. Phillips, M. and Goss, M. J., *J. Ass. Offic Agr. Chem.* **19**, 341, 350 (1936).
173. Ploetz, T., Richtzenhain, H., Deiters, W. and Giesen, G., German Patent 1,142,853 (1963).
174. Prosinki, S., Babicki, R. and Adamski, Z., *Zellstoff-Papier* **12**, 301 (1963).

175. Quaker Oats Corp., British Patent 203,691 (1923).
176. Reiferscheidt, E., *Z. Angew. Chem.* **18**, 44 (1905).
177. Riehm, Th., Ph. D. Thesis, Darmstadt 1950.
178. Riehm, Th., "FAO technical panel on wood chemistry." Tokyo, 1960.
179. Rockstroh, H., *FAO Rept.* 54/2/767 (1954).
180. Romanovskii, A. I. and Mazur, A. I., *Khim. Pererabotka Drevesiny, Ref. Inform.* No. **5**, 5-9 (1966).
181. Root, D. F., Ph. D. Thesis; see Harris *et al.* (98).
182. Root, D. F., Saeman, J. F. and Harris, J. H., *Forest Prod. J.* **9** (5), 158 (1959).
183. Saeman, J. F., *Ind. Eng. Chem.* **37**, 43 (1945); *Holzforschung* **4**, 1 (1949).
184. Saeman, J. F. Locke, E. G. and Dickerman, G. K., *FIAT Rept.* 499 (1945).
185. Sakai, Y., *Bull. Chem. Soc. Japan* **38** (6), 863 (1965).
186. Schlubach, H., Alsner, H. and Prochownik, W., *Z. Angew. Chem.* **45**, 245 (1932).
187. Schlubach, H. and Lührs, A., *Ann. Chem.* **547**, 73 (1941).
188. Schmidt, E., *Angew. Chem.* **59A**, 16 (1947).
189. Scholler, H., French Patent 706,678; *Z. Spiritusind.* **55**, 94 (1932); *Zellstofffaser* **32**, 64 (1933); *Chem. Z.* **60**, 293 (1936); **63**, 737, 752 (1939); German Patents 676,967 (1939); 704,109 (1941); US Patents 1,641,771 (1927); 1,890,304 (1932); 1,990,097 2,083,347 (1935); 2,083,348, 2,088,977 (1937); 2,108,567, 2,123,211, 2,123,212 (1938); 2,188,192, 2,188,193 (1940).
190. Scholler, H., Belgian Patent 665,689 (1965).
191. Schoenemann, K., "FAO technical panel on wood chemistry." Stockholm, 1953; *FAO Rept.* 54/2/767 (1954).
192. Schoenemannn, K. and Hofmann, H., *Chem.-Ing.-Techn.* **29**, 665 (1957).
193. Schulz, G. V. and Lohmann, H. J., *J. Prakt. Chem.* **157**, 238 (1941); *Z. Phys. Chem.* **B30**, 379 (1935); **B32**, 27 (1936); **47**, 155 (1940).
194. Schwalbe, C. G. and Schulz, W., *Z. Angew. Chem.* **26**, 499 (1913); **39**, 606 (1926).
195. Sharkov, V. I., *Chem.-Ing.-Techn.* **35** (7), 494 (1963).
196. Sharkov, V. I. and Levanova, V. P., *Gidrolizn. i Lesokhim. Prom.* **13** (1), 5 (1960).
197. Sherrard, E. C. and Blanco, G. W., *Ind. Eng. Chem.* **15**, 611 (1923); Sherrard, E. C. and Gauger, W. H., *Ind. Eng. Chem.* **15**, 1164 (1923).
198. Shilling, W. L., *Tappi* **48** (10), 105A (1965).
199. Simonsen, E. Z., *Angew. Chem.* **11**, 195, 219, 962, 1007 (1898).
200. Sjolander, N. O., Langelykke, A. F. and Peterson, W. H., *Ind. Eng. Chem.* **30**, 1251 (1938)..
201. Skoogh, C. G., Swiss Patent 405,352 (1966).
202. Skoogh, F. K., *PB Rept.* 2041 (1945).
203. Smuk, J. M., Harris, J. F. and Zoch, L. L., *US Forest Service Res. Paper FPL* Jan. 1965.
204. Sobolew, J. and Schuerch, C., *Tappi* **41** (10), 545 (1958).
205. Suida, H. and Prey, V., *Chem. Ber.* **75B**, 1580 (1942).
206. Stamm, A. J. and Cohen, W. E., *J. Phys. Chem.* **42**, 921 (1938).
207. Stamm, A. J. and Harris, E. E., "Chemical processing of wood." Chem. Publ., New York, 1953.
208. Staudinger, H. and Sorkin, M., *Ber. Chem.* **70**, 1565 (1937).
209. Sukhanovskii, S. I., Akhmina, E. I. and Milovanov, A. V., *Gidrolizn. i Lesokhim. Prom.* **16** (5), 24-6 (1963).
210. Sukhanovskii, S. I., Akhmina, E. I., Lisina, Z. I. and Oparina, L. V., *Gidrolizn. i Lesokhim. Prom.* **19** (8), 7-9 (1966).
211. Sukhanovskii, S. I. and Chadukov, M. J., *Shurn, Priklad. Chim.* **29**, 410 (1956); *Trudy Leningrad Lesotekh. Akad. S. M. Kirov*, No. 75.
212. Tai, S., Nakano, J. and Migita, N., *Nippon Mozugai Gakkaishi* **12** (2), 108 (1966).

213. Thomas, R. W. and Schuette, H. A., *J. Am. Chem. Soc.* **53,** 2334 (1931).
214. Tokarev, B. I. and Sharkov, W. I., *Tr. Leningrad Lesotekh. Akad.* **102,** 153–65 (1963).
215. Tsirlin, Y. A. and Kozlova, E. A., *Sb. Tr. Vses. Nauchn.-Issled Inst. Gidrolizn. Sulfitno Spirt Prom.* **13,** 130–39 (1965).
216. Tsirlin, Y. A. and Vasileva, V. A., *Sb. Tr. Vses. Nauchn.-Issled Inst. Gidrolizn. i Sulfitno Spirt Prom.* **13,** 122–29 (1965).
217. Tyshetskaya, O. V., Grinshtein, I. M., and Lebedev, N. V., *Sb. Tr. Vses. Nauchn.-Issled. Inst. Gidrolizn. i Sulfitno Spirt Prom.* **14,** 216–24 (1965).
218. Ungar, E., Ph. D. Thesis, Zürich 1914.
219. Urban, H., *Cellulosechemie* **7,** 73 (1926).
220. Vasyunina, N. A., Pogosov, Y. L., Balandin, A. A., Chepigo, S. W. and Barysheva, G. S., USSR Patent 168,666 (1965).
221. Vasyunina, N. A., Baladin, A. A. and Mamatov, Y., *Kinetika i Kataliz.* **4** (1), 156–62 (1963).
222. Vasyunina, N. A. Baladin, A. A. and Mamatov, Y., *Kinetika i Kataliz.* **4** (3), 443–49 (1963).
223. Vekshegonov, F. Y. and Kozlov, V. N., *Tr. Uralsk. Lesotekhn. Inst.* **18,** 84–91 (1962).
224. Vekshegonov, F. Y. and Kozlov, V. N., *Izv. Vysshikh. Uchebn. Zavadenii Lesn. Zh.* **8** (3), 134–40 (1965).
225. Vereinigte Chemische Werke, German Patents 298,593 (1915); 298,594, 298,595 (1916).
226. Vernet, G., *Chimie Industrie Spec.* No. 654, May 1923.
227. Vogel, H., *Dinglers Polytechn. J.* **1,** 235 (1820).
228. Vyrodova, L. P. and Sharkov, V. I., *Sb. Tr. Goz. Nauchn.-Issled Inst. Gidrolizn i Sulfitno Spirt Prom.* **12,** 40–8 (1964).
229. Vyrodova, L. P. and Sharkov, V. I., *Zh. Prikled. Khim.* **39** (3), 682–89 (1960).
230. Vyrodova, L. P. and Sharkov, V. I., *Sb. Tr. Vses. Nauchn.-Issled. Inst. Gidrolizn. i Sulfitno Spirt Prom.* **14,** 152–60 (1965).
231. Wenzl, H., *Papierfabrikant* **22,** 101 (1924); *Cellulosechemie* **7,** 88 (1926).
232. Wiley, A., Johnson, M. McCoy, E. and Peterson, W. H., *Ind. Eng. Chem.* **33,** 606 (1941).
233. Willstätter, R., German Patent 273,800 (1913).
234. Willstätter, R. and Zechmeister, L., *Chem. Ber.* **46,** 2401 (1913).
235. Wise, L. E., Murphy, M. and d'Addieco, A. A., *Paper Trade J.* **122** (2), 35 (1946).
236. Wohl, A., *Chem. Ber.* **23,** 2056 (1890).
237. Wohl, A. and Krull, H., *Cellulosechemie* **2,** 1 (1921).
238. Wolfrom, M. L. and Gorges, L. W., *J. Am. Chem. Soc.* **59,** 282 (1937); Wolfrom, M. L. and Snowden, J. C., *J. Am. Chem. Soc.* **60,** 3009, 1026 (1938).
239. Zechmeister, L., Ph. D. Thesis, Zurich 1913.
240. Zellstofffabrik Waldhof AG. German Patents 759,121 (1940); 905,965 (1940).
241. Zetterlund, C. G., *Wagners Jahr. Berichte* **18,** 597 (1872).

V

Further Destructive Processing of Wood

INTRODUCTION

Wood is a raw material for the chemical industry and, at the same time, an industrial material for the building of houses and for the manufacture of furniture and utensils. It was used as a fuel long before the discovery of coal, petroleum, or natural gas—which are formed chiefly by the biological decomposition of wood. So far as this book is concerned, the utilization of wood in mechanical processing industries will not be discussed.

Two completely different processes are available for using wood as a starting material in the chemical industry. These involve (a) reactions in which the structure of the fibers is preserved, and (b) reactions in which the structure is completely destroyed. In group (a) belong the processes for the production of fibers, such as the isolation of cellulose, the production of pulp, of groundwood, and of wood fibers for cardboard, fiberboard, and particleboards. These various materials differ chemically, depending upon the degree of their delignification, or mechanically, depending upon the degree of their disintegration, ranging from chips to fiber bundles or even individual fibers. In general, the mechanical processes are usually combined, to some extent, with chemical changes in the components of the wood. When purely chemical processes are involved, they will be discussed in later chapters.

In the utilization of wood by the second (b) type of reactions, the fiber structure, which characterizes the wood, is destroyed. This includes the earliest method of utilization, i.e., as fuel, in which process only the calorific value of the wood is used. On the same level as the combustion of wood

is its biological degradation in the forests, with humus being formed from the decayed substance. This rotting is, at the same time, the first step in the carbonization process.

As was shown in Chapter IV, cellulose, the main component of wood, is composed of glucose units and is converted by acids under certain reaction conditions into these units. At the same time, hexosans and pentosans are also split into low-molecular sugars. Only the lignin remains as an insoluble product. The hydrolysis of wood (as described in Chapter IV) is one process in which the destruction of its structure is involved in its utilization.

Considerably older than the hydrolysis is the pyrolysis of wood. Here, the natural process of carbonization is effected by artificial methods. The pyrolysis of wood, also known as the carbonization or dry distillation of wood, is, therefore, another destructive method of its chemical utilization in which all its components are extensively changed. Other processes that fall into this category are the pressure oxidation and the hydrogenolysis of wood. The hydrogenolysis of wood sugar lignin has been described in Chapter IV. All these methods have in common the destruction of the wood structure and the goal, in each case, is to obtain chemically interesting and economically usable products from the various wood components.

Such destructive processes, however, should be applied only to wood that cannot be used for the production of fibrous material. This restriction was applicable until a few years ago for all wood wastes from forests and from wood processing industries. Since then, however, processes for the production of fibrous materials have been improved to such an extent that even small bits of what was formerly considered to be waste wood can now be utilized. So far as forest waste woods are concerned, this applies only within certain limits, and these provide an immeasurable reservoir of wood for the chemical destructive wood utilization processes. Difficulties arise, not so much in regard to the processing itself, as in the collection of the vast amounts of waste wood, the transport of which to the wood processing plants requires greater expenditures than are returned from the finished product. It is not surprising, therefore, that countries that are especially rich in wood supplies have studied these methods of utilization extensively, while in other countries such studies have been carried out only in times of economic necessity.

1. THE PYROLYSIS OF WOOD

1.1 Effect of Time and Temperature

When wood is heated in a closed vessel with exclusion of air, it undergoes carbonization. Water vapors, organic liquids, and gases are released,

and tar, pitch, and, chiefly, charcoal remain as residues. The production of charcoal is the oldest destructive chemical process for the utilization of wood. Charcoal has recently gained increasing importance because of its special properties. The effect of the temperature on the carbonization of wood and its various components has stimulated much research, especially with regard to the reactions that take place at higher temperatures. On prolonged heating of wood, even at 100° to 105°C, significant structural changes occur, with the hygroscopically bound water being evaporated first. On still more prolonged heating, the constitutionally bound water is released. Water can be split off also from the hydroxyl groups of neighboring polysaccharide chains, with the formation of new hydrogen bridges. That chemical changes occur at these relatively low temperatures is shown by the fact that wood, treated in this way, can be pulped only with difficulty. The hygroscopicity is reduced—a fact which is utilized in wood processing industries. As is explained in Chapter IV, the pentosans are more readily degraded in an acid hydrolysis than are the cellulose and the hexosans. If the heating of the wood is continued for an extended period, the hydrolytic action of the released water and the formation of acetic acid, formic acid, and methyl alcohol can be observed. Whether the water is allowed to escape or remains in contact with the wood is an important factor in such a treatment. In the latter case, the treatment of the wood with water vapor, perhaps under pressure, can cause a hydrolytic reaction and the products formed can, in turn, participate in oxidative reactions or undergo condensation.

When the temperature of the thermal treatment is raised to 160°C for 28 days, beechwood chips lose about 10% of their weight and the holocellulose content is reduced by 48.6%. This loss in weight is due, primarily, to the release of hygroscopically bound water, while the loss in holocellulose is due to the degradation of high-molecular components to soluble low-molecular products. This loss must involve the degradation not only of the hemicelluloses but also of a part of the cellulose.

Of special interest in the carbonization is the behavior of the lignin. The wood hydrolysis shows that lignin is largely stable toward acids. Beechwood sawdust, heated for 14 days at 160°C, gave only 2.3% Klason lignin as compared to 23.8% in the original wood. If this result seems questionable, the fact still remains that this thermal treatment must have exerted significant changes in the lignin molecule. This is supported by the ultraviolet absorption spectrum, which shows the disappearance of the characteristic maximum shown by native lignin. It is obvious that such a lengthy treatment of wood at 160°C will cause the formation of low-molecular lignin decomposition products which are no longer precipitated with sulfuric acid (80).

It has long been known that the rate of decomposition of wood increases rapidly with increasing temperature. It can be said that this rate

approximately doubles for each 10° to 14°C of elevation (129). It is impossible to set a lower limit for the beginning of the decomposition of wood by thermal action because the duration of the treatment at 100° to 150°C, and even at below 100°C, affects the nature of the chemical changes. An analysis of the reactions by determining the losses in weight leads to false inferences because those that do not involve any loss in weight are not taken into account. Volatile decomposition products, such as water, acetic acid, and methanol, and noncondensable gases, such as carbon monoxide and carbon dioxide, are released at temperatures of 200°C and over. At the same time, the separation of tar occurs.

At 270°C, the decomposition occurs vigorously and, for the most part, exothermically (64,66). At what point the reaction becomes exothermic has been the subject of much discussion. Only in recent times has the differential thermal analysis (DTA) made possible a more accurate determination. In this test, the wood sample and, simultaneously, an inert substance which itself does not undergo any thermal reaction, are heated to uniformly increased temperatures and the time function of the difference in the temperatures occurring between the wood sample and the control substance is determined. These investigations show that the exothermic reactions take place in a rhythmical manner, thus indicating that the DTA curve has a series of distinct maxima (29,59,117,121). In this way it was found that the DTA curves for both sprucewood and beechwood show four distinct maxima, with the start of the exothermic reactions being at 208°C. The first maximum occurs at 240°C, the second maximum for sprucewood is at 343°C, for beechwood at 353°C, and the fourth for sprucewood at 412°C and for beechwood at 406°C (117).

Whether these maxima must be ascribed to the decomposition of certain wood components cannot be definitely decided because, in the isolation of such components, chemical changes are possible. Undoubtedly the hemicellulose portion is especially reactive, while the lignin is much more resistant. In a nitrogen atmosphere, wood and lignin are pyrolyzed in an exothermic reaction that begins at about 220°C, α-cellulose at about 275°C. As soon as the decomposition process has started, the cellulose is converted into volatile products, by a reaction that is chiefly endothermic, before a temperature of 400°C is reached. Lignin is much more slowly decomposed and loses only about half its weight in a primarily exothermic reaction when the pyrolysis is substantially terminated at 800°C. The presence of salts can reduce the threshold of the temperature for the pyrolysis of wood and facilitate the decomposition of the wood components at temperatures of below 250°C. Salts also decrease the activation energy for the pyrolysis (30).

Differential thermal analysis determinations on birchwood chips (par-

ticle size, 0.25 to 0.5 mm) in an inert gas atmosphere and at temperature gradients of 11.4°, 26.6°, and 53.3°C per minute until a final temperature of 500°C was reached showed that, at the low temperature gradients, two endothermic maxima, at 100° and 300°C, and four exothermic maxima, at 70°, 200°, 240°, and 320°C, occurred. The first endothermic maximum was due to the removal of hygroscopic water, while the second, at 300°C, was found to result from causes other than the loss of heat caused by the formation of volatile products because the changes in weight were insignificant at this temperature; one possible explanation may be intramolecular rearrangements within the residue, accompanied by the splitting off of aliphatic groups from the aromatic lignin molecule. No satisfactory explanation can be offered for the exothermic maxima; it can only be assumed that both types of thermal reactions are determined by the ratio of the energy of bond dissociation to the heat of recombination of free radicals. With increasing rate of heating, the number of peaks remained constant but their position was shifted toward the higher temperature regions. Changes in the mechanism of the decomposition of the wood become probable, as is shown by the character of the wood residue and by the increase in the overall activation energy with increasing temperature gradients (107). This activation energy for coniferous woods is 25 kcal/mole for the α-cellulose, 30 kcal/mole for the hemicelluloses, and 23.4 kcal/mole for the lignin, and 30.5 kcal/mole for cotton (129).

The effect of the oxygen in the air upon the decomposition reaction will be discussed later. The exothermic decomposition of the hemicelluloses in an inert atmosphere (nitrogen) begins at about 200°C. It may be assumed that the wood polyoses, of which the hemicelluloses are composed, behave in different ways. The decomposition of the cellulose, on the other hand, begins with an endothermic reaction at about 290°C and reaches its maximum at 315°C. In an exothermic reaction which reaches its peak at 340°C, a rapid and complete carbonization takes place. With lignin, the exothermic decomposition occurs slowly, starting at about 300°C and reaching its maximum at 425°C (117).

When the temperature is kept constant during the thermal reaction, the rate of decomposition changes with the duration of the treatment. It quickly reaches a maximum and then drops again until a complete carbonization has taken place. The composition of the volatile reaction products changes and the proportion of tarry products increases with increasing reaction time. With cellulose, a strong depolymerization can be observed, but this comes to a standstill after a degree of polymerization (DP) of about 200 has been reached. Of crucial importance in the carbonization is the type of medium in which it is carried out, especially whether the oxygen in the air is present (44,90,95).

1.2 Effect of the Surrounding Atmosphere

It is known that cellulose, on heating in air, undergoes chemical changes, including an increase in carboxyl and carbonyl groups. The removal of constitutionally bound water and the evolution of carbon dioxide are promoted. These chemical changes are combined with a depolymerization of the cellulose molecule (3). The thermal decomposition in a nitrogen atmosphere begins later, but as soon as it is in progress it occurs even more rapidly than in air (95). If the pyrolysis is carried out in a vacuum, the volatile decomposition products and the water are rapidly removed, resulting in an increased yield of tar and a decrease in charcoal (3,37).

Table V-1 shows the yield of the products obtained on the pyrolysis of birchwood under various atmospheric conditions and with increasing

TABLE V-1

DRY DISTILLATION OF BIRCHWOOD (MAXIMUM TEMPERATURE 400°C)[a] (64)

Product	High vacuum 5 hr heating	Low vacuum 5 mm Hg 5 hr	Atmospheric pressure			
			3 hr heating	8 hr heating	16 hr heating	14 days heating
Charcoal	19.38	19.54	25.51	30.85	33.18	39.44
Tar	43.66	37.18	18.0	16.94	10.1	1.8
Acids[b]	10.20	10.05	7.42	7.57	7.30	6.91
Acetic acid	7.05	7.05	6.50	6.77	6.58	6.48
Formic acid	2.40	2.30	0.71	0.61	0.55	0.33
Methyl alcohol	—	1.20	1.49	1.47	1.50	1.41
Acetone	—	0.03	0.16	0.20	0.22	0.35
Formaldehyde	1.27	1.20	1.00	0.90	—	0.80

[a] Yields in percent of weight of dry wood on ash-free basis.
[b] Acids determined as acetic acid by titration.

duration of the heating. Whereas the formation of acetic acid and methyl alcohol is practically independent of the rate of carbonization, that of formic acid shows a progressive decrease (64). If one accepts the theory of Klason that the pyrolysis occurs in two stages, in the first of which pyroligneous acids, carbon dioxide, water, coal, and, primarily, tar are formed, and, in the second of which, the tar is further pyrolyzed, partially with the formation of water, coal, carbon dioxide, and hydrocarbons, and with a very prolonged pyrolysis period even disappears completely, the values for acetic acid and methanol given in Table V-1 are difficult to explain.

When a prepyrolysis in an inert high-boiling medium, e.g., kerosene,

is carried out, the yield of formic acid seldom exceeds 3% to 4% of the total volatile acids, whereas this amount can surpass 20% when a conventional pyrolysis or dry distillation of wood is applied. In the presence of water, even higher yields are possible. From this it may be concluded that the formation of formic acid is related to hydrolytic reactions and that most probably it is caused by the degradation of furfural and hydroxymethylfurfural, and, in the dry distillation, also by oxidation by atmospheric oxygen. A prepyrolysis in a volatile medium differs fundamentally from a dry distillation because the heat-carrier (e.g., kerosene) can penetrate into the wood and can attack the inner side of the cell wall which consists mainly of hemicellulose. Water is expelled already at low temperatures and an attack on the cellulose and lignin is prevented by the timely cessation of the prepyrolysis. The prepyrolyzed wood is much more porous than fresh wood, thus the gaseous reaction products and the tar are rapidly removed in the subsequent main pyrolysis and high yields of valuable products are obtained (125).

For a rapid pyrolysis, the wood must be completely dried and preheated to 150° to 200°C. Because of technical difficulties in the gas-drying of fresh wood, the possibility of replacing the drying by a prehydrolysis in a liquid heat-carrier, such as kerosene, was investigated. The " brown wood " thus formed was used instead of fresh wood and was found to be very suitable for a rapid pyrolysis. Since it contains no hemicelluloses, it is more resistant to heat and can be preheated to 250°C without degradation. This two-stage process yields more pyrolysis products than does the conventional pyrolysis. An additional advantage in the use of brown wood is that it results in the demethylation of the wood phenols during the pyrolysis. Such phenols contain 40% less methoxyl than do the phenols from a conventional process (107a).

If the pyrolysis in its first stage is governed by hydrolytic reactions, the moisture content of the wood must play an important role. This is confirmed by the fact that the pyrolysis is less drastic when dry wood chips are used. In this case the pyrolysis products contain chiefly only slightly decomposed components, especially oxygen-containing, ether-soluble products and aliphatic, ether-insoluble compounds. Because of the low heat conductivity of the wood chips, the carbonization at a low temperature occurs only on the surface of the wood particle. The organic compounds formed therefore remain in contact with the wood for a longer time, an outer high-temperature zone is formed, and the organic components are cracked, with the formation of secondary reaction products (111). When the oxygen content of the reaction zone is increased, the yield of tar is reduced and the yield of pyroligneous products is increased. With an oxygen content of above 5%, tar is no longer formed and the yield of pyroligneous products increases linearly with the temperature (84).

2. WOOD COMPONENTS AND THEIR BEHAVIOR
IN PYROLYSIS

2.1 Cellulose

When heated to 200° to 300°C in a vacuum, cellulose forms up to 38% levoglucosan (10). Opinions as to the formation of this 1,6-anhydroglucose were, for a long time, divided. This was attributable to the fact that the yield of levoglucosan was widely dependent upon the pretreatment of the cellulose; thus, for example, hydrocellulose gave up to 40% levoglucosan (138). The assumption that the glucose chains in cellulose are first cleaved to glucose and from this, in a second stage, glucosan is formed by the splitting off of one molecule of water was therefore justified. Since cellulose and levoglucosan have the same elementary formula, $C_6H_{10}O_5$, a yield of 100% of the latter might be expected. Levoglucosan, however, is sensitive to heat and decomposes to acetic acid, acetone, phenols, and water (104). The yields of levoglucosan from the pyrolysis of celluloses after various pretreatments have provided some surprising results. Thus, cotton cellulose which had been treated for five minutes at 100°C with 1% sulfuric acid and had a hydrolyzation number of 5.21 gave 60% to 63% levoglucosan on pyrolysis in vacuum. Hydrate cellulose, regenerated from cuoxam solution and having a hydrolyzation number of 10.75, under the same conditions gave 14% to 15% levoglucosan. Finally, cotton cellulose, treated one hour at 0°C with 10% sodium hydroxide and having a hydrolyzation number of 9.55, gave 36% to 37% levoglucosan. From this it may be concluded that the changes in the structure of the cellulose, characterized by its hydrolyzability, are largely responsible for the nature of the decomposition products formed in its thermal degradation (45).

The thermal decomposition of β-D-glucose *in vacuo* gave 42.7% of a viscous distillate, many volatile products and gases, and 18.2% of a solid residue. Treatment of the distillate with an anion exchange resin gave 5.6% levoglucosan. Under the same conditions, cellulose gave 37.5%. The β-D-glucose formed in the hydrolysis of cellulose cannot therefore be considered to be an intermediate product in the formation of levoglucosan (43). When the chain length of the cellulose molecule in cotton is reduced from 1000 to 150 glucose units by a mild hydrolysis, the yield of levoglucosan in a pyrolysis *in vacuo* is not markedly reduced, but reaches an average of 59% to 63% (46). The data relating to the pyrolysis of cellulose at 200° to 400°C lead to the conclusion that its decompositon results from two competing endothermic reactions. Thus, in the first step, the cellulose is degraded to 1,4-anhydro-α-D-glucopyrosan and then, in a second step, the latter is dehydrated and forms the starting material for the exothermic pro-

cess in which carbon dioxide, carbon monoxide, water, and coal are formed
(60).

The kinetics of the reaction permit three distinct stages of pyrolysis to
be identified: in the first stage, a rapid decomposition takes place, with a
weight loss that increases with rising temperature; in the second stage, a
decomposition and volatilization occur in an essentially zero-order reaction;
and, in the third stage, a volatilization follows a first-order reaction. The
transition from the zero- to the first-order behavior occurs at a greater rate
of pyrolysis as the temperature is increased. At 288°C the initial volatilization leaves 94% of the original weight of the cellulose and the transition
from the zero- to the first-order reaction occurs at about the point where
50% of the material remains. The residual charred material amounts to
about 16% at temperatures of between 267° and 300°C and is substantially
greater at lower temperatures. The initial loss in weight is due primarily
to the decomposition of the cellulose rather than to the loss of absorbed
water. A single degree of activation energy, of 42 kcal/mole, equals that
of the decomposition and volatilization reactions in the zero-order phase
over the entire temperature range of 250° to 300°C. The reaction kinetics
of the pyrolysis of cellulose are the same, regardless of whether it is carried
out in a vacuum or in a nitrogen atmosphere (89). The number of decomposition products of the thermal decomposition of cellulose at atmospheric pressure and in the presence of oxygen is considerably larger, because
of secondary reactions, than in the pyrolysis *in vacuo* or in an inert atmosphere. Even under the most favorable reaction conditions, the theoretical
yield of levoglucosan cannot be obtained.

2.2 Hemicelluloses

The hemicelluloses, which are present in deciduous woods chiefly as
pentosans and in coniferous woods almost entirely as hexosanes, undergo
thermal decomposition very readily. It was therefore to be expected that
furan derivatives would readily be found among the decomposition products. Unlike those in the acid hydrolysis, the yields of furfural in the
pyrolysis are low. Even in a pyrolysis in a high vacuum at 300°C, only
6% furfural is formed (54). When oakwood is treated with superheated
steam at 260°C, only 4.4%, corresponding to 40% of the theoretical amount,
of furfural is obtained (92). Furfural, as an especially reactive compound,
must undergo secondary reactions under the conditions of a pyrolysis. There
is also the possibility that the pentosan molecules are split into low-
molecular fragments and, for this reason, cannot form furan rings (47). It is
also possible that furfural is partly formed by the decomposition of hexoses.
Moreover, it has been suggested that furfural can be formed in the thermal
decomposition of oxycelluloses (51).

Hydroxymethylfurfural can be formed from glucose molecules of decomposed cellulose by the splitting off of water and can form furfural by the splitting off of formaldehyde (110). The formation of furfural from hydroxymethylfurfural could be proved, but not the formation of formaldehyde (31).

Acetic acid is formed in the thermal decomposition of all three main components of wood. When the yield of acetic acid originating from the cellulose, hemicelluloses, and lignin is taken into account, the total is considerably less than the yield from the wood itself. This can be explained by the fact that the main source of acetic acid is the acetyl groups which are split off during the isolation of the single components. Undoubtedly, most of the acetyl groups are attached to the pentosans. The acetyl derivatives of xylose and arabinose are found in the products of a mild acid hydrolysis (32,93).

2.3 Lignin

The aromatic nature of lignin is shown by its pyrolysis products, of which guaiacol is that chiefly obtained from coniferous woods, and guaiacol and pyrogallol dimethyl ether from deciduous woods. Table V–2 gives a

TABLE V–2

PRODUCTS OF THE DRY DISTILLATION OF WOOD, CELLULOSE, AND LIGNIN (16)

Distillation product (%)	Wood		Cellulose		Hydrochloric acid lignin		
	Spruce	Aspen	Spruce	Aspen	Spruce	Spruce	Aspen
Reference	(66)	(52)	(66)	(52)	(53)	(48)	(52)
Charcoal	37.81	29.45	34.86	28.08	50.64	45.0	44.3
Tar	8.08	9.83	6.28	4.27	13.00	9.6	14.25
Methanol	0.96	1.48	0.07	0.00	0.90	0.7	0.87
Acetone	0.20	0.79	0.13	0.20	0.19	0.1	0.22
Acetic acid	3.19	7.37	2.79	2.66	1.09	0.6	1.28
Carbon dioxide	50.50	—	62.90	—	9.6	—	—
Carbon monoxide	32.55	—	32.42	—	50.90	—	—
Methane	9.23	—	3.12	—	37.50	—	—
Ethane	1.72	—	1.56	—	2.00	—	—

summary of the pyrolysis products obtained from the wood, cellulose, and hydrochloric acid lignin of spruce and aspen. The variations in the spruce lignin are probably due to differing conditions in the pyrolysis (16).

Lignin gives higher yields of charcoal and tar than cellulose does. The yield of methanol from lignin is lower than that from wood although lignin has a threefold higher methoxyl content than wood.

On dry distillation of Bergius lignin in a vacuum, the following results were observed: at 225°C, a white sublimate; at 335°C, a lively distillation; at 405°C, a strong evolution of gas which ceased at 450°C. In all, 56.1% aqueous distillate, 13.1% tar, and 11.9% gas were obtained (137). In the tar of the pyrolysis products of Willstätter lignin, the following saturated hydrocarbons were obtained: $C_{13}H_{26}$, $C_{14}H_{26}$, $C_{16}H_{30}$, $C_{24}H_{44}$, and $C_{30}H_{60}$, boiling at between 235° and 320°C, and the unsaturated hydrocarbons of the formulas $C_{11}H_{16}$, $C_{12}H_{16}$, and $C_{13}H_{16}$, boiling at between 200° and 260°C (105,137).

On microsublimation at 200°C, acid lignin from sprucewood is supposed to give high yields of vanillin and vanillic acid, but these results have not been confirmed (17,77). When sprucewood is heated in the presence of water for 20 hours at temperatures of 100° to 200°C and the cooled mixture is made alkaline with sodium hydroxide and then oxidized with nitrobenzene, different results are obtained. A methoxyl balance of the oxidation products shows that, on acidification of the mixture, the methoxyl-containing polymeric phenolic products are precipitated. The yield of these products increases with increasing temperature until, at 200°C, 45% of the total methoxyl is found in the precipitate. The amount of vanillin found decreases with increasing temperature and, at 200°C, is only 15% to 17% of that obtained from untreated wood. The polymeric phenolic products contain the larger part of the methoxyl and, on further oxidation with nitrobenzene, give no additional vanillin. On thermal treatment of lignin in the presence of water, condensations therefore occur. In an alkali fusion of the pyrolyzed wood, only 0.2% to 0.5% protocatechuic acid is obtained, as compared to 7% from untreated wood. This, too, confirms the fact that the lignin has undergone an autocondensation, or that condensations have taken place with the reduction products of the nitrobenzene (77).

When coniferyl alcohol is heated for up to 20 hours at up to 200°C, only 48% of vanillin is obtained on subsequent oxidation with nitrobenzene, whereas the untreated aldehyde gives 64%. Willstätter lignin, containing 15.3% methoxyl, when heated, with and without water, at 200°C, gives the same results after nitrobenzene oxidation. In this case, also, polymeric products are formed and these contain the major part of the methoxyl of the original lignin.

The autooxidation of the lignin begins already at 130°C and increases rapidly at between 140° and 160°C. Sprucewood, heated at 200°C and then oxidized, gives considerably higher yields of vanillin and less of polymeric products than when heated in the same way in the presence of water. The carbohydrates associated with the lignin obviously afford a kind of protection for the lignin. When these are removed by the hot-water treatment, the exposed lignin undergoes condensation after activation by the thermal treatment (17,77).

When lignin is subjected to a dry distillation, a considerable amount of methanol might be expected. The actual yield obtained in the pyrolysis of both coniferous and deciduous woods, however, amounts to only 0.7% to 0.9%, or not much more than 7% of the methoxyl content of the lignin. The pyrolysis of lignin from beech, birch, and oak gave 1% to 1.5% methanol, which also corresponds to only about 7% to 10% of the methoxyl content of the lignin. Thus, only about one-tenth of the methoxyl groups in the lignin participate in the formation of methanol (133).

Characteristic of the pyrolysis of lignin is the high carbon monoxide content of the gases. The pyrolysis of spruce lignin gave about 51% carbon monoxide in addition to 37.5% methane and 2% ethane. On the other hand, the carbon dioxide content was only 9.6% and that from the pyrolysis of cellulose was 62.9% (40).

Usually, for these investigations, lignins obtained in laboratory experiments were used; technical lignins, however, give practically the same results. These lignins give higher yields of charcoal (51% to 66%) and of tar (up to 14%), and the latter can be increased by a distillation under vacuum. Thus, from a technical hydrochloric acid lignin and from a soda lignin from corncobs, 13% to 28% tar is obtained. The decomposition of such tars, which are of interest for their phenolic components, will be discussed later. They contain, in addition to other substances, eugenol, phenol, *o*-cresol, guaiacol, and vinyl- and propylguaiacol (102).

3. PRODUCTS FROM THE PYROLYSIS OF WOOD

3.1 Charcoal

Charcoal is the main product from the pyrolysis of wood. The economic importance of the production of charcoal can be seen from the data for the production of wood tar and charcoal, shown in Table V–3, for Sweden, for the period from 1910 to 1948. These data show a marked increase in charcoal production during the wars because of its utilization for chemical and technical warfare application, especially in the Scandinavian countries where there was a great demand for charcoal for the operation of wood gas generators. In recent years there has again been an increased demand for charcoal for filtration, adsorption, and in the chemical industry.

For the United States, data have been gathered by the United States Forest Products Laboratory and are shown in Table V–4 (39).

In Table V–5, the products obtained from the pyrolysis of 100 kg dry coniferous wood are shown (7), while the yields of pyrolysis products from various woods are given in Table V–6 (67).

The yield of charcoal is sometimes considerably lower from deciduous than from coniferous woods, while the yield of acetic acid, methanol, and

TABLE V-3

PRODUCTION OF WOOD TAR AND CHARCOAL IN
SWEDEN, 1910–1948 (10)

Year	Wood tar[a]	Charcoal[b]
1910	8	4
1914–1918	11–15	4.0–4.5
1920	7.5	2.9
1930	7.5	2.0
1940	9.0	2.2
1944	60.0	5.2
1948	10.0	1.2

[a] In 1000 tons.
[b] In million cubic meters.

TABLE V-4

ANNUAL PRODUCTION OF CHARCOAL IN THE
UNITED STATES (39)

Year	Charcoal production[a]
1890	171,000
1909	550,000
1940	250,000
1947	213,000
1956	265,000

[a] In tons.

TABLE V-5

YIELD OF PYROLYSIS PRODUCTS FROM 100
KILOGRAMS SOFTWOOD (7)

Product	Yield
Charcoal	32[a]
A-Tar	7
B-Tar	3
Acetic acid	1.7
Acetone	0.8
Turpentine oil	0.6
Light oil	0.4
Methanol	1.0
Uncondensable gases	22.0

[a] All in kilograms.

TABLE V-6

YIELDS[a] OF PYROLYSIS PRODUCTS FROM DIFFERENT WOOD SPECIES (67)

Product	Pine	Spruce	Birch	Beech
Charcoal	37.8	37.8	31.8	35.0
Water	22.3	25.7	27.8	26.7
Gases	14.7	14.9	14.0	15.8
Acetic acid	3.5	3.2	7.1	6.0
Methanol, acetone methyl acetate	19.8	15.8	16.1	14.0

[a] All in percent of oven-dry wood.

TABLE V-7

YIELDS OF PYROLYSIS PRODUCTS FROM THE DRY DISTILLATION OF BEECHWOOD

Pyrolysis product	Yield, kg based on cubic meters wood[a] (570 kg dry wood)	Yield, based on 100 kg dry wood
Charcoal	195	34.3
Water	144	25.4
Gases	120	20.8
Acetic acid	38	6.7
Methanol, etc.	13	2.3
Tar	60	10.5

[a] For conversion of wood measurement units see Table VI-5 in Chapter VI.

acetone is higher. Table V-7 shows the yields obtained on the carbonization of beechwood in the industry; these yields are in good agreement with those shown in Table V-6.

The far-reaching effect that the carbonization temperature has on the properties of the charcoal is clearly shown in Fig. V-1. The carbon content increases sharply with increasing temperature and reaches an almost constant value at about 700°C. The same holds true for the specific weight, but to a much lesser degree. The yield, water absorbency, and the hydrogen content decrease continuously with increasing carbonization temperature. Especially impressive is the dependence of the electrical conductivity which is still 10^6 ohms at a carbonization temperature of 600°C but is even less than 1 ohm at 800°C (12). With increasing temperatures, the yield of charcoal not only decreases rapidly but its chemical composition changes increasingly to almost pure carbon; this is shown in Table V-8 (109).

Commercial wood charcoal contains about 80% carbon, 1% to 3% ash, and 12% to 15% volatile components. Because of their greatly reduced

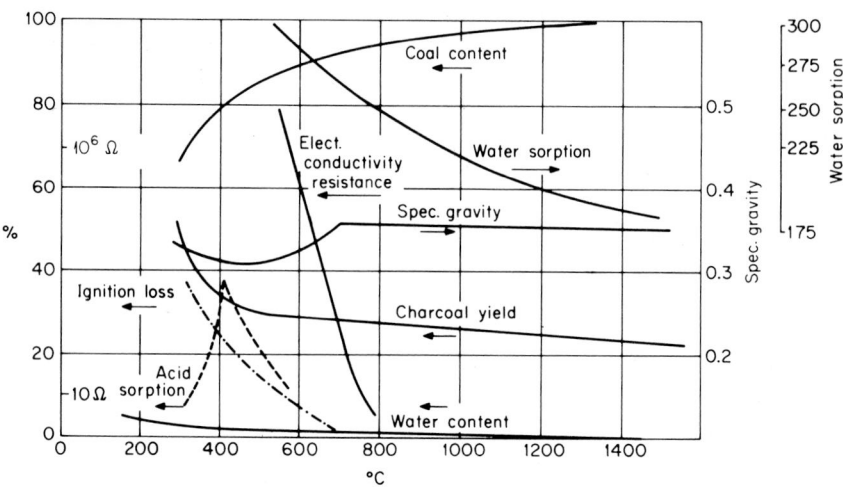

Figure V-1. Effect of temperature on coaling and charcoal quality (12). (Yield of charcoal, coal content in percent on dry wood; *idem* water content and ignition loss; acid sorption in milliliters per gram charcoal; electric conductivity resistance in ohms; water sorption in grams per 100 grams oven-dry charcoal; all figures approximate.)

TABLE V-8

ELEMENTARY COMPOSITION AND YIELD OF CHARCOAL IN RELATION TO
THE CARBONIZATION TEMPERATURE (109)

Carbonization temperature (°C)	Elementary composition (%)			Yield (%)
	C	H	O	
200	52.3	6.3	41.4	91.8
300	73.2	4.9	21.9	51.4
400	82.7	3.8	13.5	37.8
500	89.6	3.1	6.7	33.0
600	92.6	2.6	5.2	31.0
800	95.8	1.0	3.3	26.7
1000	96.6	0.5	2.9	26.5

yields, charcoals produced at a high temperature are of only theoretical interest.

When wood was carbonized in the flow of its gaseous decomposition products under conditions that prevented surface burning of the charcoal, the highest degree of shrinkage occurred in the tangential direction, the lowest in the fiber length direction. The shrinkage in all directions was dependent upon the carbonization temperature. This finding explains the

resistance of charcoal toward crushing and molding. The yields by weight and volume and the content of volatile substances and hydrogen in the charcoal decreased with increasing temperatures, while the total and non-volatile carbon contents increased. This is confirmed by the data shown in Table V-8. The type of wood used for the carbonization had no effect on these data. On the other hand, the ash content of the charcoal depends on the temperature and also on the kind of wood used. There is no correlation between the moisture content of the charcoal and the carbonization temperature. This indicates that the moisture content is affected not only by the porosity of the charcoal but also by its surface characteristics; these are the hydrophilic properties or the formation of free radicals. The mechanical resistance of charcoal toward crushing varies with the moisture content and with the nature of the wood (71).

The true density of charcoal, after milling, increases steadily with increased carbonization temperatures, while the apparent density decreases. The volume of micropores grows with the carbonization temperature up to 800° to 850°C while that of the intermediate pores increases with a temperature of up to 500° to 550°C. The volatile components have an acidic character only at carbonization temperatures of above 600°C, the acidity apparently being related to the evolution of carbon dioxide at the internal surfaces. The absorbency of the charcoal reaches a maximum at between 450° and 650°C, with its capacity to combine oxygen chemically being reached at between 420° and 450°C. The ignition temperature of charcoal increases with increasing carbonization temperature, with that of birchwood charcoal being lower, because of its more highly developed internal surfaces, than that of pinewood charcoal (72).

The absorption of oxygen by charcoal during storage depends upon the atmospheric conditions as well as the carbonization temperature. Under identical carbonization and test conditions, birchwood charcoal shows the lowest, and larchwood charcoal the highest uptake of oxygen. A postheating of the charcoal causes a considerable increase in the oxygen uptake and this, again, is dependent upon the carbonization and postheating conditions. A postheating at a low temperature and of short duration causes an increase in weight through the adsorption of oxygen. Higher temperatures (800°C) and long reaction times cause a decrease in weight due to the release of gaseous oxidation products. The resistance of the charcoal to this second oxidation reaction increases with increasing carbonization temperatures; at equal carbonization temperatures, the resistance toward oxidation increases with the hardness of the wood. The proportions of carbon dioxide and water in the oxidation products released from the charcoal are lower from that obtained at higher carbonization temperatures than from that obtained at lower temperatures; at the same time, the proportion of carbon monoxide

is greater. High temperatures in the postheating phase cause an increase in the carbon dioxide content of the oxidation products. The surface properties of charcoal are markedly affected by the postheating, as is shown by an increased hygroscopicity and a decreased absorption capacity (73).

When charcoal is heated under conditions that prevent its premature ignition, three distinct stages of oxidation can be observed. Up to a temperature of 150° to 250°C, a strong enrichment in oxygen takes place, with the formation of oxides and peroxides; after the surface of the charcoal has been saturated with oxygen, stabilization of the weight occurs in the second stage; then, in the third stage, a high loss in weight occurs. To test the properties of the surface of primary and postheated charcoal, comparable samples of pine, aspen, and birch charcoals were oxidized for 5 hours at 320°C and then reheated for short periods at 500°, 600°, and 900°C. In all cases the oxidation caused an increase in volatile compounds and a corresponding decrease in nonvolatile carbon compounds; a loss in hydrogen content was also noticed. The absorbed oxygen was split off as carbon dioxide and carbon monoxide and the yield of saturated hydrocarbons corresponded to the loss of hydrogen. For each species of wood there was a certain relationship between the content of volatile compounds and the oxidizability of the corresponding charcoal. Aspen charcoal had the highest ratio of oxidizability to volatility, while pine charcoal had the lowest loss in weight. For each species of wood there is also a relationship between charcoal activity, ash content, and porosity; but, in spite of this, charcoals from different wood species with the same or similar ash contents and porosities have completely different activities (22,74).

A different kind of charcoal production utilizes lignin from the acid wood hydrolysis process. Such lignin charcoals can be thermally activated in the same way as that described for wood charcoal. The activation can be carried out in the presence or in the absence of steam at temperatures of between 720° and 920°C and gives charcoals in approximately equal yields and having the same activation characteristics. With increasing activation temperatures, the ash content of the lignin charcoal increases, together with the adsorbency and clarification properties toward methylene blue solution; at the same time, the yield of charcoal decreases sharply. Lignin can be granulated without the addition of binders and such lignin gives a charcoal with good strength properties and porosity, suitable for the production of activated charcoal and for the manufacture of carbon disulfide (2,58,81,132).

Charcoal is useful for many purposes. Table V-9 gives a list of peculiarly characteristic uses. It may be assumed that many more uses will be found in the future, as is indicated by the increasing demand for this product (2).

TABLE V-9

COMMERCIAL APPLICATION OF CHARCOAL (5)

Domestic and specialized fuel	Metallurgical	Chemical
Recreational	Copper	Carbon disulfide
Curing tobacco	Brass	Calcium carbide
Cooking in dining cars and in	Pig iron	Silicon carbide
restaurants	Steel	Sodium cyanide
Heating, foundry, plumbing	Nickel	Potassium cyanide
equipment	Aluminum	Carbon monoxide
Heating in shipyards and citrus	Electro manganese	Activated carbon
groves	Armor plate	Black powder
	Foundry molds	Fireworks
		Rubber
		Gas adsorbent
		Crayons
		Soil conditioner
		Pharmaceuticals
		Poultry and animal feeds

3.2 Wood Tar

As can be seen from Table V-6, 15% to 20% tar, depending upon the type of process and the species of wood and based on the weight of the wood, is formed in the wood pyrolysis. From the crude wood vinegar, the A-tar, or sedimentation tar, is obtained by settling; it is phenolic in character. The B-tar, or soluble tar, is dissolved in the acetic acid. By the scrubbing of the gases from the pyrolysis plant, the B-tar can be obtained in a yield of 50%. The total amount of B-tar is, in general, about two-thirds that of the A-tar. A rapid pyrolysis, however, can produce more B-tar than A-tar. At 200°C, the B-tar decomposes in an exothermic reaction, with the evolution of carbon dioxide and the splitting off of water, to leave a water-insoluble pitch. When the acetic acid is extracted from the crude wood acid with ethyl acetate, an "extraction tar" is obtained which, in its chemical composition, resembles the B-tar. After removal of the A-tar, the crude wood acid has the following composition: 8% to 10% acetic acid or its homologs, 3% crude methanol, 7% soluble tar, and 80% water. For the separation of the tar from the crude acetic acid, a number of distillation processes have been described, of which the multiple-effect vacuum-evaporation process has proved to be the most feasible economically. The distilling acetic acid is taken up by milk of lime and recovered as calcium acetate. From the latter, the acetic acid can be recovered with sulfuric acid or, by also dry distillation, acetone can be obtained. An azeotropic distillation has been suggested for the processing of the crude wood acid (11,23,24,68,100).

Because of its phenol content, the sedimentation tar has received special attention. Attempts to isolate uniform phenols by suitable distillation processes have been numerous. The data on the yields obtained must be considered in relation to the pyrolysis process used, the species of the wood, and the type of distillation. Thus, for example, an A-tar from coniferous woods contains 60% phenol, 10% organic acids, and 30% neutral substances containing oxygen compounds and hydrocarbons. The neutral fraction from the A-tar of deciduous woods does not contain these hydrocarbons. It may be assumed that the hydrocarbons are formed by the decarboxylation of rosin acids (75).

In the fractionated distillation of an A-tar from the pyrolysis of aspenwood, the amount of phenols varied little within the temperature range of 200° to 275°C, while the amount of distillate obtained was directly proportional to the evaporation temperature. The maximum yield of carboxylic acids in the distillate was reached at 225°C. The boiling point curves showed two inflection points, the first correponding to the separation of water and low-boiling components, the second, at 150° to 152°C, being due to a rapid increase in the molecular weight and in the amount of carboxylic acids. Up to 90% of tar oils could be distilled from the A-tar at 275°C. At this temperature, the distillate amounted to 93% of the weight of the dry tar and contained 54% phenols and 12.5% carboxylic acids (4).

A vacuum distillation at 5 mm of a technical sedimentation tar from a dry distillation of wood yielded 71.7% tar 'oils and 26.3% tar pitch at temperatures of up to 230°C. From the tar oils, 47.1% phenols could be isolated, corresponding to a yield of 33.9%, based on the dry weight of the original tar. The vacuum distillation gave three fractions. A chromatographic investigation of these showed the presence of 23 components, of which the following were identified: 4.04% phenol, 4.04% o-cresol, 5.53% m- and p-cresol, 12.82% xylenols, 4.67% guaiacol, 5.81% 4-methylguaiacol,

TABLE V-10

Yield and Properties of Oils Obtained from the Pyrolysis of Various Wood Species (38)

Wood species and type of pyrolysis	Tar oil (%)	Phenol in tar oil (%)	Neutral substances (%)	Methoxyl groups (%)
Wood pyrolysis, technical	78.4	54.2	30.3	4.3
Logging waste, technical	57.3	63.2	31.3	4.6
Logging waste, experimental	66.0	56.0	35.0	5.5
Hardwood, dry distillation	86.5	42.0	52.0	4.9
Beechwood, dry distillation	57.8	50.0	33.0	6.5
Beechwood, vertical retort	64.6	35.0	24.7	7.6

8.84% guaiacol homologs, 42.2% high-boiling phenols, and 12.05% residual phenols. The average number of reactive positions in the phenols was 2.31 (99).

Table V-10 shows the yields and properties of the distillates from wood tars. That the yields of tar oil are subject to variation, depending upon the kind of pyrolysis and type of wood used, has already been mentioned in Section 1. Table V-10 shows that the phenolic fraction of the tar also varies considerably. This will be discussed further in connection with the technical installations for the wood pyrolysis (38).

The wood processing industries, such as sawmills and pulp mills, have large quantities of bark as waste material. It was therefore natural to investigate the products obtained from the pyrolysis of bark. In pilot plant and laboratory experiments, 6% to 8% A-tar, 2% to 3% B-tar, and 1.83% volatile acids could be obtained from sprucewood bark. The A-tar contained 28% to 33% neutral compounds, 23% to 27% phenols, 25% to 31% rosin and other higher acids, 4% to 5% carboxylic acids that could be extracted with ether, and 6% to 8% ether-insoluble residue. Fractional distillation of the tar yielded 49% to 51% tar oil containing 28% to 36% phenols, 23% to 24% rosin and other acids, and 35% to 40% neutral substances. At a vacuum of 4 mm, the phenols distilled within a temperature range of 75° to 190°C. The acid fraction contained chiefly fatty acids. The high lignin content of the bark indicates that the latter is a valuable raw material for the production of phenols from its A-tar (124).

The pyrolysis of wood sugar lignin has also been investigated. The yield of A-tar depended, to a great extent, on the type of lignin and its chemical composition and on the method used in the pyrolysis, including the apparatus. In the absence of gas circulation, the yield of A-tar was only about 8%, but, with circulation of the pyrolysis gases, it increased to 18%. About 40% to 50% of the tar consisted of phenols and 10% to 25% of neutral substances. The phenols were mainly methyl ethers with up to 15% methyl groups. Vacuum distillation of the tar gave up to 70% tar oil; a contact pyrolysis at 600°C resulted in an extensive demethylation of the phenol ethers (131). When the pyrolysis of the tar was carried out with superheated steam at 600°C in the presence of quartz sand as a heat transferrer, the tar contained 47.2% phenols, 28.8% neutral substances, and 3.9% volatile acids. Based on the original tar, the yield of distillate amounted to 32.6%, of gas to 18.6%, and of coke and losses to 48.8% (1).

When the pyrolysis was carried out in an externally heated reactor in which the lignin was continuously moved from the top to the bottom and the gaseous reaction products were constantly withdrawn, a maximum yield of 22% tar was obtained at a temperature of 550°C. Thus the yield was almost doubled as compared with the pyrolysis in a stationary bed. The phenol content also increased to 52%, and the yield of neutral oils in-

creased from 18% to 33% with increasing temperature, while the methoxyl content decreased. The methoxyl content at 400°C amounted to 11.8%, at 600°C to only 5%. The coke formed could be briqueted without the addition of binders (114).

The A-tars obtained from lignin do not differ fundamentally in their chemical composition from those obtained from the pyrolysis of wood. Of the 46% to 55% phenols, about 55% to 60% are guaiacols, of which about 50% is 4-methylguaiacol and the remainder is guaiacol and 4-ethyl- and 4-propylguaiacol. Other components are phenols, cresols, xylenols, pyrocatechol, 4-methylpyrocatechol, and dimethylpyrogallol and its homologs. The monohydroxyphenols and pyrocatechols can be considered to be secondary reaction products of the pyrolysis of guaiacol and methylguaiacol (123).

The most valuable products of wood pyrolysis are phenols. They are formed chiefly from the lignin and only to a small extent from the carbohydrates. On the basis of the energy of bond dissociation, the main role in the formation of volatile phenols must be ascribed to the recombination of low-molecular aryl radicals. On the other hand, a direct recombination of aromatic radicals results in the formation of heat-resistant nonvolatile products containing diphenyl and aryl ether bonds. Upon further degradation, these products are converted to carbon. To increase the yield of phenols and minimize the formation of carbon, conditions must be found which will provide a sufficient concentration of low-molecular radicals at the moment of the cleavage of primary bonds in lignin. The pyrolysis should therefore be carried out at such a rate of heating that liquid and gaseous degradation products are formed almost simultaneously. Experiments dealing with the pyrolysis of birchwood at high rates of heating showed that, as the rate was increased, the concentration of volatile products and low-molecular radicals also increased, while the formation of carbon decreased. The yield of distillable products increased with the temperature up to 500° to 600°C, then decreased, while the volume of gases steadily increased with a simultaneous decrease in the formation of carbon. The ratio of gases to liquid depended to a great extent on the duration of their retention in the reaction zone. The yield of tar and phenols increased as the retention time decreased. At a temperature of 600°C, a heating rate of 1.84 per second, and a retention time of 1.58 seconds, the yield of crude phenol was 8.6% based on the oven-dry wood (107b).

Table V–11 lists the phenolic compounds that have been found in the A-tar and its pyrolyzate. The goal of isolating single reactive phenols from the A-tar on a technical scale is complicated and can be achieved only under conditions that are not feasible from an economic point of view. In general, the objective has been limited to the isolation of certain fractions which, within a limited boiling range, have similar physical data (61,62,

TABLE V–11

PHENOLS OBTAINED IN THE PYROLYSIS OF WOOD

Phenol derivative	Reference
Phenol	99,122,123
2-Cresol	99,122
3-Cresol	99,122
4-Cresol	99,122
2,4-, 2,5-, 3,5-, 2,3-, 3,4-Xylenols	122
2-Ethylphenol	122
Guaiacol	99,122
4-Methylguaiacol	99,122,123
4-Ethylguaiacol	122,123
4-Propylguaiacol	123
Pyrocatechol	122,123
Dimethylpyrocatechol	123
3,4-Benzopyrene	28

63). The A-tar can be separated into light, heavy, and high-boiling tar oils, and pitch. The heavy oil is used for the production of creosote by treating it with sodium hydroxide in order to separate unreactive components. On acidification of the alkaline extract with sulfuric acid, the creosote separates and can be further purified by repeated distillation. Its main components are creosol (2-methoxy-4-methylphenol), cresol, and guaiacol and it has strong antiseptic properties. The high-boiling oil is used as a flotation and impregnation medium, while the pitch can be used as a cement and embedding agent.

The 3,4-benzopyrenes in pyrolysis wood tar, which have certain carcinogenic properties, occur mainly in the pyrolysis products of spruce and pine and, to a lesser extent, of birch and beech. At 350° to 400°C, the yield of 3,4-benzopyrene from the pyrolysis of wood does not depend on the presence of oxygen; at 450°C and over, the presence of oxygen markedly increases the yield as compared with that in an oxygen-free atmosphere (28a). Pyrenes and 1,12-benzoperylene can be found in the benzopyrenes. These results are of interest with regard to the smoking of meats. Such polycyclic hydrocarbons are also found in the tars that have been obtained at pyrolysis temperatures of below 300°C, and this contradicts the statement that benzopyrenes are obtained only at high pyrolysis temperatures (27).

The B-tar is dissolved in the wood acetic acid. This pyroligneous acid contains about 9% to 10% acetic acid and its homologs, about 7% soluble tar, 3% crude methanol, and 80% to 81% water. The wood acetic acid, freed of methanol, is extracted with ethyl acetate, in extraction towers filled with packing material, in a countercurrent manner. In addition to acetic acid, tar components are extracted. The acetic acid is isolated by distilling

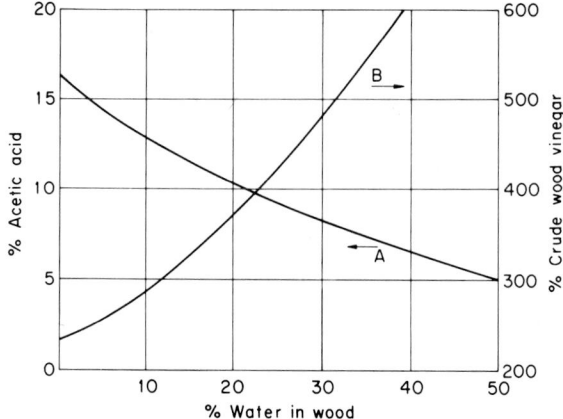

Figure V-2. Relationship between the moisture content of wood and the yield of crude wood vinegar, and the content of acetic acid therein.

off the ester, which is returned to the extraction cycle. The moisture content of the wood is of special importance for the pyrolysis; it has a decisive effect on the composition of the B-tar and the pyroligneous acid. Figure V-2 shows that, with increasing moisture content of the wood, the yield of pyroligneous acid increases sharply, with its content of acetic acid simultaneously decreasing.

When the pyroligneous acids from pyrolyses of sprucewood containing 25% and 45% moisture are extracted with ether and the extracts are deashed by means of a cation exchanger, solutions with equal amounts of soluble tar are obtained. The residues of the evaporated solutions are dissolved in water and passed through columns of basic Amberlite, the columns are eluted with formic acid, the eluates are evaporated, and the residues are extracted with ethyl acetate, giving soluble and insoluble fractions. The amount of residue from the eluate from the pyrolyzed wood with the higher moisture content is greater than from the wood with less moisture. The same is true of the residues from the filtrates from the Amberlite columns. This portion consists almost exclusively of levoglucosan, originating from the pyrolysis of the cellulose (111).

It is interesting to note that the soluble tar obtained from the pyrolysis of sprucewood bark differs considerably from that obtained from the wood. Not only is the amount of soluble tar lower, as was to be expected because of the lower carbohydrate content in the bark, but also the tar contains less levoglucosan and more phenols, hydroquinone, methylcyclopentenolone, and hydroxy acids. The ether extract of the tar pitch gives novolac resins, while the ether-insoluble residue consists of polymerized diene acids and anhydro sugars, chiefly levoglucosan (78).

3.3 Pyroligneous Acid and Methanol

The processing of pyroligneous acid to acetone has already been mentioned. It is of little technical importance today. The pyroligneous acid is extracted with ethyl acetate which is distilled off and returned to the cycle. On azeotropic distillation, the pyroligneous acid is freed of methanol, evaporated, and the wood tar is separated in a tar-removal column (127). In another column, butyl acetate is added to give an azeotropic mixture. Condensation of the vapors from this column forms an aqueous and an oily layer; the latter contains the butyl acetate, which is returned to the cycle; the former, which contains the acetic acid, is further concentrated, giving some residual tar.

When the pyroligneous acid is acidified with sulfuric acid, the soluble tar separates. The remaining solution is esterified with methanol in a distillation column by the catalytic action of the sulfuric acid. The crude ester is distilled, giving a mixture of 80% methyl acetate and 20% methanol. On fractionated distillation, pure methyl acetate can be obtained, in addition to a higher-boiling ester mixture, consisting chiefly of methyl propionate.

The first run in the distillation of the pyroligneous acid consists of about 50% methanol, 18% acetone, 7% ester, 6% aldehyde, 0.5% ethyl alcohol, 18.5% water, and small amounts of furfural. Methanol can be separated from the mixture in a wash column by extractive distillation with water.

3.4 Furfural from a Prepyrolysis

The formation of furfural in the hydrolysis of pentosans (see Chapter IV, Section 5.4) suggests combining the hydrolysis with a pyrolysis into one technical process in order to make better use of the pentosans which, because of secondary reactions, can contribute little to the formation of furfural in a pyrolysis. For such a process, there are two possible methods; (a) the acids formed in the pyrolysis can be used directly for the hydrolysis of the pentosans under certain reaction conditions; (b) the wood can be impregnated with acid or with acid salts before the pyrolysis.

To suppress the secondary reactions of the furfural formed, the wood, impregnated with acid, is subjected to a pyrolysis in high-boiling kerosene or diesel oil. The water, bound in the wood, and the main part of the furfural distill at about 240°C. The fraction distilling at 250° to 275°C contains kerosene, acid wood-decomposition products, and a small amount of furfural. With the use of acetic acid or pyroligneous acid as a hydrolyzing agent, 5% to 8% furfural can be obtained, based on the weight of the dry wood. Hydrochloric acid is considerably more effective and increases the

yield of furfural from 5% to over 10%. The yield of furfural depends primarily on the concentration of the acid, the moisture content of the wood, and the pyrolysis temperature; the highest yield is obtained with 4% to 8% hydrochloric acid at 240°C and a liquor-to-wood ratio of 2.3–2.5 to 1.

Although there is not a great difference between the pentosan contents of birchwood (21%) and aspenwood (19%), the former gives a considerably higher yield of furfural (9.3% to 9.6%) than the latter (4.9% to 5.8%). Dense woods, such as young birchwood, give low yields of furfural, probably because of incomplete impregnation of the wood with acid. Satisfactory yields can be obtained, however, at temperatures as low as 200°C. The substitution of acid salts, such as sodium bisulfate, for the hydrochloric acid does not appreciably reduce the yield of furfural.

In a pyrolysis of wood in kerosene or diesel oil as a heat-transferrer, the cellulose is drastically degraded with increasing temperatures. At 180°C, 90% of the cellulose is already degraded, and the apparent lignin content of the residue increases steadily, due to the formation of substances which are insoluble in 72% sulfuric acid. This "pseudolignin" reaches its maximum amount at a temperature of 180°C and then remains constant. It is interesting to note that this increase in lignin corresponds to the amount of degraded cellulose. Of the pentosans, 84% are degraded within one hour at a temperature of 100°C; the residual, more resistant pentosans are then degraded stepwise with increasing temperature. The distillate contains hydrochloric acid, furfural, and organic acids and amounts to about 20% of the weight of the wood. The formation of furfural, acetic acid, and small amounts of formic acid begins at about 105°C and the maximum yield of furfural is obtained at between 200° and 240°C. The splitting off of reaction water from the wood components increases with the temperature and reaches its maximum, 18% of the weight of the wood, at the maximum temperature. About 4% of the weight of the wood is converted into noncondensable gases, chiefly carbon dioxide and small amounts of carbon monoxide.

To elucidate the course of the pyrolysis reaction and the factors that control it, birchwood sawdust was subjected to a prepyrolysis, as described above, in a liquid inert heat-carrier at temperatures of between 120° and 240°C. An investigation of the residual pentosan content in the wood as a function of the prepyrolysis temperature and time was carried out; simultaneously, the residual pentose content was calculated from the furfural formed on distillation over the temperature range of 120° to 260°C. Straight line relationships were obtained between the logarithms of the pentosan and pentose contents and the heating time at each temperature; from this, reaction constants were calculated for the decomposition of the pentosans and the dehydration of the pentoses. The dependence of the

temperature on the reaction constants indicated that diffusion is the rate-controlling factor in both the decomposition of the pentosans and the formation of the furfural. In the latter reaction, three phases could be distinguished: the kinetic phase at low temperatures, the intermediate phase with increasing reaction constant, and the diffusion phase at temperatures of above 140°C. Toward the end of the prepyrolysis process, a reduction in the reaction constants could be observed; this can be explained by the increased resistance of the acid catalyst toward diffusion into residual pentosans with low accessibility (97,98,126).

When the pyrolysis of wood, impregnated with solutions of catalysts, is carried out in laboratory retorts with direct heating, higher temperatures are required. High yields of furfural can be obtained in a pyrolysis of 2 to 3 hours at temperatures of up to 420°C with wood impregnated with dilute sulfuric acid, sodium bisulfate, or mixtures of sodium chloride and sulfuric acid; such yields reach 10 to 14 times the amount obtained from nonimpregnated wood under otherwise equal conditions. Formic acid as a hydrolyzing catalyst has little effect, and an impregnation with sodium chloride has no effect. Sulfuric acid is not only the most effective, but is also the most economical catalyst; a disadvantage, however, is that the sulfur content of the charcoal is increased (70).

The addition of aluminum chloride to the hydrochloric acid as a catalyst increases the yield of furfural but it is advisable to remove the pyrolysis products from the reaction zone by a stream of carbon dioxide as quickly as possible to prevent secondary reactions (57,63). Instead of an inert gas, a fluidized-bed reactor can be used; in this, the impregnated, finely shredded wood is kept in suspension by a gas stream which simultaneously and continuously removes the reaction products. Carbon dioxide or a mixture of carbon dioxide and steam has been suggested as the carrier gas. As a catalyst, 1% to 2% sulfuric acid is used, with a temperature of between 200° and 240°C—i.e., considerably lower than that of a pyrolysis with direct heating. The moisture content of the wood is of special importance and should be between 15% and 20%. A drying after the impregnation permits the adjustment to the desired moisture content (76,113).

The effect of salts on the rate of decomposition of wood has for a long time attracted a great deal of attention for completely different reasons. The impregnation of wood with certain salts has for years been under investigation for use as protection against the danger of fire. By means of dynamic thermogravimetry, it has been found that, regardless of the pretreatment of the wood with salts, lignin undergoes a pyrolysis at a slower rate than wood or α-cellulose does. This has already been mentioned in Section 1.1. While the loss in weight at the beginning of the pyrolysis corresponds to the loss in moisture content, a loss in weight for untreated wood and lignin occurs through pyrolytic action at 220°C, but, for cel-

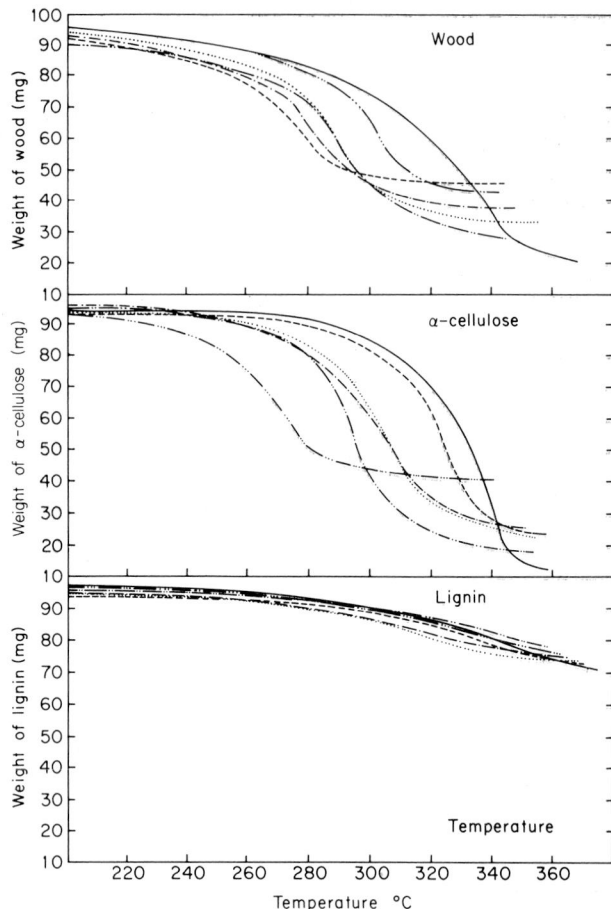

Figure V-3. Relationship of increased temperature to weight of wood, α-cellulose and lignin, with and without inorganic salts (134).

lulose, this occurs at 275°C. After this, the pyrolysis of the cellulose takes place very rapidly, while that of lignin continues slowly. The pyrolysis of cellulose is complete at 360°C, but that of lignin is only 56% complete at 420°C. On the other hand, wood is completely pyrolyzed at 360°C, with 21% charcoal remaining. The fact that the temperature threshold for the pyrolysis of lignin is lower than that of cellulose seems to be in contradiction to the reports given in Section 2. This discrepancy can be explained by the fact that only a very small amount of lignin is lost at low temperatures, through the formation of volatile compounds, whereas at such temperatures the cellulose is already completely decomposed. The loss in weight of woods impregnated with salts, with the exception of aluminum

chloride, is caused exclusively by the loss of moisture, at temperatures of up to 200°C. Aluminum chloride also has a hydrolytic effect on the wood. The effect of the impregnation with salts on the pyrolysis of wood, lignin, and α-cellulose at increasing temperatures is shown in Fig. V-3. It can be seen, for example, that the losses in weight for wood and α-cellulose which have been impregnated with sodium tetraborate correspond to the losses of untreated samples until a temperature of 300° to 350°C has been reached, at which point the decomposition of the samples is almost complete. All other salts reduce the decomposition temperature and increase the loss in weight during the pyrolysis even in the range of temperatures below 350°C. The action of monobasic ammonium phosphate is the most pronounced. The yield of charcoal is increased by the impregnation with salts, but such an impregnation has far less effect on the behavior of lignin in a pyrolysis. At temperatures of below 350°C, lignin impregnated with salts decomposes more rapidly than nonimpregnated lignin. At above 350°C, the reactions are reversed. Impregnated lignin yields more charcoal but the difference is only about 6%, whereas for wood it is 17% and for cellulose, 21%. Impregnation with salts therefore has its greatest effect on the pyrolysis of α-cellulose and its least on lignin. So far as wood is concerned, one can speak of a reciprocal action between cellulose and lignin, thus indicating the controlling role of the cellulose in the course of the pyrolysis of wood, whether impregnated or not (134).

4. TECHNICAL PROCESSES FOR THE PYROLYSIS OF WOOD

4.1 Charcoal Piles and Carbonization Furnaces without External Heating

The production of charcoal by the carbonization of piles of wood is the oldest method of utilizing waste wood. Although modern methods have replaced the classical wood carbonization method, in times of emergency the latter is still used. Thus, in Sweden during the First World War, the production of charcoal increased to about 4,000,000 cubic meters but decreased again to about 2,000,000 cubic meters after the end of the war. This increase was obtained almost exclusively by woodpile carbonization. During the Second World War, the production of charcoal increased again to over 5,000,000 cubic meters and, throughout the country, over 5,000 charcoal piles were in operation, primarily to produce charcoal for use in wood gas generators.

A charcoal pile contains, on an average, 100 to 150 rm of wood, but piles with as much as 400 rm have been used. The building of an efficiently operating pile is an art and the operation of such a pile has long since passed

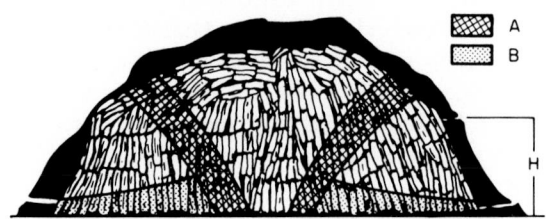

Figure V–4. Early charcoal production in earth kilns (6).

from an empirical to a more scientific undertaking. Wood blocks, chiefly from deciduous woods, are piled up in such a way that a channel remains in the center, through which the pile is ignited. The woodpile is covered with dirt, humus, or moss. The operation is started by igniting dry wood. The process is regulated by admitting air through openings at the bottom of the pile, while the smoke escapes through openings halfway up the pile. Figure V–4 shows such a charcoal pile, in which the A zone indicates the carbonization area while the B zone shows the air circulation in the pile. The air moves toward the carbonization zone and is mixed more and more with the flue gases and thus is caused to circulate within the pile. Right next to the carbonization zone is the drying zone. The difference in the specific weight of the air column, H, and the corresponding column of hot gases within the pile determines the rate of flow of the air and therefore determines the draft when a pile with a chimney is used. A charcoal pile without a chimney has many openings through which the flue gases can escape and these openings can be opened or closed in accordance with the progress of the carbonization process. The chimney facilitates better control and regulation of the carbonization.

The pyrolysis of wood in a carbonization pile does not permit the recovery of tar, and the heat required for the operation is supplied by the combustion of some of the wood, thus decreasing the yield of charcoal as compared with that from a carbonization furnace, in which the pyrolysis is started and effected by the application of external heat. It has been found that the yield of charcoal increases with an increasing content of carbon dioxide in the flue gases; thus the yield of charcoal, based on the weight of the dry wood, amounts to 19% when the carbon dioxide content of the flue gases is 22.4 vol % but reaches 29% at a carbon dioxide content of 28.6 vol % (6,65).

So-called kilns are similar in operation to the charcoal piles, and various types of construction have been designed for them. "Beehive kilns" are built from bricks and can hold 50 to 90 cords (250 to 460 rm) of wood, primarily waste wood. They are circular in shape and taper toward the top, which has a flat hood over it. Loading is achieved through a door at

the side through which the furnace is also ignited. After the wood has been lighted, the door is closed and the further course of the carbonization is controlled by regulating the admission of air. Small portable combustion kilns have also been designed from metal. In later times, large, rectangular furnaces have been built from ceramic or special concrete materials and these can hold up to 100 cords (510 rm) of wood. In all these furnaces, the wood is added at random and therefore it is easier to charge them and to control their operation. It is also easier to empty them than is the case with the charcoal pile. Such furnaces can have one or more chimneys or vents that can be opened or closed to the flue gases and to the air. All have in common the fact that the charcoal, after the carbonization is completed, requires a long time to cool before it can be removed, since, otherwise, it would be subject to spontaneous ignition. If one takes into acount the time required for the charging, the carbonization, the cooling, and the discharging of the charcoal, the cycle for the charcoal pile is 2 to 6 weeks and for the beehive kilns, about 20 to 30 days, depending upon the construction and size of the kiln. When several of such kilns are grouped together and charged in series, considerable amounts of charcoal can be produced.

4.2 Retort Furnaces

Besides the long periods required for the cycle of the furnaces described above, the impossibility of recovering the by-products has made it desirable to mechanize the process and, in this way, to make it more economical. Carbonization furnaces were therefore constructed into which the wood can be moved on carriages, or, when small pieces of wood are used, a shaftlike kiln permits the wood to be charged at the top and the charcoal to be removed at the bottom. Such furnaces operate with heat applied either externally or internally. They therefore differ from the earlier processes in that the exothermic heat is largely utilized and the by-products from the carbonization can be recovered.

Figure V–5 shows a furnace with internal circulation. By means of the blower, 1, the gases from the carbonization are kept in circulation, they pass through a heater, 2, and then strike the wood on the carriage, 3. It can be operated with wood blocks piled on the carriage, which has grated sides. The wood remains on the carriage throughout the process and the charcoal is removed on it. To save time, the carriage with the charcoal on it is moved into a cooling chamber so the retort can be refilled immediately. By means of a condensation system, 4, 5, and 6, flue gases are removed in an amount corresponding to the amount of condensable products formed in the dry distillation. The noncondensable gases are cooled, washed in the cooler, 6, and used as fuel.

In contrast to this, a shaft furnace, with external heat circulation, is

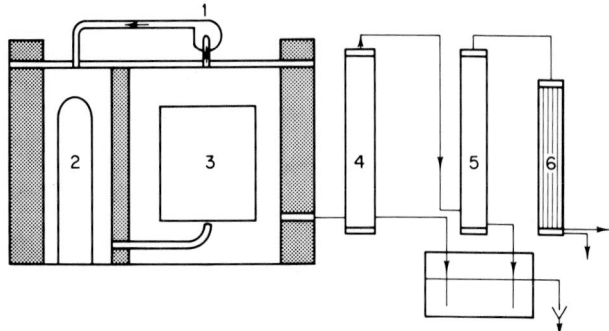

Figure V-5. Kiln with internal circulation (8). 1 ventilator; 2 heating element; 3 car; 4 washing column; 5 washing column; 6 cooling device.

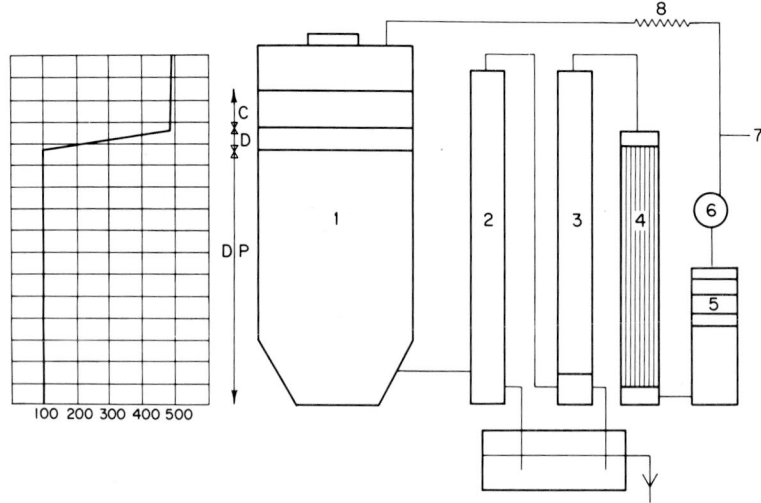

Figure V-6. Kiln with superheated, uncondensable gas in circulation (8). 1 retort; 2, 3 and 5 washing columns; 4 cooling device; 6 ventilator; 7 exit of noncondensable gas to fire place; 8 superheater.

shown in Fig. V-6. The gases coming from the furnace, 1, pass through a condensation system, 2, 3, 4, and 5, and the noncondensable gases are blown by a fan, 6, into the superheater, 8, and are returned to the furnace after an amount of gas, corresponding to that formed by the carbonization, has been taken off through the pipe, 7. The amount of gas in circulation is about 15 times greater than that formed during the carbonization. In the diagram to the left in Fig. V-6, C represents the carbonization zone, D the drying zone for the wood, and DP the zone in which the carbonization gases have a temperature that lies near the dew point.

In furnaces with inner circulation, the major part of the reaction pro-

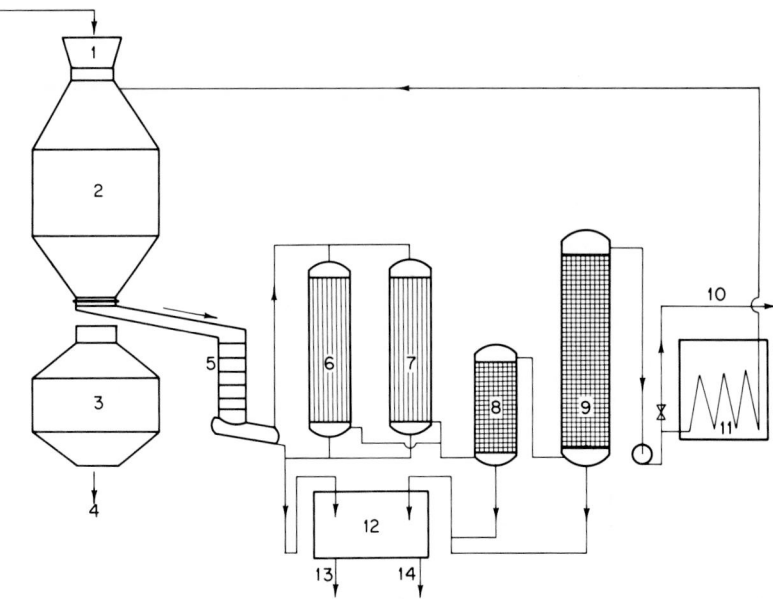

Figure V-7. The Reichert wood charring process (26). 1 wood entrance; 2 wood retort; 3 charcoal cooling bin; 4 charcoal discharge; 5 tar separator; 6 cooler; 7 cooler; 8 tar separator; 9 gas washer; 10 wood gas exit; 11 gas heater; 12 settling tank for raw wood vinegar; 13 wood tar exit; 14 raw wood vinegar exit.

ducts from the carbonization is exposed to overheating and, therefore, to partial decomposition. In furnaces with external circulation, the tar substances and the condensable products are separated prior to being returned to the cycle and thus are saved from superheating. Only the noncondensable gases and the volatile components in them are subjected to superheating. Internal circulation reduces the yield of tar because it is subject, to an increased extent, to cracking. With external circulation, the gases contain more readily volatile substances, the amount of which depends chiefly on to what extent the gases are cooled and are freed of such components by washing before being returned to the superheater, 8.

In the United States, a process has been developed in which the heat from the exothermic reaction of the wood carbonization is completely utilized and external heating is therefore unnecessary. This process, known as the Stafford process, uses a vertical retort in which a temperature of 400° to 500°C is maintained at the beginning of the carbonization by means of a fire in the retort. The wood is preheated to 100°C by waste heat and filled into the retort by a conveyor. As the carbonization begins, sufficient heat is developed by the exothermic reaction to complete the reaction without the addition of more external heat. The process can be

carried out only with dry wood and the drying is accomplished by means of hot gases from the power and steam plants. The moisture content of the wood should not exceed 0.5% (39,128).

The Seaman process is based on a similar concept and utilizes an inclined, rotating retort (119). These two processes are primarily suitable for the carbonization of small pieces of wood obtained as wood waste and sawdust from wood-processing industries. With this type of material, the problem of heat transfer is especially difficult because the densely packed state of the wood makes it a very poor heat conductor.

A further improvement is the conversion of the wood carbonization into a continuous process. Such a process is the Reichert process which is used in Germany in various wood carbonization plants. Its underlying principle is illustrated in Fig. V-7. The vertical retort, 2, has a volume of about 100 cubic meters and holds about 40 fm of wood. The heat is introduced at the top of the retort as highly heated gases. It is therefore based on a process with external circulation, such as that already shown in Fig. V-6. The rate of carbonization can be controlled by the amount and temperature of the circulating gas stream. Because the charcoal is discharged into a cooling tank, 3, after the carbonization is complete, the process can be carried out on a continuous basis, provided that the gas stream is kept constant during the charging and discharging of the retort. It is apparent that short blocks of wood should be used since, with small pieces of wood, the gas stream cannot pass through the charge (26).

The Lambiotte process is based on the principle of external circulation and is a completely continuous process. As can be seen from Fig. V-8, a vertical retort, A, which is divided into three reaction zones, is used. In the first zone, the wood is preheated, in the second zone, the exothermic carbonization reaction takes place, and, in the third, the reaction products are cooled. In contrast to the Reichert process, the heating gases are introduced in the lower part of the second zone. The gas moves upward in the retort and constantly releases its heat into the wood, which is moving downward. The gases then leave the retort at the top, pass through the condensation and cooling systems, and finally pass through the gas washer, F. From there, the excess gases are blown by a fan into the combustion chamber, P. To start the process, the gases are heated by an oil burner. Some of the returned gases are led into the third zone where they take up heat from the charcoal and are then blown into the combustion zone by the hot gas fan, B. In this process also, the dry content of the wood is of special importance for its heat economy. An increased moisture content causes not only an increase in heat consumption but also an increase in energy consumption for the circulation of the gases. This can be seen in Table V-12.

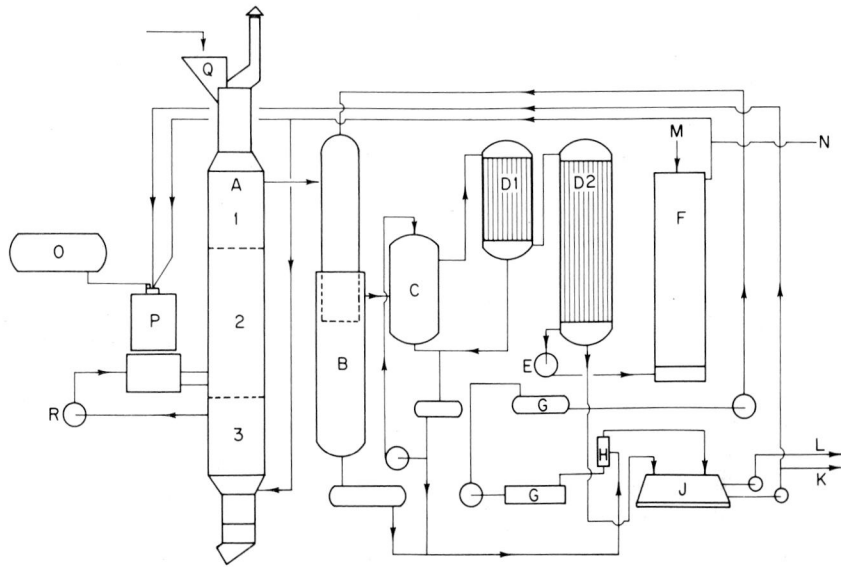

Figure V-8. Continuous wood charring plant Lambiotte system (82). A retort; 1 preheating zone; 2 charring zone; 3 cooling zone; B tar separator and precooler with tar irrigation; C tar separator with raw wood vinegar irrigation; D1 and D2 condensator and gas cooler; E ventilator; F gas washer; G filter; H tar separator; J separator bin for raw wood vinegar; K tar outlet; L raw wood vinegar outlet; M water inlet; N gas surplus exit; O bin for fuel oil; P burning chamber; Q wood entrance; R hot gas ventilator.

TABLE V-12

ENERGY CONSUMPTION IN RELATION TO MOISTURE CONTENT OF
WOOD IN THE LAMBIOTTE PROCESS (109)

Moisture content of wood (%)	Energy consumption for 1 cubic meter wood		
	kcal	Gas in circulation (m³)	kwh
5	6,000	210	2.5
10	9,500	270	3.2
15	37,000	490	4.4
20	70,000	770	5.7
25	110,000	1,050	7.2
30	155,000	1,400	9.0

A Swedish version of a continuous wood pyrolysis for the production of charcoal is shown in Fig. V-9. In all continuous processes, it is essential that the wood should be almost completely dry, to permit its thermal decomposition. Hot gases from the pyrolysis are usually used for the predrying. If the amount of gas from the exothermic reaction is insufficient,

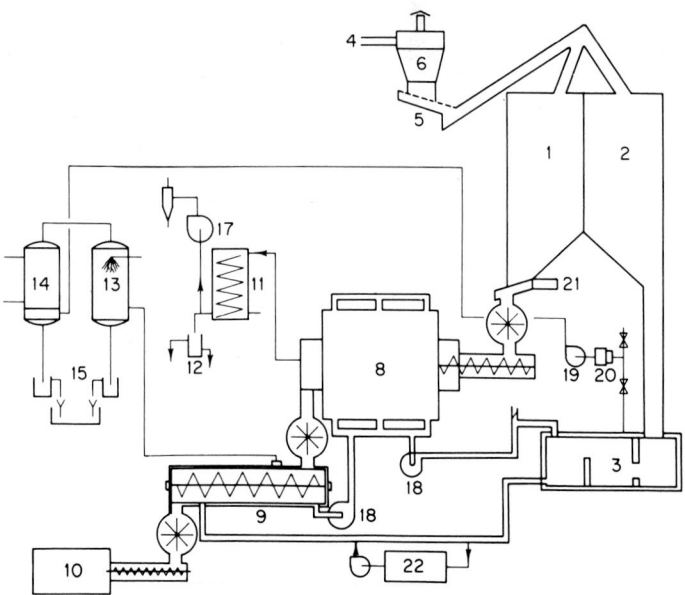

Figure V-9. Flow-sheet of a continuous wood charring plant (9). 1 charwood bin; 2 fuelwood bin; 3 furnace; 4 wood chips entrance; 5 chip screen; 6 cyclone; 8 drying drum; 9 charring chamber; 10 cooling drum; 11 spiral cooler; 12 turpentine oil separator; 13 washing tower; 14 tube collar; 15 fluid lock; 16 collecting vat; 17 drying drum ventilator; 18 flue gas ventilator; 19 ventilator for uncondensable gases; 20 flame protection; 21 measuring appliance; 22 steam boiler.

additional heat can be provided by the burning of wood wastes or by the waste gases from the power and steam plants. As Fig. V-9 shows, the wood from a wood chipper, 4, is further reduced in size and freed of dust in a cyclone, 6. After passing through the sorting sieve, 5, which removes the small fragments, the wood enters the silo, 1, while the wood which passes the sieve goes as fuelwood into a second silo, 2. From the first silo, the wood is passed into the drying chamber, 8, by a screw conveyor. The wood from the second silo is burned in the furnace, 3, and the flue gases from it are used for the heating of the drying chamber and for additional heat for the retort, 9. The charcoal drops automatically from the end of the retort over a bucket wheel into the cooling chamber, 10. The rate of flow of the wood can be regulated by the bucket wheels located before and after the drying chamber and after the retort. The other installations, 11–16, shown in the figure serve for the separation of the volatile products carried over by the gases (9).

It is understandable that in countries with plentiful supplies of wood, as in the Soviet Union, great efforts are being made to improve the processes

for the chemical utilization of wood. Recent literature shows that there, too, the procedure for the wood pyrolysis has been directed toward achieving a continuous operation that uses predried wood. In particular, multistage drying chambers precede the carbonization retorts which, in turn, are divided into several stages in order to obtain a degree of carbonization as high as possible and, at the same time, a substantial preservation of the by-products. Depending upon the dry content of the wood, which may vary between 4.8% and 30%, the total consumption of gas for the pyrolysis is 0.27 to 0.77 cubic meter per kg wood. The average yields of pyrolysis products obtained from such a process are listed in Table V–13 (74,108, 120,142).

TABLE V-13

AVERAGE YIELDS OF PYROLYSIS PRODUCTS (142)

Wood species	Wood tar (%)	Volatile acids (%)	Combustion gas (%)	Charcoal (%)
Birch	19.1	2.2	25.1	31.2
Aspen	14.1	2.5	24.0	36.6
Fir	14.7	1.8	23.8	37.8

The use of a fluidized bed for the production of furfural by pyrolysis with acid-impregnated wood has already been mentioned. In the Soviet Union a similar process has been developed. In this method, the wood particles are kept suspended in the gas stream by gravity fall; thus it has the advantage that the use of an inert gas as a heat-carrier is not required and, consequently, the equipment for the generation of such a gas can be eliminated. The material is introduced at the top into the vertical retort, which is heated electrically. The gaseous products are withdrawn from the upper part of the retort, the dimensions of which depend upon the particle size of the wood. An increase in the size of the particles requires a longer retort or an increased pyrolysis temperature to keep the degradation and the yield of reaction products constant (88).

Varying the temperature makes possible a certain selectivity of the reaction products. Thus an optimum yield of tar, of about 7.5%, is obtained at 400°C, while at 600°C a high yield (85%) of gases is obtained (85). The oxygen content of the gas also has some influence here. The total yield of liquid reaction products increases until an oxygen content of 5% is reached; at the same time, the amount of sedimentation tar decreases until it is practically zero when the oxygen content reaches 5%. Higher oxygen contents have no effect on the yield of pyroligneous acid (85,86, 87).

The fluidized-bed process has also undergone improvements. Origi-

nally conceived as a method for predrying the wood, a fluidized-bed pyrolyzator that uses the combustion gases of wood waste from a special furnace as a heat-carrier has been developed. These gases contain about 1% of oxygen. The pyrolyzed wood particles are separated from the gas in a cyclone, and the cooled gases are washed in a scrubber. The flow resistance caused by the suspended wood particles at a layer height of 50 cm is given as 15 to 22 mm water pressure. The efficiency of the apparatus per cubic meter reactor volume is 308.4 kg air-dried wood per hour. The reaction time for the pyrolysis at 400° to 600°C is about one minute (33,34).

4.3 Economic Aspects of Wood Pyrolysis

The economics underlying any industrial process requires that the final products be sold at a price that covers the cost of production and provides a certain margin of profit. The cost of production includes the cost of the raw materials—in this case, wood—of labor, power, maintenance of the technical installations, and capital investment. To this must be added the overhead cost of management and the expenses of research and development. The price of wood, in this case, plays an important role. Since the methods that destroy the wood fibers have to compete for the available wood supply with the methods that preserve the fibers, it is apparent that, in the former processes, only such woods as are unsuitable for the production of fibers should be utilized. These are primarily waste woods from industry and from forests.

A further decisive criterion for the economy of an industry is the market price of the finished product. Since the chemical industry today can produce the by-products obtained from the pyrolysis of wood, with the exception of charcoal, more cheaply than the pyrolysis process, the main emphasis in the latter is on the production of charcoal. For this reason, simple carbonization methods, similar to the original charcoal piles but in an improved form, are likely to be more economical than more complicated plants that place the emphasis on the isolation and processing of by-products. Recent developments have generally confirmed this concept. This is not to deny that, under certain economic conditions such as exist in areas where waste wood is available in great quantity, the operation of large technical plants can be justified on an economic basis. Under no circumstances should hasty conclusions be drawn from the economics of plants operating during an emergency. In the United States, small transportable carbonization furnaces were developed to produce a high-grade charcoal in good yield from a charge of only a few cords of wood. When the production of charcoal in this way can be integrated with the forest industry for utilization of waste wood, even such small units can produce charcoal economically.

The yield of charcoal from coniferous woods is 70 to 90 kg per cubic meter and, from deciduous woods, 130 to 140 kg per cubic meter, when the pyrolysis is carried out in charcoal piles or in simple carbonization furnaces. When retorts are used for the carbonization, yields of 190 to 200 kg per cubic meter, based on a volume weight of 590 kg per cubic meter for deciduous woods, are obtained. There is no appreciable difference in the yields from continuous and discontinuous processes. However, considerable differences occur with respect to the consumption of heat. In the charcoal-pile carbonization, the heat consumption is 1 to 1.5 million kcal per cubic meter of wood and this heat is provided by the burning of a part of the wood, with a corresponding loss in the yield of charcoal. When retorts are used, considerable differences are found between the continuous and discontinuous methods: for the latter, the heat consumption is 350,000 to 650,000 kcal per cubic meter, for the former, as in the Reichert process, it is 250,000 to 260,000 kcal per cubic meter, and in the Lambiotte process, 10,000 kcal per cubic meter. It must not be overlooked, however, that in the retort processes 200,000 kcal per cubic meter are produced from the carbonization gases and their heat value corresponds to about 200,000 kcal per cubic meter of wood. The low figure for the Lambiotte process is due to the use of wood with only 10% moisture content; to reduce the moisture content from 22% to 10%, 130,000 kcal per cubic meter based on wood must be added. The Stafford process, which operates without the addition of external heat, requires a wood with only 0.5% moisture content. So long as such a drastic drying of the wood cannot be carried out by waste heat, which does not cost anything, this process is no better than the others from the point of view of heat economy.

There are no reliable up-to-date data, apart from those already given, on the production of charcoal. It may be assumed that in the United States today it is close to 400,000 tons. In Sweden in 1948, the production was 1.2 million cubic meters, corresponding to 180,000 tons at an average bulk weight of 150 kg per cubic meter. In Germany, the charcoal production for 1956 was estimated at about 30,000 to 40,000 tons. For the Soviet Union there are no reliable data available, but the production figures are undoubtedly quite high (10,39,69,109).

5. HYDROLYSIS AND OXIDATION IN THE PRESENCE OF ALKALI

The alkali fusion of sawdust for the production of oxalic acid goes back to observations made by Gay-Lussac, but later investigations were needed to establish the best reaction conditions. Good yields are obtained with a ratio of alkali to wood of 2 : 1 and a temperature of 240° to 250°C.

The sodium oxalate is leached out from the melt and the solution is concentrated and allowed to crystallize. The sodium oxalate is treated with hot milk of lime to give calcium oxalate, which is decomposed with sulfuric acid to yield oxalic acid. Mixtures of potassium and sodium hydroxide for the fusion are preferred to sodium hydroxide alone. Acetic acid and methanol can also be prepared in this way. All these products, however, are made today much more cheaply by synthesis, so the alkali fusion process has lost much of its significance. Nevertheless, it seems that the alkali fusion of wood sugar lignin may still be used for the production of protocatechuic acid and oxalic acid (25,41,91,101,141).

Of greater interest from the technical and economic point of view is the reaction of wood and other vegetable fibrous materials with aqueous alkali solutions at high temperatures and pressures. Thus, for example, wood is 90% dissolved in 14% sodium hydroxide within a few hours at 5 atmospheres pressure. A 10% sodium hydroxide solution completely dissolves wood at 200° to 220°C and 20 to 25 atmospheres pressure with a wood-to-liquor ratio of 1 : 20 within 30 to 60 minutes (56,94,135). The rate of dissolution is a function not of the reaction time but of the liquor concentration; for example, wood sawdust is completely dissolved at 180°C in 5 hours with 0.9 part by weight of sodium hydroxide. The yields of acetone, methanol, and oil are, in this case, higher than the sum of their yields obtained under identical conditions from the individual wood components. This points to a chemical combination between the lignin and the carbohydrates and indicates that wood does not behave as a simple mixture of its components (139).

The oils obtained in an alkaline hydrolysis are chiefly unsaturated hydrocarbons, of which the lighter ones have a specific gravity of between 0.87 and 0.92. Sawdust from spruce, birch, and beech (100 g) was heated for 8 hours at 200°C with 48%, 50%, and 46% sodium hydroxide, respectively, based on the dry wood, at a liquor-to-wood ratio of 6 : 1. The mixtures were filtered, leaving 19.1%, 7.9%, and 6% residue, respectively. Aliquot amounts of the filtrates were analyzed for their contents of formic and acetic acid, methanol, and acetone. Other aliquot parts were heated, with and without the addition of sodium hydroxide, to 350°C and were either immediately cooled again or kept for 15 minutes at that temperature. The amounts of gas, tar and pitch, acetic acid, formic acid, methanol, and acetone formed in this way were determined, with the results given in Table V–14.

Only volatile acids, methanol, and acetone were found in the original filtrates. The strongest evolution of gas from the solutions heated at 350°C was found with beechwood and consisted chiefly of hydrogen. It was interesting to note that the formation of tar and pitch decreased with increasing reaction time at 350°C and with the addition of alkali, but formation of

TABLE V-14

DECOMPOSITION PRODUCTS OF WOOD BY AQUEOUS SODIUM HYDROXIDE
SOLUTIONS AT HIGH TEMPERATURE AND PRESSURE (49)

NaOH added (g)	Time kept at 350°C	Gas (liter/100 g wood)			Tar and pitch (g/100 g wood)		Acetic acid (g/100 g wood)		
		Spruce	Birch	Beech	Spruce	Birch	Spruce	Birch	Beech
Original filtrate		—	—	—	—	—	4.77	9.83	6.91
0	0	5.9	8.48	8.44	26.8	27.8	6.54	10.17	13.10
10	0	6.5	7.62	13.60	22.1	—	7.10	11.17	13.20
20	0	—	8.42	15.78	—	—	—	11.85	12.38
30	0	11.6	—	—	10.3	—	11.50	—	—
0	15	13.0	10.58	13.73	26.2	23.0	12.62	11.93	21.73
10	15	8.4	9.54	16.77	25.6	24.8	15.36	12.92	21.02
20	15	—	—	19.00	—	—	—	—	20.89
20	15	8.9	—	—	23.8	—	11.78	—	—

NaOH added (g)	Time kept at 350°C	Formic acid (g/100 g wood)			Methanol (g/100 g wood)			Acetone (g/100 g wood)		
		Spruce	Birch	Beech	Spruce	Birch	Beech	Spruce	Birch	Beech
Original filtrate		3.60	5.82	4.86	0.97	0.37	0.58	0.12	0.04	0.03
0	0	0.94	3.25	2.23	1.38	1.43	1.47	0.58	0.46	0.92
10	0	2.09	2.29	2.18	1.35	1.42	1.15	0.54	0.32	0.53
20	0	—	2.81	2.05	—	1.26	0.98	—	0.42	0.64
30	0	3.24	—	—	1.63	—	—	0.41	—	—
0	15	0.61	3.69	1.21	1.52	1.03	1.58	0.77	0.30	1.52
10	15	1.34	2.86	1.75	1.25	1.15	0.96	0.74	0.52	1.09
20	15	—	—	1.63	—	—	0.92	—	—	0.88
30	15	1.96	—	—	1.29	—	—	0.76	—	—

acetic acid increased, especially with beech, although the further addition of alkali had no marked effect when combined with prolonged heating. While the formation of formic acid, in each case, decreased, the yield of methanol and acetone increased.

When 3 kg wood was heated with milk of lime for 2 hours at 250°C, 27 liters of gas, 0.74% oil, and 6.31% organic distillates, based on the weight of dry wood, were obtained. In addition, large amounts of calcium compounds and an insoluble residue were obtained. When the reaction temperature was raised to 325°C and the pressure to 170 atmospheres, the yield of gas increased to 125 liters and of oils to 7.5%; distillation of the aqueous solution gave another 13.5% of oils, so that their total yield was 21% (13).

The production of gas, oils, and volatile acids by alkaline pressure oxidation and hydrolysis of wood is of no technical interest in countries having a well-developed chemical and petrochemical industry. On the

other hand, the formation of hydrocarbons may be of interest in connection with the formation of natural oil and gas (13,36,136). Thus, experiments dealing with the carbonization of carbohydrates and vegetable material in the presence of alkali at a temperature of 400°C gave phenols, phenol-carboxylic acids, methane, and other acids, in addition to bitumen. From the latter, a product resembling natural asphalt could be obtained on vacuum distillation. Hydrogenolysis of the bitumen with molybdenum compounds as a catalyst at 400°C and 70 atmospheres initial hydrogen pressure gave a mixture containing hydrocarbons similar to gasoline, naphtha, gas oil, and heavy oil. With bagasse (35% lignin) and bamboo (19% lignin), the yield of low-boiling products, such as gasoline, increased with decreasing lignin content while, on the other hand, the yield of bitumen, hydro-bitumen, heavy oils, and asphalt decreased sharply with decreasing lignin content (55,140).

6. HYDROGENOLYSIS* OF WOOD AND LIGNIN

Many of the investigations dealing with the hydrogenolysis of lignin have already been discussed in Chapter IV, Section 6. Most of these have been undertaken in an attempt to elucidate the structure of lignin. The technical processes for the hydrogenolysis of wood and wood sugar lignin are based partly on the methods used for the liquefication of coal for the production of oils and hydrocarbons to be used in gasoline engines.

The hydrogenolysis of cellulose, ground wood, and lignin at temperatures of 260° to 460°C and pressures of 200 atmospheres in the presence of nickel compounds as catalysts gave coke, distillates, and tar, the yields of which are given in Table V–15, which also shows the solubility of the tar substances in various inorganic salt solutions (35). As was to be expected, groundwood takes an intermediate position between cellulose and lignin. The yield of tar decreases with increasing lignin content of the material being hydrogenolyzed, while the amount of phenolic substances in the tar increases at the expense of the neutral substances.

The hydrogenolysis of sprucewood lignin at 240° to 350°C in a hydrogen atmosphere gave 28.7% phenols, of which 25.5% was a phenolic fraction, 2.5% was neutral phenol ethers, and was 0.7% cyclic alcohols. The phenols could be separated by fractionated distillation to give the following products: 57.1% guaiacol and homologs, 26.5% monophenols and homologs, 16.4% pyrocatechol and homologs, 0.3% formic and acetic acids, 0.5% methanol and ethanol, and 0.7% toluene. Instead of elementary hydrogen and with a metal catalyst, ethyl alcohol or isopropyl alcohol can be used

---

* See footnote in Chapter IV, Section 6.

TABLE V-15

HYDROGENOLYSIS OF CELLULOSE, MECHANICAL WOOD PULP, AND
LIGNIN AND SOLUBILITY OF TAR IN VARIOUS INORGANIC
SALT SOLUTIONS (35)

Hydrogenolysis product[a] from	Cellulose	Wood pulp	Lignin
Charcoal	1.02	03.5	15.6
Total distillate	82.0	80.2	66.8
Aqueous distillate	51.2	54.3	50.2
Tar	33.6	25.6	17.8
Solubility of the tar in	%	%	%
Sodium bisulfite	6.6	9.0	4.4
Sodium carbonate (acids)	7.5	8.2	9.7
Sodium hydroxide (phenols)	9.6	32.7	55.4
Neutral products	76.3	50.1	30.5

[a] Yields in grams from 500 g starting material

as hydrogen donors in the presence of alkali. In this way, cuoxam lignin gave 26.5% phenols and hydrochloric acid lignin gave 21% to 22% (35,40, 83).

The effectiveness of the catalyst in the hydrogenolysis is rated differently by different investigators. Raney nickel is judged to be especially effective. Other metal compounds are also mentioned as being satisfactory. The relative yields of gases and organic liquids depend not only on the amount of the catalyst but also on the reaction temperature. A controlled hydrogenolysis of carbon-containing materials requires the formation of unsaturated molecular fractions (radicals) at such a rate that these fragments can be stabilized immediately by the addition of hydrogen before they can recombine or polymerize. At the temperature used, the hydrogenation of the fragments is so rapid that no catalyst is necessary. On the other hand, a concentration as high as possible of the reaction components must be present and the use of a catalyst, in such cases, assures a sufficient supply of decomposition products at relatively low temperatures. From this point of view, the principal function of the catalyst is the promotion of the thermal decomposition reaction (15,19,35,50,116).

Certain media used for the suspension of the material to be hydrogenolyzed can be very useful because they act as solvents for the starting material and for the reaction products as well as hydrogen donors by supplying bound hydrogen, which, as compared with dissolved hydrogen, is in a large excess. Such a solvent is tetralin (1,2,3,4-tetrahydronaphthalene);

others are mixtures of ethanol and water, dioxane, and dilute alkaline solutions (15,18,19).

Of the countless products that have been obtained from the hydrogenolysis of wood and its components, only a few have thus far been identified. The prospects of carrying out the destructive hydrogenation of wood on an economical industrial basis are not good, especially since the separation of the reaction products requires complicated apparatus and the hydrogenolysis equipment, which has to withstand high pressures and temperatures, is quite expensive. To this must be added the fact that these reaction products, like those from the pyrolysis, have to compete with those produced by the chemical and petrochemical industries. In general, it may be said that wood and lignin give higher yields of phenols and lower yields of furan derivatives than cellulose does. Of the products obtained from cellulose, 30% is water and 68% are gases, consisting of about 33% carbon dioxide, 28% carbon monoxide, 38% methane, and about 1% ethane. The yields of acetic acid or acetates and methanol are about the same as obtained from the pyrolysis. The yield of cyclohexanols, calculated on the basis of the lignin content, is 60% to 70% from maple and 35% to 40% from spruce (42).

Sawdust from red spruce, when hydrogenolyzed at 170°C with a mixture of hydrogen and carbon monoxide in the presence of dicobaltoctacarbonyl as a catalyst, gives chiefly phenol and guaiacol, in addition to 4-methyl-, 4-ethyl-, and 4-propylguaiacol, 5,6,7,8-tetrahydro-3-methoxy-2-naphthol, 3-(3-methoxy-4-hydroxyphenyl)-1-butanol, and 2,2'-di-hydroxy-3,3'-dimethoxy-5,5'-dipropyldiphenyl (96). Hydrogenolysis of aspenwood meal with hydrogen and Raney nickel gave guaiacol, 4-methyl-, 4-ethyl-, and 4-propylguaiacol, dihydroconiferyl alcohol, syringol, 4-methyl-, 4-ethyl, and 4-propylsyringol, and dihydrosinapyl alcohol. The highest yields of phenolic substances were 52.2%, calculated on the basis of the lignin, obtained from a hydrogenolysis for 5 hours at 195°C and 34 atmospheres hydrogen pressure (103). Alkali lignin, acid lignin, and lignite, when hydrogenolized in the presence of various metal sulfide and metal oxide catalysts, gave phenol, protocatechol, and diesel oil. The objective of obtaining high yields of *p*-alkylphenols was not attained in this way because complicated mixtures of phenols were formed (115).

Thus it is possible to use different methods for the hydrogenation and hydrogenolysis of lignin and lignin-containing materials. The reactions can be carried out in neutral or in alkaline mediums. The hydrogenolysis can be carried out in the same medium without a catalyst when nascent hydrogen or hydrogen donors are used. Finally, a delignification can be carried out under the conditions of a hydrogenolysis. In such reactions, the various vegetable lignins behave in different ways. Bagasse lignin gives

chiefly compounds of the guaiacyl and syringyl type, as well as 4-propyl-phenol and *p*-hydroxybenzoic acid. Corncob lignin gives considerable amounts of decomposition products of alkylphenols, in addition to decomposition products of the syringyl alcohol and guaiacol series (118). The hydrogenolysis of lignin gives higher yields of phenols than its pyrolysis or the chemical treatment of fuels. Lignin from the cotton plant, when hydrogenolyzed in 5% sodium hydroxide solution in the presence of cobalt sulfide as a catalyst and phenol as an inhibitor, gave 1.3% 2,6-xylenol, 12.5% *o*-ethylphenol, and 12% of a mixture of *m*- and *p*-cresols. Under the same conditions, pinewood lignin gave 35% *m*- and *p*-cresols. The phenols, the phenolic acids, and the condensed high-molecular aromatic compounds can be sulfonated directly or can be condensed with furfural and formaldehyde and then sulfonated with oleum, to give cation-exchange resins. In this way, a complicated separation of the single components is avoided (20,21).

It is not yet known to what extent the hydrogenation and the hydrogenolysis of wood or technical lignins can be used industrially. It may be assumed, however, that this type of utilization of wood sugar lignin has a better chance of commercial application than that of waste wood.

REFERENCES

1. Akhmina, E. I., Benzmozgin, E. S., Nemchenko, A. G., Podgornaya, T. A. Sukhanovskii, S. I. and Yudkevich, Yu. D., *Khim. i Tekhnol. Topliva i Produktov ego Pererabotki Sb.* **1965**, 12–16.
2. Akhmina, E. I., Sukhanovskii, S. I., Kharlamova M. V., Andreeva, Z. N. Yumshanov, S. N. and Ilina, E. I., *Gidrolizn. Lesokhim. Prom.* **20** (4), 8–10 (1967).
3. Aschan, O., *Brennstoffchemie* **2**, 273 (1921); **4**, 129, 164 (1923).
4. Bagaev, A. N., *Gidrolizn. i Lesokhim. Prom.* **17** (2), 8–9 (1964); **18** (7), 13–15 (1965).
5. Beglinger, E. and Locke, E. G., *Economic Botany* **11** (2), 160 (1957).
6. Bergström, H., "Kolning in mila, skorstenmilor." Esselte AB, Stockhom, 1946.
7. Bergström, H., "Kolning i ugn." Stockholm, 1947.
8. Bergström, H., *Produkter ur Trä*, Vol. VII, Stockholm, 1949.
9. Bergström, H., *Produkter ur Trä*, Vol. VI, Stockholm, 1947.
10. Bergström, H., *Ingen. Vetenskap. Akad. Handl.* No. 3 (1950).
11. Bergström, H., *Kolningslaboratoriet* 1902–1952, 47–52, Stockholm, 1952.
12. Bergström, H., *Produkter ur Trä*, Vol. VIII, Stockholm, 1954.
13. Bergström, H., *Ingen. Vetenskap. Akad. Handl.* 118 (1936).
14. Berl, E. and Schildwachter, H., *Brennstoffchemie* **9**, 105, 121, 137 (1928); Berl, E. and Schmid, A., *Annalen Chem.* **461**, 192; **493**, 97, 124, 135 (1932); *Naturwissenschaften* **20**, 652 (1932).
15. Boomer, E. H. and Edwards, J., *Can. J. Res.* **13B**, 323 (1935); Boomer, E. H., Argue, G. H. and Edwards, J., *Can. J. Res.* **13B**, 337 (1935).
16. Brauns, F. E., "Chemistry of lignin," p. 581. Academic Press, New York, 1952.
17. Brauns, F. E. and Brauns, D. A., "Chemistry of lignin" Supplement Vol., p. 567. Academic Press, New York, 1960.

18. Brewer, C. P., Cooke, L. M. and Hibbert, H., *J. Am. Chem. Soc.* **65**, 1192, 1195 (1943).
19. Brewer, C. P., Cooke, L. M. and Hibbert, H., *J. Am. Chem. Soc.* **70**, 57 (1948).
20. Bronovitskii, V. E. and Kalinskaya, L., *Uzb. Khim. Zh.* **11** (2), 31 (1967).
21. Bronovitskii, V. E., Salyamova, F. and Volokovich, M. A., *Uzb. Khim. Zh.* **11** (4), 68 (1967).
22. Bronzov, O. V., Kozlov, V. N., *Khim. Pererabotka i Zashchita Drevesiny Akad. Nauch. Latv. SSR, Inst. Khim. Drevesiny* **1964**, 103–6.
23. Bugge, G., "Industrie d. Holzdestillationsprodukte," p. 101. Steinkopff, Dresden, 1927.
24. Bunburry-Elsner, "Die trockene Destillation des Holzes." Springer, Berlin, 1925.
25. Cross, C. G. Bevan, E. J. and Isaac, J. F., *J. Soc. Chem. Ind.* **111**, 966 (1892).
26. Degussa A.G., German Patents 666,387 (1932); 712,552, 713,290, 744,135 (1937).
27. Dikun, P. P., Liverovskii, A, A., Shmulevskaya, E. I., Gorelova, N. D. Parfenteva, L. N. and Vzdirnikova, R. M., *Sovrem. Probl. Unkol. Sb.* **1965**, 48–54.
28. Dikun, P. P., Liverovskii, A. A., Shmulevskaya, E. I. Gorelova, N. D., Parfenteva, L. N., Romanovskaya, L. S. and Pankina, E. I., *Vop. Onkol.* **13** (3), 80–85 (1967).
28a. Dikun, P. P., Liverovskii, A. A., Shmuleveskaya, E. I., Gorelova, N. D., Parfenteva, L. N., Romanovskaya, L. S. and Pankina, E. I., *Vop. Onkol.* **13** (11), 73 (1967).
29. Domburga, G. and Sergeeva, V. N., *Latvijas PSR Zinatun Akad. Vestis* **5**, 625 (1964).
30. Eickner, H. W., *Forest Prod. J.* **12** (4), 194 (1962).
31. Erdman, H. and Schafer, W., *Ber. Chem. Ges.* **43**, 2398 (1910).
32. Ermolaeva, S. S., *Zh. Priklad. Khim.* **21**, 543 (1948).
33. Fefilov, V. V., *Izv. Vysshikh. Uchebn. Zavadenii Lesn. Zh.* **9** (2), 130–33 (1966).
34. Fefilov, V. V. and Sokolova, A. I., *Gidrolizn. i Lesokhim. Prom.* **11** (7), 15–17 (1958).
35. Fierz-David, H. E. and Hennig, M., *Helv. Chim. Acta.* **8**, 900 (1935).
36. Fischer, F. and Schrader, H., *Brennstoffchemie* **2**, 37 (1921).
37. Fischer, F. and Tropsch, H., *Ges. Abhandl. Kenntnis Kohle* **7**, 181 (1935).
38. Foliadova, Z. I., Bystrova, O. N. and Kiprianov, A. I., *Mater. Nauch.-Tekh. Konf. Leningr. Lesotekh. Akad.* **4**, 197–203 (1966).
39. Forest Products Laboratory, Charcoal, production, marketing, use, *Forest Prod. Lab. Joint. Rept.* No. 2213 (1961).
40. Freudenberg, K. and Adam, K., *Chem. Ber.* **74** (8), 387 (1941).
41. Gay-Lussac, J. L., *Ann. Chim. Phys.* **41**, 398 (1829).
42. Godard, H. P., McCarthy, J. L. and Hibbert, H., *J. Am. Chem. Soc.* **62**, 988 (1940).
43. Golova, O. P., Andrievskaya, E. A., Pakhomov, A. M. and Merlis, N. M., *Izv. Akad. Nauk. SSSR, Otdel. Khim. Nauk.* **1957**, 389–91.
44. Golova, O. P. and Krylova, R. G., *Dokl. Akad. Nauk. SSSR.* **116**, 419 (1957).
45. Golova, O. P., Pakhomov, A. M. and Andrievskaya, E. A., *Dokl. Akad. Nauk. SSSR.* **112**, 430–32 (1957).
46. Golova, O. P., Pakhomov, A. M. and Nikolaeva, I. I., *Izv. Akad. Nauk. SSSR. Otdel. Khim. Nauk.* **1957**, 519–21.
47. Goos, A. W., see Wise and Jahn "Wood chemistry" *in* Vol. 2, p. 826 ff. Reinhold, New York, 1952.
48. Hägglund, E., *Archiv. Kemi. Mineralog., Geolog.* **7**, 8 (1918).
49. Hägglund, E., *Svensk. Kem. Tidskr.* **39**, 19 (1927).
50. Hallonquist, E. G., *Ind. Eng. Chem.* **43**, 1427 (1951).
51. Heuser, E., "Chemistry of cellulose." Wiley, New York, 1944.
52. Heuser, E. and Brotz, A., *Papierfabrikant* **23** (Festschrift), 69, 551 (1925).

53. Heuser, E. and Sjödebrand, C., *Z. Angew. Chem.* **32**, 41 (1919).
54. Heuser, E. and Scherrer, A., *Brennstoffchemie* **4**, 97 (1923).
55. Heinemann, H., *Petroleum Refiner.* **29**, 111 (1950); **33**, 161 (1954).
56. Hilpert, R. S. and Hansi, W., *Ber. Chem. Ges.* **70B**, 2209 (1937).
57. Jurkovic, J., Misovec, P. and Kosik, M., *Drevar. Vyskum* **1963** (1), 59-67.
58. Kashirskii, V. G., Raplovets, L. S. and Khotuntsev, L. L., USSR Patent 159,863 (1964).
59. Keylwerth, R. and Christoph, N., *Materialprüfung* **2** (8), 281 (1960).
60. Kilzer, F. J. and Broido, A., *Pyrodynamics* **2** (2/3), 151 (1965).
61. Kiprianov, A. I., Foliadova, Z. I. and Bystrova, O, N., *Izv. Vysshikh. Uchebn. Zavadenii, Lesn. Zh.* **8** (6), 146-50 (1965).
62. Kiprianov, A. I., Foliadova, Z. I. and Soitonen, G. P., *Izv. Vysshikh. Uchebn. Zavadenii, Lesn. Zh.* **8** (6), 145-7 (1965).
63. Klanduch, J., Kosik, M., Rendos, F. and Domansky, R., *Holzforschg.-Holzverwertg.* **17**, 1-4 (1965).
64. Klason, P., *Archiv. Kemi. Mineralog., Geolog.* **5**, 7 (1913); *J. Prakt. Chem.* **90**, 413 (1914).
65. Klason, P., *Brennstoffchemie* **1**, 79 (1920).
66. Klason, P., Heidenstam, G. and Norlin, E., *Z. Angew. Chem.* **22**, 1205 (1909).
67. Klason, P., Heidenstam, G. and Norlin, E., *Z. Angew. Chem.* **23**, 1252 (1910).
68. Klar, M., "Technololgie der Holzverkohlung." Springer, Berlin, 1923.
69. Klar, M., *Zellstofffaser* **33** (5/6), 65 (1936).
70. Koryakin, V. I. and Fursova, V. V., *Sb. Tr. Tsentr. Nauchn.-Issled. i Prochtn. Inst. Lesokhim. Prom.* **15**, 8-11 (1963).
71. Kozlov, V. N., Bronzov, O. A. and Vekshegonov, F. Z., *Izv. Vysshikh. Uchebn. Zavadenii, Lesn. Zh.* **4** (4), 136-45 (1961).
72. Kozlov, V. N, Bronzov, O. V. and Vekshegonov, F. Y., *Izv. Vysshikh. Uchebn. Zavadenii, Lesn. Zh.* **4** (6), 147-53 (1961).
73. Kozlov, V.N., Bronzov, O.V. and Utkina, E.A., *Izv. Vysshikh. Uchebn. Zavadenii, Lesn. Zh.* **5** (4), 130-5 (1962); **6** (2), 145-50 (1963); **10** (2), 130-33 (1967).
74. Kozlov, V. N. and Kozhevnikov, N. P., *Tr. Ural'sk Lesotekhn. Inst.* **18**, 92-103 (1962).
75. Kozlov, V. N. and Tishchenko, D. V., *Gidrolizn. i Lesokhim. Prom.* **18** (1), 12-13 (1965).
76. Kozmal, F., Zelnik, A., Rendos, R., Domansky, J., Pajtik, J. and Kosik, M., *Bull. Inst. Politekh. Lasi.* **11**, 364-71 (1965).
77. Kratzl, K. and Tschamler, H., *Monatsh. Chem.* **83**, 786 (1952).
78. Kromina, L. V. and Tishchenko, D. V., *Gidrolizn. i Lesokhim. Prom.* **19**, (6), 9-11 (1966).
79. Kürschner, K., "Zur Chemie der Ligninkörper." Enke, Stuttgart, 1928.
80. Kürschner, K. and Melcerova, A., *Holzforschung* **19** (6), 161, 171 (1965).
81. Kutanov, I. P. and Udarov, B. C., *Vesti Akad. Navuk Belarusk SSR, Ser Fiz.-Tekhn Navuk* **1964** (1), 44-48.
82. Lambiotte, Co. French Patent 859,741 (1936); German Patent 763,915 (1938).
83. Lautsch, W. and Piazolo, G., *Chem. Ber.* **76B**, 486 (1943).
84. Levin, E. D. and Chuprova, N. A., *Materialv. Konf. po Itogam Nauchn.-Issled Rabotza 1964 God, Sibirsk Tekhnol. Inst. Krasnoyarsk USSR* **1965**, 50-53.
85. Levin, E. D. and Chuprova, N. A., *Tr. Sibirks. Tekhnol. Inst. Sb.* **39**, 403-12 (1964).

86. Levin, E. D. and Chuprova, N. A., *Khim. Pererabotka Drevesiny, Ref. Inf.* **28**, 8-9 (1965).
87. Levin, E. D., Chuprova, N. A. and Malkov, G. A., *Gidrolizn. i Lesokhim. Prom.* **19** (3), 15-16 (1966).
88. Levin, E. D. and Malkov, G. A., *Tr. Sibirsk. Tekhnol. Inst. Sb.* **36**, 94-100 (1963).
89. Lipska, A. G. and Parker, W. J., *J. Appl. Pol. Sci.* **10** (10), 1439 (1966).
90. Madorskii, S. L., Hart, V. E. and Strauss, S., *J. Res. Nat. Bureau Standards* **56**, 343 (1956).
91. Mahood, S. A. and Cable, D. E., *Ind. Eng. Chem.* **11**, 651 (1919).
92. Merrit, R. W. and White, A. A., *Ind. Eng. Chem.* **35**, 297 (1943).
93. Minami, K. and Kawamura, K., *J. Japan Forest Soc.* **40**, (2), 61 (1958).
94. Mirlis, I. D. and Njemzowa, N. O., USSR Patent 40,567 (1934).
95. Mitchell, R. L. Seborg, R. M. and Millet, M. A., *Forest Prod. J.* **3**, (4), 38, 72 (1953).
96. Nahum, L. S., *Ind. Eng. Chem., Prod. Res. Develop.* **4**, 71 (1965).
97. Nikandrov, B. F., Slavianskii, A. K. and Sokolova, M. N., *Izv. Vysshikh. Uchebn. Zavadenii, Lesn. Zh.* **6** (2), 151-55 (1963).
98. Nikandrov, B. F., Slavianskii, A. K. and Sokolova, M. N., *Izv. Vysshikh. Uchebn. Zavadenii, Lesn. Zh.* **7** (3), 153-57 (1964).
99. Novikov, G. Y., Piyalkin, V. N. and Slavianskii, A. K., *Mater Nauch.-Tekh. Konf. Leningr. Lesotekh. Akad.* **4**, 172-73 (1966).
100. Othmer, D. F. *Chem. Metallurg. Eng.* **48**, 91 (1941).
101. Othmer, D. F., Gamer, B. and Jacobs, J. J., *Ind. Eng. Chem.* **34**, 262 (1942).
102. Panasyuk, V. G., *Zh. Priklad. Khimii* **39**, 598; 813 (1957).
103. Pepper, J. M. and Steck, W., *Can. J. Res.* **41** (11), 2867 (1963).
104. Pictet, A., *Helv. Chim. Acta* **1**, 226 (1918).
105. Pictet, A. and Gaulis, M., *Helv. Chim. Acta* **6**, 627 (1923).
106. Pictet, A. and Sarasin, I., *Helv. Chim. Acta* **1**, 87 (1918).
107. Piyalkin, V. N. and Slavianskii, A. K., *Izv. Vysshikh. Uchebn. Zavadenii, Lesn. Zh.* **9** (1), 127-29 (1966).
107a. Piyalkin, V. N. and Slavyanskii, A. K., *Tr. Leningrad. Lesotekh. Akad.* **105**, 40 (1966).
107b. Piyalkin, V. N. and Slavyanskii, A. K., *Tr. Leningrad. Lesotekh. Akad.* **105**, 31 (1966).
108. Piyalkin, V. N., Slavyanskii, A. K. and Tsyganov, E. A., *Mater. Nauch.-Tekh. Konf. Leningr. Lesotekh. Akad.* **4**, 168-72 (1966).
109. Pohl, W. *in* Ullmann, "Enzyklopadie der Techn. Chemie," Vol. 8, p. 585. Urban & Schwarzenberg, München-Berlin, 1957.
110. Ponomarjev, A. N. *in* Nikitin, N. I., "Chemie des Holzes." Akademie Verlag, Berlin, 1955.
111. Porkhorchuk, T. I. and Lyamin, V. A., *Izv. Vyssikh. Uchebn. Zavadenii, Lesn. Zh.* **8** (1), 148-56 (1965); *Mater. Nauch.-Tekh. Konf. Lening. Lesotekh. Akad.* No. **4**, 101-4 (1966).
112. Rassow, B. and Neumann, P., *Wochenbl. Papierfabr.* **66** (Special No.) 25 (1935).
113. Rendos, F., Domansky, R., Kozmal, F., Zelnik, A. and Paitik, I., *Gidrolizn. i Lesokhim. Prom.* **17** (7), 12-13 (1964).
114. Reznikov, V. M. and Morozov, E. F., *Khim. Pererabotka Drev. Ref. Inform.* **22** (10), (1966).
115. Rieche, A., Redinger, L. and Lindenhayn, K., *Brennstoffchemie* **47** (11), 326 (1966).
116. Routala, O., *Acta Chem. Fennica* **3**, 115 (1930).

117. Sandermann, W. and Augustin, H. *Holz. Roh- u. Werkstoff* **21**, 256, 305 (1963).
118. Schweers, W., *Paperi Puu* **48** (4) (1966).
119. Seaman Waste Wood Chemical Co., U.S. Patents 1,108,403 (1914); 1,115,590 (1914); 1,236,884 (1917); 1,236,885 (1917).
120. Seliverstov, N. I. and Tokishin, G. F., *Gidrolizn. i Lesokhim. Prom.* **19** (2), 20–22 (1966).
121. Sergejeva, W. N. and Waiwad, A. J., *Latvijas PSR Zinatnu Akad. Vestis* **86** (9), 103 (1954).
122. Shaposhnikov, Y. K. and Kosyukova, L. V., *Gidrolizn. Lesokhim. Prom.* **19** (8), 19–21 (1966).
123. Shaposhnikov, Y. K. Kosyukova, L. A. and Volkova, E. P., *Gidrolizn. Lesokhim. Prom.* **20** (1), 16–17 (1927).
124. Shmulevskaya, E. I. and Liverovskii, A. A., *Tr. Leningr. Lesotekhn. Akad.* **1963**, (102), 211–20.
125. Slavianskii, A. K., *Izv. Vysshikh. Uchebn. Zavadenii, Lesn. Zh.* **8** (6), 127–31 (1965).
126. Slavianskii, A. K. and Nikandrov, B. F. and Sokolov, M. N., *Izv. Vysshikh. Uchebn. Zavadenii, Lesn. Zh.* **4** (6), 133–37 (1961); **6** (7), 150–53 (1963).
127. Societé Produits Chimiques Clamecy, Etabl. Lambiotte Freres, French Patents 622,680 (1962); 760,593 (1933).
128. Stafford Retort Processing Co., U. S. Patent 1,380,262 (1919); German Patent 420,635 (1920).
129. Stamm, A. J., *Ind. Eng. Chem.* **48**, 413 (1958).
130. Staudinger, H. and Jurisch, I., *Papierfabrikant* **37**, 181 (1939).
131. Sukhanovskii, S. I., Akhimina, E. I., Podgornaya, T. A. and Lisina, Z. I., *Sb. Tr. Vses. Nauchn.-Issled. Inst. Gidrolizn. i Sulfitno-Sprit Prom.* **13**, 274–81 (1965).
132. Sukhanovskii, S. I. and Timofeeva, V. I., *Sb. Tr. Gos. Nauchn.-Issled. Inst. Gidrolizn. i Sulfitno-Sprit. Prom.* **10**, 202–14 (1962).
133. Szelenyi, G. and Gomory, A., *Brennstoffchemie* **9**, 73 (1928).
134. Tang, W. K., *US Forest Service Res. Paper FPL* 71, Jan. 1967.
135. Tauss, H., *Dinglers Polytechn. J.* **276**, 411 (1890).
136. Terres, E., *Z. Angew. Chemie.* **13** (1936).
137. Tropsch, H., *Ges. Abhandl. Kenntnis Kohle* **6**, 293 (1923).
138. Venn, J., *J. Textile Inst.* **15**, T414 (1924).
139. Wallin, H. and Oden, S., *Ing. Vetenskaps Akad. Handling.*, No. 54 (1926).
140. Wenzl, H. F. J., *Holzforschung* **10** (5), 129 (1956).
141. Zavyalov, A. N. and Frolov, S. S., USSR Patent 168,672 (1965).
142. Zhikhar, G. I. and Tsatska, E. M., *Izv. Vysshikh. Uchebn. Zavadenii, Energ.* **10** (3), 39–44 (1967).

VI

The Preparation of Wood for the Production of Pulp

1. STORAGE AND PILING

The installations and processes which have been developed for the treatment and storage of wood for the production of pulp depend largely upon regional conditions. In the Scandinavian countries, in Canada and, in part, in the United States, the pulpwood, after being felled in the forest, is delivered to the mills by dumping it in the form of 4-foot or 12- to 16-foot logs into rivers or specially designed flumes or chutes, or is towed to the mill in booms or rafts. In this operation the bark mostly remains on the log. Sometimes the tree trunks are only partially cut in even lengths. In Central Europe and also in many parts of the United States pulpwood is transported to the mills by railway flatcars or by trucks. In this case the wood is cut to a certain length, usually 4 to 6 feet or 12 to 16 feet and the transport vehicles are equipped in such a way that they can be loaded and unloaded in a minimum of time. Wood transported by rail or truck is often barked in the forest, with a large part of the bast fibers remaining on the wood. The form in which the wood is delivered to the mills has an important effect on its storage. Wood that has been delivered by floating is conveyed to the barking plant soon after its arrival and is not only freed of the bark and bast but is also cut to a certain length. The barked wood is then stored in wood stacks. In the early days, the wood was usually piled in regular rows, and this is still occasionally done today. It is understandable that when man power is lacking or is in short supply, the work connected with orderly stacking is a disadvantage. As a result, the mills have frequently resorted to storing the wood in large, irregular, conical piles, with the help of suitable conveying equipment. From these piles, it is again fed mecha-

nically to the mill. Such piles may be many meters in height and contain several thousand cords of wood.

In spite of continuous improvement in the necessary conveying installations, the dismantling of such piles still entails certain difficulties. For this reason the practice of stacking the logs has recently been abandoned in favor of storing the wood in the form of chips. This has the advantage of using considerably simpler transportation facilities. The outdoor storage of chips reduces the handling costs considerably but requires additional installations which consume a fairly large amount of power and, in addition, causes greater losses of wood. This question will be dealt with later.

The pulp mills in the southern United States obtain their wood in cordwood length by trucks or rail, and store a supply sufficient for only three to six weeks. Other pulp mills receive 12- to 16-foot logs which are barked mechanically and converted into chips. Both types of mills obtain the wood during the whole year in regular deliveries and keep only a small amount of unbarked wood on hand. The trucks and railway cars are of standard design in order that they may load and unload in the shortest time and with the minimum amount of labor.

One of the most noteworthy developments in the wood yard is undoubtedly the outdoor storage of wood chips. This type of storage had its origin in the use of sawmill waste as raw material for pulp mills. The development has gone even further and has recently led to an attempt to provide the mills wherever possible with wood chips direct from the forest. Pipe lines through which the chips can be transported play a decisive role. However, such pipe lines are economical only if the mill is fully and continuously in operation. Before such lines are installed, a series of technical problems and local prerequisites must be checked. Some of these problems are: sufficient water supply at the starting point of the pipe line; returning of the water used for the transportation; difficulties in the resumption of the operation of the line after a stoppage; protection of the line against freezing; distribution of the flow of chips during a cessation of operations in the mill; and storage of the wet chips, especially during the winter months. Longer seasonal delivery periods are of decisive importance, regardless of the form in which the wood is delivered to the mill (58).

It is comprehensible that the question of the effect of damage of the wood during its storage on the yield and quality of the pulp should have been carefully studied. As already mentioned, a large part of the wood is delivered to the mills by floating it on water. It is well known that wood that has been in water for a long time produces a high splinter portion in the pulping process. These splinters lead to a loss in yield of 2% to 5% of the pulp as compared with that produced from wood that had not been floated or that had been barked before floating. The outer splint layer in the wood, which is about 5 mm in thickness and directly attached to the

bark, often appears to be yellow in color. This layer exhibits a pronounced resistance toward sulfite pulping. It has been assumed that this resistance is the result of an enrichment of extractives, which, during the floating or during the storage in water, migrates by diffusion from the bark into the sapwood. It has been found that sound wood can be made resistant to pulping by the sulfite process by impregnation with an acetone extract of wood that had been floated or had been stored in water. When such wood is treated with a mixture of methanol and sulfuric acid, a typical color reaction is obtained, caused by the oxidation of leucoanthocyanidines by the oxygen of the air. Besides leucoanthocyanidines, tannins of the catechol type may diffuse from the bark into the sapwood. The above color reaction corresponds to the lignin-phloroglucinol color reaction and is the result of a condensation of coniferyl aldehyde groups in the lignin with the polyphenolic catechol tannin. A better sulfite cook is obtained when the cook is first carried out with a sodium sulfite cooking liquor and is then finished with a calcium bisulfite liquor. The blocking of the reactive groups of the phenolic compounds in the wood by a pretreatment with formaldehyde in alkaline solution before the cook also results in a considerable improvement in the sulfite cooking process (1,36).

Prolonged storage causes losses in pulp yield. In this connection, the moisture content in the wood pile, especially of floated wood, plays an important role. Differences in moisture content of between 26% and 47% within the wood pile have been found and, as a result, the moisture content of the sapwood varies between 49% and 52%, that of the heartwood between 28% and 35% (66). Deciduous woods are, in general, more susceptible to storage damage than coniferous woods (125). A comparison of the storage in river water, seawater, and on land shows that storage in seawater is the most advantageous, while storage in river water is more favorable than storage on land. The storage period for wood chips from deciduous woods should not be longer than half a year, that for chips from coniferous wood, not longer than one year. So long as there is no great damage by fungal attack, no harmful effects upon the mechanical properties of the pulp prepared from stored wood have been observed. On the other hand, a considerable decrease in the brightness of the resulting pulp has been found. The losses of wood substance from stored roundwood of unseasoned peeled pine during one and two warm seasons have been found to be $1.1 \pm 0.6\%$ and $1.7 \pm 0.6\%$, respectively, and were much smaller than those of chips of green pine stored outside; the latter usually lose 2% to 4% after a storage of 2 to 4 months (2,62a,83,135).

Losses in yield also result from other causes. The storage of wood chips from fresh sprucewood in piles of from 8000 to 9000 cubic meters for 4 to 13 months resulted in considerable wood losses after the longer storage periods. After 14 months of storage, wood losses of up to 4% have been

observed. A deacetylation of carbohydrates and the formation of acetic acid accompany this wood loss and cause a discoloration of the wood. The possibility of additional losses due to the action of microorganisms should not be ruled out (4,104).

Damage through fungus attack is not necessarily disadvantageous to yields in the sulfite pulping process. However, when the density of the wood is markedly decreased, considerable losses in pulp strength and a decrease in yield occur. Barked wood is less affected by fungus attack and its ill effects than is unbarked wood. Jack pine, after storage for 26 months, showed a decrease in density from 0.414 to 0.378, with the decrease occurring chiefly during the first 12 months. The critical moisture content lies at 23%, i.e., below this point no substantial damage by fungi and no substantial decrease in density take place (105).

The reports on wood losses caused by storage vary considerably. There can be no question that storage and climatic conditions are of considerable significance. On storage of logs, wood losses of up to 20% have been observed with piles of small logs suffering higher losses than bigger logs. Spraying with fungicides allegedly is of no avail (96). The environmental conditions certainly play an important role. Insects usually are the direct cause of only minor destruction of wood. However, they carry a variety of fungi into the wood and, therefore, lead to earlier and more rapid fungus deterioration than would otherwise occur, particularly in raw wood (70). It has been generally observed that deciduous woods are more readily susceptible to fungal attack than coniferous woods are. On storage in big piles the outer layers are naturally more readily attacked by fungi than the inner layers. Fungal attack on the wood causes damage to the quality of the pulp; even the increase in yield occasionally observed does not refute this statement because the increase may be the result of a specific attack on the lignin in the wood (76).

As already mentioned, temperature and moisture conditions play a significant role. It is therefore important to observe and to study these conditions as they occur within a large pile of wood chips. Accurate measurements show that the temperature within the pile increases rapidly and so does the moisture content. The loss of density of the wood caused by fungal attack is greater in the outer aerated layers than in the inner layers which are in a compressed state. A not unimportant factor in wood-chip storage is also the higher degree of soiling of the chips (15).

A comprehensive investigation of the temperature, moisture, and density changes in a wood-chip pile during long periods of storage—one, two, three, five, and six months—has been reported. Inserted thermocouples showed that the temperature in the dense inner layers of the pile increases rapidly and remains high for several months. At the same time, the moisture content also increases constantly. The temperature course within

the pile is controlled by a series of factors, especially by the type of wood, its moisture content before storage, the temperature of the wood, the type of piling, and by local conditions. Piles of birch, pine, and spruce wood chips, built up during the summer, reach relative maximum temperatures of 69°, 63°, and 58°C, respectively. Birch chip piles reach their maximum temperature after only 10 to 14 days, whereas pine and spruce chip piles require 14 to 20 days. The heat development resulting from the biochemical processes is estimated to be at least 70 kcal/kg of absolutely dry wood substance in the cooler bottom layers, and 120 kcal/kg in the warmer upper layers. It has been calculated that the air within the pile changes at least ten times every day. When such a chip pile is built up by successive horizontal layers forming a symmetrical, blunt conical pile, it is found that the zone of maximum temperature is around the central vertical axis. The isotherms are ellipsoidal and include the zone of maximum temperature, which moves slowly upward (15,72).

The effect of variations of temperature and oxygen and carbon dioxide content within the chip pile on the growth of wood-destroying microorganisms has been studied by laboratory investigations. In order to delay the growth, very great variations have to occur and these are very difficult to obtain in a chip pile. After a 15-month storage the wood chips in the middle zone of the pile are mostly dry, whereas in the upper part they always take up water. Below the saturation point the wood chips undergo no substantial change in their moisture content regardless of their position in the pile. Rot fungi are found predominantly in the center of the pile, while *Ascomycetes, Fungi imperfecti*, and bacteria are generally found in the outer zones. Wood chips which, during the drying, have passed through the moisture zone especially favorable for a fungal attack gave a very low pulp yield after 15 months of storage. It can therefore be said that a short storage period for chips under favorable atmospheric conditions causes no damage and no decrease in the yield of pulp; however, a longer storage period, especially of moist spruce and pine wood chips, has an unfavorable effect (14).

The isolation of numerous microorganisms was achieved by an analysis of the microflora. An especially destructive type of mold, *Chrysosporum pruinosum*, was frequently found. The number of different microorganisms per wood unit is considerably greater in the storage of chips than in stacked roundwood storage. Wood-destroying *Basidiomycetes* are limited by the temperature factor and are normally found only in the outer part of the pile or stack. Temperatures of above 60°C are required to sterilize wood chips. It is probable that the microorganisms contribute to the heat formation by their metabolism (85).

The changes found during storage in the extractives are of special significance and affect, in the first place, the yield of turpentine and tall oil.

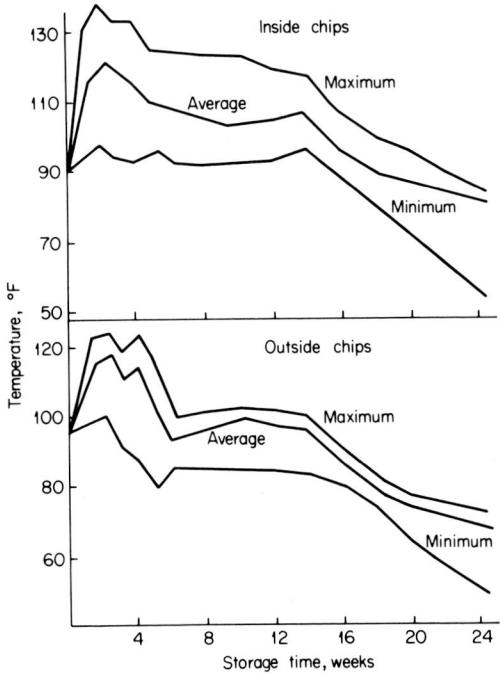

Figure VI-1. Chip pile temperatures (118).

Since it may be assumed that the development of heat within the pile or stack has a detrimental effect, measurements obtained with thermocouples are shown in Fig. VI-1.

The temperature curve of the chips in the innermost part of the pile shows a rapid rise in the maximum curve during the first two weeks, followed by a general decline to about 80°F (37°C). The minimum temperatures of the chips were, in general, only a few degrees higher than the surrounding temperature. These minimum temperatures were measured at the bottom and the top of the pile. The temperature changes in the outer layers, in general, followed the course of the temperatures measured in the center, but were essentially lower (118).

The type of storage also has an effect upon the resin troubles later found in the processing of the pulp. This effect is especially pronounced when the wood has been stored under water. Floated wood which is later stored on land shows an increase in resin troubles with increasing storage time. The storage of wood chips reduces the resin content of the wood, and the velocity of the enzymatic hydrolysis and the oxidation of the lipidlike material are accelerated as compared with the roundwood storage in piles. This affects the behavior of the resin during the cooking process. The total

content of extractives in the sapwood decreases considerably more rapidly than that in the heartwood. The fat content decreases without a simultaneous increase in petroleum ether-insoluble components, i.e., oxidized fatty acids. Only the chips near the cool zones at the surface show an increase of substances insoluble in petroleum ether; this corresponds approximately to the conditions which have been observed in the storage of roundwood. These differences may be attributed to the different course of the autoxidation of the fats at high and low temperatures. The oxidized fatty acids, in contrast to those not degraded by autoxidation at low temperatures, act as dispersing agents for neutral substances in alkaline media. They are therefore not desired in the pulp production, especially in the cooking of birch. Spruce contains resin acids and less neutral substances and therefore causes less trouble (5,10,28,31,33).

The storage of wood chips lowers the yield of tall oil and turpentine to a considerable degree. This is especially important for the kraft pulp industry. Losses of up to 80% of these products have been observed with woods stored for 30 weeks. The loss of tall oil is accompanied by a strong carbon dioxide evolution in the chip pile. The changes in amount and composition of the extractives during storage are the result of two competing reactions, i.e., enzymatic degradation and autoxidation. At moderate temperatures and high moisture content the enzymatic degradation predominates, i.e., the products soluble in ether and acetone decrease rapidly. Fats are probably degraded to water-soluble or volatile compounds, possibly to carbon dioxide, and are removed during the storage. With dry wood at low temperatures and also with moist wood at high temperatures, autoxidation takes place. The content of extractives then remains constant. Under otherwise equal storage conditions, autoxidation is more predominant in roundwood storage than in chip pile storage. The reason for this may be that the enzymatic degradation depends upon the ready removal of the carbon dioxide which is formed but which is not produced during autoxidation (29,109).

The yield of tall oil from fresh pinewood decreases from 50 kg originally to about 20 kg per ton 90% dry sulfate pulp after the wood has been stored for four months. The decrease occurs very rapidly in the case of green wood, and, after a storage period of one month, the yield amounts to only 55% to 60% of that from fresh and nonstored wood. Wood stored on land, regardless of whether or not it had previously been floated, yields 30 kg per ton of pulp and this decreases to 20 kg after a storage period of four months. Roundwood storage results in a considerably smaller decrease in the yield of tall oil, especially when the wood stack has been built up during a cold season (87).

The course of the loss in turpentine from roundwood and chips during storage in the open is shown in Fig. VI-2. This loss is relatively low during

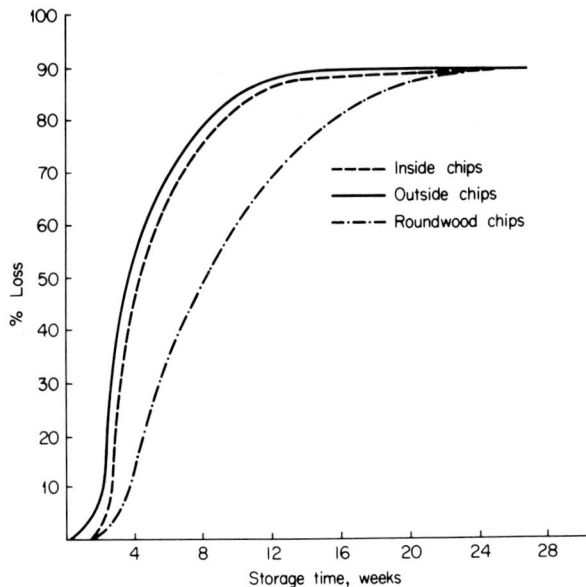

Figure VI-2. Turpentine loss during chip storage (118).

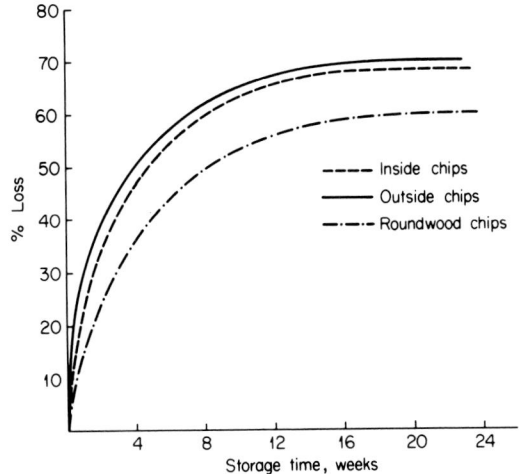

Figure VI-3. Tall oil loss (paper pulp cooks) (118).

the first two weeks but increases rapidly during the next six weeks. Chips stored outdoors show a slightly increased loss of turpentine as compared with that of chips stored indoors. The turpentine loss from roundwood increases uniformly and reaches its maximum at the end of 22 weeks.

The percentage of losses of tall oil in a paper pulp cook are shown in Fig. VI-3. Although these experimental results do not necessarily correspond to

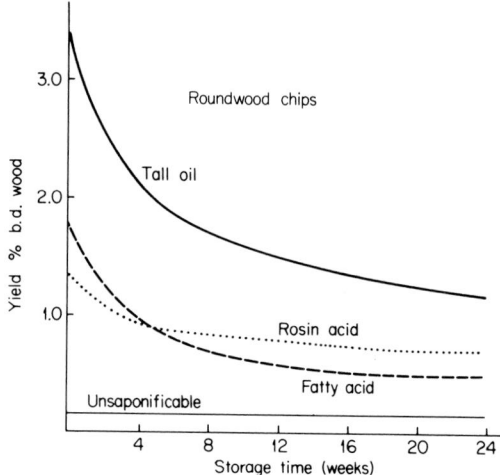

Figure VI-4. Tall oil and tall oil composition (paper pulp cooks) (118).

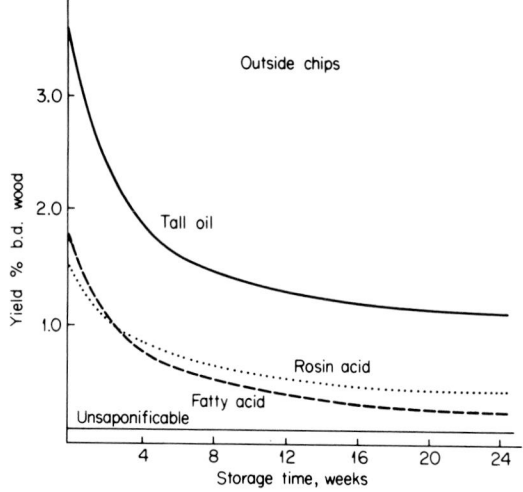

Figure VI-5. Tall oil and tall oil composition (paper pulp cooks) (118).

those experienced in the mill, they show the fundamental course of the basic reactions. The percentage losses increase rapidly within the first 4 to 6 weeks and then rise uniformly with increasing storage time. The tall oil losses from stored logs are smaller than those from chips stored indoors, whereas chips stored outdoors always show the greatest losses. To show the effect of storage upon the individual components of the tall oil, the single losses related to dry wood and a paper pulp cook are shown in Figs. VI-4, VI-5, and VI-6. The wood used was slash pine from Florida (118). The yields of tall oil, resin, and fatty acids from three types of chips (chips from the

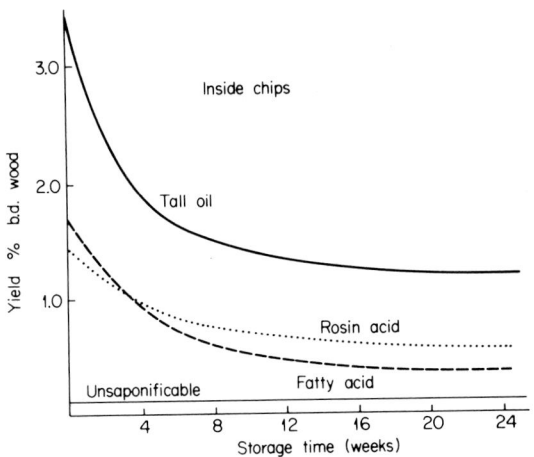

Figure VI-6. Tall oil and tall oil composition (paper pulp cooks) (118).

outer layer of the chip pile, from the inner layer, and from stored round-wood) all decrease with the time of storage. The fraction of unsaponifiable components does not change appreciably with storage.

The question has been raised as to whether the storage of chips in large piles does not considerably increase the fire hazard. Investigations in this direction have shown that it is possible that the layers on the surface of wood-chip piles can be ignited by an outer fire source. The two principal factors influencing such an ignition are the moisture content and the wind velocity. Experience has shown that the spread of such surface fires is very slow under the moisture conditions usually prevailing and that it occurs only in the uppermost, relatively dry layer. It must be taken into consideration that the chips lose their moisture much less rapidly than they take it up from highly humid air. The extinguishing of such surface fires can be accomplished with a minimum of water, especially by the use of sprayers. Any loosening or disturbing of the chips should be avoided since the fire will die out in the inner part of the pile from lack of oxygen (94).

2. THE BARK AS A CHARACTERISTIC COMPONENT OF THE WOOD

2.1 Growth Factors

Generally the term "bark" means the outer part of the trunk; the bark surrounds the wood. There are various types of bark tissues and two principal layers can be distinguished: the inner bark and the outer bark. The inner bark includes all those tissues which are situated between the cambium and

the most recently formed periderm. These tissues form a secondary phloem region, while the outer bark surrounds the tissues which are outside the last-formed periderm. It is impossible to draw a sharp line between the inner and outer bark, and the terms which are in use for the various types of tissues are not universally or officially recognized. In the following, the nomenclature as proposed by Chang (24) has been used.

The function of the inner bark is essentially the transportation of the products formed by assimilation and the storage of these nutrients. The outer bark, on the other hand, consists of dead tissue and is therefore physiologically inactive. It acts as a protective cover against chemical and mechanical damage to the tree and the wood. The first step in the growth involves the primary tissue which, in ordinary barks of seed plants, consists of the epidermis, the cortex, and the primary phloem. The second stage of growth involves two types of secondary tissues which differ from each other by two special meristems; one is generally called the cambium, the other, the phellogen or cork cambium. The secondary phloem is derived from the cambium and is often called "bast." The periderm or cork originates from the phellogen. The bark of a young twig or stem consists of both the primary and secondary tissues. The bark of the main trunk can contain a part of the primary tissues or can be devoid of them. The location of the various bark tissues in ordinary stems at various stages of growth is shown in Fig. VI-7. In order to avoid confusion about inner and outer bark, these designations should be made according to the anatomical structure and the physiological activity of their types of tissue. According to Chang, the inner and outer bark can be defined as follows: the inner bark comprises the region of the secondary phloem from the cambium to the last-formed periderm. The outer bark includes the tissue layers from the last-formed periderm to the outer surface of the bark. The tissues of the inner bark are continuously formed by cell division of the cambium and a part of these tissues are physiologically active. The tissues of the outer bark are chiefly physiologically inactive and, because of the visible change in the cells, are morphological in nature, in contrast to the inner bark. Chang called the part of the outer bark which is composed of alternating layers of periderm and dead secondary phloem "rhytidome."

The inner bark consists of sieve elements, the phloem parenchyma, the sclerenchyma, and the medullary-ray parenchyma. The most important tissue is composed of the sieve elements. These are sieve cells in the gymnosperms and sieve tubes in the angiosperms. The sieve elements are formed by longitudinal rows of cells. The protoplasm of the adjacent elements is connected by innumerable fine strands. This cell-wall region which is permeable through such strands is also called the sieve zone. In the sieve cells the sieve zones differ very little but in the sieve tubes the sieve zones vary greatly in shape. In general, the sieve elements function only one year, then

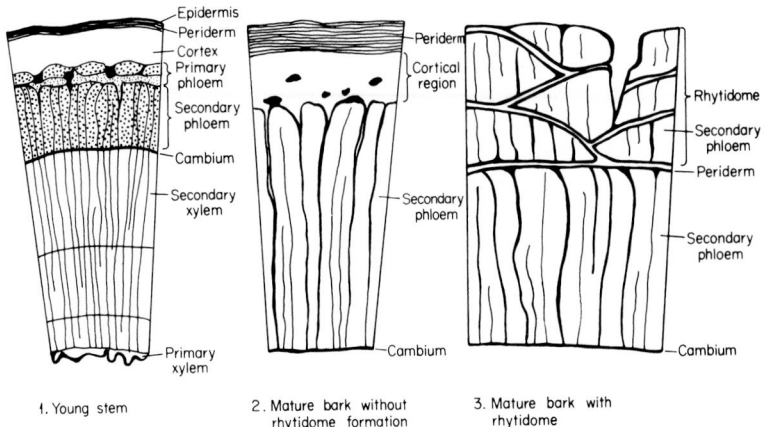

1. Young stem   2. Mature bark without   3. Mature bark with
                              rhytidome formation   rhytidome

Figure VI-7. Diagrammatic drawings showing the main tissues in different types of bark. (1) Cross-section of young branch of stem. (2) Cross-section of bark having persistent cortex, such as that in middle-aged balsam fir and quaking aspen. (3) Mature bark with rhytidome formation (23). (Reproduction by permission of Technical Association of the Pulp and Paper Industry from Tappi Monograph No. 14.)

they shrink, probably because of the dissolution of some substances from their walls. The walls then become elastic and are compressed through the continuous growth of adjacent parenchymatic tissues.

Horizontal rays and vertical parenchyma, which can be divided into strips or ribbons, are formed between the sieve elements and tissues of the inner bark by parenchymatic cells. The walls of these cells are thin and not lignified. The horizontal cells, the phloem medullary rays, are direct continuations of the xylem medullary rays and both are formed from the vascular cambium. Because of a secondary thickness growth, a strong pressure is exerted outwards toward the bark, causing a tangential stretching of the tissue and a general adjustment of the cells through the expansion, or causing new cell divisions to be started. The medullary rays of the xylem are longer than those of the phloem, due to the fact that more xylem cells than phloem cells are formed by the division of cambium cells. The sclerenchyma cells, i.e., the bast fibers and sclereids, form the supporting tissue of the inner bark. The portion of bast fibers varies considerably and sometimes amounts to as much as 35% to 48% of the inner bark. The thickness of the bast amounts, in general, to between three and ten millimeters. Pine and larch have a narrow bast area, fir and beech a wide one. The thick cell walls are more or less permeated by lignin. They are normally arranged in tangential rows, the formation of which is regularly continued. The lignified tissues of the inner bark consist chiefly of sclerenchymatic tissue. The portion of the secondary phloem in the total bark varies between 72% and 82% in the gymnosperms, and between 60% and 88% in the angiosperms.

The primary tissue of the outer bark, the epidermis, is formed during the first stage of growth of the tree trunk in the outermost tissue layer. This tissue is the outermost cell layer and consists exclusively of one single layer of cells. It is soon replaced by a periderm in which an established cork cambium, or phellogen, deposits dead cork cells outward and bark parenchyma inward. The cell wall of the cork is comparatively thin and without marked intercellular spaces. The cork cells soon die. Some wood species form a thick cork which shows lamination. This lamination is due to the fact that the cork cells which are formed during the late summer are thicker, i.e., they have thicker walls, than the cork cells formed during the spring. The surface or primary periderm, the cork, and the periderm layers can also be formed in the deeper layers of the bark, i.e., in the bast. In cross-sections of such inner or secondary periderms, the single periderm layers can be seen as curved bands in which the chronological interval between the formation of the periderm layers amounts to several years. The cork cells are dead and filled with air and consequently the sap flow from the inner parts of the stem toward the outer layers of the periderm is interrupted. The rhytidome (bark) does not consist of a separate tissue but contains the residue of all dead tissues outside the periderm layers, such as cork, parenchymatic cells, bast fibers, sclereids, and so on. Since the rhytidome consists entirely of dead tissue, it cannot adjust to the growth in thickness of the stem and therefore breaks up and partially drops off. This process takes place in a specific way for each tree species, depending upon the anatomical structure of the rhytidome.

For the production of pulp the complete removal of the bark is required. This means that not only the outer bark must be removed but also the inner bark, which consists of the region of the secondary phloem of the cambium and the most recently formed periderm. It is clear that the removal of the bark from the adjacent cambium can be readily achieved so long as the cells of the secondary phloem are still carrying out their functions, that is, in the spring. The possibility of barking the stem while it is still fresh is slight, as will be shown later. Barked wood is also more susceptible to decomposition (17,24,94,121).

2.2 Chemical Composition

Data on the chemical composition of bark must be given with reservations. In the first place, they depend upon the wood species and also upon the time when the sample was taken. Certain agreements in the chemical composition, not from the qualitative but from the quantitative standpoint, have been found when the results obtained with various barks are examined. The various components of barks will be discussed in connection with the possible utilization of the bark obtained in the barking of wood for the pulp industry.

The components of bark are obtained mostly by extraction with various solvents. Table VI–1 shows the solubility or insolubility of the single components of the extracts and makes possible a compilation of these components into certain groups. From the table it can be seen that the principal components can be isolated as groups by selective extraction. On successive extraction with several different solvents the order in which the solvents are used is of significance.

The celluloses from the extractive-free barks of *Abies amabilis, Picea*

TABLE VI–1

SOLUBILITY OF EXTRACTIVES FROM BARKS IN VARIOUS SOLVENTS (62)

Group	Type of substances	Soluble in	Insoluble in
I	Volatile terpenes, aldehydes, etc., aliphatic oils, fats, waxes, higher acids and alcohols, hydrocarbon resenes, plant sterols	Petroleum ether Ethyl ether Alcohol	Water
II	Amorphous resins and resin acids, substances as in group I, but more highly hydroxylated	Ethyl ether Alcohol	Water Petroleum ether
III	Phlobaphenes, some glycosides	Alcohol	Water Petroleum ether Ethyl ether
IV	Tannins, simple sugars, glycosides, etc.	Alcohol Water	Petroleum ether Ethyl ether
V	Polysaccharides, gums, and pectins	Water	Petroleum ether Ethyl ether Alcohol

engelmanni, and *Pinus contorta* were obtained in 38.1%, 30.9%, and 30.4% yield, respectively. The nitrates prepared from them gave weight average degrees of polymerization of 7200, 7100, and 10,300. On hydrolysis, the celluloses yielded D-glucose and traces of mannose and xylose (103). In addition, the bark contained other carbohydrates such as pectin, polyuronic acids, gums, and mucilages, and hemicelluloses such as pentosans and hexosans. Arabino-4-O-methylglucuronoxylan and galactoglucomannan have been isolated from the bark of *Picea engelmanni.* This bark also contained considerable amounts of pectinlike material (95). A water-soluble arabinomethylglucuronoxylan was also isolated from the bark of *Abies amabilis.* This bark xylan is similar to, if not identical with, that contained in the wood (120). A general view of the chemical composition of extractive-free barks of spruce, fir, and pine is given in Table VI–2.

TABLE VI-2

CHEMICAL COMPOSITION OF EXTRACTIVE-FREE BARKS (119)

Component (summative data)	Fir	Spruce	Pine	
			Inner	Outer
"Lignin"	39.4	38.1	24.6	46.9
Ash	4.0	2.1	3.3	2.2
Acetyl	0.5	0.8	0.2	0.8
Uronic anhydride	8.0	5.6	9.9	7.7
Residues of				
Galactose	2.4	1.6	4.3	4.2
Glucose	35.7	37.4	40.9	26.8
Mannose	2.9	8.0	2.5	2.5
Arabinose	3.3	3.2	10.6	5.5
Xylose	3.8	3.2	3.7	3.4
Other data				
Pentosan	11.3	9.1	19.3	13.0
Cold-water-soluble material	9.0	4.8	13.7	5.4
Hot-water-soluble material	20.3	8.4	22.7	12.7
Material soluble in NaOH (1%)	51.6	35.4	—	—

A more comprehensive list of the chemical components of the bark from 24 species of American woods is given in Table VI-3. As compared with the wood, the barks contain relatively large amounts of ash, extractives, "lignin" and, to a smaller extent, carbohydrates. With regard to any connection between chemical composition and anatomical structure, no definite statements can be made. There seems to be a certain relationship between the amount of periderm and cork and the benzene- and alcohol-soluble components. The relationship between the significant tissue elements in the secondary phloem and cellulose and "lignin" is not unequivocal. High proportions of phloem fibers in the barks of deciduous woods, on total hydrolysis, give proportionally higher yields of reducing sugars than of "lignin." The ratio of sugar yield as compared with that of lignin becomes larger after extraction with 1% sodium hydroxide. In spite of the fact that the barks of coniferous woods are simpler in their structure they show no uniform relationships between chemical composition and structure.

The chemical utilization of bark is probably most nearly possible through its extractives. The part which can be extracted with 1% sodium hydroxide corresponds approximately to that which is obtained by exhaustive extraction with benzene. However, it must be mentioned that during the barking process of the wood for the pulp industry a part of the extractives is lost, especially when a wet barking operation is used. The yield

TABLE VI-3

ASH CONTENT, EXTRACTIVES, "LIGNIN,"[a] AND METHOXYL OF "LIGNIN" OF BARKS FROM VARIOUS WOOD SPECIES[b] (24)

	Ash	Material dissolved by successive extractions with				"Lignin" from extractive-free bark		"Lignin" from NaOH-extracted bark		MeO in unextracted bark
		Benzene	95% EtOH	Hot H$_2$O	1% NaOH	Yield[c]	MeO[d]	Yield[c]	MeO[d]	
Balsam fir	2.3	13.2	3.3	2.7	30.6	27.7	8.5	15.0	12.7	3.30
Western larch	1.6	1.3	14.8	3.8	22.7	30.0	8.9	19.6	10.7	3.14
Engelmann spruce	2.5	5.2	25.9	10.9	22.2	17.9	7.2	8.7	9.7	2.90
Black spruce	2.0	5.0	14.6	4.4	28.0	25.3	8.2	14.4	10.0	3.20
Jack pine	1.7	8.0	12.4	3.0	41.3	42.2	4.7	14.4	10.1	3.07
Lodgepole pine	2.0	28.7	10.9	5.6	29.8	14.8	5.1	5.4	8.1	1.99
Slash pine	0.6	3.4	10.6	3.7	28.9	49.9	6.3	26.1	10.2	3.95
Sugar pine	0.6	1.5	21.7	3.2	36.0	49.9	3.9	20.1	8.2	2.45
Eastern hemlock	1.6	2.8	21.2	3.3	24.6	35.8	8.0	20.1	10.5	3.61
Box elder	6.2	2.4	6.3	6.2	23.7	30.1	11.1	22.8	12.4	4.03
Sugar maple	6.3	1.2	3.9	2.4	19.2	37.3	11.2	27.4	11.9	5.05
Red alder	3.1	2.3	3.9	3.7	27.5	40.9	7.9	28.5	9.2	3.85
Yellow birch	1.7	4.3	10.8	2.3	28.4	40.6	7.6	26.3	9.2	3.46
Paper birch	1.5	9.4	10.5	2.5	25.1	37.8	8.4	22.7	12.1	4.04
Sweet pecan	7.5	0.8	18.4	5.4	25.3	24.9	7.5	16.0	9.5	2.69
Sweet gum	5.7	1.5	17.7	7.4	21.3	25.3	10.7	18.7	12.9	3.37
Black gum	7.2	2.5	4.6	5.3	27.8	38.3	10.3	25.0	12.5	4.97
Sycamore	5.8	2.1	6.0	3.6	22.0	26.6	15.0	21.3	16.8	5.53
Swamp cotton wood	4.0	1.9	8.0	4.8	20.2	33.4	12.0	27.2	13.7	5.36
Quaking aspen	2.8	4.0	11.6	4.7	22.0	31.2	10.3	21.0	13.6	4.75
White oak	10.7	2.7	4.4	5.8	26.5	31.8	7.3	20.9	9.3	3.28
Northern red oak	5.4	4.8	7.9	3.6	22.3	34.8	9.1	23.3	11.3	4.32
Black willow	6.0	1.6	3.8	4.8	23.8	29.0	9.2	20.9	12.0	3.74
American elm	9.5	0.5	10.1	6.0	27.0	27.5	6.9	16.9	9.7	2.91

[a] Acid-insoluble residue from bark by 72% H$_2$SO$_4$.
[b] Percentages based on oven-dry unextracted bark
[c] Yield based on oven-dry weight.
[d] Based on oven-dry weight of lignin residue.

of extractives from the bark so obtained is therefore not necessarily the same as that from the original bark (24).

As shown by the analytical data in Table VI–2, the water extract contains a large number of soluble substances. This is especially important with regard to the fact that wood often comes in contact with water in an unbarked state so that water-soluble compounds can diffuse into the wood during the floating, during the storage of the wood in water, and during hydraulic and friction barking. These processes will be dealt with in detail.

An important component of the bark is the tannins. These are substances that are capable of converting animal hides into leather. In general, two groups of tannins can be distinguished: the hydrolyzable tannins and the condensed tannins, the latter of which are converted by acids into phlobaphenes. The majority of the tannins capable of making hides into leather belong to the second group. The structure of the phlobaphenes is still largely unknown although it can be assumed that the tannins, phlobaphenes, and the phenolic acids in the bark are polymers of catechin and related substances. Other polyphenols in the bark with ligninlike properties may be copolymers of substances which contain catechol and guaiacyl groups (44).

The water extract of the newly formed inner bark of Douglas fir contains numerous amino acids, as is shown by paper chromatography. The presence of tyrosine and phenylalanine is probable. Whereas sucrose, fructose, and glucose have been found in variable quantities throughout the entire year, raffinose seems to be present only during the growing period. The presence of catechols in the living bark suggests that they are the precursors of the polyphenolic components of the bark and the wood. The presence of amino acids seems to be of decisive importance, especially for the formation of proteins and, to a lesser extent, for the synthesis of structural elements of the wood (57). In the hot-water extract of poplar bark (*Populus grandidentata*), after clarification with basic lead acetate, the following compounds were identified: salicin, tremuloidin, salireposide, populin, salicyl alcohol, and vanillic, ferulic, *p*-coumaric, and *p*-hydroxybenzoic acids. With regard to the isolation of populin (salicin-6-monobenzoate) and tremuloidin (salicin-2-monobenzoate) it could be shown that their formation depended upon the alkalinity of the solvent used for the extraction (89,90).

The extraction of bark, already extracted with organic solvents, with alkaline solutions gives a series of compounds which are usually called "phenolic acids." These substances, which are insoluble in water, are found in the barks of conifers in amounts of up to 40%. Obviously a remarkable similarity in structure of the building units exists between these phenolic acids, the phlobaphenes, and the water-soluble tannins. Numerous investigations by various authors make it appear probable that monomeric fla-

vanoids and leucoanthocyanins are the precursors which alternately form the tannins, phlobaphenes, and the high-molecular polyphenols, the last of which resist extraction with neutral solvents. On extraction with 3% sodium hydroxide at room temperature an amorphous phenolic acid has been isolated from slash pine bark after neutralization of the extract. This acid was free of tannin and lignin but containted about 5% carbohydrates as impurities. It could be methylated and was identified by various reactions such as alkali fusion, pyrolysis, oxidation, and reduction, as a polymer consisting of substituted catechin-type units (37).

By means of organic solvents a series of substances, generally to be designated as waxes, have been extracted from bark. Among these substances were considerable amounts of rosin acids, triterpenes, nonaliphatic alcohols, and others. When the bark of Douglas fir was boiled with an alcoholic solution of sodium hydroxide, additional amounts of wax were obtained from the lignin complex of the bark. This wax had different properties from those of the wax obtained by extraction with organic solvents (20).

From the methanol extract of the inner bark of *Populus tremuloides*, pyrocatechol and benzoic, vanillic, *p*-hydroxybenzoic, and *p*-coumaric acids were isolated and obtained in a crystallized state. Paper chromatography showed the presence also of salicin, populin, tremuloidin, and ferulic acid. Some of these compounds were already mentioned as appearing in the water extract. From the seasonal variations in the appearance of these products in the extract, the hypothesis was advanced that activity of the cambium and the formation of methoxyl groups are interconnected (38).

Suberin is a characteristic component of the cell wall of the cork. The constitution and formation of suberin are still to a large extent unknown although the assumption that it is an ester of an aliphatic hydroxy acid and a phenolic acid has gained wide credence. According to this, the cork cell wall consists of a lignocellulosic network which is impregnated by these polyesters (55).

The anthocyanins are glycosides, the aglycones of which, the anthocyanidins, probably do not exist in a natural free state. Leucoanthocyanin from the inner bark of black spruce gives anthocyanidin on acid treatment. The leucoanthocyanidins lie, chemically, between the flavonols and the anthocyanidins. The flavonols are closely related to D-catechin (56,73).

The sterols, mostly saturated alcohols with 27 to 31 carbon atoms, have recently gained increasing interest. From the bark of pine could be isolated chiefly β-sitosterol and, in smaller amounts, also campesterol (99).

This short and somewhat incomplete presentation of the constituents of bark can give only a suggestion of their complicated but meaningful structure. The possibility of the chemical and pharmaceutical utilization of some of the components will be dealt with later. However, it must be

mentioned that a close connection exists between the chemical composition and anatomical structure of the bark. For example, the high caloric value of some barks is due to their high suberin content. On the other hand, an especially low moisture content of bark even on the living tree is the result of the formation of the cell from predominantly hydrophobic material. The power to resist destruction by microorganisms can be ascribed to the presence of hydrophobic and antiseptic components in the bark. The bark is therefore a natural protector of the tree against mechanical, chemical, and biochemical damage (61).

2.3 The Bark Portion of the Trunk

The bark portion of the trunk varies according to location, age, and species of the tree and shows appreciable differences with increasing height of the tree. In spruce, such differences due to the location amount to from 7% to 18%. In northern Finland, for example, the bark amounts to 18%, in southern Finland to 14%. Spruce grown in the higher mountain regions of Germany has an average of 7% to 12% bark, in the lower regions, 9%

TABLE VI-4

BARK CONTENT OF VARIOUS AMERICAN WOODS (112)

Wood species	Bark content[a]
Engelman spruce	11.1
White spruce	12.4
Grand fir	9.1
Silver fir	15.9
Douglas fir	10.6
Jack pine	9.8
Loblolly pine	10.5
Lodgepole pine	7.5
Long-leaf pine	11.5
Short-leaf pine	11.9
Slash pine	15.6
White Eastern pine	12.5
Eastern hemlock	18.9
Western hemlock	9.7
Western larch	8.8
Paper birch	13.2
American elm	9.6
Black gum	12.4
Sugar maple	13.7
Quaking aspen	18.4
Eastern cottonwood	14.7

[a] Percent of the weight of the unpeeled wood.

to 11% bark. In a survey made by the Forest Products Laboratory at Madison, Wisconsin, the data given in Table VI–4 were obtained (112).

The bark portion increases with the height of the tree. With a south German spruce, for example, the amounts of bark were: 8.2% at 1.3 meters, 7.2% at 9.7 meters, 12.8% at 20.3 meters, and 18.9% at 28.7 meters. With pine this fluctuation is even higher. According to measurements made in south German pine groves, trees varying in age from 25 to 280 years gave variations of from 8.2% to 28.4%, while single trees varied as much as from 3.7% to 37.4%. At equal ages, trees in poor locations had a higher percentage of bark. With spruce, on the other hand, the lower parts of the stem had a higher proportion of bark, averaging 16%, whereas parts of the upper tree had a lower proportion, averaging 7.5%. As can be seen from the table, the various pine species show considerable differences in their bark content. Red beech and grove beech contain an average of 9% bark in the young wood, and 6% in the old wood.

3. BARKING

3.1 Measurement Units and Measurement

Pulp wood can be measured by volume and by weight. In forestry circles, measurement by volume is preferred because it is difficult to weigh the tree trunks in the woods. Pulp mills calculate their pulp yields on the basis of the weight of the dry wood; they therefore want to know the weight of the wood. In the United States, various measurement units are used, as, for example, the cord, the cubic unit (or cunit), the board foot measure (or log scale), and, finally, as the only weight measure, the ton.

The cord is that amount of wood which is contained in an 8-foot long and 4-foot high stack of 4-foot long sticks. The sticks must be arranged in parallel layers, but even then this is not a measure of the volume of solid wood. The spaces between the sticks in a cord amount to about 128 cubic feet. The volume taken up by the wood is considerably lower because there are cavities between the single sticks and the sticks themselves can differ in thickness at their two ends. Whether the wood is barked or unbarked also makes a difference.

The cubic unit is that amount of unbarked wood contained in 100 cubic feet of solid wood. The dimensions of the cubic foot correspond to those of the cord. As a matter of fact, a cord seldom contains more than 100 cubic feet of solid wood.

The board foot measure is generally used in the wood processing industry. A board foot (bd ft) is that amount of wood which is contained in a board one foot long, one foot wide, and one inch thick. The one ton

unit equals 2000 pounds and is used for the calculation of the yield of pulp, provided that the bark content and the moisture content of the wood are known. The pulp and paper industry deals generally with air-dried wood and air-dried pulp. Precisely speaking, this is a very uncertain term which cannot be used technically without a more exact definition. By air-dried is meant that the dry content is 90% absolutely dry, i.e., 90% wood and 10% water.

In Europe the "Festmeter" (fm) and "Raummeter" (rm) are used as measurement units. The Festmeter is that amount of solid wood which is contained in a one-meter cube; it is therefore identical with one cubic meter of solid wood. A Raummeter is the amount of wood that is contained in a stack of logs one meter long, deep, and high. It is evident that the content of solid wood in a Raummeter depends to a great extent on the average diameter of the logs. The solid content of 1 rm sprucewood of an average diameter of 8 to 18 cm, for example, varies between 0.70 and 0.78 fm. The conversion from Festmeters to weight units, i.e., tons, requires a knowledge of the specific gravity of the wood. In addition to Festmeter and Raummeter, still other units are in use, as for instance: ster (southern Germany), standard (U.S.A. and England), ton (U.S.A.), and stere (France). Table VI-5 gives the conversion factors.

As already mentioned, the pulp industry is interested in knowing the true weight of the wood that goes into the digesters. The volumetric measurement is made according to the xylometric method, in which the wood is dropped into a tank filled with water and the amount of water displaced is measured. In order to obtain the true wood weight, the density of the wood and the moisture content of samples are determined. At first glance the method seems to be troublesome, but it can be extensively automated (64). The actual solid volume of stacked wood can also be estimated by

TABLE VI-5

CONVERSION FACTORS FOR UNITS OF MEASUREMENTS OF WOOD[a]

Unit of measurement	Fest- meter (fm)	Raum- meter (rm)	Ster (st)	Cord (cd)	Standard (std)	Ton	Stere (str)
Festmeter	1.0000	1.4285	1.3471	0.2803	0.2943	0.8826	1.0000
Raummeter	0.7000	1.0000	0.9430	0.1962	0.2060	0.6178	0.7000
Ster (Ger.)	0.7423	1.0604	1.0000	0.2081	0.2185	0.6552	0.7423
Cord	3.5678	5.0968	4.8064	1.0000	1.0500	3.1490	3.5678
Standard	3.3980	4.8543	4.5776	0.9525	1.0000	3.0000	3.3980
Ton	1.1330	1.6185	1.5263	0.3175	0.3334	1.0000	1.1330
Stere (Fr.)	1.0000	1.4285	1.3471	0.2803	0.2943	0.8826	1.0000

[a] These factors are valid for roundwood only.

determining the volume of the single logs and adding these found values to the mass unit. This method also requires a knowledge of the density of the wood and its moisture content if it is to lead to usable results with regard to the weight. It must be added that single logs vary in thickness along their length. The wood stack must therefore be looked at from both the front and the back and the average of the two must be taken.

All these methods are time-consuming and require much man power. A photographic method has therefore been proposed in which the stack is photographed from a certain distance, the picture is enlarged and provided with a screen with dots all the same distance apart. By counting the dots which fall on the surface of the cuts and those which cover the intermediate spaces, the content of the wood in the stack can be calculated. It is assumed that the solid content of the logs has the same relationship to the Raum-meter as the sum of the cut surface of the logs has to the square meter of the front wall of the stack; the prerequisite is that the logs are equally thick along their whole length. This may be correct for short logs, but with long ones considerable errors in the calculation may occur (16,46,67,82).

The bark content of stacked wood can be readily calculated by means of a bark gauge. The diameters of the logs are measured once with the bark and once without it. The ratio between the inner diameter, d, without the bark and the outer diameter, D, with the bark gives the average value $k = d/D$, which is determined for 20 logs. The following equation is then obtained:

$$k = \frac{\Sigma d}{\Sigma D} = \frac{\Sigma D - \Sigma 2B}{\Sigma D}$$

in which $2B$ signifies the sum of the measured double bark thickness. The bark volume (%) of the unpeeled wood is then calculated from the equation:

$$\text{bark volume } (BV) \text{ in } \% = 80 \, (1 - k^2)$$

The factor 80 takes into consideration the fact that the ridges and the fissures of the bolts of wood fit into each other to some degree. This factor, obtained experimentally, is not universally valid but must be determined specifically for woods of the same species and the same locality (22).

3.2 Mechanical Barking

As shown in Section 2, the bark is separated from the wood by a layer of living cells which is called the cambium. The chemical composition of the cambial zone differs considerably from that of the newly formed wood. The cell walls of the newly formed cells consist mainly of a mixture of pectic material which forms the middle lamella of the fully grown cells. In the course of further development of the primary wall, cellulose and

lignin are embedded in the cell wall. The cambial zone is rich in extractives and contains considerable amounts of protein, but only small amounts of lignin. During the growing period, growth-promoting substances and hormones are present, the origin of which has not yet been completely elucidated. During stationary growing periods the bark is attached to the wood very firmly and even strong traction does not give a clean separation in the cambial zone. Mostly the separation occurs within the layers of the bark itself and parts of the bark and the bast stay with the wood. These conditions of the strong attachment of the bark to the wood prevail during most of the year (from August–September until April–May), depending upon the geographical location, the altitude, the weather conditions, and the species of tree. It has been assumed that, at the end of the growing period, hydrophobic colloids, especially fats and lipoids, concentrate at the surface of the protoplasm of the living cambium cells and, in this way, the capacity to swell is lessened and the absorption of water is prevented. On the other hand, at the end of the dormant period a physiological change in the lipoid layers is to be expected and thus the swelling capability of the protoplasm is increased. This reactivation of the cambium after the dormant period has a direct effect on the ease of removal of the bark.

Before cell division, the cells expand strongly in a radial direction and the radial cell walls become thinner, swell, and soften. During the cell division and until the end of the cambial activity, a break between the bark and the wood readily occurs as the result of outside action, mostly in the zone of the cell expansion on the side of the cambium turned toward the wood, i.e., in the young xylem cells before the formation and lignification of the secondary cell wall. It is interesting to note that the ease of removal of the bark increases with the beginning of the growing period and proceeds from the crown of the tree downward. The downward migrating substances which initiate the activity of the cambium are not assimilation products. The anatomical structure of the formed wood cells has no decisive effect on the barking capacity. Trees felled during the winter are most easily barked during the following summer, but somewhat later than freshly felled trees. The period also ends somewhat earlier than that for growing trees. Rapid drying of the felled tree prevents the formation of such a barking period. If the tree is kept in the forest with its crown attached it loses its water more rapidly than does the tree from which the top has been removed (91).

The adhesion of the bark to the wood is therefore closely connected with the growing process of the cambium and there are physiological conditions, primarily swelling processes, which affect the removal of the bark disadvantageously. A successful attempt has therefore been made to measure the adhesion of the bark to the wood and it was found that the treatment of the unbarked wood with hot water considerably facilitates the removal

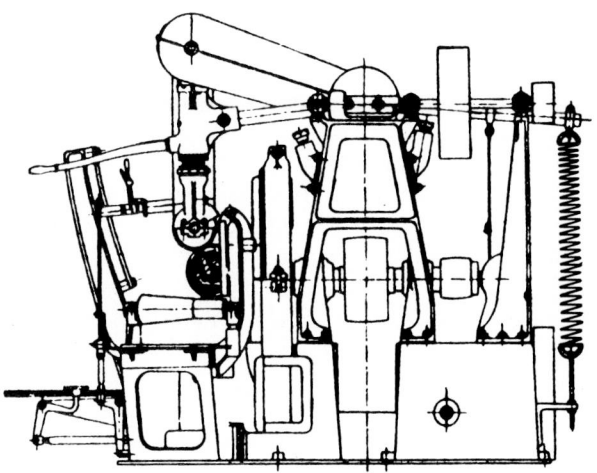

Figure VI-8. Knife barking machine system Fresk.

of the bark. This fact has been known for a long time and the way in which it has been used technically in barking by pretreatment of the wood will be discussed in a later section (124).

Commercial barking machines use essentially the following operational procedures, often with several of them being combined: (*a*) stationary or rotating knives, (*b*) mechanical friction by rubbing or pounding action, and (*c*) hydraulic barking by water jets.

The simplest method is undoubtedly the use of a peeling knife with which, even today, much wood is barked in the forest. Wood thus peeled still contains large amounts of inner bark and bast and has to be further peeled in the mill, to an extent depending upon the quality requirements of the pulp to be produced from it. Such bark residues are especially difficult to remove. Barking in the forest has the advantage that the loss of wood is low and the waste consists almost entirely of bark, which remains in the forest. On the other hand, barking by hand requires more man power and is confined chiefly to middle-European forests where the wood, in the form of cut logs, is transported by rail or truck.

Barking machines are based, in general, on the use of rapidly rotating knives which peel the moving stems. There is a great variety, depending upon whether the peeling knives are arranged in a stationary or a moveable position. In the same way, the pressure of the log against the peeling disk may be flexible. The barking machines can be designed for long or short logs. Assuming that the wood has already been barked in the forest, the bark waste amounts to about 5% to 10% and consists chiefly of wood. Such a barker, for the barking of wood up to 2.5 meters in length and up to 30 cm in diameter, is shown in Fig. VI-8.

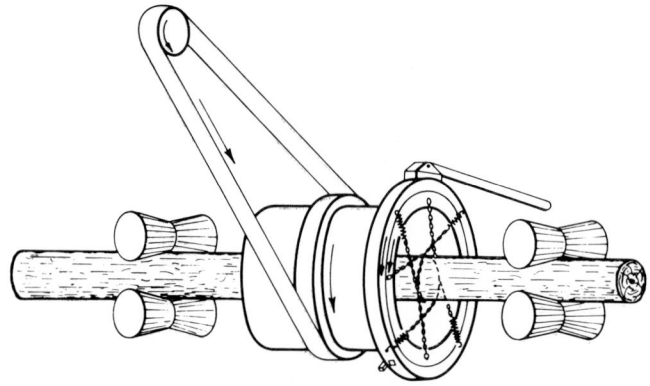

Figure VI-9. Barking machine system Aström. Swedish Patent 130,000, 1944.

The knife plate is equipped with six knives, the transporting rollers are mechanically pressed against the wood, and the angle of the knives can be adjusted automatically. The power consumption of the knife disk is about 30 hp, that of the feeding rollers about 4 hp. The efficiency of the machine depends upon the diameter of the logs and varies between 11 and 20 cubic meters per hour. One man is sufficient for the operation. This type of barker serves mainly for supplementary peeling of prebarked wood.

There is also a series of machines now in use which remove all the bark and bast by purely mechanical means in a single process. In order to increase the efficiency of such machines and to utilize them to the full extent, every effort is made to keep the wood passing continuously through the machine. The mechanical installations which accomplish the removal of the bark by friction consist of rotating rollers, spike rollers, rotating fluted rollers, and dull peeling irons which are pneumatically or hydraulically pressed against the logs, causing them to pass through the barker in a rotating or nonrotating condition. It is natural that the highest performance is obtained when stem after stem or log after log passes through the machine in uninterrupted sequence. The barking process is therefore closely linked to a well-designed conveyor system. The output of such machines, which are also portable so that they can be used in the forest, too, is mostly reported in running meters of barked wood. Chains and wire ropes which are rigged crosswise to a rotating ring are also used for the mechanical barking. All these installations demand that the wood should still have almost its entire bark, so that the bark and the bast are removed together.

The possibilities of variation are so manifold and the suggestions, in part patented, often differing only in minor details of construction, are so numerous that it is impossible to present an exhaustive description. In Fig. VI-9 an arrangement is shown schematically which uses a rotating ring

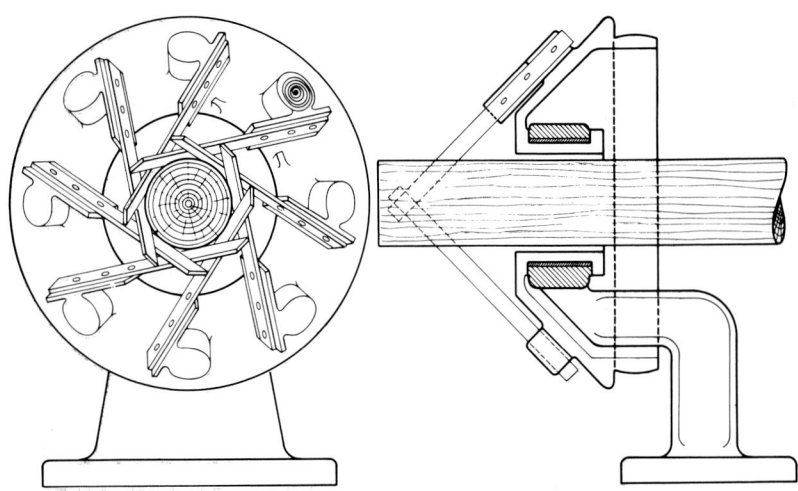

Figure VI-10. Barking machine system Carlson. Swedish Patent 118,829, 1946.

formed by spring-tightened chains that automatically surround the log to remove the bark.

Such chain barking machines are available in numerous designs. That shown in Fig. VI-9 should be considered only as a prototype. As already mentioned, peel irons, spike rollers, wire ropes, etc. are used instead of chains. Figure VI-10 shows a typical example.

It has been proved, particularly in the barking of deciduous woods, that knocking or pounding can remove the bark more easily than can be done by peeling. For this reason chains have been attached to a rotating shaft in such a way that the ends of the chains, as the result of centrifugal force, exert an impact action on the logs as they are passed through this arrangement in a longitudinal direction. In this way the bark is loosened (11,59,65,79,116).

3.3 Hydraulic Barking

The term "hydraulic barking" derives from the fact that the wood is barked in a wet state by the use of the mechanical shearing force of high-powered water jets. Hydraulic barking therefore uses a mechanical force for the removal of the bark. The installations for the technical operation of such hydraulic barking processes may be divided into three categories: (a) trough or pocket barkers, (b) drum barkers, and (c) water jet barkers. In all three types the wood is barked in a wet state, in some cases even with water being constantly added. One of the oldest examples of this type is shown in Fig. VI-11. It is a multipocket barker, characterized by the fact that the logs are kept in motion in three pockets by revolving double-ended

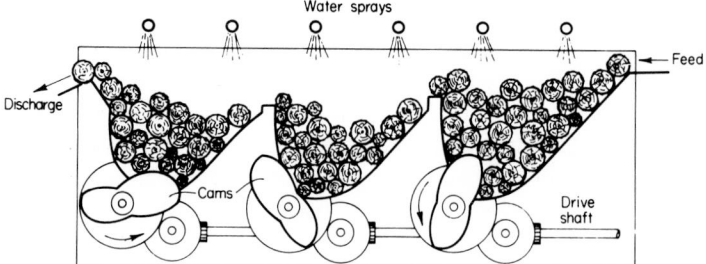

Figure VI-11. Pocket barking machine system Thorne. German Patent 619,649, 1935.

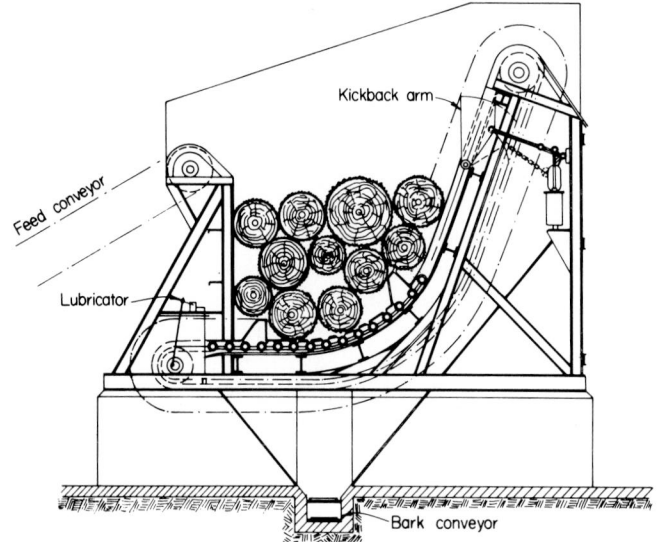

Figure VI-12. Pocket barking machine system Hillbom. Swedish Patent 119,814, 1935.

cams at the bottom of each pocket. The cams raise the logs and give them a rolling motion as they are dropped from one pocket into the next. The logs with bark remaining on them are then returned to the first pocket for another passage. Because of the great power consumption and the relatively heavy wear on such machines, they have become outmoded. They are suitable only for logs one to two meters in length (68,84,92,100,127).

In order to bark longer logs, especially those which have been delivered to the mill by floating, a one-pocket barker has been designed for a continuous process. This is shown schematically in Fig. VI-12. With this type, also, many variations in the apparatus design are possible and Fig. VI-12 illustrates only the working principle of the machine without giving further construction details. The logs are thrown from a conveyor into the trough and are moved around from the lower part of the trough to the upper by a moving endless chain equipped with carrying attachments; then they are

thrown back by a moveable arm or deflector. In this way the logs are kept rolling and rubbing against each other. This process is repeated until the desired debarking has been achieved. During this operation the woods are continuously sprayed with water, warm, if necessary. Such installations are built to take logs four to eight meters in length. The bottom of the trough is perforated to allow the bark to be constantly carried away by a conveyor.

This principle is often used as a continuous process in which several of such troughs are connected in a series whereby the logs are conveyed from one trough to the next. The number of troughs depends upon the wood species and on the difficulty with which it is barked. This continuous process is similar to that shown in Fig. VI–11, but requires a considerably lower power consumption. All the pocket barkers require floated wood, i.e., wet wood; otherwise, the wood has to be pretreated in a suitable way.

The drum barkers for mechanical barking have found by far the widest acceptance. The idea of removing the bark by rubbing the logs against each other in rotating drums, totally or partially immersed in water, is very old and was mentioned in a patent as early as 1891. Since then improvements have been made in the construction of the drums themselves, in their arrangement and their reinforcement to prevent deformation by the forces involved, in the hump-shaped channel sections, and in the drive. While originally the wood was put in parallel order into the drum and the latter was closed, today the wood is fed into the drum irregularly and moves continuously through it. The maximum length of the logs depends upon the diameter of the drum. Drums up to five meters in diameter are in use. The length of the drum depends upon the time the wood is intended to remain in it and this again depends upon the speed of rotation and the desired degree of barking. The drums are submerged in a trough partially filled with water, which also fills the drum about one-third or one-half. The higher the water level the greater the tendency of the wood to float. The rubbing of the logs against each other and against the drum wall decreases as the water level rises. To accelerate the barking and to facilitate the discharge of the bark, many structural changes have been proposed which cannot be discussed here in detail. In general, it has been found that sharp-edged barking irons in the inner wall of the grated drum constructed of profile steel are unsatisfactory because they cause severe slicing and splintering of the logs, especially at the ends. Such broomlike fringed ends on the wood pick up sections of bark and lead to objectionable impurities in the pulp. In addition, the splintering causes losses of wood.

Really contradictory reports have been made about the efficiency of barking drums. This is not surprising because the efficiency of such a barking device depends upon many external factors which differ from place to place. First of all, the condition of the wood is a determining factor for the degree of efficiency to be expected. Furthermore, whether the wood

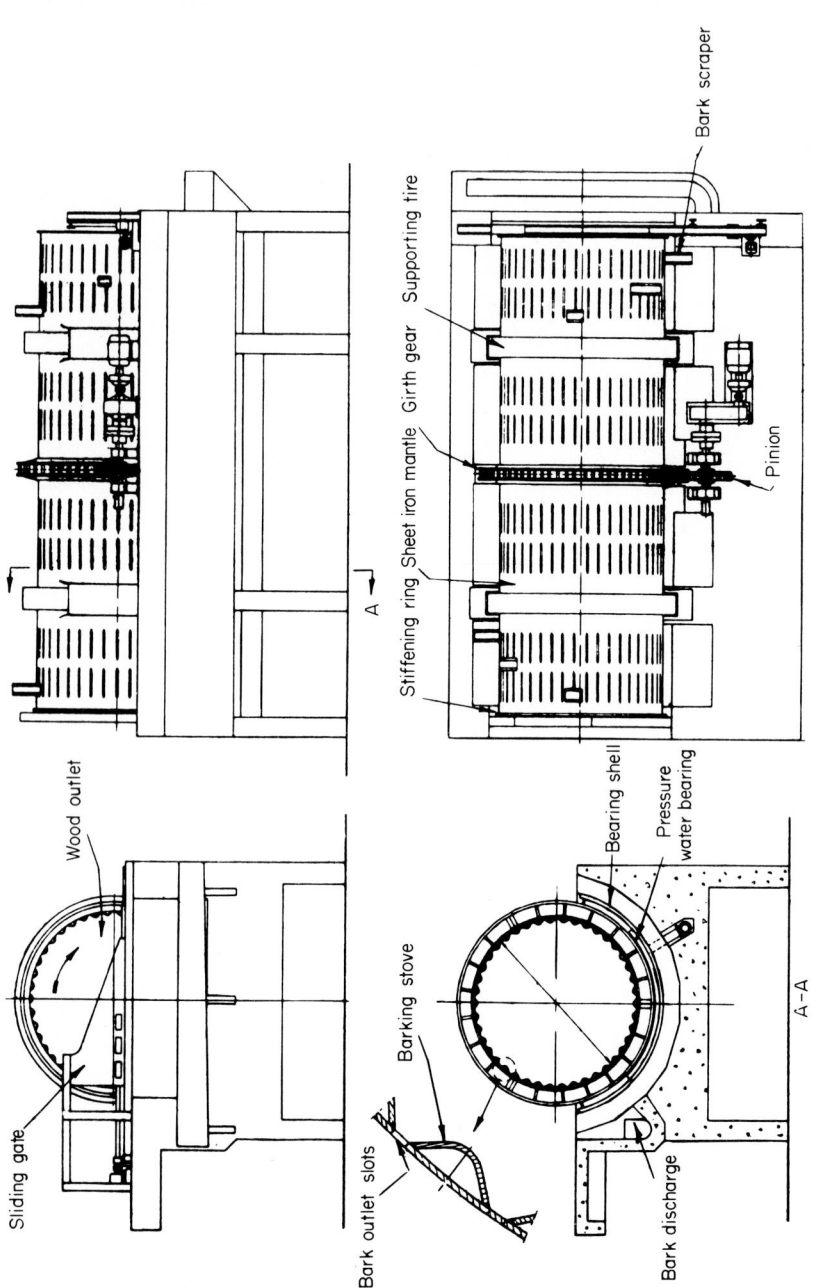

Figure VI-13. Barking system Waplan-KMW-Myren. Hydraulically supported wet barking drum.

Figure VI-14. Wood preparation plant system Billerud A.B. (13). Right side—wood storage, overhead crane with wood bundle; center—feeding and sorting tables; left—damping tables.

still retains all its bark, only a part of it, or only bast, whether the wood is dry or wet, and whether it was transported by floating or by rail are all factors of fundamental importance. In a continuous operation of the drums the length of the logs plays a decisive role. With pocket barkers the continuous or almost uninterrupted feeding of the wood into the barker is the goal to be striven for. According to experiences in Scandinavia (127) barking drums are most efficient when the drum contains one quarter of its volume of wood. A drum of 4.20 meters in diameter and 10 meters in length has a volume of 138.5 cubic meters and should be charged with about 34.5 fm wood. With a 3-hour barking time scheduled, such a drum would bark about 277 fm or 367 rm of wood in 24 hours. A barking time of 3 hours is considered sufficient for wood that has been felled during the summer, has been floated, and then stacked for one year. According to more recent Canadian reports, a drum of 3.65 meters in diameter and 20.5 meters in length barks 40 to 45 cords of freshly floated wood per hour, corresponding to 200 to 225 rm. These, by European experience, astonishingly high results can be explained only as resulting from especially favorable wood conditions. Calculated for a drum charge of about 76.5 rm, the barking time is about 25 minutes. A barking drum of Swedish construction is shown in Fig. VI-13.

Such drums are also suitable for hardwoods (80). A new barking plant, which as a whole may be determinative, has recently been put into opera-

Figure VI-15. Wood preparation plant system Billerud A.B. (13). In the foreground, the sorting table; in the background, right-hand, the sorter's cabin; left-hand the damping tables.

tion in Sweden (13). Whereas in regions of intensive forestry activity where wood is produced exclusively for the pulp industry, every endeavor is made to convert the wood as quickly as possible into a valuable industrial product, i.e., if possible, to deliver to the pulp mill, wood chips ready for the digester—where the forestry regions are widely distributed and the delivery routes are relatively short, the conditions are fundamentally different. In this case the cost of transporting unpeeled wood to the mill is compensated for by the fact that the barking in up-to-date central plants is cheaper and the storage periods are considerably shortened. To this must be added the value of the bark as fuel for the mill. Because various wood species are used (fir, spruce, and deciduous woods) sorting can be done satisfactorily

Figure VI-16. Wood preparation plant system Billerud A.B. (13). The barking drums operator's cabin with two monitors surveying the operation of the drums and the feeding of the wood.

only when the wood is in an unpeeled state. In the plant shown in Figs. VI-14, VI-15, VI-16, and VI-17, 800,000 fm wood are processed annually. The wood, in the form of unpeeled and unsorted logs three meters long, is delivered to the plant partly by truck and partly by rail. The shipments consist to about 10% of deciduous wood. The unloading is carried out by two cranes which move along a 185-meter long steel structure. Below the travelling crane there is a storage place for 5000 fm wood and this is the only storage place between the forest and the mill. The cranes deposit the wood on three feeding tables, so-called chain tables, which are 4 meters wide and 20 meters long. Three chain sections and one stepwise feeder permit three different speeds for each section, with a capacity of 40 logs per minute. The chain tables transport the wood to the sorting plant where it is sorted into fir, spruce, and deciduous woods. The sorting is carried out by a checker and by means of pneumatically operated openings through which the wood is tossed onto one of the three conveyors, according to

Figure VI-17. Wood preparation plant system Billerud A.B. (13). Two barking drums in operation.

species. The checker, or sorter, sits in a heated cabin and operates the openings by means of pneumatic steering gears.

The deciduous wood travels on the conveyor through a chute onto another conveyor and from there is dropped into two pocket-shaped troughs where it is bundled. The bundles are then lifted by means of a crane into a pocket barker which, during the winter, can be closed by an insulating roll-top. During the winter the wood is warmed by steam and the barking is carried out as described above. The volume of the pocket barker is 22 cubic meters, the output 25 to 35 fm per hour, and the steam consumption 20 kg per fm.

The coniferous woods, fir and spruce, are transported through conveyors to the steaming tables on which the wood is moved laterally by means of chains. During the winter the tables are heated by steam to facilitate the barking. The length of time that the wood remains on these tables is 4 to 10 minutes for fir and 2 to 5 minutes for spruce, with the steam consumption being 15 to 25 kg per fm at 3 atmospheres. After the steaming, the logs are dropped onto conveyors which carry them to the drum barkers. These conveyors have an initial velocity of 1 m per sec and, shortly before arriving at the entrance to the drum, a speed of 5 m per sec. Two drums are used, each 34 meters long. The wood is barked without the addition of water. The drum is 2.85 meters at the entrance and 3.35 meters at the outlet, i.e., the cone at the outlet is wider. The number of revolutions of

the drums is 9.6 rpm and their efficiency is 90 fm per hour for fir and 110 fm per hour for spruce. The power consumption for both drums is 200 hp. The peeled wood is then sent to the chippers.

The water-jet barker uses high-powered water jets directed toward the unbarked stems or logs, and the removal of the bark is effected by shearing force. The first experiments in this direction were carried out a few years before the Second World War. A Swedish company obtained a patent for such a process, which was simultaneously combined with a thermal pretreatment of the wood. The thermal pretreatment will be dealt with later. The process itself has been described elsewhere (127,129). In the United States, work on a water-jet barking system had already been carried out in the 1930s (3). Simultaneously, similar experiments were also conducted in Europe and a Norwegian company developed a ring of nozzles through which the stems or logs are forced so that the sheetlike water jets which hit the unbarked logs at certain variable angles peel the bark off (39,40). In this type of barker, too, the wood must have been floated to soften the bark, or else the wood has to be subjected to a special pretreatment.

After the end of the war, the very efficient American water-jet barkers became more and more well-known and were introduced into wood processing plants. Quite a variety of such installations have been described and all have in common the fact that high-powered water jets are used instead of knives, barking irons, rubbing rolls, scrapers, or other mechanical devices. The various models differ, however, in the method of feeding in the wood; some make the logs rotate during the barking, others keep them stationary, while still others let the nozzles rotate. The angles at which the water hits the wood also vary in the individual designs. There are arrangements with only one nozzle, others with several. The pressure of the water jet is usually very high; while some models use 25 atmospheres, the newer ones have pressures of 85 to 100 atmospheres and over. It is understandable that the continuous feeding of the stems or logs is of decisive importance for the efficiency and power consumption of such a plant.

From reports received and from the structural details shown in Figs. VI–18, VI–19, and VI–20 of a water-jet barker, the following data are presented. The so-called "package barker" is suitable for logs 3 to 6 meters long. By the addition of more aggregates, stems of up to 12 meters in length can be processed. The stem to be barked is placed on two rotating rollers and is itself rotated. Specially designed swivel cranes are used to deposit and remove the logs. A carriage with jet nozzles runs above the log, with its high-powered water jets directed axially along the log. As shown in Fig. VI–18, these nozzles can be turned 360° and bent vertically 45° to each side. Figure VI–19 shows the package barker which is suitable for barking pulp wood of the usual length of 1 to 3 meters and 15 to 120 centimeters in diameter. The bark thickness varies from 6 mm for hemlock up to 5 cm

Figure VI-18. Sumner package barker mechanism to revolve the nozzle by 360° (115).

for old spruce stems. Except with frozen wood where warming with hot water is required, barking is carried out without pretreatment. The performance is one cord per minute with logs 2.5 meters in length and up to 40 centimeters thick, and one-third cord with logs of 20 centimeters diameter. The water consumption varies between 2.27 cubic meters per minute with package barkers and up to 5.7 cubic meters per minute with sawmill and woodroom barkers. The standard nozzle has 29 openings, 3/16 of an inch in diameter. The power consumption is about 1 hp per gallon, or about 600 hp for the package barker (115).

The reported figures for power and water consumption are subject to numerous fluctuations because some firms give the consumption only for the barking equipment and others include the consumption for the transport

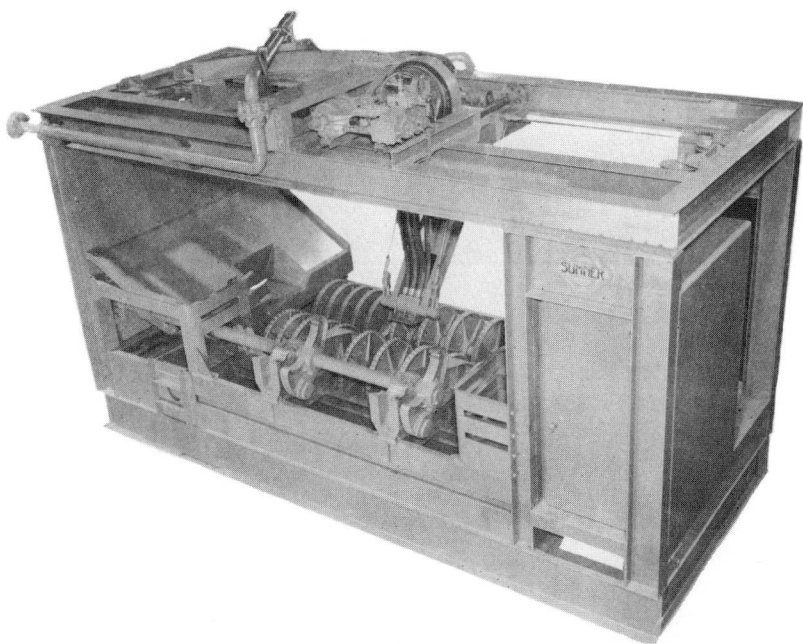

Figure VI-19. Package type of barker. System Sumner iron works (115).

Figure VI-20. Sumner package barker in operation (115).

section. Efficiency and energy expenditure depend to a great extent upon the condition of the wood, the thickness and length of the stem, the proportion of bark, and the local conditions (71). Comparisons of the efficiency and reports on the losses in wood are valid only when the operational conditions are defined in detail. It is understandable that the attack on the wood itself and, with it, the loss of wood, increase with increasing power expenditure and the resulting increase in water pressure (133).

3.4 Preparation of the Wood for Barking

Barking by the rubbing of completely dry or dried-out wood is not possible. As is shown by the anatomical structure and the biosynthesis of the bark, the cambial layer must be in a swollen state to facilitate the removal of the bark. The force which effects the separation of the bark from the wood must be sufficiently great to overcome the adhesion of the bark to the wood on the one hand, and, on the other hand, it must not be so strong that the cohesive strength of the wood is surpassed, thus causing loss of wood. It has already been shown how this has been achieved with the various barkers. The barking of completely dry wood requires a water treatment which causes a reswelling of the inner layers of the bark and thus gives them a certain plasticity. The action of water or steam is therefore used commercially for the pretreatment processes. Many methods have been proposed but it is impossible to discuss them all here. In general, they mostly have in common the immersion of the wood either under pressure by its own weight, or with the help of mechanical devices such as drums or chains, in troughs, vats, or channels filled with water of different temperatures and for varying periods of time. In this way the desired rubbing of the logs against each other is achieved (63,67,114). In another version of the preparation of the logs, the wood is kept in chambers or pits filled with warm water and then is transported in a warm wet state to the drum barker. Figure VI–21 shows such a plant schematically. At the right is a cross-section of the soaking installation. The properly cut wood, delivered by truck, is placed in the cells by means of a gripper. The wood lying under the grate, IV, is in the process of being soaked. The wood, I, in the opposite chamber, II, is raised by means of the float trough, III, in the rising water, dropped onto the conveyor, V, and transported to the drum barker (35).

Another method which has undergone various changes treats the wood in hermetically sealed steaming chambers. Figure VI–22 shows schematically such a structure. The wood passes first, by means of a circulating double chain, through the soaking bath, D, and is then transported by the elevator, E, into the steam chamber, F. After the trap, E', is closed, the wood lying parallel on chains is treated with steam for periods depending upon the wood species and its dry content. During this treatment the chains remain

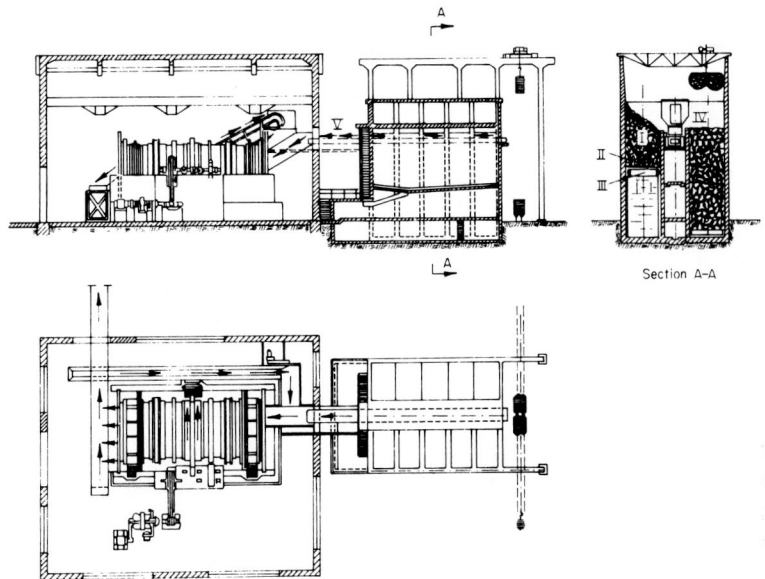

Figure VI-21. Drum barking with thermal pretreatment of wood system J. W. Enqvist A.B. German Patent 708,790, 1938.

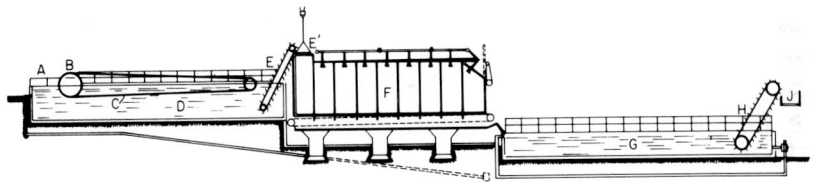

Figure VI-22. Thermal pretreatment of wood system Wikmanshytte Bruks (128).

motionless, but afterwards the chains are set in motion and bring the wood into the conditioning bath, G. While the chamber is being emptied, it is simultaneously being filled again at E'. From G the wood is transported by means of the elevator, H, through the conveyor, J, to the drum barker. The disadvantage of this structure is the difficulty of storing the logs in parallel order. After the steam treatment, the wood is quenched in cold water. This treatment tends to prevent the drying out of the hot wood and the shrinkage of the inner bark layer (129).

To overcome the difficulties involved in the storing of the wood in the steam chamber, it has been suggested that the prepared wood stack should be covered by a kind of moveable hood under which it can be steamed (128).

3.5 Chemical Barking

The term "chemical barking" is misleading. What is meant is the treatment of the growing tree in the forest with certain chemicals which later make it possible to remove the bark easily from the felled tree. The removal of the bark has to be prepared for by the infusion of suitable chemicals into the sap stream of the living tree. According to Canadian experiments, arsenic compounds, when fed to the tree in June or July, facilitate the removal of the bark after cutting. With coniferous woods a considerable loss of wood substance has been observed, whereas deciduous woods did not show this loss (6). Experiments with 2,4-dichlorophenoxy-acetic acid, 2,4,5-trichlorophenoxyacetic acid, and ammonium sulfamate in combination with a solution of volatile esters or diesel oil had little or no effect with regard to the loosening of the bark, but showed distinct insecticidal action (9). According to other reports, a 40% solution of sodium arsenite proved especially effective. To carry out the treatment a ring of bark is removed at a certain distance from the root of the tree and the chemical solution is brushed on the bark-free zone. The tree dies within a few days. Trees treated in this way can be readily barked 8 to 10 months after cutting, whereas with untreated trees this is possible only within 8 to 12 weeks. The toxic hazards involved should not be underestimated. Damage to deer is presumably possible only within a few days after the inoculation of the tree because the sap soon carries the poison into the inner stem. This process is welcomed by the forester because it facilitates the work in the forest and leaves the bark there to form humus. Thus far, however, there is insufficient evidence available to decide whether or not the accumulation of the poison in the bark by a general application of this process will cause any damage to the humus formation (7,60,75,111,130).

That the arsenic content of the pulp produced from such treated wood can reach sufficient amounts to be injurious to health is denied (107). Various investigations on the action of the chemicals used are available. As mentioned already, the chemicals, when applied in the right way at the right place, pass readily into the sap stream. The velocity of this movement depends upon the width of the ring, the wood species, its age, and the surrounding conditions. The upward movement of the sap of the ring zone occurs within the evaporation stream of the sapwood, the downward motion in the assimilation stream of the bast. If the bast has been removed during the preparation of the ring, no downward motion takes place. With the upward movement a radial movement occurs simultaneously. The diffusion velocity depends upon the ratio of sapwood to heartwood. In trees with narrow rings of sapwood the upward movement in the outer layers is limited. With wide sapwood rings the movement of the poisonous solution pene-

trates into deeper layers and to a greater distance from the cambium. This may be the reason that some wood species react more slowly to the chemical infusion than others do. In the case of damage to the sapwood during the ring cutting, the movement of the solution is likewise deeper and to a greater distance from the cambium. The cambium and the neighboring tissues are killed. When the chemical inoculation is carried out while the sap is running in the tree, tissues are destroyed when the newly formed xylem cells start to form a weak bond between the wood and the bark. The tissues shrink through the action of the chemical before they die (101). So far as is known, the chemical treatment of living trees has not yet found much application. The idea of using less toxic, but plant-physiologically active substances such as enzymes is still under investigation.

4. PROCESSING AND UTILIZATION OF THE BARK

If the wood is barked in the forest, the bark remains there and is used in the formation of humus for the soil. Wood which is barked at the pulp mill yields large quantities of bark. It is understandable that these mills are interested in finding a suitable means of utilization of the bark. Recent Scandinavian investigations have shown that the amount of bark, expressed in kilograms of dry bark per Festmeter of barked wood, varies, depending upon the diameter of the logs, from 33 to 67 kg or an average of 47.6 kg with spruce, and between 20 and 45 kg or an average of 38.2 kg with pine. These figures give some information on the amount of bark, by weight only, on a dry basis. The diameter of the logs varies in this case between 18 and 21 centimeters. Depending upon the type of barking, the bark is obtained with a more or less high moisture content. This moisture content, however, is of great importance when the bark is used as fuel. The relationship between the effective calorific value per kilogram of wet bark in kcal/kg and the water content in percent of the total wet weight is shown in Fig. VI-23.

Figure VI-24, on the other hand, gives the relationship between the calorific value of the dry bark in kcal/kg with reference to the percent of water content of the total weight of the bark. According to Scandinavian reports, the effective heat value of absolutely dry bark is 4500 kcal/kg for spruce, 4750 for pine, and 5450 for birch. Table VI-6 shows American figures for barks of various origin with a dry content of 5% to 6.7%.

It is apparent that the moisture content of the bark is of decisive importance for its use as fuel. The highest permissible water content for the bark in especially designed fireboxes with corresponding air preheating is about 60% (122,123). The ash content of the bark is considerably higher

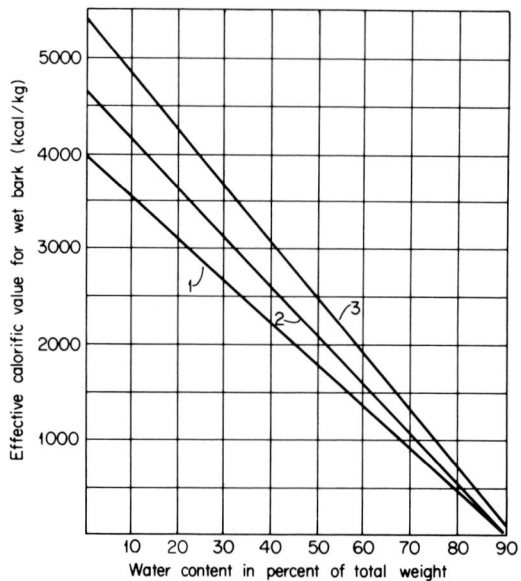

Figure VI-23. Effective calorific value of wet bark (123). 1 spruce; 2 pine; 3 birch.

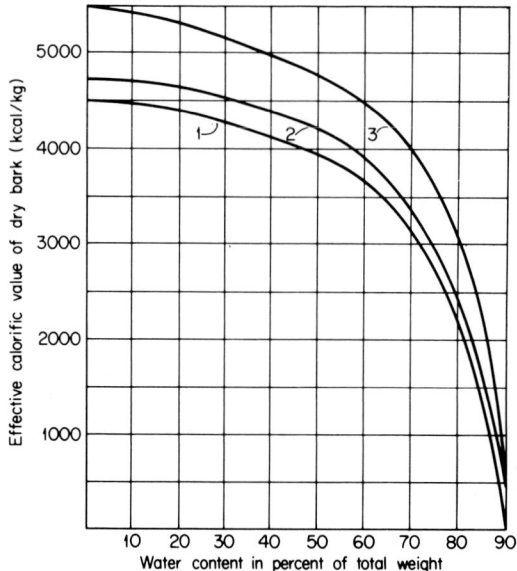

Figure VI-24. Effective calorific value of dry bark (123). 1 spruce; 2 pine; 3 birch.

than that of the wood. Some ash values have already been mentioned, others are found in Table VI-6. Depending upon the transportation of the wood, whether by floating or by rail or truck, the bark contains greater or

lesser amounts of sand. American sources report up to 7% ash contents of barks if the wood has been shipped overland (34).

The moisture content depends to a large extent upon the type of barking process. Oven-dried bark contains about 10% water, air-dried between 15% and 20%. Drum barking in a nearly dry state gives bark with 35% to 50% moisture content, in a wet state, bark with 45% to 65% moisture content, and barking with water jets increases the moisture content to 60% to 75%. With increasing moisture content the heat value decreases and, with it, the efficiency of the combustion chamber. For the simultaneous combustion of bark and gas or oil, various kinds of boiler constructions are in operation. With a moisture content of 50% to 65% it is recommended that the bark be fed into the combustion chamber in batches so that the part of the bark that is burning will cause a certain amount of drying of the rest of it. With

TABLE VI-6

CALORIFIC VALUES OF BARKS FROM VARIOUS WOODS (122)

Wood species	Ash^a	Moisture content (%)	Calories per gram^b	British thermal units per pound^b
Balsam fir	2.3	6.5	4.923	8.861
Western larch	1.6	6.7	4.558	8.204
Engelmann spruce	2.5	5.5	4.644	8.359
Black spruce	2.0	6.5	4.581	8.246
Jack pine	1.7	6.6	4.867	8.761
Lodgepole pine^c	2.0	5.6	5.661	10.190
Slash pine	0.6	6.4	5.001	9.002
Sugar pine	0.6	—	—	—
Eastern hemlock	1.6	6.2	4.890	8.802
Box elder	6.2	—	—	—
Sugar maple	6.3	6.0	4.056	7.301
Red alder	3.1	5.8	4.415	7.947
Yellow birch	1.7	5.2	5.042	9.076
Paper birch	1.5	4.8	5.241	9.434
Pecan	7.5	—	—	—
Sweet gum	5.7	6.2	4.139	7.450
Black gum	7.2	6.0	4.409	7.936
American sycamore	5.8	6.4	4.113	7.403
Swamp cottonwood	4.0	—	—	—
Quaking aspen	2.8	5.5	4.685	8.433
White oak^d	10.7	6.5	3.886	6.995
Red oak	5.4	4.4	4.461	8.030
Black willow	6.0	6.7	3.982	7.168
American elm	9.5	6.7	3.845	6.921

^{a} Based on weight of oven-dry wood.
^{b} Values are for samples of the indicated moisture content.
^{c} High heat value probably due to high content of benzene extractives.
^{d} Low heat value probably due to high ash content.

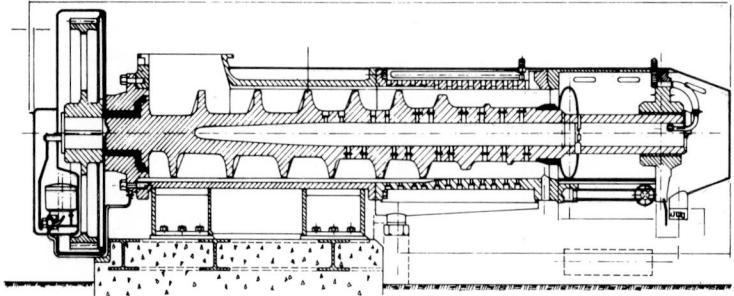

Figure VI-25. Screw press for bark. (Sunds Verkstäder A.B.)

moisture contents of below 50%, boilers are used that have a travelling grate on which the bark is spread out in a thin layer. In this way simultaneous ignition and combustion are assured (21,34).

The ideal procedure would be to send the bark directly from the barker to the boiler. This is possible, however, only in rare cases because the barker is operated only during one shift while the boiler is in operation continuously. Moreover, it is necessary to bring the bark into a form suitable for combustion, with a suitable moisture content. The crushing of the bark is usually carried out in hammer mills which cut it into as uniformly sized pieces as possible. If the bark is taken from large piles in which it has been stored for some time, it has to be freed of foreign bodies. It has been suggested that the bark should be partially dried in drying plants which use the heat of the flue gases from the boiler. For a reduction of the water content of the bark, especially that from hydraulic barkers, presses with vertically arranged grooved rollers, screw presses, and chain-band presses are used (88,93,123). Figure VI–25 shows a screw press, Fig. VI–26 a double-roll press with two vertical rollers, and Fig. VI–27 a chain conveyor press of Finnish construction. Screw and roller presses handle up to 20 cubic meters of bark per hour; chain conveyor presses between 17 and 70 cubic meters per hour (92).

Combustion of the bark means destruction, so far as material is concerned, even if it may be justified from an economic point of view. The wood industry, of which the pulp industry is an associate of considerable importance, has an enormous amount of refuse in the form of waste wood, which includes the bark. In the United States, the amount of such refuse amounts to around 40,000,000 cords per annum; this about corresponds to the yearly consumption of wood in the pulp industry. The amount of bark obtained yearly in Sweden is 4,100,000 cubic meters, in Finland 3,500,000, and in Norway 900,000. The total amount of bark obtained yearly in Europe, without including European Russia, is estimated to be 25,600,000 cubic meters. It is understandable that efforts are being made to utilize these wood residues in a more substance-preserving way.

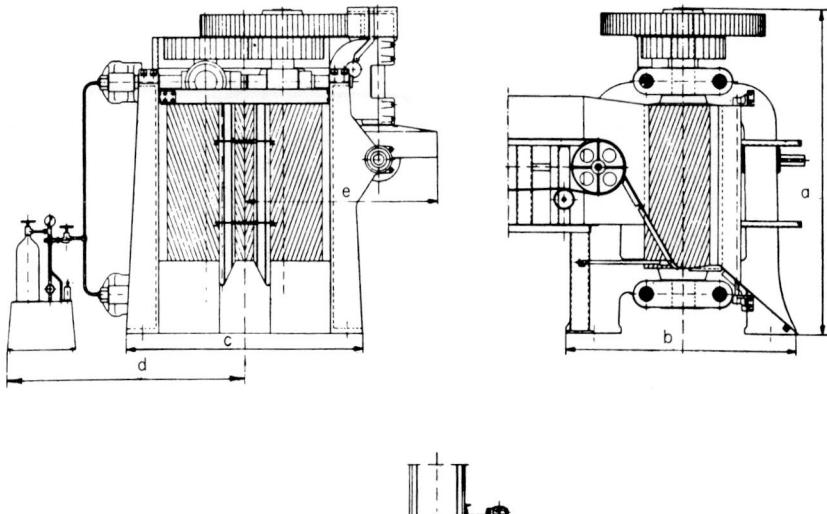

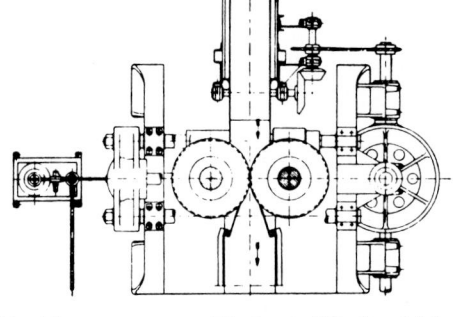

Figure VI-26. Roll barking press system Waplan. (Waplans Mekaniska Verkstad A.B. Waplan, Sweden.)

What are the prospects for utilization? The chemical composition and the anatomical structure of bark have been discussed already. A very valuable and large part of the bark, but varying in amount according to the wood species, is the cork. To obtain the cork, the bark has to be processed mechanically by a kind of fractionation. This can be done by milling, by chemical treatment, by sorting, by flotation, or by explosive defibration. Cork can be used, apart from its classical use as bottle corks, chiefly as filling material for floor coverings. The fibrous portion of the bark, although short-fibered and brittle, can be applied advantageously for floor tiles, board, filter material, in oil drilling, as filling material in ceramics, and in the concrete-cement industry. Finely ground bark is used in phenolic molding compositions. In this process the bark, because of its chemical composition, undergoes reactions with other molding additives during heat-pressing. The properties of the bark in each wood species are different and its uses there-

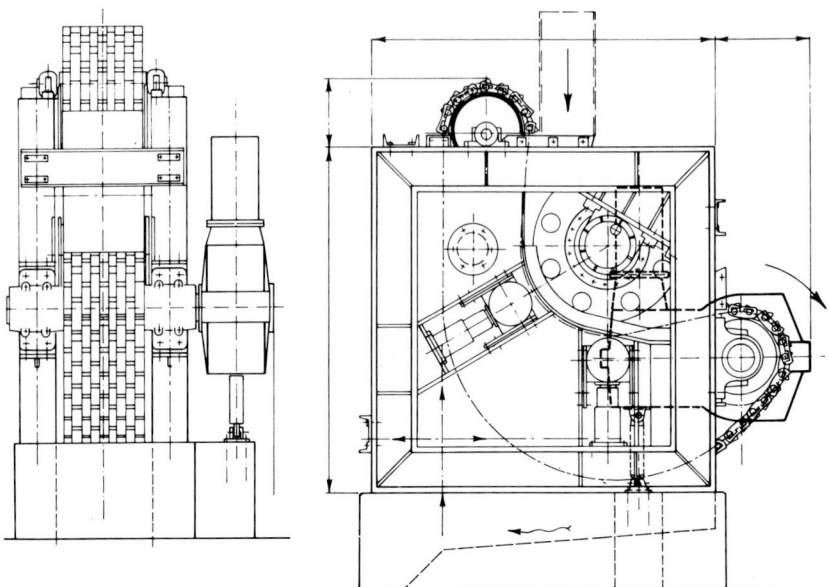

Figure VI-27. Warkaus bark press with electric drive. (A. Ahlström O. Y. Warkaus Mekaniska Verkstad, Warkaus, Finland.)

fore differ completely, depending upon the origin of the bark and the degree of its physical fractionation.

As far as the chemical utilization of bark is concerned, as mentioned before, a large number of organic compounds can be isolated from it, mainly by extraction. To mention only some of these, there are the flavanoids, alkaloids, carbohydrates, inositole, terpinoids, glycosides, saponins, steroids, esters, fats, waxes, and complex phenols. In general, the isolation of pure organic compounds from bark is economically feasible only if other salable components can also be isolated by physical fractionation. Further difficulties, especially in the preparation of wood for the pulp industry, arise from the fact that generally barks from different wood species are obtained mixed together and must first be converted into a form suitable for further processing. Among the flavanoids, quercitin undoubtedly plays an important role because it can be used in the pharmacological and dyestuff industries and as an antioxidant. The tannins have gained considerable importance on the market but they have to compete with the more effective tannins obtained from quebracho wood. However, there is some demand for the bark tannins for adhesives. The literature on proposals for the utilization of bark is extensive, but, in general, its utilization seems largely to be still in its initial phase (74,110,122,134).

5. CUTTING AND CHIPPING OF THE WOOD

5.1 Equipment for Chipping

For any kind of chemical degradation of wood it has to be converted into a form in which it can be impregnated rapidly, uniformly, and completely by chemicals in aqueous solution. The penetration of the liquids into the wood is a complex process which takes place partly by capillary action and partly by diffusion. Since this movement of the liquid is of major importance for the complete course of the reaction it will be treated in detail in one of the next chapters. It is obvious that the thickness of the chip which has to be penetrated, as well as its moisture content, will influence the movement of the liquor. The logs therefore must be chopped into chips as uniform as possible. The chippers used for this purpose have originated from three different concepts. The knife disk chipper, which can almost be designated as classical, receives the wood in the form of logs of suitable length through a chute at an angle of 35° to 45° to the knife disk. The disk has a diameter of 220 to 280 centimeters and is equipped with 4, 8, 10, and sometimes 12 knives. Figure VI–28 shows such a chipper. The wood is fed from a conveyor into the chute and slides down into the chipper spout; there it comes in contact with the knives, which are mounted on the rotating disk. The chips are cut by the impact of the knives and leave the chipper through the slots behind the disks.

The chipping process can be divided into three stages: (*a*) the cutting, (*b*) the thrust, and (*c*) the breaking action. The chief energy consumption is that used by the thrust action of the knife disk. The shearing strength of the wood against a force acting at a *right angle* to the fiber direction is about one-third of that which resists a force acting *parallel* to the fiber direction. A compression acting parallel to the fiber direction decreases the strength of a wood pulp considerably more than occurs with a pulp prepared from a wood that has been subjected to a compressing force vertical to the fiber direction. With a chipper constructed on the basis of these experiences, the force of the knife edge is adjusted in such a way that only a small force, sufficient for the pull-in of the log, is used parallel to the fiber direction. The main part of the force is applied at a right angle to the fiber direction so that only relatively little is required for the removal by chipping (51).

The energy consumption of the chipping increases with the angle of the knives and is greater for conventional chipping than for that done approximately parallel to the fiber direction. An axial deformation of more than 1% in the fiber direction can be noted from the fringed ends of chips and sometimes in the chips themselves obtained by all conventional methods, but, with almost parallel chipping, no axial deformation can be observed although there may be considerable deformation vertical to the fiber direction (18).

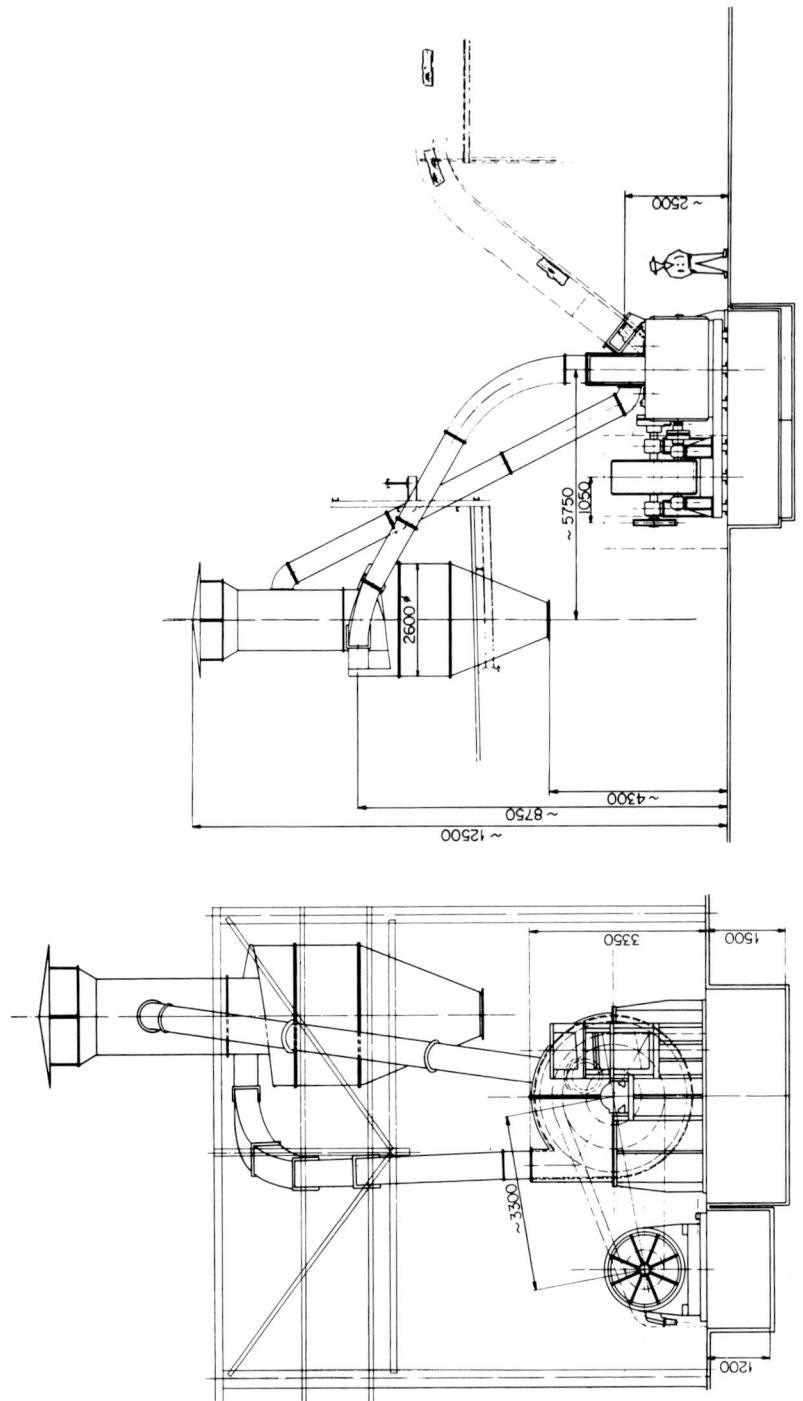

Figure VI-28. Multiknife chipper system H. Wigger, Unna, Germany.

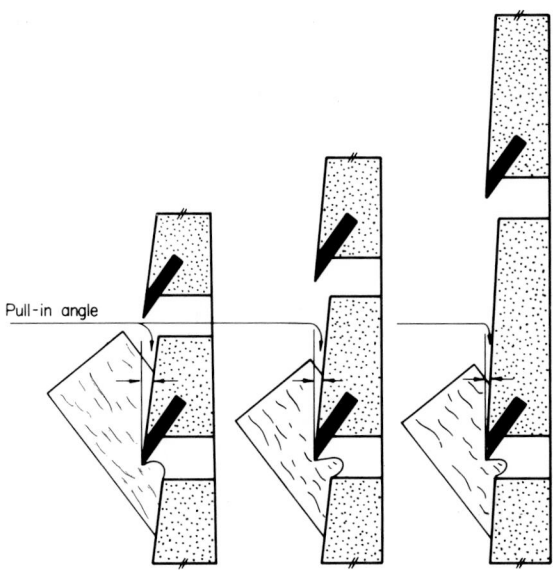

Figure VI-29. Norman chipper. The pull-in angle decreases with increasing distance of the knives (27).

A wide uniformity in the wood chips is of great importance for a uniform cook. In the alkaline cooking process the liquor penetrates almost uniformly in both directions. The thickness and width of the chips are therefore almost as important as the length (131). Differences in thickness of between three and eight millimeters do not greatly affect the yield of screened pulp and its lignin content; however, chips of over eight millimeters or less than three millimeters in thickness have a distinct effect on the selectivity of the cooking liquor. For the acid bisulfite process, 25 mm is the minimum chip length still acceptable. With constant chip length the thickness increases with the increasing angle of inclination of the chute toward the knife disk (25,132). While the knife of the rotating disk slices through the wood, the cut face of the log follows as far as possible the adjacent surface of the knife. The log must be pressed against the surface of the rotating disk so that when it arrives at the next chip slot it is ready for another cut. The angle formed by the vertical line and the adjacent knife surface (the pull-in angle) is of preponderant importance. A pull-in angle that is too small is as bad as an angle that is too great. As is shown in Fig. VI-29, the cutting angle decreases with increasing space between the knives. The "control surface" of the knife must aim at the front edge of the next chip slot to obtain the best results. This forms the principle of the well-known Norman chipper (27).

The importance of the uniformity of the chips is shown in Table VI-

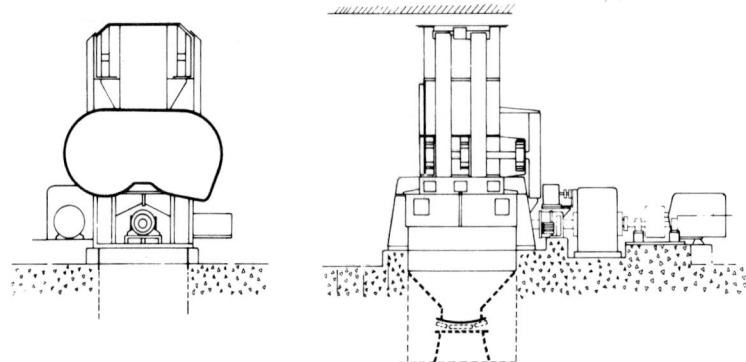

Figure VI-30. Drum chipper system Söderhamns Verkstäder (108).

7, which compares the results of cooking experiments by the sulfate process of wood chips obtained with various chippers. A conventional 10-knife chipper and a 12-knife Norman chipper were used and some of the chips so obtained were further shredded. The great differences in yield of screened pulp result from the different proportions of oversized chips. The chips from the 10-knife chipper contained 27% oversized chips, those from the 12-knife chipper only 7%, while the shredded chips did not contain any oversized specimens. The strength properties of the original and the shredded chips from the same chipper showed no noteworthy differences (86).

TABLE VI-7

YIELD OF SULFATE PULP AND SCREENINGS FROM CHIPS FROM VARIOUS
CHIPPERS WITHOUT AND WITH CHIP SHREDDING (86)

	10-Knife chipper	12-Knife chipper	12-Knife chipper with chip shredding
Total yield (%)	49.7	48.5	48.7
Screened pulp (%)	39.3	44.9	48.6
Screenings (% of total yield)	20.9	7.4	0.2

Further improvements in chippers with ten and more knives were provided primarily by steadying the logs at the entrance into the chipper to prevent them from bouncing and "fishtailing." A helicoidal surface of the multiknife disk has proved to be advantageous (19).

A principal deviation from the orthodox chipper construction is the drum chipper. Figure VI–30 shows such a machine schematically. At first glance this machine more closely resembles a magazine grinder than a chipper. The principal part is a large drum, 1.8 meters in diameter and 1 to 2.5 meters long. The wood is fed onto the drum through a shaft through

which it is moved downward and pressed against the drum by endless chains. The logs are placed in the shaft parallel to the drum axle. The drum itself is equipped with a large number of knives, 33 mm in width and 33 mm apart. These small knives cut 33 mm-wide bands or strips parallel to the log axis while a knife, the full width of the drum, then cuts the shavings peeled from the log by the short knives. The drum makes about 30 revolutions per minute and a drum 1.5 meters wide produces about 400 cubic meters of chips per hour. Because the production is continuous and the chipping takes place parallel to the fiber axis the power consumption is much lower than is the case with a conventional chipper of comparable output. A further advantage is the marked reduction in the volume of noise. The main advantage, however, is that the drum chipper produces unusually uniform chips which make a subsequent screening practically unnecessary. There is hardly any dust or fine splinter formation. The length of the logs is determined by the width of the drum.

For long stems a spiral chipper has been developed. It consists of two truncated cones which are arranged with their tapered ends downward, thus forming a V-shaped space above them. These cones are equipped with a number of spirally arranged knives in such a way that the diameter of the spirals increases with each rotation. The number of knives depends upon the desired thickness of the chips and the dimensions of the logs. In general, the action of the spiral chipper is the same as that of the drum chipper (24a,67a,108,125).

A chipper which comes between the drum and spiral chippers has been developed at the Swedish Central Laboratory of the Pulp Industry for the chipping of logs of varied lengths. The goal of this chipper was to maintain the principle of chipping along the fiber axis, but the problem was the continuous feeding of the long stems into the machine because it had no provision for moving them forward. This problem was solved without an increase in power consumption or impairment of the quality of the chips by feeding the stems at an angle of 20° to the drum axis (52,125).

5.2 Effect of Chipping on the Properties of the Chips and Pulp

The effect of the chip thickness is more marked in the acid sulfite cook than in the alkaline cooking processes. The Roe number of the pulp increases with increasing chip thickness under equal cooking conditions. The strength properties and the behavior of the pulp on beating show qualitative losses with increasing chip thickness. These losses are lower in cooks with soluble bases than in those with insoluble bases (30,41,69). The knives, especially when their edges have lost some of their sharpness or when the cutting angle is too great, cause a compression of the wood fiber. When small blocks of wood are exposed in axial direction to the pressure

of a hydraulic press they undergo a compression and, with otherwise equal pulping conditions, the compressed blocks are less readily delignified in an alkaline cook than are noncompressed blocks. The pulp obtained from compressed wood has diminished strength properties. Chips damaged by sawing give a sulfite pulp with markedly lower strength properties. The hydrolytic attack by the cooking liquor upon the cellulose increases as the result of the mechanical damage to the fiber structure. Dissolving pulps prepared from chips damaged by compression show a decrease in viscosity and a considerable increase in 10% sodium hydroxide-soluble material (26, 47,54,106).

Two types of wood damage can be distinguished: (*a*) mechanical damage, which causes a shortening and fragmentation of the fibers, and (*b*) latent damage, appearing exclusively after the action of the cooking liquor. The mechanical damage affects the properties of any pulp, no matter by what cooking process it has been produced. Latent damages occur only in sulfite pulps and are caused by the hydrolytic action of the acid cooking liquor; they are traced to an axial compression and displacement of the microfibrils of the cellulose in the cell wall. The degree of damage is not proportional to the energy used for the deformation of the wood, but warm wood is damaged more than cold wood. The fiber strength of a sulfite pulp from spruce summer wood shows a marked decrease, around 20%, when the wood has been compressed 10% in the axial direction. A comparison of a sulfite pulp prepared from undamaged chips with one obtained from compressed wood showed that neither the specific energy of the fiber bonding nor the binding area, as measured by the optical contact method, is affected. The damaging of chips affects the strength properties of the fiber itself (48,53,113). A strong compression of the wood causes dislocations and disorder in the outer and inner secondary layers of the secondary wall, which is thereby weakened. In swelling agents the fibers swell rapidly and practically without resistance. Compression damage in chipping is more pronounced with springwood than with summerwood. Microtome cross-sections under polarized light show that damage in springwood is more uniform and more consistent than in summerwood (49,50).

5.3 Control of Operations

The importance of controlled wood preparation for the quality of the pulp has led to a comprehensive automation of the control systems. Special installations which have been developed for the weighing of the wood are the filled system force cell method, the pneumatic force balance, the strain gauge force cell measuring, and the simple water-displacing methods. The control of the transport of the wood, the barking process, and the conveying of the chips are handled through a central office. The use of television

equipment for the control is no longer a rarity in modern pulp mills (13,32).

The determination of the moisture content of chips is of special importance because it more or less impedes the penetration of the cooking liquor into the wood structure. A hydrodensometer determines the moisture content by radiation from a radium-beryllium source. The radium emits γ rays which excite the beryllium to the emission of rapid neutrons, the weight of one of which equals the weight of one hydrogen atom in the water molecule. When these rapid neutrons strike a hydrogen atom of the water in the chips they are slowed down and reflected as slow neutrons. The number of slow neutrons per time unit are counted and, from this number, the moisture content of the chips is calculated with the help of an experimentally determined factor. The method requires a considerable sample of chips from the conveyor and their preparation for the measurement. The determination itself requires only a few minutes (42,45).

This method is not too satisfactory because it is discontinuous and the continuous cooking procedure requires a similar control of the chips. One is thus confronted with a determination of the moisture of a stream of semisolid materials. The size of the particles, the speed of the flow, and the density and arrangement of the chips are the most important factors influencing the data. The moisture determination alone is therefore not sufficient and must be combined with a weight determination. By a combination of a weight-cell measurement unit and a capacity moisture determination, the amount of wood fed into a continuous digester can be estimated on a dry wood basis (12).

Continuous measurement of wood chip moisture requires the determination of the amount of water present and also of the amount of wood associated with the water. One of these measurements alone will not be sufficient unless the bulk density is constant. A two-measurement system based on nuclear radiation is capable of measuring the percentage of moisture in wood chips continuously, with an average error not exceeding 0.5%. The Nuclear-Chicago Qualicon system utilizes nuclear radiation to make separate measurements of bulk density (total mass) and total hydrogen in wet wood chips. The resulting signals are combined in a ratio computer to yield a single electrical signal which is proportional to percent moisture. The signal can be displayed, recorded, or used for control. The bulk density channel uses gamma radiation to measure total mass independently of composition. Radiation which penetrates a fixed cross-section of material is detected by an ionization chamber which develops a small current proportional to the radiation. Since the principle is that absorption of radiation is proportional to mass, the current is inversely proportional to bulk density. The hydrogen channel uses high-energy neutrons to measure the total hydrogen in the product. The fast neutrons lose energy by collision with the nuclei of matter. The number of slow neutrons resulting from

these collisions is proportional to the hydrogen content of the material. The slow neutrons are detected by an ionization chamber which develops a small current directly proportional to hydrogen density (97).

To summarize, it may be said that the progressive development of control methods assures a uniform charging of continuous digesters and in this way increases the great economic advantages resulting from the continuous cooking process as well as from a uniform and improved pulp quality.

5.4 Chip Shredding and Transportation

The importance of chip size has already been mentioned repeatedly. It is clear that the cooking process can be accelerated by further shredding of the chips so long as the shredding does not cause qualitative damage in the pulp. This chip shredding is especially useful if a high yield of pulp, chiefly from the kraft process, is desired. Because the yield is primarily responsible for the quality of the pulp, the goal of high yields usually means a certain loss in quality in favor of economic advantages. The goal of higher yields can be achieved more readily with shredded chips. Screening of the crushed chips is superfluous and knot mills are not required or can be operated with low power consumption. The pulp from shredded chips contains fewer splinters and impurities, the cooking time can be reduced, and the production increased. The reduced strength properties that were feared for pulp from crushed chips can be avoided by suitable cooking conditions, as has been shown by recent investigations. It is therefore to be expected that further information will lead to an increased use of crushed chips, especially in continuous cooking processes (43,78,102).

In pulp mills with a wood preparation plant, the transport of the chips takes place by conveyors or through pneumatically operated pipe lines. In large wood-processing plants with widespread wood delivery sources, the wood-processing plant and the pulp mill are often locally separated. Some pulp mills obtain from large sawmills and wood-working plants wood and sawmill wastes which are already in small pieces or are converted at their source into chips. In such cases, chips, ready to be cooked, are delivered to the pulp mill by truck or rail or on barges and are stored in large piles or in chip bins and from there are taken into the mill. This transportation within the mill area again usually takes place by conveyors or pipe lines. Chips stored in high piles often pack together during storage and must be loosened by a specially designed apparatus before they go into the transportation lines. Hydraulic transport through pipes has been considered and might be economical for greater transportation distances if the amount of chips reaches 900,000 cubic meters or more per year; the water used can then be utilized in the mill (8,98,117).

REFERENCES

1. Adler, E., *Svensk Papperstidn.* **54** (13), 445 (1951); **54** (14), 477 (1951).
2. Alhojarvi, J., Alm, A. A., Jarvela, O. and Virkola, N. E., *Paperi Puu* **46** (4) 209 (1964).
3. Allis Chalmer Corp., *Paper Trade J.* **111** (Aug. 15) 31 (1940).
4. Annergren, G., Bengtsson, B., Dillner, B., Hagglund, A. and Jaegerud, G., *Svensk Papperstidn.* **68** (9), 309 (1965).
5. Annergren, G., Dillén, S. and Vardheim, S., *Svensk Papperstidn.* **67** (4), 125 (1964).
6. Anon., *Pulp Paper Mag. Can.* **53** (1), 113 (1952).
7. Anon., *Pulp Paper Mag. Can.* **54** (8), 158 (1953).
8. Anon., *Pulp Paper Mag. Can.* **67** (8), 96 (1966).
9. Arend, J. L., *Pulp Paper Mag. Can.* **53** (7), 159 (1952).
10. Assarson, A., Croon, I. and Donetzhuber, A., *Svensk Papperstidn.* **66** (22), 940 (1963).
11. Bell, G. E., *Pulp Paper Mag. Can.* Woodland Rev. 130 (Dec. 1953).
12. Benton, D., *Tappi* **49** (9), 56A (1966).
13. Billerud A.B., Personal communication (1965).
14. Bjoerkman, E. B. and Haeger, G. E., *Svensk Papperstidn.* **66** (15), 558 (1963).
15. Bois, P. I., Flick, R. A., and Gilmer, W. D., *Tappi* **45** (8), 609 (1962).
16. Brabender, G. J., *Tappi* **32** (8), 337 (1949).
17. Brown, H. P. *in* Wise and Jahn, "Wood chemistry" Vol. 1. Reinhold, New York, 1962.
18. Buchanan, J. G. and Tuchnicki, T. S., *Pulp Paper Mag. Can.* **64** (5), T.235 (1963).
19. Buechler, R. M., Duncan, L. L. and Haller, J. R., *Tappi* **44** (8), 178A (1961).
20. Burgon, W. J., *Tappi* **47** (5), 124A (1964).
21. Bush, Ch. C. and Tribble, J. J., *Tappi* **46** (6), 160A (1963).
22. Chamberlain, E. B. and Meyer, H. A., *Tappi* **33** (11), 554 (1950).
23. Chang, Y. P., *Tappi* Monograph Series No. 14, New York, 1954.
24. Chang, Y. P. and Mitchell, R. L., *Tappi* **38** (5), 315 (1955).
24a. Clark, J. d'A., *Tappi* **50** (6), 136 (1967); U.S. Patents 2,735,762; 2,849,038.
25. Colombo, P. Corbetta, D., Pirotta, A. and Ruffini, G., *Svensk Papperstidn.* **67** (12), 505 (1964).
26. Cowan, W. F., *Tappi* **42** (2), 152 (1959).
27. Crowley, E. L. and Wardwell, N. P., *Tappi* **44** (8), 175A (1961).
28. Dahm, H. P., *Norsk Skogind.* **18** (10), 362 (1964).
29. Dahm, H. P., *Norsk Skogind.* **20** (1), 10 (1966).
30. Dahm, H. P. and Loeschbrandt, F., *Norsk Skogind.* **14** (11), 419 (1960).
31. Donetzhuber, A. and Swan, B., *Svensk Papperstidn.* **68** (11), 419 (1965).
32. Dykes, J. T., *Pulp Paper Mag. Can.* **62** (5), T-274 (1961).
33. Ellefson, O. and Langselmo, O., *Norsk Skogind.* **14** (11), 474 (1960).
34. Elmore, C. P. and Rochford, R. S., *Tappi* **46** (6), 157A (1963).
35. Enqvist, A. B., German Patent 708,790 (1938).
36. Erdtman, H., *Svensk Papperstidn.* **43** (13), 241 (1940).
37. Erman, W. F. and Lyness, W. I., *Tappi* **48** (4), 249 (1965).
38. Faber, H. B., *Tappi* **43** (5), 406 (1960).
39. Faserchemie Gesellschaft mbH., German Patent Appl. French Patent 92,359 (1942).
40. Faserchemie Gesellschaft mbH., German Patent Appl. French Patent 94,292 (1943).

41. Feiner, J. H. and Gallay, W., *Pulp Paper Mag. Can.* **63** (9), T-435 (1962).
42. Folsom, J. H., *Paper Trade J.* **140** (46), 54 (1964); *Tappi* **48** (6), 100A (1965).
43. Freedman, H.; *Paper Trade J.* **149** (31), 36 (1965).
44. Fuji, M. and Kurth, E. F., *Tappi* **49** (2), 92 (1966).
45. Gibson, E. J. and Rusten, D., *Norsk Skogind.* **18** (10), 351 (1964).
46. Glaeser, H., *Papier* **6** (23/24), 524 (1952).
47. Hartler, N., *Svensk Papperstidn.* **65** (1), 6 (1962).
48. Hartler, N., *Svensk Papperstidn.* **66** (10), 412 (1963).
49. Hartler, N., *Svensk Papperstidn.* **66** (11), 443 (1963).
50. Hartler, N., *Svensk Papperstidn.* **66** (14), 526 (1963).
51. Hartler, N., *Svensk Papperstidn.* **66** (17), 650 (1963).
52. Hartler, N., *Svensk Papperstidn.* **68** (18), 618 (1965).
53. Hartler, N., Kull, G. and Stockman, L., *Svensk Papperstidn.* **66** (8), 309 (1963).
54. Hartler, N. and Sundberg, O., *Svensk Papperstidn.* **63** (8), 263 (1960).
55. Hergert, H. L., *Forest. Prod. J.* **8** (11), 355 (1958).
56. Hergert, H. L. and Kurth, E. F., *Tappi* **36** (3), 137 (1953).
57. Holmes, C. W. and Kurth, E. F., *Tappi* **44** (12), 893 (1961).
58. Hurtler, A. M. and Jelinek, V., *Pulp Paper Mag. Can.* **67** (3), T-179 (1966).
59. Industriekompaniet A.B., Swedish Patent 118,829 (1964); Korsnäs, A.B., Swedish Patent 121,038, (1944); Gustafson, A., Swedish Patent 132,652 (1948); Swedish Patent 132,734 (1946); Svenska Cellulosa A.B., Swedish Patent 128,197 (1944).
60. Jahn, E. C., *Pulp Paper Mag. Can.* **55** (8), 154 (1954).
61. Jensen, W., *Paperi Puu* **31** (7), 113 (1949).
62. Jensen, W., Fremer, K., Sierilä, and Wartiavaara, W. *in* B. L., Browning, "The chemistry of wood," Wiley (Interscience), New York 1963.
62a. Jerkeman, P. *Svensk Papperstidn.* **70** (18) 587 (1967).
63. Johansson, A. G., German Patents 657,216, 657,217 (1934).
64. Jones, J. B., *Tappi* **32** (2), 58 (1949).
65. Juŝtŝuk, E., *Holz Roh-u. Werkstoff* **10** (6), 229 (1952).
66. Kaijser, Ch., *Svensk Papperstidn.* **55** (5), 173 (1952).
67. Keepers, C. H., *Paper Trade J.* **117** (9), 93 (1943).
67a. Larson, L. E., *Tappi* **50** (2) 61A (1967).
68. Lassberg, J. v., *Papierfabrik.* **37**, 243 (1939).
69. Lea, N. S. and Siverstone, S., *Tappi* **45** (10), 157A (1962).
70. Lindgren, R. M. and Eslyn, W. E., *Tappi* **44** (6), 419 (1961).
71. Ljunggren, B., *Svensk Papperstidn.* **48** (23), 567 (1945).
72. Lungqvist, K. H., *Svensk Papperstidn.* **68** (16), 527 (1965).
73. Manson, D. W., *Tappi* **43** (1), 59 (1960).
74. Marian, J. E. and Wissing, A., *Svensk Papperstidn.* **59**, 751, 800, 836, (1956); **60**, 45, 85, 124, 170, 255, 348, 522 (1957).
75. McIntosh, D. C., *Tappi* **36** (8), 150A (1953).
76. McNeel, W., Shenefelt, R. D., Pascoe, T. A. and Scheffer, T. C., *Tappi* **43** (4), 323 (1960).
77. Miag, A. G., German Patent 871,365 (1941).
78. Miller, R. L. and Rothrock, C. W., *Tappi* **46** (7), 174A (1963).
79. Moqvist, J. W., Swedish Patent 117,459 (1945); Zellstofffabrik Waldhof, German Patent 878,703 (1950).
80. Morgan, B. G. and Wilson, J. C., *Pulp Paper Mag. Can.* **65** (5), T-228 (1964).
81. Mosher, J. P., *Pulp Paper Mag. Can.* **65** (10), T-437 (1964).
82. Mountain, H. S., *J. Forestry* **47**, 627 (1949).

83. Nickerson, R. E., *Pulp Paper Mag. Can.* **65** (C), T-123 (1964).
84. Niethammer, W., *Papierfabrikant.* **27,** 600 (1929).
85. Nilsson, Th., *Svensk Papperstidn.* **68** (15), 495 (1965).
86. Nolan, W. J., *Tappi* **46** (8), 458 (1963).
87. Nordin, B. and Selleby, L., *Svensk Papperstidn.* **68** (1), 1 (1965).
88. Olson, D. J., *Pulp Paper Mag. Can.* **65** (C), T-88 (1964).
89. Pearl, I. A. and Darling, St. F., *Tappi* **47** (6), 377 (1964).
90. Pearl, I. A., Justman, O., Beyer, D. L. and Whitney, D., *Tappi* **45** (8), 663 (1962).
91. Perem, E., *Pulp Paper Mag. Can.* **59** (9), 109 (1958).
92. Possanner, J. v., *Wochbl. Papierfabrik.* **28,** 645 (1930).
93. Price, F. A., *Pulp Paper Mag. Can.* **60** (1), 80 (1959).
94. Pryce, J. N. and Cole, A. H., *Pulp Paper Mag. Can.* **64** (9), T-389 (1963).
95. Ramalingam, K. V. and Timell, T., *Svensk Papperstidn.* **67** (12), 512 (1964).
96. Reid, H. A., *Appita J.* **13** (1), 41 (1959).
97. Richeson, M. A., Clapp, L. L. and Miller, R. L., *Paper Trade J.* **150** (39), 58 (1959).
98. Ross, J. H., *Tappi* **47** (7), 153A (1964).
99. Rowe, J. R., *Phytochemistry* **4** (1), (1965).
100. Schmid, W., *Osterr. Zentralbl. Papierind.* 435, 1930.
101. Schuett, Ch., *Paperi Puu* **42** (11), 591 (1960).
102. Scott, J. B., *Norsk Skogind.* **20** (9), 320 (1966).
103. Segall, G. H. and Purves, C. B., *Pulp Paper Mag. Can.* **47** (3), 148 (1946).
104. Selleby, L., *Svensk Papperstidn.* **68** (14), 477 (1965).
105. Sheridan, T. G., *Pulp Paper Mag. Can.* **59** (C), 228 (1958).
106. Sihtola, H., Saarinen, A., Wigren, G. and Ulmanen, T., *Paperi Puu* **37** (11), 511 (1955).
107. Smith, R. F. and Mathiesen, K., *Tappi* **37** (10), 451 (1954).
108. Soederhamns Verkstäder A.B., Private communication.
109. Somson, R. A., *Tappi* **45** (8), 623 (1962).
110. Sproull, R. C. and Pierce, G. A., *Tappi* **46** (8), 175A (1963).
111. Stegmann, E., *Papier* **5** (9), 162 (1951).
112. Stephenson, J. N., "Pulp and paper manufacture." McGraw-Hill, New York, 1950.
113. Stone, J. E., *Tappi* **44** (8), 166A (1961).
114. Suckdorff, B., *Svensk Papperstidn.* **61** (9), 279 (1958).
115. Sumner Iron Works, Private communication.
116. Svenska Cellulosa A.B., Swedish Patents 130,000 (1944); 132,509 (1945); 124,273 (1945); 122,987 (1945).
117. Thiessmeyer, L. R., *Unasylva* **19** (1), 19 (1965).
118. Thornburg, W. L., *Tappi* **46** (8), 453 (1963).
119. Timell, T. E., *Svensk Papperstidn.* **64** (18), 651, (19), 685 (1961).
120. Timell, T. E., *Svensk Papperstidn.* **64** (20), 651, 744 (1961).
121. Trendlenburg, R. and Mayer-Wegelin, H., "Holz, Roh Werkstoff." C. Hanser, München, 1958.
122. *U.S. Forest Prod. Lab.* Ann. Rept. No. 1666–5, June 1961.
123. Velle, A., *Norsk Skogind.* **18** (11), 441 (1964).
124. Vogt, H., *Norsk Skogind.* **19** (5), 181 (1965).
125. Vogt, H., *Paper Trade J.* **149** (51), 26 (1965).
126. Watson, A. J., *Appita J.* **13** (1), 49 (1959).
127. Wenzl, H. F. J., *Holzforschung* **1** (4), 112 (1947); **2** (1), 12 (1948).
128. Wenzl, H. F. J., German Patent 941,452 (1954).
129. Wikmanshytte Bruks A.B., German Patent 610,954 (1933); 664,799 (1936).

130. Woodfin, R. O., *Tappi* **46** (2), 73 (1963).
131. Wretne, A., *Paper Trade J.* **148** (46), 56 (1964).
132. Wultsch, F. and Holtappels, H., *Wochbl. Papierfabrik.* **93** (3), 77 (1965).
133. Wultsch, F. and Salzer, H., *Papier* **6,** 75, 115, 169 (1952).
134. Young, H. E., Gammon, C. B. and Ashley, M., *Tappi* **47** (9), 555 (1964).
135. Zak, H. and Krauthauf, E., *Papier* **18** (11), 691 (1964).

VII

The Sulfite Pulp Cooking Process

REVIEW AND OUTLOOK

A hundred years have passed since B. C. Tilghman received a patent in which the decomposition of wood with sulfurous acid or its salts was used for the production of fibrous pulp for the manufacture of paper. The technical use of this process, however, first occurred eight years later, in 1874, when Ekman, a Swedish chemist, marketed the first sulfite pulp. At that time pulp was not an unknown product because wood could also be decomposed with sodium hydroxide solution; the pulp obtained in this way was dark in color and had no outstanding qualitative properties. The process was expensive and the recovery of the chemicals was still largely undeveloped, expensive, and time-consuming. The first digesters were small units, a few cubic meters in volume; they were in a horizontal position and were heated by the direct injection of steam. The charging and discharging of the digesters was cumbersome and time-consuming. The use of sulfurous acid with its corrosive properties required the use of corrosion-resisting materials. After long and expensive experiments this problem was finally solved by using digesters lined with a ceramic layer and heating coils of hard lead.

Almost simultaneously with Ekman, A. Mitscherlich in Germany worked on the sulfite cooking process of wood and, in 1880, after many difficulties and legal controversies, he started a sulfite pulp mill in Zell in south Germany, the results of which attracted great attention. Mitscherlich's cooking process was marked by the use of heating coils, i.e., by an indirect heating of the digester charge with steam. In 1822, C. Kellner obtained a patent for a sulfite cooking process, first with sodium bisulfite

solution, later with calcium bisulfite solution as cooking reagent. Unlike Mitscherlich, Kellner used vertical digesters and heated them directly by the injection of steam. The pulp thus produced was quite light in color and cottonlike in texture and could be readily bleached.

The essential advances of the following years consisted chiefly in the improvement and development of the plants for the preparation of the cooking liquor. Initially, iron pyrites were used for the production of sulfur dioxide but even at that time elementary sulfur was occasionally used. The increasing demand for such sulfite pulp necessitated the building of new pulp mills and the enlargement of existing ones. To utilize the digester capacity as completely as possible, charging equipment was designed which, according to the type of operation, accomplished a certain degree of deaeration of the wood chips. A denser charging of the digester and a larger digester capacity required an improvement in the circulation of the cooking liquor in the digester, the turbulence caused by the injected steam being insufficient. The cooking liquor was therefore circulated by pumps and heated by heat exchangers outside the digesters. One of the principal difficulties which had to be overcome in connection with increasing digester capacity was the uniform impregnation of the wood chips by cooking liquor. Within the last twenty years many proposals have been made and these will be discussed in more detail in the following sections of this chapter.

Chemical research has contributed considerably to the understanding and improvement of the sulfite pulping process by the study of the alternate action of the cooking acid on the different wood components, i. e., the cellulose, hemicellulose, and lignin. This knowledge of the course of the reaction has put the process upon a secure and sound basis. Especially the role of the hemicelluloses has been clarified and it has been found that their importance can vary widely, depending upon the later use of the pulp. Whereas the manufacture of paper requires a certain amount of hemicelluloses in the pulp, the rayon industry demands a pulp from which the hemicelluloses have been removed as completely as possible. Processes have therefore been developed for removing the hemicelluloses from the sulfite pulp and giving it a content as high as possible in "alkali-resistant" cellulose.

With the development toward the end of the nineteenth century of the so-called kraft pulp cooking process—erroneously also called the sulfate pulp cooking process—the sulfite pulping process faced strong competition. Sulfite pulp was light in color and could be used for paper without further bleaching, while kraft pulp was dark in color and required bleaching when used for writing paper, but it surpassed the former by its excellent strength properties. The sulfite process was suitable chiefly for spruce- and firwoods, whereas with the kraft pulping process practically all types of wood and

waste wood could by processed. It must be mentioned here that bleaching processes have been tremendously improved during the past twenty-five years. They have been divided into several bleaching stages, and new bleaching agents, such as chlorite and chlorine dioxide, have been used. The kraft cooking process has the disadvantage of being accompanied by the formation of the obnoxious odors of mercaptans and hydrogen sulfide. It has, however, the great advantage that its spent liquors can be evaporated and the chemicals recovered from the residue, with heat being generated by the burning of the organic matter, all of which contributes to the economy of the process.

The spent sulfite liquor presents a serious problem to the sulfite pulp industry because the release of the liquor into rivers affects the biochemical oxygen demand (BOD) to an inadmissible degree. Investigations have therefore been carried out to utilize the liquor in an economical way. The spent liquor contains certain hexoses and pentoses which can be fermented to produce marketable products, such as alcohol and yeast, but only a small percentage of the organic matter in the liquor can be utilized in this way. The use of soluble bases, such as magnesium, sodium and ammonium, instead of calcium has helped to solve the problem by the regeneration of these bases in a way similar to that used in the kraft pulping industry, with the burning of organic substances and utilization of their heat value. It is true that the calcium sulfite spent liquor can be evaporated and burned, but this is not economically feasible. With soluble bases, recovery of the chemicals by burning the liquor solves the problem in an economical way. Pulping with soluble bases has also made it possible to apply the sulfite process to woods high in resin content, such as pine and most of the deciduous woods.

The use of soluble bases is one of the most important advances made in the field of the sulfite pulping process in recent years. Of course a complete regeneration plant must be added to the mill and this requires large operating units and a considerable capital investment. For small pulp mills which have a low daily production and are usually combined with a paper plant, as they are occasionally found in Europe, compulsion to evaporate and burn the spent liquor would mean a definite shutdown.

A further important improvement in sulfite pulping is the continuous process which has already been used successfully in the kraft cooking process for over 15 years. Its introduction in the acid sulfite process at first met with great difficulties in the apparatus and operation. The introduction of soluble bases has here also facilitated the use of the continuous cooking method. Whereas in earlier days several digesters were housed in great buildings, today one continuous digester, which can be located outdoors, can produce 300 tons of pulp daily. In this connection it must be mentioned that extraordinary developments in construction types were

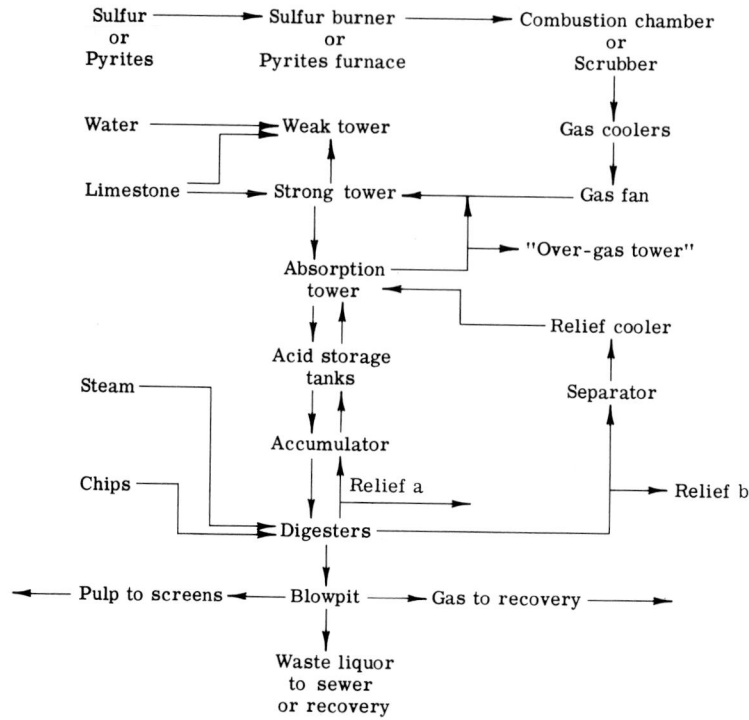

Figure VII-1a. Flow-sheet of the sulfite tower process.

required to build dependable equipment for such highly productive capacities. The same holds true for the use of construction materials, in particular for highly resistant special steels which have made the use of ceramic linings in the digesters unnecessary.

During the last two decades the sulfite pulping industry has made every effort to overtake the kraft cooking process again. It is therefore interesting to speculate as to what other developments may be expected. The continuous process will certainly undergo further constructive and operative improvements. A combination of the continuous process and the use of soluble bases will undoubtedly be a future result. Multistage cooking processes with uniform bases but under various cooking conditions, such as temperature, pH, and cooking time, and cooking with different bases under the same or different reaction conditions may be developed. As far as suggestions for new processes are available or are already in use, these will be discussed in detail in the following sections. Before this is done, the course of the sulfite cooking process will be explained briefly by a flow-sheet, regardless of whether the process is carried out batch-wise or continuously. In the flow-sheet it is assumed that calcium base is used.

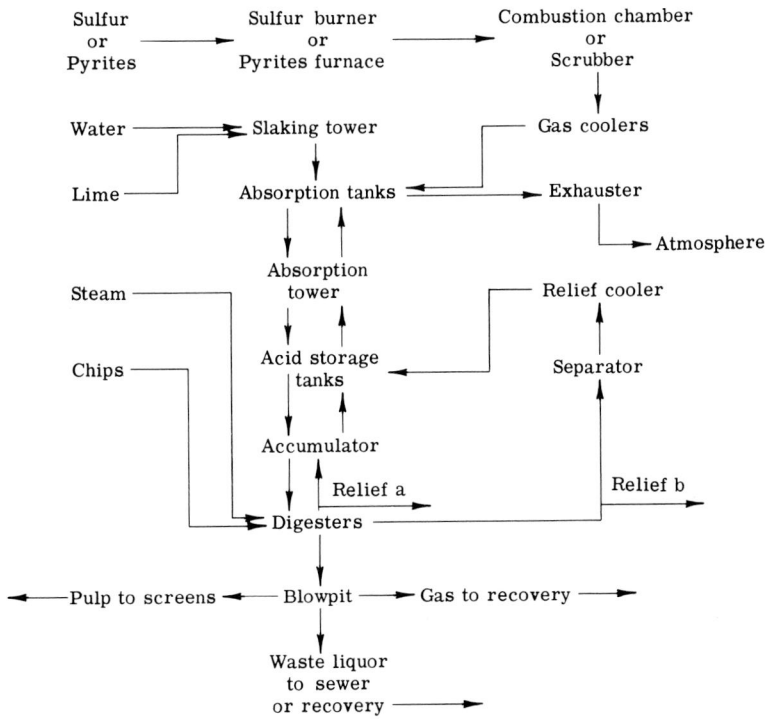

Figure VII-1b. Flow-sheet of the milk of lime system.

Deviations from the process by the use of soluble bases will be dealt with in the relevant sections. The scheme contains no data on the processing of the spent liquor, which will be discussed in another section. Figure VII-1 shows the flow-sheet. It begins with the preparation of the cooking liquor. Two systems are used: the tower process and the milk of lime process. It may be mentioned here that the cooking liquor preparation from soluble bases is very similar to that with milk of lime. This will be dealt with later.

Sulfur dioxide for the preparation of the cooking liquor is produced from sulfur ore, pyrites, or elementary sulfur, either in pyrite burners or in sulfur kilns. Whereas the sulfur dioxide from elementary sulfur is used directly, that from the pyrite burner is passed through a scrubber to free the gas from retained dust. The gas is then cooled and passed into the bottom of the first tower, filled with limestone, over which water is trickled in a countercurrent system. The nonabsorbed gas is passed from the top of the tower into a similar second tower. The weak sulfurous acid from the latter is fortified by repassing it through the first tower. The concentrated acid from the first tower is drained off at the bottom and clarified by al-

lowing it to settle. In many mills the acid is passed into the cooking liquor storage tank. Other mills, which operate with a more concentrated cooking liquor, high in free sulfur dioxide, have a third absorption tower into which concentrated sulfur dioxide is passed under a specified pressure. Other tanks are inserted for the separation of sludge and colloidal sulfur before the liquor goes into the storage tank. Before the cooking liquor is passed into the digester it is enriched with relief gas from the digesters in an accumulator which is kept under a certain pressure and at 50° to 70°C. The relief gases are passed first into the accumulator, then into the acid storage tank and, finally, into the first tower. If there is no accumulator available, the relief gases are led into a separator and from there into a gas cooler and a special absorption tower. The enriched acid is then passed into the digester, which is filled with wood chips. The course of the cook will be discussed later. The digester contents, containing the pulp, are blown and the relief gases which still contain a considerable amount of sulfur dioxide are, as mentioned above, used to enrich the cooking liquor. The spent sulfite liquor is separated and subjected to special processing.

The milk of lime process differs from the tower process in that, in lieu of limestone, a sludge of hydrate of lime (milk of lime) or an aqueous suspension of calcium carbonate is used. The sulfur dioxide is passed directly into the suspension or into a tower in which the sludge trickles over contact surfaces. Further enrichment is carried out in the same way as in the tower process.

The relief gases from the digester contain a considerable amount of heat which has to be removed either by cooling or condensation if the gas absorption takes place under atmospheric conditions. Pressure reservoirs for the acid have been introduced and these must resist not only the usual pressures and temperatures used in the sulfite process but also the effect of the acid itself. Usually old digesters are used as such reservoirs and have proven very useful for this purpose. New reservoirs are constructed in a spherical shape. This pressure storage process permits the utilization of the heat and also the charging of the digester with hot cooking liquor, whereby a certain hydraulic pressure facilitates impregnation of the chips by the cooking acid.

1. THE PREPARATION OF THE SULFITE COOKING LIQUOR

1.1 Raw Materials for the Preparation of the Cooking Liquor

As mentioned in the introduction, sulfurous acid is used as the cooking agent. The classical sulfite pulp cooking process uses a calcium bisulfite solution, containing a greater or lesser proportion of free sulfur dioxide.

Other cooking liquors may contain a certain amount of magnesium, especially when limestone containing magnesium, such as dolomite, is used. In recent times, soluble bases such as sodium, magnesium, and ammonium are being used on an increasing scale.

Sulfur dioxide is prepared from sulfidic ores, pyrites, or elementary sulfur. Pyrites are iron sulfides with a theoretical content of 46.6% iron and 53.4% sulfur. Their burning temperature is between 400° and 500°C. The pyrites burn with the addition of air to sulfur dioxide and iron oxide. The commercial pyrites contain, in addition to iron, still other metal sulfides such as copper, zinc, arsenic, and lead sulfides. The theoretical amount of sulfur dioxide therefore is often not reached although in some cases it is even surpassed. The chemical composition of some of the common pyrites are given in Table VII–1.

Elementary sulfur is found in nature in a free state with a sulfur content of 99.5% to 99.9%. The most important production centers are in Louisiana and Texas, in Italy, and in Japan. It is obvious that the countries which have elementary sulfur at their disposal will use it for the preparation of sulfur dioxide; countries rich in pyrites will use them, often in combination with free sulfur. Sulfur melts at 112°C to a brown liquid and with increasing temperatures it becomes more liquid and dark red in color. At temperatures of above 160°C the viscosity greatly increases, with further darkening. The boiling point is at 445°C. The preferred temperature for use is 140° to 160°C because of the low viscosity and favorable flow properties at this temperature.

In its purest form, limestone is marble and, as such, it is crystalline and pure white. All other limestones, the colors of which vary from white to yellow to gray, contain more or less large proportions of magnesium and small amounts of iron. Dolomite usually contains 54% calcium carbonate and 46% magnesium carbonate. When heated at a high temperature, dolomite loses carbon dioxide and is converted into oxides which, when suspended in water, form a sludge of the calcium and magnesium hydroxides; these are used directly for the absorption of the sulfur dioxide.

The use of soluble bases for the preparation of sulfite cooking liquor will be discussed later. However, it must be mentioned here that suspensions of finely milled limestone have been used successfully in absorption towers (20). Large amounts of calcium-containing sludge are obtained in sulfate pulp mills which use large quantities of lime in their caustification plants. This sludge can be used for the preparation of cooking liquor without any qualitative damage. A certain retardation at the beginning of the cook has obviously no detrimental effect on the qualities of the pulp. The silica content, which is almost ten times higher in a cooking liquor from such a caustification sludge than in a sludge from limestone, is due to the fact that this silicic acid is soluble in the aqueous sulfur dioxide solution. Whereas

TABLE VII-1

CHEMICAL COMPOSITION OF SOME COMMERCIAL PYRITES[a]

Pyrite sample	S	Fe	Cu	Zn	Pb	As	Se
Germany							
Meggen	39–40	32–34	0.05	0.9–7.3	—	0.06	—
Spain							
Rio Tinto	46–47	39–40	0.6–1.3	1.5–3.1	1.1–1.2	0.4–0.6	0.008
Tharsis	47–48	41–42	0.7–1.7	0.7–2.0	0.5–0.7	0.4–0.5	0.004
San Telmo	45–47	39–40	1.9–2.5	1.0–2.8	0.1	0.1–0.2	0.01
Portugal							
Aljustrel	45–46	39–40	0.7–1.8	3.2–4.2	1.2–1.4	0.5–0.6	0.01
Sweden							
Falun	41–42	39	0.3	4.6	1.20	0.01	—
Boliden[b]	50–51	45–47	0.09–0.15	0.3	0.05	0.05	0.03
Norway							
Orkla	41–46	37–40	2.2–2.7	1.6–2.0	0.06	0.06	0.001
Killingdal	42–45	35–38	1.6–1.7	6.5	0.5–1.1	0.1	—
Björkaasen	46–48	45–46	0.3–0.5	0.01–0.2	0.8–1.8	0.01–0.03	—
Finland							
Outukumpu	41–43	40–44	0.5–1.4	0.1–2.1	0.3	0.06	0.001
Italy							
Montecatini	47–48	42–44	0.02–0.03	0.01	0.01	0.04	—
Greece							
Kassandra	46–48	41–43	0.1–0.3	0.3–0.5	0.3–0.5	0.3–0.8	0.003–0.008
Cyprus							
Kalavassos	47–48	41–45	0.8–1.2	0.3–0.6	0.02	0.01–0.05	0.01
Yugoslavia							
Trepca[b]	48–49	44–45	0.1	0.1–0.2	0.2–0.3	0.3–0.4	—
Maydanpek	40	34.5	0.4	0.8	0.1	0.08	—
France							
Saint Bel.	49	43	Trace	Trace	1.0–1.1	0.02	—
Canada							
Pyrite[b]	47–48	43–44	0.4–0.5	1.0–1.3	—	0.02	0.008
Cuba							
Cienfugos	48–49	43–44	1.3	1.2–1.3	0.03	0.03–0.2	0.02
Japan							
Dowa Mining	47	?	0.5	1.0	0.2	0.05–0.06	—

[a] Ullmanns "Enzyklopädie der techn. Chemie" Third Ed., Vol. 15, p. 390. Urban and Schwarzenberg, München, 1964.
[b] Flotation product.

the high silica content apparently has no damaging effect upon the cooking process, troubles may arise if the spent sulfite liquor is later to be evaporated, or the silica content may lead to scaling of the heat exchangers in the indirect cooking process. The use of lime sludge is especially economical when a sulfite and a sulfate mill are combined or at least located not too

far apart. Coal particles in the sludge, however, may cause impurity of the pulp; to avoid this, the sludge would first have to be burned (144,377).

Liquid sulfur dioxide is used in some mills, especially when a high-grade cooking acid is used. The economy in this case depends on the cost of the liquid sulfur dioxide and this cost is warranted only if, for example, the sulfur dioxide is obtained as a by-product in the chemical industry.

1.2 Combustion of Sulfur

The combustion of pure sulfur is certainly the most efficient method for the preparation of sulfur dioxide. For this purpose two types of special sulfur burners are used, the rotary, and the spray type. In the rotary burner, molten sulfur is vaporized at 310° to 370°C. The molten sulfur is distributed by slow rotation of the drumlike oven as a thin film on the inner wall and comes in contact with primary air admitted and regulated by a damper. Behind the rotating oven is the combustion chamber into which secondary air is admitted. The combustion of the sulfur occurs at 850° to 950°C. A rotary sulfur burner is show in Fig. VII–2. It consists of a cylindrical drum, A, with two conical ends, B. The cylinder, A, rests on rollers, K, and makes about 1.5 rpm. The oven is charged with sulfur in either molten or solid form, in the latter case by means of a transport worm. With molten sulfur the feeding is achieved through a pipe from the tank, D, in which the sulfur is melted. The regulation of the primary air occurs through the damper, at the front cone behind the sulfur entrance. The heat of the sulfur burner is often used to melt the sulfur. The actual combustion takes place in the combustion chamber, E, where the secondary air is admitted through the damper, H, between the rotary back cone and the combustion chamber. It is relatively easy to start a rotary oven which is fed with powdered sulfur since this type of sulfur ignites very readily. To put the oven in operation with molten sulfur, the latter must be melted in the tank, D, and the pipes.

The spray-type burner obtains the sulfur by spraying it in a molten state mixed with the air required for its combustion. The combustion temperature in this case is at 650° to 850°C. The melting of the sulfur is done by the use of waste heat. The combustion heat, in turn, is used to heat the primary and secondary combustion air. The greatest possible accuracy in the regulation of the combustion air is very important to obtain the maximum yield of sulfur dioxide and to prevent the formation of sulfur trioxide. One volume of oxygen will give one volume of sulfur dioxide. Air, as the source of oxygen, contains 21% oxygen. The volume of sulfur dioxide in the gas leaving the burner can therefore never be higher than 21%. Well-regulated sulfur burners usually deliver a gas with a sulfur dioxide con-

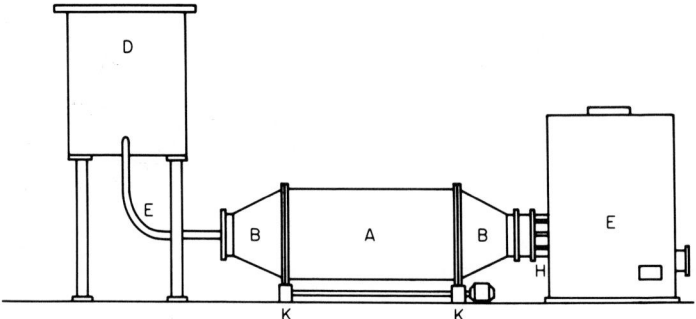

Figure VII-2. Diagram of a rotary sulfur burner.

centration of between 14% and 19%. The formation of sulfur trioxide according to the equation

$$SO_2 + O = SO_3$$

depends, first of all, upon the reaction temperature. The reaction is accelerated catalytically by metals. Because the gas in all burners comes in contact with metals, especially iron, it is important to know that the maximum formation of sulfur trioxide occurs at a temperature region of between 600° and 900°C. At below 400°C and above 1000°C, the sulfur trioxide formation is insignificant. It is therefore necessary to keep the temperature high, i.e., above 1000°C, and to cool the gas as quickly as possible as it leaves the combustion chamber.

The spray-type burner is especially well-suited for this purpose. The burner, with its wide combustion room where the sulfur enters and its much narrower main combustion chamber, allows a high intake of sulfur and a high gas velocity. The strong turbulence caused by the tangential admission of the secondary air causes an intimate and quick mixing of the vaporized sulfur and the oxygen of the air. Figure VII–3 shows a cross-section of a spray-type sulfur burner in which 1 is the spray nozzle, 2 the tangential nozzles for the secondary air, 3 the combustion room, 4 the chamber lining, 5 the main combustion chamber, and 6 the burner wall; 8 are elbows, 9 is an observation window, and 10 is the first spray cooler. The flame temperature of the burning sulfur reaches 1260° to 1320°C. The gas temperature immediately before the first cooler is 980° to 1050°C and is therefore outside the danger zone for the formation of sulfur trioxide. The gas concentration reaches 18% to 19% sulfur dioxide and 1.1% to 3.6% sulfur trioxide, calculated on the weight of the sulfur charge (47,130,413):

In this connection it is interesting that in recent times the delivery of sulfur in a molten state is also possible. The sulfur is melted at the place of origin and filled into tank cars. Even after a travel time of several days

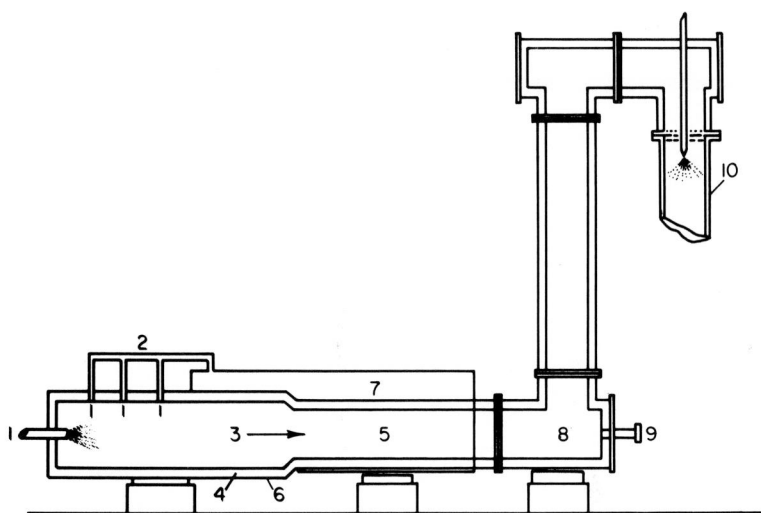

Figure VII-3. Diagram of a tubular high-mass sulfur burner.

it arrives at the mill still in a liquid state at its center and, after further heating and complete melting, it is transferred into the storage tanks. This process simplifies the reloading, saves heat for melting, and prevents the infusion of impurities (168).

1.3 Pyrite Roasters

There are several types of pyrite roasters which differ in details of their construction, but which are basically identical in their reactions. The mode of action of a Herreshoff pyrite roaster is described here without the expression of any preference for it over other types. As Fig. VII-4 shows, this oven consists of a steel cylinder, which is fitted out with a fireproof ceramic lining. A large number of arches covered with firebricks divide the cylinder into several floors, one above the other. The bottom of these floors forms the hearth on which the ore burns. The oven has a vertical axle attached with scrapers. The ore is fed into the oven by the hopper, and is moved by the upper scraper from the center to the edge of the first floor; from there it drops to the outer edge of the second floor, and from the center there to the center of the next floor, and so on. The burnt ore leaves the oven at the bottom. The air is admitted at the bottom and passes the ore in a countercurrent direction. The central axle and the scraper are cooled by air (47,130).

The operation of such an oven requires an exact regulation of the air supply and of the temperature of the various compartments in order to ob-

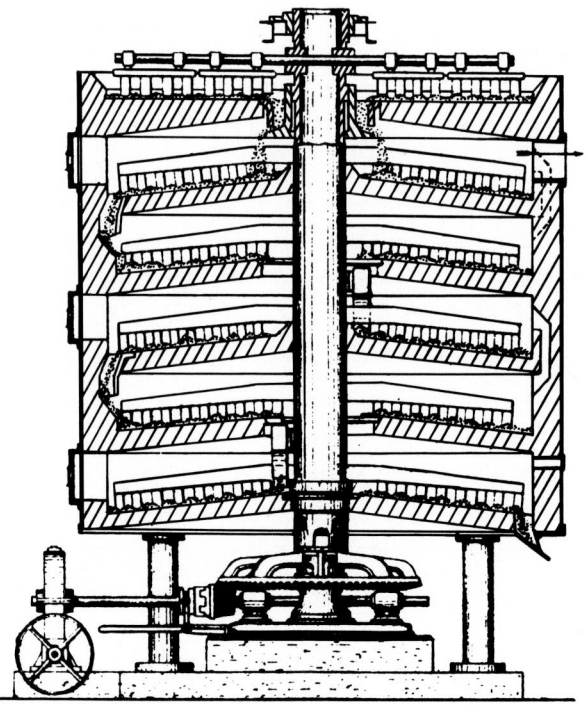

Figure VII-4. Pyrite furnace system Herreshoff.

tain a complete roasting of the ore, with a high yield of sulfur dioxide. Also the formation of sulfur trioxide and sintering of the pyrite have to be avoided. To set such an oven in operation is more complicated than that of a sulfur oven. The formation of sulfur dioxide from pyrite takes place by the following equation:

$$2\,FeS_2 + 11\,O = Fe_2O_3 + 4\,SO_2$$

Part of the oxygen of the air is therefore consumed by the oxidation of the iron to iron oxide. For the formation of a sulfur dioxide of 16 vol % from sulfur, the theoretical amount of oxygen required is 3.86 cubic meters per kg air; with pyrite and a 10 to 14 vol% sulfur dioxide the amount of oxygen is 2.5 cubic meters per kg air under atmospheric conditions. Metal catalysts and selenium further the formation of sulfur trioxide, as can be seen from Fig. VII-5. This catalytic action is stronger with lower concentrations of sulfur dioxide in the gas stream. Besides the shelved ovens, rotary kilns in operation are like those used in the roasting of ores or in the drying of granular materials. Such rotary ovens, in order to be economical, require large throughputs of ore and two of such ovens have to

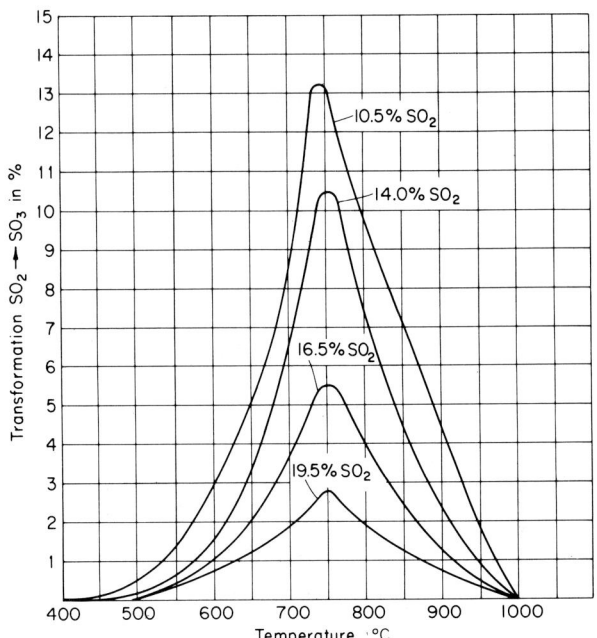

Figure VII-5. Conversion of SO₂ into SO₃ in contact with iron and dependence on temperature.

be available in case of a breakdown or needed repairs to avoid any interruption in the preparation of the cooking liquor (74).

Recent investigations have shown that the rotary oven still can stand improvements, especially with large volumes of up to 140 tons daily, in order to be economical (410). The cleavage of sulfur from pyrite occurs in two stages: the first is the dissociation of iron disulfide into monosulfide and sulfur, which occurs already at a low temperature; the second is the conversion of the monosulfide into ferric oxide and the oxidation of the sulfur to sulfur dioxide. Occasional disturbing sintering occurs when the temperature in the front zone of the furnace is kept too high. The installation of cooling coils prevents such sintering and simultaneously produces steam.

Experience gained in the burning of powdered coal has led to the construction of the pyrite flash roaster, which is shown schematically in Fig. VII-6. It is suitable for the burning of floated pyrites of very fine grain. To prevent the formation of lumps, the moisture content must be kept at about 1%. The dustlike pyrite is blown from the vibrating hopper, A, by the fan, C (which, in some cases, may be attached to a ball mill, B), through the burner nozzle, D, into the combustion chamber, E. The dust flame moves down the center of the chamber, up again at the side walls,

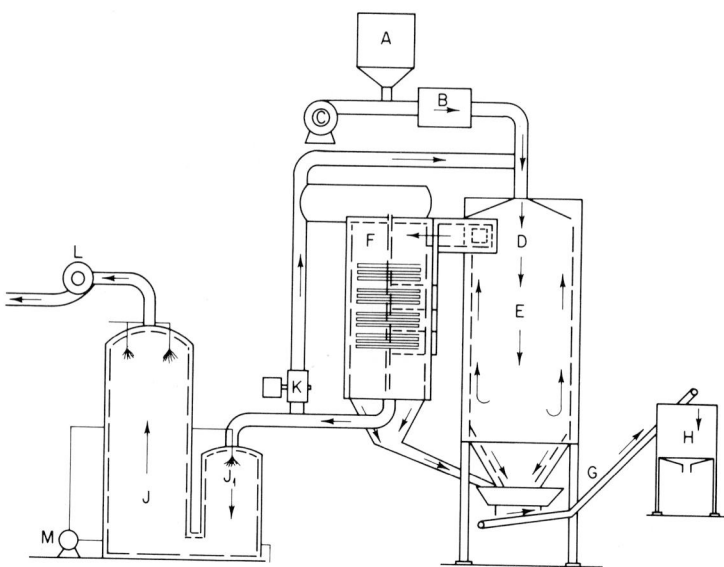

Figure VII-6. Pyrite flash roaster.

and from there into the dust chamber, F, which is connected with the boiler. The flame temperature is high enough to prevent the formation of sulfur trioxide and the gas is cooled quickly. The cinders are transported by the conveyer, G, to the silo, H. The temperature in the dust chamber is kept so high that no condensation can occur. The boiler tubes are protected by cast iron jackets from corrosion and dust erosion. If necessary, a part of the gases is returned through the fan, K, to the combustion chamber to keep the combustion, temperature, and gas concentration constant. Acid wash water from the scrubber, J, is returned by the pump, M, to the scrubber, J_1, and discharged through an outlet, into the sewer. The fan, L, blows the washed and flue-dust-free gas to the absorption tower. At the same time, the fan blows air into the combustion chamber when the cinders are to be taken out. The gas contains 11% to 12% sulfur dioxide. At the entrance to the dust chamber it has a temperature of about 1000°C and leaves it at 350°C. The roasted pyrites contain, on an average, 65% iron, 2.5% silicium dioxide, 1.75% aluminum oxide, and 0.5% sulfur. The sulfur efficiency from pyrite is over 94%. From one ton of pyrites, 1.5 to 2 tons of steam are produced. Arsenic and selenium are almost completely removed by the scrubber (11,118,196,197,269,297).

The essential problem of the flash oven is the afterburning of the pyrite dust in an atmosphere which already contains sulfur dioxide. An intensive study of the combustion course and combustion time has shown that the operation of such an oven is uneconomical when operated without after-

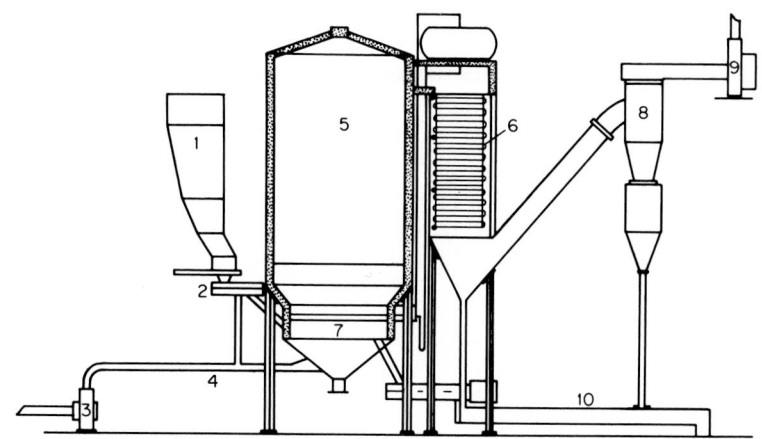

Figure VII-7. Nichols-Freeman pyrite furnace modified as pyrite flash roaster. 1 pyrite funnel; 2 weight control; 3 primary air fan; 4 air pipe; 5 flash roaster; 6 steam boiler; 7 cooling pipes; 8 cyclones; 9 flue gas fan; 10 cinder conveyor.

combustion of the pyrite dust which is carried over with the gas when there is an overloading of the flow bed. The combustion velocity cannot be increased by the admission of secondary air. The only advantage of the secondary air is its cooling action in the afterburning zones. Instead of secondary air, a partial recirculation of the combustion gases is recommended. It has also been proposed that the turbulence bed should be divided into two compartments, into one of which fresh pyrite is fed and from which the loosely bound sulfur is expelled by admission of roasting gas cooled to 400°C after passing through a heat exchanger. In the other compartment, which is fed with fresh air, the remaining part of the sulfur in the pyrite is burned. The gas admission to the separate compartments can be measured out in such a way that a burning temperature of about 1300° to 1400°C is maintained in the combustion room while the gas in the heat exchanger is cooled to 400°C (106,184).

Floor combustion ovens have been successfully converted into turbulence ovens. Figure VII-7 shows a Nichols-Freeman oven which has been remodeled to a whirlpool oven as developed by the Badische Anilin and Soda Fabrik. After removal of the central axle and the floors, the bottom of the oven is converted into a whirl-bed. The pyrite runs from the hopper, 1, over a volumeter, 2, into the oven. Air is blown by means of the primary air fan, 3, through the pipe line, 4, into the oven, 5. The hot combustion gases pass through the heat exchanger, 6, and are sucked off through the fan, 9, after they have been freed of carried-over dust by the cyclone, 8. The whirl-bed is cooled by cooling coils. The cinders are carried away by a conveyer, 10 (131).

The "Fluo-Solids" system of pyrite burning of the Dorr-Oliver Company is very similar to the above-mentioned whirlpool ovens. The pyrite is introduced through a feeder into a cylindrical brick-lined reaction chamber and deposited on a perforated bottom. The pyrite may be dry or may contain up to 20% water. Air is blown in under slight pressure through the bottom with sufficient force to keep the upper layers of the pyrite in suspension. The sulfur dioxide gas is sucked off from the upper part of the reactor, passed through a cyclone to separate the flue dust, and from there goes into the absorption plant. The cinders are taken out by a kind of gas-tight overflow at the bottom of the oven. An oil burner is used to ignite the pyrite to start the oven; after that the roasting process is independent of external addition of heat and is continuous (198).

1.4 The Cooling of the Sulfur Dioxide Gas

Cooling of the hot sulfur dioxide gases is necessary for several reasons. The gases must be removed by cooling from the temperature region which is particularly favorable for the formation of sulfur trioxide. A proper absorption of the gas in the absorption towers also requires a low gas temperature. Further, there is the endeavor to utilize the heat of the gases as far as possible. For the cooling, various systems are used. Originally the gas was passed through lead pipes which were placed horizontally in a tank filled with running cold water. Such cooling systems were not very efficient and were replaced by vertically arranged pipes which were cooled by trickling them with cold water. Much more efficient than these pipe coolers is the spray-type cooler in which the very finely divided water comes in direct contact with the hot gas. This water immediately vaporizes and the gas is saturated with steam. The temperature decrease caused in this way is not accompanied by any loss of heat because, by the saturation with steam, the latent heat of water vapor is merely added to the gas. In the second stage the latent heat of the steam and the heat of the gas are removed. This is achieved by a double cooling-tower system which is shown schematically in Fig. VII–8. The hot gas enters the first tower, A, through the pipe, J, and is saturated with water vapors by spraying it with cold water. The water-saturated gas then enters the second, larger tower, C, at the bottom and passes upward against a water spray, then leaves the tower through pipe, K. The acid water from tower C is pumped from the bottom through the heat exchanger, D, into tower C at the top. A part of the water is removed before it reaches the heat exchanger and is pumped through pipe E into tower A. The heat exchanger is fed with cold water through the inlet and outlet pipes, F. The water level in the towers is kept constant through the overflow vessel, B. The safety valve, H, regulates the fresh water supply in case of an operation stoppage and to prevent any damage to the system from overheating (124,173).

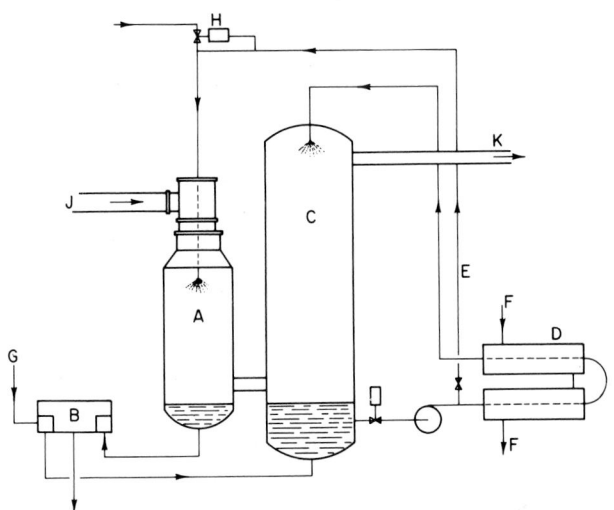

Figure VII-8. Chemipulp gas spray cooling system (124).

As will be shown later, the temperature of the water should be as low as possible in order to achieve a high concentration of sulfur dioxide in the cooking liquor. Water for industrial use often is subjected to seasonal variations in temperature. To avoid these changes, suggestions have been made for keeping the cooling water at a constant temperature. The proposed steam-jet vacuum-cooling system is based on the fact that the boiling point of a liquid, in this case water, depends on the pressure. When water is sprayed into an evacuated chamber, it evaporates when the vacuum matches the pertinent boiling point. During the evaporation, the water takes up heat from its environment. In this way it is possible to cool relatively large amounts of water by small amounts of evaporated water. The required vaccum can be produced by steam ejectors (84).

With the use of spray coolers, washing of the gas takes place simultaneously with the cooling, whereas with pipe coolers a washing must follow the cooling (225). For the separation of the carried-over sulfur trioxide mist, which has to be removed from the sulfur dioxide gas as soon as possible, electro filters are used. It has been found that stationary sound waves can precipitate the sulfur trioxide mist in a relatively simple way. Such sound waves can coagulate the mist droplets of 0.01 to 0.2 micron to give an 80 to 800 thousandfold mass enlargement. The difficulty in applying these acoustic precipitation systems is that the room in which the acoustic irradiation takes place has to be insulated to prevent sound-wave losses, otherwise no stationary wave can be produced. A solution to this problem has been found in a pipe system of special arrangement. At present it is not known to what extent this interesting system has been introduced into the industry (183).

2. ABSORPTION OF THE SULFUR DIOXIDE FOR THE PREPARATION OF THE COOKING LIQUOR

2.1 The Absorption Mechanism

The absorption process of the gases coming from the sulfur or pyrite oven is a typical example for a gas mixture; one of its components, sulfur dioxide, is readily soluble in water, whereas the other components, nitrogen and oxygen, have such a low degree of solubility that they may be considered to be practically water-insoluble. The absorption of the sulfur dioxide can be compared with an extraction process in which the soluble portion of a mixture is dissolved, leaving the insoluble portion as a residue. The process which takes place between the liquid and the absorbed gas is accompanied by a separation process of the gas mixture. The gas to be absorbed is dissolved out from the gas mixture at the dissolving zone; consequently its concentration, C_{gs} (partial pressure), decreases, in contrast to the average concentration in the mixture, C_g. The concentration difference thus formed causes the migration of the gas from the center of the mixture to the dissolving zone, to be absorbed there. This process is indeed the reverse of the dissolving process, in which the dissolved compound migrates away from the dissolving zone, but otherwise it conforms with the same law of solution. The more favorable the transition of the gas, the smaller is the difference, $C_g - C_{gs}$, between the concentration of the gas to be absorbed in the center of the gas mixture and that in the dissolving zone. These transition relationships are primarily determined by the degree of motion in the gas mixture.

The concentration of the gas to be absorbed at the liquid side, C_s, depends upon the concentration of the gas in the dissolving zone at the gas side, C_{gs}. The specific absorption velocity, b, per square meter of dissolving surface depends on the transition conditions at the gas side for which the equation

$$b = \delta_g (C_g - C_{gs}) C_g \text{ kg/m}^2\text{hr}$$

is valid. In this equation, g is the transition number at the gas side for the transition of the gas to be absorbed from the center of the gas mixture to the dissolving zone. In the same way the transition conditions influence the liquid side, as expressed by the equation

$$b = \delta_g (C_s - C_f) C_g \text{ kg/m}^2\text{hr}$$

in which the saturation concentration, C_s, depends only on the concentration, C_{gs}, of the gas to be absorbed at the gas side of the dissolving zone. Both equations must result in the same value for b, therefore

$$\delta_g (C_g - C_{gs}) = \delta (C_s - C_f)$$

Henry's law, which gives a simple relationship between the two concentrations at the dissolving zone, is no longer even approximately valid for the great solubility of the gas, sulfur dioxide, under discussion. The enlarging of the dissolving surface by an extensive distribution of the liquid and gas streams causes high absorption velocities.

In the absorption of a uniform gas in a continuously operating apparatus it is of no importance whether it operates as a direct current or as a countercurrent. The pressure and saturation concentration, C_s, are universally equal. Countercurrent can effect a better distribution of the gas and the liquid and, with it, an improvement in the transition conditions. With the absorption of gas mixtures the direct and countercurrent directions result in fundamentally different effects, whereas the saturation concentration, C_s, at the dissolving zone decreases constantly with the passage through the tower. With direct current, the concentration difference, C_s–C_f, effective for the absorption, decreases through the increase in the average concentration, C_f, of the solution and through the decrease in saturation concentration, C_s, along the gas and liquid currents. Whereas with a direct current the final highly concentrated sulfur dioxide solution is in contact with a gas which is already largely freed of absorbable gas, with a countercurrent the highly concentrated solution comes into contact with a fresh gas mixture with a high content of absorbable gas.

The absorption coefficients of sulfur dioxide in water have been determined with the help of an absorption tower filled with Raschig rings. Rates of flow for water at between 0.09 and 0.69 ft per minute and gas velocities of between 0.037 and 0.30 ft per sec were used. The absorption coefficient for sulfur dioxide in a sulfur dioxide-air mixture varies with the velocity of the gas at a constant velocity for the absorption liquid, and, vice versa, with the velocity of the liquid at a constant gas velocity. An absorption resistance is observed primarily in the liquid film when the velocity of the liquid is low. This resistance, however, changes to the gas film at high liquid velocities (191a).

The relationships between sulfur dioxide concentration, temperature, and total gas pressure of the system SO_2/H_2O have been examined repeatedly. Figure VII–9 shows these relationships up to a temperature of 110°C and up to the formation of a two-phase system. It is known that at 25°C and a sulfur dioxide concentration of about 25% by weight, two liquid phases are formed (241,298,418).

The concentration equilibriums between the liquid phase and the vapor phase were measured over a liquid-concentration range of between 0.59 and 4.48 g per 100 g solution, and at a temperature range of from 30° to 80°C. The values obtained were related to each other, taking into consideration the dissociation of the various types of molecules in solution and the activity coefficient of the types of ions. A table can be set up to re-

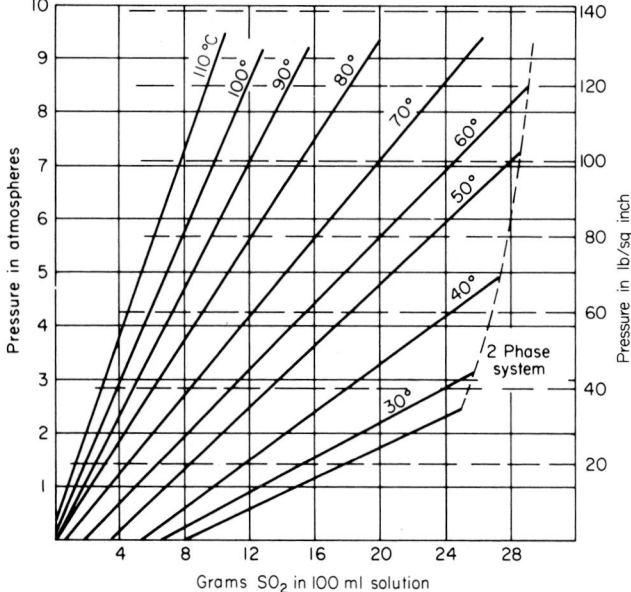

Figure VII-9. Gauge pressures for system SO_2/H_2O (418).

present the partial vapor pressure of sulfur dioxide above the aqueous SO_2 solution for the concentration region of 0.01 to 20.0 g $SO_2/100$ g H_2O and for temperatures of 0° to 130°C (Table VII–2) (299).

A great many variables are of importance in an absorption tower. For a well-synchronized operation, they permit an equal degree of absorption to be reached. Such variables are the packing material, the ratio of the liquid throughput (lb-mole per hr per sq ft) to gas throughput (lb-mole per hr per sq ft), the gas velocity, the temperature, and the composition of the absorption liquid so far as this is not specified by the process to be used (373).

If the sulfur dioxide absorption takes place in an absorption tower and the absorption medium is water, the absorption mechanism depends, according to the above-mentioned two-film theory, essentially on the properties of the absorption medium. The absorption medium in the preparation of sulfite cooking liquor, however, is usually not water but a more or less alkaline aqueous solution. In a strongly alkaline absorption medium, for instance, the greatest absorption resistance is shifted toward the side of the gas layer. More weakly alkaline mediums, such as sodium bicarbonate and sodium sulfite, shift this absorption resistance more toward the liquid layer. Depending upon the base used, the dissolution of the gas takes place in three stages. In the first stage the hydroxide or carbonate is converted into monosulfite or bicarbonate. In the second stage, these are

TABLE VII-2

PARTIAL VAPOR PRESSURE OF SULFUR DIOXIDE OVER AQUEOUS SOLUTIONS, mm of Hg (299)

Grams SO₂ in 100 g H₂O	Temperature (°C)													
	0	10	20	30	40	50	60	70	80	90	100	110	120	130
0.01	0.02	0.04	0.07	0.12	0.19	0.29	0.43	0.62	0.87	1.21	1.63	2.16	2.82	3.61
0.02	0.08	0.14	0.24	0.39	0.60	0.91	1.33	1.89	2.62	3.55	4.71	6.13	7.86	9.92
0.03	0.16	0.28	0.47	0.76	1.16	1.73	2.49	3.49	4.78	6.41	8.41	10.9	13.8	17.3
0.04	0.26	0.46	0.75	1.19	1.81	2.67	3.81	5.29	7.19	9.56	12.5	16.0	20.2	25.2
0.05	0.38	0.66	1.07	1.68	2.53	3.69	5.24	7.24	9.78	12.9	16.8	21.5	27.0	33.6
0.10	1.15	1.91	3.03	4.62	6.80	9.71	13.5	18.3	24.3	31.7	40.7	51.4	63.9	78.6
0.15	2.10	3.44	5.37	8.07	11.7	16.5	22.7	30.6	40.3	52.2	66.6	83.6	104.0	127.0
0.20	3.17	5.13	7.93	11.8	17.0	23.8	32.6	43.6	57.1	73.7	93.5	117.0	145.0	177.0
0.25	4.34	6.93	10.6	15.7	22.5	31.4	42.8	57.0	74.5	95.8	121.0	151.0	186.0	227.0
0.30	5.57	8.84	13.5	19.8	28.2	39.2	53.3	70.7	92.3	118.0	149.0	186.0	229.0	279.0
0.35	6.85	10.8	16.4	24.0	34.1	47.2	63.9	84.7	110.0	141.0	178.0	222.0	272.0	331.0
0.40	8.17	12.8	19.4	28.3	40.1	55.3	74.7	98.9	129.0	164.0	207.0	257.0	316.0	384.0
0.45	9.53	14.9	22.5	32.7	46.2	63.9	85.7	113.0	147.0	188.0	236.0	293.0	360.0	437.0
0.50	10.9	17.0	25.6	37.1	52.3	72.0	96.8	128.0	166.0	211.0	266.0	329.0	404.0	490.0
1.00	25.8	39.5	58.4	83.7	117.0	159.0	212.0	278.0	358.0	454.0	567.0	701.0	856	
1.50	42.0	63.6	93.2	132.0	184.0	249.0	331.0	433.0	555.0	703.0	877.0			
2.00	58.6	88.5	129.0	183.0	253.0	342.0	453.0	590.0	756.0	955.0				
2.50	75.7	114.0	165.0	234.0	323.0	435.0	576.0	749.0	958.0					
3.00	93.2	139.0	202.0	285.0	393.0	530.0	700.0	908.0						
3.50	111.0	166.0	240.0	337.0	464.0	625.0	825.0							

4.00	129.0	192.0	277.0	389.0	535.0	720.0
4.50	147.0	218.0	315.0	442.0	607.0	816.0
5.00	165.0	245.0	353.0	496.0	679.0	
6.00	202.0	299.0	430.0	602.0	824.0	
7.00	238.0	353.0	507.0	710.0		
8.00	275.0	407.0	585.0	818.0		
9.00	313.0	462.0	663.0			
10.00	351.0	517.0	741.0			
11.00	389.0	573.0	819.0			
12.00	427.0	628.0				
13.00	465.0	684.0				
14.00	504.0	740.0				
15.00	542.0	796.0				
16.00	581.0	852.0				
17.00	619.0					
18.00	658.0					
19.00	697.0					
20.00	735.0					

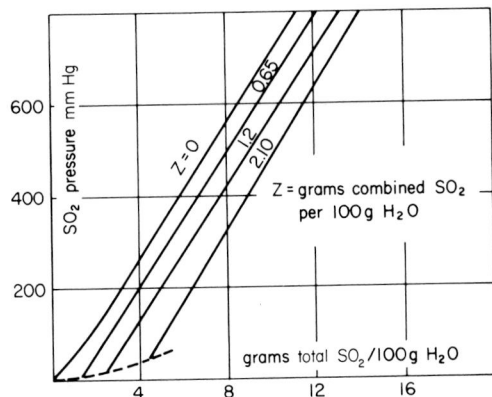

Figure VII-10. Solubility relations in the $SO_2/H_2O/CaO$ system (423).

converted into bisulfite, while in a third stage free sulfurous acid is formed in the solution. The absorption coefficients, calculated for the gas layer, refer to the first stage, with sodium and ammonium being used as base. With calcium and magnesium as base, these values have been found to be too high because of too low solubility of the various components of the system. The effect of the temperature in this stage is small. Only a few experimental facts are available for the second reaction stage. In this case only an average absorption coefficient, which is probably valid for all bases, can be used. For the third stage experimental data for the SO_2/H_2O system can be used (191a), but the solubility equilibriums must be taken into account (201, 254,423,425).

Investigations on the solubility of sulfur dioxide in calcium bisulfite solutions show that, within the concentration regions investigated, the gas pressure is a linear function of the concentration. By extrapolation the straight lines obtained cut the abscissa about at double the value of the concentration of bound acid. In this way the pressure equilibrium of the sulfurous acid within the two-phase zone will be directly proportional to the concentration of actually free sulfur dioxide in the solution. Figure VII-10 shows the relationships between the concentration of total sulfur dioxide in the solution and the pressure, depending on the concentration of bound sulfur dioxide in the solution (423).

A statistical evaluation of results obtained by six different authors on the relationships of the equilibrium constants of the systems $SO_2/H_2O/CaO$ showed the best agreement for a concentration of bound sulfur dioxide of 0.5 g $SO_2/100$ g H_2O and within a concentration region for the total acid content of 1.0 to 6.0 g $SO_2/100$ g H_2O. Figure VII-11 shows graphically the equilibrium conditions for concentration, vapor pressure, and temperature, in which P_m is the vapor pressure of the sulfur dioxide and water over a

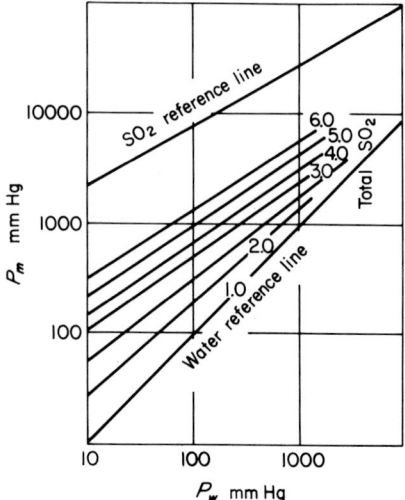

Figure VII-11. Modified Othmer-Cox chart showing equilibrium concentration-vapor pressure-temperature relationships for the system $SO_2/H_2O/CaO$; combined SO_2 concentration 0.5 g SO_2 per 100 g H_2O (75).

solution of constant composition at a given temperature, and P_w is the vapor pressure of the water at the same temperature (75).

Recent investigations have shown that an apparently simple system such as SO_2/H_2O/base can, at closer study, be fairly complicated. According to this, the understanding of the behavior of each component of this system is facilitated by the elimination of the concept of sulfurous acid, H_2SO_3, as a component of the system. The two principal anions in the sulfite system, i.e., HSO_3^- and SO_3^{2-}, are formed by direct dipole association of SO_2 with H_2O or OH^- and exist beside each other at room temperature in solution, although the distribution ratio varies strongly and depends on the pH. The reaction of sulfur dioxide remains incomplete and free SO_2 is always present, but only in minute amounts at higher pH. In sulfite solutions small parts of thiosulfate ions, $S_2O_3^{2-}$, and polythionate ions, $S_xO_6^{2-}$, are present as intermediate products of a spontaneous disproportionation reaction:

$$3\,SO_2 = 2\,SO_3 + S$$

The equilibriums between the anions mentioned are dependent, even at room temperature, on the concentration and the basic strength of the cation. The thermal stability of the bisulfite ion is considerably greater than that of the sulfite ion. At 160°C and in the presence of sodium ions the bisulfite ion remains stable, whereas the sulfite ion is markedly weakened. In the

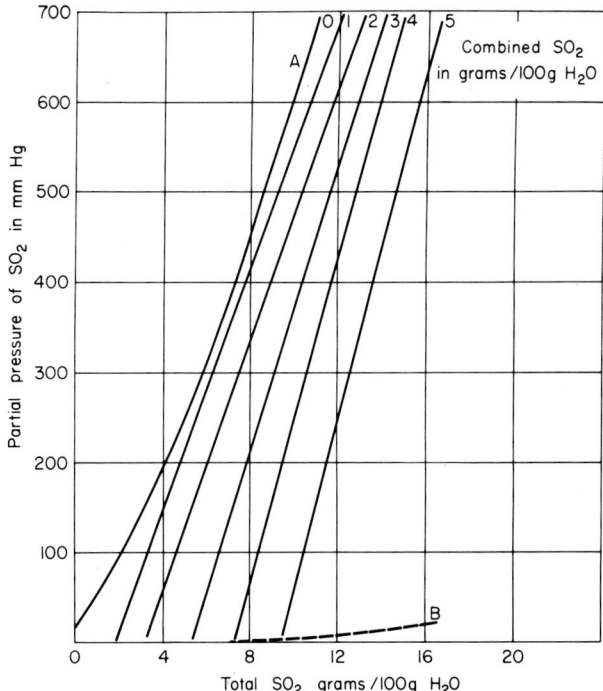

Figure VII-12. *P-T-Z* Diagram for 25°C (215). *P* = SO₂ pressure above solution; *T* = total SO₂ per 100 grams H₂O; *Z* = grams combined SO₂ as MgSO₃ per 100 grams H₂O.

presence of ammonium ions the sulfite ion disappears at between 25° and 80°C and the bisulfite ion exhibits a weakening at 150°C. Regardless of the great differences in their solubility at higher pH values, the ions of the metallic bases, sodium, calcium, and magnesium, remain thermally stable, whereas the ammonium ion, deriving from the hydrated gas, is just as unstable as the bisulfite ion (180,182).

   Magnesium bisulfite is much more soluble at equal partial pressure of the sulfur dioxide than is calcium bisulfite. Also the ratio between total acid and bound acid is, in all cases studied, below 2; it is almost constant for all saturated solutions; and it increases slightly with a strong increase in the partial pressure of the sulfur dioxide. In Fig. VII–12, the straight line, OA, represents the solubility of SO₂ in water. The curve, OB, is the saturation line, *P* is the pressure above the solution, *T* the total acid concentration in grams per 100 grams water, and, finally, *Z* is the concentration of the bound SO₂ as magnesium bisulfite in grams per 100 grams water. The value for the reference ratio, *T/Z*, is clearly different for the magnesium and calcium systems. The value for the latter varies between 2.07 and 3.24 for the temperature range of 15° to 25°C over the same pressure region. The

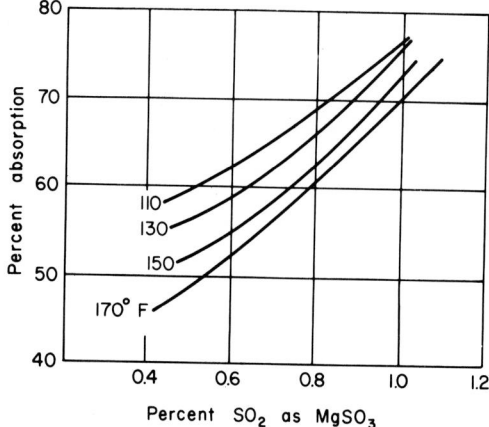

Figure VII-13. The effect of MgSO₃ on Venturi absorption efficiency (251).

straight lines for constant bound sulfur dioxide lie above the curve, OB, in the region of unsaturated solution, while below the curve OB is the supersaturated zone, which is unstable. The technically used cooking acids are not saturated and contain gas in addition to that necessary for the formation of bisulfite. The solubility equilibriums were therefore calculated from recalculated data of earlier investigations (215,369).

To check the efficiency of an absorption plant, the sulfur dioxide vapor pressure above a magnesium bisulfite-magnesium monosulfite solution was studied at two total acid concentrations of 4.4% and 6.1% SO₂ and temperatures of 38°, 49°, and 60°C. The result was that the experimentally found vapor pressures in the region of a low magnesium sulfite concentration (0.5 to 0.2% SO₂ as MgSO₃) are considerably higher than those empirically calculated. With a concentration of above 0.6% sulfur dioxide as magnesium sulfite, the experimentally found and the calculated values approach each other and become almost equal at 1% or higher sulfur dioxide content as magnesium sulfite. The importance of these results lies in their applicability to the general investigation of the sulfur dioxide absorption. The difference between the vapor pressure of the sulfur dioxide in the gas and in the liquor phase is the motivating force for the absorption of the sulfur dioxide. The marked influence of the magnesium sulfite concentration of the spray acid on the efficiency of a Venturi absorption system is shown in Fig. VII-13. There the relationships between absorption efficiency and magnesium sulfite concentration are given for different temperature parameters. This shows that with an increase in the magnesium sulfite concentration from 0.5% sulfur dioxide as magnesium sulfite to 1% sulfur dioxide the efficiency increases by 15% to 20%. This will be dealt with again later (251).

TABLE VII-3

DATA ON AQUEOUS SODIUM BISULFITE SOLUTIONS (218)

Specific gravity	Be	% NaHSO₃	% SO₂
1.006	1	1.6	0.4
1.038	5	3.6	2.2
1.068	9	6.5	3.9
1.084	11	8.0	4.8
1.116	15	11.2	6.8
1.152	19	14.6	9.0
1.190	23	18.5	11.5
1.240	27	23.5	14.5
1.275	31	28.9	17.8

TABLE VII-4

RELATIONSHIP BETWEEN pH AND PERCENT AMMONIA
IN AQUEOUS AMMONIA SOLUTIONS (27)

% NH₃ in solution	pH of solution
0.036	10.49
0.051	10.61
0.132	10.89
0.358	11.38
0.609	11.57
0.920	11.63
1.410	11.70
2.100	11.81
2.820	11.89
4.000	11.99
9.670	12.06

TABLE VII-5

VARIATIONS WITH pH IN THE SYSTEM NH₃/SO₂/H₂O (27)

% Free SO₂	% Total SO₂	% Combined SO₂	% Free / % Total SO₂	% Total / % Combined SO₂	pH
1.66	6.80	5.14	0.244	1.20	6.39
2.26	7.19	4.93	0.315	1.45	6.17
3.18	8.05	4.87	0.396	1.81	5.87
4.35	9.31	4.95	0.468	1.88	5.27
5.77	10.69	4.92	0.541	1.85	2.82
6.69	11.56	4.87	0.579	2.37	2.47
8.87	13.64	4.77	0.650	2.85	2.11

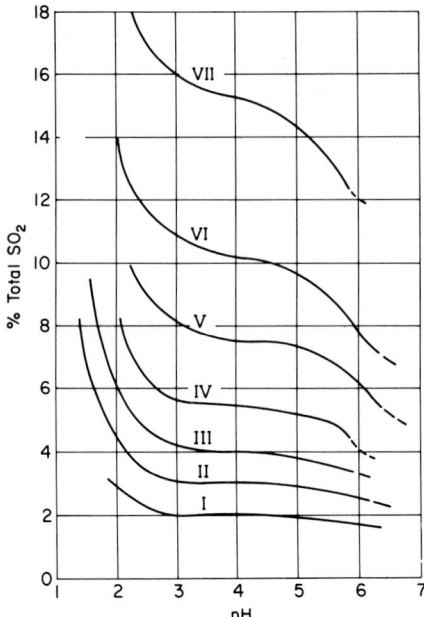

Figure VII-14. Variation of pH value in solutions of different NH_3 concentration with the concentration of total SO_2 (27).

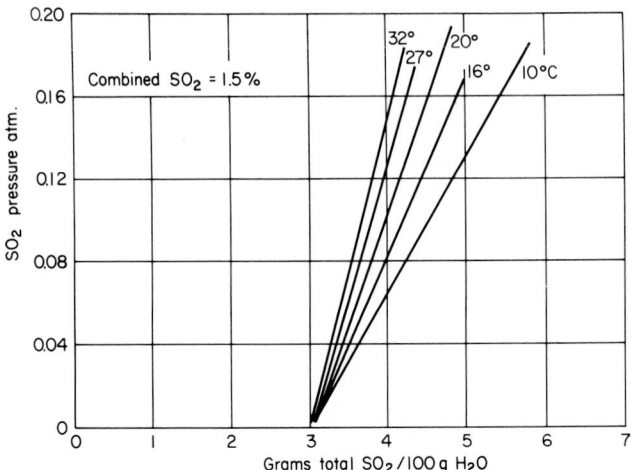

Figure VII-15. Relationship between sulfur dioxide pressure and total sulfur dioxide concentration in an ammonium-base cooking liquor at various temperatures and constant bound sulfur dioxide concentration.

The sodium salts of the sulfurous acid are well-known in the industry and their preparation and the properties of their solutions are so familiar

that only a few data are to be found in the literature. Table VII-3 shows the sodium bisulfite and sulfur dioxide content of aqueous sodium bisulfite solutions (218).

Table VII-4 shows the pH values of aqueous solutions of ammonia with increasing concentrations (27). The common effect of ammonia and sulfur dioxide on the pH value of the solution is interesting in this connection. Table VII-5 shows these relationships. The figures in this table have been calculated from curve VI in Fig. VII-14. The curves in Fig. VII-14 represent the following ammonia concentrations in percent: I, 0.358; II, 0.609; III, 0.920; IV, 1.41; V, 2.10; VI, 2.28; and VII, 4.00.

Figure VII-15 shows the relationship between sulfur dioxide pressure and total sulfur dioxide concentration in an ammonium-base cooking liquor at various temperatures and a concentration of 1.5% of bound sulfur dioxide (27).

2.2 The Absorption Installations

So-called acid towers are generally used for the absorption of sulfur dioxide for the preparation of calcium bisulfite cooking liquor. They are known as Jenssen towers after their inventor. These acid towers have replaced the old wooden absorption towers. Figure VII-16 shows such a Jenssen tower. It consists of a series of two connected towers of vertical concrete tubes, 2 to 3 meters in diameter and 40 meters or more in height. These towers are filled with limestone. The gas coming from the sulfur or pyrite burners (1-7) is cooled in the gas cooler, 8, freed of dust in the gas purifying plant, 9, and passed by means of the fan, 10, through the distributor, 12, into the tower, 13. This tower is closed at the top and from there the nonabsorbed gas is led to the bottom of the second tower, 14. The limestone in tower 14 is trickled with cold fresh water and absorbs the remaining nonabsorbed sulfur dioxide from the first tower. The relatively weak acid from tower 14 is used for trickling the limestone in tower 13 and is enriched by the concentrated gases from the burner to give the desired cooking liquor concentration. The third tower, 24, houses the elevator for the limestone. Tower 13 is also called the strong acid tower, where about 75% to 95% of the sulfur dioxide of the gas is absorbed. The gas distributor, 12, makes it possible to change the passage of the gas from one tower to the other. Because the second tower, 14, also called the weak acid tower, can be charged with limestone during its operation, an auxiliary third absorption tower is unnecessary. There are mills with three or even four absorption towers, with one, in the first case, serving as a reserve tower, and, in the second case, two being used as strong acid towers.

The concentration of total sulfur dioxide in the acid coming from the strong acid tower is regulated by the gas stream and the water supply

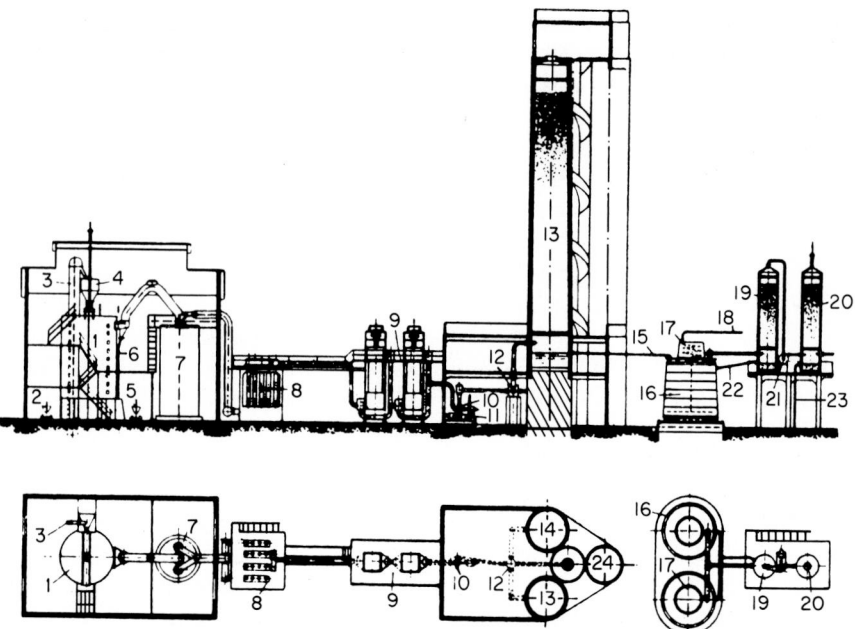

Figure VII–16. SO$_2$-absorption plant system Jenssen. (Reproduced from F. Müller, "Die Papierfabrikation und ihre Maschinen" by permission Güntter Staib Verlag, Biberach, Germany.) 1 pyrite roaster; 2 pyrite transporting car; 3 pyrite elevator; 4 pyrite inlet; 5 pyrite cinder car; 6 pyrite cinder exit; 7 washing tower; 8 gas cooler; 9 Cottrell wet precipitator; 10 ventilator; 11 pump for weak acid; 12 gas distributor; 13, 14 absorption towers; 15 pipe line for strong acid; 16 acid storage tank; 17 cooler; 18 flue gas pipe line; 19 Glover tower; 20 absorption tower with limestone; 21 ventilator; 22, 23 pipe lines for acid; 24 elevator tower.

from the weak acid tower. The amount of sulfur dioxide bound to calcium or magnesium, expressed as monosulfite, can be regulated by the water supply and the water temperature. The reaction of sulfur dioxide with the limestone takes place more readily at higher temperatures because the temperature of the water plays a decisive role in the solubility characteristic of sulfur dioxide, as was already mentioned in the previous section.

The reaction in the absorption tower can be represented by the following equations, in which calcium carbonate may be replaced by magnesium carbonate or dolomite:

$$CaCO_3 + SO_2 + H_2O = CaSO_3 + H_2O + CO_2$$
$$CaSO_3 + SO_2 + H_2O = Ca(HSO_3)_2$$

In cases where soluble bases are used for the absorption of sulfur dioxide, the Jenssen towers can be used by replacing the limestone by wooden blocks

or other filling material; however, special absorption towers are much better. Such installations are also suitable when milk of lime is used instead of limestone. In this case, a suspension of slaked lime is trickled over the contact area and passes the sulfur dioxide gas in a countercurrent through the contact tower. Various types of construction and equipment have been suggested for contact towers, all having in common the fact that the absorption liquid, spread in a very finely divided form or in a very thin layer over large surfaces, is thus exposed to the sulfur dioxide gas in a manner designed to achieve a very rapid and complete chemical reaction. Some absorption towers are divided into several sections which are outfitted with baffle surfaces or filling bodies to provide large reaction surfaces. Such absorption towers are also connected in a series so that here, also, both strong and weak acid towers are obtained. The course of the reaction with calcium and magnesium hydroxide suspensions is very similar. No carbon dioxide is formed in a first reaction stage as no carbonates are present. Whereas in the Jenssen tower operation, magnesium is not desired with the calcium, in the milk of lime absorption tower the presence of some magnesium is considered advantageous. The heat of reaction in the milk of lime process is greater than in the tower process and can cause an increase of as much as $8°C$ in the temperature in the absorption plant.

The concentration of the cooking acid can be considerably increased, independently of the temperature and the concentration of the bound sulfur dioxide, by the application of pressure. In this case, the acid is led into a specially equipped pressure absorption tank at a pressure of 22 psi. This absorption system has been found especially suitable when milk of lime is used as an absorbent.

The Turbulent Contact Absorber (82) is an interesting further development. It consists of an absorption tower of relatively small dimensions and is divided into four sections which are separated by a layer of hollow polyethylene plastic spheres. The milk of lime or, better, the magnesium oxide slurry is introduced into the top section of the tower and passes downward through the four sections, while the gas enters the tower in a countercurrent stream at the bottom. Another improvement was achieved with complete absorption, even of the residual gas, by the introduction of cold water into the first section and the magnesium oxide sludge into the second. In this way the carrying over of liquid by the fan is avoided and a further cooling of the gas is achieved, together with the recovery of heat. The basic principle of such an absorber consists of a high gas velocity—up to 1200 ft per minute—and a high liquor velocity—up to 60 gal per minte— per sq ft of the tower cross-section. Through the turbulence an intimate contact between gas and absorption medium is effected. The draining acid has a temperature of $65°$ to $70°C$. This temperature, unusual for a calcium bisulfite cooking liquor, is obtained through the supply of hot, saturated gas

of 70°C and an absorption medium of 45°C. The final liquor has a concentration of 7% to 8% total sulfur dioxide. The absorption coefficient and the reaction velocity increase with increasing temperature.

A three-phase fluidization has been suggested for the sulfur dioxide absorption of fluidized limestone suspensions. Such a process promises a number of advantages, in particular, an increased capacity of the absorption installation with a small apparative volume. On the other hand, the crushing of the limestone and the power required for the fluidization would cause increases in costs and an additional filtering plant would be required (412).

Another type of acid gas absorber, called the flooded-bed absorber, uses glass marbles as bed packing. Each bed is relatively shallow and the marbles are supported on stainless steel wires. Irrigation of the bed is accomplished by introducing the absorbing medium by means of spray nozzles operating at 15 to 18 psi. The gas entering at the bottom section of the tower flows upward, transports the absorbent liquor into the marble beds, filling the voids in them, and forms a deep turbulent, bubbly layer of gas and absorbent liquor. The depth of the layer is controlled by overflow weirs which drain the absorbent from the layer at the same rate as it is supplied by the spray nozzles below the bed. Substantial quantities of droplets and mist are carried from the turbulent layer by the gas which passes a demisting section before leaving the tower. This flooded-bed absorber was designed originally as a dust collector or scrubber. For the gas-liquid absorption or gas-liquid chemical reactions the unit provides the equivalent of several feet of conventional packing operated at below flooding velocities; it provides a simple but very efficient method for regenerating cooking acid from magnesium-base waste pulping effluents (146a).

2.3 The Recovery of Sulfur Dioxide from the Relief Gases

Modern installations which have to take up the pressurized relief gases and condensates from the digester room require an increased concentration of bound sulfur dioxide in order to counteract the dilution of the cooking acid. For this purpose pressure storage reservoirs for low and high pressure are used. If the digester gases are passed into the acid tower or into an absorption plant for milk of lime they must be cooled first. Often older digesters are used as pressure reservoirs. When several storage vessels are used, the relief gases at the start of the cook are used for reinforcement and warming of the cooking liquor from the low-pressure reservoir. Gas and acid are mixed by means of an injector and completion is attained. Gas from the high-pressure reservoir is used for strengthening the cold crude acid which is pumped from the absorption tower through a 'low-pressure injector into the low-pressure reservoir. Relief gases from the low-pressure storage tank are sent directly into the crude acid tank. The recovery of

sulfur dioxide and heat from the relief gases and spent liquor plays an important role in the heat economy of the sulfite pulp mill. This will be discussed further in a later section. If the heat used as cooking steam is taken as 100, about 50% of the applied heat can be recovered with a suitable recovery system (411). The pressure storage tank process acts according to the principle of stepwise absorption (63). Numerous other processes have been suggested; these operate partially with heat exchangers or with special absorption media from which the absorbed gas can be expelled by heating (164,213,237,238,249,396). The increasing use of the magnesium bisulfite cooking process makes the recovery of the chemicals by burning the spent sulfite liquor feasible. The problems connected with this will be dealt with in detail in Section 9. Here it should by mentioned only that the sulfur dioxide gases, which come from the spent liquor burner in a concentration of 1%, are freed of fly ash (MgO) and cooled in a heat exchanger and are then passed into two Venturi absorbers which recover 95% of the sulfur contained in the evaporated spent liquor (251).

The recovery of liquid sulfur dioxide from the digester relief gases has been described and patented. The gases, which are under the pressure of the digester, are cooled in several stages after separation of the carried-over mist and pulp particles. At first a sulfur dioxide-containing condensate is obtained and the almost dry sulfur dioxide gas, cooled to 20° to 30°C, is liquefied under the excess pressure present. The yield of liquid sulfur dioxide amounts to up to 90%. A yield of 100% cannot be obtained because noncondensable gases originating from the wood are present during the cook (213,371).

3. EFFECT OF THE COOKING ACID ON THE WOOD COMPONENTS

3.1 Lignin

In technological-economics terms, the goal of pulping is the removal of the lignin and noncellulosic components from the wood for the production of cellulose in a fibrous form. Depending upon the later use of the pulp, a partial removal, only of the lignin and the hemicelluloses, may be satisfactory. It is not the purpose of this book to give a presentation of the chemistry of lignin. Several books are available for reference on the subject (40,138,328,434). The reactions between the cooking liquor and the lignin are of interest, however, in connection with the sulfite cooking process. According to an early theory, the process of dissolving the lignin occurs in two steps: in the first, the lignin is sulfonated to form a solid lignosulfonic acid which, in the second stage, is rendered soluble by hy-

drolysis and is extracted from the wood structure. According to this theory, the degree of delignification is proportional to the acid concentration and depends upon the presence of other cations (139). In the opinion of other authors, this theory is no longer valid. The delignification velocity depends on the product of the hydrogen ion and bisulfite ion concentrations and the sulfonation reaction is therefore the determining factor for the reaction velocity (55). Klason's theory suggests an addition reaction of sulfurous acid to an ethylene group in the lignin molecule. In this case, one carbonyl group would be present for each sulfonic acid group. A lignosulfonic acid with one sulfonic acid group for each lignin building unit must behave like a coniferyl aldehyde sulfonic acid (208). Freudenberg and co-workers, on the other hand, assume that the entrance of the sulfurous acid, in the first stage, takes place by the splitting of a ring, with the formation of a phenolic and a secondary hydroxyl group and with the latter being partially replaced by a sulfonic acid group either directly or after the splitting off of water (120,330).

The ultraviolet absorption of the lignin is not affected by the sulfonation; the latter therefore cannot be due to a reaction involving an unsaturated system (18,19,97). Thus it can be assumed that the sulfonation takes place by a reaction involving alcoholic hydroxyl groups or carbonyl groups in the lignin molecule (94). The action of lignosulfonic acid as an ion exchanger is well-known and has been studied extensively (222,223). When wood is heated with bisulfite cooking liquor an insoluble lignosulfonic acid is formed first. This acid acts as a cation exchanger and takes up calcium ions from the cooking liquor and in this way increases the degree of acidity. In the second stage, the solubility of the solid lignosulfonic acid develops (95). The formation of the solid lignosulfonic acid in a first step can readily be shown by a cook of wood chips with a bisulfite solution—a mixture of sodium monosulfite and bisulfite—at a pH of 6. At a normal cooking temperature of about 110°C only a very small part of the lignin is dissolved. The main part remains in the wood and is rapidly sulfonated there until one sulfur atom per three or four methoxyl groups has been introduced. A bisulfite cook at pH 4.5 leads to a further sulfonation until one sulfur atom per two methoxyl groups is present. Even at this pH value the dissolution of lignin is small and the higher degree of sulfonation does not increase the solubility of the lignosulfonic acid. When this sulfonated wood is heated with an acid buffer solution the lignosulfonic acid is dissolved. The reaction velocity in the second stage, which leads to the dissolution of the lignosulfonic acid, is controlled by the hydrogen ion concentration (137).

The phenylpropyl units which form the lignin molecule are combined with each other or with the carbohydrates in the wood through benzyl ether groups. All, or at least most, phenylpropyl units contain a benzyl alcohol

or benzylalkyl ether group. The analogy which exists between the sulfonation of β-guaiacyl ethers of veratrylglycerol and the benzyl alcohol and benzyl ether groups in lignin makes it appear probable that at least a considerable part of the phenylpropane monomers are formed from phenylglycerol derivatives (235). With regard to the dissolution velocity of the lignin, it can be said that at high sulfurous acid concentrations the dissolution of the lignosulfonic acid is determined by the hydrolysis, while at low concentrations the sulfonation reaction also has a decisive influence on the velocity of the reaction (141).

There are still differences of opinion regarding the type of linkage of the lignin in wood. Such bonds undoubtedly are present, but the question is whether all the lignin is combined to carbohydrates (260). Investigations on variations in the average molecular weights of hemlock, spruce, and maple lignosulfonic acids and on changes in these weights during the sulfite cooking process have been carried out, with the goal of obtaining additional information on the sulfonation process of the lignin. Lignosulfonic acid from gymnosperms shows first a low molecular weight which, during the course of the cook, passes through a maximum and, after a temporary decrease, rises again toward the end of the cook. The bonds between the lignin units show a variation in behavior: one type is readily hydrolyzable by aqueous acid solutions, another type is more resistant. If the bonds within the lignin molecule are cleaved by hydrolysis and subsequent sulfonation, the low-molecular fragments soluble in water diffuse first out of the wood structure, and with progressing delignification, the higher-molecular portions also diffuse. While the hydrolysis of the dissolved lignosulfonate is progressing, the higher-molecular portion becomes enriched until the molecular weight reaches a maximum. When all the hydrolyzable bonds are cleaved, the average molecular weight passes through a minimum, but increases again as the result of condensation reactions (275). When hemlock chips are treated with sodium bisulfite-sulfur dioxide solutions with acidity increasing from pH 4.8 to 2.9, only a third of the lignin can be extracted. Such lignosulfonates show a molecular weight of about 3000. The removal of the residual lignin fraction requires stronger acid solutions and the lignosulfonic acids thus formed have average molecular weights of about 30,000. A subsequent hydrolysis causes a marked decrease in the molecular weights. Condensation reactions apparently prevent a further decrease after a minimum is reached (108). Changes in the composition, the acidity, and the reaction temperature of the cooking liquor show no direct relationship to the changes in the molecular weights. The molecular weight increases gradually, reaching a maximum when 75% to 80% of the lignin is dissolved. At the same time, a stepwise decrease in the methoxyl content of the lignosulfonate occurs. The increase in the molecular weight is not the result of a lignin condensation and the decrease in methoxyl is not caused by a

complex formation with sugars in the spent liquor. The hydrolytic depolymerization does not seem to play an important role in the dissolution of
the lignin. When wood is heated for a longer time (15 hours) at 164°C
with a solution of lignosulfonic acid, 26% of the wood substance is dissolved, with the splitting off of formic and acetic acids, and the apparent
lignin content of the undissolved residue increases. Such prehydrolyzed
wood, after a subsequent sulfite cook, gives only 6.8% of hydrolyzed carbohydrates and the residue now contains 42% lignin, as compared with 40%
before the cook and 28% in the original wood. The heating of wood with
lignosulfonic acid therefore leads to a deactivation of the lignin and to
a hindering of the delignification of the wood in the subsequent sulfite
cook. The increase of free lignosulfonic acid in the cooking acid thus has
a detrimental effect on the yield of pulp. In the first stage of the sulfite
cook, sulfonic acid and sulfonic acid esters are formed in the reaction with
the lignin. These sulfonic acid esters lead to a combination of lignin molecules and cause an increase in the molecular weight of the lignosulfonic
acid formed and a decrease in its solubility. Probably a linkage through
ester groups with the polysaccharides takes place. In a later stage of the
cook these ester bonds are cleaved by acid hydrolysis and a soluble lignosulfonic acid is liberated (271,272,273).

The degree of sulfonation depends extensively upon the temperature.
Sulfonation at a low temperature requires a period of up to 16 months. The
presence of bound sulfur dioxide, in this case, acts disadvantageously. A
prehydrolysis of the wood reduces the sulfonation, both at low temperatures and in a subsequent sulfite cook. The hydrolysis of presulfonated
wood with cooking liquors which do not contain free sulfur dioxide causes
only a partial delignification except when the presulfonation has been carried out with free sulfurous acid and this has led to a sulfur content of over
4%, based on the lignin. The sulfonation of wood with bisulfite or monosulfite at high temperatures also causes a noticeable retardation of the delignification (305).

The exchange reaction of the sulfur between the sulfonated lignin and
bisulfite solutions which contain radioactive ^{35}S increases during the first
hours and then remains constant. In highly sulfonated lignins, 7% to 38% of
the original sulfur could be exchanged. There seems to be no connection
between the amount of exchanged sulfur and the original sulfur content of
the lignosulfonic acid. On the other hand, a linear relationship exists
between the initial sulfur content and the time required to dissolve a lignin-
sulfonate fraction (303).

The concept of the velocity-determining reaction in the two-stage
dissolution process of the lignin from the wood structure, which is described
above, has undergone a certain change in that the second step is now considered to be a mixture of the reactions, i.e., hydrolysis and sulfitolysis of

the sulfonated lignin. In both cases, the primary reaction is a proton addition to a benzyl ether oxygen atom and a cleavage of the C–O bond. The observation that in technical sulfite cooks, in contrast to laboratory cooks, which are usually carried out with a large excess of cooking liquor, the dissolution of the lignin takes place rapidly after the sulfonation, leads to the conclusion that the sulfonation is the primary velocity-determining factor. In the event that the dissolution is carried out by sulfitolysis, the whole delignification process can be considered to be a one-step process. The process is then a continuously increasing sulfonation of the lignin in which the dissolution takes place within the range of a certain degree of sulfonation (314).

It has been mentioned many times that the cations which are combined with the lignosulfonic acid exert an influence on the course and velocity of the reactions. The lower the valency of the cation the greater should be the diffusion velocity of the lignosulfonate (310,327). A limited sulfonation of 0.2 sulfur atom per methoxyl group leads to a relatively rapid and complete delignification in a cook with aqueous sulfur dioxide solution. A like result cannot be obtained with mineral acids. It must therefore be assumed that there is an additional sulfonation by the sulfurous acid, followed by a hydrolysis or sulfitolysis, or both. The velocity with which these reactions occur must depend upon the acidity in the reaction zone, i.e., the interface between the wood fiber and the cooking liquor. The velocity is a function not only of the acidity of the cooking liquor but also of the existing Donnan equilibriums. The various actions of the different cations can therefore be explained rather as resulting from a shift of these Donnan equilibriums than from a different solubility effect. Bivalent cations are preferably absorbed by the solid phase, thus causing the acidity to be lowered much more strongly than is the case with monovalent cations. A roughly estimated calculation of the acidity conditions in the solid phase according to Donnan's law gave $0.18 N$ for sodium and $0.007 N$ for calcium. It is surprising that this considerable difference does not have a stronger effect; on the other hand, this confirms again that, in this case also, the hydrolysis cannot be the determining factor for the reaction velocity of the delignification process (314).

Comparative cooks of sprucewood meal with cooking liquor containing 7% total sulfur dioxide and 0% to 7% base (Ca, Mg, Na, and NH_4) gave a yield of 67.7% residue with 25% lignin (as compared with 28.2% in the wood) and 0.77% sulfur, after a cook with 7% sulfur dioxide solution without a base. At base contents of 0.5%, 1%, and 2% and equivalent concentrations, the degree of sulfonation of the lignin is not dependent upon the nature of the base used. The degree of sulfonation increases only slightly with the concentration of the base. The solid residue had a sulfur content of around 1%, in spite of the fact that the yield increased noticeably with

increasing concentration of the base. From this it may be concluded that the presence of a low base concentration results in a hindrance of minor importance to the delignification but of major importance to the hydrolysis of the hemicellulose portion. The highest degree of sulfonation, 0.82% sulfur content, was obtained with a sodium base, and the lowest, 0.67%, with an ammonium base (268).

As mentioned repeatedly, delignification can also be achieved with sulfur dioxide solutions without a base. Methanol and other water-soluble organic solvents have proved to be favorable additives. The presence of such organic solvents seems to prevent the danger of lignin condensation and a subsequent burnt cook (Schwarzkochung) (344). In connection with the lignin condensation, it must be mentioned that this danger can be lessened considerably if high sulfur dioxide concentrations and a low temperature are used. In this case, no organic solvents are required. The subsequent hydrolysis of the lignosulfonate then occurs at temperatures which are above those used in normal sulfite cooks. With regard to the sulfite and hydrogen ion concentrations, approximately the same reaction conditions can be maintained as are used in a normal sulfite cook on addition of sodium sulfate to the sulfur dioxide solution. The danger of condensation increases with increasing hydrogen ion concentration and temperature, but it decreases with an increasing degree of sulfonation of the lignin (380).

As mentioned earlier, the extraction of lignin is strongly retarded when the wood has been prehydrolyzed. This phenomenon has been attributed to a coalescence of the lignin, thus causing a lowering of the phase boundary surfaces. This coalescence also reduces the accessibility of the functional groups in the lignin. Only reactions within the lignin molecule itself and with other substances should be meant by "lignin condensation," in which a chemical bond is formed and an increase in molecular weight takes place (209). First of all, polyphenols, which are present in the bark and bast of softwoods in considerable amounts, have to be considered for such reactions. Comparative cooks with the addition of resorcinol in sodium-base sulfite cooks at pH 4.5 and pH 2.5 showed that the hindrance of the lignin dissolution and darkening of the pulp are functions of the hydrogen ion concentration and the resorcinol concentration. The least effect was found to be at pH 4.5. These condensations can be avoided with resorcinol concentration of up to 1% and with a sufficient concentration of hydrogen ions during the sulfonation stage (110).

A determination of the lignin in the sulfite cooking liquor can be carried out spectrophotometrically by measuring the light absorption in the region of 200 to 205 or 280 mμ. The disturbing effect of foreign substances must be taken into account in the choice of wavelengths. The formation of such substances and their presence in the cooking liquor depend

on the cooking conditions. Substances which can disturb through absorption in the shorter wave region are sulfur dioxide and acetic and other weak acids, those which interfere in the longer wave region are furfural and other decomposition products of the carbohydrates. In general, satisfactory results in acid sulfite cooks are obtained at 280 mμ; however, this is not the case with bisulfite cooks that are carried out at a high temperature (364). The absorption curves obtained at 280 μm are not a direct measure for the dissolved lignin when the cooking conditions have been chosen in such a way that a considerable hydrolysis of carbohydrates occurs. Cooks which have been carried out with a cooking liquor containing some spent sulfite liquor show, at a certain delignification region, an approximately linear relationship between the absorption data taken at 205 and at 280 mμ, so that an estimation of the lignin content in the cooking liquor can be made on an empirical basis (210).

A fractionation by gel filtration of the aromatic compounds in the spent liquors from four bisulfite cooks which had been carried out under different cooking conditions showed that 14% to 24% of the original (28.6%) lignin of spruce is dissolved, regardless of the pH of the cooking acid, and that yellow methoxyl- and sulfur-containing ligninlike substances are formed, the ultraviolet spectra of which differ in alkaline solution from the spectrum of lignosulfonic acid. On the other hand, the dissolution of the lignin depends to a great extent on the pH. On redissolution of the low-molecular lignosulfonic acids, it was possible to analyze them for their sulfur content and reducing groups. It was found that the degree of sulfonation increases with decreasing molecular weight of the lignosulfonic acids and that the sulfonation is accompanied by the formation of a proportional amount of reducing groups. Lignosulfonic acids of equal molecular weight had a constant degree of sulfonation, which was independent of the cooking conditions. The differences in the elementary composition of lignosulfonic acids of different molecular weight show clearly that the sulfonation of lignin includes the addition of hydrogen, oxygen, and sulfur atoms in a ratio of 4:4:1 $(SO_2 + 2 H_2O)_n$. As found earlier, the formation of monosaccharides decreases with increasing pH, while the formation of nonaromatic acids reaches a maximum in a cook with a cooking liquor of pH 3 (117,117a).

3.2 Cellulose

Cellulose can be hydrolyzed to glucose under the effect of temperature, depending upon the concentration and the type of acid. This acid hydrolysis is used technically as described in Chapter IV.

The goal of the sulfite pulping process is, of course, not the production of low-molecular sugars from the carbohydrates of the wood, but the prep-

aration of undamaged fibers of pure cellulose. The cellulose is present in wood with other carbohydrates and lignin so that the isolation of the cellulose deals with a hydrolysis of carbohydrates and a delignification, reactions which occur either simultaneously or successively. The isolation is controlled by three factors: time, temperature, and acidity, whereas the delignification depends also on the bisulfite concentration, as was shown in the preceding section.

For the characterization of wood cellulose and the process of its decomposition during the sulfite cook, the determination of the average chain length of the cellulose proved to be of value. A mild dissolving out of the lignin from the wood structure, the subsequent nitration of the isolated cellulose, and the determination of the viscosity of the nitrate in acetone permit an estimation to be made of the average degree of polymerization (DP). The average DP of wood cellulose is between 2300 and 2400, but it can decrease in quick cooks to 1400; in comparison, cotton has a DP of around 4000. The essential difference between wood and cotton celluloses lies in the far greater uniformity of the molecularity of cotton cellulose and the pronounced polymolecularity of wood cellulose. In the sulfite pulping process, long molecular chains are cleaved into shorter chains and, in this way, a homogeneity of the chain-length differences is obtained, which is especially important for the chemical processing by industry of wood cellulose (15,157).

Greater damage to the cellulose occurs when the cooking process is combined with a mechanical action (134,135). Even minor amounts of oxygen seem to have a pronounced effect on the hydrolysis (415). Lignin, at least at the start of the cook, exerts a certain protective effect on the fiber. The danger of damage to the cellulose by hydrolysis increases with progressive delignification, regardless of other factors which act in the same direction. The effect of topochemical factors will be discussed later. The cellulose in deciduous woods is protected by lignin to a lesser degree so that, here, especially mild cooking conditions must be applied.

Comprehensive investigations on the effect of sulfite cooks over the total pH range, from 1.3 to 12, upon the yield and properties of the pulp have shown that, of all the carbohydrates in wood, cellulose, at pH 7, is the least resistant component. It shows a maximum resistance in the strongly acid region and another at near pH 11 (400). Figure VII–17 shows the influence of the pH conditions of the sulfite cook on the viscosity of cellulose-Cuene solutions. The cellulose degradation at neutral pH values depends upon the temperature, cooking time, and the hydrogen ion concentration of the cooking liquor. The strong decrease in viscosity of wood pulp which has been treated at temperatures of above 140°C in liquors of pH 8 to 10 cannot be traced to the presence of hydrogen ions in the cooking liquor because arabofuranose and galactopyranose, which are especially sensitive

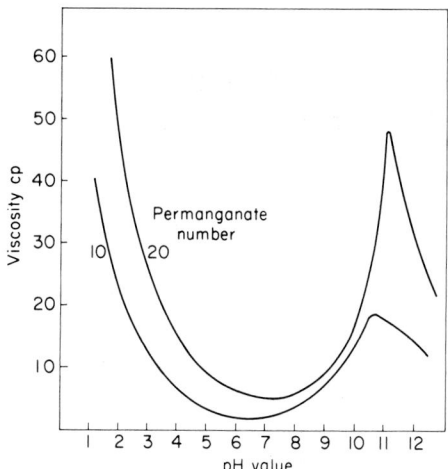

Figure VII-17. Variation of viscosity of pulps with average pH of cooking liquor (399).

to hydrolysis, remain in the pulp. The degradation of cotton linters and of hemicelluloses from wood pulp shows that the hemicelluloses are subject to a different degradation mechanism than that of cellulose. The hemicelluloses undergo an increasing loss in yield and viscosity with increasing pH value of the cooking liquor. Glucans, dispersed in alkali, when treated with cooking liquor of varying pH values at 180°C, undergo a decomposition which is similar to that of hemicellulose but not to that of cotton linters. This suggests that the degradation of insoluble cellulose in the neutral region of the cooking liquor can be traced to the formation of acid decomposition products as the result of a pyrolysis of the cellulose. With increasing alkalinity of the cooking liquor, an increasing penetration of the insoluble cellulose by alkali, a neutralization of the acid decomposition products and, with it, a decrease in cellulose degradation take place. First at high pH values is the favorable buffer action of the alkali replaced by its own hydrolysis action and then further decomposition takes place (399).

When methyl β-D-glucopyranoside, as model substance for cellulose, is treated with sulfite cooking liquor under the conditions of a neutral sulfite cook at pH 6 to 7 and 150° to 180°C, the decomposition occurs considerably faster than in cooks in the absence of sulfite under otherwise equal experimental conditions (147).

3.3 Hemicelluloses

A considerable portion of the polysaccharides of the wood are hydrolyzed during the sulfite cook. Not only the yield but also the quality

of the pulp depend to a large extent on the course of the hydrolysis. A general idea of the types of resistance of the carbohydrates against the hydrolyzing cooking liquor can be gathered when the various kinds of sugars formed during the hydrolysis are determined in the cooking liquor at different cooking times. At the end of the cook the same types of sugars, i.e., mannose, galactose, glucose, xylose, and arabinose, are always found, regardless of the wood species used. However, there are great differences between the quantitative yields of the various sugars, especially of galactose and mannose, in the spent liquors of coniferous and deciduous woods. The wood species, itself, obviously is not responsible for the sequence in which the various sugars are dissolved. Arabinose is found in the liquor before a temperature of 100°C has been reached in the digester. Shortly afterward, or simultaneously, the hydrolysis of xylan and galactan starts. Mannan is hydrolyzed considerably later and not before a temperature of 130°C is reached. Glucose, as a degradation product of cellulose, appears last, and usually not before a temperature of 140°C or a prolonged period at 130°C. Paper chromatographic investigations of the cooking liquor taken from the acid accumulator of cooks of spruce have shown the presence of, for example, 34.3% arabinose, 25.5% xylose, 9.2% galactose, 26.5% mannose, and 4.4% glucose. On the other hand, 3% to 5% arabinose, 15% to 25% xylose, 9% to 12% galactose, 50% to 60% mannose, and 5% to 15% glucose were found in the sulfite spent liquor itself. Large amounts of hemicelluloses can be isolated from the cooking liquor already in the early stage of a sulfite cook by exchange of the metal ions by hydrogen and precipitation with ethanol. In addition to the above sugars, uronic acids and small amounts of xylan units are found. In cooking liquor samples taken at 110°C and treated with a cation-exchange resin, the following sugars were isolated by precipitation and chromatography: 20.8% galactose, 15.8% glucose, 59.3% mannose, and 4.1% xylose. Ultraviolet absorption spectra confirmed that these sugars were practically free of lignin. Carbonyl groups were present as end-groups (98,99,358,391).

The characteristic differences between sulfite and sulfate or kraft pulps can be explained by the presence in various amounts of noncellulosic carbohydrates in the latter. The absence of galactose, arabinose, rhamnose, and galacturonic acid in sulfite and in prehydrolyzed kraft pulp depends on the acid-catalyzed hydrolysis of the polymers which contain these sugars. In general, these polymeric carbohydrates are either degraded to short-chain fragments or are dissolved in the cooking liquor. Some types of sugars, such as arabinose, are bound as acid-unstable end-groups in different polymers and are quickly split off under the conditions of the cook. This behavior is decisive for the conversion of 4–*O*-methylglucuronoaraboxylan in coniferous woods into 4–*O*-methylglucuronoxylan. This new polysaccharide, to some extent a transformation product in coniferous woods but

an original polymer in deciduous woods, is degraded by acid hydrolysis to products of lower degrees of polymerization. But not all these polymers are dissolved in the cooking liquor (145). When a normal bisulfite cook is preceded by a neutral cook, a pulp with a lower α-cellulose content is obtained than in a one-stage sulfite cook. The effect therefore depends on the pH in the first stage, and a pulp with normal α-cellulose content is obtained at a low pH. Pulps with low α-cellulose content contain less xylan but more glucomannan. The attachment of linear molecule fragments on the surface area of cellulose fibrils seems to be possible. This may explain a lesser accessibility for hydrolysis in the acid cooking stage. A minor stabilization of the glucomannan in a normal acid sulfite cook can be expected when the cook is not preceded by a sufficiently long impregnation period at low temperature. This stabilization of glucomannan found with spruce was also observed with pine, but it was not found with birch, which contains little glucomannan. The presence of sulfite or bisulfite ions during the first cooking stage is of no importance. The reactions of the lignin also seem to be without effect. The stabilization reaction occurs fairly rapidly in an alkaline medium and thus has led to the assumption that the cleavage of ester bonds precedes the glucomannan stabilization which, in turn, is connected with a deacetylation of the wood. The combined action of the pH and the temperature in the deacetylation and the glucomannan stabilization in such two-stage cooks is marked by a shift toward the acid region. Thus an increase in temperature from 125° to 160°C gives a shift in the pH of about 1.5 units. The resistance capability of glucomannan in pretreated holocellulose shows the same dependence on pH and temperature as was found with a two-stage cook of wood (8,9,70).

Hemicelluloses of the xylan type were subjected to sulfite cooks under technical cooking conditions and at various pH of the cooking liquors. It was found that the uronic acid content in the xylans, isolated from pulps, was only small. Xylans isolated from birchwood holocelluloses and treated in a homogeneous medium at pH 1.6 to 13.4 showed strong fluctuations in uronic acid content. Only xylan which was isolated from a pulp cooked at a pH of between 11.4 and 9.5 showed a uronic acid content of about 7%. Xylans, isolated from pulps obtained from cooks at other pH ranges, had uronic acid contents of only about 4%. The degrees of polymerization of the xylans obtained under acid conditions were considerably lower than those from pulps cooked under alkaline conditions. Uronic acids are readily split off in alkaline cooks because the glycosidic bond between the uronic acid and the polysaccharides is strongly alkali-unstable (145,258, 400).

During the sulfite cook not only the lignin but, to a certain degree, the hemicelluloses and the sugars formed by hydrolysis also react with the cooking liquor. The reaction takes place not merely until sugar sulfonic

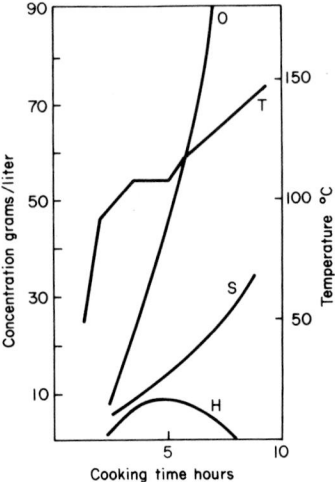

Figure VII-18. The hemicellulose content in the cooking liquor as a function of time (laboratory cooking) (99). O = organic substance, grams per liter; H = hemicellulose, grams per liter; S = reducing sugars, grams per liter; T = temperature, °C.

acids are formed but it can also proceed to the formation of irreversible conversion products and, in this way, can cause a loss of fermentable sugars. This reaction occurs partially via an oxidation of aldoses to aldonic acids. Arabonic acid is formed during a very early stage of the cook and its concentration reaches a practically constant value toward the end of the cook. Xylonic acid is also formed relatively early, whereas mannonic acid appears in small amounts at the beginning of the cook, only to surpass all other aldonic acids toward the end of the cook. Gluconic and galactonic acids are also found in the cooking liquor. The formation of the sugar sulfonic acids is not yet completely elucidated. Only gluconic acid is formed on heating of glucose with cooking liquor of low calcium-base content, whereas a similar treatment of holocellulose leads to the formation of aldonic and sulfonic acids. Of the sugars destroyed during the sulfite cook, 15% to 30% were identified as stable sulfonic acids (1,278,322,334).

Figure VII-18 shows the dependence of the hemicellulose content in the cooking liquor on the cooking time. It is apparent that the content of hemicelluloses increases during the impregnation period and at temperatures of between 100° and 110°C and decreases again rapidly on a further rise in the temperature. In contrast to this, the amount of reducing sugars increases continuously until the end of the cook (98). Figure VII-19, on the other hand, shows the dependence of the hemicellulose content of the pulp on the pH during the cook. The curves of the pH region of only 1 to 6 fall practically within the framework of a so-called sulfite cook. A com-

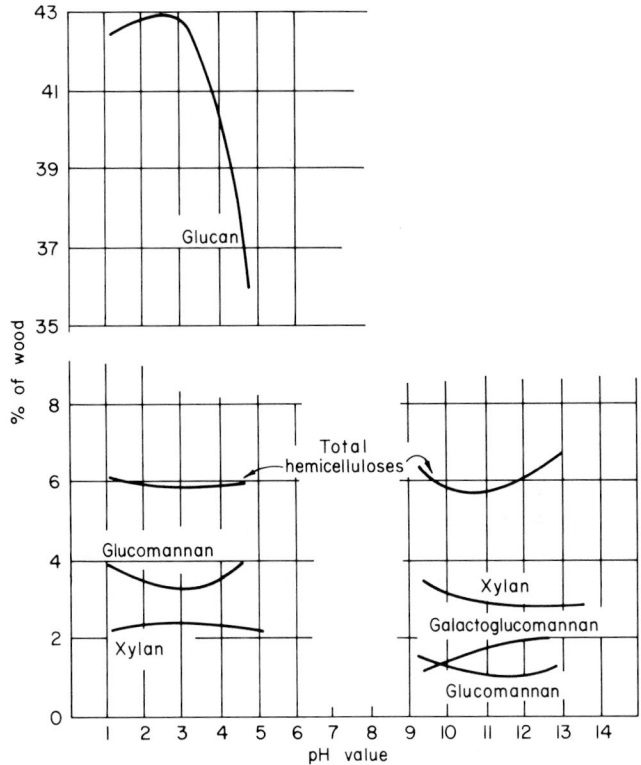

Figure VII-19. A comparison of carbohydrate content with cooking pH (400).

parison of the curves at the alkaline side of the diagram allows one to draw conclusions regarding the different chemical and physical behavior of the pulps cooked under acid, neutral, or alkaline conditions so far as this behavior is influenced by their hemicellulose contents (98,400).

3.4 Reaction with Various Other Wood Components

Relatively great differences exist in the decomposition reactions of different wood species during the sulfite cooking process, although appreciable progress has been achieved in recent years with almost all wood species by the use of soluble bases such as sodium, ammonium, and magnesium. Extractives in the heartwood are chiefly responsible for the strong resistance of pinewood toward the sulfite pulping process. Pinosylvin and its monomethyl ether are examples of such disturbing phenols. Some groups in lignin react readily with phenols in a strongly acid reaction medium, other groups react only weakly or not at all. The more highly the lignin is

sulfonated in the absence of phenols, the less the further pulping of the wood is hindered by the presence of phenols. The different reactivities of the sulfonated lignin groups toward sulfite and phenols explain the possibility of pulping pinewood by a two-stage process in which, in the first stage, the sulfonation is accomplished in a less acid reaction medium. Even relatively small amounts of phenol are sufficient to prevent the sulfite cook. Sulfonation and phenolation reactions are strongly temperature-dependent, with the phenolation obviously starting only at higher temperatures, and the sulfonation occurring already at lower temperatures (93,96).

Pinosylvin is a dihydroxystilbene or flavanone, and a 3,3',4',5,7-pentahydroxy derivative has been found in the heartwood of Douglas fir. This flavanone exhibits the same inhibiting properties that pinosylvin does (294). A series of investigations confirmed the necessity of a low maximum temperature, a slow temperature increase to this maximum, and the dependence of the phenolation and sulfonation upon the acid concentration. A sodium-base cooking liquor will lead to a uniform pulping of the heartwood; in the same way, a sulfonation with sodium-base cooking liquor and subsequent cook with calcium bisulfite cooking liquor will lead to the same result when the liquor is low in bound sulfur dioxide. Ammonium-base cooking liquors behave in about the same way as sodium-base liquors, but magnesium-base liquors take up a middle position between calcium and sodium (66,67). A sulfonation at a special stage of the cook and under mild temperature and acidity conditions is therefore advantageous in every case (304).

Instead of a retarding effect by the phenols on the sulfite pulping process, some authors assume that a decomposition of the bisulfite is caused by these phenolic compounds. Dihydroquercitin has been isolated from Douglas fir and was found to be a mild inhibitor in the sulfite cook without, however, forming an insoluble condensation product with the lignin. Dihydroquercitin reduces bisulfite to thiosulfate which, in turn, accelerates the decomposition of the bisulfite in an autocatalytic reaction of the first order. The strongly retarded dissolution of lignin in a calcium bisulfite cook of Douglas fir is the result of a bisulfite decomposition, by means of which considerable amounts of sulfuric acid are formed before the dissolution of the lignin is completed. Sulfuric acid causes the precipitation of calcium sulfate and consequently all the dangers associated with the depletion of the base. The outstanding advantage of cooking liquors from soluble bases is the extensive availability of the base throughout the cooking period (2,165).

The reddening of sulfite pulps has been discussed frequently. Several investigators assume that the compound responsible for the reddening is present in a leuco form and first becomes visible on oxidation and after removal of reducing constituents. Metal ions are believed to exert a catalytic

effect, while complex ion formation prevents the reddening. The complex causing the reddening is contained in the spent liquor and is difficult to wash from the pulp. Also the fluorescence under ultraviolet radiation, which goes parallel with the reddening, is ascribed to the same cause (3,333).

3.5 Topochemical Aspects

The properties of the pulp fiber isolated from wood are connected with the structure of the wood itself. It was not long ago that a specific investigation of the cell wall was carried out and this has led to surprising and valuable results. Details have been discussed in Chapter II. Here, only those observations relative to the sulfite cooking process will be dealt with. Microphotographs of wood cross-sections obtained at various stages of the cook have shown that the intercellular material consists chiefly of lignin and, possibly, some hemicelluloses. The cambial wall is the integrating part of the pulp fiber, which consists mainly of cellulose. The secondary wall, on the other hand, consists of two components, of which the outer wall contains considerable amounts of lignin and the inner, chiefly only cellulose. When, during the cook, fairly large amounts of lignin are removed from the outer wall, a hollow space is formed between the secondary wall and the cambial wall. Alkaline solutions have a selective action on the lignin-containing components of the wood and consequently the intercellular lignin is dissolved rapidly without any noteworthy attack on the cell-wall lignin. The sulfite cooking liquor does not have this selective action and the intercellular and cell-wall lignins are attacked more or less simultaneously. The dissolution of the lignin is appreciably slower than with alkaline liquors; this is partly due to the swelling action of the alkaline solution on the secondary wall (28,204).

The cooking liquors can penetrate into the microscopic and submicroscopic capillary spaces only by capillarity and diffusion, whereupon an exchange of material takes place by sorption and desorption (419). During the delignification process, the capillary system is subjected to constant changes. In an acid cook the middle lamella (see Fig. VII–20) is delignified more rapidly than the fiber wall. In the middle lamella, bonds between the fibers are assumed to be present and to be more greatly developed in the tangential middle layers than in the radial layers. An acid cook strongly attacks the primary wall and completely decomposes it. The outer layer adjacent to the primary wall, too, is attacked to an appreciable extent, causing a decrease in, or the complete disappearance of the double refraction of this wall constituent. In an alkaline cook, or on a mild and careful delignification, the secondary wall swells as the result of the hydration and swelling of the embedded wood polyoses. The polyoses are hydrolyzed

Figure VII-20. Section of adjacent walls of a tracheid (204, see also Fig. II-7). a = truly isotropic intercellular material; b = cambial or primary wall; c = outer layer of secondary wall; d=central layer of secondary wall; e = inner layer of secondary wall.

during an acid cook and their removal causes a shrinkage of the swollen layer. With the help of color reactions, a controlled partial hydrolysis of the tertiary wall in sprucewood could be accomplished. The hydrolyzate contained 45% xylose, 24% mannose, 16% arabinose, 13% galactose, and 2% glucose (all as mole %). The presence of uronic acids in the tertiary wall could be shown by staining, but they were not detectable in the hydrolyzate. Figure VII-21, parts A to C, clearly show the attack on the middle lamella after 4 hours cooking time (A), the period during which the cell walls are still held together at the corners by incrusted material (B), and the collapse of the cells after 9 hours cooking time (C) (49,50,51).

Sprucewood chips were partially delignified by the acid-sulfite and sulfate processes. After this, the lignin content of the cooked chips was compared with that of the original sprucewood; the chips from the sulfite cook were found to contain 27.5%, and the chips from the sulfate cook, 29.8% of their original lignin. Microtome cross-sections were prepared from the cooked chips that could not be defibered and were investigated under the ultraviolet microscope at a wavelength of 280 mμ. After the partial sulfite cook, the lignin content of the cell wall was practically unchanged, whereas the lignin of the middle lamella was almost completely removed. The delignification started at the middle lamella and proceeded from there toward the lumen. On the other hand, the partially pulped wood from the sulfate cook showed that a considerable amount of lignin was removed from the cell wall as well as from the middle lamella, indicating that the delignification of the cell wall occurred from both sides, from the middle lamella and from the lumen (186a).

Similar investigations after a 50% delignification by the acid-sulfite and sulfate pulping processes and also by a neutral sulfite cook of sprucewood confirmed the fact that, in kraft and acid-sulfite pulping, lignin is preferentially removed from the secondary wall, while, at the same time, the middle lamella and the cell corner areas are strongly attacked. Little topochemical preference was found in the removal of lignin by the neutral sulfite process at up to 50% delignification. No concentration gradients of lignin were produced in the secondary wall by any of the cooking processes. The middle lamella was resistant to solution even when it came into direct

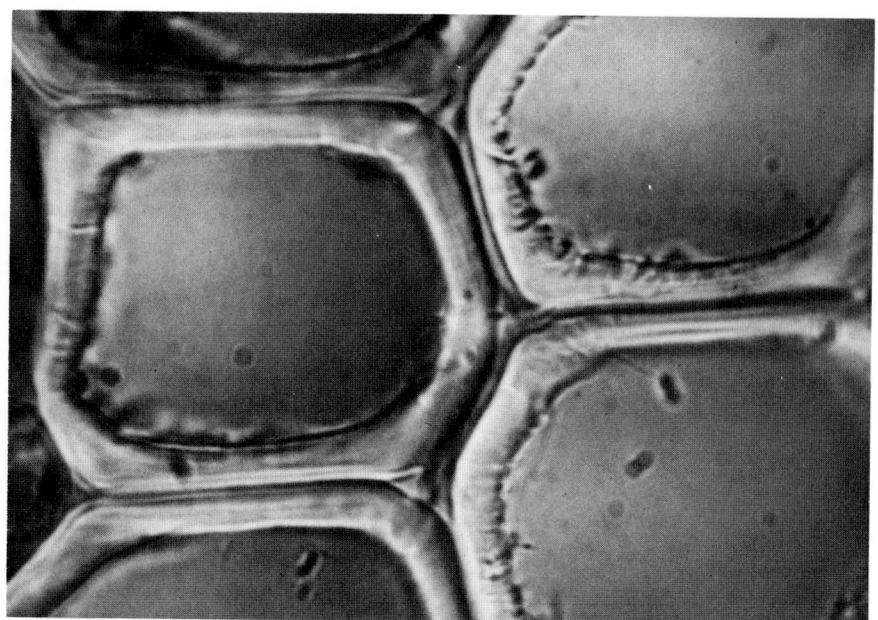

Figure VII-21A. Cross-section of sprucewood. Middle lamella, after 4 hours sulfite cooking, heavily damaged (49). Magn. 1600:1. (Courtesy Cellulosefabrik Attisholz.)

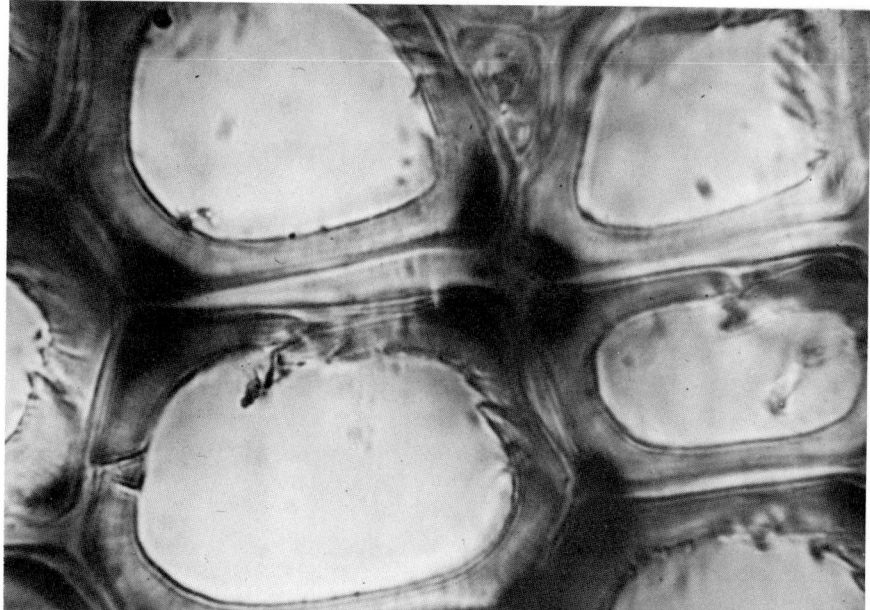

Figure VII-21B. Latewood cells of sprucewood with incrustations in the edges at beginning of separation, after 7 hours sulfite cooking (49). Magn. 1600:1. (Courtesy Cellulosefabrik Attisholz.)

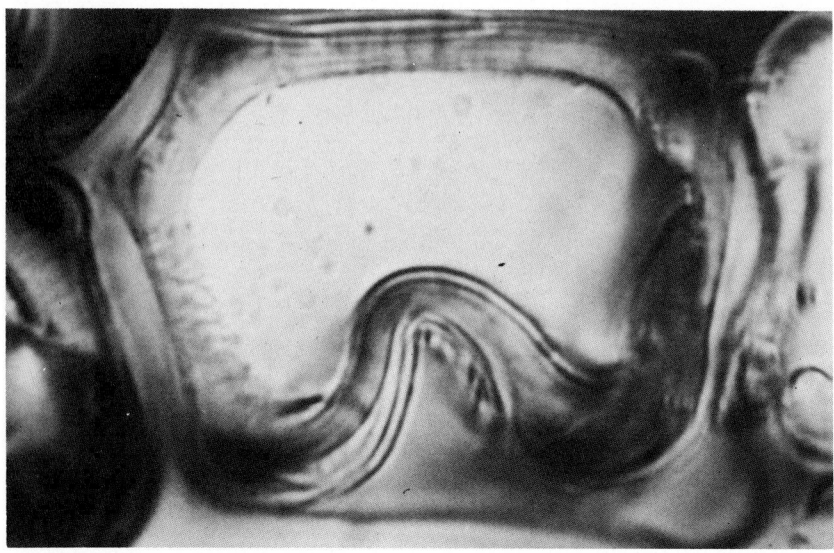

Figure VII-21C. The same cells, after 9 hours of sulfite cooking, begin to fold (49). Magn. 1600:1. (Courtesy Cellulosefabrik Attisholz.)

contact with the liquor at the margin of the pit cavity. This suggests that diffusion of the chemicals into, or of lignin macromolecules out of, the secondary wall does not play the dominant role in controlling the rate of the pulping reaction (298a).

When wood chips saturated with cooking liquor were dyed with malachite green and *p,p'*-azodimethylaniline, it could be shown by microscopic tests that the middle lamella in aspenwood is the first zone to be sulfonated in an acid or neutral sulfite cook, while the lignin in the cell wall is dissolved first. This dissolution perhaps takes place without a previous sulfonation. Guaiacyl and syringyl nuclei are fractionated by the cooking liquor. It is possible that they are distributed heterogeneously across the cell walls of aspenwood, with the larger amounts of syringyl groups being present in the middle lamella and the larger amounts of guaiacyl groups being in the lignin within the cell wall (254a). Investigations with Jack pine (*Pinus banksiana*) showed that wood chips which had been stored for 6 years in sulfite cooking liquor at 25°C had lost almost half of their original lignin content. As shown under the microscope, the middle lamella was partly dissolved and this led to a partial separation of the fibers, chiefly in the radial direction. The secondary wall showed a strong swelling which could be traced to the presence of highly sulfonated lignin. An electron microscopic investigation showed that the middle lamella either was in a state of dissolution or was completely removed. x-Ray investigations of the lattice

structure showed no changes within the cellulose molecule during the cold sulfite treatment. The crystalline orientation appeared to have increased slightly, probably due to the hydrolysis of amorphous cellulose units and crystallization of broken chains (211).

The distribution of hemicelluloses, especially of xylan and mannan, within the cell wall of spruce tracheids was studied chromatographically with easily hydrolyzable polysaccharides in finely milled wood. About 30% to 35% of the total xylose portion was found in the inner layers, 25% to 30% in the outer layers, and 35% to 40% within the central secondary wall. Galactose- and arabinose-containing polyoses are limited chiefly to the inner layer of the secondary wall, while the principal part of the glucomannan is found in the outer layers (5). The inner surface of the wood does not increase greatly at the beginning of the cook but it increases strongly during the cook at temperatures of above 110°C and reaches its maximum of 300 to 400 square meters per gram after a cooking time of 0.5 hour at a pulp yield of 65% to 75%. On continuation of the cook the surface area decreases again and amounts to 150 to 300 square meters per gram toward the end of the cook. The enlargement of the inner surface is connected with the formation of new capillaries and is the result of the destruction of transverse bonds between the elements of the wood structure in the cell wall. The resistivity of these submicroscopic capillaries decreases strongly with increasing delignification (100).

3.6 Wood Quality and Effects of Growth

The influence of the quality of the wood upon the properties of the pulp has been recognized for a long time. In spite of this, the classification of pulp wood with regard to its technical suitability encounters difficulties. The density, the fibril angle, fiber length, and the proportion of spring- and summerwood can be ascertained by taking drilling cores. In addition, the determination of the growth rate, the chemical composition, the fiber diameter, the cell-wall thickness, the fiber strength in the wood, the strength of the fiber-to-fiber bond, and the portion of knots and of compression and tension wood are of importance. By selective cultivation it is hoped that trees with greater density, longer fibers, lower lignin content, and higher cellulose content can be grown (71,113,142,263,395).

The damage occurring during the preparation of the wood, especially during the chipping operation or during the storage of the wood in the form of logs, stems, or chips, has been reported in Chapter VI, therefore only specific growth damages will be discussed here. Tracheids of coniferous woods show a weakened structure at the places where they have been in contact with medullary rays. The tracheids are cracked on defibering of the cooked chips and damaged places, known as nodules, remain; when

the fibers are swollen, they break at these places. On the other hand, a tree which, during its growing period, is exposed to a crushing stress, e.g., by wind pressure, tries to increase its stability by the formation of compression fibers. Such fibers show radial cracks in the secondary wall and have no tertiary wall. They have higher lignin and galactan contents. The yield of pulp from such compression wood is generally lower. Deciduous woods behave differently, through the formation of tension wood. Such tension wood fibers differ chemically from normal angiosperm fibers by a lower lignin content, with the secondary wall being almost completely lignin-free. The pulp yield and the pulp properties from tension wood are improved (114,154,188,295).

The yield of pulp, free of splinters, per volume unit decreases markedly with pine as the diameter of the trunk increases. The pulp yields from the middle parts of the tree trunk are definitely higher than from very fast- or very slow-grown stems. On a weight basis, the yield decreases from the top downward and, vice versa, on a volume basis, increases from the top downward. On the basis of these facts, a choice of the wood could lead to an increase in the yield and an improvement in the quality of the pulp (64, 65). There is no direct relationship between the average tree diameter and the solid volume, but the average density increases with decreasing average log diameter. Slash pinewood with a growth rate 2.5-fold above average showed no noteworthy deviations with regard to the density of the wood. That the density, which is important for the yield and quality of the pulp, can be influenced by changing the surrounding factors and by genetic action is considered possible and seems to be promising, but the realization of the necessary actions may not be so easy (38,52,293).

The differences between sapwood and mature wood lie mainly in the lower density of the former. Every tree has some sapwood and a pulp from this wood has inferior properties. Only minor relationships exist between growth velocity and density. On the other hand, there are direct relationships between the water-resistant carbohydrate and the α-cellulose contents in sapwood and mature wood. A low yield of pulp from sapwood, therefore, also means a low yield from mature wood and, vice versa, a high yield from sapwood means a high yield from mature wood. With loblolly pines from five different stands, those from only one showed a certain relationship between pulp yield and density of the wood, whereas, on a volume basis, a clear relationship between yield and density is evident (356,439). A study of the effects of surrounding factors on the density and volume of a tree showed that more favorable surroundings gave a larger amount of wood with a low density, and, under poor growing conditions, a wood with higher density resulted. The distance between the trees in a wood stand had little effect on the wood volume, but the effect was intensified when the distance was small. With all trees, the density of the wood de-

creased from the root to the top and, in the same way, from the cambium toward the center of the trunk (185,440).

4. THE PERMEABILITY OF THE WOOD

4.1 The Movement of Gases and Condensable Vapors in Wood

Gases can penetrate wood only through the permanent capillary structure and above the fiber saturation point, since free water must be displaced from the capillaries. Condensable vapors penetrate the wood under pressure in the same way as gases do if the relative vapor pressure is below that pressure formed by the condensation of the vapors in the openings of the porous membrane. Vapors are mostly adsorbed and condensed within the cell-wall structure. The swelling is thereby directly related to the volume of the condensed liquid, but it has a negligible effect on the dimensions of the fiber lumens. Swelling lowers the permeability of the pores. Any kind of liquid movement within the cell wall takes place by diffusion and is independent of the hydrostatic pressure used. A hydrostatic pressure is therefore active only when it increases the solubility of the gas in water. The diffusion of condensable vapors into the permanent structure of dry wood therefore proceeds similarly to that of gases, but it is supplemented by a diffusion of the condensed vapor as "bound" liquid within the cell wall. It is evident that great importance should be attached to the relationship between the gas, vapor, and liquor movements during the sulfite cooking process.

4.2 The Movement of Liquids in Wood

Wood is a complex porous material. As already mentioned, the movement of liquids within the cell-wall structure is possible only by diffusion. Diffusion is characterized by the migration of molecularly or colloidally dissolved substances in the direction of a concentration gradient. On dissolution of a substance in a solvent, an infinitely thin boundary layer is formed at the contact surfaces of both substances. The migration of the dissolved substances in the solvent proceeds by the irregular movement of the molecules. When a visible hydrodynamic movement is imparted to the liquor particles, a convection is formed. Each dissolving process is connected with changes in density and thus causes buoyant forces and inner movement. Diffusion and convection interchange with each other. Analogous conditions were discussed in Section 2, with reference to the dissolution of sulfur dioxide in water or bases.

When diffusion takes place through a membrane which, under special

conditions, is permeable for crystalloids but not for colloids, the process is called dialysis. The passage of the liquor through the membranes takes place through microscopic openings or cracks. In this way a capillary sifting action can take place, allowing small molecules to pass through the openings but retaining large molecules. The presence of electrolytes determines the state of charge of the membranes. The greater the permeability of the membrane, the smaller the electromotive effect.

The wood structure consists of a complex capillary system. The flow of liquids through capillaries conforms to special laws. A stationary boundary layer is formed at the wall of the capillary. In direct contact with this boundary layer, a laminar flow prevails, with a turbulent flow joined to it. The movement of substances from the porous capillary wall (cell wall) into the boundary layer and, from there, into the layer of laminar flow (cooking liquor), takes place by diffusion and, further, by means of convection. The mechanism of the penetration of liquids into wood becomes more complicated because of the variability of the flow movements to which the liquids are subjected, and the variability of the forces which influence these actions. To this must be added the complex structure of the wood. In the submicroscopic region there still exists, in addition to the permanent structure, a transient one, which appears only as the swelling occurs. The liquor flow within this submicroscopic structure can take place simultaneously or, at times, successively with that within the permanent structure.

Liquids penetrate the wood in the axial direction 50 to 200 times faster than in the transverse direction. Enclosed air greatly reduces the velocity of the penetration. Under hydrostatic pressure, liquids penetrate the wood until the formation of a liquid-air meniscus. Further penetration can take place when the capillaries are not circular or when they possess hygroscopic properties. Evaporation can take place from the meniscuses and vapors of swelling liquids can enter the cell walls by diffusion and can condense there. In this way the cell walls can be saturated with liquid by condensation before the liquid itself has penetrated there. Nonswelling liquids require certain pressures which must be greater than the surface tension force in the pore openings before they can penetrate into the capillary structure. Wetting agents may have an advantageous effect by decreasing the surface tension in the pores, but their action is still disputed (256,419).

The air enclosed in the pores and hollow spaces has a special effect on the permeability of wood. Reactive liquids which react with one or several components of the wood show a strong dependence on time and temperature in their influence upon the permeability. The dissolution of lignin and other accessory components of the cellulose causes a continuous alteration in the macroscopic and submicroscopic structure of wood. In this way new openings are constantly formed or existing ones enlarged. Only about 14% of the total water uptake is due to water vapor diffusion, the other 86%

being due to capillary ascent. The capillary penetration which occurs first is, therefore, of prime importance, especially since the diffusion into the cell walls takes place relatively rapidly because only small distances have to be traveled. All means that facilitate the capillary penetration of the liquid act favorably upon the course of the cooking process.

The cooking process, however, is not completed with the impregnation and the chemical reaction of the cooking liquor with the components of the wood. The products which have been rendered soluble must diffuse out from the still intact wood structure. The velocity of this dissolving out process depends to a large extent upon three factors: the size of the contact surface, the flow velocity, and the solvent gradient. The contact surface can be enlarged by an extensive mechanical degradation of the cooked material. This process is limited, as already mentioned in Section 3.2. A high flow velocity helps the dissolving out by convection. The technical procedures to do this are still to be discussed. The formation of a concentration gradient as great as possible requires a multistage process because, in a single-stage process, equilibriums at the contact surface will be established after a certain reaction time.

A series of investigations has been carried out, dealing with a general determination of the permeability of wood by liquids. Long experience has confirmed the fact that wood chips, placed in a closed vessel of water, will float on the surface; when the air in the vessel is removed, some of the chips sink, while others remain on the surface of the water. Chemical dyeing easily shows that the chips that sink consist mainly of sapwood, whereas the floating chips consist of heartwood. This is especially true with coniferous woods. With deciduous woods, the poor penetration is caused, first of all, by the formation of tyloses (bladderlike growths) in the vessels. The above-mentioned method of evacuation of wood chips suspended in water can be used for the determination of the permeability of wood. A more reliable method is to measure the flow of air through the capillary system of the wood under certain specified conditions and to calculate the penetration factor found with the help of the Poiseulle equation. A wooden dowel is connected parallel to a calibrated glass capillary. According to the equation

$$r_w^{\ 4} = \text{const}\left(\frac{P_g}{P_w}\right)$$

the permeability factor is proportional to the gradient ratio between the glass capillary and the dowel. The permeability of the wood is calculated under the assumption that all capillaries are combined to form a single capillary which corresponds to 1 cm² cross-section of wood. In the equation, P_g is the pressure drop of the capillary and P_w that of the wood, r_w is the

radius of this hypothetical capillary, and the penetration factor is the fourth power of this radius.

Table VII-6 shows the differences in such penetration measurements with various coniferous and deciduous woods. It must be borne in mind that a wood with a penetration factor of, for example, 100×10^{-10} is ten times more difficult to penetrate than a wood with a penetration factor of 1000×10^{-10}. As mentioned above, 86% of the liquid penetration of wood is by capillary ascent. This must not lead to the conclusion that the term "penetration" is the same as capillary infiltration. The remaining 14% of the liquid, which enters the wood by diffusion, is a part of the penetration, which is therefore a collective term for any kind of infiltration of liquid into wood. The expression "capillarity" seems to be a better term for the

TABLE VII-6

VARIATION OF PENETRABILITY BETWEEN WOOD SPECIES

Wood species	Penetration factor $\times 10^{10}$	
	Sapwood	Heartwood
Black spruce (*Picea mariana*)	—	2
Engelmann spruce (*Picea engelmanni*)	5	3
Slash pine (*Pinus caribaea*)	6000	10
White pine (*Pinus monticola*)	100	10
Short-leaf pine (*Pinus echinata*)	120	5
Long-leaf pine (*Pinus palustris*)	4000	2·
Lodgepole pine (*Pinus contorta*)	300	—
Douglas fir (*Pseudotsuga taxifolia*)	70	5
Elm (*Ulmus americana*)	400	70
Box elder (*Acer negundo*)	1300	400
Hickory (*Hicoria cordiformis*)	4000	400
Sycamore (*Platanus occidentalis*)	4000	4000
Willow (*Salix nigra*)	4000	—
Birch (*Betula papyrifera*)	1300	450
Beech (*Fagus grandifolia*)	1000	0.5
Sweetgum (*Liquidamber styraciflua*)	1200	850
Ash (*Fraxinus nigra*)	80	—
Maple (*Acer rubrum*)	400	120
Red oak (*Quercus falcata*)	4000	5
Scarlet oak (*Quercus coccinea*)	1000	400
Post oak (*Quercus stellata*)	67	5
White oak (*Quercus alba*)	0.7	—
Cottonwood (*Populus deltoides*)	4000	500
Balm of Gilead (*Populus tacamahaca*)	800	300
Aspen (*Populus tremuloides*)	2500	1
Aspen (Finland) (*Populus tremula*)	5000	4
Triploid aspen (Finland) (*Populus tremula*)	1000	2000

movement of the liquid under the effect of a hydrostatic pressure gradient. What is measured and denoted as the penetration factor is, therefore, the capillarity of the wood structure. This basic difference between the two kinds of impregnation is of special importance for deciduous woods. Here the penetration of the wood is completely dependent on the presence of vessel cells, while the fibers themselves contribute little or nothing at all. The penetration velocity varies within wide limits, depending upon whether the vessel cells are open or closed. Diffusion, on the other hand, can take place also through the cell walls and is independent of the presence of vessel cells. Moreover, the usually obstructive tyloses have no influence on it (381,382,383,384,385).

4.3 Impregnation and Deaeration

The importance of a complete and uniform impregnation of the wood chips with cooking liquor has already been discussed in detail. The processes which cause the penetration of the wood with liquor were outlined in the preceding section. Since the hollow spaces in wood are filled with air and the porous volume is extremely large, the removal of the air is a major problem. Several possible methods for achieving this are available. They are (*a*) steaming with excess vapor pressure, with or without rapid pressure relief; (*b*) evacuation; (*c*) hydrostatic pressure treatment; and (*d*) introduction of hot cooking liquor into the digester.

A process developed in Canada and known as the "Va-Purge-Process" builds up a rapid vapor pressure in the digester, filled with chips and cooking liquor, by a jerkily percussive introduction of steam, with the pressure being relieved by an equally rapid release. This vapor thrust technique, which gave the process its name, is repeated several times at short intervals. To some extent, this process is a combination of some of the above-mentioned possibilities for an improvement of the impregnation (151,242).

On the other hand, the evacuation of the digester when it is filled with chips causes the removal of the air from the hollow spaces of the wood through the creation of a vacuum. This process is carried out in the calcium bisulfite cooking process in the absence of the cooking liquor. Such an evacuation of the digester, however, is a time-consuming process, and a digester with an acid-proof lining cannot be evacuated. After the evacuation and before the restoration of the atmospheric pressure in the digester, the latter is filled with cooking liquor. It has also been suggested that the chips and cooking liquor should be mixed and the evacuation carried out in a special tank (101). According to another suggestion, three tanks are used: in the first, the chips are moistened with cooking liquor; in the second, the chips are evacuated; and in the third, the chips are completely impregnated and then placed in the digester (261). A Finnish process re-

duces the pressure in the presence of the cooking liquor and in this way achieves a rapid impregnation (408). All these suggestions have in common the removal of the air from the wood pores by reducing the pressure.

In contrast to these processes, the use of hydrostatic pressure means a compression of the enclosed air rather than its removal. The application of higher pressures, of course, is not a solution to the problem, but raising the partial vapor pressure of the sulfur dioxide by heating the cooking liquor has a favorable effect (332). A rhythmic hydrostatic pulsation process, by pressure and pressure release, has been recommended for promoting penetration. After such an impregnation process has led to a quick and uniform distribution of the cooking liquor within the chips, the cook can be carried out relatively quickly and, under certain circumstances, even at the vapor phase. Besides a uniform impregnation, a high proportion of bisulfite and a content of not more than 3% free sulfur dioxide are necessary (259,355). Not only the impregnation but the whole cooking process can be greatly accelerated when preheated cooking liquor from a pressure storage tank is used (153). It has also been suggested that the air in the pores should be displaced by carbon dioxide (17).

The dimensions of the wood chips are of special importance for the impregnation process. In particular, a far-reaching uniformity in the dimensions must be striven for. The more rapid the impregnation process is to be, the shorter must be the route to be traveled by the cooking liquor for the capillary penetration. The comminution of the wood chips, of course, is limited because of possible damage to the fibers. The temperature also plays an important role because the penetration of the cooking liquor does not occur in all directions with the same velocity; it takes place much more rapidly in the fiber direction than in the radial or tangential direction. With coniferous woods, the penetration occurs through the bordered pit from one tracheid to the next; the medullary rays are the transportation routes in the radial direction; resin ducts can facilitate the penetration in the axial direction. In deciduous woods, the vessels are filled first and from there the liquor penetrates the medullary rays and into the sclerenchyma fibers. The tyloses in the heartwood especially retard the penetration (79,91,194,417).

4.4 Forced Circulation

The pulping process basically can be considered to be a continuous process. In a first stage, the active chemicals react with some wood components which are converted into soluble products, partially by hydrolysis, partially by reducing their molecular size. In a second stage, these soluble components are leached out of the wood structure. Taking the dimensions of modern digesters into account, it is understandable that the extraction

process of the second stage can take place uniformly only when the acid concentration and the temperature within the digester are approximately constant. The circulation caused by heat buoyancy or by the direct introduction of steam is insufficient. The economic efficiency of the digester's volume requires a charge as dense as possible and a low liquor-to-wood ratio. The forced circulation of the liquor during the cook was therefore one of the first steps taken in the progressive development of the pulping process. This forced circulation also made it possible to use indirect heating of the cooking liquor by continuously taking out liquor from the digester by a pumping system, passing it through a heat exchanger, and returning it in a reheated state to the digester. The direct introduction of steam for heating into the circulating liquor, of course, is also possible if desired. The take-out and the input of the liquor in the digester take place at different locations, depending upon the system used. Some systems withdraw the liquor at the top and return it at the bottom, others use a reverse circulation. Still others withdraw the liquor from the middle of the digester and return it at the top and bottom. Finally, there are systems in operation which reverse the circulation from time to time, i.e., they take off the liquor alternately from the top or bottom and inject it again in the opposite direction. The advantages of the indirect cooking are, in the first place, the maintenance to a large extent of a high acid concentration because dilution by condensation of steam is avoided. The cooking time can be curtailed. The condensate from the heat exchanger can be used in the boiler, thus effecting a saving of steam. Disadvantages of indirect cooking are an increased power consumption for the pumping system and the considerable cost of the upkeep of the circulation system. In addition, the formation of scales in the circulating system and in the heat exchangers is a disadvantage, but this can largely be prevented by the right composition of the cooking liquor, by a low content of combined sulfur dioxide, and by effective filtration of the cooking liquor.

A circulation system without pumping was designed in Sweden. The cooking liquor is withdrawn through an ascending pipe which is installed on the inner wall of the digester and has a perforated bottom at the conical base of the digester. The pipe has two steam nozzles in the lower part and two nozzles for steam and two injectors for sulfur dioxide at the top. It is debatable whether direct or indirect cooking is preferable, but it is generally agreed that direct contact between steam and wood chips should be avoided. In any case, a circulation system requires a sufficiently large perforated bottom to the digester for drainage of the spent liquor so that the sieve is not clogged even with dense digester charges or when the chips start to defiber (105,245,307,428).

5. ACIDITY, PRESSURE, TEMPERATURE, AND COOKING TIME, AND THEIR EFFECT ON THE SULFITE PULP COOKING PROCESS

5.1 Measurement of pH and Degree of Dissociation

The withdrawal of acid samples from the digester and their potentiometric measurement after being cooled do not suffice for the determination of the actual acidity conditions during the cook. Direct measurement of the hydrogen ion concentration in the hot cooking liquor while it is under pressure have first been made possible by suitable instruments during the past 15 years. An indirect determination of the acid conditions had been proposed before and had been carried out by determining the saccharification velocity of soluble starch by the cooking liquor under the conventional reaction conditions. The extensive dependence of the acidity on the temperature and the formation of dissociable reaction products could be shown (140). The glass electrode was found to be especially suitable for pH measurements under extreme experimental conditions. Such a special electrode differs from the normal type in that the potential-determining reaction is not a redox reaction but is probably a diffusion or ion migration phenomenon. Because the glass electrode is actually a chain, its combination with a simultaneously used reference electrode serves only for a potential determination. Water and dilute acids cause the hydrolysis of the silicates on the glass surface. In a permutoid reaction they replace the metal ions by H_3O^+. The swollen silica-gel layer, charged with H_3O^+ ions, is the site of the potential formation. Only after this layer has been formed does the electrode give constant values for the electromotive force (349, 350).

The pH measurements within the digester require special measuring instruments. Electrodes have been made which allow measurements at temperatures of up to 200°C and corresponding pressures. A built-in cooling device makes a separation of the high- and low-temperature parts possible and allows the temperature gradient to be limited to the electrolyte solution. The arrangement of the half-cell within the pressure zone requires no special adjustment for a reduction of the pressure. The effect of the pressure on the electrochemical behavior of the electrolytes can be estimated from the shift of the activity coefficient, and measurements with potassium chloride solutions have shown that these shifts at pressures of up to 100 atmospheres are without significance. Flows in the electrolyte solution due to convection, however, can have a disturbing effect (174,176).

To avoid penetration of the cooking liquor into the measuring cell, the latter is under an excess nitrogen pressure. The connection with the digester contents is formed by a capillary which is filled with asbestos. The reference calomel electrode is some distance away from the digester and is kept at room temperature; direct readings are made possible by means of a temperature compensator and no corrections are necessary (30). Three measurement sites, for example, are installed inside the digester, one in the upper part, one at the bottom cone, and another in the circulation pipe. In confirmation of earlier results, it was found that the pH value at the start of the cook is between 2.2 and 2.5, depending upon the composition of the cooking liquor; after 3 hours cooking time at 140°C the pH rises to 2.7 and then drops slowly to its previous value. As soon as the circulation is interrupted, strong deviations are found at the various measuring sites; this again clearly proves the importance of the circulation on the course of the cook (374).

The effect of the acidity upon the delignification process and on the other wood components is of great importance with regard to the properties of the pulp obtained under specific cooking conditions. According to the law of mass action, the reaction velocity depends on the concentration of the reacting materials. An increase in the concentration of the sulfurous acid and the resulting simultaneous increase in pressure, even at a constant temperature, lead to an acceleration of the cook. However, the maximum temperature is of greater importance than the higher sulfur dioxide concentration for the quality of the pulp. Direct relationships have also been found between the degree of cooking and the acid concentration. With pulp cooks with constant sulfur dioxide content but with different pH values and without the presence of any other chemicals, a marked retardation of the cook could be noticed with increased initial pH values for the cooking liquor. Sulfite cooks over the total pH region of from 1.5 to 11 showed that all the pulps obtained at pH 1.5 to 6 contained a 4-O-methylglucuronoxylan, whereas those obtained at pH 9 to 11 contained a 4-O-methylglucuronoaraboxylan. In contrast to this, kraft pulps contain an araboxylan. A comparison of the physical properties of the pulps obtained showed that those with a higher lignin content, i.e., those which were less cooked, possessed greater fiber strength. The xylan content seems to have a special influence upon the physical properties of the pulps (167,291,292,296,400).

With calcium- or sodium-base cooking liquors the pH value increases with increasing temperature. Lignin derivatives which are formed during the cook are only slightly acid under conventional cooking conditions and consequently the hydrolysis of carbohydrates and polymerization of lignin are minimal (56,174).

The acids contained in the cooking liquor and those formed during the cook are partly nondissociated, partly completely dissociated, while the

sulfurous acid is partly dissociated. Some strong acids, such as the α-hydroxysulfonic acids, are not stable. The acidity of the cook at given temperature conditions can be calculated from the pH determinations of the cooled liquor if the changes in the concentration of the α-hydroxysulfonic acids and the dissociation conditions of the sulfurous acid are known. With cooks with indirect steam heating and with cooks of a moderate degree of pulping—i.e., with so-called hard cooks—the pH value is above 2, in good agreement with direct pH determinations. Cooks with a high degree of decompositon—so-called soft cooks—show an increase in the pH toward the end of the cook. This increase is connected with a drop in the viscosity of the pulp. The dissociation constant of the sulfurous acid decreases sharply with increasing temperature; at 90°C only about 8%, and at 130°C only about 2% of the SO_2 is hydrated. In the sulfonation of the lignin the free sulfurous acid may not be the decisive reaction component because its actual concentration at the high temperature and its activity are small as compared with that of the bisulfite ion. The bisulfite ion also plays a role in the formation of organic acids, in spite of the fact that some authors claim that no strong acids, but only their anions, are formed. The bisulfite ions are continuously used up during the cook and exert no buffer action. The ratio between the hydrogen ion and the bisulfite ion concentrations for different types of cooks is shown in Fig. VII–22. Whereas the bisulfite ion concentration decreasess continuously during the cook, the hydrogen ion concentration first decreases somewhat and then increases again toward the end of the cook. When the pressure in the digester is released, a further rise in the hydrogen ion concentration occurs, probably as the result of the decomposition of α-hydroxysulfonic acids. Cooks with a low base content reach this reversal point earlier than those with a high base content. The minimum amount of hydrogen ion is at 0.002 to 0.003 grams per liter, the maximum at 0.03 to 0.06 grams per liter, or at pH 2.5 to 2.7 for the minimum and 1.2 to 1.5 for the maximum (56,180,312,313).

The three metallic bases, sodium, calcium, and magnesium, regardless, of considerable differences in their solubility, form temperature-resistant ions, whereas the ammonium ion is just as unstable as the bisulfite ion. Because of the absence of a specific acid and the volatility of sulfur dioxide, a strong dependence of the ionization on pressure, temperature, and concentration exists for the sulfite system in the presence of a base and, to a greater extent, in its absence (SO_2/H_2O). It is therefore of little use to determine the ionization for a narrow range of reaction conditions. A considerable number of physicochemical measurements are required before a thermodynamic solution for each base can be attempted. This should be carried out at temperature and pressure ranges which lie within technically significant cooking conditions. Another complication arises from the fact that decomposition reactions of the sulfite ions and reactions with the wood

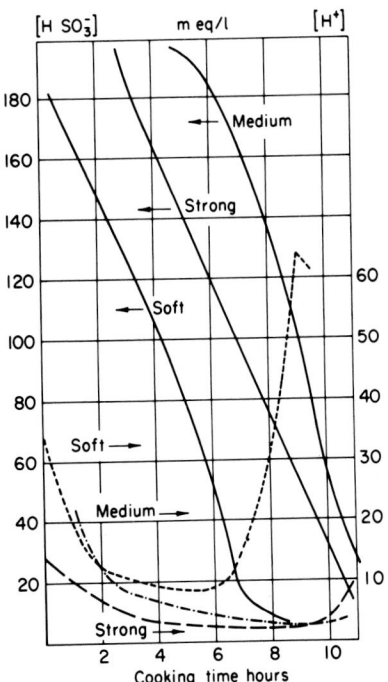

Figure VII-22. Changes in hydrogen ion and bisulfite ion concentration in three types of sulfite cooks (190).

components must be taken into account. The state of the alkali and alkaline earth ions is affected only a little, whereas ammonium reacts with lignin and other wood components. Bisulfite and sulfite ions, on the other hand, react with lignin; carbonyl groups in the reaction products show an increased acidity (lignosulfonation) or a considerably diminished acidity (carbonyl-bisulfite complex). Formic, acetic, and oxalic acids which are formed during the cook have a marked effect on bisulfite and alkaline sulfite cooks. Carboxyl groups become anionic in alkaline solutions. Such an anionic activity is not confined only to ions in solution but occurs also on the surface of the inside of the polyelectrolytes which make up the solid phase of the wood (181).

A comprehensive study of the sodium-base system in the sulfite pulping of black spruce, at between 115° and 175°C and pH 1 and pH 12, was achieved through the control of both the temperature and the pH in cooks with varying base concentrations to various yield levels. It was found that the pH is the controlling variable in all sulfite pulping and is followed in importance by the temperature. The bound sulfur dioxide and the yield show a limited influence, but only in certain areas. Small changes in the

pH may give rise to drastic changes in all significant pulp characteristics. The pH cannot be allowed to follow its natural course in a sulfite cook if control over the quality of the pulp is desired. Hence, pulps with any desired combination of properties can be obtained from sulfite cooks at controlled pH levels, including cooks in two and more stages. The difference between alkaline sulfite and kraft cooking liquors is apparent in the sulfur-containing chemicals, Na_2S and Na_2SO_3, which differ by three oxygen atoms (182a).

As was shown earlier, the damaging effect of large amounts of oxygen in the free gas space of the digester decreases with the increasing pH value of the sulfite cooking liquors used. This leads to an increased sulfate formation and, with it, to an increase in acidity which, in turn, interferes with the sulfonation reaction. The main effect of the oxygen lies in the change it causes in the cooking liquor, not in its direct influence on the cellulose (415,416). From these considerations it is obvious that the sulfur dioxide concentration cannot have any integrating effect on the quality of the pulp. On the other hand, the hydrogen ion concentration has a considerable effect. By varying the pH value during the cook, pulps of very different qualities can be produced. In general, tensile and burst strengths increase with a higher pH value of the cooking liquor during the pulping stage, while the breaking strength decreases. With a constant base-to-chip ratio, when the initial pH value varies and the cooking temperature has been chosen so that the delignification velocity remains as constant as possible, an increasing pH value gives a softer pulp when cooked for a high pulp yield, but the opposite is the case with a cook for a low yield. It has been found that the maximum strength properties of such pulps are obtained at pH 4. With sodium-base cooking liquors, the pH value can be kept constant by the continuous injection of base or acid throughout the cook. Experiments in this direction, carried out at a constant pH of 2.8, 3.0, and 3.2, showed strong variations in the composition of the cooking liquors and in the pressure in the digester, but the pulps themselves were of almost identical quality; only their viscosity showed a certain dependence on the acidity. Within the acidity range tested, the cooking time at a certain temperature to obtain constant yields was a linear function of the pH value. The amounts of base or acid required to maintain a constant pH gave a good indication of the buffer action of the digester contents, whereas, on the other hand, the formation of acids during the cook could be quantitatively determined by a continuous measurement of the pH (158,175,177,179).

5.2 Pressure, Temperature, and Time

Based on investigations of the kraft pulping process, it has been suggested that the reaction time and the temperature should be expressed as

a single variable. To do this, the well-known Arrhenius equation, in a modified form, has been used:

$$\ln k = B - A/T$$

in which k is the reaction velocity, T the temperature in absolute degrees, and A and B are constants. The reaction velocity at 100°C was chosen arbitrarily as the unit and all other values were calculated on the basis of this unit. When the reaction velocity, k, is plotted against the cooking time, the space below the curve represents the degree of decomposition reached at a given cooking period. Corresponding values for k for various temperatures were calculated from the Arrhenius equation, using 21,500 cal as the value for the activation energy and $k = 1.00$ at 100°C. When the reaction velocity at constant temperature is directly proportional to the partial pressure of the sulfur dioxide, then the velocity is expressed simply by $k(pSO_2)$. Relating this term to time then gives the calculated degree of cooking (414,437).

It is obvious that the effect of the acid concentration, of the degree of dissociation, of the pressure, the temperature, and the reaction time on the course of the cook cannot be strictly separated. The temperature affects not only the diffusion processes but also the dissociation and equilibrium conditions between the gaseous, dissolved, and combined sulfur dioxide. The effects of an increase in the sulfur dioxide content of the cooking liquor on the course of the cook were investigated in a series of experiments which comprised practically the total pH range of from 1 to 13. For this purpose a sodium hydroxide solution, of about 0.6 M, was increasingly enriched with sulfur dioxide. Starting with a purely alkaline cook, the regions of a neutral sulfite, a bisulfite, and an acid bisulfite cook were covered. The cooking time varied between 3 and 4 hours and the temperature maximum from 125°, 140°, 155°, to 170°C. Yields of 80% and higher were obtained, but these did not represent pulps in the usual sense. However, the results were interesting with regard to the production of semi-pulps and high-yield pulps. Figure VII–23 shows the pH range of a 0.6 M sodium hydroxide solution with increasing sulfur dioxide content; the curve A shows the conditions before the cook, curve B those at the end of the cook. All pH values were measured at room temperature and therefore give no direct indication of the acidity conditions during the cook under pressure and increased temperature. While a sharp decrease occurs in curve A on addition of about 0.3 mole SO_2 (formation of neutral sulfite), this drop takes place in curve B after an addition of only 0.18 mole SO_2. A second break appears in curve A after the addition of about 0.6 mole SO_2. This break does not appear in curve B, where a continuous decrease in the pH curve can be observed until the formation of sodium bisulfite. This phenomenon can be explained only by the formation of acids from the

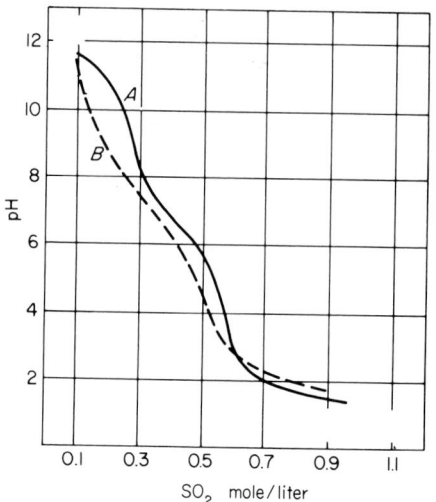

Figure VII-23. Change of pH of a 0.6 M NaOH solution containing increasing amounts of SO$_2$ (308). A = original solution; B = solution after sulfite cook.

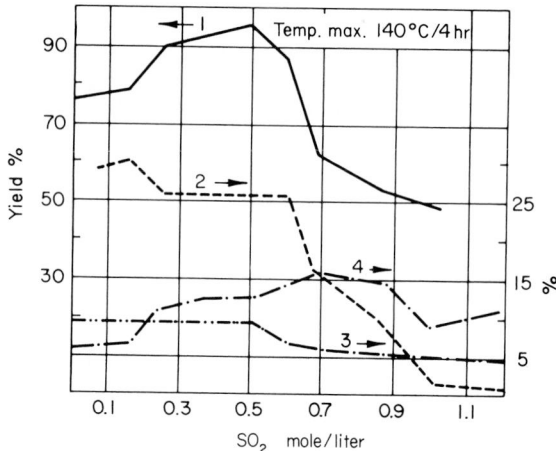

Figure VII-24. Yield and extraction of various wood components by cooking liquors with increasing SO$_2$ content (308). 1 = yield; 2 = lignin; 3 = pentosan; 4 = alkali-soluble components.

wood during the cook. So far as the pH values obtained at room temperature permit conclusions to be drawn, the cooking time plays only a minor role in the development of the pH. The temperature has a strong effect on the yield and delignification. Figure VII-24 shows that a considerable dissolution of wood substance occurs even in the still alkaline region. With decreasing pH, and especially in the region of the mono- and bisulfite for-

mation, this dissolving process comes to a temporary standstill and then increases strongly with the appearance of free sulfur dioxide in the cooking liquor. Here, again, the marked dependence of the delignification process upon the temperature is apparent (308,386).

In the prepartion of a Ross diagram, which is especially instructive for the description of the relationships just mentioned, the following figures are used for the composition of the wood, based on investigations of Swedish sprucewood:

Carbohydrate portion		65.8%
Difficultly hydrolyzable polyoses	8.3%	
Readily hydrolyzable polyoses	16.0%	
Cellulose	41.5%	
Lignin		28.0%
Various components		6.2%
Acetyl	1.4%	
Resins, ash, proteins, residue	4.8%	
		100.0%

The carbohydrates and lignin together amount to 93.8%. Within the diagram, the percentage relationship between carbohydrate and lignin contents in the pulp can be marked off in a "decomposition zone" which is valid, when the composition of the wood is the same, for all pulps and semipulps regardless of the method used for their production. The basic concept of the Ross diagram arises from the assumption that the pulping process is an extraction process. Its goal is the dissolution of all noncellulosic components, and this can be attained completely or only partially. It is obvious that a certain simplification of this very complicated process underlies this assumption, but such a diagram reveals a series of interesting concepts if one guards against generalized conclusions. Table VII–7 contains the figures which have led to the Ross diagram shown in Fig. VII–25 for the cook of sprucewood at various temperatures and pH values. Table VII–7 shows that the series of experiments starts far in the alkaline region; it enters the neutral zone, according to Fig. VII–25, at 0.3 mole sulfur dioxide, and, at 0.6 mole, reaches the conditions of an acid sodium bisulfite cook. The cooking time in all cases was 4 hours. The Ross diagram clearly shows the relationships. The curve constructed for each of the four temperature regions for the relationship between the lignin/carbohydrate (x/y) portion in the pulp and the yield $(x + y)$ gives an approximate idea of the alkalinity and/or acidity during the cook, and thus, from column 2 of Table VII–7, the mole proportion of sulfur dioxide in the cooking liquor can be found; it is shown graphically in Fig. VII–25. Each curve starts in the upper left or alkaline region, then turns to the right into the neutral region while a reversal at the point of the highest yield shifts it, with in-

TABLE VII-7

EFFECT OF SO_2 CONCENTRATION AND pH IN THE COOKING OF SPRUCE-
WOOD WITH SULFITE SOLUTIONS (308)

Time at max. temp.	SO_2 (mole/ liter)	Yield (%)	Pen-tosan (%)	Sul-fur (%)	TAPPI lignin (%)	CH$_3$O lignin (%)	NaOH soluble (%)	S in TAPPI lignin (%)	Residual Pen-tosan (%)	Residual TAPPI lignin (%)	Residual CH$_3$O lignin (%)
4 hr 125°C	0.0	78.4	11.2	0.00	30.2	35.0	5.3	0.0	8.8	23.7	27.4
	0.173	81.0	10.0	0.30	30.6	34.2	6.7	—	8.1	24.7	29.7
	0.275	94.5	9.5	0.71	25.3	24.5	12.4	2.8	9.0	23.8	23.1
	0.345	94.7	9.3	0.72	28.2	27.3	11.8	2.5	8.8	26.7	25.8
	0.425	95.8	9.1	0.72	28.5	24.5	16.3	2.5	8.7	27.3	23.5
	0.628	84.4	7.4	0.61	27.6	23.5	9.3	2.2	6.3	23.3	19.8
	0.641	85.5	6.7	0.89	29.2	24.8	19.8	3.0	5.7	25.0	21.1
	1.414	63.6	7.0	0.98	17.8	14.5	18.6	5.5	4.5	11.3	9.2
4 hr 140°C	0.00	76.6	9.4	0.00	29.3	34.4	6.3	0.0	7.2	22.4	26.4
	0.150	79.2	8.5	0.40	30.3	34.8	6.7	—	6.7	24.4	27.6
	0.263	90.6	8.9	0.85	27.0	26.9	11.1	3.1	8.0	24.5	24.4
	0.328	91.7	8.8	0.91	27.3	26.4	11.5	3.3	8.0	25.0	24.2
	0.369	94.0	8.9	0.94	27.0	26.0	12.6	3.5	8.3	25.3	24.5
	0.403	92.7	9.0	0.86	27.0	25.4	11.3	3.2	8.3	25.0	23.6
	0.506	93.6	9.0	0.85	26.9	25.0	12.0	3.2	8.4	24.8	23.4
	0.590	87.5	7.2	1.08	25.5	23.3	18.5	4.3	6.3	22.3	20.4
	0.688	62.4	6.4	0.83	16.2	19.9	15.3	5.1	4.0	10.1	12.4
	0.869	54.1	6.1	0.61	8.5	12.7	14.7	7.2	3.3	4.6	6.9
	1.075	48.8	5.0	—	1.8	4.2	8.7	—	2.9	0.6	2.0
	1.270	45.4	4.7	—	1.3	1.2	10.8	—	2.1	0.6	0.5
4 hr 155°C	0.00	69.6	8.5	0.00	27.8	32.6	7.9	0.0	5.9	19.3	22.7
	0.173	73.3	6.8	0.30	29.4	32.2	6.8	—	5.0	21.5	23.6
	0.278	86.0	9.1	0.74	26.1	24.6	9.3	2.8	7.8	22.5	21.2
	0.345	88.4	8.7	0.86	25.6	23.9	9.8	3.3	7.7	22.6	21.1
	0.455	89.6	8.7	0.84	24.2	22.4	9.7	3.4	7.8	12.7	20.1
	0.511	81.9	5.7	0.92	23.4	20.2	16.9	3.9	4.7	19.2	16.6
	0.617	54.2	7.5	—	7.4	10.0	9.1	—	4.0	4.0	5.4
	0.972	45.1	6.5	—	0.2	1.1	10.3	—	2.9	0.9	0.5
4 hr 170°C	0.00	65.5	8.7	0.00	24.3	30.4	7.8	0.0	5.7	16.0	19.9
	0.169	73.4	8.1	0.50	28.8	32.2	7.6	—	5.8	21.2	23.6
	0.275	81.0	9.0	0.72	24.6	31.0	9.3	2.9	7.3	20.0	25.1
	0.369	81.7	7.5	0.97	21.0	28.4	10.4	4.6	6.1	17.2	23.2
	0.445	82.3	7.1	1.01	20.9	26.4	12.1	4.8	5.9	17.2	21.8
	0.481	74.6	5.8	0.98	18.9	19.4	16.5	5.2	4.3	14.1	14.5
	0.681	43.7	5.3	—	0.9	1.5	10.0	—	2.3	0.4	0.7

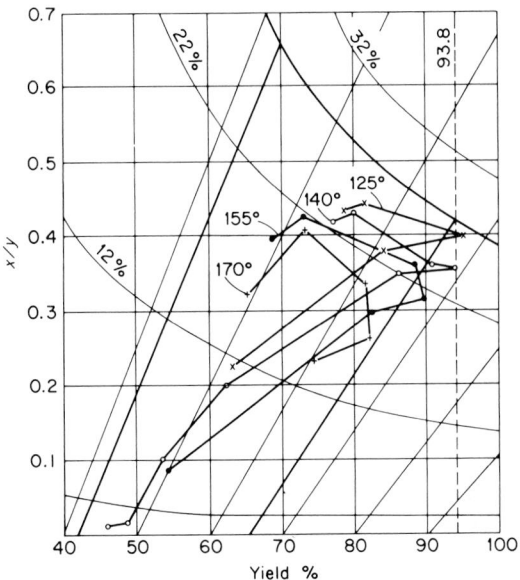

Figure VII-25. Ross diagram for the sulfite pulping of a spruce-balsam mixture at various temperatures and pH levels (386).

creasing temperature, steadily more into the acid region. An increasing dissolution of the hemicelluloses is connected with this reversal, but the dissolution of the lignin occurs later, in the region of greater acidity. Indeed the region of the semipulps and the high-yield pulps is passed through until an almost completely lignin-free pulp emerges (308,386).

A pulp obtained in a high yield was used in a subsequent bisulfite cook to study the effect of time, temperature, and the composition of the cooking liquor. The velocity of the delignification and the dissolution of some roughly estimated carbohydrate fractions were determined at temperatures of 135° and 180°C, pH regions of 1.5 and 4.0, and a sodium ion normality of 0.5 N and 2 N. Whereas the temperature and the pH value, determined with the cold cooking liquor, had a marked effect on the reaction velocities, no effect from the variation of the normality of the Na ions could be found within the regions of the experiment. In the bisulfite cooking process, the reactions between the wood constituents and the cooking liquor components are obviously not influenced by the final degree of decomposition. The total yield of pulp at any stage of cooking is, therefore, a function of the initial hydrogen ion concentration and the temperature of the cook. When the latter is raised, the effect of the initial hydrogen ion concentration becomes less, so that the assumption that, at higher temperatures, the composition of the original cooking liquor loses its significance seems to be justified. This general finding is valid for the dissolution of all wood components and equally so for the total yield. The

temperature and the composition of the cooking liquor showed no effect upon the carbohydrate/lignin ratio at any time in the cook. Over the wide range of conditions used, no indications were found that the experimental conditions of a bisulfite cook could be changed to effect a better dissolution of the lignin (436).

The pH measurements on fresh cooking liquor prepared with different bases and on the spent liquor, at temperatures of up to 120°C, show approximately the same development with increasing temperatures. This is true for cooking acids with calcium, magnesium, and sodium base and, with certain limitations, also for fresh cooking liquor with ammonium base. The difference in the initial sulfonation velocity at 90°C can be ignored even if, at this temperature, certain differences in pH have been found. The hydrolysis velocity for the lignosulfonic acids and hemicelluloses and, with it, for the pulping velocity in general is governed solely by the acidity of the cooking acid (214). Ross diagrams of quick laboratory cooks at high cooking temperature and with an acid bisulfite, a bisulfite, and a sulfite process showed that, during the 15 minutes required to heat the digester up to 140° to 150°C, the sulfonation and the dissolution of the part of the lignin that is not combined in the lignin-carbohydrate complex were complete. During the next 10 minutes of cooking time while the temperature was raised to 160° to 170°C, the course of the delignification varied, depending upon whether the cook was carried out with an acid bisulfite cooking liquor or with one with a higher pH value. In the acid bisulfite cook, the cleavage of the lignin-polysaccharide bonds and dissolution of the hemicelluloses occurred, while the lignin content remained constant. Further quick sulfonation and dissolution of the lignin occurred only after complete splitting of the complex bonds. From this it appeared that quick acid bisulfite cooks at high temperatures are not suitable for the production of high-yield or semipulps. On the other hand, with bisulfite cooks and sulfite cooks in the higher temperature region, and with acid bisulfite cooks at lower temperatures (140°C maximum), the liberation of the lignin from the lignin-polysaccharide complex takes place simultaneously with the delignification and, subsequently, the hydrolysis of the hemicelluloses and the solution of the lignin also occur at the same time (34).

6. EFFECT OF OTHER DEGRADATION REACTIONS ON THE COURSE OF THE SULFITE PULP COOK

6.1 Reaction Mechanism of the Formation of Thiosulfate

The role of the thiosulfate in a sulfite pulp cook has already been briefly discussed in connection with the hindrance of the sulfite pulping

by a prior phenolation of the lignin (165). Thiosulfate is formed when an alkaline barium lignosulfonate solution is heated at 150°C (319). It can be concluded from model experiments that the thiosulfate formed disappears again as the result of its reaction with lignin during the cook; it is assumed that this reaction consists in a transformation of benzyl alcohol and benzyl ether groups into benzyl sulfide groups (127). The reaction of the thiosulfate with the sulfonatable groups of the lignin causes a retardation of the dissolution of the latter, which, in the first stage of the cook, also acts as a retarder to the dissolution of the carbohydrates. The reaction of the lignin with thiosulfate takes place more rapidly than that with the bisulfite of the cooking liquor. The sulfidic groups thus formed do not render the lignin hydrophilic to the same degree as do the sulfonic groups (155). Thiosulfate affects the stability of the cooking liquor; formic acid and terpenes react with bisulfite with the formation of thiosulfate. Because a certain amount of thiosulfate is always formed in the modern regeneration plants for sodium-base spent liquor, the initial content of thiosulfate in the cooking liquor is of great importance. The sulfur dioxide content decreases through the decomposition of the cooking acid, with an increase in sulfate and the separation of elementary sulfur. Calcium sulfate can precipitate from calcium-base cooking liquors.

Investigations dealing with the changes in cooking liquors with various contents of thiosulfate have shown that the decomposition of acid sulfite cooking liquor is accelerated by increasing temperature and thiosulfate content. The addition of sugars has no marked effect on the stability of the cooking liquor. The storage of bisulfite and acid bisulfite liquors under the same conditions showed that the former is considerably more stable than the latter. When a sodium bisulfite liquor (pH 4.5) is stored for 2 days at 105°C the amount of base and of thiosulfate remain unchanged. The reason for this must be that the formation of thiosulfate as well as the initial decomposition of the cooking acid depend on the hydrogen ion concentration (155). The effect of thiosulfate and of tetrathionate upon the decomposition of sulfurous acid is practically the same. These two components are always found in solutions which contain sulfur dioxide. The presence of these substances in small quantities may even be advantageous for the course of the cook. In larger amounts, however, they cause an increased condensation of the lignin (368). Model experiments with vanillyl alcohol and thiosulfate have shown that the reaction mechanism is very similar to that of the sulfonation with sulfite (127,128).

Investigations on the reaction of the thiosulfate in the cooking liquor with the wood were carried out with inactive and radioactive thiosulfate. In this way, a decrease in the thiosulfate content in the cooking liquor was found right at the beginning of the cook; however, there was no increase in the polythionate content. The decrease in thiosulfate was extensively

dependent on the base content of the cooking liquor. At pH 4 no decrease was found. An increase in sulfate content went parallel with the decrease in the thiosulfate so that, at high base content, the decrease in thiosulfate was equivalent to the increase in the sulfate content. The reason for the low thiosulfate content is perhaps that the formation of excess organic sulfur takes place considerably more rapidly than the reaction between aldoses and bisulfite. On the other hand, it may be assumed that the formation of thiosulfate and sulfate via polythionate and sulfite continues slowly throughout the sulfite cook. Investigations into the decomposition of acid sodium bisulfite solutions with an increase in temperature from 110° to 148°C and with a different base, sulfur dioxide, and different thiosulfate concentrations showed that the reaction constant is strongly temperature-dependent, since it increased fourfold at a temperature increase of 10°C. Deviations from the kinetic equations were found at high thiosulfate ion and high H ion contents; this may be traced to the fact that, in these cases, the course of the reaction is of second or higher order. The formation of thiosulfate from bisulfite and aldoses is also dependent on the temperature, and with an increase of 10°C the value for the reaction constant almost doubles (339,340). With acid sodium bisulfite and sodium bisulfite cooks with various additions of thiosulfate and at different temperatures it could be confirmed that the thiosulfate is formed during most of the cooking time via aldose and bisulfite and that, first toward the end of the acid bisulfite cook, the formation of thiosulfate via polythionate and bisulfite begins to be preponderant, while that via aldose and bisulfite becomes slower. With increasing thiosulfate content in the cooking acid, its formation via polythionate and bisulfite gains in importance. Sulfate ions are formed in the reaction between thiosulfate and the components of the wood. This sulfate formation corresponds to the decrease in thiosulfate and justifies the assumption that sulfide groups are formed in the reaction between wood and thiosulfate (341).

The effect of thiosulfate on the sulfite cook has been studied for one-stage and two-stage cooks. In one-stage cooks, the effect of the thiosulfate was examined in the production of pulp for paper (1.36% Na_2O, 6% total SO_2, 135°C), of dissolving pulp (1.00% Na_2O, 6% total SO_2, 148°C), and of bisulfite cooks (1.92% Na_2O, 3.96% total SO_2, 155°C). Two-stage cooks for dissolving pulps were carried out with pinewood (pH 5.6 in the first stage, followed by an acid second stage). The addition of thiosulfate to the cooking liquor resulted in a retardation of the delignification in the shortest cooking times and in one-stage cooks. In bisulfite cooks, on the other hand, the delignification velocity increased during the final stages when thiosulfate was added to the cooking liquor. In the final stages of the cook, therefore, the variation in the lignin content of the pulps was always less, regardless of whether the cook was carried out with or without the addition

of thiosulfate. No such uniformly rapid decrease in variation of the lignin contents was found in bisulfite cooks. In two-stage cooks, the addition of thiosulfate resulted in a decrease in the cooking time in the second stage if the cook was terminated when a constant viscosity of the pulp was attained. The Roe number, however, was not affected.

During the first stage of such a two-stage cook, the thiosulfate content of the cooking liquor decreases slowly. While the formation of thiosulfate can be ignored under the prevailing cooking conditions, the small decrease in thiosulfate content in the first stage of the cook indicates that the formation of sulfide and similar groups in the lignin takes place very slowly, certainly much more slowly than the formation of sulfonic acid groups. In the second stage of the cook a fast reaction takes place between the thiosulfate and the lignin and dissolved lignosulfonic acid. This is shown by the rapid decrease in the thiosulfate content in the second stage. In this stage a rapid decomposition of the bisulfite results from the high thiosulfate content in the cooking liquor, and this results in a rapid increase in the H ion concentration. The increasing H ion concentration, in turn, leads to a rapid dissolution of the lignin, as in a one-stage cook (339,340,341,342, 367).

The reactions of the bisulfite ion in an acid bisulfite cook were studied by labeling the sodium bisulfite in the cooking liquor with radioactive isotope ^{35}S and gel filtration of the radioactive spent sulfite liquor. Only three-quarters of the sulfur in the cooking liquor reacted with the formation of compounds which contained so-called loosely combined sulfur dioxide. About 13% of the sulfur reacted with the formation of organic compounds which contained firmly combined sulfur. About 66% of the sulfur was combined to the lignin, while the remainder was attached to about a dozen ligninlike and nonaromatic compounds. In an acid bisulfite cook, the bisulfite ions took part in numerous unknown side reactions (193).

6.2 Other Sulfur Compounds and Degradation Products; Ion-Exchange Effects

The decomposition of the cooking liquor can also be brought about by formic acid, which reacts as a hydroxyaldehyde with the bisulfite. At a low lime content in the cooking liquor, even a small amount of formic acid is sufficient to exert a noticeable effect (378). The stability of the cooking liquor can also be affected by selenium, reducing sugars, terpenes, and by finely divided sulfur (155,266). Other sulfur-containing compounds in addition to sulfonic acid groups are formed in sulfite cooks. The amount of this so-called "excess sulfur" is, to a great extent, dependent on the decomposition of the cooking liquor. From the ratio of the sulfuric acid formed to the excess sulfur, conclusions can be drawn as to the oxidizing

and reducing action of the compounds which accelerate the decomposition of the cooking liquor; it also shows that reducing compounds are chiefly responsible, the action of these being more pronounced in buffered than in unbuffered solutions. At a low pH, of about 1.4, a slow dehydration of hexoses occurs with the formation of levulinic acid from glucose and compounds of sulfonic acid-like character from xylose (302,368).

The ion-exchange capacity of lignosulfonic acids has already been mentioned. Dilute solutions of simple and completely dissociated acids and bases are originally unbuffered, i.e., their ion reserves are exhausted. When wood chips, cellulose, or lignin is added to such solutions, a marked change in the ion concentration occurs. This change takes place in a few seconds when these ion-exchangeable substances are present in a sufficiently finely distributed state, regardless of whether the solution used has a high or low pH (178). The ion-exchange capacity of sulfite pulp is a function of the kappa number of the pulp, and the degree of sulfonation of the lignin remaining in the pulp stays constant over a wide kappa number region. Sulfur determinations of pulps have shown that those with a low kappa number contain, in comparison with their exchange capacity, more sulfur than corresponds to their sulfonic acid group content. The amount of excess sulfur is particularly high in bisulfite pulps and, as mentioned before, depends on the decomposition of the cooking liquor (363).

7. EFFECT OF VARIOUS BASES ON THE SULFITE PULP COOKING PROCESS

7.1 Comparison of Pulps from Cooks with Different Bases

The effect of cation exchange upon the solubility of sulfurous acid in cooking liquors has been described in Section 2, while the effect of the acidity has been treated in Section 5. To avoid repetition, only the specific action of cations on the sulfite cooking process will be discussed here.

It should be stated first that it is also possible to carry out a sulfite cook without the presence of a base. This, however, is not successful under the usual cooking conditions, but requires low temperatures and long cooking periods to avoid burnt cooks which would result at high acid concentrations. Such cooking conditions affect the economy of the process (156). It is interesting to note that alcohols and even acetone can replace the base. A cooking liquor consisting of a 1:1 mixture of methanol and water and containing 5.5% sulfur dioxide gives a good result in 5 hours at 130°C (343).

A comparison of the various bases used shows, first of all, a greater variability in the use of soluble bases, such as sodium and ammonia (345). Sodium- and ammonium-base cooking liquors are quite similar in their

action, whereas magnesium occupies a middle position between them and calcium. The range of application with regard to the various wood species is wider in the case of water-soluble bases than with calcium-base liquor. Comparative cooks with different wood species have been carried out with calcium-, magnesium-, sodium-, and ammonium-base cooking liquors. The goal was to produce pulps for artificial fibers (dissolving pulps). The unbleached pulps obtained from the cooks with sodium- and ammonium-base liquor had more favorable properties with regard to permanganate number, viscosity, degree of delignification, brightness and alpha cellulose content than those from calcium- and magnesium-base liquors. The pulps prepared with sodium- and ammonium-base cooking liquors had similar viscosities and brightness. When the unbleached pulp viscosity was used as the criterion, the relative cooking time with the four bases increased in the following order: calcium, magnesium, ammonium, and sodium; when the bases were arranged according to the pulps with decreasing α-cellulose and on the basis of equal viscosity, the following order was obtained: ammonium, sodium, magnesium, and calcium (149,345). Sodium- and ammonium-base cooking liquors gave fewer splinters in the pulp, although no marked difference in the course of the delignification reaction could be observed. The greatest dissolution of pentosan was obtained with magnesium-base cooking liquors. No basic differences were found in the strength properties of the various pulps (387). In agreement with statements reported earlier, laboratory cooking experiments with sodium-base cooking liquors at a pH range of from 1.4 to 11.0 showed that pulps obtained in the neutral sulfite region had the highest strength properties. Sodium bisulfite cooks and those with excess sulfur dioxide gave pulps with the highest degree of brightness (100 a,230,308).

Comparative cooks with calcium-, sodium-, and magnesium-base liquors to equal yields of 60% to 65% showed that the soluble bases gave pulps with considerably better physical properties in a 20% to 38% shorter cooking time. The sulfite spent liquors contained 0.5% to 0.6% more sugar, calculated on the basis of dry wood. It is assumed that the shorter cooking time is due to a better impregnation of the wood chips by the sodium- and ammonium-base liquor (54). Whereas in the early days of sulfite pulping it was necessary to cook spruce- and pinewood by different pulping processes, today, by modification of the process, as, for example, by the use of pure bisulfite or neutral sulfite cooks, it is possible to pulp pinewood, too, by a sulfite process. Because pinewood has a higher density (0.41) than spruce (0.40), a higher yield of pulp could be expected. Pine generally gives pulps with higher tearing strength but with lower tensile and burst strengths (69).

Deciduous woods, especially birchwood, can be cooked advantageously

by the sulfite process. Comparative cooks with calcium-, magnesium-, and sodium-base acid bisulfite liquors and with magnesium and sodium bisulfite liquors with an initial pH of 3 to 6 gave pulp yields of 50% to 70%. Pulps from birch, prepared by the acid bisulfite process with cooking liquors of different bases, showed no substantial differences. The selectivity with regard to delignification is lower with calcium-base cooking liquors than with the soluble-base liquors. The bisulfite cook with magnesium-base proceeds somewhat more rapidly than with sodium-base liquor within the comparable kappa number region. With sodium bisulfite cooks at different pH, an increase of the initial pH value causes a clear retardation of the delignification. Bisulfite pulps obtained at pH 6 have a lower tearing strength, but, in general, they are markedly stronger and lighter in color than neutral sulfite pulps if the yield is above 53%. If a considerable improvement of the pulp properties in the bisulfite cook as compared to the sulfite cook is desired, the pulps must be cooked to a higher yield (about 60%). The neutral sodium sulfite cooking process (or NSSC) gives the highest yield of bleached pulp (about 60%), followed by the kraft pulp process (53%), the bisulfite (51%), and the sulfite (49%). The bleached kraft pulps of birch are visibly superior to the birch pulps obtained by the other processes (195).

In a series of investigations with Norwegian hardwoods (birch, aspen, and alder), the effects of various cooking processes were tested with regard to yield and quality of pulps obtained. For this purpose the following cooks were carried out: normal calcium bisulfite cooks at maximum temperatures of 118° and 135°C; a one-stage sodium bisulfite cook at an initial pH of 4 and at a maximum temperature of 155°C (the so-called Arbiso process); a magnesium bisulfite cook (Magnefite process); finally, a two-stage sodium sulfite cook was carried out at a pH of 9.7 in the first stage, then, after 1 hour at 125°C, the cooking liquor was drawn off and the cook was finished in the second stage with a conventional calcium bisulfite cook at 130°C. For comparison, a kraft cook at a maximum temperature of 160°C was performed. For pulps with the same chlorine number, alder required the longest cooking time at the maximum temperature, while aspen was delignified most readily and birch took a middle position. The kraft process gave the highest yield with all three woods. The bisulfite and the two-stage cooks gave the highest pulp yields of the sulfite processes. The chlorine consumption in the bleaching and the alkali consumption in the extractions between the bleaching stages were higher than with spruce, in agreement with earlier findings. All pulps had good drainage properties. The beatability was better with aspen than with alder pulp. The kraft pulps of all three woods had considerably higher tensile strength than the corresponding sulfite pulps (407).

7.2 Sodium-Base Sulfite Cooks

Sodium-base cooks require considerably shorter cooking times than the conventional calcium bisulfite cooks. With regard to yield, conflicting results have been reported. In evaluating the quality of the pulp, the cooking conditions used must be taken into consideration. Undoubtedly, sodium-base sulfite cooks can produce pulps of any desired degree of pulping and over a wide range of yields. The viscosities of the pulps of comparable degrees of cooking are higher and the filterability of the viscoses from these pulps is better than of those from calcium-base cooks. The cooks can be heated up more rapidly since the danger of burnt cooks is lessened. The removal of pentosans is generally smaller than in calcium-base sulfite cooks. Mixed cooks with dolomite calcium-base liquor and sodium-base liquor have been carried out successfully (89,187,267,346,347,366).

The wide range of pH values possible with sodium-base sulfite cooks is shown in a Ross diagram for poplar in Fig. VII–26. A similar diagram for spruce is shown in Fig. VII–25, in Section 5.2. The diagram for poplar shows that the total yield reaches its maximum within the pH region that includes the presence of mono- and bisulfite in the cooking liquor. The lignin content decreases at first in the alkaline region, increases again in the neutral region, and drops continuously again in the acid region (205,230). The high-yield pulps are therefore chiefly obtained in the region of the neutral zone (80,81,257,308).

An increase in the ratio of chemicals to wood results in a more rapid delignification, whereas the dissolution of the carbohydrates remains unaffected. In this way, higher yields, shorter cooking periods, and pulps with better strength properties are produced; at equal cooking times, a softer pulp is obtained. An increase in the concentration of the chemicals in the cooking liquor gives the same results at the critical liquor-to-wood ratio of 2:1. Below this ratio, a burnt cook readily results (158). A comparison of an acid calcium bisulfite, a sodium-base bisulfite, and a sulfate cook of various wood species and their mixtures showed that the calcium-acid bisulfite cooks and sodium bisulfite cooks gave the same yields. The "defibering point" of the cooked chips was higher in the sodium bisulfite cooks than in the corresponding calcium-base cooks. Under equal cooking conditions, brightness and strength properties were better with the pulps from the sodium bisulfite cooks. The sulfate cooks, on the other hand, gave pulps with higher strength properties. A disadvantage of the sodium bisulfite cook was the greatly increased separation of rosin. The content of reducing sugars was lessened by 30% in the spent liquor of the cooks with the sodium-base cooking liquor (159). A pulp with 91% α-cellulose and a viscosity of 25 cp was obtained in a yield of 41% to 42% in a cook

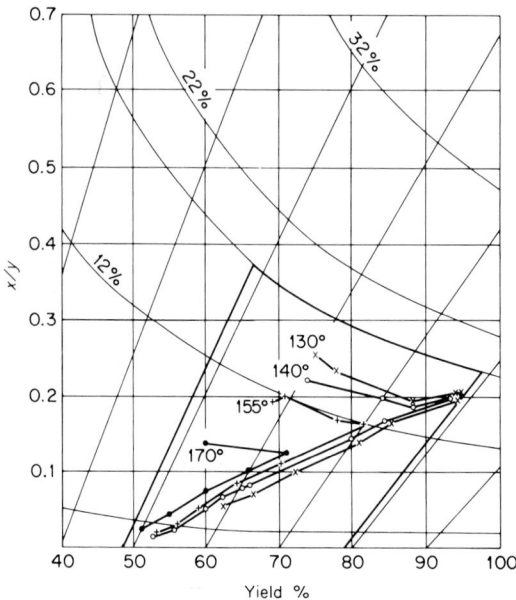

Figure VII-26. Sulfite cooking of poplar at various temperatures with varying pH levels (386). x = lignin content in pulp; y = carbohydrate content in pulp.

with a cooking liquor with 5% total sulfur dioxide, about 1% sodium-base-combined sulfur dioxide, and at a maximum temperature of 145°C for 9 hours. An increase in temperature and curtailed cooking time resulted in an inferior pulp (219).

The delignification in an acid sodium bisulfite cook at pH 1.35 and in a neutral cook at pH 6.0 was most successful; within the pH region of 3.0 to 5.0 the delignification occurred more slowly, but in this region the pentosan content of the pulps was highest and the pulps showed the best strength properties (409).

Monterey pine (*Pinus radiata*) was pulped with sodium bisulfite cooking liquor at pH 4 and 18% to 21% bisulfite, on the basis of oven-dry wood, for 2.5 hours at 165°C and gave yields of 47% to 50% screened pulp. The pulps had kappa numbers of about 30; they had good brightness and tensile strength but low tearing strength. Cooks of the same wood with calcium bisulfite cooking liquor for periods of 8 to 9 hours gave yields of 47% to 50% screened pulp, with kappa numbers ranging from 20 to 30 (404a).

Sitka spruce was pulped by the single- and two-stage processes and the advantage of the latter over the former in retaining a higher proportion of glucomannan was found to be relatively small at kappa numbers of below 30. An advantage of the two-stage cook of greater practical importance, however, was the completely uniform delignification throughout

the chips at kappa numbers of as high as 78. No uncooked centers remained even at this low degree of delignification (437 a).

The Arbiso process is a cooking process which operates with a sodium bisulfite liquor without free sulfur dioxide at a pH of 4 or higher. This process has proved to be a good one, especially for deciduous woods and mixtures thereof (80,81,257).

7.3 Ammonium-Base Cooking

A comprehensive documentation of the use of ammonium as base for a cooking acid for the period of 1900 to 1953 has been published (161, 171). From this it can be seen that the idea of using ammonia as base for a sulfite pulping liquor is at least as old as the use of sodium or magnesium. With regard to the problems concerning the apparative device for the preparation of the cooking liquor, reference is made to Section 2 (379). In general, no major changes need be made in the installation for conversion to ammonia (226). The heat liberated during the absorption of ammonia by water requires control of the temperature in order to obtain the required strength of the cooking liquor (129). Generally speaking, the replacement of sodium by ammonia promises an increased production and an improvement in the quality of the pulp from coniferous woods. One exception is in the brightness of the pulp, which is usually lower (226,375,387). The cooking velocity of an ammonium bisulfite liquor of pH 2.8 is considerably greater than that of a similar monosulfite solution. Complete pulping is obtained in 4 hours at 140°C with a cooking liquor containing more than 5% total sulfur dioxide. At 150°C maximum temperature, the total sulfur dioxide content can be lowered to 2.5% to obtain the same result. Pulps cooked under acid conditions have a lower holocellulose content than pulps cooked under neutral conditions, but they have a higher lignin and a higher mannan content. Pulps cooked under acid conditions to a yield of approximately 50% are definitely superior in strength properties to those obtained from neutral cooks (281). The special advantages of the ammonium-base cooking liquor are the faster impregnation of the chips with the cooking liquor and the ready solubility of the ammonium salts. The variation in the course of the pulping in accordance with the reaction conditions and with the choice of the type of wood is obvious. The determination of the end-point of the cook is difficult because the spent liquor quickly turns dark toward the end of the cook. Resinous woods consume a considerable amount of the ammonium ions (200).

The conditions for the cooking of deciduous woods, especially beech, are less favorable. The pulps contain a large proportion of splinters. Larch, on the other hand, is readily cooked and, when impregnated long enough (2 to 2.5 hours at 105°C), even its heartwood can be pulped without diffi-

culty (274,282). Under the same cooking conditions, the four bases show no essential difference with regard to yield. In spite of reports to the contrary, no effect of the base on the α-cellulose and strength properties has been observed (387). An increased consumption of chlorine during the bleaching may be connected with the poor initial brightness of the ammonia base pulp and the strong discoloration of the cooking liquor toward the end of the cook (375).

7.4 Magnesium-Base Cooking

The use of magnesium-base cooking liquors, as such, suggested itself already as a result of the occasional use of dolomite for the preparation of the liquor. The dissociation and solubility data for a total sulfur dioxide content of from 0.5 to 1 M permit the calculation of the ion equilibriums of a magnesium-base cooking liquor at 25°C and allow the ion distribution in the liquid phase to be predicted as a function of pH and concentration (206). The better solubility of the magnesium bisulfite as compared with calcium bisulfite permits a lowering of the acid/chip ratio to 2.5:1, so long as a sufficient liquor circulation in the digester has been provided for. In this way, the concentration of organic substance in the spent liquor is increased, and this increase results in a decreased steam consumption in the further processing of the spent liquor. Magnesium sulfate, as opposed to calcium sulfate, is readily soluble and this precludes the precipitation of incrustants so often experienced in the customary sulfite process. In the regeneration process, magnesium sulfate is converted into magnesium oxide. Whereas the sulfur consumption of a pulp mill operating by the magnesium-base process without chemical recovery amounts to 90 to 100 kg per ton pulp and the magnesium oxide consumption to 65 kg per ton, their consumption in a mill with complete recovery (which is required today because of water pollution regulations) amounts to not more than 11 to 12 kg per ton and 4 kg per ton, respectively (72,160). Questions concerning the processing of the spent liquor and the recovery of the chemicals will be dealt with in Section 9. Pulp yields of 60% based on the wood can be obtained within a pH range of 3 to 5 with cooking temperatures varying correspondingly. The pulps have a high brightness and improved strength properties as compared with pulps obtained from acid calcium bisulfite cooks. The increasing use of magnesium-base cooking liquor is due to the fact that the magnesium-base sulfite spent liquors can be processed relatively simply, with the recovery of the chemicals, and that, in this way, the problem of the disposal of the waste liquor is solved in an economical way. Some magnesium bisulfite cooking processes, with recovery of the chemicals, are known by the term "Magnefite" processes (32,62,125,306,401).

It must be noted that this type of process is the result of the efforts of

many pulp mills. All have in common the use of magnesium as base, but the procedures may vary in the different mills. The acid magnesium bisulfite process was developed not only to overcome the difficulties involved in the disposal of the spent liquor of the conventional calcium-base cooking process but also because it made possible the recovery of the chemicals. The acid magnesium bisulfite cook is limited to use with certain wood species. The chief advantage of the Magnefite process, on the other hand, is that it can be used for practically all wood species, such as pine, and even for dense deciduous woods which had previously been considered unsuitable for the sulfite process. The special characteristic of this process is the use of an almost pure magnesium bisulfite cooking liquor; in other woods, it contains practically no free sulfur dioxide. Such magnesium bisulfite solutions have a pH of between 3 and 4, in contrast to calcium bisulfite cooking liquor which, because of the presence of free sulfur dioxide, has a pH of between 1 and 2. This pH value of 3 to 4 is maintained during the cook by strict control of the pressure in the digester. The cooking temperature is higher than in the acid calcium bisulfite cook. In this way, the sulfonation of the lignin is accelerated and the cooking time can be shortened. The yields are generally higher, and the strength properties of the pulp are markedly better when compared with those of the pulp obtained by the calcium bisulfite process from the same wood and at the same degree of pulping. The pulps can be bleached to a higher brightness with the same amount of chlorine because of the higher brightness of the unbleached pulps.

The southern pine species, which do not respond to the conventional sulfite cooking process because of the phenolic extractives of their heartwood, were readily pulped with magnesium bisulfite cooking liquor. The pitch problem commonly observed in nonalkaline pulping processes was avoided by storing the logs for several months before the cook. Such storage often results in blue staining of the wood but, in spite of this, the unbleached sulfite pulp was usually much brighter than kraft pulp from the same wood. The pulp was readily bleached in three stages by the use of chlorine, caustic extraction, and hypochlorite. Under favorable experimental cooking conditions, no sticky pitch deposits were noted. The pulps from pinewood were not consistently either stronger or weaker than the acid sulfite pulps from sprucewood, and the southern pine pulps should be competitive with northern coniferous pulps wherever the coarser nature of the fibers of the pinewood is not undesirable (200a).

A two-stage magnesium cooking process has been developed in order to produce pulps with particularly good strength properties. This two-stage process operates at pH values of between 3 and 4 in the first stage, and at pH 5.5 to 6.5, through the addition of magnesium oxide, in the second stage. The sulfonation takes place chiefly during the first stage, whereas the increased pH in the second stage contributes to the preservation of the

TABLE VII-8

COMPARISON OF VARIOUS COOKING PROCESSES—INVESTIGATIONS OF UNBLEACHED PULPS (44)

Type of cook	Cooking conditions						Yield		Cl no.	C.E.D. viscosity	G.E. brightness	Strength properties at 450 Can. Standard Freeness					
	First stage			Second stage			Screened	Rejects				Beating time (min)	Tensile (km)	Burst factor	Tear factor	Burst +1/2 tear	Fold
	pH	Max. temp. (°C)	Time at temp. (hr)	pH	Max. temp. (°C)	Time at temp. (hr)											
Two-stage acid	5.33	166	2	1.5	140	1.75	55.8	0.1	6.5	30.0	64.7	16	11.5	64	66	97	650
	5.35	166	2	1.5	140	2	57.05	0.1	7.4	37.4	65.5	20	12.2	67	66	100	790
Magnefite	3.2	166	2.75	—	—	—	51.7	0.1	5.2	38.5	68.1	31	12.1	78	86	121	840
	3.55	166	2.5	—	—	—	52.9	0.1	7.0	41.4	65.1	27	12.8	82	73	119	1000
	3.55	166	2.5	—	—	—	53.4	0.1	8.7	51.4	66.9	27	12.3	75	88	119	980
Two-stage Magnefite	3.72	166	1.5	6.6	175 (Max. temp.)	2 (Time at temp.)	51.2	0.1	6.1	27.0	60.8	29	13.1	92	85	135	1720
	3.50	166	1.5	6.5	175	1.75	52.2	0.03	7.6	29.3	54.3	28	13.1	83	78	122	1380
	3.55	166	1.25	6.7	175	2	53.4	0.2	8.3	29.3	55.1	28	13.0	87	87	131	1400
Kraft	14 (Active alkali)	31.7 (Sulfidity)	177 (Max. temp.)	1.25 (Time at temp.)			47.3	0.1	4.7	25.3	30.2	54	13.8	101	118	160	2560
	13	30.8	177	1.25			47.9	0.3	7.3	28.7	24.8	62	13.6	111	114	168	1730
	13	30.7	177	1.25			46.8	0.2	5.7	32.1	27.4	61	14.0	98	117	157	3000

pulp by reducing the hydrolysis of the hemicelluloses and certain cellulose portions. The pulp produced by this two-stage process has excellent strength properties and the yield is not lower at the same degree of pulping than that from a one-stage process.

Another two-stage cooking process, which is a kind of reverse of that described above, is one in which the cook is carried out in the first stage with a magnesium monosulfite-bisulfite cooking liquor at a pH of between 5 and 6, and, in the second stage, with an acid magnesium bisulfite solution. This two-stage process gives especially high yields (44,45).

The two-stage magnesium-base processes have considerably broadened the quality range for pulps from the same wood. In Tables VII–8 and VII–9 the results of a series of comparative cooks are given to show the different effects of one-stage and two-stage cooking processes on the yield, chlorine number, viscosity, brightness, and strength properties of the unbleached pulps (Table VII–8), and of bleached pulps of some selected cooks (Table VII–9) (44). Table VII–8 shows that, at first, two-stage magnesium-base cooks were carried out, the first stage at pH 5.3, the second at pH 1.5. In a second series, one-step Magnefite cooks at pH 3.2 and 3.55 were performed, and in another set, two-stage cooks at pH 3.5 to 3.7 and then at pH 6.5 to

TABLE VII-9

Comparison of Cooks Selected from Table VII-8—Investigation of
Bleached Pulps Obtained in Four- and Five-Stage
Bleaching Processes (44)

Cook No.	Un-bleached yield	Cl no.	Bleached yield	Bleached brightness	Bleached viscosity	Beating time	Burst + 1/2 Tear	
							Un-bleached	Bleached
A. Pulps bleached in five bleaching stages								
Two-stage acid	57.0	7.4	52.75	90.1	28.5	19	100	105
Magnefite	53.4	8.7	48.9	89.6	38.2	31	119	123
Two-stage Magnefite	52.2	7.6	48.1	90.8	23.6	35	122	127
Kraft	46.7	5.7	43.2	88.6	23.5	39	157	152
B. Pulps bleached in four bleaching stages								
Magnefite	51.7	5.2	47.6	88.9	32.7	31	121	105
Magnefite	52.9	7.0	48.5	89.3	30.5	29	119	110
Two-stage Magnefite	51.2	6.1	48.1	89.7	25.4	30	135	137
Two-stage Magnefite	53.4	8.3	49.0	90.2	26.0	28	131	125
Kraft	47.3	4.7	45.5	87.7	13.7	41	160	147
Kraft	47.9	7.3	44.8	86.5	17.9	43	168	156

6.7 were carried out. For comparison, some kraft cooks were included, with the cooking conditions being given in the table. Especially striking are the variations in yield from process to process. The yields in the magnesium-base cooks varied from 52% to 57%, those in the kraft cooks from 47% to 48%. The yields of the one-stage Magnefite cooks amounted to 52% to 53%, those of the two-stage magnesium-base cooks to 56% to 57%. The brightness of the pulps from magnesium-base cooks was much better than that of those from the kraft cooks. All the cooks were carried out with sprucewood. The strength properties and beating data to reach a comparable degree of beating showed that those of the kraft cooks had the highest values, those of the two-stage acid magnesium sulfite cooks the lowest, while those of the Magnefite cook lay in between.

Pulps from selected cooks of this series were bleached by multistage bleaching processes, with the results listed in Table VII-9. The pulps from magnesium-base cooks required a chlorine consumption of 114% to 125% of the corresponding chlorine numbers in order to obtain a G. E. brightness of 89% to 91%. The kraft pulps, of the corresponding chlorine numbers, on the other hand, required 178% and 234% of chlorine to obtain a brightness of 86.6 to 88.6 G. E. Not only the total bleach consumption but also the chlorine dioxide consumption was lower for the magnesium-base cook pulps than for the kraft pulps. The sequence in the four-stage bleaching was chlorination, alkaline extraction, chlorine dioxide, and hypochlorite; that in the five-stage process was chlorination, alkaline extraction, chlorine dioxide, alkaline extraction, and hypochlorite (44,45).

It is said that magnesium-base sulfite pulps are specially suitable for the production of newsprint. The increase in the yields is considerable and may amount to 10% to 12%. The amount of splinters decreases by over 60%. The higher strength properties and the better brightness permit a reduction to be made in the pulp fraction in the production of newsprint (53).

White oak and two mixtures of deciduous woods, cooked with magnesium-base sulfite cooking liquors containing ratios of bound and total sulfur dioxide of between 50% and 100% based on the wood, gave pulp yields of between 70% and 80%. Corrugated material produced from these pulps was somewhat stiffer than that from neutral sulfite semipulp. At a slightly higher pulp yield, the corrugated paper had the same quality as that of a NSSC semipulp (324).

The vapor-phase cooking process can also be applied to the magnesium-base sulfite cook. The impregnation has to be carried out at 50°C. Spruce chips absorb between 1.37 and 1.57 grams of liquor per gram of wood; then enough cooking liquor is added to keep the acid-to-chip ratio constant at 1.68. The best results are obtained with a magnesium-base cooking liquor containing 8% total sulfur dioxide, with an impregnation

period of 5 minutes, a heating-up time of 45 minutes, and a cooking time of 8 to 14 minutes at a maximum temperature of 180°C. Under these conditions, pulps of excellent strength properties are obtained in 78% yield (186).

Vapor-phase Magnefite pulping of sprucewood was performed in a pulp-yield range of 45.7% to 71.0%, with combined sulfur dioxide contents of the impregnation liquor of 2% to 6%, a wood-to-liquor ratio of 1:2.35, and cooking temperatures of 160° and 170°C. Increasing the cooking liquor concentration decreased the cooking time needed to obtain a certain yield, and twice as much time was required when a cooking temperature of 160°C instead of 170°C was used. Although the pentosan contents of pulps cooked under various conditions did not differ, more hemicelluloses were retained in pulps cooked at a higher chemical concentration. The rates of delignification in vapor-phase and standard liquid-phase cooking were nearly the same, but the yield from vapor-phase pulping for a given cooking time was lower than that from the standard Magnefite process at yields of over 50%. The vapor-phase process tended to give a burnt cook sooner than the standard process in the low-yield range, especially at higher concentrations. In the range of 2% and 4% combined sulfur dioxide the strength properties of vapor-phase Magnefite pulps were similar to those of liquid-phase pulps, but the former, when cooked at 6% combined sulfur dioxide at 170°C, showed lower strength properties (280a).

Finally, it must be mentioned that aluminum-base bisulfite cooking liquors can be used for pulping. In this case, the aluminum content is 0.2% and the total sulfur dioxide content is 6% to 8%. The highest temperature is 120° to 130°C and the pulps obtained have, at the same degree of pulping, properties similar to those of calcium bisulfite pulps. Because aluminum sulfite decomposes readily even at 500° to 600°C, simple methods for the recovery of the base are to be expected (351).

8. TECHNOLOGICAL STEPS FOR THE OPERATION OF THE SULFITE COOKING PROCESS

8.1 The Digester

The technical construction of sulfite pulp digesters will not be discussed since that is beyond the scope of this book. However, it seems appropriate to discuss some technological problems, especially corrosion problems. During the last two decades, the use of acid-proof steels for the lining of digesters has gained considerable importance. In particular, steels containing chromium-nickel-molybdenum alloys have proved very satisfactory. In general, calcium- and magnesium-base sulfite cooking liquors have almost the same corrosive properties. In special cases, a strong enrichment

of chlorides in the digester can cause the dissolution of the inactive protective layer and, with it, greater corrosion (111,329). The cause of this has been found to be woods which had been floated through salt water. Special care must be taken when welding is carried out on the steel sheets of the lining so that no places poor in carbon are formed through diffusion; such places are especially susceptible to corrosion. One of the principal causes of corrosion is the occurrence of precipitates in the digesters. These consist largely of gypsum and, to a smaller extent, of calcium sulfite. When the precipitates form thick layers, especially at places which cannot readily be reached and are protected from the circulation of the liquor, the liquid beneath these layers becomes devoid of oxygen. Anodes of oxygen-concentration elements are formed, the cathodes of which are the places not covered by the layers that are in contact with the cooking liquor rich in oxygen (354).

Ceramic linings are still being used, as well as linings with acid-proof steel sheathing. Whether soluble or insoluble bases are used in the cooking liquor plays an important role. The pores of the ceramic material are filled by the precipitates from cooking liquors from insoluble bases and in this way a dense protective layer against the cooking acid is soon formed on the digester jacket. Cooking liquors from soluble bases do not form such precipitates and two ceramic linings, one on top of the other, must be applied. The upper layer, which comes in direct contact with the liquor, must be thicker, and is applied by using an acid-resistant synthetic plastic cement. With welded digesters a rubber or synthetic resin layer instead of the cement layer is applied to the digester wall, followed by two layers of ceramic lining, which are separated by a synthetic resin layer (403). A two-layer lining, with the inner one of carbon bricks, has been proposed. The swelling and shrinking properties of the synthetic resin cements have to be taken into consideration so that the tensions and movements inside the lining will remain within permissible limits and no cracks will occur (317,398,421).

8.2 Impregnation and Forced Circulation

The importance of the impregnation of the chips with cooking liquor and the physical-chemical prerequisites for the liquor movement within the wood have been discussed in Section 4. Because the volume of the pores in the chips is very large and the pores are filled with air so long as the wood is not completely wet, one of the most important steps for a rapid and complete impregnation of chips with cooking liquor is the removal of the air from the pores. This can be accomplished in different ways: (a) by steam pressure, with or without sudden pressure relief; (b) by evacuation; (c) by an artificially produced hydrostatic pressure; and (d) by injecting hot cooking liquor into the filled digester.

Investigations over many years in Canadian research laboratories

showed that a pretreatment of the wood chips with hot steam gave the best results. Such steaming not only expels the air from the pores but also forms a certain equilibration of the moisture content of the chips. The differences in the penetrability of the heartwood and sapwood are reduced when clogged or luted bordered pits are exposed. Such a presteaming must be applied with caution. A maximum temperature of 110°C should not be surpassed for long. The use of superheated steam, especially, requires caution. The hydrolysis of the hemicelluloses starts readily at temperatures of above 120°C and the acids formed can have a detrimental effect on the sulfonation of the lignin. When steaming and impregnation by pressure have been carried out successively, a considerable portion of the cooking liquor can be drained off, provided that sufficient amounts of bound sulfur dioxide are present. In this case, it is possible to carry out the cook in the steam phase, especially when soluble bases are used. That the expulsion of the air from the pores in the wood is necessary is shown by the fact that a certain proportion of the sulfur dioxide is oxidized to sulfur trioxide and the sulfuric acid formed affects the lignin unfavorably at temperatures of above 110°C, causing the formation of so-called "burnt centers" which often appear in forced conventional calcium bisulfite cooks. The air content of the wood chips depends on the moisture content. Chips of black spruce with a 50% moisture content contain a third of their volume of air. The composition of the air in the chips is not exactly identical with that of the surrounding air; in general, the chips contain less oxygen and more carbon dioxide, which can be 90 times as great as that in the surrounding air. The solubility of air in water must also be considered, and this holds even truer for the solubility of carbon dioxide in water. At 100°C the solubility of these gases is practically zero, and this explains the efficiency of steaming to expel the dissolved gases (170).

The steaming of the chips contributes not only to the expulsion of the air but also to the lessening of the harmful rosin content, as has already been mentioned. The terpenes are almost completely removed on steaming of the chips for 20 minutes at 130°C, whereupon the rosin becomes brittle and harmless. This holds true for both spruce- and pinewood (243).

8.3 Multistage and Two-Base Cooking Processes

There are various possibilities for dividing the sulfite pulping process into several stages or for using different bases in the various stages. In a two-stage cook, the pH value may be lower in the first stage than in the second. In this case, sodium hydroxide, for example, may be added before the start of the second stage; or the pH in the second stage may be lower than in the first, which is accomplished by the addition of sulfur dioxide before the start of the second stage. Whereas the pH value of the hot cook-

ing acid in an acid bisulfite cook at first increases, regardless of the kind of base used, and then, toward the end of the cook, decreases again due to the formation of acid reaction products, the pH value of the hot cooking liquor in the first group of two-stage processes is higher than that of the cold liquor. These two-stage processes are characterized by the constancy of the pH of the cold liquor by buffering in the second stage. The pH value of the hot liquor in cooking processes with a high pH in the first stage is higher in both stages than in the cold liquor. In this case the main goal is the protection of the hemicelluloses against acid hydrolysis during the first stage, whereas in the second stage the delignification must be accomplished. On the other hand, a two-stage cooking process, in which the pH value in the second stage is higher than in the first, may be termed an "alkaline" sulfite process. In this case the delignification occurs chiefly in the first or acid stage, and hydrolysis of the hemicelluloses takes place simultaneously. These partially degraded hemicelluloses are then dissolved in the second stage by the alkali. In this way pulps can be produced with good strength properties, especially from hardwoods (244,406).

The "Stora" cooking process is a sulfite cooking process with sodium base. It was developed in Sweden and, in common with other similar processes, it can be used extensively for rosin-containing woods such as pine. As mentioned earlier, the acid components of the calcium bisulfite penetrate the heartwood faster than the base does and this causes an increase in the hydrogen ion concentration, which, in turn, causes condensation reactions in the lignin. In the first stage of the Stora process, a lower hydrogen ion concentration of about 4 is applied; in the second stage, sulfur dioxide is added and thus the hydrogen ion concentration is increased (102,228,370, 430). The pulps obtained in this way possess a high hemicellulose content because of the the low hydrolysis in the first stage, and they are readily beatable. They are especially suitable in combination with hardwood pulps, which require a high degree of beating, for the production of papers. Figure VII–27 shows the course of delignification in the two-stage pulping of pinewood at various pH levels in the first stage, while Fig. VII–28 shows the yields of carbohydrates in percent, based on the weight of wood, for different pulping processes. Figure VII–27 clearly shows that the delignification in the first stage decreases with increasing pH, whereas Fig. VII–28 indicaes the increase in the yield of carbohydrates, especially the hemicelluloses, in the two-stage soda-base sulfite process as compared with the single-stage sodium bisulfite, calcium bisulfite, and kraft pulping processes (227,397). The two-stage process, with a slightly acid liquor in the first stage and a more strongly acid liquor in the second, can also be carried out with two different bases in the two stages by using sodium-base cooking liquor in the first stage and calcium bisulfite in the second. In this way a more selective delignification and better homogeneity of the pulp are

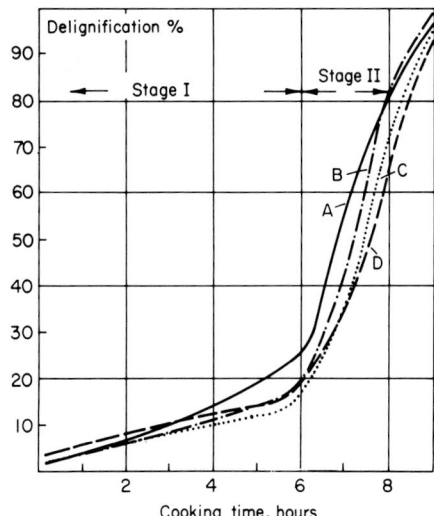

Figure VII-27. Delignification of pinewood in two-stage cooking at various pH levels in stage one (227). A = pH 4; B = pH 6; C = pH 7; D = pH 8.

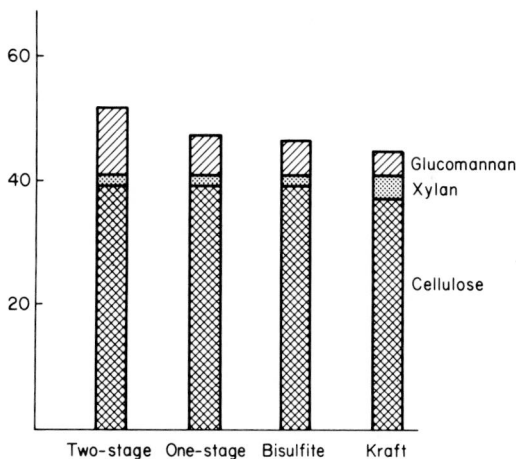

Figure VII-28. Carbohydrate yield in percent of wood, for different pulping processes (227).

achieved. The defibration point—i.e., the stage at which the fiber bundles can readily be separated from each other—is reached in a one-stage calcium bisulfite process at a 53% or lower pulp yield, and in a two-stage, two-base cook, already at a yield of 60% to 61% (33).

The "Sivola" process, developed in Finland, is a two-stage pulping process carried out at pH 4 in the first stage and finished in a second stage

at pH 7.5 to 9 by the addition of sodium carbonate. The process uses sodium as base in both stages and is, in the second stage, an alkaline cook, because the cooking liquor contains a considerable amount of sodium carbonate. In this way an attempt is made to combine the advantages of a sodium-base bisulfite cook with a soda cook (48,289). Although the pulps thus obtained appear to be especially suitable for rayon, they may also have many possibilities for the manufacture of paper.

Recent investigations have had as their goal the study of the use of sodium, magnesium, calcium, and ammonia base in the first stage and sodium hydroxide, sodium carbonate, or sodium sulfite, or calcium, magnesium, or ammonium hydroxide in the second stage. In this way, two-stage and two-base cooking processes were developed. It was found that, in the first as well as in the second stage, the cation has a decisive influence on the strength and brightness of the pulp. The effect of the hydrogen ion concentration, on the other hand, in a pH range of 6.3 to 10 is of secondary importance. The temperature, the concentration of chemicals in the cooking liquor, and the yield of the pulp in the second stage also have an influence on the strength properties of the pulp. The degree of brightness of all pulps produced by two-stage processes was relatively low, but it could be improved by lowering the temperature in the first stage and by shifting the major part of the pulping operation to this stage.

Such two-stage, two-base cooks can be carried out technically by two different processes: (a), the cooking liquor of the first stage is completely drained off and, without intermediate washing, an alkaline cooking liquor is added; or (b), the cooking liquor of the first stage remains in the digester and is adjusted to the desired pH value by direct addition of alkali. It was found that a two-stage pulp produced according to the first method had higher tensile strength than a pulp obtained from a one-stage cook. A subsequent sodium sulfite cook did not result in an improvement of the strength properties of the pulp. An ammonium bisulfite-alkali cook was only partially successful. Ammonium or calcium hydroxide in the second stage gave unfavorable results. The differences in pH are chiefly responsible for the difference in the efficiency of the alkalis used. It is known that the ionic product and, with it, the hydrogen ion concentration of the water increases with the temperature. This contributes to a lowering of the pH values of the solutions. There are still other factors that affect the temperature-pH relationship, such as, for example, the dissociation of the sodium carbonate, the solubility of the calcium hydroxide, and the dissociation constant of the ammonia-water system.

Pulps produced by the two-stage cooking process by the second method are distinguished by a high burst strength and high tensile strength, but they have a lower tear strength. In general, the average strength properties are superior to those of one-stage bisulfite pulps but inferior to those of

kraft pulps. Ammonium and calcium hydroxide in the second stage of an ammonium bisulfite cook give pulps that are inferior to those obtained by the sodium bisulfite process. The same holds true for pulps which are pro- duced in the first stage with magnesium bisulfite and are then treated, in the second stage, with sodium hydroxide, sodium carbonate, or magnesium hydroxide. The pH value at cooking temperature first increases and then decreases again. The dependence of the pH value of the ammonium and sodium bisulfite solutions on the temperature in the second stage has al- ready been mentioned (180,181). The poor quality of the pulps produced by the ammonium bisulfite alkali process is generally attributed to the effect of the pH. This contradicts the fact that excellent pulps can be pro- duced by the magnesium bisulfite-magnesium hydroxide process and the sodium bisulfite-sodium hydroxide process. It may be possible that still unknown but important reactions take place between the ammonium base and the prevailing cation in the second stage (88,234).

Jack pine, balsam fir, spruce, oak, and sweet gum woods were sub- jected to three two-stage cooks: bisulfite-neutral sulfite, bisulfite-acid bisul- fite, and neutral sulfite-acid sulfite. Without draining off the original cooking liquor of the first stage from the digester, the pH was adjusted in the second stage either with liquid sulfur dioxide or with anhydrous sodium carbonate. The pulps obtained in the lowest yield had the best strength properties; these were highest from the bisulfite-neutral sulfite cook which approached those usually found with kraft pulp. The resistance of the pulp to the hydrolytic action of the acid increased with increasing pH value of the cooking liquor in the second stage. With oak and sweet gum the yields were lower than with corresponding kraft cooks. The noticeable difference between deciduous and coniferous woods with regard to the efficiency of the various cooking processes can be explained, at least par- tially, by the low glucomannan content of the deciduous woods. A test for sugars in the acid spent liquors showed that the main reason for the increased yield is the greater retention of the glucomannan in the neutral- acid sulfite process.

The well-known difficulties in the pulping of heartwood from Douglas fir in a one-stage calcium bisulfite cook are caused by the decomposition of the cooking liquor by the dihydroquercitin, present in the heartwood; this can be overcome by a two-stage bisulfite-alkali sulfite cook. The pulps thus obtained possess almost the same strength properties as those of kraft pulps. However, if the second stage is carried out in a neutral or an acid medium, an increase in yield of 10% to 20% is obtained, but the strength properties suffer a loss which is considerably higher than that of a like cook with other coniferous woods. This is explained by the presence in Douglas fir of thick cell walls which are especially sensitive to hydrolysis. These thick cell walls originate in the compression wood, the formation of which is

quite common in this species. The addition of ammonia in the second stage after a magnesium bisulfite cook in the first stage did not give any better results (325,326).

A variation between 4 and 8 in the pH values in the first stage in cooks of pine- and birchwood showed that the delignification velocity of the latter was considerably greater than that of the former. A fractionation of the lignosulfonic acids formed showed a continuous increase in the average molecular weight during the cook, with those of the lignosulfonic acids from birch being lower than those of pine. The degree of sulfonation also differed. At higher pH values a considerable amount of the dissolved carbohydrates is converted into acids. At pH 6 or higher, only minor portions of the carbohydrates go into solution. This shows the effect of the hydrogen ion concentration upon the stabilization of the glucomannan toward acid hydrolysis (365).

Combinations of sodium bisulfite cooks and kraft cooks have also been suggested. Pulps produced in this way are distinguished by their high α-cellulose content (236). As mentioned before, drainage of the cooking liquor after the first stage can be carried out, or even a washing step can be introduced before the second stage. This stage is then carried out with a pure soda liquor. Several advantages are attributable to such a process: the capacity of the digester can be fully utilized; the spent liquor from the first stage can be fermented to alcohol, which is impossible when soda is added to the cook without draining off the liquor from the first stage; in addition, the alkali consumption is lower because none is consumed by the neutralization of the acid. The pulps thus produced possess a low resin content and have good bleachability (4).

"Mixed-base" cooking liquors have been used, too. A calcium bisulfite cook containing sodium ions gives pulps with noticeably improved properties. Such mixed cooks can be carried out by the addition of relatively concentrated sodium sulfite or bisulfite solutions to the calcium bisulfite liquor during the charging of the digester, or the wood chips may be impregnated with sodium sulfite solution prior to the cook. The sulfite consumption amounts to about 2% to 5%, based on the weight of the wood. The heating-up period and the cooking time at the maximum temperature can be considerably shortened. The pulps thus obtained have good strength properties, and any lignin condensation can be extensively minimized by the presence of sodium ions (35). Pulps produced by means of mixed cooking liquors correspond approximately to those obtained by the pure sodium or ammonium bisulfite cooking process with regard to their degree of pulping, their α-cellulose content, and their splinter content (433). Because pure ammonium bisulfite spent liquors cannot be fermented, spent liquors from mixed calcium-ammonium cooks have been used for this purpose, with the result that the presence of 40% to 50% of ammo-

nium spent liquor has been found to be sufficient to provide the nitrogen required for the fermentation (217).

Magnesium salts are adequately soluble over a wide range of pH values and this makes it possible to convert the magnesium bisulfite cooking process, which has already been discussed in detail, into a two-stage process. For this purpose, a magnesium oxide slurry is added to the acid liquor after the first (acid) stage until the pH in the digester is adjusted to 6.0 to 6.5. In this way the strength properties and the bleachability of the pulp are favorably affected and the pulp yield remains the same (46). A reversed sequence can also be used by carrying out the cook with a magnesium-base cooking liquor of pH 6 in the first stage, and with a pure sulfur dioxide solution in the second. The pulp obtained in this way is especially readily hydrated and has a high hemicellulose content. According to another suggestion, an acid magnesium bisulfite solution of pH 3.7 may be used in the first stage, followed by an almost neutral cook at pH 6.0 to 6.5 in the second. This last process has already been mentioned, and it is claimed that it gives pulps with high α-cellulose content (21). Finally, an almost saturated magnesium sulfite solution with a pH of 7 to 9 may be used in the first stage and a magnesium bisulfite cook in the second stage. An especially high retention of glucomannan is attributed to this process (270).

8.4 The Continuous Cook

Whereas the continuous kraft cooking process has found wide acceptance—several hundred such mills are already in operation—the introduction of a continuous acid pulping process required a great deal of special preliminary work to bring it to technical maturity. The problems involved were manifold and differed fundamentally from those which had to be solved for the application of the continuous kraft pulping process. Any continuous process must be based, in general, on a constant temperature if the investment is not to become too high to make the process economically feasible. The sulfite pulping process distinguishes itself by a series of reactions in which each single step requires special reaction conditions to become fully effective. It is therefore difficult to maintain a uniform temperature for such a varied reaction process. The chief advantage of a continuous cook is that one or more digester aggregates can take over the work of a series of digesters. This requires that the reaction should take place relatively rapidly; in other words, short cooking periods must be successfully attained. The impregnation of the chips with cooking liquor requires a certain amount of time. The penetration of the various cooking liquor components is not uniform; the acid component penetrates the wood chips more rapidly than the base does. To obtain a rapid and complete penetra-

tion, an attempt has been made to pulp normal and shredded sprucewood chips with an acid ammonium bisulfite liquor (5.2% total, and 1.2% combined sulfur dioxide) and to vary the cooking time and maximum temperature as follows: 9 hours at 135°C, 6 hours at 142°C, and, finally, 0.5 hour at 148°C. The first two conditions correspond roughly to the classical bisulfite cook, whereas the short cook fulfills more nearly the requirements of a continuous cooking process. It was found that even an increase in the temperature to 148°C within 3 minutes after the chips had been mixed with the liquor did not result in a burnt cook. In all cases, shredded chips were quickly impregnated and could be pulped in a cooking time of 2 hours at a maximum temperature of 148°C to give a pulp in 50.6% yield with a lignin content of 8.2%. This was not possible with conventional chips. Wood chips impregnated for 6 hours with a cooking liquor of the above composition showed a decrease in the reaction velocity and in the yield of screened pulp, together with an increased attack on the cellulose. Prior impregnation with cooking liquor without any free sulfur dioxide had no effect on the reaction velocity or on the yield of screened pulp, but it also caused slight damage to the cellulose (6,216,276,277).

In the discontinuous sulfite cooking process, the autocatalytic decomposition of the sulfurous acid, according to the equation

$$3\ SO_2 \cdot H_2O \longrightarrow 2\ H_2SO_4 + S + H_2O$$

does not play a major role because the heating-up period is relatively slow and the greater part of the reactions between the wood and the cooking liquor is already terminated before the maximum temperature is reached. In contrast to this, the autocatalytic reaction reaches a dangerous point in a continuous cooking process with a constant temperature and, in this way, increases the danger of a burnt cook. Cooking liquors containing free sulfur dioxide develop very high pressures at the maximum cooking temperatures. Another problem is the strongly corrosive properties of the cooking liquor, especially in the zones between liquor and gas phase. Ceramic linings cannot be used in a continuous digester in which the material to be cooked is continuously in movement. Even stainless steel shows signs of corrosion. The auxiliary installations which are necessary for the continuous feeding of the chips into the digester and the take-off of the pulped chips from the digester are especially susceptible to corrosion (216,277,309,311).

A large American pulp and paper mill put a continuous magnesium-base sulfite cooking plant into operation at a plant in Ontario, Canada, in March 1964. Figure VII–29 shows an outline of the plant, and, because such a mill is typical for the continuous Magnefite cook, it will be described briefly. The mill uses four-fifths black spruce from its own forests, with the remainder being white fir of other origin, together with about 4%

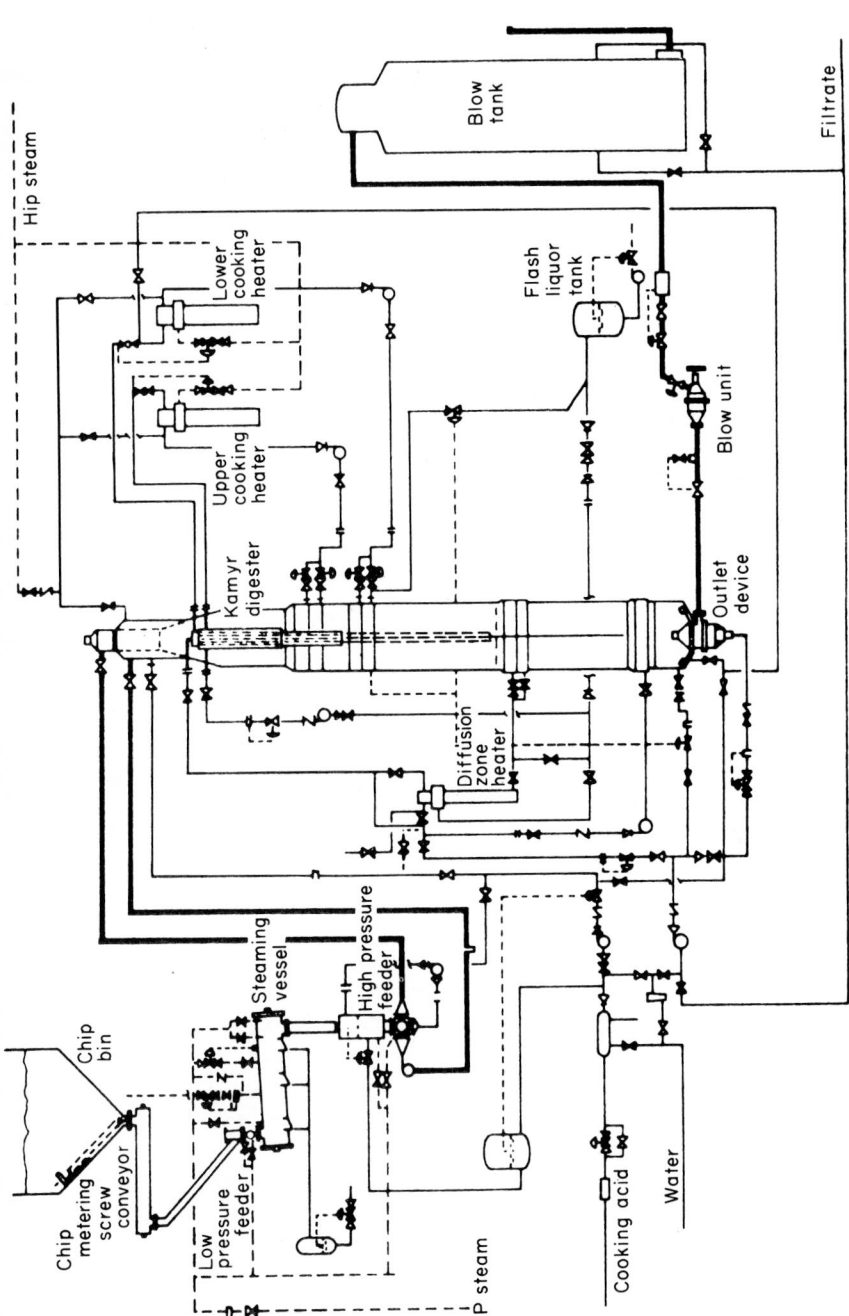

Figure VII-29. Kamyr continuous Magnefite pulping system as installed in Spruce Falls Power and Paper Co., Kapuskasing, Ontario (103). (Reproduced from "Sulphite pulping technology" by permission of *Paper Trade Journal.*)

balsam fir and minor amounts of Jack pine. The wood has an average moisture content of 42%. The chips are somewhat smaller than those conventionally used. Metallic impurities are carefully removed on a conveyor band to protect the measuring instruments and the transportation units to the digester from damage. The measurement of the wood is carried out by an integrator which indicates the revolution of the screw conveyor and, with it, the volume of the fed chips. The volumetric chip-metering screw conveyor moves the chips through a low-pressure feeder into a steaming vessel. This steaming chamber is under 14 to 19 psi pressure and the pressure feeder acts simultaneously as a pressure seal toward the outside. At the same time, the steam heats the chips. This steaming of the chips is important for their subsequent uniform impregnation with cooking liquor.

In contrast to the principle used in the conventional kraft pulp cook, the steam for steaming is taken, not from the digester, but from a separate tank. The steam penetrates the chips and, so far as it is not condensed, is blown off with the expelled air. One of the most important devices of the plant is the high-pressure feeder that carries the steamed chips from the low-pressure zone to the high-pressure zone. The condensate and surplus acid which escape by leakage from the high-pressure feeder are recovered and recycled via a niveau tank. In the same way, the acid escaping from the outlet device at the bottom of the digester is returned to the low-pressure feeder. The pressure in the digester is kept constant at between 185 and 200 psi. This pressure considerably exceeds the steam pressure of the cooking liquor at the maximum cooking temperature of 175°C. In this way boiling up and the formation of steam bubbles are avoided. A pressure pump takes care of the circulation of the liquor and returns it via the high-pressure feeder to the digester at a constant flow of 2500 gpm and thus simultaneously adds to the movement of the chips in the digester.

The digester itself is 120 feet high and has a diameter of 10 feet. It is designed for a pressure of 250 psi. Concentrically arranged feed pipes for the cooking liquor and ring-shaped sieve zones for the withdrawal of acid samples are arranged to allow 45 minutes for the impregnation, 2 hours for the cooking, and one hour for the countercurrent washing. Most of the cooking liquor which enters the digester with the chips is drawn off at the top, through a perforated pipe. All parts that come in contact with the cooking acid are made of a special stainless steel. The fresh cooking liquor enters the digester through a jacket around the upper separator. This cooking liquor is made up of regenerated chemicals together with some additional ones, but contains no spent liquor as the classical calcium bisulfite process often does. The total sulfur dioxide content of the magnesium-base acid amounts to 6% to 7% and the pH value varies between 3.7 and 4.0. Cooking acid flow and chip passage are synchronized and can be

varied within certain limits to achieve a predetermined liquor-to-wood ratio. Under normal conditions, this ratio is 3.25 : 1. The maintenance of a constant temperature is of special importance for the production of a uniform pulp. A Rosenblad spiral heat exchanger keeps the final temperature at about 82°C. The fresh cooking liquor is stored in outside tanks and is injected, through a flow meter and a heat exchanger lined with Teflon, into the pressure system by means of a pressure pump. A niveau tank, which assures a constant liquor level in the inlet opening of the high-pressure feeder, conducts the overflow cooking liquor to the suction pipe of the pressure pump. Under normal conditions, 16% to 25% sulfur dioxide, based on the weight of the dry chips, is consumed in magnesium bisulfite cooks during one cooking cycle. In the continuous Magnefite plant described here, a consumption of 16% sulfur dioxide has been calculated, although the actual consumption is about 14%.

The digester itself has the usual two heating zones: a cooking zone and a separate countercurrent washing zone. The mixture of chips and cooking liquor moves from the impregnation zone at the top of the digester into the upper heating zone. Here some liquor is drawn off through two ring strainers, heated indirectly with steam at 160 psi through a heat exchanger, and then injected directly through nozzles into the same zone. The cooking mass then moves through a quiet zone in which the impregnation of the chips is completed. From there the wood chips enter the second heating zone where the maximum temperature is reached. This is achieved by drawing off, reheating, and returning the cooking liquor in the same way as in the first heating zone. The maximum temperature is 170°C to obtain a daily production of 200 tons of pulp. Supplementary heat exchangers are installed in a parallel position in order to prevent any interruption of the operation if the first heat exchangers have to be cleaned of incrustants. Magnesium bisulfite cooks cause only minor incrustations as compared with continuous kraft cooks. After the second heating zone, the wood enters the actual cooking zone in which the pulping reactions are completed and are then stopped by cooling with liquor at room temperature. During the cook, the volume of cooked wood and liquor decreases, since solid constituents are dissolved. This phenomenon, together with the recirculation of the cooking liquor, forms a danger point, because variations in the rate of flow can occur and, in turn, can cause the formation of thiosulfate. As already mentioned, thiosulfate is an undesirable factor in the cooking process.

The cook is interrupted by the withdrawal of hot acid and the addition of cold liquor. The hot liquor is taken off through the upper sieve rings at a rate of 545 gpm, cooled, and 300 gpm are added again at the

same point. The residual 245 gpm are transferred under digester pressure into a relief tank and from there into a storage tank of the regeneration plant. The difference between the withdrawn and the returned cooking liquor is made up by replacing it with thin liquor (washing liquor), which is injected at the bottom of the digester and in this way causes a countercurrent washing in the wash zone. The outlet device at the bottom of the digester disintegrates the cooked chips, cooled wash liquor is added, and the chips leave the digester under digester pressure. The consistency of the pulp is lowered in this way from 19% to 8% (16,103,309).

It has been reported that, with a Swedish continuous experimental digester with a daily production of 14 tons of pulp, sodium- and magnesium-base bisulfite, two-stage sodium-base-sulfite, and acid bisulfite cooks can be carried out. The digester is, in general, similar to the conventional Kamyr digester except that the separator is installed in an inclined position at the side of the digester, beside the digester dome. This arrangement permits the formation of a steam phase in the digester dome, but the usual feeding system, including the Kamyr high-pressure feeder system, is maintained. Doubts were quickly dispelled as to whether the short impregnation period, possible with this type of feeder system, would be sufficient to permit the wood then to be subjected directly to the high cooking temperature. The amount of chemicals which penetrate the chips is sufficient for the pulping reactions and a quick replenishment of these chemicals occurs in the digester proper by diffusion. Pulps in yields of between 53% and 75% can be obtained, depending upon the operational conditions, with a quality ranging from bleachable pulp to semichemical pulp.

Two-stage acid bisulfite cooks can also be carried out successfully with such a Kamyr digester. The first stage, based on earlier results, is carried out in a practically neutral medium at a high temperature and for a longer cooking time, or in an alkaline solution at a moderate temperature and for a shorter cooking time. This first cooking stage serves primarily for the deacetylation of the hemicelluloses (9). The second (acid) stage is carried out by the injection of liquid sulfur dioxide into the circulating cooking liquor. At the same time, the cooking temperature is adjusted. The washing of the pulp takes place by the countercurrent process, as already described. Some difficulties are encountered in the zone that separates the two cooking zones when there is a low liquor-to-chip ratio and a low temperature in the first stage, but they are overcome by a slight increase in the temperature in the upper zone. Figure VII–30 shows the operational scheme of such a two-stage continuous cook, and Table VII–10 gives the cooking conditions and data on the properties of pulps from spruce- and pinewoods (10).

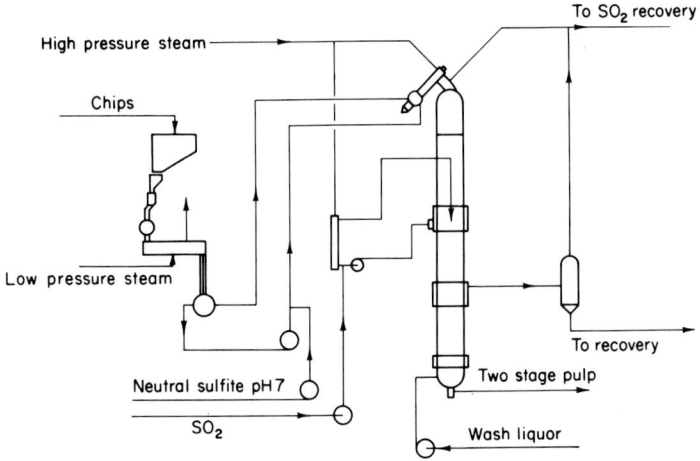

Figure VII-30. Kamyr digester with inclined top separator used for two-stage neutral sulfite/acid sulfite pulping.

TABLE VII-10

COOKING CONDITIONS AND PULP PROPERTIES FOR A CONTINUOUS, TWO-STAGE
NEUTRAL SULFITE-ACID BISULFITE COOK OF SPRUCEWOOD
AND PINEWOOD (10)

	Spruce	Pine
Production, short tons per 24 hr	15	15
Cooking conditions		
Neutral sulfite stage		
Digester pressure, psi	140	140
Combined SO_2 per a.d. wood	4.9	5.2
Cooking liquor pH at start of the cook	7.0	7.0
Cooking liquor:wood ratio	2.0:1	2.1:1
Temperature maximum, °C	155	155
Cooking time, hr	2	2
Acid bisulfite stage		
Liquid SO_2, % per a.d. wood	14.0	13.0
Total SO_2, grams per liter	34	34
Cooking liquor:wood ratio	2.4:1	2.4:1
Temperature, °C	142	143
Cooking time, hr	1.5	1.5
Pulp properties		
Kappa no.	23	32
Rejects (%) of the pulp	1.4	3.3
Brightness, % SCAN	71.1	63.5
Glucose, %	81.1	82.4
Mannose, %	13.2	13.1
Xylose, %	5.8	4.5

8.5 High-Yield Sulfite Pulp

Special efforts have been made in recent years to increase the yield of pulp from wood without impairing the specific properties of the pulp fiber. These efforts have therefore been made in the direction of the semipulps without being identical with that process. To differentiate them from the semipulps, these are called "high-yield pulps." It is understandable that the demarcation line between the two types of pulp is not sharp and some "high-yield pulp" may also be called "semipulp." It has already been pointed out in previous sections which factors affect the yield in the sulfite pulping process. Certain operational problems arise when the yield is increased much above 50%. The chief problem is an adequate impregnation of the wood chips with cooking liquor in as short a time as possible. Equally important are the right composition of the cooking liquor and the choice of the best reaction conditions such as the acidity, time, and temperature. A further problem is the defibering of the high-yield pulps. The power consumption required increases with increasing lignin and hemicellulose contents. The efficiency of the disintegrators usually used in the pulp industry is not adequate to separate the fiber bundles from the chips and the fibers from the fiber bundles in such lignin-rich pulps.

The problem of impregnation has already been discussed—the last time in connection with continuous cooking, which requires very short impregnation periods. Impregnation with preheated cooking liquor and subsequent cooking in the vapor phase have also been mentioned before (152, 153,283,355,388,432). Figure VII–24, in Section 5.2, shows the effect of acidity and sulfur dioxide concentration upon the dissolution of the various wood components (308). The Ross diagrams, shown in Figs. VII–25 and VII–26, show the effect of the variation in temperature for each combination of cooking acids. From these curves, the reaction conditions for high-yield pulp cooks can be predetermined, provided that an adequate and especially uniform impregnation has been achieved.

The required range in the composition of the cooking acid can be reached by the use of soluble bases, or a two-stage cooking process must be employed (280,362). The addition of 0.2% to 2.0% of borohydride in a weakly or strongly alkaline sulfite cook of Douglas fir or spruce gave a 15% to 20% higher yield, but of slightly inferior pulp (323). Hardwood impregnated with sodium sulfite is supposed to give a groundwood which approaches in quality a sulfite pulp (406). This shows how the borders between pulp, semipulp, and chemigroundwood are more or less blotted out.

The defibering problem of chips from a relatively hard cook required the development of a special apparatus to prevent the presence of fiber bundles and splinters. Under the name "Curlator," devices are in use which not only disintegrate the fiber bundles but also crinkle the fibers, and this

leads to a better sheet formation in the paper industry (360). Pumps are not suited to the transportation of nondefibered chips, but screw conveyors have proved effective for this purpose (58). Disk refiners achieve a disintegration of fiber bundles by rubbing them without cutting the individual fibers (286,376). Such disintegration operations consume energy, and attempts have therefore been made to measure the cohesive forces that keep the fibers in the fiber bundle together (435).

Great efforts have been made during the past 15 years to increase the yield of sulfite pulp and hence to make better utilization of the wood. As a result of these endeavors, a great many new details have been found, leading to the discovery that sulfite pulps can be produced in 65% to 70% yield by the use of soluble bases, and that these pulps can be used to good advantage in the paper industry (229). For comparison, it may be mentioned that kraft cooking liquors penetrate the chips much more rapidly, and that the air enclosed in the chips affects the kraft pulping process much less than the sulfite process. The production of high-yield pulp by the kraft cooking process is handicapped by the greater loss of carbohydrates. This loss amounts, at equal lignin content, to about 4% more than in the sulfite process. The Ross diagram shown in Fig. VII–31 clearly indicates the difference between the sulfite and kraft pulping processes (150,386).

Within arbitrarily defined limits, the two-stage pulping conditions for red alder to produce a high-yield pulp with the best overall strength properties were determined. The cooks were carried out in the first stage with a bisulfite, and in the second stage, with a slightly alkaline cooking liquor, and gave pulps with the highest strength characteristics. Only chips between 0.5 and 0.75 inch in size and with a moisture content of 47.7% were used. The cooking liquor contained 5.25% (by weight) of sodium bisulfite or sodium sulfite. The pH desired for the liquor was reached by buffering it with sodium bicarbonate or sodium carbonate. The initial pH's were 4.5, 6.0, 7.5, 9.0, and 10.5. All these liquors were used in the second stage and all except that with pH 10.5 were used in the first stage. The liquor-to-wood ratio was 4 : 1. Before the liquor was added to the digester, the chips were presteamed for 10 minutes at 50 psi. Immediately after the steaming, the cold first-stage liquor was injected and the temperature was increased within 10 to 15 minutes to the desired cooking temperature. At the end of an hour the liquor was drained off, the second-stage liquor was injected, and the desired cooking temperature was maintained for one hour. The pulping temperatures were 150°, 155°, 160°, and 165°C. After the cook, the chips were drained and then defibered in a Bauer double-disk refiner at 65° to 75°C. In Table VII–11 the strength properties of selected pulps are compared with those of other pulps produced from red alder by the neutral sulfite semichemical (NSSC) and kraft semichemical processes within the same yield range.

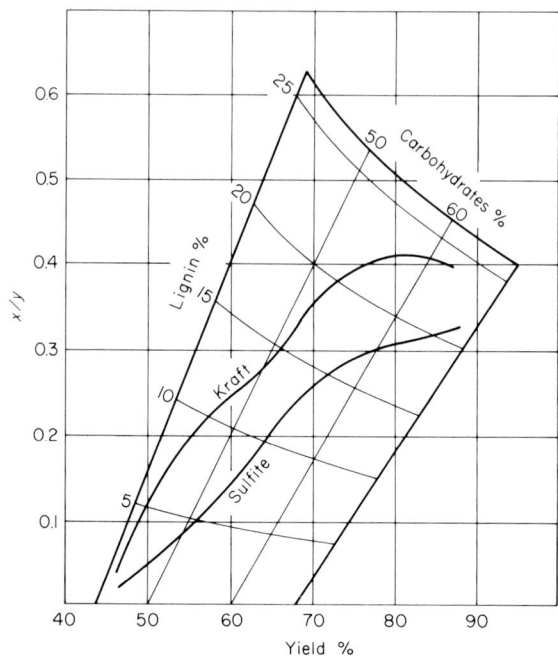

Figure VII–31. Ross diagram comparing high-yield sodium bisulfite pulp with kraft pulp (386).

The pulping conditions used in the two-stage process were much milder than those used in either the NSSC or the kraft process and the resulting pulps had better overall strength properties. The total pulping time in the two-stage process was 2.75 hours, whereas that in the NSSC process was 7 hours, for a 70% pulp yield. The pulps from the two-stage process were analyzed, with particular attention being devoted to the xylan constituent. The principal hemicellulose, *O*-acetyl-(4-*O*-methylglucurono)-xylan, had most of its acetyl groups removed and its reducing end-groups were partially oxidized, while the uronic acid groups were completely retained during the pulping. The cellulose also appeared to be retained almost quantitatively and, coupled with a relatively high hemicellulose content, this produced a high-yield pulp with desirable properties. The good retention of the carbohydrate material by the pulp was ascribed to the oxidative effect of the first-stage bisulfite liquor, resulting in protection against alkaline degradation in the second stage of the cook. This protection was shown by the identification of xylonic acid groups as one of the more important acidic end-groups in the xylan (200b,344a).

TABLE VII-11

COMPARISON OF STRENGTH PROPERTIES OF RED ALDER PULPS
FROM VARIOUS PULPING PROCESSES (10)

Pulping method and conditions used		CSf (ml)	Yield (%)	Burst (Mullen) factor	Tear factor	Breaking length	Density (g/ml)
Two-stage sulfite process (10)							
First-stage pH	Second-stage pH						
4.5	8.0	600	75.9	28.0	59.6	4.89	0.68
6.0	10.0	600	76.5	23.4	62.9	4.52	0.63
4.5	8.0	600	66.1	25.6	73.1	4.94	0.63
6.0	10.0	600	66.0	25.4	71.4	4.84	0.66
4.5	8.0	400	75.0	39.9	59.8	6.45	0.75
6.0	10.0	400	76.5	36.9	57.9	5.95	0.76
4.5	8.0	400	66.1	41.5	66.4	7.31	0.77
6.0	10.0	400	66.0	39.4	67.6	7.42	0.76
Neutral sulfite semichemical process (200b)							
Na_2SO_3 % of wood	Pulping temp.[a] (°C) / time (hr)						
13.8	170 / 1.2	450	76.8	30.0	60.6	—	0.59
20.1	170 / 4.5	450	69.8	36.0	67.7	—	0.64
27.5	175 / 5.0	450	63.8	37.5	64.2	—	0.64
Kraft semichemical process (200b)							
Active alkali % of wood	Pulping temp.[b] (°C) / time (hr)						
7.8	170 / 0	450	76.4	18.0	46.3	—	0.58
9.8	170 / 1.5	450	68.3	22.3	59.8	—	0.63
11.7	170 / 1.5	450	58.3	35.5	73.5	—	0.74

[a] Heating to pulping temperature in 2 hours.
[b] Heating to pulping temperature in 1.5 hours.

9. THE SPENT SULFITE LIQUOR

9.1 Properties and Composition

In studies of the sulfite cooking process, the spent sulfite liquor must occupy a prominent place. A review of the literature shows that a separate

presentation would be justified. However, it is not the goal here to give an exhaustive survey of the pertinent literature. Only the technological importance of the spent liquor for the sulfite pulping process, the possibilities of its processing, its economic utilization, and its disposal will be discussed here.

With limestone as the basis for the preparation of the cooking liquor, there is no economic interest in a processing of the spent sulfite liquor. Recovery of the chemicals used for the preparation of the cooking liquor is both troublesome and uneconomical. The partial utilization of the organic matter in the spent liquor is limited to the use of the sugars for the production of alcohol and yeast, and these sugars constitute only a relatively small portion of the organic products in the spent liquor. Assuming a 47% yield of a normal pulp, based on the weight of the wood, it is clear that for each ton of pulp more than one ton of organic matter is present in the spent liquor. These large amounts of organic material form a tremendous load on the oxygen content of the rivers into which the spent liquor is discharged. Increasingly urgent demands for the prevention of stream pollution have made an intensive study of the spent liquor problem imperative. Disposal of the spent liquor by burning is possible only after it has been concentrated. This involves a considerable heat demand which is economically justified only when it can be provided, to a large extent, by the burning of the organic matter, and such is not the case with calcium bisulfite spent liquor.

Another development which has attracted a great deal of attention in recent years is the recovery of chemicals from spent sulfite liquor. As has been known for many years, the kraft pulping industry supplies a great part of its heat demand by burning its black liquor, with the simultaneous recovery of chemicals. The use of soluble bases in the sulfite cooking process has led to a new kind of disposal process, with the goal not only of destroying the organic material by combustion but also of recovering the chemicals used in the preparation of the cooking liquor. The replacement of the calcium by magnesium, sodium, and ammonium made the evaporation and combustion of the spent liquor interesting and economically necessary, and has proved the practicability of these bases, the advantages of which have been discussed above. In this way, after years of development, the cycle has been completed in which the sulfite pulping industry, through a complete regeneration plant, recovers its chemicals. The organic matter is utilized with economy of heat and, at the same time, with a contribution toward the necessary reduction of stream pollution.

This type of processing of the spent liquor requires a knowledge of its properties and its composition. The first question was: what is the molecular size of the lignosulfonic acid salts in the spent liquor? By fractionated dialysis with membranes of different permeability, ammonium

lignosulfonates with molecular weights of 300 to 400, 1500 to 2000, and about 20,000 were found. In spent sulfite liquors from spruce, lignosulfonates with molecular weights of between 1650 and 2000 were found. When such liquors were heated again, e.g., for 10 hours at 180°C, the molecular weight increased to 3250. These data agree with those of a dialyzed spent liquor which were determined by diffusion, pycnometer, and viscosity determinations and gave molecular weights of 3400 (23,172,348, 352).

During the sulfite cooking process, the molecular weight of the lignosulfonates in the neutral and acid regions varies between 100 and 100,000; however, values of over 15,000 led to the conclusion that the material under investigation had been subjected to a subsequent treatment, e.g., drying at a higher temperature. The surface activity of lignosulfonates, such as their dispersing action on kaolin suspensions, increases up to an optimum molecular weight and then decreases again with a further increase in the molecular weight by polymerization (24).

The viscosity of the spent sulfite liquor depends on its concentration and temperature and plays an important role in the processing. These relationships form the basis of the TAPPI Data Sheet 156. The flow properties correspond to those of Newtonian fluids. Density and solid content show an almost linear decrease with increasing temperature at constant solid content. The density of a magnesium-base spent liquor from beech is somewhat higher than that of spruce at the same concentration, but, in general, sodium and ammonium spent liquors show no essential differences from calcium-base spent liquors (132,172).

The average thermal conductivity of a spent liquor with 60% solid content, according to the physical method of measuring, amounts at 70°C to 0.00132 cal cm^{-1} sec^{-1} degree^{-1}, whereas that of water at the same temperature is 0.00157 cal cm^{-1} sec^{-1} degree^{-1}. The heat transfer coefficients for spent sulfite liquor have been determined for a temperature range of 50° to 135°C. The processes of heat conduction and convection are represented by the heat transfer coefficient. This number is not a specific figure for each substance but rather a function of several properties of the substance, the form and nature of the neighboring areas, and the flow velocity. Through the application of the laws of similarity, general solutions are given for the processes which take place in the heat transfer areas. The resulting coefficients are without dimension and are known as Reynolds, Prandtl, and Nusselt numbers. They correspond in varying degrees to the velocity of the flow, to the density, dynamic or kinematic tenacity, to temperature conductivity, heat transfer, and heat conductivity numbers, and to general characteristic measurements such as, for example, the pipe diameter (132,133,172).

Comparative investigations of the calorific value of spent sulfite and

spent kraft liquors showed 4170 to 4370 kcal per kg and 3250 to 3450 kcal per kg, respectively. These figures refer to a moisture-free substance. The burning of spent calcium sulfite liquor is thus uneconomical even in regions with high costs for fuel (76,264,315).

In accord with the wood components, the composition of the spent liquors varies greatly. By a combination of dialysis, precipitation with solvents, ion exchange, and paper chromatography, the polysaccharides can be separated from the monomeric sugars. Another method of separation is by precipitation with organic bases and fractionation with activated carbon. By after-hydrolysis with sulfuric acid followed by separation by paper chromatography, mannose, galactose, glucose, xylose, and, in some cases, rhamnose were found as building stones of the polysaccharides. These sugars, after they had been separated by paper chromatography and elution, were also identified and determined by oxidation with bichromate and sulfuric acid (14,357). By the use of suitable experimental conditions and appropriate ion-exchange resins, chiefly either sugars or lignosulfonic acids can be obtained. Experiments with known substances have shown that this separation depends upon the different types of exchange resins, on adsorption conditions, and on molecular sieve action (108). By extraction of a 50% spent sulfite liquor with hot ethyl alcohol, carbohydrate and lignosulfonate fractions can be separated (92).

The extraction of a technical ammonium-base spent sulfite liquor from an aspen cook with ether and subsequent fractionation by ion exchange yielded a neutral, a weakly acid, and a strongly acid fraction. These fractions were then subjected to an acid and an alkaline hydrolysis, an alkaline oxidation with nitrobenzene, and a countercurrent distribution. The neutral fractions contained chiefly simple sugars or low-molecular oligosaccharides and some phenolic and ligninlike materials, mainly of the syringyl type. The fractions from the weakly acid portion contained considerable amounts of carbohydrates in glycosidic linkage with phenolic compounds. The countercurrent distribution of this fraction affected the fractionation of the lignosulfonates. The strongly acid fraction consisted chiefly of lignosulfonates. This fraction also gave sugars on acid hydrolysis. An investigation of the original spent liquor for aldonic acids gave xylanic acid as the principal component, in addition to small amounts of arabanic, mannanic galactanic, glucanic, and some unidentified acids (290).

The high-molecular lignosulfonic acids can be separated by dialysis after removal of the mineral acids and the calcium by ion-exchange resins. Other fractions with molecular weights of 3700 to 58,000 can be obtained by ultrafiltration. The lignosulfonates are considered to be flexible polyelectrolytes of which those with molecular weights of below 5000 associate themselves in a way similar to the micelle formation of colloidal electrolytes. The dispersion and adhesion properties of the lignosulfonates are the result

of their polyelectrolytic nature (121). Neutralized and demineralized spent sulfite liquor, after being repeatedly extracted with ethyl ether and amyl alcohol, shows ultraviolet absorption curves which correspond to those of lignosulfonates obtained in high yield and of good purity. The latter was confirmed by the methoxyl and sulfur contents (148). For the determination of the lignin content by ultraviolet absorption, the region of 200 to 205 mμ can be used under the condition that the spent liquor contains less than 5 grams per liter of total sulfur dioxide. Larger amounts of hydrolysis products, such as furfural, do not disturb the result. For spent liquors with more than 5 grams per liter of sulfur dioxide and less than 0.5 gram per liter of furfural, the wavelength at 280 mμ is preferred. For spent liquors that contain larger amounts of interfering substances, the measurement at 200 to 205 mμ can be used after prior oxidation and precipitation-elimination of the sulfites and the sulfur dioxide. This type of control of the course of delignification has been applied successfully in the pulping process of a two-stage sulfite cook according to the so-called Stora process (78,284).

Low-molecular organic acids, which, as sulfurous acid, interfere in the determination of the lignosulfonates can be removed with weakly basic anion exchangers which do not react with lignosulfonates. Concentrated spent sulfite liquor was passed through a column of a strong cation-exchange resin and then eluted with water. The eluate could be separated into several fractions. The major portion of the first fractions consisted of lignosulfonic acids. No difference in the chemical composition of these lignosulfonic acids could be observed; the differentiation between α- and β-lignosulfonic acids seems, therefore, to be unjustified, although there is a considerable difference in their molecular size. About 52% of the dry substance of the spent liquor under investigation consisted of lignosulfonic acids. Small amounts of α-conidendrin, hydroxymatairesinolsulfonic acid, and carbohydrate sulfonic acids were found. The second series of fractions probably contained 4-O-methyl-D-glucuronic acid to which xylose seemed to be attached. The third group consisted chiefly of glucuronic acid, and the fourth group contained polysaccharides which, on acid hydrolysis, gave glucose, mannose, galactose, and xylose. The fifth group consisted of D-glucose, D-xylose, D-mannose, D-galactose, and L-arabinose. Aldonic acid may also have been present. The final group contained chiefly acetic acid and minor amounts of low-molecular products (115,116).

The lignosulfonic acids can be further separated by gel filtration into fractions with various molecular weights. All fractions show a uniform ultraviolet absorption spectrum, with an absorption maximum at 280 mμ, and all have the same methoxyl content. Ligninlike substances can be separated into different components and can be differentiated by ultraviolet absorption (192).

9.2 Evaporation

To prevent pollution of the public water supply, the disposal of spent sulfite liquor into rivers is, in many cases, forbidden. The only solution possible is to evaporate it or even to dry it out completely. Public interest must supersede economic considerations. As mentioned already, the evaporation and burning of calcium-base spent liquors is uneconomical. The solution of the problem of evaporation of sulfite spent liquor has been greatly facilitated by the wide experiences of the kraft pulping industry, which has had to deal with this problem ever since its inception. Especially since the Second World War, great efforts have been made to improve the evaporation processes. The introduction of soluble bases into the sulfite pulping process gave new impetus to these efforts since the problem of the recovery of the chemicals used in the preparation of the cooking liquor was simultaneously connected with the processing of the spent liquors. The disposal of the spent sulfite liquors from soluble bases, therefore, approaches, from a technical point of view, the chemical recovery in the kraft pulp industry (76,231).

The economical utilization of the heat from the organic matter in the spent liquor necessitates three steps to guarantee its economy. These are: (*a*) the separation of the waste liquor from the pulp in a state as undiluted as possible; (*b*) the evaporation of the liquor to a concentration that makes burning possible; and (*c*) as complete as possible utilization of the heat generated. Under normal conditions, the solid content of a spent sulfite liquor amounts to 10% to 14%. At an average pulp yield of 47% of the weight of the wood, about 1025 kg solid material, with a heat value of 4100 to 4400 kcal per kg dry substance, is present in the spent liquor per ton of pulp. For the calculation of the amount of heat, it is important to know how much combustible material is available in the liquor. For this purpose, the amount of liquor, its concentration, and composition must be known. These values should be as high as possible—in other words, as much spent liquor as possible in a high degree of concentration should be recovered. Figure VII–32 shows the gross and net steam production from concentrated calcium-base spent sulfite liquor under the following assumptions: (*a*) the total amount of dry substance, including the chemicals, is taken as being 1170 kg per ton of pulp at an assumed 100% recovery; (*b*) the heat value of the solids amounts to 4400 kcal per kg dry substance; (*c*) the solid content of the evaporated liquor from a five-unit evaporator is 55%; (*d*) the efficiency of the evaporator plant is 3.8:1; and (*e*) the overall efficiency of the boiler plant is 65%. These figures also show the importance of the amount, in percent, of recovered solids and of the solid content of the

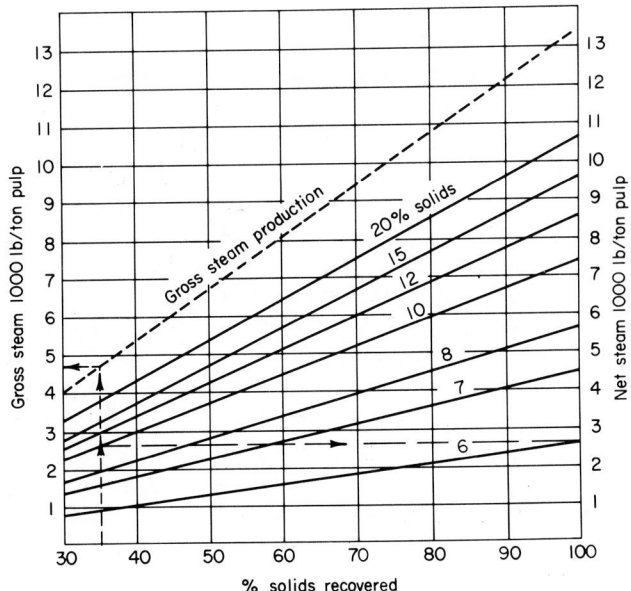

Figure VII-32. Gross and net steam production from sulfite spent liquor in relation to solids content and percent solids recovery (73).

liquor before evaporation to ensure the heat economy of the whole process (73).

For the technical evaporation, various methods are available: (a) the multistage vacuum evaporation; (b) back-pressure evaporation, and (c) evaporation by thermocompression. Under normal conditions a concentration of the liquor to a solid content of about 55% is desired, as is indicated by the above-mentioned properties of the liquor.

The back-pressure process is notable for its great heat economy, but it makes necessary the profitable utilization of the back-pressure steam. Incrusting of the equipment occurs in greatly increased strength. Because of the high temperatures during the evaporation process, thermal decomposition of the liquor may occur (404).

The thermal compression method is especially advantageous when cheap power is available. The steam formed during the evaporation is compressed and passed along the outer wall of the evaporator where it is condensed and its latent heat transferred to the boiling liquor. The total amount of heat consumed is large in comparison with the amount of energy needed for the operation of the steam compressor. The rise in the boiling point is low because of the high molecular weight of the lignosulfonate, and the efficiency of the plant is high with regard to the conversion of fuel to energy and of energy to compressed steam, while, at the same time, the temperature gradient at the heat transfer areas is small (212).

TABLE VII-12

Heat Consumption and Fuel Savings with Different Types
of Evaporation Plants (404)

Type of plant	Spent liquor used in evaporation plant	Amount of heat (kcal/kg evaporated water)				Net coal saving (kg coal/ton 90%)	
		Primary heat require-ments for evap. alcohol plant	Deduc-tion for normal alcohol plant	Change in secon-dary heat[a]	Net heat require-ment	Rayon pulp	Paper pulp
Mill with alcohol plant							
Vacuum 5 effects	Fermented	215	105	−10	120	330	265
Back pressure	Fermented and stripped	140	100	+ 5	35	445	360
Thermo-compression	Fermented and stripped	80	100	−10	− 10	520	420
Mill without alcohol plant							
Vacuum 5 effects	Acid liquor	140	—	− 5	145	410	310
Back pressure	Acid liquor	40	—	0	40	555	435
Thermo-compression	Acid liquor	10	—	+10	0	610	480

[a] (+) = increase, (−) = decrease

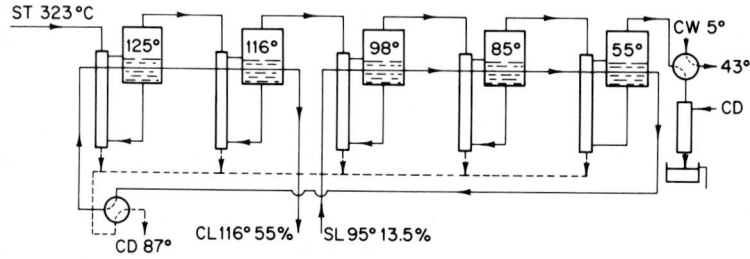

Figure VII-33A. Flow diagram of multiple effect vacuum liquor evaporation system (42). ST = steam; CW = cooling water; CD = condensate; CL = concentrated liquor; SL = spent liquor.

Vacuum evaporators are less efficient. Pulp mills with a subsidiary alcohol plant can combine the evaporation of the liquor with the distillation of alcohol. With a three-stage evaporation for the processing of the condensate formed, an alcohol yield of 98% can be obtained. This yield decreases to 91% with a two-stage unit (318,404).

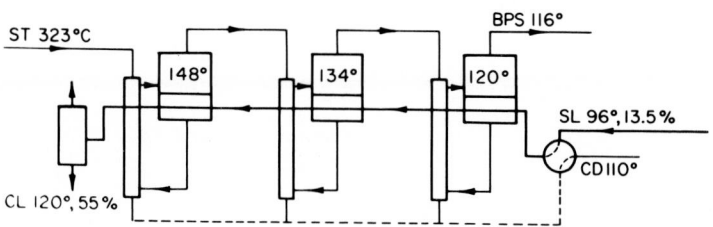

Figure VII-33B. Flow diagram of back-pressure liquor evaporation system (42). ST = steam; BPS = back-pressure steam; SL = spent liquor; CL = concentrated liquor; CD = condensate.

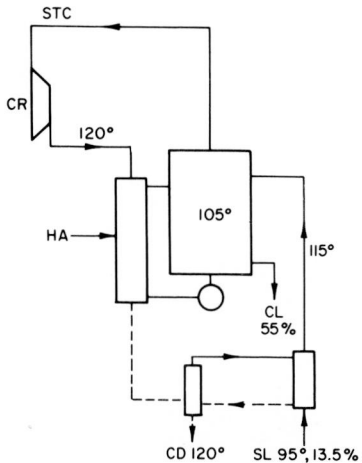

Figure VII-33C. Flow diagram of a thermocompression waste liquor evaporation system (42). STC = steam to compressor; CR = compressor; HA = heat addition; CL = concentrated liquor; CD = condensate; SL = spent liquor.

Table VII-12 shows the heat requirement and the fuel savings with various evaporator plants, while Table VII-13 gives a survey of the fuel and power consumption for a sulfite pulp mill with and without an attached alcohol plant, for dissolving pulp and paper pulp, and for different evaporation systems (44). Figure VII-33, parts A, B, and C, show flow-sheets of three evaporation processes. The vacuum and back-pressure methods can be combined with an alcohol distillation plant. In this way, the conventional alcohol plant can be smaller in size and is also safer in operation. The concentrated spent liquor can be burned both with and without the addition of other fuel (42).

During the evaporation, calcium sulfite and calcium sulfate are present partly in the form of supersaturated solutions. A spent liquor with 10% to 14% solid content may contain between 2 and 3 grams per liter of calcium

TABLE VII-13

FUEL AND POWER CONSUMPTION FOR A PULP MILL WITH LIQUOR EVAPORATION AND COMBUSTION[a] (404)

Pulp grade alcohol plant evaporation system	Dissolving pulp with alcohol plant			Dissolving pulp without alcohol plant			Paper pulp with alcohol plant			Paper pulp without alcohol plant		
	V	B	T	V	B	T	V	B	T	V	B	T
Liquor for evaporation plant	NF	NFS	NFS	Acid	Acid	Acid	NF	NFS	NFS	Acid	Acid	Acid
Heat consumption for mill without evaporation												
Cooking Mcal/ton 90%	—	1300	—	—	1300	—	—	1000	—	—	1000	—
Bleaching (no chlorine dioxide) (Mcal/ton 90%)	—	400	—	—	600	—	—	300	—	—	500	—
Drying (Mcal/ton 90%)	—	850	—	—	850	—	—	850	—	—	850	—
Alcohol distn. (Mcal/ton 90%)	—	850	—	—	—	—	—	700	—	—	—	—
Heating, etc. (Mcal/ton 90%)	—	400	—	—	400	—	—	400	—	—	400	—
Total (Mcal/ton 90%)	—	3800	—	—	3150	—	—	3250	—	—	2750	—
Fuel for mill without evaporation												
Fuel requirements (kg coal/ton 90%)	—	690	—	—	570	—	—	590	—	—	500	—
Fuel available												
Pyrite (kg coal/ton 90%)	—	35	—	—	35	—	—	30	—	—	30	—
Bark and wood refuse (kg coal/ton 90%)	—	70	—	—	70	—	—	60	—	—	60	—
Total (kg coal/ton 90%)	—	105	—	—	105	—	—	90	—	—	90	—
Fuel deficiency (kg coal/ton 90%)	—	585	—	—	465	—	—	500	—	—	410	—
Fuel for mill with evaporation												
Fuel addition from evaporation (kg coal/ton 90%)	330	455	520	410	555	610	265	360	420	310	435	480
Fuel deficiency (−) or surplus (+) (kg coal/ton 90%)	−225	−130	−65	−55	+90	+145	−235	−140	−80	−100	+25	+70
Power for mill with evaporation												
Increase in power requirements (kWh/ton 90%)	40	40	430	40	40	390	30	30	350	30	30	320
Increase (+) or decrease (−) in back-pressure energy (kWh/ton 90%)	+230	−30	−10	+230	−50	0	+190	−60	−10	+190	−50	0

[a] V = vacuum 5 effects; B = back pressure; T = thermocompression; NF = neutralized and fermented liquor; NFS = stripped NF liquor.

sulfate in solution, but this solubility decreases in an evaporated solution of 50% to 55% solid content to about 1 gram per liter (393). The difficulties in combatting incrustations on the evaporator walls have led to the construction of an apparatus which prevents the formation of crusts or eliminates them by suitable flow velocities and changes in the flow direction. It has also been suggested that crystals of gypsum should be added as crystallization nuclei and, in this way, form coarse-grained precipitations (12).

A spray film evaporator, originally developed for the processing of sea water, was used successfully for the evaporation of spent sulfite liquor. The special advantages of this type of evaporator are: (a) a considerable limitation in pipe lines and valves, and thus an appreciable saving in installation costs and maintenance expenses; (b) undisturbed operation of the plant even if the liquor is inclined to foam; (c) relatively low construction costs due to smaller space requirements than those of conventional vertical evaporators; (d) the steam-compression cycle can be used also with the spray film evaporator, thus securing very low total energy costs. Evaporation plants with a three-, four-, and five-fold effect can be built with this system, thus allowing the use of low-pressure waste steam. Precipitations of calcium sulfate can be removed by periodic washing with condensate or fresh water. The spray film evaporator has the same heat transfer economic value as evaporation aggregates with forced circulation or jet feeding but requires considerably less energy consumption for the pumps. Ventilating preheaters take care of the removal of noncondensable gases (126).

The suggestion has recently been made that the evaporation should be carried out by direct contact of the liquor with the flue gases of the boiler. This principle plays an important role in the concentration of the kraft black liquor and will be discussed in the next chapter. The incrustation of the evaporation aggregates during the concentration of spent sulfite liquor has increased interest in such a direct-contact evaporation process. Although it has been said that this type of evaporation has an inferior heat efficiency factor, more recent developments have shown that such a process can also be carried out in a "multiple effect" manner. The number of heat exchangers, which give rise to incrustation, can be greatly reduced. What further technical developments this promising evaporation process will bring remains to be seen (107). In this connection, it should be mentioned that the freezing out of the sulfite liquor has been suggested for the concentration, in lieu of evaporation (265,279). The submerged combustion evaporation, which is used in the concentration of wood sugar solutions, has also been recommended for the evaporation of neutral spent sulfite liquor (285).

The tabulation of a sulfur balance in an evaporation plant may be of interest. When the amount of sulfur in the liquor as it leaves the digester is assumed to be 100%, the figures given in Table VII-14 are found for the

TABLE VII–14

SULFUR BALANCE IN MILLS WITH EVAPORATOR SYSTEMS (337)

Mill	Aa	Ab	B	C	D
Pulp	Rayon	Paper	Rayon	Rayon	Strong[a]
Liquor	NFS[b]	Acid	Acid	NFS	Acid
Evaporator type	Back press.	Back press.	Back press.	Vacuum	Vacuum
Liquor temperature, °C max.	155	155	150	100	123
Liquor temperature, °C min.	117	117	138	40	56
Percent sulfur in spent liquor	100	100	100	100	100
Concentrated liquor	87	80	80	98	83
Condensate	9	9	9	2	4
Steam (6th stage)	3	9	3	0	13
Vent steam	1	2	8	0	—

[a] Hard or high chlorine number pulp.
[b] Neutralized, fermented and stripped liquor.

different cooking processes, depending upon whether the liquor had been fermented or not. The table shows that the loss of sulfur in the condensate, in the waste steam of the final stage, and in the ventilation steam increases with increasing temperature and decreasing pH of the spent liquor (337).

In cases where the spent liquor is further processed, e.g., for the production of chemical products, it must be obtained, in many instances, in a completely dry state. To achieve this, the liquor is first concentrated to a solid content of 20% to 25% and then dried by flue gases in a spray drier (85,112,224,231,331).

9.3 Combustion of Spent Liquor

With the burning of the organic matter in the spent liquor, the heat value of the liquor should be utilized as completely as possible, with the goal of supplying the maximum amount of the heat required for the evaporation. In addition, the inorganic chemicals in the liquor must be recovered. Depending upon the base used in the pulping process, very different problems arise. With calcium sulfite spent liquor, the sulfur is the only chemical worth recovering. With the soluble bases, the recovery of the base itself is an economic necessity. Figure VII–34 shows the amount of heat that can be recovered on evaporation and burning of calcium-base sulfite liquor. The regenerated heat is expressed in kilograms of coal, having a heat value of 6500 kcal per kg. The importance of the dry content of the spent liquor, and the difference in the evaporation in the multistage vacuum and back-pressure processes can be seen from the curves. The concentration of the original spent liquor before evaporation is of great importance for the heat

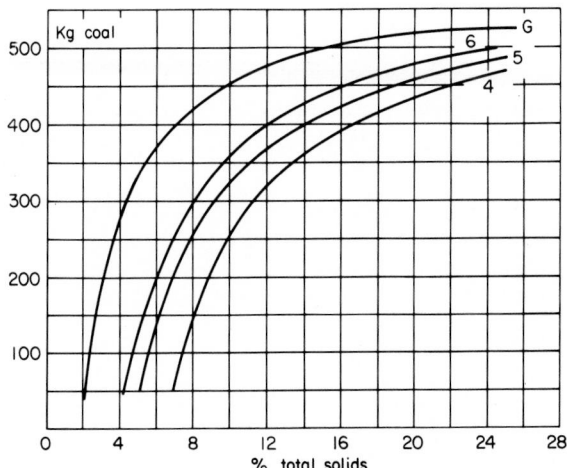

Figure VII-34. Heat production from sulfite spent liquor (as kilograms of coal with 6500 kcal per kg) in relation to dry content and evaporation system used (86). G = back-pressure evaporation; 4, 5, 6 = vacuum evaporation system with respective number of effects.

economy and for a possible fermentation of the liquor to alcohol. This concentration depends, in turn, to a great extent on how far the liquor has been drawn from the pulp—in other words, to what extent the liquor can be taken off without surpassing the limits of economy. Results of Swedish investigations along these lines are listed in Table VII-15. In this table, the "economic solid content" T_e is mentioned. This is the organic matter content of the spent liquor and, on its economic utilization in the form of heat produced by its burning, in the form of saleable alcohol, or by combining these uses, it just covers the costs of the processing of the liquor (evaporation and alcohol production).

The table shows the importance of the dry content of the liquor for all further processing. The economic solid content for alcohol production plus evaporation is below that solely for evaporation, but above that solely for alcohol production. This means that, with a combined alcohol production and evaporation, the same amount of liquor gives an economic surplus in the alcohol production but a deficit in the evaporation. The table further shows the effect of the price of coal on the economy of the operation (86,335).

Completely different results are obtained with the evaporation of a liquor to include the recovery of chemicals. Table VII-16 gives the combustion products and the regenerated chemicals from spent sulfite liquors with different bases. The amount of liquor varies with the different cooking processes and amounts, in a sulfite process, to 8 tons, in a kraft process, to 5 tons, and in a neutral sulfite cook, to 2 tons per ton of air-dried pulp.

TABLE VII-15

ECONOMICAL SOLID CONTENT, REQUIRED IN THE VARIOUS
EVAPORATION STAGES (335)

	T_e	
Type of pulp	Rayon pulp	Paper pulp
Price of coal A[a]		
Alcohol plant only	27.4	30.4
Evaporation only (65)[b]	69.9	69.9
Evaporation only (155)	89.5	89.5
Alcohol production + evaporation (65)	54.2	56.5
Alcohol production + evaporation (155)	66.0	68.9
Price of coal B[c]		
Alcohol plant only	44.6	49.3
Evaporation only (65)	46.1	46.1
Evaporation only (155)	70.6	70.6
Alcohol production + evaporation (65)	52.1	53.7
Alcohol production + evaporation (155)	69.7	71.6

[a] Coal price 40 sKr/ton.

[b] Heat consumption in kcal/kg water evaporation.

[c] Coal price 85 sKr/ton. Efficiency of the boiler plant: 0.78. Steam consumption in alcohol plant: 142 kg steam/ton of fermented liquor. Sugar yield × degree of fermentation × fermentation efficiency = 0.75. Spent liquor from rayon pulp: dry solid content 155 kg/ton, fermentable sugar 32.5 kg/ton. Spent liquor from paper pulp: dry solid content 140 kg/ton, fermentable sugar 26.6 kg/ton.

TABLE VII-16

COMBUSTION PRODUCTS AND CHEMICALS RECOVERED IN DIFFERENT
SULFITE PULPING PROCESSES (420)

Process	Combustion products		Recovered chemicals
	Ash or smelt	Flue gas	
Ca-sulfite	CaO, $CaCO_3$, $CaSO_4$	SO_2	none
Mg-sulfite	MgO	SO_2	$MgO + SO_2$
Na-sulfite	Na_2S, Na_2CO_3	SO_2	$Na_2S + Na_2CO_3$
NH_4-sulfite	none	SO_2, N_2	none

Steam generation and requirements for the SCA-Billerud recovery process (25) are given in Table VII-17. The net production of electric power has been calculated to be 450 kWh per metric ton a.d. pulp, with a back pressure of the turbine of 3.5 atm. Calculations show that the heat and power requirements for the entire mill will be very closely balanced by its own production of steam and electrical power, when steam generation from burning bark is included. The fairly large amount of fuel oil required for

TABLE VII-17

Steam Generation and Requirements for the SCA-Billerud
Recovery Process (162)

	High-pressure steam 64 atm 450°C		Low-pressure steam 10 atm saturated	
	metric	tons/ton	metric	tons/ton
Steam generation from	tons/hr	a.d. pulp	tons/hr	a.d. pulp
Gas cooler	—	—	20	1.6
Sulfur burner	—	—	2	0.1
Steam boiler	50	4.0	—	—
Total	50	4.0	22	1.7
Steam consumption in the recovery plant	—	—	9	0.7
Net steam production	50	4.0	12	1.0

the recovery process therefore does not add to the total operation cost of the mill (162).

The economic data for the amount of heat and the chemicals that can be recovered from the various pulping processes are shown with special clarity in Fig. VII-35. This clearly illustrates why the optimum recovery of the solids from the magnesium- and sodium-base cooking liquors is higher than from the calcium- and ammonium-base liquors. The data in Fig. VII-35 are shown on the assumption that the simultaneous production of alcohol is not involved.

With regard to conditions for the burning of the spent liquor, the simultaneous production of alcohol by neutralization and fermentation of the spent liquor can be considered to be economically feasible only if the heat value of the hexoses in the liquor surpasses the calorific value of the liquor. This consideration, in some cases, has led to the fact that, with high fuel costs, some pulp mills which had combined the calcium bisulfite cooking process with the production of alcohol have abandoned the latter for economic reasons. On the other hand, the economy of the process can be improved by a suitable combination of fermentation and evaporation when, for example, the fermentation is preceded by a partial concentration, or when the fermented liquor is distilled in the evaporation plant (420).

An evaporated spent sulfite liquor with 50% solid content contains an average of 5% to 8% calcium oxide, 5% to 8% sulfur, and 7% to 15% ash. The sulfur content of the coal is almost completely converted into sulfur dioxide but that of the spent liquor only to the extent of 75%, with the remainder left in the ash. Fly-ash and sulfur dioxide cause damage and incrustations in the boiler tubes. Attempts to recover the sulfur dioxide

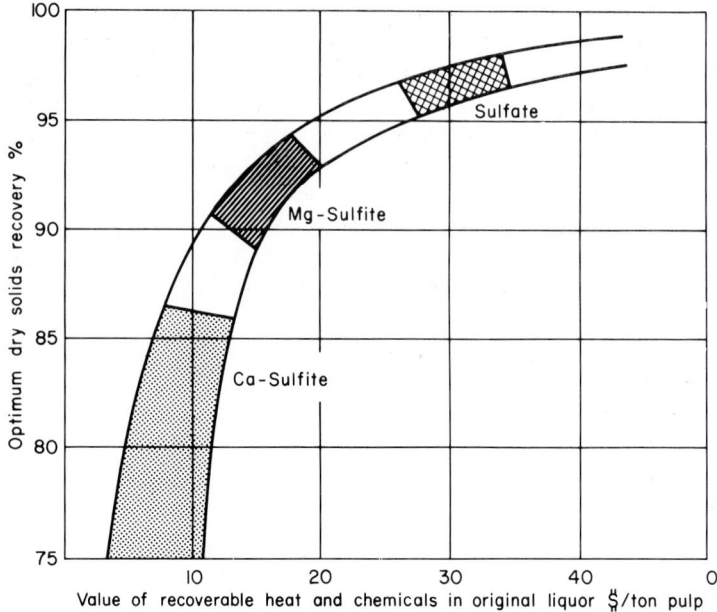

Figure VII-35. Optimum sodium sulfite liquor recovery as a function of the heat and chemical value (420).

and the fly-ash from the flue gases with scrubbers showed that 30% of the sulfur dioxide is taken up by the ash sludge and 60% is oxidized to sulfate (336). The physical properties of the fly-ash, especially its sedimentation velocity, play an important role. Its chemical composition varies with the various combustion conditions such as the combustion temperature and the temperature of the spent liquor and its solid content, and the load in the combustion oven. A maximum combustion temperature of 1400°C can be reached by increasing the temperature of the liquor and its solid content. At this temperature, the ash contains about 47% calcium oxide and 30% calcium sulfate. When the temperature is increased by additional oil-firing to 1550°C, the calcium oxide content is increased to 63%, while the calcium sulfate content decreases to 12%. Simultaneously, the sulfur dioxide content of the flue gases increases to 0.75% (90,199).

The recovery of the sulfur from the flue gases of the boiler depends on the reaction equilibriums of the single gas components under the various reaction conditions for the formation of the corresponding oxides, sulfides, and sulfates from the calcium- and magnesium-base liquors. With calcium spent sulfite liquors, the recovery of the sulfate is simple, but it is of no economic or technical interest. Sulfur can be recovered as sulfur dioxide from the flue gases, or as sulfide from the ash. The maintenance of constant

reaction equilibriums and the occurrence of side reactions during the cooling of the gases and the ash cause difficulties. With magnesium-base spent liquors, all the sulfur is present in the gas as sulfur dioxide or hydrogen sulfide and, as a result, the reaction conditions can vary within a wide range. No magnesium sulfide is obtained in the burning (359,424).

The direct injection of the concentrated liquor into the combustion chamber of the boiler results in incomplete combustion. A cylindrical combustion chamber was installed in front of the boiler and here the liquor was blown, under steam pressure, in the form of a fine spray while the combustion air was introduced tangentially, with the resulting turbulence causing a good mixing of the liquor spray and the air. Especially efficient combustion has been obtained when the liquor spray has a solid content of between 52% and 58% and is injected at a velocity of 2.5 to 3 kilograms per minute. The amount of air required reaches 115% of the theoretical value. Such a combustion chamber is known as a Loddby chamber and has become a pertinent part of many liquor-combustion installations (29,264, 272,361).

The fluidized-bed process has also been applied to the combustion of spent sulfite liquor. The process has already been discussed in connection with sulfide roasting and will be met again in Chapter VIII in the fields of drying, calcination, lime mud reburning, and incineration. This spent liquor combustion system represents an economical method for the abatement of water and air pollution by converting the spent liquor into a granular inorganic solid. Weak spent liquor is pumped at 14% solids content into a two-effect evaporator where the solids concentration is increased to 20.5%. The liquor is then metered into a Venturi evaporator-scrubber and is there concentrated to 41% solids by direct contact with the reactor exit gases which enter the throat section at about 1200°F and exit from a cyclone at about 220°F. This concentrated liquor is continuously recycled to the Venturi throat section at a rate determined by a density recorder-controller in the recycle loop. A bleed-off from this loop conveys the feed liquor to the reactor and meters the liquor directly into the combustion bed through a series of feed guns. A bed temperature of 1350°F is maintained by the autogenous combustion of the organic fraction of the solids. The inorganic ash, consisting of a mixture of sodium sulfate and sodium carbonate (in the case of a NSSC process), adheres to pellets already in the bed. The space above the fluid bed, called the freeboard, is expanded in diameter to decrease the velocity of the rising gases and to alloy fine solids to fall back into the bed to act as nuclei for new pellet growth. The exit gases pass through a hot cyclone collector to remove the entrained dust. Ash pellets, which range from 10 to 100 mesh in size, are discharged from the combustion bed through an internal discharge pipe. The rate of this flow is automatically controlled to maintain a predetermined depth in the combustion bed. The reactor is provided with a gas-fired burner to preheat

the bed at start-up. This spent liquor combustion system is not limited to the disposal of NSSC spent liquor; other pulping waste liquors lend themselves to equally effective disposal. However, operating conditions with respect to combustion temperatures and minor flow-sheet modifications change for the various bases (92a).

The so-called wet combustion of the spent liquor is based on a completely different principle and should, more correctly, be called a flameless oxidation. This type of process was originally developed for the disposal of municipal sewage. As the so-called Zimmermann process, it has also been applied in the processing of spent sulfite liquor. Air is used as an oxidizing agent and it oxidizes the organic matter at temperatures of 300° to 320°C and pressures of 180 to 200 atmospheres. The pressure in the oxidation vessel has to be high enough that the main part of the water is still present in a condensed form. The heat development corresponds to the upper calorific value of the dry substance in the liquor; this solid content should amount to between 3% and 15%. Higher concentrations are undesirable because the reactor may "cook itself dry." Sawdust, peat, and sludges of other organic materials suspended in water may be added to increase the calorific value. The high pressure in the reactor requires an efficient compressor and a corresponding consumption of power. The steam and gas coming from the reactor can be utilized in a turbine for the production of power. Such a reactor reaches, under the most favorable conditions, an efficiency of 85% as compared with 62% for a back-pressure evaporation plant (43,146).

A different type of wet combustion is wet carbonization. In this case, also, a relatively high temperature, of about 225°C, and correspondingly high pressures are used. In contrast to the flameless oxidation, only a part of the organic material is oxidized, while the other part separates as a carbonaceous residue. For the complete oxidation of low-molecular organic acids, such as acetic acid, which escape with the steam-gas mixture, copper-manganese-chromite catalysts are needed. This principle of wet carbonization was investigated decades ago, but at that time the necessary high temperatures and pressures could not be technically applied. Even today, flameless pressure oxidation and wet charring depend on a satisfactory solution to the apparative and heat-economy problems for an economic processing of spent sulfite liquor. To this must be added the fact that the increasing use of soluble bases demands the simultaneous solution of the problem of the recovery of the chemicals (13,31,59,60,353,389,438).

9.4 Recovery of the Chemicals from the Spent Liquors

The recovery of the chemicals from the spent liquor can be divided operationally into three stages: (*a*) concentration or evaporation of the liquor, whereby liberated sulfur dioxide is led to the cooking liquor prep-

aration plant; (*b*) combustion of the concentrated liquor, in which step the freed sulfur dioxide again goes into the liquor preparation plant; and (*c*) chemical processing of the basic components in the melt. Although the methods for the working up of the spent liquors from the sodium-, ammonium-, and magnesium-base cooking processes are largely similar, they will be treated separately here.

9.4.1 SODIUM-BASE SPENT SULFITE LIQUORS

The processing of this type of liquor can be accomplished in two different ways: either the hydrogen sulfide is liberated from the sodium sulfide-containing ash in the combustion chamber, or the sulfide is oxidized directly to sulfite.

The so-called "Mead" process is based on the utilization of a sodium sulfite solution in the cooking process. The spent liquor is evaporated to a solid content of about 50% and is then further concentrated by bringing it into contact in a gas scrubber with boiler gases, whereupon the concentration is increased to 65%, with the simultaneous absorption of the gases and flue ash. The combustion of this liquor is then carried out in a conventional kraft black liquor combustion chamber. The melt is dissolved and clarified. Thus far the process does not differ essentially from the operation of the spent liquor recovery plant of a kraft mill. Further processing then takes place according to the following reaction scheme:

1. Precarbonization:

$$Na_2S + H_2S \longrightarrow 2\ NaHS$$
$$Na_2S + CO_2 + H_2O \longrightarrow Na_2CO_3 + H_2S$$

2. Carbonization:

$$2\ NaHS + H_2O + CO_2 \longrightarrow Na_2CO_3 + 2\ H_2S$$
$$2\ NaOH + CO_2 \longrightarrow Na_2CO_3 + H_2O$$

3. Final carbonization:

$$Na_2CO_3 + H_2O + CO_2 \longrightarrow 2\ NaHCO_3$$

4. Sulfitation:

$$SO_2 + H_2O \longrightarrow H_2SO_3$$
$$H_2SO_3 + 2\ Na_2CO_3 \longrightarrow Na_2SO_3 + 2\ NaHCO_3$$
$$H_2SO_3 + 2\ NaHCO_3 \longrightarrow Na_2SO_3 + 2\ H_2O + 2\ CO_2$$

During the carbonization steps, sodium sulfide and sodium carbonate are converted chiefly into sodium bicarbonate and the sulfur escapes as hydrogen sulfide and is oxidized under the boiler to sulfur dioxide. The sulfur dioxide formed is absorbed in the sulfitation step. According to this

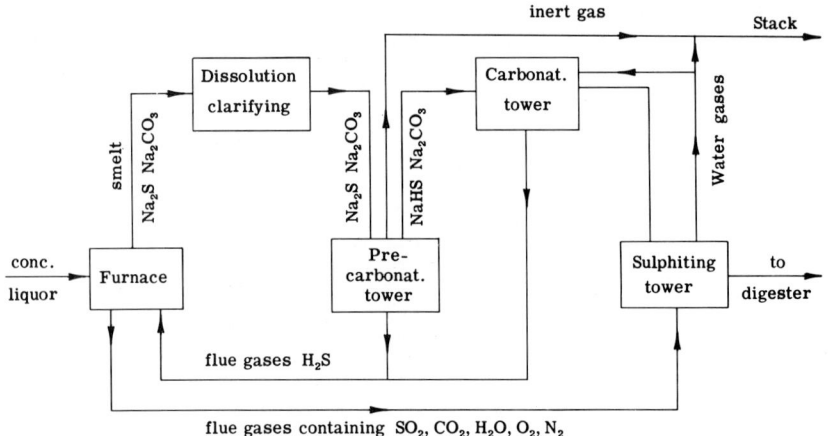

Figure VII-36. Mead NSSC recovery system (250).

scheme, a yield of chemicals of up to 95% can be obtained. When this procedure is applied to the bisulfite pulping process, additional sulfur must be added in the sulfitation step. Some difficulties may occur in obtaining complete absorption of the carbonate dust from the flue gases, and as a result of the relatively high thiosulfate content of the cooking liquor prepared from the regenerated chemicals (250) (see Fig. VII-36).

The direct sulfitation of the dissolved melt by sulfur dioxide has been applied for many years in the neutral sulfite cooking process. This direct sulfitation of the dissolved melt by sulfur dioxide-containing gases has been studied, particularly with regard to the formation of thiosulfate. An improved system, which uses a bisulfite-monosulfite solution, has been developed for this sulfitation process. Under the most favorable reaction conditions this bisulfite method gives a cooking liquor which contains only 2 to 3 grams of thiosulfate per liter. The advantage lies in the simple sulfur recovery (425). In the process, the melt is mixed with bisulfite solution and, as soon as it reaches its boiling point, is injected into a chamber. The solution flows down an absorption column while steam simultaneously moves upward in a countercurrent stream. Hydrogen sulfide and carbon dioxide are expelled as reaction products and are removed as acid gas. A part of the reaction solution is used for the preparation of the bisulfite solution by absorption of sulfur dioxide gas. The acid gas, after a partial condensation, is burned in a sulfur burner and in this way forms the sulfur dioxide required in the sulfitation cycle (427).

The so-called "Western" precipitation process, shown in the flow-sheet in Fig. VII-37, separates the sodium carbonate from the sodium sulfide in the melt by crystallization. The acid spent sulfite liquor from the digester

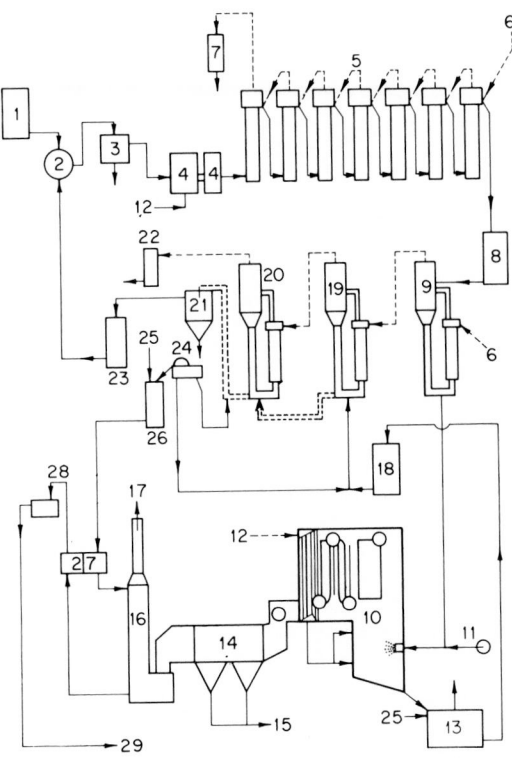

Figure VII-37. Flow diagram of the Western precipitation recovery system (39). 1 spent liquor tank; 2 pump; 3 thickener; 4 Bradley reactor; 5 multiple effect evaporator (kraft type); 6 steam; 7 condenser; 8 concentrated liquor tank; 9 concentrator; 10 boiler; 11 fan; 12 air to furnace; 13 green liquor; 14 Cottrell precipitator; 15 sodium sulfate; 16 sulfur dioxide; 17 waste gas; 18 storage tank; 19, 20 crystallizers; 21 separator; 22 condenser; 23 sodium sulfide tank; 24 soda filter; 25 water addition; 26 dissolution of soda; 27 liquor preparation; 28 filter; 29 acid bisulfite liquor to sulfite mill.

is mixed with sodium sulfide mother liquor by means of a pump, with the admixture of the sodium sulfide solution being controlled by an automatic pH meter. Calcium carbonate, flue gas residues, and pulp fibers in the liquor are removed in a settling tank, which allows two hours time for the settling. The liquor then goes to a Bradley reactor in which it develops a large reaction surface through the formation of foam, thus causing the sulfide and sulfite ions to be oxidized to thiosulfate and polysulfates. This reaction requires 15 to 30 minutes. The liquor is then passed through a defoaming chamber where it is deaerated and then is led into a storage tank. From there it goes to the evaporation plant where it is concentrated to a 45% to 50% solid content. The concentrated liquor then goes into a multiple effect concentrator and from there into the crystallization plant.

In the former, the solid content is increased to 65% to 70% and, in this state, the liquor is burned under the boiler. A part of the concentrated liquor is used for the dissolution of the sodium sulfate precipitated in the Cottrell gas precipitator and is then combined with the liquor which is injected under the boilers. The melt, flowing out from the burner, consists of an almost equimolecular mixture of sodium sulfide and sodium carbonate. The flue gases from the boiler contain some sodium sulfate which is precipitated in the electric gas purifier, i.e., in the Cottrell unit, and about 0.7% to 1.5% sulfur dioxide, which is absorbed by a solution of sodium sulfite and sodium bisulfite in a countercurrent in a tower. The solution leaving the absorption tower consists of sodium bisulfite to the extent of 90%. The melt is dissolved in water to a controlled concentration and the solution is passed, via a storage tank, to the crystallization plant (steps 2 and 3). The sodium carbonate crystallizes out, is filtered off, washed, and placed in a storage tank near the absorption tower. There it is mixed with the solution coming from the tower and is passed over the tower for further enrichment. The enriched solution is filtered and used in the digester room. The sodium sulfide mother liquor goes with the fresh spent liquor, as described above, to the Bradley reactor (36).

More water has to be evaporated than in the normal evaporation process because the melt must be dissolved for the crystallization of the sodium carbonate and then must be reconcentrated. As long as the two evaporation plants are well synchronized, the apparently higher steam consumption is extensively compensated for by equivalent heat recovery. The absorption tower works with an efficiency of 97% and without odor annoyance. The flow-sheet in Fig. VII–37 explains the process (26,37,39,255,288).

Another regenerating process, which is especially suitable for a two-stage cook, is the Sivola recovery process. This two-stage cooking procedure has been described before (48,289). Figure VII–38 shows a simplified scheme of this recovery process. The dilute spent liquor is first evaporated to a solid content of 53%, then concentrated in a Cascade evaporator to 62%, and, finally, is burned under the boiler. The melt is clarified and dissolved as usual. The green liquor is then reacted with the flue gases in the precarbonization stage. The carbon dioxide converts the sodium sulfide into sodium carbonate and sodium hydrosulfide. In the main carbonization step, the sodium hydrosulfide is converted into hydrogen sulfide, with the formation of sodium bicarbonate. The latter is thermally decomposed to sodium carbonate and carbon dioxide, and a small part of the carbonate is used for the preparation of cooking liquor. The main part is transported to the Claus reactor and is treated there with bisulfite, with the formation of sodium sulfite and carbon dioxide. Part of the sodium sulfite is used in the cooking process, the other part is used for the formation of sodium bisulfite. The sulfur dioxide required for the sulfitation is produced by burning the

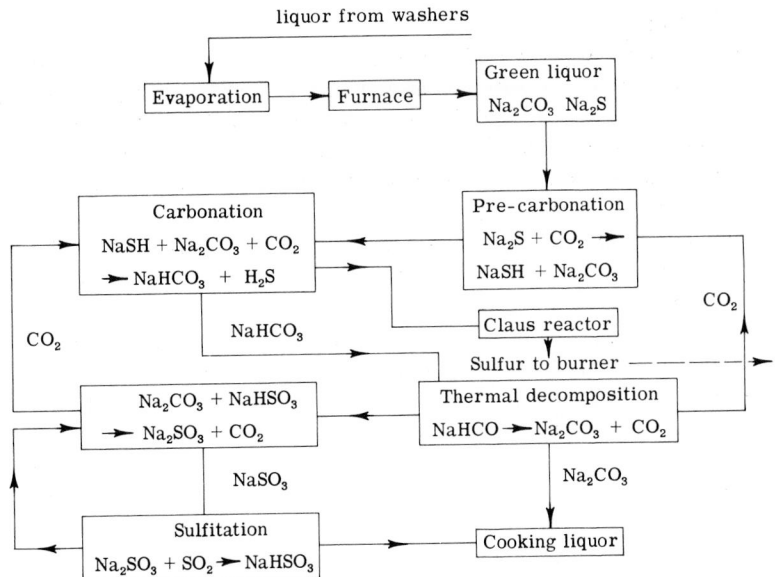

Figure VII-38. Simplified diagram of the Sivola recovery process (316).

hydrogen sulfide, formed in the carbonization step, in the sulfur burner where additional sulfur is burned. The whole process is self-supporting with regard to heat consumption and also produces additional mechanical energy (316).

In contrast to the processes already discussed, in the "Storafite" process of the Stora Kopparberg Bergslag, the sulfur is precipitated in its elementary form. For this purpose the melt is dissolved to give a concentration of 100 to 130 grams per liter of active sodium oxide and a sulfidity of about 75%. In the precarbonization tower, the sulfide is decomposed by passing in carbon dioxide in a countercurrent stream, thus liberating hydrogen sulfide. Steam, carbon dioxide, and 16% to 18% of hydrogen sulfide escape from the tower; the steam is condensed and the gas absorbed in a Claus reactor where the hydrogen sulfide reacts with sulfur dioxide to give elementary sulfur. The exothermic reaction takes place by the catalytic action of bauxite, according to the following equation:

$$2 \, H_2S + SO_2 \longrightarrow 2 \, H_2O + 3 \, S$$

The reactor consists of two towers, one acting at a higher, the other at a lower temperature. Following each tower there is a gas scrubber in which the gas is washed with liquid sulfur at 135° to 140°C. The sulfur leaves the scrubber at 150°C and cooling then takes place in a heat exchanger, with the formation of steam which is used for the heating of the carboni-

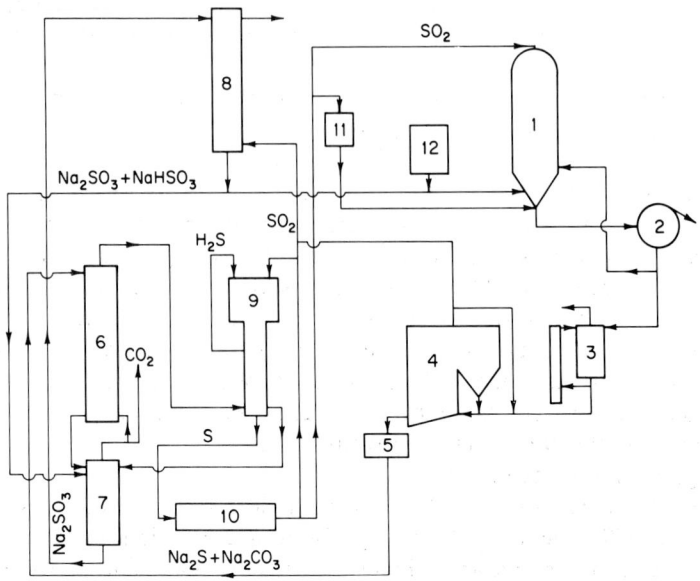

Figure VII–39. Stora Kopparberg recovery system for sodium bisulfite liquors (104). 1 digester; 2 pulp washers; 3 evaporators; 4 recovery unit; 5 smelt dissolver; 6 carbonation tower; 7 decarbonation tower; 8 absorption tower; 9 Claus reactor; 10 sulfur burner; 11 compressor; 12 first-stage "drawn" liquor tank.

zation tower. The excess sulfur goes into the sulfur burner. The concentration equilibriums between hydrogen sulfide, sulfur dioxide, and sulfur in the Claus reactor are regulated by proportional composition of the components, by the steam content, and by the temperature. The reaction can be carried out only in steps since the excess of sulfur dioxide and the temperature have to be held as low as possible. The gas mixture, consisting of carbon dioxide and hydrogen sulfide from the carbonization tower, is mixed with an adequate amount of sulfur dioxide and is passed at 300°C into the first reaction tower. Because of the washing with liquid sulfur, the gas leaving the tower is considerably poorer in hydrogen sulfide and enters the second tower at a temperature of 240°C. The sulfur goes into the sulfur burner, the carbon dioxide into the carbonization tower. Figure VII–39 shows a flow-sheet of the Storafite regeneration process (61,104,338).

The regeneration processes hitherto discussed have in common, to a certain degree, a prior concentration of the liquor and subsequent combustion, followed by processing of the melt. It has also been suggested that the original dilute spent liquor should be processed directly. Two different methods are available for this purpose: the organic matter in the liquor can be converted to gas by pyrolysis, or the base can be recovered by ion

exchange. The pyrolysis process also requires an extensive concentration of the liquor, but the ion-exchange process can be carried out without prior concentration.

The pyrolysis process has become known as the "atomized suspension technique" or, in short, the AST process. In this process, a sodium bisulfite spent liquor is sprayed into a reactor that is kept at two atmospheres pressure and is heated to a wall temperature of 650° to 800°C. This causes an almost instantaneous pyrolysis through the contact of the finely suspended solid particles with the hot walls of the reactor. The reaction time amounts to only 15 to 20 seconds. In this way, the sodium salts are precipitated to an extent of 90% as carbonate and 10% as sulfate. The gases formed possess a high calorific value and consist of hydrogen sulfide, carbon monoxide, hydrogen, and hydrocarbons. After separation of the solids in a cyclone and a washer, the noncondensed gases are burned under the boiler. The heat of the flue gases is recovered in an economizer, and the cooled gases go to the sulfitation plant.

For the elucidation of the most suitable reaction conditions, a concentrated NSSC spent liquor was pyrolyzed under various temperature and pressure conditions and at a higher rate of flow than is normally used. It was found that the conversion of the organic components into a gaseous form and the formation of carbonate were complete at a temperature of 800°C and with a limited input. The pressure in the reactor had little effect on the type of reaction products, but a higher concentration of the injected liquor caused a slight reduction in the gas formation. When the reaction temperature was lowered and the input was increased, a large proportion of carbon remained in the residue. In spite of this, the formation of carbonate amounted to an average of 85% to 90%. The proportion of sulfate in the residue was independent of the sulfate content of the original spent liquor. A build-up of sulfate, therefore, was not to be feared in the closed system. The heat of the pyrolysis gases surpassed that required, thus the whole process was self-supporting with regard to heat economy when the liquor concentration was more than 50%. A special advantage of the process is the low proportion of sulfate and thiosulfate in the solid residue. A problem still remaining is the strong corrosion of the walls of the reactor at the high temperatures required (22,122,233,332).

When concentrated spent sulfite liquor is heated rapidly, it is converted into a powder which contains all the sodium as carbonate and produces a gas that contains all the sulfur as hydrogen sulfide. Two Swedish mills, the Svenska Cellulose A.B. and the Billerud A.B., have developed a process known as SCA-Billerud shock pyrolysis. This shock pyrolysis is carried out by the addition of limited amounts of air, with part of the heat required for the pyrolysis being provided by partial combustion of the organic substances. The reaction products, which can be readily separated, consist of

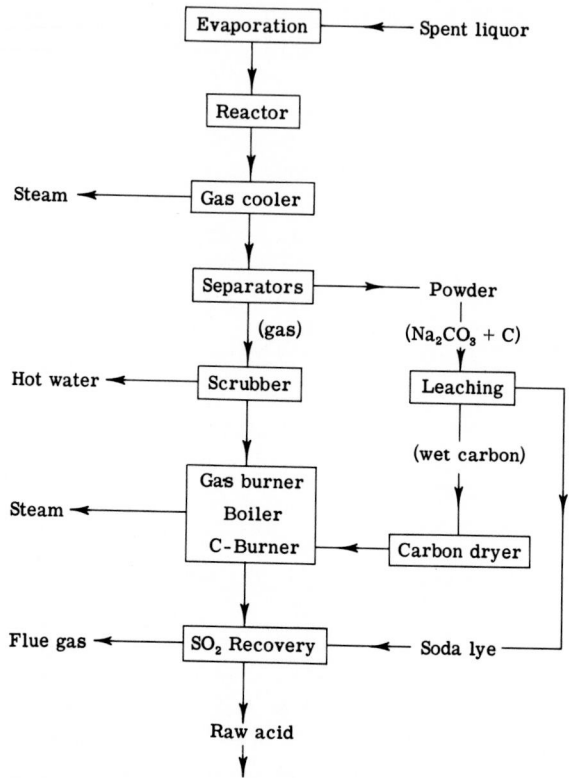

Figure VII–40. Principle of the Svenska Cellulosa A.B.-Billerud recovery process (162).

sodium carbonate, which is free of sulfide, a combustible gas, and various amounts of carbon. The gas and carbon are burned in a conventional boiler and the flue gases are reacted in a scrubber with sodium carbonate solution to form crude acid for the sulfite cook (25).

The main aggregate of the whole plant is the reactor. The concentrated liquor, together with combustion gases and an excess of air, is injected into the reactor where the amount of oxygen is insufficient for the complete combustion of the organic matter. The direct transformation of all the sodium salts into carbonate is unique for such a regeneration system. The almost complete separation of the sulfur is also typical for this process. Since these transformations take place in a single reaction sequence, the simplicity of the procedure is striking. A conventional apparatus can be used for the operation of the process. Figure VII–40 shows a flow-sheet of the method. The only novelty is the reactor itself. The spent liquor is sprayed into the reactor, together with a large excess of air, by means of steam and hot flue gases from an oil burner. The carbon content of the ash

decreases with increasing temperature in the reactor, but it is very difficult to obtain a powdery residue completely free of carbon without surpassing the melting point of the carbonate. For this reason the temperature in the reactor is limited to about 750°C. Data on the possible steam production and the power consumption are given in Table VII–17. From this it can be seen that the heat and power requirements for a whole pulp mill can be covered almost completely by the steam which is produced when that generated by the burning of the bark is included. About 95% of the sodium and 85% of the sulfur originally used can be recovered. The cooking liquor prepared from the regenerated chemicals contains about 21% of the total sodium as sulfate and less than 1.5% as thiosulfate. The relatively high proportion of sulfate has been found not to be detrimental in the bisulfite cook (162).

A prior concentration of the spent liquor is unnecessary for the ion-exchange process. Sulfonic acid-type exchange resins are used as cation exchangers and the regeneration of the exchanger is carried out with a sulfurous acid solution. The eluate from the exchanger contains about 1% of bound sulfur dioxide, and about 65% of the sodium can be recovered. The remaining liquor has a very low ash content and is well-suited for evaporation and burning or for the production of by-products (87,246). Although evaporation of the liquor is no longer necessary for the recovery of the chemicals, the problem of stream pollution remains unsolved because of the organic matter still present in the liquor. Consequently, even in this process, evaporation and burning, with simultaneous heat recovery, are unavoidable and are required for technical reasons relating to waste water.

An ion-exchange process, known as the "Abiperm" recovery process, is supposed to be suitable also for magnesium- and ammonium-base spent sulfite liquors. The ion-exchange cell is charged with a sulfonic acid exchange resin through which the liquor is passed and from which the base is recovered by eluting the cell, in a countercurrent system, with a regeneration liquor rich in sulfur dioxide (20%) under low pressure. Since the sulfur dioxide content is much higher than the degree of concentration of the cooking liquor, the excess sulfur dioxide is returned by an isobaric method into the cycle of fresh regeneration liquor. The liquor, free of excess sulfur dioxide, is mixed with additional base for the preparation of fresh cooking liquor. The process is suitable for the preparation of a cooking liquor with 0.9% to 1.4% bound sulfur dioxide. For stronger cooking acids, the addition of more chemicals is necessary.

The Abiperm process converts the lignosulfonates into free ligno-sulfonic acid which is a suitable starting material for the production of by-products and which also has a high calorific value and a low ash content (166).

Electrodialysis is a physico-chemical process in which the ion transport is made through a membrane by an electromotive force. The mass-transfer

process differs from that in which a chemical change in the material occurs. The demineralization by electrodialysis plays a special role in which anion- and cation-selective membranes are alternately used. This type of process has become known especially in the desalting of sea water (262).

For the processing of spent sulfite liquor, a special cell has been designed which uses an ion-conducting barrier cell and permits the maintenance of separate reaction process streams without the need of recombining the separate components. Laboratory experiments for processing ammonium-, sodium-, magnesium-, and calcium-base spent sulfite liquors from acid bisulfite, bisulfite, monosulfite, semichemical, Magnefite, and various other modifications of sulfite pulping processes have been carried out. The yield, based on the power consumption of the bases, amounts to about 100% and the current density runs about 10 to 20 times as high as in the salt water desalination. The chemicals of the pulp mills are recovered by three main process streams: (*a*) the first stream produces a strong aqueous sulfite solution free of organic components and with up to 70 grams of sodium sulfite per liter; (*b*) the second gives solutions of free lignosulfonic acids of high molecular weight; and (*c*) the third contains a solution of low-molecular acids, i.e., formic, acetic, and sulfurous acids. Numerous problems still must be solved before this electrodialysis process is ready for technical operation, but there is a possibility that it may be useful for the regeneration of the bases from the spent sulfite liquors and also for the production of saleable organic by-products (83).

9.4.2 AMMONIUM-BASE SPENT SULFITE LIQUORS

When ammonium spent sulfite liquor is evaporated in vacuo no nitrogen is split off, although a certain amount is present in a firmly bound state and is not split off even on boiling with weakly alkaline solutions. The release of sulfur dioxide on steam distillation occurs much more slowly than with calcium-base spent liquors. Such a distillation leads to a considerable decrease in the loosely combined sulfur dioxide and a decrease in the acidity, but no essential change occurs in the content of reducing sugars. Almost all ammonium ions can be replaced by hydrogen ions by ion exchange (252). Although the recovery of the sulfur dioxide is not difficult, the recovery of the ammonium base is possible neither by evaporation nor by steam distillation (287). By mixing ammonium-base spent sulfite liquor with kraft black liquor, 60% to 100% of the ammonia can be recovered, depending upon the ratio in which the liquors are mixed. This process, however, requires the combination of a sulfite mill with a kraft pulp mill (189,191).

The pyrolysis process has also been suggested for the regeneration of the chemicals. In this process, the sulfur is obtained in a form which makes its direct utilization impossible, and this also complicates the re-

covery of the base (253). These difficulties have stimulated the use of ion-exchange resins for the processing of ammonium-base spent sulfite liquors. The study of the recovery of monovalent bases by means of ion exchangers has shown that the efficiency of the ammonium regeneration is controlled by the limited volume of weakly ionized sulfur dioxide solutions as regenerate, and is also affected by the presence of other multivalent cations which originate from the wood and the water used. Two methods have been developed for the elimination of the multivalent cations. The lignosulfonic acid obtained as a by-product is unsuitable for this purpose, but ammonium bisulfite solution has been found to be satisfactory as a means of regeneration (87). The fundamental difference between the Abiperm process discussed above and the Pritchard-Ontario Research Foundation process is the requirement of the elimination of the multivalent cations before the exchange of the monovalent base. For this purpose, the process is carried out in two steps: in the first step, the multivalent cations are removed with a column charged with an exchange resin in an NH_4^+ form, then the NH_4 is absorbed in a column charged with a resin in an H^+ form (203). The modified Pritchard-Fraxon process uses a sulfur dioxide solution to which some acetone has been added to increase the hydrogen ion concentration of the eluent. It is claimed that about 95% of the ammonia can be recovered in this way. The recovered acid can be used directly in the sulfite cooking process and permits the preparation of an acid with a low or high degree of bound sulfur dioxide. The acid is free of polyvalent cations and the eluent can be obtained from the burner gases without the addition of liquid sulfur dioxide. The ion exchange takes place in only one column (431).

Another interesting modification that may lead to a satisfactory solution to these problems is the distillation of the ammonium-base spent sulfite liquor in the presence of magnesium oxide. In the distillation at atmospheric or elevated pressure, losses in nitrogen occur, but this can be avoided by carrying out the distillation in vacuo at 30° to 35°C. The highest yields of ammonia are obtained by steam distillation. The magnesium oxide ashes obtained on burning are less active than many commercial brands of magnesium oxide. The activity of the magnesium oxide obviously plays an important role in the vacuum distillation but is of practically no importance in the steam distillation (320,321). However, the processing of ammonium-base spent liquors involves strong corrosion problems which cannot be eliminated by the use of stainless steel (190).

9.4.3 Magnesium-Base Spent Liquors

The use of magnesium as base affords numerous advantages in the regeneration. The sulfur is burned exclusively to sulfur dioxide while the magnesium, at the temperatures prevailing in the combustion chamber, forms no volatile products and no colloidal smoke, but only an ash that

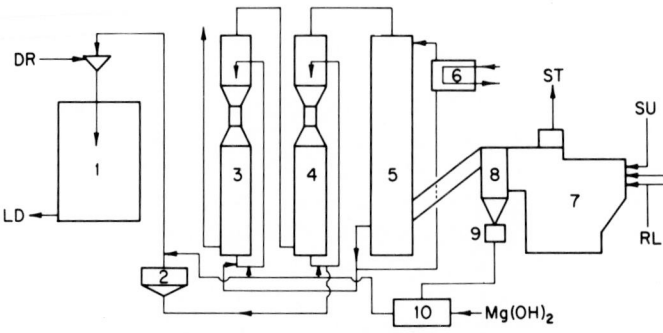

Figure VII–41. Recovery system for magnesium bisulfite spent liquor. 1 liquor storage; 2 liquor filter; 3, 4 absorption towers; 5 cooling tower; 6 water cooler; 7 boiler; 8 mechanical dust collector; 9 magnesium oxide; 10 magnesium hydroxide slurry; DR digester relief line; ST steam for process and power; SU make-up sulfur; RL red liquor from evaporators; LD liquor to digesters.

separates readily. Although the proportion of bound sulfur dioxide in the cooking liquor is twice as high as in the conventional calcium-base cooking liquor, the amount of ash formed is only 6 to 7 kg per 100 kg of wood while the ash content from a kraft cook amounts to 32 to 35 kg per 100 kg of wood. The regeneration process of the magnesium-base spent liquor is technically self-supporting with regard to heat and power economy, with low heat and chemical losses. Stream and air pollution are eliminated. The ratio of wood to acid and the concentration can be increased and in this way the steam consumption can be reduced. The introduction of a continuous process is therefore considerably simplified. A flow-sheet of a process for the regeneration of a spent liquor of the magnesium-base sulfite cooking process is shown in Fig. VII–41 (41,402).

In a modern continuous Magnefite plant with a production of 200 tons per day, the pulp receives its first washing already in the digester. Further washing is carried out in two one-stage vacuum washers. The spent liquor is circulated in a countercurrent process through the filters and in the digester for increased efficiency for the washing. In this way the pulp is simultaneously cooled before it leaves the digester. The drained liquor contains about 13% solids; it is filtered, passed into a four-stage evaporator, and concentrated to 50% solid content. It is then further concentrated in a cyclone evaporator, simultaneously utilizing the burner flue gases as a heat source. The concentrated liquor is then burned in an specially designed combustion oven, with the formation of magnesium oxide, sulfur dioxide, and surplus steam. Additional amounts of magnesium oxide are obtained from magnesium hydroxide. The sulfur dioxide is converted in twin absorption towers into cooking liquor, with additional sulfur dioxide being obtained by the burning of elementary sulfur (103,309).

9.4.4 COMBINED PROCESSES

In cases where a sodium sulfite pulp mill is close to a kraft pulp mill, the sulfite spent liquor can be processed together with the black liquor for the recovery of the chemicals. This demands that the two mills be so closely located that not only the recovery of the chemicals but also the total heat economy can be adjusted for both plants. This represents an operational partnership, not a combined process.

Sulfited black liquor from a soda cook forms a valuable starting material for the production of sulfite cooking liquor. In addition, the black liquor from a sulfur-free alkaline semipulp or the spent liquor from a cold soda pulping process can be used. It has been suggested that neutral sodium sulfite liquor could be prepared by direct sulfitation of kraft black liquor. The sulfitation can also be carried out in connection with the carbonation. It has been found, however, that sulfite cooking liquor which has been prepared by direct sulfitation of conventional kraft black liquor causes poor delignification. This may be the result of the presence of thiosulfate which considerably surpasses the amount usually present. The sulfide therefore must be expelled as hydrogen sulfide during the carbonation. A sulfite cooking liquor prepared from black liquor contains all the acid-soluble components of the black liquor, including saccharinic acids, formic and acetic acids, and methyl alcohol, etc. As is known, formic acid accelerates the decomposition of sulfite cooking liquor, with the formation of thiosulfate. In so far as volatile organic compounds are freed on acidification, they can be partially removed by distillation. Only a part of the sodium present in the black liquor is available for the formation of bound sulfur dioxide. About 40% of the total sodium is present as carbonate, sulfide, and hydroxide and forms sodium bisulfite stoichiometrically; about 50% is attached to decomposition products of the wood; the remainder is present as inorganic compounds, e.g., sulfate. The amount of alkali in the kraft black liquor that can be used as base in the sulfite cook has not yet been definitely established. It should be mentioned that about 70% of the consumed alkali in the kraft cooking process is consumed by the carbohydrates, especially the hemicelluloses, and only 20% to 25% is required for the delignification; the remaining 5% to 10% is consumed by the acids formed in the cook (247).

The Sivola cooking process constitutes a special case. It is a two-stage process using the same base, with a sodium bisulfite cook in the first stage, followed by an alkaline cook in the second stage (48,289). There is no washing between the two stages. The regeneration process of the alkaline spent liquor can be adapted to the variability of the cooking process. In its basic procedure, it does not differ from the regeneration processes for sodium-base spent liquors (202,240,247,289).

9.5 The Chemical Utilization of the Organic Components Present in Sulfite Spent Liquor

The composition and the properties of sulfite spent liquor have been discussed in general in Section 9.1. It is known that, in a normal cooking process, about half the wood substance goes into solution. Some cellulose and the hemicelluloses are converted by partial or complete hydrolysis into low-molecular sugars. Part of these sugars react further with bisulfite, forming sugar sulfonic acids, or are converted by oxidation into aldonic acids. Lignin is converted into lignosulfonates. The methoxyl groups of the glucuronoxylan form methanol, the acetyl groups of the xylan give acetic acid. The hydroxymethyl groups of the lignin yield formaldehyde, by dehydration of the pentoses furfural is obtained, and bisulfite oxidation of terpenes gives cymene. All attempts at an economical utilization of the large amounts of organic compounds present in the spent sulfite liquor have led to only partial success, such as the conversion of the sugars into ethyl alcohol and yeast. But this utilization is of only minor importance considering the great problem caused by the presence of oxidizable matter resulting from the drainage of the liquor into rivers and lakes. Utilization of the lignosulfonates which make up the major part of the organic components would be much more important and valuable. Some other possibilities for utilization will be discussed later, but these play only a minor role in view of the tremendous amounts of spent liquor that are obtained daily in sulfite pulp mills. To this must be added that not only the sugars but also the products which may be produced from the lignosulfonates meet with strong competition from synthetic products of the chemical industry. Similar problems have been discussed already in connection with the hydrolysis of wood and the processing of the wood sugar solutions and the hydrolysis lignins.

The use of soluble bases, especially of sodium and magnesium bases, in the sulfite cooking process promises a fundamental change in this situation. The forced recovery of the chemicals with these bases requires the use of the major part of the organic matter as fuel. Under certain conditions the utilization of the sugars may be possible, but the separation and processing of the lignosulfonates may be advisable only in exceptional cases. However, local operational conditions may be such as to prevent a separate processing of the spent liquor components, and the spent liquor as a whole would have to be used as fuel.

9.5.1 Hexoses and Pentoses of the Spent Sulfite Liquor

The fermentation of spent sulfite liquor by means of yeast (*Saccharomyces cerecisiae*) has been carried out for almost 60 years and today is an

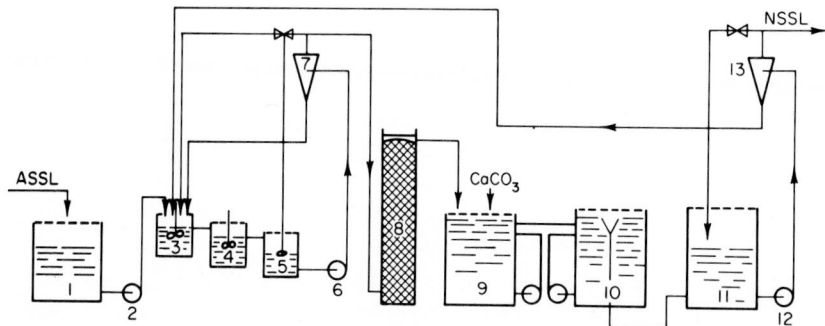

Figure VII-42. Flow-sheet of the neutralization, clarification and sulfate precipitation process of sulfite spent liquor (57). 1 acid liquor storage; 2 acid liquor pump; 3, 4 seeding tanks; 5 buffer tanks; 6 pump; 7 seeding cyclones (4–8 units); 8 limestone towers (4 units); 9, 10 final neutralization tanks; 11 buffer tanks; 12 pump; 13 clarification cyclones (5–8 units); ASSL acid spent sulfite liquor; NSSL neutralized spent sulfite liquor.

integral part of the plant in most European sulfite pulp mills. Glucose and mannose are directly fermented by yeast; galactose can also be fermented under certain reaction conditions. Pentoses, on the other hand, remain unfermented. To obtain a yield of sugar as high as possible, the cooking conditions must be varied accordingly. The further the hydrolysis proceeds, the higher, in general, the yield of sugar. The demand for a high yield of sugar is limited because it is the quality of the pulp, not the yield of sugar, that is the deciding factor for the type of cook. It is important that a spent liquor, as concentrated as possible, should be drawn off from the pulp and should, generally speaking, be between 10% and 15%. About a third of the sugars are destroyed by the reaction with bisulfite and by oxidation. The yield of sugars can be expected to amount to 100 to 140 kg per ton of pulp, depending on the type of cook needed for the production of a specific quality of pulp. The fermentation of the hexoses stoichiometrically allows a yield of 51% ethyl alcohol to be expected. The alcohol yield anticipated from the above sugar yields varies, therefore, between 100 and 130 liters from a rayon pulp cook, and between 50 and 100 liters from a paper pulp cook.

Figure VII-42 shows a flow-sheet of a plant of a Swedish pulp mill for the neutralization, clarification, and separation of the sulfate from a sulfite spent liquor for fermentation (57). The acid spent liquor is pumped from a storage tank into two settling tanks, from there it passes into a buffer tank, and then, depending upon the throughput, into from 4 to 8 cyclone separators. The sludge from the separators, together with that from the clarification cyclones, passes into the settling tanks. From the cyclone separators the liquor is led through a four-unit limestone tower. For complete neutralization, calcium carbonate is added. The neutralized liquor then flows

through a buffer tank into five to eight cyclone separators and, from there, into the fermentation vats. The lime sludge from the separators goes back into the settling tanks.

The neutralized and clarified spent liquor is cooled to 30° to 35°C and mixed with the required nutrient salts and yeast reclaimed from the cycle. The liquor is generally passed into the fermentation vats at the bottom, and the overflowing liquor is led to the bottom of the next vat. The fermented wort is then led into the after-fermentation vat which serves simultaneously as a buffer tank. The yeast contained in the wort is centrifuged off and returned to the first fermentation vat, while the liquor goes into the distillation plant. The yeast in the vats is kept in suspension and thus built-in fittings in the vats are unnecessary and settling of the yeast is prevented. The fermentation period is usually 20 to 30 hours. The higher the concentration of yeast the more rapid the fermentation.

The distillation of the alcohol is normally carried out in two columns: in the first, the alcohol is distilled off, in the second, the alcohol is rectified. In the latter column, methanol, aldehydes, and certain fusel oils are obtained and are, normally, returned to the fermentation vats. Experiments to prevent the formation of aldehydes by varying the temperature, acidity, or concentration of the yeast have been unsuccessful. However, a 5% increase in alcohol yield is obtained when the total first run is added at the start of the fermentation instead of being continuously added while the fermentation is in progress (392).

Certain complications occur when sodium-base sulfite cooking liquor is used. When evaporation and recovery of chemicals are intended, the liquor, after the removal of the alcohol, must be passed into the evaporation plant as quantitatively as possible, and without loss of heat, in order to prevent the need of added heat. Further, in the neutralization of the liquor, an excess of alkali occurs in the regeneration system, surpassing the total loss of alkali in the regeneration. Experiments to carry out the neutralization by using alkali from the regeneration system, itself, have shown that about 20% of the alkali required can come from this source, containing about 75% Na_2S and 25% Na_2CO_3. Larger proportions of regeneration alkali will cause damage to the yeast, the damage being caused chiefly by the formation of colloidal sulfur during the neutralization. This formation of sulfur can be prevented when the neutralization is carried out to a pH of 5.3 to 6.0. In this way, the fermentation is considerably improved and the use of large amounts of regeneration liquor for the neutralization is possible. The biological action of sulfite, present in its different forms in the spent liquor, on the fermentation process is especially pronounced in the case of free or loosely combined sulfite, and its presence in the acid spent liquor from sodium-base cooks should be kept below 100 milliequivalents per liter, i.e., 3.2 g SO_2 per liter (77).

As mentioned before, the production of alcohol from spent sulfite liquor has to meet strong competition from alcohol produced by other processes. For this reason, some pulp mills have applied other chemical reactions for the utilization of the alcohol. Here the alcohol is converted to butanol, via acetaldehyde, to acetic acid, or to various types of esters. The production of glycol via ethylene oxide, or the condensation of ethylene oxide with phenol or other reactive compounds has also been considered. Other types of fermentations have been carried out by the use of specific fermentation organisms, as, for instance, the production of polyalcohols, of butanol-acetone, of acetic and/or lactic acid. But thus far the economic aspects of these processes have not been proven. The amount of pentoses and furfural in spent sulfite liquor is not sufficient to make possible an economic recovery of the latter from the liquor (429).

Before and during the Second World War, the production of fodder and nutrient yeast from wood sugar and from spent sulfite liquor gained special attention. The production of yeast from wood sugar has already been discussed in Chapter IV in connection with the wood saccharification process; consequently, a repetition need not be given here.

9.5.2 LIGNOSULFONATES

The technical utilization of the lignosulfonates in the spent liquor has been the goal of pulp mills for a long time. The enormous amount of spent sulfite liquor available and the more or less complicated processes used for the production of useful technical products have not as yet led to an economical solution of the problem. This does not mean that lignosulfonates as such, or after conversion into other organic compounds, have not found, to a limited degree, a profitable use. However, the preparation of such lignosulfonates and conversion products is limited to only a few mills. Lignosulfonates, as such, of various degrees of purity are used as tannins, dispersion agents, as binding materials for road construction, in the ceramic industry, in the preparation of animal fodder, and in the plywood industry. The properties of such commercial products are described in the respective company pamphlets (7). The preparation of lignosulfonates is carried out chiefly by the Howard process of the Marathon Division of the American Can Company. In this process, the main part of the sulfite and sulfate in the spent liquor is precipitated by the addition of lime until a pH of 10.5 has been reached. Then more lime is added to the filtered solution to precipitate the basic calcium lignosulfonate. The formation of this precipitate occurs at a pH range of 10.5 to 12.2. Somewhat more than 50% of the lignosulfonates present are precipitated and isolated by filtration. On treatment of a sludge of this precipitate with sulfuric acid or magnesium or sodium sulfate, the calcium is converted into gypsum which is removed by filtration or centrifugation. The neutral magnesium or sodium

lignosulfonate in the filtrate is then obtained as a light brown powder on evaporation and drying, and can be used for various chemical reactions (169).

The reaction taking place in this precipitation with lime has been studied in detail. It has been found that, at pH 8.5, calcium sulfite is precipitated and the sugars are converted into saccharinic acids. At pH 10.8, the main part of the calcium lignosulfonate is precipitated. At pH 11.8, an additional amount of basic calcium lignosulfonate and calcium saccharate separates. The sodium, ammonium, and magnesium lignosulfonates prepared from the calcium salt do not differ essentially in their dispersing action on suspensions of precipitated calcium carbonate, finely ground limestone, anatase, and kaolin, but they are inactive toward rutile dispersions (143).

Because the dispersing action of the lignosulfonates is affected to a considerable extent by the presence of sugars, attempts have been made to find processes for the preparation of sugar-free lignosulfonates. For this purpose, a liquid-liquid extraction process with acetone and methanol has been used (422). Other processes apply a polymerization of the lignosulfonic acids and, if necessary, of the sugars, to improve the dispersing action. Here, the spent liquor, after being freed of cations by ion exchange, is concentrated to 30% or higher and is then heated at temperatures of 100° to 250°C (390). The polymerization is also accelerated by heating the liquor after addition of sulfuric acid to a pH of 0.2 to 4 (123). The addition of a soluble sulfite salt, corresponding to 2% to 4% SO_2 based on the dry content, to the concentrated liquor and heating it at 120° to 200°C also accelerates the polymerization (300). The precipitation of lignosulfonates can be facilitated by the addition of basic metal salts, such as basic aluminum chloride (394). A selective fractionation of lignosulfonates can be carried out by precipitation with high-molecular quaternary ammonium bases (301). A fractionated extraction and concentration of lignosulfonic acid has been described in a patent, by which a diluted solution of free lignosulfonic acid is neutralized with a small excess of a suitable alkyl amine which is practically insoluble in water, and the amine salt formed is extracted with a higher aliphatic alcohol, with 8 to 16 carbon atoms (207).

All these processes have as their goal an improvement in the properties of the isolated lignosulfonic acids without affecting their general usefulness.

It has long been known that vanillin can be obtained from lignin, from spent sulfite liquor, and from kraft black liquor. When spent sulfite liquor is treated with alkali under certain reaction conditions, vanillin is formed (220). The yield varies between 0.7 and 2.39 grams per liter of liquor, depending upon the origin and the composition of the latter (221). The

basic calcium lignosulfonate obtained in the Howard process is especially suitable for the technical preparation of vanillin. This sulfonate is mixed with sodium hydroxide and heated under pressure in the presence of air for one hour at 160° to 165°C. The mixture is then acidified and the vanillin extracted with butanol. The yield amounts to between 5% and 10%, based on the lignin content of the spent liquor. This process has later been supplemented by an extraction of the butanol extract with an aqueous alkali solution (248).

Since the formation of vanillin from lignin and lignosulfonates is obviously an oxidation process in an alkaline medium, numerous experiments have been carried out by adding a mild oxidizing agent to the alkaline solution. Nitrobenzene has been proposed for this purpose (119,134). In this way the yield of vanillin can be increased considerably and yields of 17.7% to 19.7%, based on the lignin content of the liquor, have been obtained. Vanillin can be obtained not only from spent liquor lignin but also from other types of isolated lignin, with the highest yields having been obtained from alkali lignin and the lowest from sulfuric acid hydrolysis lignin. Nitrobenzene oxidation of lignin from gymnosperms gives vanillin; that of lignin from angiosperms gives vanillin and syringaldehyde (68). Although numerous derivatives and conversion products, such as vanillic acid, protocatechuic acid, vanillyl alcohol, and vanillin esters, can be prepared from the vanillin, the production of these products offers no solution to the spent sulfite liquor problem because the amount of lignosulfonates obtained daily from sulfite mills far surpasses the demand for these products.

REFERENCES

1. Adler, E., *Svensk Papperstidn.* **49** (15), 339 (1946).
2. Adler, E. and Häggroth, S., *Svensk Papperstidn.* **53** (11), 287 (1950).
3. Adler, E. and Häggroth, S., *Svensk Papperstidn.* **53** (12), 321 (1950).
4. Ahlen, L., *Tappi* **46** (11), 143A (1963).
5. Akim, L. E. and Geles, I. S., *Bumazhn. Prom.* **18**, 26 (1965).
6. Allgulander, O., Rydholm, S. and Willberg, S., *Svensk Papperstidn.* **57** (15), 542 (1954).
7. American Can Co., Marathon Div., Bulletin no. 130.
8. Annergren, G. E. and Rydholm, S. A., *Svensk Papperstidn.* **62** (20), 737 (1959).
9. Annergren, G. E. and Rydholm, S. A., *Svensk Papperstidn.* **63** (18), 591 (1960).
10. Annergren, G. E. and Backlund, A., *Pulp Paper Mag. Can.*, **67** (4), T-220 (1966).
11. Anon., *Svensk Papperstidn.* **42**, 629 (1939).
12. Anon., *Pulp Paper Mag., Can.* **59** (2), 89 (1958).
13. Anon., *Svensk Papperstidn.* **64** (14), 528 (1961).
14. Ant-Wuorinen, O. and Halonen, A., *Paperi Puu* **40** (10), 481 (1958).
15. Ant-Wuorinen, O. and Visipää, A., *Makromol. Chem.* **30** (1), 1 (1959).
16. Arnold, G. C., *Paper Trade J.*, **149**, 41 (Oct. 12, 1965).
17. Aschaffenburger Zellstoffwerke A.G., German Patent 1,049,220.

18. Aulin-Erdtman, G., Björkman, A., Erdtman, H., and Hägglund, S. E., *Svensk Papperstidn.* **50** (11B), 81 (1947).
19. Aulin-Erdtman, G., *Tappi* **32** (4), 160 (1949).
20. Babushkina, M. D., Babajew, E., Kirjakov, M. F. Karasik, S. and Scharapova, S., *Bumazhn. Prom.* **32** (2), 2 (1951); **34** (9), 13 (1959).
21. Bailey, E. L., *Tappi* **45** (9), 689 (1962).
22. Barclay, H. G., Prahacs, S. and Gravel, J. O., *Pulp Paper Mag. Can.* **65** (12), T-553 (1964).
23. Bardany, K. and Guba, F., *Faserforsch. Textiltech.* **8** (1), 27 (1957).
24. Benko, J., *Tappi* **44** (11) 766, 771, (12) 849 (1961).
25. Bergholm, A., *Svensk Papperstidn.* **66** (4), 125 (1963).
26. Berry, L. A. and Larsen, A. D., *Tappi* **45** (11), 887 (1962).
27. Bishop, F. F. and Honstead, J. F., *Tappi* **34** (7), 318 (1951).
28. Bixler, A. L. M., *Tech. Assoc. Papers* **21**, 181 (1938); *Paper Trade J.* **107** (15), 29 (1938).
29. Black, H. H., *Ind. Eng. Chem.* **50** (10), 95A (1958).
30. Blackmore, K. A. E. and Markham, A. E., *Tappi* **41** (7), 138A (1958).
31. Blikstad, F., *Norsk Skogind.* **10** (5), 172 (1956).
32. Bobrow, A. I., Turanova, A. I., Popow, A. D., Cherepanow, B. F. and Khorshev, V. M., *Bumazhn. Prom.* **39** (2), 5 (1964).
33. Borisek, R., *Zellstoff Papier* **12** (9), 267 (1963).
34. Borisek, R. and Minarik, F., *Sb. Vyskum. Prac. Odboru Celluozy Papiera* **7**, 47 (1962).
35. Borisek, R., Schmied, J. and Enderst, V., *Svensk Papperstidn.* **64** (9), 341 (1961).
36. Boyer, R. Q., *Tappi* **43** (8), 688 (1960).
37. Boyer, R. Q., *Tappi* **42** (5), 356 (1959).
38. Brabender, G. J., *Tappi* **32** (7), 337 (1949).
39. Bradley, R. Q., U.S. Patent 2,792,350; May 14, 1957.
40. Brauns, F. E., "The chemistry of lignin." Academic Press, New York, 1952.
   Brauns, F. E. and Brauns, D. A., "The chemistry of lignin" Supplement Volume. Academic Press, New York, 1960.
41. Brown, J. V. and Roger, C. E., *Tappi* **49** (11), 819 (1956).
42. Brunes, B., *Svensk Papperstidn.* **57** (9), 317 (1954).
43. Brunes, B., Jarnberg, T. and Jonsson, S. E., *Svensk Papperstidn.* **58** (9), 332 (1955).
44. Bryce, J. R., Lamed, S. and Tomlinson II, G. H., *Pulp Paper Mag. Can.* **63** (3), T-155 (1964).
45. Bryce, J. R., Lamed, S. and Tomlinson II, G. H., *Paper Trade J.* **148** (15), 36 (1964).
46. Bryce, J. R. and Tomlinson II, G. H., *Pulp Paper Mag. Can.* **63** (7), T-355 (1962).
47. Bryde, O., *Svensk Papperstidn.* **55** (19), 734 (1952).
48. Buchanan, J. S. and Kennedy, E. H., *Tappi* **42** (2), 153A (1959).
49. Bucher, H., "Morphologie und Struktur von Holzfasern." Cellulosefabrik, Attisholz, 1947.
50. Bucher, H., *Chimia* **13** (12), 397 (1959).
51. Bucher, H., *Papier* **14** (10a), 542 (1960).
52. Buijtenen, J. P. V., *Tappi* **41** (4), 175 (1958).
53. Burrows, D., *Paper Trade J.* **148** (21), 30 (1964).
54. Butko, Y. G. and Pelevin, Y., *Khim. Pererabotka Drevesiny* **3**, 24 (1962).
55. Calhoun, J. M., Yorston, F. H. and Maass, O., *Can. J. Research* **15B** (11), 457 (1937); **17B** (4), 121 (1939).
56. Campbell, W. B. and Maass, O., *Can. J. Research* **2**, 42 (1930).

57. Carlson, G., Moden, H. and Scholander, A., *Svensk Papperstidn.* **61** (18B), 815 (1958).
58. Carufel, G. de, *Pulp Paper Mag. Can.* **60** (9), T-272 (1959).
59. Cederquist, K. N., *Svensk Papperstidn.* **58** (5), 154 (1955); **61** (1), 38 (1958).
60. Cederquist, K. N., *Svensk Papperstidn.* **61** (2), 38 (1958).
61. Cederquist, K. N., Ahlborg, N. K., Lunden, B. and Wentworth, T. O., *Tappi* **43** (8), 702 (1960).
62. Chao, Sh. Sh., Ku, Y. Ch. and Huang, L. Y., *Taiwan Lin Yen Shi Yen* No. **90** (1963).
63. Chemical Sulfite Mill Operation, Chemipulp Process Inc., Watertown, N.Y., 1939.
64. Chidester, G. H., Bray, M. W. and Curran, C. E., *Tech. Assoc. Papers* **23**, 661 (1940).
65. Chidester, G. H. and Curran, C. E., *U.S. Forest Prod. Lab. Rept.* 1286, Nov. 1959.
66. Chidester, G. H. and McGovern, J. W., *Paper Trade J.* **110** (10), 39 (1940); **113** (16), 32 (1941).
67. Chidester, G. H. and McGovern, J. W., *Tech. Assoc. Papers* **24**, 226 (1941).
68. Creighton, R. H., McCarthy, J. J. and Hibbert, H., *J. Am. Chem. Soc.* **63**, 312 (1941); **66**, 32 (1944).
69. Croon, J., *Svensk Papperstidn.* **67** (11), 467 (1964).
70. Croon, J., Enström, B. F. and Rydholm, S. A., *Svensk Papperstidn.* **67** (5), 196 (1964).
71. Dahm, H. P., *Norsk Skogind.* **14** (3), 92 (1960).
72. Dahm, H. P., *Pulp Paper Int.* **3** (1), 30 (1961).
73. Darmstadt, W. and Richter, F. H., *Tappi* **36** (6), 171A (1953).
74. Debusch, C. B., *Wochbl. Papierfarik.* **25** (24), 365 (1927).
75. Dickens, W. A. and Plummer, A. W., *Tappi* **40** (11), 895 (1957).
76. Dickerman, G. K., *Pulp Paper Mag. Can.* **61** (3), T-200 (1960).
77. Dillen, S., *Svensk Papperstidn.* **64** (8), 283; (15), 545; (22), 819 (1961).
78. Dioszegi, O. and Kleinert, Th. N., *Pulp Paper Mag. Can.* **61** (7), T-374 (1960).
79. Dorland, R. M., Leask, R. A. and McKinney, J. H., *Pulp Paper Mag. Can.* **58** (6), 135 (1957).
80. Dorland, R. M., Leask, R. A. and McKinney, J. H., *Pulp Paper Mag. Can.* **59** (5), 236 (1958).
81. Dorland, R. M., Leask, R. A. and McKinney, J. H., *Pulp Paper Mag. Can.* **60** (2), T-37 (1959).
82. Douglas, M. R., Snyder, I. W. and Tomlinson, G. H., *Pulp Paper Mag. Can.* **66** (6), T-316 (1965).
83. Dubey, G. A., McElhinney, T. R. and Wiley, A. J., *Tappi* **48** (2), 95 (1965).
84. Duhs, E. F., *Tappi* **33** (7), 333 (1950).
85. Edling, G., *Tech. Assoc. Papers* **29**, 615 (1946).
86. Edling, G., *Papier* **4** (23), 438 (1950).
87. Effer, W. R., Hopper, W. E. and Marshall, H. N., *Pulp Paper Mag. Can.* **62** (10), T-447 (1961).
88. Enderst, V., Kubelka, V., Jurkovicowa, D. and Gaydos, J., *Zellstoff Papier* **10** (4), 134 (1961).
89. Engelhardt, G. and Bergmann, G., *Zellstoff Papier* **10** (4), 125 (1961).
90. Engström, M. and Stockman, L., *Svensk Papperstidn.* **66** (13), 513 (1963).
91. Enomoto, S., Okada, M. and Koshigawa, T., *Tappi* **41** (9), 552 (1958).
92. Enkvist, T. and Turunen, K., *Paperi Puu* **42** (4a), 157 (1960).
92a. Erdman, A., *Tappi* **50** (6), 110A (1967).

93. Erdtman, H., *Cellulosechemie* **18**, 53 (1940); *Svensk Papperstidn.* **43**, 241 (1940).
94. Erdtman, H., *Tappi* **32** (2), 75 (1949).
95. Erdtman, H., *Tappi* **32** (8), 346 (1949).
96. Erdtman, H., *Tappi* **32** (7), 303 (1949).
97. Erdtman, H., Aulin-Erdtman, G. and Lindgren, B., *Svensk Papperstidn.* **49** (9), 199 (1946).
98. Eriksson, E. and Samuelson, O., *Svensk Papperstidn.* **64** (4), 138 (1961).
99. Eriksson, E. and Samuelson, O., *Svensk Papperstidn.* **65** (16), 600 (1962).
100. Erins P. and Odincovs, P., *Celuloza si Hirtie* (Bucharest) **14** (7/9), 447 (1965).
100a. Ernest, F. M. and Harman, S. M., *Tappi* **50** (12), 110A (1967).
101. Escher, Wyss GmbH, U.S. Patent 2,999,748.
102. Evans, J. C., *Paper Trade J.* **143** (36), 42 (1959).
103. Eyken, H. K. V., *Pulp Paper Mag. Can.* **65** (9), 103 (1964).
104. Eyken, H. K. and Price, F. A., *Pulp Paper Mag. Can.* **64** (1), 71 (1963).
105. Fahlgren, S., *Tech. Assoc. Papers.* **23**, 369 (1940).
106. Farbenfabriken Bayer A.G., German Patent 1,000,841 (1957).
107. Farin, W. G., *Paper Trade J.* **150** (47), 46 (1966).
108. Felicetta, V. F., Lung, M. and McCarthy, J. L., *Tappi* **42** (6), 157 (1960).
109. Felicetta, V. F. and McCarthy, J. L., *J. Am. Chem. Soc.* **79** (16), 4499 (1957).
110. Fellegi, J. and Janci, J., *Sb. Vyskum. Prac. Odboru Celulozy Papiera* **7**, 73 (1962).
111. Finsen, J. E., *Tappi* **42** (2), 104 (1959).
112. Fogler, H. H., Herbolzheimer, F., Stichfield, R. M. and Jenness, L. C., *Tappi* **32** (9), 389 (1949),
113. Forest Biology Subcommittee, *Tappi* **43** (11), 40A (1960).
114. Forgacs, O. I., *Tappi* **44** (2), 112 (1961).
115. Forss, K., *Meddel. Ind. Centrallab.* No. 263, Helsinki 1961.
116. Forss, K., *Paperi Puu* **43** (11), 676 (1961).
117. Forss, K. and Fremer, K., *Tappi* **47** (8), 485 (1964).
117a. Forss, K., Fremer, K. and Stenlund, B., *Papper och Trä* **48** (11), 669 (1966).
118. Freeman, H., *Chem. Met. Eng.* **44** (6), 311 (1937).
119. Freudenberg, K., Lautsch, W. and Engler, K., *Chem. Ber.* **73B**, 167 (1940).
120. Freudenberg, K., Sohns, F. and Jansen, A., *Ann. Chem.* **518** (1), 62 (1935).
121. Gardon, J. L. and Mason, S. G., *Can. J. Chem.* **33**, 1477 (1955); **37**, 1491 (1955).
122. Gauvin, W. H. and Gravel, J. O., *Tappi* **43** (8), 678 (1960).
123. Georgia Pacific Corp., U.S. Patent 3,138,555, June 23, 1964.
124. Gerace, Th., *Tech. Assoc. Papers* **31**, 604 (1947).
125. Giertz, H. W., *Norsk Skogind.* **14** (10), 369 (1960).
126. Goeldner, R. W. and Leitner, G. F., *Tappi* **47** (11), 185A (1964).
127. Goliath, M. and Lindgren, B. O., *Svensk Papperstidn.* **64** (4), 109 (1961).
128. Goliath, M. and Lindgren, B. O., *Svensk Papperstidn.* **64** (12), 469 (1961).
129. Gordon, L. I., *Tappi* **39** (4), 172A (1956).
130. Grangard, G., *Svensk Papperstidn.* **57** (17), 605 (1954).
131. Grangard, G., *Svensk Papperstidn.* **62** (24), 920 (1959).
132. Grimsrud, L. and McCarthy, J. L., *Tappi* **42** (6), 503 (1959).
133. Grimsrud, L., *Norsk Skogind.* **11** (11), 455 (1957).
134. Gustavson, T., Swedish Patent 82,886 (1935).
135. Hägglund, E., *Cellulosechemie* **16**, 41 (1935).
136. Hägglund, E., *Papierfabrikant* **34**, 313 (1936).
137. Hägglund, E., *Tappi* **33** (10), 520 (1950).
138. Hägglund, E., "Chemistry of wood." Academic Press, New York, 1951.

139. Hägglund, E., *Svensk Kemisk Tidskr.* **37,** 116 (1925); **38,** 177 (1926); *Papierfabrikant* **34** (34), 313 (1936); Hägglund, E. and Arnold, S., *Papierfabrikant* **36** (24), 266 (1938).
140. Hägglund, E. and Johanssen, A., *Svensk Papperstidn.* **35,** 475 (1932); *Cellulosechemie* **13** (8/9), 139 (1932).
141. Häggroth, S., Lindgren, B. O. and Seaden, U., *Svensk Papperstidn.* **57** (17), 660 (1953).
142. Hale, J. D., *Tappi* **45** (7), 538 (1957).
143. Hall, L., *Svensk Papperstidn.* **59** (20), 716 (1956); **60** (6), 199 (1957).
144. Hallgren, P. A. and Stockman, L., *Svensk Papperstidn.* **54** (4), 136 (1951).
145. Hamilton, J. K. and Thompson, N. S., *Pulp Paper Mag. Can.* **61** (4), T-272 (1960).
146. Hanssen, J., *Norsk Skogind.* **8** (12), 454 (1954).
146a. Hanway, J. E., Henby, E. B. and Smithson, G. R., *Tappi* **50** (10), 64A (1967).
147. Hardell, H. L. and Theander, O., *Svensk Papperstidn.* **68** (14), 282 (1965).
148. Harris, E. E. and Hogan, D., *Ind. Eng. Chem.* **49** (9), 1393 (1957).
149. Harris, G., *Pulp Paper Mag. Can.* **58** (3), 284 (1957); Harris, G. R. and Wayman, M., *Pulp Paper Mag. Can.* **57** (3), 231 (1956).
150. Hart, H. S., *Tappi* **41** (5), 218A (1958).
151. Hart, H. S., *Tappi* **37** (8), 331 (1954).
152. Hart, J. S. and Woods, J. M., *Pulp Paper Mag. Can.* **56** (9), 95 (1955).
153. Hart, J. S. and Woods, J. M., *Pulp Paper Mag. Can.* **57** (4), 158 (1956).
154. Hartler, N., *Papier* **16** (5) 181 (1962).
155. Hartler, N., Lind, L. and Stockman, L., *Svensk Papperstidn.* **64** (5), 160 (9), 336 (1961).
156. Hartler, N., Rönström, P. and Stockman, L., *Svensk Papperstidn.* **64** (19), 699 (1961).
157. Hartler, N. and Samuelson, O., *Svensk Papperstidn.* **55** (22), 851 (1952).
158. Hartler, N., Stockman, L. and Sundberg, O., *Svensk Papperstidn.* **64** (2), 33 (1961).
159. Hartler, N., Sundberg, O. and Stockman, L., *Svensk Papperstidn.* **64** (3), 67 (1961).
160. Hatch, R. S., *Tech. Assoc. Papers* **29,** 485 (1946).
161. Helleur, D. E., *Tappi* **37** (1), 177A (1954).
162. Hellström, B., *Pulp Paper Mag. Can.* **66** (5), T-289 (1965).
163. Heuser, E., *Tappi* **33** (3), 118 (1950).
164. Hoar, F. J., *Tech. Assoc. Papers* **31,** 81 (1948).
165. Hoge, W. H., *Tappi* **37** (9), 369 (1954).
166. Holder, D. A., Mindler, A. N. and Manchester, D. F., *Pulp Paper Mag. Can.* **66** (2), T-55 (1665).
167. Holzer, W. F., *Tech. Assoc. Papers* **27,** 276 (1944).
168. Hopkins, A. B., *Pulp Paper Mag. Can.* **64** (4), T-213 (1963).
169. Howard, C. C., U.S. Patents 1,848,292 (1932); 1,856,588 (1932).
170. Howard, E. J., *Pulp Paper Mag. Can.* **52** (7), 91 (1951); *Can. Pulp Paper Assoc. Tech. Sect. Proc.* 384 (1954).
171. Hull, W. Q., Smitz, B. C., Hull, J. H. and Holzer, W. F., *Ind. Eng. Chem.* **46** (8), 1546 (1954).
172. Hunter, R. E., Tracer, J., Cutts, R., Young, R. E., Olin, J. and McCarthy, J. L., *Tappi* **36** (11), 493 (1953).
173. Inderdohnen, J. F., *Tech. Assoc. Papers* **31,** 111 (1948).
174. Ingruber, O. V., *Pulp Paper Mag. Can.* **55** (10), 124 (1954).
175. Ingruber, O. V., *Pulp Paper Mag. Can.* **58** (10), 161 (1957).
176. Ingruber, O. V., *Pulp Paper Mag. Can.* **58** (13), 131 (1957).
177. Ingruber, O. V., *Tappi* **41** (12), 764 (1958).
178. Ingruber, O. V., *Pulp Paper Mag. Can.* **59** (11), 135 (1958).

179. Ingruber, O. V., *Pulp Paper Mag. Can.* **60** (11), T-346 (1959).
180. Ingruber, O. V., *Svensk Papperstidn.* **65** (11), 448 (1962).
181. Ingruber, O. V., Comm. Int. Sulfite Pulping Conference 1964.
182. Ingruber, O. V., *Pulp Paper Mag. Can.* **66** (4), T-215 (1965).
182a. Ingruber, O. V. and Allard, G. A., *Tappi* **50** (12), 597 (1967).
183. Jahn, R., *Papier* **16** (9), 433 (1962); Austrian Patent 200,121.
184. Javorski, B., *Svensk Papperstidn.* **59** (22), 787 (1956).
185. Jayne, B. A., *Tappi* **41** (4) 162 (1958).
186. Jayme, G., Broschinski, L. and Matzke, W., *Papier* **18** (7), 308 (1964).
186a. Jayme, G. and Forgersen, H. F., *Holzforschung* **21** (4), 110 (1967).
187. Jayme, G., Gasche, U. and Dubach, M., *Papier* **15**, 538 (1961).
188. Jayme, G., Harders Steinhäuser, M. and Mohrberg, W., *Papier* **5** (19/20) 411; (21/22) 445; (23/24) 504 (1951).
189. Jenness, L. C., *Tappi* **37** (8), 137A (1954).
190. Jenness, L. C., *Tappi* **45** (5), 404 (1959).
191. Jenness, L. C., Durst, R. E. and Thode, E. F., *Tappi* **36** (8), 337 (1953).
191a. Jenness, L. C. and Gaulfield, J. G. L., *Tech. Assoc. Papers* **23**, 654 (1940).
192. Jensen, W., Fremer, K. and Forss, K., *Tappi* **45** (2), 122 (1962).
193. Jensen, W., Fogelberg, B. C., Forss, K., Fremer, K. and Johanson, M., Comm. Int. Sulfite Pulping Conference 1964.
194. Jensen, W., Fogelberg, B. C. and Johanson, M., *Paperi Puu* **42** (7), 393 (1960).
195. Jensen, W., Virkola, N. E. and Alm, A., *Pulp Paper* **45** (8), 391 (1963).
196. Jentz, C. D., *Pulp Paper Mag. Can.* **36** (2), 61 (1935).
197. Jentz, C. D., *Pulp Paper Mag. Can.* **42** (2), 101 (1941).
198. Jull, N. A. and Marston, W. T., *Pulp Paper Mag. Can.* **52** (7), 83 (1951).
199. Jullander, I. and Olsson, B., *Svensk Papperstidn.* **53** (17), 518 (1950).
200. Kajimo, H., *J. Japan Tappi* **9** (3), 107 (1955).
200a. Keller, E. L. and Fahey, D. J., *Tappi* **51** (2), 98 (1968); *US Forest Service Res. Paper FPL* **78** August 1967.
200b. Keller, E. L., Martin, J. S. and Kingsbury, R. M., *Forest Products Lab. Rept.* 1912 (1956).
201. Kenetti, A., *Svensk Papperstidn.* **55** (14), 477 (1952).
202. Kennedy, E. H., *Tappi* **43** (8), 683 (1960).
203. Kerr, W. R., *Pulp Paper Mag. Can.* **62** (10), T-455 (1961).
204. Kerr, T. and Bailey, I. W., *Arnold Arboretum* **15**, 327 (1934).
205. Kerr, W. D. and Harding, S. A., *Pulp Paper Mag. Can.* **56** (9), 102 (1955).
206. Kesler, R. B. and Han, S. T., *Tappi* **45** (7), 534 (1962).
207. Kimberley Clark Corp. U.S. Patent 3,251,820 (1966).
208. Klason, P., *Chem. Ber.* **58**, 1761 (1925).
209. Kleinert, Th. N., *Holzforschung* **18** (5), 139 (1954).
210. Kleinert, Th. N., *Pulp Paper Mag. Can.* **65** (12), T-565 (1964).
211. Kleinert, Th. N. and McKinney, J. W., *Tappi* **45** (7), 529 (1962).
212. Kleinschmidt, R. V. and Bowen, E. C., *Tappi* **39** (5), 295 (1956).
213. Klinga, K. J., U.S. Patent 2,891,886 June 23, 1959.
214. Koch, P., *Norsk Skogind.* **18** (10), 366 (1964).
215. Konrad, F. H. and Brice, D. B., *Tappi* **32** (5), 222 (1949).
216. Kosaja, G. S., *Bumazh. Prom.* **32** (4), 4 (1957).
217. Kretschmar, G. and Martin, R., *Zellstoff Papier* **9** (7), 256 (1960).
218. Kubelka, V., *Zellstoff Papier* **8** (9), 351 (1959).
219. Kubelka, V., Enderst, V. and Gajdos, J., *Zellstoff Papier* **10** (6), 290 (1961).

220. Kürschner, K. J., *Pract. Chem.* 118 (2), 238 (1928).
221. Kürschner, K. and Schramek, W., *Tech. Chem. Papier Zellst. Fabr.* 28 (5), 65 (1931).
222. Kullgren, C., *Svensk Chem. Tidskr.* 42, 179 (1930); 44, 15 (1932).
223. Kullgren, C., *Svensk Papperstidn.* 55 (1), 1 (1952).
224. Kuhles, W., *Papierfabrikant* 28 (38), 604 (1931).
225. Kusminikh, I. N., Babaev, E. V., Babushkina, M. D. and Skvortsov, K. A., *Bumazh. Prom.* 32 (2), 2 (1957).
226. LaFond, L. A. and Holzer, W. F., *Tappi* 34 (6), 241 (1951).
227. Lagergren, S., *Svensk Papperstidn.* 67 (6), 238 (1964).
228. Lagergren, S. and Lunden, B., *Pulp Paper Mag. Can.* 60 (11), 338 (1959).
229. Lambert, J. E., *Paper Trade J.* 114 (43), 36 (1960).
230. Lang, A. R. G., *Pulp Paper Mag. Can.* 63 (6), T-331 (1962).
231. Lassberg, J. v., *Papierfabrikant* 39, 262 (1941); 40, 193, 201 (1942).
232. Lee, G. and Gauvin, W. H., *Tappi* 41 (3), 110 (1958).
233. Lee, G., Themelis, N. J. and Gauvin, W. H., *Pulp Paper Mag. Can.* 59 (3), 140 (1958).
234. Lightfood, R. G. and Sepall, O., *Pulp Paper Mag. Can.* 66 (5), T-279 (1965).
235. Lindgren, B. O., *Svensk Papperstidn.* 55 (3), 78 (1952).
236. Litwinova, W. B., Litwinov, A. B., Demtschenkow, and Nepenin, I. N., *Bumazh. Prom.* 33 (12), 4 (1958).
237. Lockman, C. G., German Patent 968,088, Jan. 2, 1958.
238. Lockman, C. J., U.S. Patent 2,847,043, Feb. 17, 1959.
239. Lorant, M., *Chem. Rundschau* 6, 153 (March 15, 1960).
240. Lüthgens, M. W., *Tappi* 45 (11), 837 (1962).
241. Maass, C. M. and Maass, O. J., *J. Am. Chem. Soc.* 50, 1352 (1928).
242. Maass, O., *Pulp Paper Mag. Can.* 54 (8), 98-134 (1953).
243. Mahdalik, M. and Slavik, I., *Paperi Puu* 47 (9), 503 (1965).
244. Makkonen, H., Järvelä, O. and Virkola, N. E., *Paperi Puu* 47 (2), 59 (1965).
245. Malik, A., *Zellstoff Papier* 4 (9), 264 (1955).
246. Manchester, D. F. and Termini, J. P., *Pulp Paper Mag. Can.* 62 (9), T-415 (1961).
247. Mannbro, N., *Svensk Papperstidn.* 54 (1), 19; (2), 61 (1951); 58 (15), 525, (16), 571 (1955); 66 (2), 25 (1963); *Pulp Paper Mag. Can.* 62 (2), T-66 (1961).
248. Marathon Corp., U.S. Patent 2,399,607.
249. Marcheguet, H. C. L. and Gandon, L., U.S. Patent 2,994,585 (Aug. 1, 1961).
250. Markant, H. P., *Tappi* 43 (8), 699 (1960).
251. Markant, H. P., McIlroy, R. A. and Matty, R. E., *Tappi* 45 (11), 849 (1962).
252. Markham, A. E. and McCarthy, J. L., *Tech. Assoc. Papers* 31, 236 (1948).
253. Markham, A. E., Peniston, Q. P. and McCarthy, J. L., *Tech. Assoc. Papers* 31, 407 (1948); Markham, A. E. and McCarthy, J. L., *Tappi* 37 (8), 355 (1954); Felicetta, V. F., Markham, A. E. and McCarthy, J. L., *Tappi* 37 (10), 431 (1954); Bolme, D. W., Li, P. S., Markham, A. E., Johnson, L. N. and McCarthy, J. L., *Tappi* 42 (5), 379 (1959).
254. Marriner, D. E. and Whitney, P. P., *Tech. Assoc. Papers* 31, 143 (1949).
254a. Marth, D. E., *Tappi* 42 (4), 301 (1959).
255. Martin, A., Felicetta, V. F. and McCarthy, J. L., *Tappi* 42 (6), 510 (1959).
256. McGovern, J. W. and Chidester, G. H., *Tech. Assoc. Papers* 24, 579 (1941).
257. McKinney, J. W., *Tappi* 42 (2), 153A (1959).
258. Meier, H., *Svensk Papperstidn.* 65 (8), 299 (1962).

259. Melnikow, S. F., *Bumazh. Prom.* **33** (8), 11 (1958).
260. Merewether, J. W. T., *Holzforschung* **11** (3), 65 (1957).
261. Miag Mühlenbau Industrie, German Patent 1,065,716.
262. Mintz, M. S., *Ind. Eng. Chem.* **55** (6), 18 (1963).
263. Mitchell, H. L., *Tappi* **41** (4), 150 (1958).
264. Molsberry, M. V., *Pulp Paper Mag. Can.* **62** (12), T-532 (1961).
265. Moorhead, G. R. *Pulp and Paper* **25** (11), 46 (1951).
266. Nelson, P. F., *Holzforschung* **17** (5), 134 (1963).
267. Nepenin, I. N., *Bumazhn. Prom.* **32** (7), 5 (1957); **33** (10), 11 (1958); **33** (5) 7.
268. Nepenin, I. N. and Buevskaya, A. D., *Bumazhn. Prom.* **1965** (12), 3.
269. Nichols-Freeman, *Svensk Papperstidn.* **42** (24), 629 (1939).
270. Nilson, O. and Stockman, L., *Svensk Papperstidn.* **65** (18), 711 (1962).
271. Nikitin, Y. N. and Eliashberg, M. G., *Bumazhn. Prom.* **1965** (1), 4.
272. Nikitin, M. N. and Nikitin, Y. N., *Bumazhn. Prom.* **1966** (3), 5.
273. Nikitin, V. N. and Vasilev, N. I., *Bumazhn. Prom.* **1965** (9), 3.
274. Nokihara, E., *J. Japan Tappi* **10** (7), 347 (1956).
275. Nokihara, E., Tuttle, M. J., Felicetta, V. F. and McCarthy, J. L., *J. Am. Chem. Soc.* **79** (16), 4499 (1957).
276. Nolan, W. J., *Tappi* **44** (7), 484 (1961).
277. Nolan, W. J., *Tappi* **44** (11), 753 (1961).
278. Nord, S. I., Samuelson, O. and Simonson, R., *Svensk Papperstidn.* **65** (19), 767 (1962).
279. Oeman, E., "Cellulosaindustrien." Bonnier, Stockholm, 1944.
280. Ogait, A., *Papier* **17** (11), 639 (1963).
280a. Ogawa, E. and Gorbatsevich, S. N., *Tappi* **51** (4), 171 (1968).
281. Ogiwara, Y., *J. Japan Tappi* **9** (2), 65 (1955).
282. Ogiwara, Y., *J. Japan Tappi* **9** (4), 157 (1955).
283. Onisko, W., *Przeglad Papier* **19** (12), 380 (1963).
284. Orsler, R. J. and Packman, D. F., *Svensk Papperstidn.* **67** (21), 855 (1964).
285. Owen, J. J. and Moggio, W. A., *Tappi* **38** (2), 144A (1955).
286. Palmen, L., *Svensk Papperstidn.* **63** (17), 550 (1960).
287. Palmrose, C. V. and Hull, J. H., *Tappi* **35** (5), 193 (1952).
288. Parsons, S. R. and Boyer, R. Q., *Tappi* **42** (7), 565 (1959).
289. Pascoe, T. A., Buchanan, J. S., Kennedy, E. H. and Sivola, G., *Tappi* **42** (4), 265 (1959).
290. Pearl, I. A. and Beyer, D. L., *Tappi* **47** (8), 458 (1964).
291. Peckham, J. R. and Drunen, V. v., *Tappi* **44** (5), 374 (1961).
292. Peckham, J. R. and Drunen, V. v., *Tappi* **48** (3), 193 (1965).
293. Perry, Th. O. and Chi Wu, W., *Tappi* **41** (4), 178 (1958).
294. Pew, C. J., *Tappi* **32** (1), 39 (1949).
295. Pillow, M. Y., Schaefer, E. R. and Pew, C. J., *Tech. Assoc. Papers* **19,** 178 (1936).
296. Pittam, W., *Tech. Assoc. Papers* **29,** 613 (1946).
297. Pitz, R., *Arch. Waermewirtsch.* **20** (4), 107 (1939).
298. Plummer, A. W., *Chem. Eng. Progress* **46,** 369 (1950).
298a. Procter, A. R., Yean, W. Q. and Goering, D. A. I., *Pulp Paper Mag. Can.* **68** (9), T-445 (1967).
299. Rabe, A. E. and Harris, J. F., *J. Chem. Eng. Data* **8** (3), 333 (1963).
300. Rayonier Corp., French Patent 1,321,648, Feb. 11, 1963.
301. Rayonier, Corp., U.S. Patent 3,271,382, Sept. 6, 1966.

302. Regestad, S. O. and Samuelson, O., *Svensk Papperstidn.* **61** (18B), 735 (1958).
303. Rezanowich, R., Allen, G. A. and Mason, S. G., *Pulp Paper Mag. Can.* **58** (11), 153 (1957).
304. Richter, G. A., *Tappi* **32** (12), 553 (1949).
305. Richter, G. A. and Pancoast, L. H., *Tappi* **37** (6), 263 (1954).
306. Robinson, M. D. and Harris, D. W., *Pulp Paper Mag. Can.* **60** (8), T-243 (1959).
307. Rosenblads Patenter, German Patent 972,469.
308. Ross, J. H., Hart, J. S., Strapp, R. K. and Yean, W. Q., *Pulp Paper Mag. Can.* **52** (10), 116 (1951).
309. Rowlandson, C., *Pulp Paper Mag. Can.* **66** (2), T-65 (1965).
310. Rozenberger, N. A., *Bumazhn. Prom.* **32** (10), 4 (1957).
311. Rozenberger, N. A., *Bumazhn. Prom.* **32** (10), 10 (1957).
312. Rydholm, S., *Svensk Papperstidn.* **57** (12), 427 (1954).
313. Rydholm, S., *Svensk Papperstidn.* **58** (8), 273 (1955).
314. Rydholm, S. and Lagergren, S., *Svensk Papperstidn.* **62** (4), 103 (1959).
315. Sadler, H., Bellak, F. and Roeder, H., *Papier* **7** (21), 416 (1953).
316. Saiha, E., *Paper Trade J.* **148** (23), 30 (1964).
317. Sallinen, K. A., *Paperi Puu* **5** (4), 53 (1954).
318. Samuelson, O., Gabrielson, G. and Hartler, N., *Svensk Papperstidn.* **55** (16) 613 (1952).
319. Samuelson, O. and Lindgren, B. O., *Svensk Papperstidn.* **51**, 179 (1948).
320. Samuelson, O. and Schoon, N. H., *Svensk Papperstidn.* **59** (21), 743 (1965).
321. Samuelson, O. and Schoon, N. H., *Svensk Papperstidn.* **60** (7), 259 (1957).
322. Samuelson, O. and Simonson, R., *Svensk Papperstidn.* **65** (18), 685 (1962).
323. Sanyer, N., Itho, T. and Keller, E. L., *Tappi* **47** (6), 323 (1964).
324. Sanyer, N. and Keller, E. L., *Tappi* **48** (2), 99 (1965).
325. Sanyer, N. and Keller, E. L., *Tappi* **48** (10), 545 (1965).
326. Sanyer, N., Keller, E. L. and Chidester, C. H., *Tappi* **45** (2), 90 (1662).
327. Sapotnitzki, S. A., *Bumazh. Prom.* **32** (8), 5 (1957).
328. Sarkanen, K. V. *in* B. L. Browning, "The chemistry of wood." Wiley (Interscience), New York, 1963.
329. Scheil, M. A., *Tappi* **36** (6), 241 (1953).
330. Schenk, W. A., *Tech. Assoc. Papers* **27**, 593 (1944).
331. Schippel, H. W., *Papier* **13** (17), 416 (1959).
332. Schmied, J., *Zellstoff Papier* **8** (6), 222 (1959).
333. Schmidt, E., *Svensk Papperstidn.* **55** (4), 134 (1952).
334. Schmidt, U., *Holzforschung* **15** (3), 79 (1961).
335. Scholander, A., *Svensk Papperstidn.* **53** (4), 95 (1950).
336. Scholander, A., *Svensk Papperstidn.* **53** (2), 35 (1950).
337. Scholander, A., *Svensk Papperstidn.* **55** (16), 620 (1952).
338. Scholander, A., *Tappi* **43** (8), 706 (1960).
339. Schoon, N. H., *Svensk Papperstidn.* **64** (17), 624 (1961).
340. Schoon, N. H., *Svensk Papperstidn.* **65** (19), 729 (1962).
341. Schoon, N. H., *Svensk Papperstidn.* **65** (23), 965 (1962).
342. Schoon, N. H., *Svensk Papperstidn.* **66** (17), 659 (1963).
343. Schorning, P., *Faserforsch. Textiltech.* **8** (12), 487 (1957).
344. Schorning, P., *Zellstoff Papier* **7** (1), 8 (1958).
344a. Schroeder, H. A. and Hansen, E. D., *Tappi* **51** (1), 1 (1968).
345. Schur, M. O. and Baker, R. E., *Tech. Assoc. Papers* **24**, 405 (1941).
346. Schur, M. O. and Baker, R. E., *Tech. Assoc. Papers* **25**, 453 (1942).
347. Schur, M. O. and Ingalls, E. G., *Tech. Assoc. Papers* **26**, 296 (1943).

348. Schwabe, K., *Cellulosechemie* **20** (3), 61 (1942).
349. Schwabe, K., "Fortschritte der pH-Messtechnik." Akademie Verlag, Berlin, 1953.
350. Schwabe, K., *Chem. Ing. Techn.* **29** (10), 656 (1957).
351. Schwabe, K. and Abdelmoniem, S. A., *Zellstoff Papier* **9** (7), 291 (1960).
352. Schwabe, K. and Hasner, L., *Wochbl. Papierfabr.* **4**, 115 (1934).
353. Schwalbe, C. G., *Papierfabrikant* **22** (16), 169 (1924).
354. Scott, W. M., *Can. Pulp Paper Ind.* **13** (3), 26 (1960).
355. Seaman, S. E., U.S. Patent 2,698,234.
356. Seborg, C. O. and Simmonds, F. A., *Tech. Assoc. Papers* **25**, 639 (1942).
357. Shaw, A. C., *Pulp Paper Mag. Can.* **57** (1), 95 (1956).
358. Sihtola, H., Anthoni, B. and Rosqvist, E., *Paperi Puu* **40** (10), 493 (1958).
359. Sillen, L. G. and Andresson, T., *Svensk Papperstidn.* **55** (16), 622 (1952).
360. Silver, F. P. and Beath, L. R., *Tappi* **36** (7), 305 (1953).
361. Simmons, T., *Svensk Papperstidn.* **56** (4), 121 (1953).
362. Sjöstrom, E. S., *Norsk Skogind.* **18** (6), 212 (1964).
363. Sjöstrom, E. and Haglund, P., *Svensk Papperstidn.* **65** (9), 370 (1962).
364. Sjöstrom, E. and Haglund, P., *Tappi* **47** (5), 286 (1964).
365. Sjöstrom, E., Haglund, P. and Janson, J., *Svensk Papperstidn.* **65** (21), 855 (1962).
366. Skoggard, C. O, and Libby, C. E., *Tech. Assoc. Papers* **29**, 479 (1946).
367. Slavik, I., *Zellstoff Papier* **6** (11), 331 (1957).
368. Slavik, I., *Svensk Papperstidn.* **64** (11), 427 (1961).
369. Smith, W. T. and Parkhurst, R. B., *J. Am. Chem. Soc.* **44**, 1819 (1922).
370. Söderquist, R., *Svensk Papperstidn.* **58** (19), 706 (1955).
371. Somer, V., *Papier* **17** (10a), 568 (1963).
372. Sorensen, H. A. and Douty, C. J., *Tappi* **40** (8), 658 (1957).
373. Spalding, C. W. and Han, S. T., *Tappi* **45** (3), 192 (1962).
374. Stanners, L. N. and Ingruber, O. V., *Pulp Paper Mag. Can.* **59** (1), 89 (1958).
375. Stevens, R. J., *Pulp Paper Mag. Can.* **59** (1), 96 (1958).
376. Stewart, D. L. and Crotogino, H. F., *Pulp Paper Mag. Can.* **60** (9), T-284 (1959).
377. Stockman, L., *Svensk Papperstidn.* **54** (18), 621 (1951).
378. Stockman, L., *Svensk Papperstidn.* **54** (18), 641 (1951).
379. Stockman, L., *Svensk Papperstidn.* **56** (1), 11 (1953).
380. Stockman, L. G., *Tappi* **43** (2), 112 (1960).
381. Stone, J. E., *Tappi* **38** (8), 449, 452 (1955).
382. Stone, J. E., *Tappi* **40** (7), 539 (1957).
383. Stone, J. E., *Pulp Paper Mag. Can.* **57**, (7), 139 (1956).
384. Stone, J. E. and Foerderreuther, C., *Tappi* **39** (10), 679 (1956).
385. Stone, J. E. and Green, H. V., *Pulp Paper Mag. Can.* **59** (10), 223 (1958).
386. Strapp, R. K., *Pulp Paper Mag. Can.* **56** (3), 179 (1955).
387. Strapp, R. K., Kerr, W. D. and Vroom, K. E., *Pulp Paper Mag. Can.* **58** (3), 277 (1957).
388. Strapp, R. K. and Woods, J. M., *Pulp Paper Mag. Can.* **57** (5), 124 (1956).
389. Strehlenert, R. W., *Svensk Kem. Tidskr.* **25**, 28 (1913).
390. Sulphite Products Corp., U.S. Patent 3,148,177 Sept. 8, 1964.
391. Sundman, J., *Paperi Puu* **32B** (12), 379 (1950).
392. Sundman, J. and Sundman, V., *Paperi Puu* **32B** (12), 379 (1950).
393. Svanoe, H., *Tappi* **40** (3), 152 (1957).
394. Taki Fertilizer Mfg. Co., U.S. Patent 3,270,001 Aug. 30, 1966.
395. Tappi Monograph Series No. 4, New York 1947.
396. Tarbutton, G., U.S. Patent 2,984,545 May 16, 1961.
397. Tenescu, S. and Cardi, C., *Atip Bull.* **18** (1), 15 (1964).

398. Thomas, B., *Pulp Paper Mag. Can.* **58** (12), 139 (1957).
399. Thompson, N. S. and Kaustinen, O. A., *Tappi* **49** (12), 550 (1966).
400. Thompson, N. S. Peckham, J. R. and Thode, E. F., *Tappi* **45** (6), 433 (1962).
401. Tomlinson, G. H., Tomlinson, G. H., II, Bryce, J. R. and Tuck, N. G., *Pulp Paper Mag. Can.* **59** (C), 252 (1959).
402. Tomlinson, G. H., Tomlinson, G. H., II, Bryce, J. R. and Tuck, N. G., *Pulp Paper Mag. Can.* **59** (5), 247 (1958).
403. Tucker, E. F., *Tappi* **33** (1), 29 (1950).
404. Ulfsparre, S., *Svensk Papperstidn.* **61** (18B), 803 (1958).
404a. Uprichard, J. M., *Appita J.* **21** (5), 164 (1968).
405. Utaka, G., *Tappi* **48** (1), 78A (1965).
406. Utaka, G., Oku, K., Matsuura, H. and Sakai, *Tappi* **48** (5), 273 (1965).
407. Vethe, A., Loras, V. and Loeschbrandt, F., *Norsk Skogind.* **16** (11), 458 (1962).
408. Vilamo, E., Aho, O. and Auino, K., *Svensk Papperstidn.* **58** (12), 452 (1955).
409. Virkola, N. E. and Alm, A., *Papier* **15** (10A), 522 (1961).
410. Vogt, G., *Chem. Ing. Tech.* **35** (1), 48 (1963).
411. Vollprecht, M. and Müller, G., *Wochbl. Papierfabrik.* **87** (24), 1051 (1959).
412. Volpicelli, G. and Massimilia, L., *Pulp Paper Mag. Can.* **66** (10), T-512 (1965).
413. Vos, J. W. de, *Tappi* **39** (12), 833 (1956).
414. Vroom, K. E., *Pulp Paper Mag. Can.* **58** (C), 228 (1957).
415. Wacek, A. v. and Schlegl, F., *Papier* **13** (7/8), 125 (1959).
416. Wacek, A. v., Grablowitz, O. and Oberbichler, W., *Papier* **20** (7), 376 (1966).
417. Wardrop, A. B. and Davies, G. W., *Holzforschung* **15** (5), 129 (1961).
418. Wells, F. L. and Mac Claren, R. H., *Tappi* **38** (11) 668 (1955).
419. Wenzl, H. F. J., *Papier* **11** (19/20), 435 (1957).
420. Westerberg, E. N., *Pulp Paper Mag. Can.* **68** (2), T-40 (1967).
421. Wester, A. and Aggeryd, B., *Svensk Papperstidn.* **58** (21), 788 (1955); 59 (3), 80 (1956).
422. Weyerhaeuser Co., British Patent 1,024,136 March 30, 1966.
423. White, C. K., Vivian, J. E. and Whitney, R. P., *Tech. Assoc. Papers* **31**, 141 (1948).
424. Whitney, R. P., Elias, R. M. and May, M. N., *Tappi* **34** (9), 396 (1951).
425. Whitney, R. P., Han, S. T. and Davis J. L., *Tappi* **36** (4), 172 (1953).
426. Whitney, R. P., Han, S. T. and Davis, J. L., *Tappi* **40** (7), 587 (1957).
427. Whitney, R. P., Han, S. T., Kesler, R. B. and Bakken, J. F., *Tappi* **48** (3), 1 (1965).
428. Wilcox, J. M., *Tappi* **33** (6), 272 (1950).
429. Wiley, A. J., Whitmore, L. M. and Boggs, L. A., *Tappi* **42** (5), 14A (1959).
430. Wilson, A. W., *Pulp Paper* **34** (4), 92 (1960).
431. Wilson, W. J., Comm. Int. Sulphite Pulping Conference 1964.
432. Wilson, J. W. and O'Meara, D., *Pulp Paper Mag. Can.* **61** (4), T-259 (1960).
433. Wilson, J. W. and Worster, *Pulp Paper Mag. Can.* **62** (C), T-98 (1961).
434. Wise, L. E. and Jahn, E. C., "Wood chemistry" Vols. I and II. Reinhold, New York, 1963.
435. Woods, J. M. and Hart, J. S., *Tappi* **38** (9), 548 (1955).
436. Yean, W. Q., Ross, J. H. and Vroom, K. E., *Pulp Paper Mag. Can.* **58** (7), 197 (1957).
437. Yorston, F. H. and Liebergott, N., *Pulp Paper Mag. Can.* **66** (5), T-272 (1965).
437a. Young, W. D. and Packman, D. F., *Appita J.* **21** (5), 144 (1968).
438. Zimmermann, F. J. and Diddams, D. G., *Tappi* **43** (8), 710 (1960).
439. Zobel, B. J. and McElwee, R. L., *Tappi* **41** (4), 158 (1958).
440. Zobel, B. J. and McElwee, R. L., *Tappi* **41** (4), 167 (1958)·

VIII

The Production of Pulp by Alkaline Reagents

INTRODUCTION

The production of pulp by means of alkaline reagents is considerably older than pulping with sulfurous acid and its salts. It has been reported that, in China, as early as 2000 years ago, bast fibers were isolated from mulberry trees by treatment with alkalis, presumably potash. This, of course, was not a production of pulp according to today's meaning, especially when wood is used. The first experiments along these lines may have been carried out toward the end of the 18th century, but they were not very successful. It was only after the large-scale production of alkalis, such as sodium carbonate and sodium hydroxide, had made their use economically possible that they were used for the pulping of wood. The pulp obtained in this way was dark brown and was not well-suited for the production of paper for printing or writing. Together with the industrial manufacture of soda, it was necessary to produce chlorine and hypochlorite of lime, by means of which the pulp could be bleached and thus be made suitable for varied uses. As it happened, these important chemical discoveries occurred at almost the same time and thus permitted the production of pulp from wood by means of alkaline reagents.

Almost simultaneously, Burgess in the United States and Watt in England succeeded in decomposing wood with alkali. In the years from 1853 to 1858, the first patents dealing with these processes appeared in these countries. In contrast to earlier unsuccessful proposals, these used closed vessels under pressure, with caustic soda as the reagent. In the first patents, a cooking temperature of 150°C, corresponding to 4.2 atm pressure, was mentioned, while in later patents the temperature was increased

to 180° to 190°C, corresponding to 11 to 12 atm. The strength of the alkaline solution was given as 6.5° to 7°Bé. It is interesting to note that, in a supplementary claim, the evaporation and combustion of the waste liquor for the recovery of the chemicals were specified.

The first mills in the United States (Royresford and Manayunk) and in England (Conemills) operated partly with vertical and partly with horizontal digesters. The heating was effected either by the direct introduction of superheated steam or by means of heating coils. Direct heating by gas flames was also used. The digesters were relatively small units of about 10 cubic meters. In the 1870s, the first mills were built in Germany and the one in Dalbke deserves special mention because it used, for the first time, a liquor circulation system. The liquor was heated in heating coils outside the digester by direct heat and was introduced into the digester at the top and withdrawn at the bottom. Dresel, who constructed this plant, also considerably improved the regeneration of the chemicals and the chipping of the wood by the introduction of ingenious ideas and designs.

Ungerer had found that the dissolution of the noncellulosic components varied, i.e., some were dissolved readily, others only with difficulty. His cooking process applied the diffusion cooking principle for the first time to the production of pulp. In this way, the wood first comes into contact with a liquor enriched with organic substances but having only a small amount of free alkali, which is completely consumed by the readily soluble hemicelluloses. The largely decomposed wood chips are then treated with fresh alkali cooking liquor and in this way a light-colored, readily bleachable pulp is obtained. Although Ungerer's ideas aroused great interest and pointed toward today's developments, his process itself was not basically successful because the recovery of the chemicals was found to be too expensive.

It was early recognized, especially in the United States, that it was necessary for the recovery of the chemicals to be carried out economically. To accomplish this, a separation, as complete as possible, of the pulp from the spent liquor had to be attempted. Dahl introduced the principle of diffusers into the alkaline pulping process. By this means the contents of the digester are blown under pressure into so-called diffusers, thus causing the separation of the fibers from the softened wood chips and, in this way, facilitating the separation of the liquor from the pulp. Because the diffusers are closed vessels, the excess steam can be utilized economically with regard to heat. For the evaporation of the spent liquor and calcination of the residue, experience gained from the soda industry was available. However, it was soon found that the economical operation of a mill under these conditions was impossible. The consumption of coal was too high to permit the evaporation to be carried out in this way. As a consequence, the sugar industry, with its multiple effect evaporating systems, served as

a model for the pulp industry. It would take too long to discuss in detail the numerous improvements that have been carried out in the decades following on the evaporation of the liquor and, with it, on the recovery of the chemicals. An early attempt was made to utilize the hot furnace gases for the evaporation of the liquor. Disk evaporators were constructed and exposed large surfaces of the liquor to the flame and furnace gases. The original open-hearth furnaces in which the organic substances of the liquor were burned were later converted into rotating furnaces, and, most recently, the concentrated spent liquor, injected in a finely divided form into ultramodern spent liquor boilers, serves directly as fuel and thus makes the alkaline pulping process self-supporting as regards heat economy.

This process as applied to wood underwent a fundamental change by the addition of sodium sulfide to the sodium carbonate and sodium hydroxide cooking liquor. The first experiments in which sodium sulfide, exclusively, was used as pulping reagent were carried out by Poole (1853), Henri (1861), and Eaton (1871). In 1879, Dahl compensated for the losses of alkali occuring during the pulping process by the addition of sodium sulfate to the black liquor in the recovery of the chemicals. During the combustion process, the sodium sulfate was reduced to sodium sulfide and the cooking liquor prepared from the regenerated chemicals contained a small amount of sodium sulfide, together with the main body of sodium hydroxide. This type of alkaline compensation by the addition of sodium sulfate gave this particular cooking process the name "sulfate cooking process," although this designation is misleading. The introduction of sulfur into the cooking liquor gave rise to other problems, especially that of the creation of obnoxious odors, with the mercaptans formed during the course of the cook causing a high degree of air pollution. In Sweden, with its sparsely populated areas and extensive forests, this process found wide application. By the end of the 1880s, however, the sulfite pulping process underwent an unexpectedly rapid growth and began to surpass the production of alkaline pulp. The sulfite process was practically odorless and the pulp produced was light in color and could be used for many types of paper even in an unbleached state. The chief disadvantage of the process was the tremendous amount of sulfite spent liquor formed which could be discarded into rivers only if the mill was located near a sufficiently large vorfluter (draining ditch). Another disadvantage was that the process was limited to certain species of wood, whereas the alkaline pulping process could be adjusted to woods rich in resins and hence unsuitable for the sulfite process. The further development of the sulfite process has been discussed in the preceding chapter.

The alkaline pulping process, in turn, underwent an unprecedented growth in the Scandinavian countries, the United States, and Canada and, during the last decade, also in Japan, and considerably surpassed the pro-

duction of sulfite pulp in these countries. There are various reasons for this. Alkali pulp possesses mechanical properties that favor the manufacture of paper, especially of wrapping paper. Furthermore, the alkaline pulping process can utilize practically all wood species, even waste woods from forests and wood-processing industries, and this is especially advantageous in countries having an extensive forest and wood economy. Finally, as will be discussed later, the processes involved in the pulping, in the regeneration of the chemicals from the spent liquor, and the bleaching of the originally dark-colored pulp have been improved to such an extent that alkali pulp can be used for practically all kinds of paper products. Another aspect that should not be underestimated is the fact that the odor annoyance has been almost completely eliminated and the major problem of the disposal of spent liquor from sulfite mills has never existed for alkali pulp mills because of their recovery of the chemicals. As has been mentioned before, with the use of soluble bases the operation of the sulfite pulping process more nearly approaches that of the alkaline process, and it will be interesting to observe whether this will lead to a trend toward the building of new sulfite pulp mills.

Before the statistical data for the production of pulps by the alkaline methods are discussed, it is necessary to make some observations regarding the nomenclature. As mentioned above, the term "sulfate pulp" is misleading. In spite of this, the term is widely used and can hardly be avoided. Originally, unbleached sulfate pulp was used mainly for the manufacture of high-strength wrapping paper, or so-called kraft paper. This term was also used for the pulp, i.e., kraft pulp. Today, however, there is no longer a distinction between sulfate and kraft pulps because there are highly bleached sulfate pulps that are also exceptionally strong. According to the "Dictionary of Paper," published by the American Paper and Pulp Association, the terms "sulfate" and "kraft" are synonymous. Because "sulfate" with reference to the pulping of wood is a misnomer, the term "kraft pulping process" will be used for the pulping of wood with alkali in the presence of sodium sulfide, and the pulp obtained by this method will be called "kraft pulp." A pulp obtained by a cook with sodium carbonate or sodium hydroxide, alone, will be termed a "soda pulp."

A survey of the world's production of pulp shows an unusually large increase in the production of kraft pulp, especially since the Second World War. According to the U.S. Pulp Producer's Association, 33,000,000 short tons of kraft pulp were produced in the Free World in 1964, as compared with about 4,500,000 in 1937. In the U.S., 5,600,000 tons of sulfite pulp and 23,000,000 tons of kraft pulp were produced in 1964. In the same year, Canada produced 2,900,000 tons of sulfite and 3,500,000 tons of kraft pulp. These figures do not include the pulp (dissolving pulp) used

TABLE VIII-1

Pulp Production of Some Principal Countries

Country	Production in millions metric tons			
	Total	Dissolving	Sulfite	Kraft
United States	27,722	1,745	5,093	20,884
Canada	6,258	427	2,628	3,203
Finland	3,545	273	1,196	2,076
Sweden	5,016	446	1,697	2,873
Norway	828	130	896	
Japan (1963)	2,952	4.3	372	2,167

for the manufacture of artificial fibers. The statistics of the FAO of the United Nations for 1965 do not differentiate between the types of pulp according to the process used. Table VIII-I includes figures indicating the marked increase in kraft pulp production in Japan. In other countries the production of sulfite and kraft pulps is approximately equal. It must therefore be assumed that this increase in the U.S. and Japan is chiefly the result of the supply of pulp wood. Accurate data on the world's production of pulp are not readily obtainable because that of both the European and Asiatic parts of Russia can only be estimated approximately.

It is difficult, if not impossible, to make definite statements on the future developments of the two competing processes. Sulfite and kraft pulps have their specific properties and hence their specific industrial uses. It must not be overlooked that, in the production of kraft pulp, improvements in the cooking process, in the recovery of the chemicals, in heat technology, and in its economy in general, the adoption of the continuous cooking process long before its application to the sulfite process, and the possibility of varying the properties of the pulp by different cooking and bleaching procedures have constantly widened the field of its application. From the very beginning, kraft pulp mills were forced to process their waste liquors and therefore they have the equipment and experience necessary to do this. With sulfite pulp mills, the overloading of the vorfluter with spent liquor has made its processing absolutely indispensable. Such processing is economical only when it is combined with the simultaneous recovery of the chemicals and utilization of the calorific value of the spent liquor. These prerequisites are met only through the use of soluble bases. It is therefore doubtful that small pulp mills are suitable for the application of the modified sulfite process. The introduction of the continuous pulping process also shows that, in the future, only large mills will be able to compete. Under these conditions, a promising future can be predicted without reservations for both pulping processes.

1. THE ACTION OF THE CHEMICALS USED IN THE KRAFT PULPING PROCESS

1.1 The Alkali

First, some generally used terms for the reactive agents must be defined. The terms used here are those specified by TAPPI (356), as follows:

1. Total alkali: $NaOH + Na_2S + Na_2CO_3 + 1/2 Na_2SO_3$, all expressed as Na_2O.
2. Active alkali: $NaOH + Na_2S$, expressed as Na_2O.
3. Effective alkali: $NaOH + 1/2 Na_2S$, expressed as Na_2O.
4. Causticity: the percentage ratio of NaOH, expressed as Na_2O, to active alkali.
5. Sulfidity: the percentage ratio of Na_2S, expressed as Na_2O, to active alkali.
6. Reduction: in green liquor, the percentage ratio of Na_2S to $Na_2SO_4 + Na_2S +$ any other soda sulfur compounds, all expressed as Na_2O.

According to Scandinavian standards, these terms are based on NaOH.

This comparison shows that, at least in the kraft pulping process, it is difficult to separate the action of the alkali from that of the sulfur compounds because the latter, in the form of Na_2S, are included in the total alkali. Pulping experiments in the absence of sulfur showed that only about 3% to 4% of the sodium hydroxide used is consumed by the dissolution of the lignin, about 1.5% for the neutralization of the formic and acetic acid groups, and 10.5% to 11% for the dissolution of carbohydrates. Only after a considerable amount of hemicelluloses is dissolved does the reaction with the lignin proceed. On the other hand, the sulfur component, when present, furthers primarily the dissolution of the lignin.

In the reaction between sodium hydroxide and hemicelluloses, sugar acids, lactones, simple organic acids, and humic substances are formed. In the latter stages of the cook the cellulose itself, after having been freed of hemicellulose and lignin, is attacked. In general, lowering the amount of active alkali in the cooking liquor by about 12% does not markedly affect the yield; on the other hand, an increase of this size results in a pulp with improved mechanical properties. Even at temperatures of below 100°C, 30% to 40% of the alkali is consumed, but the dissolution of the lignin and the carbohydrates at this temperature is insignificant. At between 100° and 150°C, about 70% of the glucomannan originally present is dissolved, while at 140° to 150°C the dissolution of the lignin begins and, at 170°C, about 50% of the latter has gone into solution. When the amount of active alkali is drastically decreased, the yield of xylan increases to a considerable

extent, while the yield of glucomannan is not affected. At the same degree of pulping (kappa number) an increase in the alkali content of the liquor results in a small decrease in the yield of pulp and a greater loss in the yield of xylan; at the same time, the brightness of the pulp increases, but the pulps thus obtained require a greater consumption of power during their beating (18,48,63).

The dissolution of the lignin undoubtedly occurs in several stages. In the first, alkali is adsorbed at the lignin-liquor boundary surface. The adsorption takes place rapidly and reaches an equilibrium after saturation. The influence of the concentration of the alkali is small up to a temperature of about 140°C and after that increases two-fold with an increase of 10°C in temperature. No impregnation of the wood chips is necessary before the actual reaction between the lignin and the cooking liquor takes place because the reaction area follows the penetration of the liquor into the interior of the chips. A continuous "lignin gradient" exists between the outer surface of the wood chip and the reaction zone, with very low values at the outer surface and up to about 20% lignin in the immediate vicinity of the reaction zone. Temperature, liquor concentration, and the dimensions of the wood chips are reaction-determining factors for the advance of the reaction zone into the wood structure. The reaction is actually a diffusion process in which the boundary-reaction zones move slowly into the interior of the wood chips, while the dissolved reaction products diffuse out into the surrounding liquor. In the second stage, therefore, the pulping process becomes an extraction process (49,54,223,272,277).

According to Nernst's equation,

$$\frac{dx}{dt} = kS\,(a - x) = \frac{D}{\delta}S\,(a - x)$$

in which S is the contact surface, $(a - x)$ is the concentration of the reagent at any instant, and k is a constant, the magnitude of which depends upon the diffusion coefficient, D, and the thickness of the interfacial layer, S — an equilibrium is formed at each separating layer between two phases, with almost infinitely great velocity as compared with the diffusion velocity. The latter therefore becomes the time-determining factor with regard to the course of the reaction (324,387).

For each degree of pulping there is obviously a precisely defined minimum of effective alkali requirement and a corresponding minimum value of pH in the black liquor; the effective alkali rather than the active alkali expresses the actual alkali consumption in the kraft cooking process. An increase in temperature of about 20°C causes almost a tripling of the delignification velocity. Rapid cooks (of up to 30 minutes) of sprucewood chips with caustic soda to a yield of 50%, by rapid heating to a maximum temperature, resulted in an alkali consumption of 12% to 12.5% of sodium

hydroxide (9.3% to 9.7% Na_2O), whereas a consumption of 16% of sodium hydroxide (12.4% Na_2O) is reported in the literature as a normal alkali consumption for an alkali cook. Such pulping experiments have been carried out in a liquor phase and a vapor phase, in the latter case with chips previously impregnated with the liquor. At constant sulfidity the active alkali of the cooking liquor increases with increasing concentration, but the consumption of active alkali and sulfide remains fairly constant within the range of the concentrations investigated. During the impregnation step only a little sulfur is bound to the wood in irreversible form even under drastic reaction conditions (20 minutes at 140°C) (173,204).

1.2 The Sulfur Components

It is generally agreed that the sulfur components facilitate the dissolution of the lignin in the alkaline pulping process, but there is considerable disagreement as to the required degree of sulfidity. On the one hand, it has been found that an increase in sulfidity from 15% to 31% results in only a slight improvement in the pulping, while, on the other hand, a sulfidity of even 100% has been considered to be the optimum. An older theory maintained that a retardation of the delignification is caused when

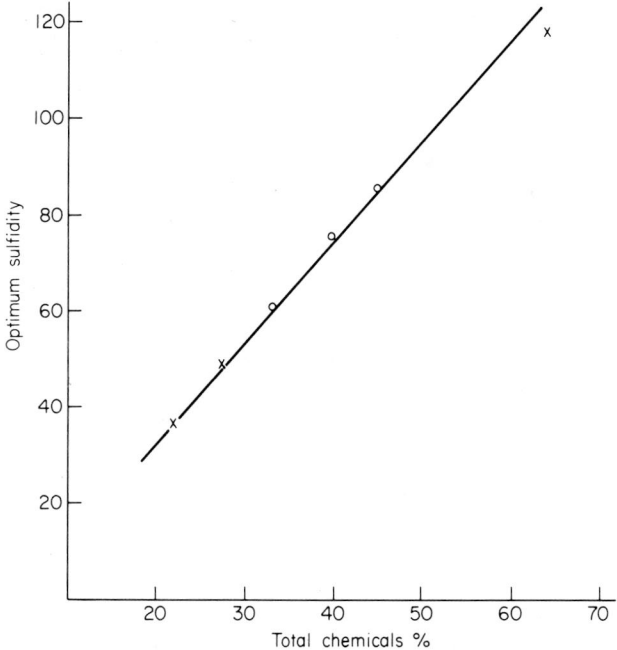

Figure VIII-1. Relashionship between total chemical consumption and optimum sulfidity in kraft pulping (155).

a certain degree of sulfidity is surpassed (149,212,258). This disagreement can perhaps be explained as resulting from a misunderstanding of the action of the sulfide in the cooking liquor. The sodium sulfide represents an integral part of the total alkalinity; when the latter is increased, the sulfidity also can be increased. The hydrolysis of the sulfide plays an important role here. It has been found that, at the start of the cook, about 40% of the sulfide is hydrolyzed, whereas toward the end of the cook, over 90% may be hydrolyzed. These figures are valid only for mixtures of sodium hydroxide and sodium sulfide and within the usual ranges of concentration in the kraft pulping process. There can be no doubt that the alkali component in the cooking liquor must have a certain numerical relationship to the sulfur component. Figure VIII-1 shows the relationship between the total chemical content and the optimum sulfidity (155,236).

A pure alkaline cook without the presence of a sulfur component and within the normal alkali concentration range, at 160°C, results in a pulp with about 5% residual lignin in 10 hours, whereas with a sulfidity of 5.25% the reaction time is reduced to 8 hours and the residual lignin content in the pulp is lowered to 3%. This fact proves that there exists an optimum relationship between the two alkaline components. At a sulfidity of 15% the reaction time is reduced to 6 hours, but the lignin content in the pulp is not further reduced. This reaction apparently occurs in two stages, in the first of which the lignin molecule takes up sulfur. It is assumed that the hydroxyl in substituted benzyl alcohol groups is replaced by a mercaptan group; the latter is unstable and is converted into a sulfide group by the reaction with a hydroxyl group. In the second stage, the alkali causes a hydrolytic cleavage of the phenyl ether group, free phenolic groups are formed, and the lignin dissolves in the alkali. Disulfide groups, formed by the oxidation of sulfide groups in lignin, are split into the corresponding mercaptans and sulfenic acids. The concept that the sulfur groups entering the lignin molecule increase its acidity and, with it, its solubility in alkali (155) is contradicted by the suggestion that the sulfidation, instead, causes the blocking of a group sensitive to condensation, and the role of the sulfur component in the cooking liquor therefore consists primarily in the hindrance of such condensation reactions. The addition of hydrogen sulfide to a carbonyl group in the lignin molecule is improbable because no mercaptan sulfur can be detected in thiolignin (5, 55,116,146).

The negligible effect of the sulfidity on deciduous woods is attributed to the high methoxyl content of their lignin. To produce a good pulp from beech, for example, a definite minimum amount of alkali is required (about 8.6% sodium hydroxide and 1% sodium sulfide, based on the absolutely dry wood). Although an average optimum could be shown for a sulfidity of 20%, even very small amounts of sodium sulfide caused a con-

siderable change in the properties of the pulp obtained. The beech lignin, with its high methoxyl content, is less susceptible to condensation reactions (193). The sodium sulfide exerts a pronounced buffer reaction. The organic salts in the black liquor also exhibit a buffer action. A prior impregnation of the wood chips by black liquor has therefore been reported to be advantageous. In any case, a reduction in the sulfidity is desirable because it results simultaneously in a reduction of the formation of mercaptan and, hence, of air pollution (386).

As mentioned above, opinions concerning the optimum composition of the cooking liquor vary to a considerable degree, but the reaction-promoting effect of the sulfide on the delignification and its buffer action are unanimously acknowledged. An increase in sulfidity and active alkali in the cooking liquor accelerates the pulping velocity up to a maximum value above which a further increase affects the qualitative properties and the yield of the pulp. In comparing these different results with regard to the efficiency of the liquor components, it must be taken into consideration that the results were obtained in experiments with quite different woods. Pure sodium sulfide cooks have shown that practically all the sulfur remains in the cooking liquor and only a very small amount reacts with the lignin. At below 120°C a reaction of the sulfur with the lignin is practically nonexistent. The effect of the sulfide in a kraft cook in the presence of alkali is therefore not limited only to a reaction with the lignin but also extends to its oxidation, in which the sodium sulfide effects the dissolution of the lignin and decreases the dissolution of the carbohydrates (227,319).

Furthermore, the heterogeneous reactions taking place between the inorganic components of the cooking liquor must be taken into consideration. Potentiometric titrations with silver nitrate solutions seem to indicate the occurrence of the reactions below (*a* to *g*), but the sequence in which they are listed does not denote the order of their occurrence in the digester under the conditions of the kraft cook. The following reactions are possible between sodium hydroxide and the sulfide component:

(*a*) $Na_2S + H_2O = NaOH + NaHS$, then further, via (*b*) to (*c*);

(*b*) $NaHS + H_2O = NaOH + H_2S$, then to (*c*);

(*c*) $NaHS + H_2O = NaOH + H_2S \uparrow$, on heating;

(*d*) $NaHS + NaOH = Na_2S + H_2O$, then further, via (*b*) to (*c*);

(*e*) $NaOH + H_2S = NaHS + H_2O$, with excess of H_2S, then via (*b*) to (*c*);

(*f*) $2NaOH + H_2S = Na_2S + H_2O$, with excess of NaOH, then to (*c*);

(*g*) $H_2S + Na_2S = 2NaHS$, at about 150°C.

Some of these reactions could convert sodium sulfide extensively and rapidly into hydrogen sulfide. At lower temperatures, especially at the beginning of the cook, and in the presence of a sufficient amount of sodium hydroxide, the hydrogen sulfide is converted into sodium sulfide again (ac-

cording to *f*) or to sodium hydrogen sulfide (according to *e* and *g*). At higher temperatures, reaction (*g*) may proceed preferentially. The above reactions do not deal with the formation of organic salts, thiolignin, mercaptans, or inorganic oxidation products such as thiosulfate or sulfates. It is certain that the presence of carbonate plays a role in the sequence of these reactions, and, on this basis, one can visualize the following reactions as being possible:

(*h*) $H_2S + Na_2CO_3 = NaHCO_3 + NaHS$ (at 100° to 125°C); then to (*b*), (*c*), (*d*), and (*i*);

(*i*) $2NaHCO_3 + H_2O = Na_2CO_3 + 2H_2O + CO_2 \uparrow$ (upon heating); then to (*h*) and (*j*);

(*j*) $CO_2 + Na_2S + H_2O = Na_2CO_3 + H_2S \uparrow$;

(*k*) $NaOH + CO_2 = NaHCO_3$; then (*i*).

In these reactions, sodium carbonate, hydrogen sulfide, and carbon dioxide are formed, with the latter contributing to the formation of the hydrogen sulfide according to equation (*j*) (13,43,44).

Whether the sodium sulfide, sodium hydrosulfide, or hydrogen sulfide is the reaction-determining component for the formation of thiolignin cannot be definitely decided. The hydrolysis of sodium sulfide to sodium hydrosulfide is probably far advanced during the first stage of the cook, but a reaction between the sodium sulfide and lignin cannot be excluded either. Unlike sodium hydrogen sulfide, sodium sulfide is a strong base and therefore is a better pulping reagent. The presence of sufficient amounts of free hydrogen sulfide is shown by gas analysis. Although hydrogen sulfide is very soluble in water, its solubility is reduced by the presence of numerous electrolytes. The hypothesis mentioned above, that there is no preimpregnation of the wood chips by the cooking liquor but that a reaction zone directly follows the diffusion-limit zone, does not exclude the possibility that hydrogen sulfide can participate directly in the formation of thiolignin.

1.3 Polysulfide

The addition of sulfur to the sodium hydroxide cooking liquor was mentioned in a patent in 1911 as a means of improving the yield of pulp (113). Recent investigations have intensified interest in the polysulfide cooking process. There are various methods for enriching the cooking liquor with polysulfide: (*a*) by the addition of elementary sulfur to the alkaline sodium sulfide-containing cooking liquor; (*b*) by the separation of the green liquor into sulfide and carbonate solutions and electrolytic oxidation of the former, or expulsion of the hydrogen sulfide by means of carbon dioxide and its oxidation into sulfur; (*c*) finally, by the direct addition of alkali polysulfide (32,379). Aqueous polysulfide solutions can

also be obtained by boiling sodium sulfide liquors with sulfur (141,247).

The possibility of producing polysulfides by processing, at an elevated temperature, the sodium-sulfur compounds available in the recovery system of a kraft pulp mill has recently been considered. It is known that polysulfides can be obtained by treating sodium sulfide with sulfur at over 400°C. Furthermore, some claims have been made that polysulfides can be formed when sodium sulfate is reacted with carbon in an electrode furnace or when sodium thiosulfate is heated to 300°C. There are three ways in which polysulfides may be formed from the aforementioned sodium-sulfur compounds: the thermal decomposition of thiosulfate, the partial oxidation of sulfide, and the partial reduction of sulfate. These reactions take place at elevated temperatures in the absence of water, at atmospheric pressure, and with limited air contact. The reaction product must be cooled before its dissolution in water, otherwise the polysulfides will partly decompose into thiosulfate and sulfide.

When thiosulfate is heated to above 450°C it decomposes to polysulfide and sulfate, probably according to the following equations:

$$4S_2O_3^{2-} \rightarrow 4S + 4SO_3^{2-}$$
$$4SO_3^{2-} \rightarrow 3SO_4^{2-} + S^{2-}$$
$$4S_2O_3^{2-} \rightarrow 3SO_4^{2-} + \text{``}S_4S^{2-}\text{''}$$

The "S_4S^{2-}" represents a mixture of polysulfides with the stoichiometric composition, Na_2S_5. On heating thiosulfate to 550°C for 15 minutes, the yields of sulfide and polysulfide excess sulfur obtained are 11.3% and 47.2%, respectively. This is close to the theoretical values of 12.5% sulfide and 50% excess sulfur (4a).

A study of the chemistry of the preparation of polysulfide cooking liquor from white liquor and sulfur showed that, in order to obtain a higher action rate, it is essential to keep the sulfide sulfur concentration at a high level and to recycle part of the polysulfide formed so that the ratio of polysulfide-surplus sulfur to sulfide sulfur is greater than 1 ($X_s > 1$). An intimate contact between the sulfur and the liquid phase is necessary. The temperature has comparatively little effect on the reaction rate. To obtain the highest possible polysulfide yield, X_s should be kept below 2.5 and the temperature below 80°C. Contact with air and addition of sulfite should be avoided (160a,228a).

Although a polysulfide-containing cooking liquor, with 26 grams per liter Na_2S_2 shows no change upon heating to about 130°C, the reaction of the polysulfide with wood takes place already at temperatures of 80° to 90°C and is almost completed at 140°C. It may be assumed that the polysulfide, through the consumption of alkali, very probably decomposes into sulfide and sulfite ions according to the following scheme:

$$3S_2^{2-} + 6OH^- \rightarrow 5S^{2-} + SO_3^{2-} + 3H_2O$$

The sulfite ions then can react further with polysulfide ions with the formation of thiosulfate:

$$SO_3^{2-} + S_2^{2-} \rightarrow S_2O_3^{2-} + S^{2-}$$

The reaction between polysulfide ions and the wood components occurs with the consumption of alkali as follows:

$$S_2^{2-} + 3OH^- + RCHO \rightarrow RCHOO^- + 2S^{2-} + 2H_2O$$

and

$$2S^{2-} + 2H_2O \rightarrow 2SH^- + 2OH^-$$

In polysulfide cooks it is advantageous to keep the alkali content of the cooking liquor somewhat higher than in normal cooks (215).

The velocity of the decomposition of the polysulfide into sulfide and thiosulfate increases with increasing hydroxide concentration and with increasing sulfur atoms in the polysulfide ion. Thiosulfate has no noticeable effect on the decomposition velocity. It can be concluded from the changes in the concentration of polysulfide, thiosulfate, and sulfide during the course of the cook that about 60% of the polysulfide decomposes. The remainder of the polysulfide sulfur is consumed by reactions with organic substances. The thiosulfate plays an active role in the reactions during the cook, as was shown when thiosulfate was added to a normal kraft cook and it was found that the concentration of thiosulfate had decreased (282). The advantages of the addition of polysulfide are shown, first by an increase in yield, then by a better bleachability and improved mechanical properties of the pulp (51,145,326,340).

A higher degree of delignification has been observed within the temperature range of 115° to 170°C. The increase in yield depends upon the species of wood used and may amount to about 8% based on the dry wood. The advantage of this increase in yield, however, is partially counteracted by the necessary changes in the chemical-regeneration system during the processing of the spent liquor. The special qualities of the polysulfide pulp must therefore justify, economically, the utilization of this type of cook (290).

These qualitative improvements in the polysulfide pulps are due, first of all, to their high carbohydrate content. When pinewood pulps prepared by the kraft process, with and without the addition of polysulfide, are hydrolyzed, the former are found to contain chiefly gluconic acid end-groups, while the latter contain only a very minute quantity of these groups. Some of the aldehyde end-groups that are formed by the cleavage of carbohydrate chains are oxidized to aldonic acid end-groups under the conditions of the polysulfide cook. This may, in part, explain the increase in yield (9).

Cellobiose and hydrocellulose were treated with polysulfide cooking liquor at 100°C. The reducing end-glucose moieties were thereby transformed into alkali-stable acid moieties which, together with glucometasaccharinic acids formed by an alkaline rearrangement, formed aldonic acids (4b). The pulping of wood with kraft liquor containing sodium polysulfide gives pulps with higher carbohydrate-to-lignin ratios than normal kraft pulps. Commercial application has been hindered because the single-stage polysulfide process suffers from several disadvantages, e.g., the polysulfide ion is largely destroyed at a high temperature before it can exert its full stabilizing effect on the polysaccharides. One way to avoid this is to separate the polysaccharide stabilization stage from the alkaline delignification stage. As a result, a lower consumption of both polysulfide and alkali has been observed. But some decomposition of polysulfide still occurs at the comparatively high temperature (130°C) necessary to achieve the stabilization reaction at a reasonable rate. Another method is to subdivide the first stage into individual stages for impregnation and stabilization. Optimum conditions for the stabilization can then be chosen without the necessity of protecting the polysulfide against decomposition (75a).

Undoubtedly the dissolution of the lignin is accelerated, but the increase in yield and, with it, the preservation of the carbohydrate fraction are more important advantages of this process. About half the increase in yield can be ascribed to the doubling of the yield of glucomannan while the other half can be explained by a ten percent higher yield of cellulose and xylose. At constant yield, polysulfide pulps have lower lignin and higher hemicellulose contents (89,315).

Another advantage of the polysulfide cook is the fact that, in comparison with a normal kraft cook, it is much more easily controlled. In conventional kraft cooks, the active alkali is the decisive variable for the delignification and the carbohydrate yield, but they also depend upon the sulfidity. At a high degree of sulfidity, the effective alkali seems to be the controlling variable, whereas at a low sulfidity, which is more generally used, the delignification is controlled by the active alkali. Of the 4.5% increase in yield, based on the wood, of the polysulfide cook, about 5% can be ascribed to lignin, 40% to cellulose, and 55% to hemicelluloses, consisting only to a small extent of pentosans (109).

The introduction of polysulfide pulping in a kraft mill entails a certian operational change-over. The preparation of the cooking liquor is best carried out by the dissolution of sulfur in the white liquor during the regeneration process. For satisfactory control of the cooking process an exact knowledge of the two parameters—effective alkali and amount of excess sulfur—is important. The polysulfide does not affect the delignification to any extent. Without the addition of supplementary alkali only relatively small amounts of carbohydrate polymers are dissolved in the cooking

liquor. The increase in yield corresponds roughly to that obtained through the addition of borohydride. Odorous organic sulfur compounds are produced in greater amount in the polysulfide cook, but they cannot be attributed directly to the presence of the polysulfide. The polysulfide is split into thiosulfate and sulfide even before the temperature required for the formation of these organic sulfur compounds is reached. The increased formation of sulfide is probably responsible for the stronger odor (160).

The major obstacle to the commercial use of the process is the lack of an effective recovery system. The present kraft recovery method can be used only after major modifications have been made because of the high sulfidity of the smelt that would result from the burning of polysulfide liquor. The sulfur or polysulfide added to the cooking liquor must either be taken out of the system or be converted into polysulfide from the white or green liquor by an oxidation process. Methods for the oxidation of black liquor are discussed in Section 5.4. In regard to the formation of polysulfide, it has been observed that it is formed not only during oxidation but also during storage after the oxidation has been completed. The oxidation process occurs in several steps and oxidized organic substances are intermediate products. In black liquor it is the sulfate lignin which will be oxidized first, and the oxidized products of this will, in turn, be capable of oxidizing sulfide to polysulfide. Much polysulfide is formed during storage and, at the same time, the amount of sulfide decreases. The polysulfide concentration reaches its maximum value when about 50% of the sulfide is oxidized. This oxidation process should be well suited for the preparation of a cooking liquor in the polysulfide process. Normally, white liquor and black liquor are mixed before or when they are in the digester in proportions determined by the required ratio of wood to liquor. If the mixture, before it is charged into the digester, is oxidized with air by means of simple oxidation equipment, it is possible to produce polysulfide. The higher the sulfide concentration of the liquor the greater the amount of polysulfide that can be produced. The sulfide concentration in this mixture is, in turn, dependent upon the sulfidity, and the increases in yield obtainable are therefore a function of the sulfidity of the smelt (221, 314a).

1.4 Stabilizing Additives

The purpose of the use of stabilizing additives, primarily those with reducing action, in the cooking liquor is mainly to preserve the amorphous carbohydrate fractions of the wood, i.e., the hemicelluloses. In connection with the decomposition of these carbohydrates, the alkaline hydrolysis of glucosidic linkages is of considerable importance. Phenylglucosides and glucosides of β-substituted alcohols, the β-substituent of which is an elec-

tron acceptor, are especially readily attacked by alkali (229). Sodium
hydrosulfite is a strong reducing agent. Aspen sawdust, heated for two
hours at 160°C with a mixed liquor of sodium hydroxide and sodium hydrosulfite, showed no sign of decomposition in spite of the fact that the
lignin content had decreased from 16.4% to 3.5% after the treatment.
Better results were obtained when the temperature was increased to 170°C
and the reaction time to 24 hours. The pulps thus obtained were light in
color, had a high polysaccharide content, and showed good strength properties (196,259).

Ammonium hydroxide is a good alkaline pulping agent when used at
high concentrations and high temperatures. A mixture of ammonium hydroxide and ammonium sulfide gives good results even at lower temperatures and concentrations. These pulps are distinguished by their high polysaccharide content (346).

Sodium formate has been suggested as a stabilizing agent but has not
proved a success (204). The use of borohydride as an additive in the
cooking liquor has recently been advocated, but thus far from a theoretical rather than from a practical point of view. Borohydride ($NaBH_4$) is a
sodium tetraboranate. When the aldehydic end-groups of the carbohydrate
polymers are reduced with borohydride, they are stabilized toward alkali.
An increase in yield by the stabilization of the carbohydrate fraction in the
solid phase is made possible through a selective delignification (158).

A 10% increase in yield occurs with the addition of only 1% sodium
borohydride, based on the weight of the wood. This increase in yield is
mainly the result of an extensive preservation (80%) of the glucomannans
present in the wood. In a normal kraft cook, only about 30% of the glucomannans are preserved. The amount of xylan in a pulp from a cook to
which sodium borohydride has been added is somewhat lower because of
the lower resorption of the xylan from the cooking liquor. When borohydride is added to the cooking liquor before the start of the cook, it is decomposed already in the first phase of the cook. The reducing action of
this reagent in an alkaline medium takes place according to the following
scheme:

$$BH_4^- + 8OH^- \rightarrow BO_2^- + 6H_2O + 8e^-$$

and, at higher temperatures, according to the scheme

$$BH_4^- + 4H_2O \rightarrow 4H_2 + OH^- + B(OH)_3$$

A reduction in the decomposition of the carbohydrates is paralleled by a
reduction in the consumption of alkali, as was already shown in Section
1.1. It must also be remembered that the addition of borohydride reduces
the reactivity of the lignin toward the alkaline components in the cooking
liquor, thus nullifying the reduction in the consumption of alkali through

the decrease in the delignification capacity. On the other hand, a decrease in alkali consumption is desirable for economic reasons to counteract the cost of the borohydride. An alkaline solution of borohydride can also be applied to the wood chips before the kraft cook. In this way, a 1% solution of borohydride, based on the weight of the wood, is applied for 30 minutes at 80°C, with a liquor-to-wood ratio of 5:1. By this method, the increase in yield is double that obtained when the borohydride is added directly to the digester (17,18,157,292).

Another reducing agent which seems to increase the yield of pulp from kraft cooking is hydrazine. It has been found that the higher pulp yield—about 10% for spruce and pine and 2% to 4% for birch, calculated on the basis of dry wood—originates mainly from a stabilization of the glucomannan in spruce and pine, and of the cellulose in birch. Much higher quantities of hydrazine are required to give results comparable to those obtained with borohydride (267a).

In the first phase of the alkaline cook, when the polymer carbohydrates are dissolved and the chain-splitting reactions begin, an environment that stabililizes the carbohydrates, through either reduction or oxidation, must be established. Neither borohydride nor polysulfide is an ideal agent for this purpose. The former is too expensive, and the protection exerted by the latter is inadequate. During the further course of the cook, it is especially important that a rapid reaction should take place at the α-carbon atoms of the lignin molecule, which contain free phenolic hydroxyl groups in paraposition, in order to prevent the occurrence of condensation reactions. During this stage of the cook, the presence of active chemicals capable of forming thio groups is desired in as great a concentration as possible. Such a chemical is, above all, hydrogen sulfide. It may be assumed that the condensation velocity decreases with decreasing hydroxyl ion concentration. For this reason, a low pH value in the cooking liquor is desirable during this stage. The dissolution of the lignin by the splitting of aryl alkyl ether linkages and the formation of new phenolic groups, on the other hand, requires a hydroxyl ion concentration as high as possible. Toward the end of the cook, the carbohydrates, which by now are dissolved in the cooking liquor, are stabilized and are removed from the cycle when the liquor is drained off; after the pH has been adjusted to 12, the carbohydrates in the liquor are returned to the digester. In this way, a large part of the carbohydrates is recovered by resorption into the pulp. These considerations clearly show that the kraft pulping process, also, is a stepwise process and that, by manipulation of the operational conditions of each individual step, advantages can be gained such as have not been achieved before. The continous cook, which will be discussed in detail later, seems to be destined to accomplish this selective stepwise cooking process in an economical way (159).

2. THE EFFECT OF THE COOKING LIQUOR
ON THE WOOD COMPONENTS

2.1   Lignin

Because lignin and carbohydrates are closely associated with each other in the wood—the type of linkages cannot be discussed here—it is obvious that the reaction of one of these components will, to a certain extent, have a direct effect on the other. As already mentioned, the component in the cooking liquor that is responsible for the formation of thiolignin cannot be definitely identified. There is some indication that the hydrogen sulfide, which is formed according to the above reaction sequence, participates directly in the thiolignin formation. A commercial thiolignin (Indulin) was dissolved in dilute sodium hydroxide, the solution was acidified stepwise, the precipitated lignin was extracted with acetone, and the residues of the extracts were chromatographed over aluminum oxide to give a series of substances which may be designated as thiolignins, although the various fractions differ fundamentally from each other. Upon reaction of the thiolignin by heating it with Raney nickel, aqueous alkali, and water, at least four different types of sulfur linkages appeared to be present in it. After oxidation of the methylated product, infrared spectroscopy seemed to indicate the presence of sulfoxide and sulfone linkages. Only traces of inorganically bound sulfur were present (120).

When eucalyptus wood is treated with buffered solutions of sodium sulfide ranging from a pH of 7.2 to 11.2, the solubility of the lignin in the alkali increases with the increase in pH. At near the neutral point, only the action of the sodium hydrosulfide is apparent. As already mentioned, sodium hydrosulfide is not a suitable pulping agent. On the other hand, a partial sulfidation of the lignin is also possible with sodium hydrosulfide or hydrogen sulfide within a pH range of 7.0 to 8.5. The lignin remains undissolved, however, but can be extracted with alkali or pyridine. Apparently different sulfur components take part in the sulfidation of the lignin during the course of the pulping process and various types of sulfur linkages are formed (211).

In order to learn more about the reaction of lignin in a kraft cook, experiments with model substances have been carried out. In this way, it could be shown that aryl alkyl ethers, which have a hydroxyl group at the neighboring carbon atom in the alkyl group (I), are split by alkali via the epoxide (II) into the corresponding phenols (III) and 1,2-glycols (IV) (134). A repetition of these experiments with finely ground lignin from *Picea abies* resulted in the same reaction sequence as with the model substance; here, also, aryl alkyl ether linkages (V) were split via the corresponding

epoxy intermediate (VI) into the 1,2-glycol structures (VII) and the corresponding phenols (VIII) (132,133,137).

Although β-aryl ether linkages are completely cleaved by white liquor containing sodium sulfide, only about 30% is split by $2N$ sodium hydroxide; this is elucidated by experiments with model substances which showed that the partial cleavage with sodium hydroxide takes place via the corresponding methylene quinone and enol ether structures, while the cleavage with white liquor is the result of the formation of sulfhydryl linkages (135).

It has also been suggested that the sulfur in thiolignin is present in its elementary form and not in a chemically bound state. In this way, the hydrogen sulfide in the cooking liquor would cleave the hydroxydiphenylmethane structures in the lignin between the substituted methylene bridges and the phenolic nuclei, with the formation of the corresponding mono-

cyclic phenolic structures and of elementary sulfur. 2,4′-Dihydroxydiphenylmethane and 4,4′-dihydroxydiphenylmethane structures may be formed in lignin during the alkali cook by the condensation of *p*-hydroxybenzyl alcohol or *p*-hydroxybenzylalkyl ether with phenolic units and other *p*-hydroxybenzyl alcohol structures (69, 405). Dihydroxydiphenylmethanes, however, are completely stable on treatment with alkaline sulfide solutions under the conditions of a kraft cook, which indicates that the above theory, which also contradicts the hitherto acknowledged hypothesis on the importance of sulfhydryl ions in the formation of thiolignin, can hardly be considered to be valid (136).

The mechanism of delignification by means of alkaline liquors may be summarized as follows: in liquors that are free of sulfur, the cleavage of alkyl aryl ether linkages takes place at high temperatures and between monomeric compounds which contain no free phenolic hydroxyl groups. The cleavage products dissolve through the formation of phenolates. By the cleavage of methoxyl groups, additional phenolic hydroxyl groups may be formed. Benzyl alcohol groups can condense through the activation of benzene rings or through the formation of dibenzylmethane structures. The formation of formaldehyde also may contribute to such condensations. In the presence of sulfide and hydrosulfide ions in sulfur-containing cooking liquor, mercaptan groups are formed even at low temperatures from *p*-hydroxybenzyl alcohol and alkyl ether groups of the lignin and, on further heating, they form dibenzylsulfide structures and can contribute to the cleavage of alkyl aryl ether linkages. Sulfide and hydrosulfide ions are constantly reproduced by alkaline hydrolysis of the thiolignin. Thus the decomposition of the lignin molecule by the cleavage of alkyl aryl ether bonds is accelerated by the sulfur components, and the formation of formaldehyde, which is activated by free phenolic hydroxyl groups, is prevented through a substitution by sulfur. In this way and through the substitution of the benzyl alcohol groups, the condensation is largely prevented and the delignification is facilitated (117).

As mentioned above, about 40% to 50% of the alkali present is consumed during the first part of the cook and at temperatures of below 100°C. During this phase of the cook, the dissolution of lignin is negligible. At between 100° and 150°C appreciable amounts of cellulose and xylan are dissolved. In the meantime, the remaining hemicelluloses have almost completely disappeared. At 140° to 150°C, the dissolution of the lignin is preponderant and, at 170°C, up to 75% has gone into solution. An increase in effective alkali causes primarily an increase in the dissolution of xylan and thus a decrease in the yield of pulp (15).

A quantitative determination of the dissolution velocity of the various wood components in an alkaline cook is made more difficult by the complexity of the physical and chemical structure of the wood and by the lack

of an exact detection of the liquor equilibriums. It may be assumed that the two factors which cause the movement of the liquor in the wood become less important when very thin wood chips are used and the alkali in the cooking liquor is not exhausted. If the movement of the liquor plays an important role, the diffusion can be held responsible for the inequalities of the pulp. Undoubtedly there are sorption and desorption reactions, the importance of which must still be clarified. Sulfides and hydroxides play a decisive role in the delignification, but their action in the dissolution of carbohydrates is less clear. The velocity of the delignification is related to the temperature in the Arrhenius equation. Lignin is an amorphous substance and is susceptible to alkaline reagents. The consumption of alkali is of prime importance for the degree of dissolution of the carbohydrates, whereas the consumption of sulfide is a function of the degree of delignification. In a pure soda cook, the velocity of the delignification is proportional to the product of the undissolved part of the lignin and the alkaline concentration of the cooking liquor. The proportionality constant depends on the temperature. In the kraft cook the mode of action of hydroxide and sulfide is additive; therefore the sulfide rather than the hydrosulfide may be the active delignification reagent. A model of such a kinetic reaction agrees very well with experimental results. The deduced reaction equation, however, is not universally applicable to experimental conditions with time-related changes in liquor concentration and with variations caused by high sulfidity or low liquor concentration (394).

A study of the reaction kinetics of a soda cook of sulfidized (with sodium hydrosulfide or hydrogen sulfide) or sulfonated wood, on the one hand, and of a kraft cook of wood pretreated with alkali, on the other hand, showed that the delignification of sulfonated wood occurs very rapidly at first, but slows down very drastically after a short time. During the first stage of the cook, the activation energy of the delignification was relatively small and amounted to 21 to 23 kcal after which it increased rapidly to 38 to 39 kcal. From this it may be concluded that, in the first stage of a soda cook of sulfonated wood, the reaction mechanism is similar to that of a kraft cook, but, in the second stage of the cook, it more closely resembles that of a soda cook. In the kraft cook, a continuous sulfidation of the lignin is important to achieve a mild dissolution. The reaction velocity in a kraft cook of wood pretreated with alkali is considerably lower than that of untreated wood. This difference became very apparent in a cook of wood from red pine, whereas a cook of beechwood behaved almost the same in both cases. This indicates that lignin from coniferous woods lends itself much more readily to condensation by alkali than does lignin from deciduous woods. The delignification velocity, however, increases for both wood species with increasing sulfidity. A maximum was observed at 60% to 100% sulfidity. Thus, the delignification of pinewood is less,

especially at low sulfidity, than of beechwood. The ratio of the delignification between the two wood species varies with the sulfidity and amounts to approximately 3.9, 3.1, 2.7, 2.5, and 2.3 at sulfidities of 0%, 10%, 27%, 60%, and 100%. The activation energy of the delignification, as mentioned above, was not constant, but decreased with increasing sulfidity in about the following order:

Sulfidity (%)	Activation energy of delignification (kcal)	
	Pine	Beech
0	38.6	37.9
27	32.1	29.8
100	30.4	28.7

The effect of sulfidity on the delignification velocity is low at higher temperatures (314).

Investigations on the alkaline delignification of sprucewood led to the concept that two different mechanisms can be observed: the rapid "bulk delignification" and the slowly occurring "residual delignification." The bulk delignification is ascribed to a complicated reaction mechanism of

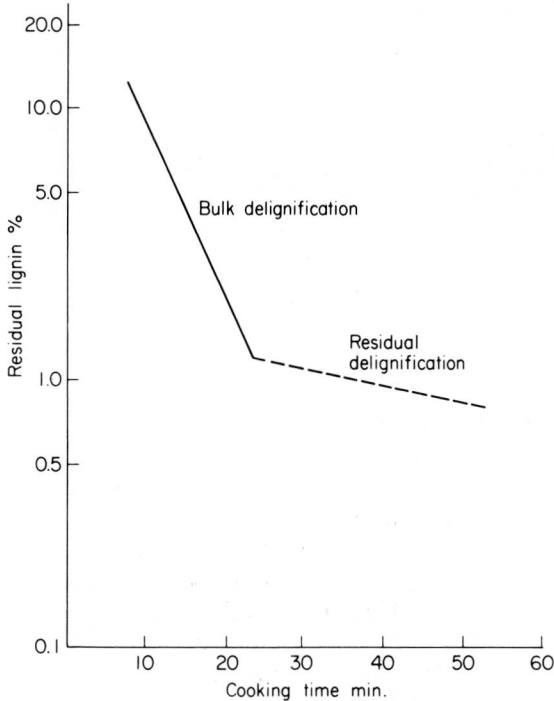

Figure VIII-2. Alkaline delignification (202).

liberated radicals of lignin fractions, whereas the residual delignification is attributed to an alkaline decomposition of cellulose fractions to which lignin is grafted in the course of the cooking process. Both reaction mechanisms show the general scheme of first-order reactions, but they have different reaction constants. This can be seen in Fig. VIII-2. Here the delignification is shown in a sprucewood cook in which the wood is first impregnated at a low temperature with a kraft cooking liquor, and then, within three minutes, is heated to 180°C. Within 25 minutes, the bulk delignification leads to a lignin content of the pulp of 1.3%, based on the weight of the wood; then a sudden retardation sets in, with a change to the residual delignification. Within the temperature range of 130° to 185°C, the effect of the reaction temperature results in a lower lignin content in a kraft pulp than in a soda cook at the point where the change from the bulk to the residual delignification occurs. The energy of activation in the kraft cook for the bulk delignification was found to be 32 kcal per mole and for the residual delignification, about two-thirds of this value. When the reaction temperature is lowered, the portion of residual lignin increases at the transition point. During the residual delignification, a considerable decrease in yield, based on lignin-free pulp, occurs. Based on the weight unit of dissolved lignin, this loss in yield is about ten times greater in the residual than in the bulk delignification. It is therefore recommended that the kraft cook should be limited to the bulk delignification phase and not be extended to the residual delignification (202).

The decomposition of lignin to monomeric phenols by alkaline reagents has been the subject of many investigations. The acidity of the phenolic hydroxyl groups of lignin depends upon the nature and the position of the substituents in the ring. The determination of the dissociation constants of such groups can reveal the changes that the lignin molecule undergoes during the alkaline cook. A differentiation of these acidic groups forms a "lignin spectrum" when the parts of the various acidic groups are represented as a function of the dissociation constants. In this case, it is to be assumed that lignin does not contain acidic groups with pK 5 to 7. The pK value of carboxyl groups lies below 5, that of the phenolic hydroxyl group is greater than 7. The lignin spectra obtained in this way from various steps of soda and kraft cooks with 10% and 30% sulfidity show an accumulation of weak acidic groups (pK 12), especially in the absence of sodium sulfide, i.e., in soda cooks, whereas kraft cooks, especially at high sulfidity, provide the lignin with more acid properties, and this can be explained by the accumulation of carboxyl groups and phenolic hydroxyl groups with pK 8 to 10 (39).

Model experiments with lignins from kraft and soda cooks showed that the dark color of such pulps is caused, above all, by chromophoric *o*-quinoid groups. These are formed by hydrolysis of methoxyl groups

neighboring the phenolic hydroxyl groups, followed by oxidation. In order to obtain lighter pulps, either the hydrolysis of the methoxyl groups must be suppressed or the chromophoric groups must be destroyed during the cook (70).

When lignin is treated with solutions of sodium hydroxide and sodium sulfide at temperatures of between 270° and 290°C, up to 6% catechol can be obtained. In order to elucidate the reaction mechanism, which is important for an understanding of the reactions taking place in the kraft cook, experiments have been carried out with simple model substances which may be considered to be the building stones of lignin. The majority of the demethylation experiments were carried out with guaiacol, those for the cleavage of C–C linkages, with dihydrodehydrodiisoeugenol and its derivatives. Sodium methylmercaptide was an especially effective demethylating agent, somewhat less effective was sodium hydrosulfide, and the least effective was sodium hydroxide (371). The fact that, on cooking of kraft black liquor, only half the theoretic amount of dimethylsulfide is obtained can possibly be explained by reactions of the mercaptide ions with carbon bridges and decomposition products of the lignin. A complete demethylation of the lignin could not be achieved by cooking it at atmospheric or higher pressure. Catechol is therefore formed by the splitting of the aliphatic carbon chain and demethylation, and lignin is probably decomposed by similar reactions. The variety of building stones which make up the lignin molecule is responsible for the difference of its decomposition products from those obtained from dihydrodehydrodiisoeugenol, from which considerably larger amounts of acids are obtained. A considerable part of the acids formed undoubtedly results from the decomposition of the hemicelluloses, but the larger portion is aromatic in nature. Experiments with dimeric substances as model compounds do not explain all the reactions that take place with lignin in alkaline cooking liquors. Propylguaiacol and propylcatechol are obtained from lignin in only minor amounts, but, in addition to catechol as the main product, larger portions of ethylcatechol and methylcatechol are obtained and, therefore, the yield of catechol and its simple derivatives is lower from lignin because of its larger molecule than the yield from its dimeric substances (371).

2.2 Hemicelluloses

The distribution of the hemicelluloses in wood is shown in Fig. VIII–3. The total amount of hemicelluloses in plant material varies between 15% and 30%. In the mechanical degradation of the wood structure, e.g., by grinding, the chemical composition is not taken into consideration. The goal is to obtain a pulp as uniform as possible. In the chemical degradation of wood, the production of a cellulose as pure as possible, in the form

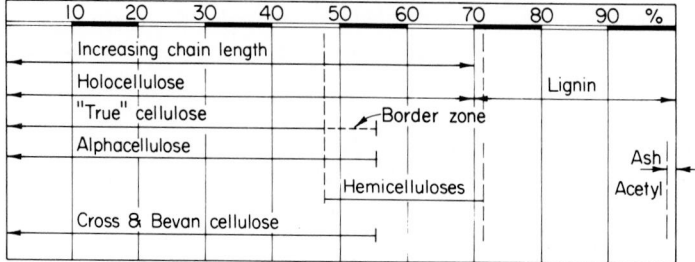

Figure VIII-3. Schematic diagram showing the components of the extractive-free cell walls of wood.

of fibers, is the goal. Because of the chemical and anatomic structure of the wood, this goal can be reached only to a limited extent, which means that the production of pure cellulose fibers is achieved only by a reduction in the yield. Cellulose, hemicelluloses, and lignin are so closely associated that the isolation of the total amount of one component is impossible without affecting another component. In the processing of pulp for paper it has been shown that the hemicelluloses are of special importance in the beating and the sheet formation. Only the artificial fiber industry, which dissolves the cellulose by chemical reactions and, by preciptiation or other methods, produces fibers or films, is interested in a pulp with the greatest possible cellulose content. Kraft pulps have a relatively high hemicellulose content and, in order to increase the yield of pulp, precautions are taken to preserve the hemicelluloses during the pulping process.

Depending upon the wood species used, the kraft process produces araboxylan from coniferous woods and xylan from deciduous woods. Hemicelluloses containing xylose in the various pulps therefore originate either from 4-*O*-methylglucuronoaraboxylan or 4-*O*-methylglucuronoxylan. When the cook is preceded by a hydrolysis the pulp from both wood species contains only xylan. The basic skeleton of the hemicelluloses is formed by a chain of anhydroxylopyranose with β-1,4-glucosidic linkages. Kraft pulps contain araboxylan which is formed from the original polymers in the wood by selective alkaline hydrolysis of uronic acid units. This araboxylan consists of arabofuranose units in which each twelfth to eighteenth xylose unit is combined in the basic skeleton by 1,3-glucosidic linkages. The average degree of polymerization (DP) varies between 76 and 90. Prehydrolyzed kraft pulp contains a xylan with a DP of 22 to 60 (151).

Of special importance is the fact that a part of the xylan formed during the cook and dissolved in the liquor is resorbed by the cellulose. Here the type of cellulose plays an important role. Cotton resorbs twice as much xylan as cold-refined sprucewood pulp does. This adsorption is proportional to the xylan concentration in the cooking liquor and is irreversible. The hexosans dissolved from the wood are rapidly destroyed, whereas the pento-

sans show much greater resistance. Although the xylan content in the cooking liquor reaches a maximum at the start of maximum temperature (173°C) and decreases within 2 hours at this temperature to about one-sixth of its original amount, it remains, in general, proportional to the xylan content of the wood (313, 404).

The sorption of the dissolved lignin and carbohydrate fractions on kraft pulp from pinewood and on cotton linters shows a certain dependence on the concentration of the organic components dissolved in the liquor and on the pH value. The sorption increases with increasing concentration and decreasing pH and reaches a maximum at pH 6. Certain portions of the adsorbed compounds adhere very firmly to the cellulose fibers and behave almost like native lignin with regard to the action of strong mineral acids and mild oxidation. This can simulate a higher lignin content in analyses. The lack of a certain excess of free alkali in the final stage of the cook can cause a shift of the dynamic equilibrium between the decomposition of the wood and the sorption and, with this, can lead to an increased resorption of the organic substances dissolved in the cooking liquor. In this way, the pulping process is retarded, if not even prevented. The liquor thus rendered deficient in alkali and enriched in organic material should therefore be removed and replaced by small amounts of fresh liquor in order to obtain well-cooked pulps, light in color and low in pentosan content. The desirability of a two-stage cooking process is thus indicated (351).

The redeposition of hemicelluloses during the cook has been proven with birch xylan labeled with tritium. In this way it has been shown that this redeposition can amount to as much as 3% of the weight of the pulp, regardless of how much hemicellulose is already on the fibers (76).

The presence of hemicelluloses in the cooking liquor reduces their extractability, especially under conditions when the concentration of active alkali in the liquor is lower and the temperature is higher. Extraction and resorption of hemicelluloses of the xylan type are to be considered, in a kinetic sense, as a pair of reversed reactions. The extraction is caused by the reciprocal action and the attractive force between the hemicellulose molecules and the solvent, on the one hand, and the presence of ions of electrolytes, on the other. The resorption is governed by the attractive force between the hemicellulose molecules and the cellulose chains. The two opposing reactions occur at different velocities, depending upon the concentration and temperature, with the coefficient of the latter being higher for the resorption than for the extraction (111).

The redeposited xylan may be present in the pulp in three different forms: (*a*) in loosely combined form, either as a precipitate or adsorbed on the surface, (*b*) precipitated in a crystalline form on the cellulose fraction, and (*c*) chemically combined with the cellulose by transglucosidation. How large each of these fractions of redeposited xylan is depends upon the

reaction conditions of the cook, the nature of the resorbing cellulose, and the molecular size and compositon of the xylan in the solution. The maximal redeposition of xylan during a kraft cook of birchwood does not parallel the maximal concentration of xylan in the spent liquor. Prehydrolyzed xylan from birchwood and straw shows little tendency toward redeposition on cotton cellulose on heating in alkaline solution, but xylan preparations which have been treated with warm dilute sulfuric acid show no loss in their redeposition properties. About half of the xylan present in a normal pinewood kraft pulp may be the result of redeposition, whereas in a soda cook this amount is considerably lower (250).

Experiments with beechwood confirmed the fact that a part of the dissolved xylan is readsorbed by the cellulose from the cooking liquor during an alkali cook. This is not the case if the wood has been subjected to a prehydrolysis. The greater part of the xylan remains in the pulp. These xylans are largely resistant and cannot be removed even in a cold-refining process. Obviously the hemicelluloses are rendered resistant to hydrolytic decomposition by the cooking liquor at the beginning of the cook because of the mild reaction conditions. A mild prehydrolysis, followed by a strong alkaline cook does not result in pulps with a low pentosan content (181).

The proportion of 4-*O*-methyl-D-glucuronic acid groups in xylan from deciduous woods varies between 6.4% and 13.8%. Thus the uronic acid content in the pulp is lower than in the wood. The xylan isolated from a kraft pulp is never completely free of uronic acid groups. The hydrolysis of the uronic acid groups is most pronounced at 160° to 170°C. The hydrolysis is incomplete because, at this stage of the cook, a large part of the alkali has already been consumed. In any case, the diminution of the uronic acid portion causes a stabilization of the xylan at temperatures of above 160°C. The hemicellulose fractions isolated from a kraft pulp of Scots pine could be separated into subfractions rich in mannose and rich in xylose. The arabinoglucuronoxylan lost, during the cook, 60% to 70% of its 4-*O*-methylglucuronic acid groups and about 10% of its arabinose units, whereas the degree of polymerization (DP) dropped from 130 to between 80 and 90. The alkali content of the cooking liquor determined, to a large extent, the degree of modification of the arabinoglucuronoxylan during the cook (98).

Relatively pure polysaccharide fractions can be isolated during the cook from the cooking liquor by neutralization and precipitation with alcohol. About 8% of the weight of the wood can be isolated as polysaccharides at 150°C, corresponding to a concentration of 20 grams per liter in the cooking liquor. With increasing duration of the cook and increasing temperature, these fractions decrease again and only 2 grams per liter are found in the spent liquor. While these fractions show the same composi-

tion of their neutral carbohydrates at up to about 160°C, the amount of xylan drops at 170°C, and, as a result, the content of arabinose and galactose increases. The carbohydrates, consisting of galactose, arabinose, rhamnose, and uronic acids, show a remarkable stability toward alkaline decomposition (19).

A practically lignin-free hemicellulose fraction can be precipitated from the liquor with alcohol when the lignin, which is insoluble in acid, has been previously removed by the addition of a cation-exchange resin charged with protons. From the lignin, an additional hemicellulose fraction can be obtained by reprecipitation and treatment of the lignin with chlorine dioxide at pH 4. The concentration of lignin-free hemicelluloses is low at the beginning of the cook and reaches its maximum after a cooking period of one hour at 170°C. The amount of hemicelluloses that can be isolated from the lignin is also low at the beginning of the cook but increases strongly during the time that the temperature is being raised, and then remains almost constant. For this reason, lignin-carbohydrate compounds must be assumed to be present in the cooking liquor. In a birchwood kraft cook, a hemicellulose concentration amounting to 8 grams per liter was found in the early stage of the cook. The maximum, 10.7 grams per liter, was found toward the end of the temperature-raising period although the concentration of hemicelluloses was still relatively high, 7 grams per liter, toward the end of the cook. The hemicelluloses thus obtained consisted, to a large extent, of xylan together with small amounts of polysaccharides consisting of galactose and arabinose units. From the spent liquor of a polysulfide cook, considerably larger amounts of the hemicelluloses could be isolated. The maximum amount that could be precipitated in this case was 5.7%, based on the weight of the wood. Even at the end of the cook, the total concentration of hemicelluloses was still 10 grams per liter, or 3.3% of the weight of the wood. The fractions consisted chiefly of xylose units. The higher xylan content of the cooking liquor and the greater xylan yield make it seem probable that during the polysulfide cook a stabilization of the birch xylan takes place (332).

The treatment of wood chips from *Pinus radiata* with borohydride at room temperature causes a far-reaching stabilization of the hexoses, but the pentoses are not affected by this treatment. The increase in yield from kraft cooks in the presence of borohydride can be attributed exclusively to the greater retention of the hexoses. The addition of 5% to 10% sodium metaborate results in a similar increase in yield. Kraft pulps produced in the presence of borohydride have a considerably lower pentosan content, a lower permanganate number, a higher hexosan content, and a slightly lower α-cellulose content. The lignin content of these pulps is slightly higher at high alkali concentrations than that of comparable kraft pulps. With lower alkali concentrations the reverse of this statement may be true.

The alkali consumption in the conventional kraft cook is higher than in such a cook with the addition of borohydride. Borohydride kraft pulps are more readily beaten than pure kraft pulps (249,251).

When birch xylan is heated with alkali at 100°C, the glucuronic acid bond remains stable so long as the glucuronic acid is not combined to a reducing xylose end-group. However, when the substituent is attached to a reducing end-group, the latter is decomposed even at temperatures of below 100°C. The reaction occurs via a series of still unknown intermediate steps and the probable final result is a cleavage of the bridge between the glucuronic acid and the xylose unit. This indicates that a 4-*O*-methyl-D-glucuronic acid substituent at the C-2 position in the xylose does not stop the decomposition reaction from taking place in the alkaline medium when the substituted glucose unit has been reached. If the stability toward alkaline reaction is great, as is the case with coniferous wood xylans which have arabinose substituents that can prevent the decomposition reaction, additional stabilizing measures are ineffective. On the other hand, when the stability toward alkali is low, as is the case with glucomannans which have no substituents or with xylans from deciduous woods which have only a few reactive glucuronic acids, stabilizing measures may be quite successful (162).

The reprecipitation of dissolved hemicelluloses can be explained as resulting from a decrease in the alkalinity of the pulping liquor through alkali consumption, from an increase in the concentration of organic salts and dissolved lignin, and from a decrease in the solubility of the original xylan through liberation of side chains which bear carboxyl groups (189a, 250a). The ability of polysulfide solutions to stabilize carbohydrates at a given temperature was found to increase with the X_S value (the polysulfide excess sulfur concentration) and the hydroxyl ion concentration, as well as with the excess sulfur concentration. This ability to stabilize can be predicted by means of a semiempirical parameter containing the redox potential (358a).

A modification of kraft pulping consists in the drawing off of all free cooking liquor when the concentration of dissolved, relatively undegraded hemicellulose is high. This liquor is cooled and added to a new cook. A gain in yield in the range of 1% to 1.5% (lignin-free yield calculated on the wood basis) can be achieved in birch cooks, and 0.5% to 1.0% in pine cooks. The additon of polysulfide or borohydride to the recirculated liquor can increase the yield through a stabilizing effect. Polysulfide or borohydride does not, however, have any important influence upon the dissolved hemicelluloses in the liquor. If the liquor is drawn off at an early stage of the cook a considerable amount of unconsumed polysulfide is recirculated and this means a more efficient use of polysulfide than can be achieved in a normal one-stage cook. Furthermore, the recirculation technique gives

a lower alkali consumption and results in an increased retention of hemicelluloses in the pulp (109a).

2.3 Cellulose

The objective of obtaining a yield as high as possible in as short as possible a cooking time suggests that the concentration of chemicals and the cooking temperature should be increased. Although the delignification can be accelerated by these means, the attack on the carbohydrates is simultaneously increased. An increase in the concentration of the chemicals from 30 to 60 grams per liter total alkali, and then from 60 to 90 grams per liter, at times reduces the cooking time by a half, but the attack on the cellulose is greater than that on the lignin. Variations in the concentration of the cooking liquor within the range of 70 to 100 grams per liter have little effect on the delignification velocity when the ratio of chemicals to wood is 15%; however, when the ratio is 30%, the increase in concentration of the liquor causes a great decrease in the yield and the α-cellulose content of the pulp (49,323).

Cooks carried out over the total pH range should indicate especially the behavior of the hemicelluloses and the stability of the cellulose toward the attack of the cooking reagent. Both components show a pronounced maximum in the alkaline region of pH 11, while the cellulose, at the neutral point, shows its lowest resistance, as is shown by the low yield and low viscosity of the pulp (360).

A study of native and mercerized cotton celluloses in alkali cooks at 170°C, and comparative determinations of the number of carboxyl groups in the decomposed cellulose and of the degree of polymerization showed that this cellulose contains an average of 0.66 carboxyl groups per chain molecule. In this case, the carboxyl groups at which the cleavage reaction stops in the chain molecule may be end-groups, whereas the end-groups which are formed at the start of the reaction may be hydroxyl groups. The carbonyl groups in the degraded cellulose are of no importance. Small amounts of oxygen in the cooking liquor play no significant role in the decomposition. The number of residual glucose groups which go into solution at each cleavage of the cellulose chain is widely independent of the degree of polymerization; it is smaller for the mercerized than for the native cellulose. In the decomposition of the mercerized cotton cellulose in an alkali cook at 120°C, the number of carboxyl groups per molecule decreases with decreasing DP. A comparison of the results obtained in a cook at 170°C with the one at 120°C shows a good agreement; the reaction mechanism, therefore, appears to be the same in both cases. To characterize the carboxylic end-groups formed in the alkali cook, alkaline cooked cotton was subjected to an acid hydrolysis and the acids in the hydrolyzate were

isolated by ion-exchange chromatography. In addition to the bands of the original cellulose obtained in a control experiment, two other bands for acids were found. These acids have been indentified as α- and β-glucometasaccharinic acids. The cleavage reaction, which occurs after the degradation of the cellulose chains during the alkaline cook, is obviously terminated by a rearrangement in the cellulose phase which leads to the formation of glucometasaccharinic acid end-groups (7,8,72,126).

2.4 Wood Structure and Growth Factors

Alkaline solutions exert a lasting effect on the crystalline structure of the cellulose. The macroscopic properties of the fibers caused by the modifications of the fine structure become apparent in the form of changes in the moisture and dyestuff absorption, and in the swelling, softness, and elongation of the fibers. The crystalline order not only shows variations in the wood but also may undergo changes during the cooking process. Thus, for instance, it has been found that sulfite pulps possess small, orderly, crystalline areas having excellent orientation, whereas kraft pulps show medium-sized areas with very different orientations. Cotton linters have larger areas with a strongly pronounced orientation. The degree of orientation in woods proceeds according to the following, diminishing, order: Douglas fir, spruce, hemlock, and pine. The chain length in the cellulose in spruce and Douglas fir are similar, whereas pine, hemlock, and cedar show group similarities (73,170,286).

The wood from the crown of a tree contains proportionally more springwood and, in general, gives pulps with a higher burst and tension strength, whereas the wood from the trunk contains chiefly summerwood and gives pulps with a higher tensile strength. The springwood contains thin-walled cells. Rapidly grown sapwood from Douglas fir gives pulps of low density; slowly grown sapwood has the lowest proportion of springwood. On the other hand, mature, rapidly grown wood has a higher density and requires less chemicals in a kraft cook; the pentosan content and the tensile strength of the pulp are increased. Of the different parts of the tree trunk, pine gives the highest yield from the upper part, about 1% lower yield from the middle, and about 3% lower yield from the bottom part. The middle parts of the stem are more suitable for the production of bleached pulp; burst and tension strengths of the pulps increase from the bottom to the top, while the tensile strength increases in the opposite direction (50, 56, 239).

Pulps obtained in various yields from loblolly pine and completely delignified with peracetic acid and sodium tetraboranate were studied for the properties of their single fibers. Springwood fibers were superior to summerwood fibers in regard to their tensile strength as well as their bonding

strength per available unit of their surface of contact. Single fibers from summerwood showed, with increasing pulp yield, a remarkable decrease in strength; with regard to the cross-section surface of the fiber wall, no significant changes in strength were found except in pulps of very low yields. The bonding strength at a higher yield was more strongly affected than the fiber strength. Fibers from springwood underwent a gradual decrease in fiber strength with decreasing yields and had a relatively uniform bonding strength in the different yields of pulp. The chemical analysis of pulps from summerwood showed a decrease in the lignin content, in the degree of polymerization, and in hemicellulose content with decreasing yields. A clear relationship between the fiber strength and any of the wood components could not be established, but a good agreement was found between the fiber strength (in grams) and the mannan content and degree of polymerization. The chemical composition and the structure of the fiber surface are not sufficiently well known to deduce relationships between the bonding strength and the chemical composition of the pulp (245).

Morphological changes in the cell wall are closely related to the respective chemical treatment. An acid treatment attacks the cell wall uniformly without causing a strong swelling effect and decomposes especially the hemicelluloses in the primary wall. An alkali treatment, on the other hand, causes a strong swelling and attacks, primarily, the inner part of the secondary wall, with the primary wall being less affected. Hemicelluloses with acid groups in the side-chain (polyuronides) are degraded in an acid treatment to low-molecular fragments, whereas in an alkaline treatment the side-chains are split off and an unbranched xylan chain is formed, which, in a close linkage with the cellulose chain, is resistant to acid hydrolysis. The structural and chemical differences caused in a two-stage cook are of special importance. When an alkaline treatment is carried out in the first stage, the residual pentosan content is about five times as high as when an acid treatment is first applied. Electron microscope investigations have confirmed that, when poplarwood is treated with 4% sodium hydroxide solution before an acid hydrolysis, the fibril bundles are preserved and hemicelluloses are found to be embedded between the microfibrils and are bound to the cellulose, apparently by hydrogen bridges. By a prehydrolysis with acid the hemicelluloses between the microfibrils are degraded and are dissolved in the subsequent alkaline treatment. In this way the cellulose skeleton is set free. These facts are of special importance for the kraft pulping process with a prehydrolysis (182).

Kraft cooks at 150°C and within a yield range of from 90% to 45% show that the dissolution of the various wood components is widely independent of the physical structure of the wood. Glucomannan is rapidly dissolved to the extent of about 40% of its original amount when the cook is carried out to a yield of about 75%. The remaining glucomannan is then

mostly resistant to further degradation. At the beginning of the cook, lignin is dissolved more slowly than the glucomannan, but the degree of delignification remains fairly constant throughout the cooking process. In contrast to this, the degree of polymerization of the cellulose decreases markedly at the beginning of the cook and then remains substantially constant. Xylan is dissolved more slowly than lignin toward the end of the cook. No differences in the dissolution rates of the various components in spring- and summerwood could be found. The stability of the glucomannan and the xylan toward the end of the cook must be attributable to stabilizing structural changes occurring in connection with the alkaline decomposition (6).

The significance of the morphological characteristics of the woods and the strength of the individual fibers on the properties of the pulps produced from them has been emphasized repeatedly. Thus, the thickness of the cell wall and the ratio between the width of the lumen and the diameter of the fiber influence not only the behavior of the pulp on beating but also the density and the tensile and burst strengths of the sheet formed from the fibers. When pulps from pinewood were separated into spring- and summerwood fractions and sheets were formed from the different fibers, the springwood fractions gave dense, relatively nonporous sheets which had higher mullen and lower tear strength, but better smoothness than those obtained from the original nonfractionated pulp. Summerwood fractions formed bulky and porous sheets with lower mullen and higher tear strength (198).

Extensive investigations of pulps from loblolly pine showed that, except for the yield, the eventual properties of the paper produced could be predicted from the morphological properties of the wood and the fibers. With regard to the distribution of the various chemical wood components within the tree, the content of holocellulose and α-cellulose increases radially from sapwood to the outside, whereas, in general, the opposite is true for lignin, hemicelluloses, and extractives. Little is known about the distribution of the components throughout the length of the tree, but tree trunks with the highest specific weight also have the highest content of holo- and α-cellulose. The specific weight of the extracted wood, the cell-wall thickness of the summerwood, and the proportion of summerwood in all the trees investigated were higher in the sapwood than in the heartwood (23).

With regard to the relationship between the various components of the wood and the properties of the pulp and of the paper produced from it, both α-cellulose and hemicellulose show a far-reaching interrelationship with the properties of the paper. The amount of hemicelluloses was calculated in these investigations from the difference between holo- and α-cellulose; it could therefore be expected that a good agreement between the properties of the paper and α-cellulose would result in a similar agree-

ment between the hemicellulose content and the properties of the paper. Wood pulp with a high α-cellulose and low hemicellulose content required a shorter beating period to achieve a desired degree of beating (freeness). Beating signifies hydration of the fibers and an increase in their plasticity. The lignin content has no visible effect on the course of the beating, but pulps with a low α-cellulose and a high hemicellulose content require a longer beating time. The density of the papers from such fibers depends largely upon the plasticity of the fibers which is produced by the beating of the fibers in a wet state.

Burst and tensile strengths are widely dependent upon the quantity and quality of the fiber-to-fiber bonds. These bonds are strongly influenced by the holocellulose, i.e., the α-cellulose and hemicellulose contents. From a purely chemical point of view it can be assumed that a high hemicellulose content facilitates the hydration and, in this way, strengthens the fiber-to-fiber bonding. The tensile strength is influenced primarily by the strength of the individual fiber. Pulp fibers with the highest cellulose content and the lowest hemicellulose and lignin contents should therefore show the greatest individual strength. This hypothesis has been confirmed by experiments. Since the yield is, to a certain extent, a criterion for the dissolution of the noncellulosic components of the wood, and thus, indirectly, also for the purity of the pulp, a relationship between the yield and the properties of the paper could be expected here, too. On the other hand, no clear relationships were observed between any of the chemical components of the wood and the total yield, but such relationships could be deduced for the screened pulp. The total yield and the screened yield obviously depend upon some factors other than the chemical composition of the wood (60).

With regard to the relationship between chemical composition and the morphology of the fibers, it was found that a high content of α-cellulose and a low content of hemicellulose with a high specific weight of the wood bear a relationship to the thick cells of the summerwood, the high proportion of this wood, the low proportion of tension wood, and the "Runkel factor." The Runkel factor (R) is a measurement for the properties of the fiber which collapses during the processing, and is represented by the equation

$$R = \frac{\text{cell-wall thickness} \times 2}{\text{cell-lumen diameter}}$$

A low R factor therefore means an extensive collapse of the fibers. Sixty to ninety percent of the total variations in the duration of the beating time, the apparent density of the fibers, and their burst, breaking, and tensile strengths can be related to the dimensions of the cells of the summerwood. In this case, the density (thickness) of the cells of the summerwood plays a major role, but sometimes the length of the tracheids is an important secondary factor (24,381,382).

The differences often observed between kraft pulps from southern and northern American woods cannot be explained by the usual physical methods; nor do chemical analyses offer an explanation. Because the proportion of summerwood is greater in southern woods, it must be assumed that the actual bonded areas per gram of paper from this pulp are considerably smaller than those in pulp from northern woods (74,214). Douglas firwoods between 40 and 350 years old, when pulped in kraft cooks under identical conditions and to the same degree of pulping, showed that the yield varies in close relationship with the proportion of sapwood. Burst strength, folding endurance, and density vary with the proportion of springwood. Young woods give readily bleachable pulps which are delicate and suffer a loss in strength on bleaching. Summerwoods give pulps with an initially low specific surface which is greatly increased by beating and considerably surpasses that of springwood (153,176).

The degree of swelling of the wood also plays an important role in pulping by the kraft process. Prolonged drying of the wood causes a certain amount of inner hornification which can nullify the reversibility of the swelling. Prolonged soaking of the wood, on the other hand, is favorable. A suitable variation of the cooking conditions produces a new type of spruce kraft pulp of which the biostructure and state of swelling correspond approximately to those of freshly cut wood. The ribbonlike and swollen springwood tracheids are therefore considered to be the principal carriers of the biological properties. The drying of wood causes a loss in strength of about 8% in sapwood and about 22% in heartwood. A pulp produced from air-dried summerwood from spruce has only about 63% of the strength of that from freshly cut springwood (189,190,192).

Like the differences in the wood's growth, the chemical and mechanical treatments which it undergoes before its conversion into pulp affect the quality of the pulp. For this reason, wood which has been extracted for the production of rosin is not well-suited for pulping. This is partly due to a dehydration of the cell walls and partly to the presence of residual solvents in the wood. Sawmill wastes mixed with sound wood chips also impair the properties of the pulp in proportion to their stage of comminution. The effect that attacks by fungi on wood have on the quality of its pulp has been discussed in Chapter VI (53,213,238).

The wide range of variations which can be applied to the kraft cooking process is especially favorable for the pulping of deciduous woods. Woods such as eucalyptus—of which more than 600 species are known—and scrub oak (*Quercus laevis*) can be pulped satisfactorily by suitable liquor combinations. The tannins contained in such woods do not cause major difficulties. Beechwood can be pulped more readily than, for instance, birch or maple, but less readily than poplar. Pulps for chemical conversion require the removal of the high hemicellulose portions, usually by means of an acid prehydrolysis. A great many tropical woods can be

pulped by the kraft cooking process. Many of them produce pulps with relatively short fibers but which can be used to advantage in mixtures with kraft pulps from coniferous woods for printing paper (22,52,125,163,224, 283,370).

2.5 The Movement of the Liquor in the Wood

The movement of the liquor in the wood and the significance of capillarity and diffusion have been discussed in detail in Chapter VII, Section 4. Only the basic concepts, especially those relating to the alkaline pulping of wood, will be considered here. Coniferous woods consist of discontinuous, long fibers or tracheids of tubelike structure, with closed ends. Their average length is 3.5 mm, their diameter 30 to 35μ, and that of the lumen 20 to 30μ. The flow of the liquor between the tracheids takes place through the wood pits which are braced by a membrane with a thick center. The torus consists of microfibrils, arranged in a parallel direction and running in a circular form (129). The pores are closed by a primary wall. The membrane contains pores of 0.03μ in diameter. Numerous pits in the heartwood may be inactivated. The rosin ducts contribute only a little to the liquor movement. The medullary rays, running from the center to the outside, are discontinuous tubes which are connected with each other by pits. They facilitate the radial penetration more than the transverse.

Deciduous woods consist chiefly of long fibers which have about one-third the dimensions of the tracheids of coniferous woods. They also contain pores which are, however, considerably wider than those of the tracheids. They are less effective since they are closed by tylose membranes which contain very fine openings.

In the submicroscopic region there exists a "transient" structure of the cell wall which becomes apparent only in a swollen state and is therefore of special significance for the pulping with alkaline reagents. The movement of the liquor through this submicroscopic structure can take place simultaneously or successively with that through the permanent capillary structure (341,387).

Most woods have a capillary volume of 50% to 80% of their total volume. Gases penetrate the wood through the permanent capillary structure above the fiber saturation point (30% moisture) only with great difficulty. Free water must be displaced from the capillaries and this requires pressures of over 20 kg per cm². Condensable vapors behave similarly when their relative vapor pressure is below the pressure that is formed by the condensation of the vapors in the openings of the porous membranes. In the case of water vapor, the critical condensation point lies at over 90% relative humidity.

The swelling is directly related to the volume of the condensed liquid.

It has a negligible effect upon the dimensions of the hollow spaces of the fibers but a reducing effect on the permeability of the pores and, thus, on the pressure permeability (342,343). Each type of liquor movement within the cell wall takes place by diffusion and is therefore independent of the hydrostatic pressure. Thus the latter is effective only when it increases the solubility of vapors in water. The diffusion of condensable vapors into the permanent structure of dry wood is followed by a diffusion of the condensed vapors as "bound" liquid into the cell wall.

Liquids penetrate the wood in an axial direction 50 to 200 times more rapidly than in a transverse direction. Occluded air offers great resistance to the penetration. During the penetration of liquid into dry wood, a liquid-air meniscus is formed in the capillaries; to this point, the capillary ascent occurs readily. A further capillary rise can occur when the capillaries are not circular. Swelling liquids can evaporate at the meniscuses and can condense in the adjacent cell walls because the reduction of the vapor pressure in the cell walls is greater than in the meniscuses. In this way the cell walls can be saturated even before they are reached by the liquid. Liquids with swelling capacity, in general, penetrate more rapidly into the wood than nonswelling liquids. Concentrated sodium hydroxide solutions penetrate the wood in both directions at approximately the same rate. The penetration rate increases with the temperature. The specific gravity of the wood has no effect on the rate of penetration of sodium hydroxide. Liquids that react with one or several components in the wood exhibit a strong dependence on time and temperature with regard to their penetrative ability. It is obvious that new hollow spaces are formed by the dissolution of lignin and other cellulosic constituents. About 86% of the absorbed liquid results from capillary ascent and only 14% from water vapor diffusion. Because diffusion and swelling run parallel and have to travel only very small distances in the thin cell walls, the diffusion occurs relatively rapidly.

With the impregnation and the chemical reaction of the cooking liquor with the wood components, the pulping process is not completed; on the contrary, the already dissolved components and those that have been rendered soluble must be removed from the still intact wood-chip structure. The second stage, therefore, is an extraction process which can be accelerated and facilitated by increasing the contact surface (dissolution area) between the liquid and the solid to be extracted as much as possible. Moreover, the flow rate must be kept high to achieve the extraction through convection. Finally, the dissolution gradient — i.e., the difference between the concentration at the dissolution border and that of the circulating cooking liquor — should be as great as possible. The first of these requirements can be completely fulfilled only by an extensive defibering of the wood chips. The flow velocity can be adjusted at will for all digesters

having forced circulation. To fulfill the last requirement, a multistage cooking process should be applied because a concentration equilibrium is formed in the digester after a relatively short time.

The proposals that have as their goal an acceleration of the penetration of the wood by the cooking liquor at the beginning of the cook have been dealt with in detail in Chapter VII, Section 4, and consequently a repetition of these proposals, which are the same for acid and alkaline cooks, is unnecessary.

3. THE REACTION PARAMETERS OF THE KRAFT COOKING PROCESS

3.1   The Dimensions of the Wood Chips

If the cooking process is regarded as a series of reactions — in the first of which the impregnation of the wood by cooking liquor takes place, in the second, the wood components react with the chemicals and, in the third, the components that have been rendered soluble are extracted — it is apparent, from the previous section, that the dimensions of the wood chips will have a considerable effect on the course of the reactions, so far as time is concerned. The rate of the penetration is influenced by the dimensions of the wood chips, by differences in their morphological structure, and by the density and moisture content of the wood. The course of the impregnation is controlled essentially by the rate of deaeration of the wood, by the hydrostatic pressure, by the ratio of wood to cooking liquor, and by the temperature of the latter. The diffusion is influenced only by the temperature and the concentration of the cooking liquor. The sorption of the alkali by the wood is primarily a function of the lignin.

The importance of obtaining a deaeration of the wood that is as complete as possible has already been mentioned. A temperature of between 100° and 150°C and a hydrostatic excess pressure facilitate and accelerate the impregnation. The maximum temperature is about 150°C and this should not be surpassed in order to avoid retarding the dissolution of the lignin. If the impregnation is carried out with a considerable excess of liquor, the impregnation period with coniferous woods should not exceed 20 minutes and the ratio of liquor to wood should be decreased to 2.0–2.2:1. A prolongation of the impregnation period under the above conditions leads to a partial dissolution of the hemicelluloses and involves an increase in alkali consumption; in addition, difficulties in the dissolution of the lignin occur. Uniform distribution of the alkali in the wood has a lasting effect on the residual lignin and on the splinter fraction of the pulp. The addition of wetting agents to the liquor has no positive effect on the impregnation (203,204,205).

The importance of the dimensions of the wood chips increases with the curtailment of the cooking time, which is especially necessary in the continuous cooking process. For this reason, conventional wood chips have been shredded, not only in a dry but in a presteamed state. The shredded chips were then separated into several fractions by screening. Such chips gave a screened pulp in a yield of 50% in a cooking time of 1.5 hours, as compared with a 43.4% yield in 2.5 hours from normal chips under equal cooking conditions. Other wood chips, before shredding, were subjected to a steaming process with dilute sodium hydroxide solution for 10 minutes at 170 psi (about 185°C). The pulps from the shredded chips without previous steaming showed a somewhat smaller tensile strength but otherwise were equal to the pulps from normal chips. Pulps from presteamed and shredded chips showed improved tensile and folding strengths but otherwise were equal to those from conventional wood chips. With regard to the various fractions after shredding, those from the 4 to 8 mesh gave pulps with constant strength properties, that from 10 mesh showed a minor decrease, and that from 16 mesh showed a distinct drop in strength properties. Shredding of dry wood chips in a hammer mill is recommended for the conventional batch cook, whereas shredding combined with steaming is advantageous for the continuous cook (274).

Within 25 to 30 minutes cooking time at an initial alkali concentration of 20 to 35 grams Na_2O per liter and 150 to 170 psi (182° to 185°C), shredded slash pinewood chips give pulps whose properties equal those of the best pulps obtained in a discontinuous cook of 2 to 3 hours. When the cooking pressure is reduced to 125 psi (about 174°C), high-quality pulps are obtained in a high yield, but the duration of the cook is increased beyond that required for a continuous cook. A preimpregnation with the cooking liquor reduces the cooking time but, at the same time, it reduces the yield and quality of the pulp. Conventional wood chips are not suitable for a quick cook either with or without a prior impregnation. As regards the effect of the moisture content of the wood chips on the yield and quality of the pulp, it can be said that dry chips give pulps in lower yield and of inferior quality, whereas chips that have been impregnated with water give pulps similar to those from freshly cut chips. A preimpregnation with dilute alkali to avoid an acid hydrolysis during the steaming proved to be disadvantageous (270, 273).

The thickness of the chips unquestionably has a marked effect on the course of the cook and the quality of the pulp. Increasing the cooking temperature and thickness of the chips gives a higher amount of screenings. The viscosities of the holocelluloses of the unbleached pulps from pinewood do not differ when the thickness of the chips is increased from 2 to 7 mm. From this it may be concluded that the strength properties are also practically constant. Special attention should therefore be paid to the size of the

wood chips. With chips of 8 to 15 mm in thickness, the selectivity of the cooking reaction is affected unfavorably. The same holds true for very thin chips of less than 1.5 mm in thickness (83,161,299). With cooks of mixed birch and pine chips, differences in their dimensions have an especially unfavorable effect, in addition to their different chemical compositions (14).

An enlargement of the reaction surface has been attempted by a mechanical after-treatment of the wood chips. The chips can be compressed in a parallel or perpendicular direction to the fiber axis. The greatest damage to the fibers is caused by compression parallel to the fiber; the effect of this damage is greatest in the acid cooking process and least in the alkaline. There are reports, however, claiming that a compression in axial fiber direction shortens the cooking time and facilitates the delignification (93, 345). The effect of the moisture content has already been discussed, but here, also, opinions are contradictory, although the majority of authors believe that fresh, moist wood chips or presteamed wood chips are best with regard to the yield and quality of the pulp (21,273,372).

3.2 Temperature and Time

The effect of temperature and time on the course of the cook has already been indicated in the discussions of the movement of the liquor in the wood and the effect of the dimensions of the wood chips. It has been suggested that the reaction time and the temperature should be expressed by a single variable. By means of a modified Arrhenius equation the following formula is obtained:

$$\ln k = B - A/T$$

in which k is the reaction rate, T is the temperature in absolute degrees, and A and B are constants (380). The reaction velocity at 100°C was taken arbitrarily as a unit and all other data are calculated on the basis of this unit. The equation then becomes

$$O = B - 16.113/373$$

and the reaction rate at any other temperature is calculated as

$$\ln^{-1}(43.20 - 16.113/T)$$

From this formula, tables can be set up for the relative reaction velocity for each desired temperature and a curve of the rate vs time (in hours) can be plotted for any cooking cycle; the area below this curve — also designated as H factor — can be employed as a means to adjust deviations of the temperature-time scheme of any arbitrarily chosen cook in order to obtain the desired and predetermined degree of pulping.

In order to express the course of the reaction of a cook numerically, the most important variables of the reactions involved must be known. The

variables are primarily temperature, concentration of the components taking part in the reaction, and the time. For a calculation of the single reaction sequences, one variable must be systematically changed while the others remain unchanged. From the results of experiments with a neutral sulfite and a kraft cook with slash pine, the effect of the temperature at constant composition of the cooking liquor on the delignification velocity and on the dissolution of the carbohydrates can be determined quantitatively (218). The data thus obtained, and used in conjunction with the Ross diagram, will illustrate the essential similarities and differences of the two cooking methods. All these reactions are strongly temperature-dependent, because an increase in temperature accelerates the reaction rate. The neutral sulfite process is shown to be more selective and removes more lignin at an equal yield of pulp. This selectivity is especially apparent with short cooking times. Figure VIII–4 shows a Ross diagram comparison of neutral sulfite and kraft pulping. From this the following principal differences between the two processes can be deduced: (*a*) if a high pulping velocity and a relatively low pulp yield are desired, this can best be achieved by the use of a

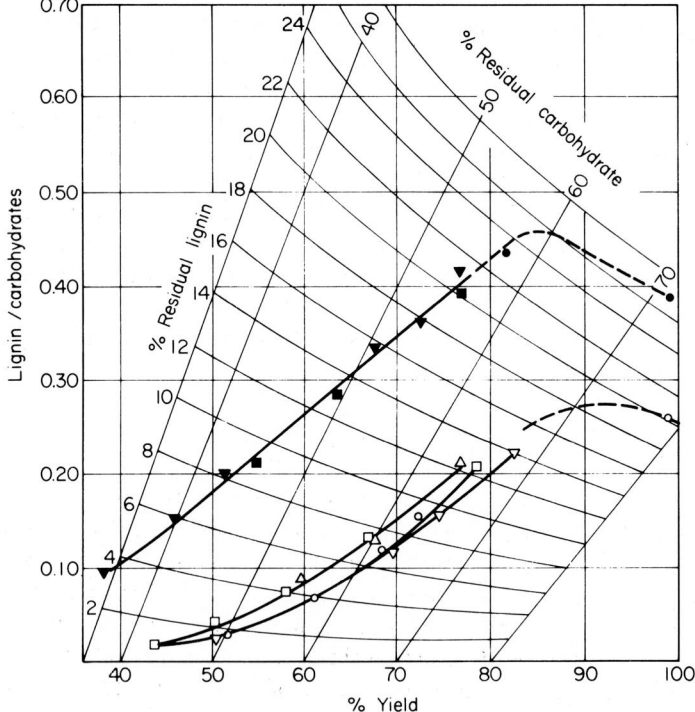

Figure VIII–4. Ross diagram comparison of neutral sulfite and kraft pulping (395). Neutral sulfite pulping: □ 210°C, △ 200°C, ○ 156°C, ▽ 173°C. Kraft pulping: ▼ 180°C, ■ 170°C, ▲ 160°C, ● 140°C.

high-temperature kraft cook; (*b*) if the primary objective is the production in a high yield of a pulp low in lignin content, the neutral sulfite process is more suitable than the kraft process; (*c*) although, in both cases, an increase in cooking temperature increases the velocity of the reactions, the lignin-to-carbohydrate ratio is more adversely affected by an increase in temperature in the neutral sulfite process (395).

With pure soda cooks at increasing temperatures and liquor concentration, the temperature has the greater effect on the dissolution of the lignin and the pentosans and on the decomposition of the cellulose. At constant temperature, the delignification velocity is primarily a function of the alkali concentration and the liquor-to-wood ratio (175). When the reaction temperature in a kraft cook, with 25.8% active alkali and 26% sulfidity, is increased from 175° to 235°C, the cooking time is shortened from 3 to 5 hours to a few minutes. In this case, however, the yield drops from 47% to 33%, and the pulps obtained show unfavorable mechanical properties and low viscosities (362).

The great significance of the diffussion processes, not only in the impregnation of the chips with cooking liquor but also in the subsequent removal of the dissolved wood components, was indicated in experiments that showed that temperature and time can be greatly reduced if the cooking liquor is circulated through the digester in a continuous flow. Sprucewood chips could be completely pulped with a liquor containing 90 grams per liter Na_2O in 35 minutes at 170°C, and aspenwood chips with a liquor containing 45 grams per liter Na_2O in the same time at the same temperature, to give a pulp in 50% yield with an α-cellulose content of 88.6%. The technical problem of such a method is the economical processing of the great volume of spent liquor if it cannot be reused after being fortified with fresh chemicals (325).

The delignification follows a first-order reaction and can be represented by the equation

$$\ln \frac{L_0}{L} = k_0 t$$

in which L is the undissolved lignin (in grams per 100 grams wood), L_0 is the apparent initial lignin content (as obtained from the first-order reaction approximation in grams of lignin per 100 grams wood), t is the cooking time (in minutes), and k_0 is the rate constant (a function of the liquor composition, the liquor concentration, and the digestion time in minutes^{-1}). The great interest displayed in reaction times as short as possible in continuous cooks has led to the initiation of a series of kraft and soda cooks that were carried out at constant active alkali content and a sulfidity of 25%, at temperatures of 165°, 180°, and 195°C. The yields of pulp obtained at the various temperatures varied between 45% and 65%. The results of these

experiments, with regard to the effect of the temperature, can be summarized as follows: (*a*) the delignification velocity follows a first-order reaction (see above) and is related to temperature by the Arrhenius equation; this holds for both processes; (*b*) the activation energies found were in agreement with those found by other investigators; (*c*) the kraft process exhibited twice the rate of lignin removal that the soda process did at the same temperature; (*d*) the viscosities of the pulps decreased as the pulping temperature was increased, in both cases; the kraft pulp viscosities were about double those of the soda pulps at the same temperature and lignin content; (*e*) the carbohydrate retention at a given lignin content was higher in the kraft pulps than in the soda pulps and was independent of the reaction temperature; this higher carbohydrate retention can be attributed to the higher cellulose content and to a higher retention of hemicelluloses containing mannan; (*f*) at equal lignin contents, the cellulose content decreased with increasing reaction temperature, with the kraft pulps at a given lignin content and temperature containing about 5% more cellulose than the soda pulps; the retention of the hemicelluloses was independent of the reaction temperature and was about 15% higher in the kraft pulps; (*g*) at constant lignin content, the retention of methoxyl was higher at higher temperatures; (*h*) although the mechanical properties of the kraft pulps had double the strength of the soda pulps, they were strongly temperature-dependent, whereas those of the soda pulps were not affected by the temperature (100).

3.3 Concentration of the Chemicals, and the Liquor-to-Wood Ratio

Increased amounts of alkali in kraft cooks of birchwood, to the same degree of pulping, lead to a decrease in yield and in cooking time. This decrease is mainly due to a lower retention of xylan. Figure VIII–5 shows the relationships between effective alkali and cooking time at various degrees of pulping (kappa number) and to various yields. Figure VIII–6 clearly shows that the yield of xylan and the total yield decrease uniformly with increasing amounts of effective alkali. The brightness of the pulps increases 0.5 point with each 1% increase in alkali, and the beating time to a specified freeness increases at the same rate (16).

Extensive investigations of the pulping of beechwood with kraft cooking liquors of various compositions showed not only the effect of the variation of the total alkali content but also the action of the principal liquor components, such as sodium hydroxide, sodium carbonate, and sodium sulfide, on the yield and the physical, chemical, and mechanical properties of the pulps. To summarize, it can be said that there is a reduction in the influence of the concentration of the sodium hydroxide in the cooking liquor upon lignin content of the pulp, upon the yield, on a lignin-free basis, of

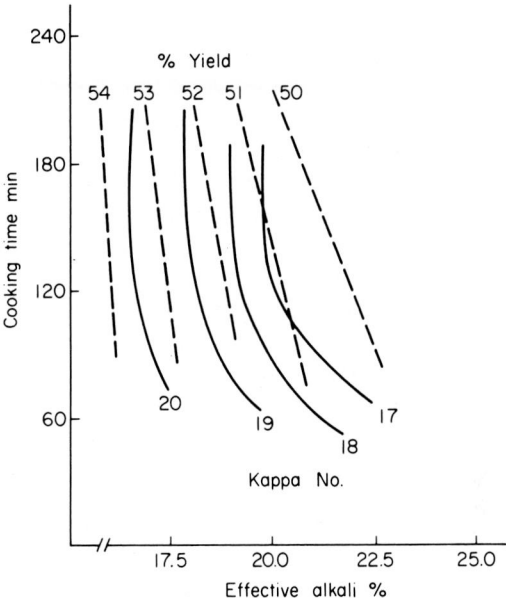

Figure VIII-5. Kappa number and yield for different charges of effective alkali and different times at the cooking temperature 170°C (16).

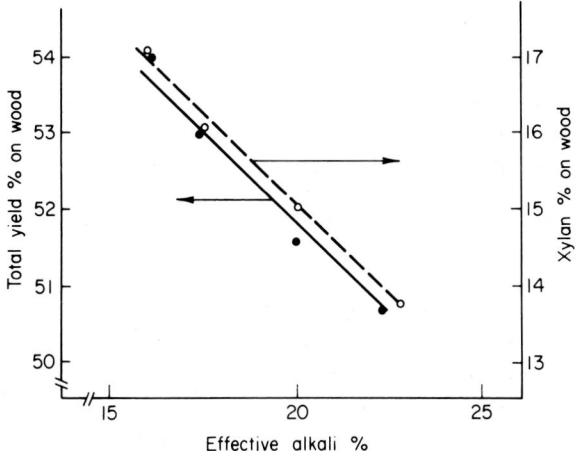

Figure VIII-6. Pulp yield and xylan yield versus effective alkali; kappa number approx. 19 (16).

pulp which is resistant toward 17.5% sodium hydroxide, upon the hemicelluloses as characterized by furfural determinations, and upon more or less all the physical and mechanical properties of the pulps. On the other hand, the sodium sulfide and sodium carbonate react, in general, in opposite

directions; in some cases the action of the former greatly surpasses that of the latter, and vice versa. To obtain pulps with certain optimum chemical and physical characteristics, liquor compositions and concentrations can be defined, probable changes in the properties of the pulps can thus be predicted, and desired modifications can be obtained by suitable alterations in the liquor (195).

In addition to deliberate modifications in the composition of the liquor and in the liquor-to-wood ratio, the moisture content of the wood chips can also cause such changes. It has been found that the liquor concentration required for a certain degree of pulping decreases with increasing sulfide content in the white liquor. The charge of the digester with wood chips, calculated on a water-free basis, remains constant regardless of the moisture content and weight of the wood chips. This conclusion is confirmed by the fact that the dimensions of the chips remain unchanged even if the saturation point of the fibers, which, for most woods, is 23%, is exceeded. This statement is based on the presumption that the conditions under which the digester is charged remain unchanged (59). The method of heating the digester also plays a role. The direct injection of steam, which has proved to be satisfactory with deciduous woods, causes a considerable loss in quality, which can be attributed to the change in the liquor-to-wood ratio by the formation of condensate, in the case of coniferous woods (180).

3.4 The Addition of Black Liquor to the Cooking Liquor

The addition of black liquor to the cooking liquor is of interest from both the chemical and the economic point of view. An enrichment of the cooking liquor with organic matter can be advantageous in the regeneration of the chemicals from the spent liquor. Opinions about the advantages and disadvantages of such additions are not unanimous. According to some authors, the addition of black liquor will not only act as a substitute for an equivalent amount of alkali in the white liquor but will also have a favorable effect on the permanganate number, yield, and screenings; moreover, the strength properties of the pulps cooked with the additon of black liquor will be improved. It is assumed that alkali is released from organic compounds in the black liquor (95). According to other investigators, the addition of black liquor to the cooking liquor results in an immediate loss of at least half of the titratable sulfide. Since this loss occurs before the cooking liquor reacts with the wood, the latter cannot be involved in this reaction. In contrast to the above-mentioned improvement, an increase in screenings was reported in this instance. The latter experiments were carried out with scrub oak, the former with balsam fir (310). An occassional disadvantage of the addition of black liquor was noticed in the occurrence of

an increase in the lignin content of the pulp and a decrease in brightness and α-cellulose content. With straight soda cooks, i.e., without any sulfur compound, unfavorable effects on the quality of the pulp and the consumption of chemicals and bleaching agents were also found (102,241).

To summarize, it can be said that the active alkalis in the black liquor can more or less replace an equivalent amount of alkali in the white liquor. The active chemicals contribute to the pulping reaction in approximately the same ratio as they are present in the liquors. The activity decreases with repeated use (244). The overloading of the cooking liquor with organic matter is obviously not without its consequences when the content of effective alkali is low and the amount of black liquor added surpasses a certain limit of tolerance, which lies at about 40%.

3.5   Graphic Representation of the Course of the Cook

There are several possible ways to present graphically the pulping reaction and the changes occurring in the lignin and carbohydrate fractions. The Ross diagram uses a rectangular coordinate system in which the ordinate represents the ratio of lignin to carbohydrate in the pulp and the abscissa is formed by the sum of lignin and carbohydrate, in other words, the yield. Over this system a curved net is laid which shows the coordinates for the percentage proportion of each of the two major components, i.e., lignin and carbohydrates.

This system is based on the concept that the pulping of plant material is primarily an extraction process. The material to be extracted consists

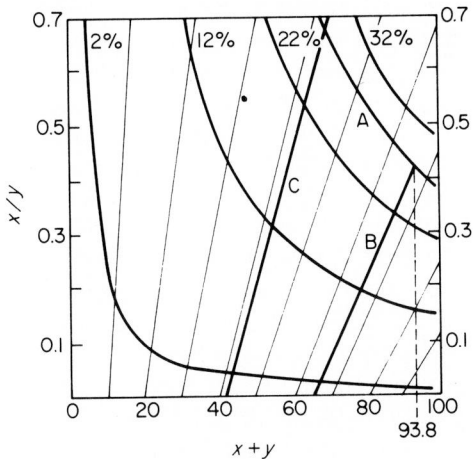

Figure VIII-7. Ross diagram: sulfate pulping of Swedish spruce (334)

$$\frac{x}{y} = \frac{\text{Lignin in pulp}}{\text{Carbohydrates in pulp}}$$

$x + y = $ Lignin + Carbohydrates = Yield.

chiefly of carbohydrates and lignin; the purpose of the extraction medium, i.e., the cooking liquor, is to dissolve the lignin and, depending upon the experimental conditions used, also a part of the carbohydrates from the wood structure. This carbohydrate portion also contains some of the cellulose. In setting up a Ross diagram, the amounts of the main components of the wood must be known. Swedish sprucewood contains about 93.8% carbohydrates and lignin (see also Chapter VII, Section 5.2). A Ross diagram can also be set up, of course, for any other lignocellulose material so long as the composition of their major components is known. In such a diagram the relationship between the carbohydrate and lignin portions in the pulp can be defined by a kind of "cooking zone" which is valid for all pulps and semipulps and for all pulping processes. Proceeding from the composition of Swedish sprucewood, on which the diagram shown in Fig. VIII–7 is based, the highest lignin content, if none is dissolved, is 28%. Curve A in the diagram therefore represents the boundary line for lignin, and line B, the boundary line for the maximum carbohydrate portion of 65.8%. Line C represents the maximum cellulose portion, with 41.5%; it delineates the fourth boundary line, the abscissa, and indicates the yield. Figure VIII–8 shows a Ross diagram for sprucewood for the soda, kraft,

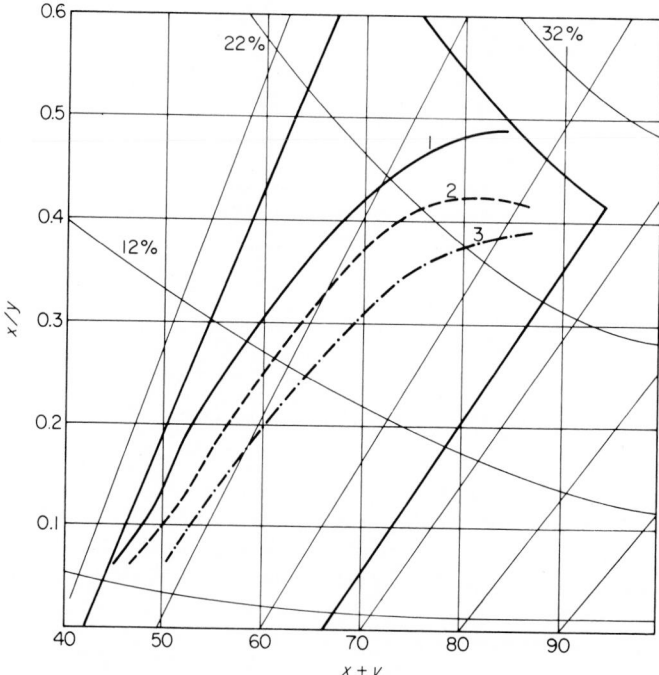

Figure VIII–8. Ross diagram (334). (1) soda pulping; (2) sulfate (kraft) pulping; (3) sulfite pulping.

and sulfite cooking processes. A comparison of these processes shows that, at a pulp yield of about 50%, the delignifying action of the kraft cook drops strongly, whereas that of the soda cook remains approximately constant (147,257,347).

Another type of graphic representation is the construction of a lignin-hemicellulose diagram (LH diagram). It is based on the assumption that the cook is carried out in such a way that the cellulose is not attacked to such an extent that it goes into solution. The variations in the pulp are, therefore, attributed exclusively to changes in the lignin and hemicellulose content. The yield of pulp and the lignin and hemicellulose contents are expressed in percentage of the wood. The rosin content is disregarded. For sprucewood, the cellulose content is then 42%, the lignin 28%, and the hemicellulose 30%. If the yield and the lignin contents of a number of pulps from the same cooking series are known, a "delignification curve" can be inserted into the diagram. Figure VIII-9 shows such a diagram

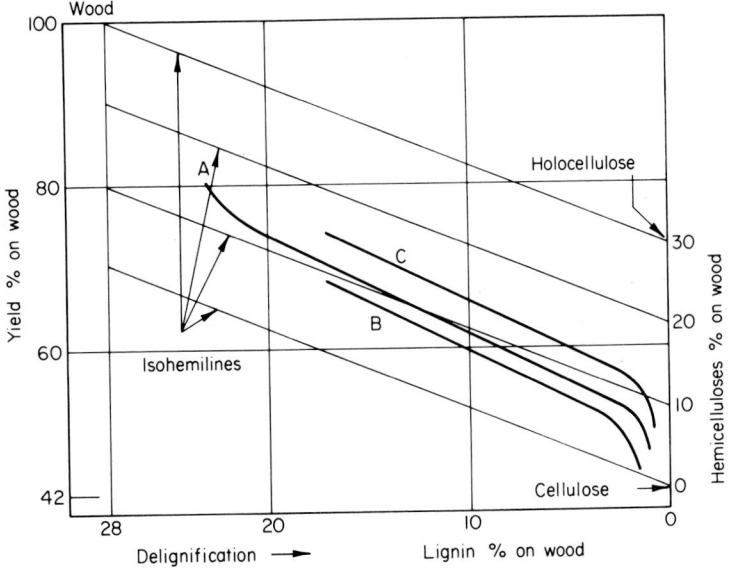

Figure VIII-9. LH diagram: lignin-hemicellulose diagram (388).

with the delignification curve A. The lignin content can be calculated from the kappa number. As curve A indicates, in the present case more hemicellulose and only a little lignin are dissolved at the beginning of the cook. The curve shifts, however, in a horizontal or vertical direction, depending upon the cooking conditions. A horizontal shift (curve B) shows an increase or a decrease in the amount of condensed lignin, whereas a vertical shift (curve C) indicates an increased retention of hemicelluloses

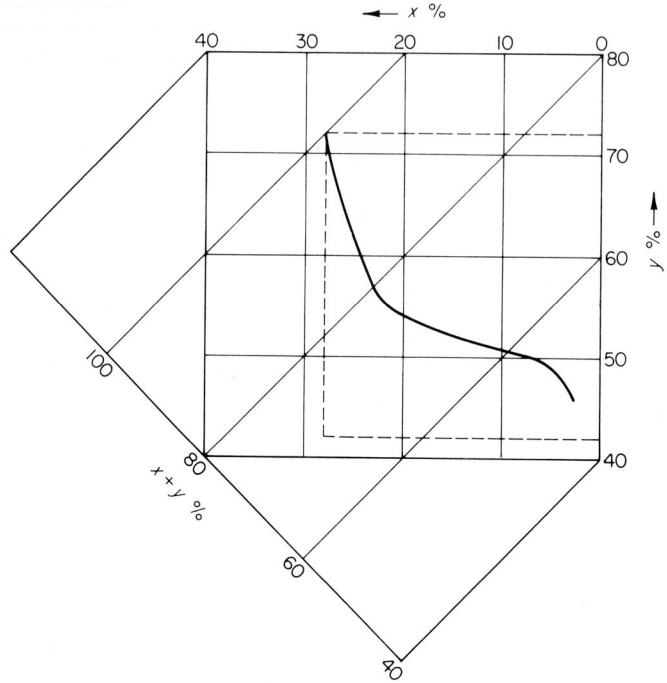

Figure VIII-10. Schmied diagram (320).

(138,232). The diagram shown in Fig. VIII-10 permits a direct reading of the values for lignin (x), carbohydrates (y), and the yield $(x + y)$ to be made. This system uses a rectangular coordinate system with uniform distances for the x and y values and a second linear net with uniform scales for $x + y$, turned 45° toward this coordinate system. From this diagram, the delignification velocity, d_x/d_t, the decomposition rate of the carbohydrates, d_y/d_t, and the relationship between the two velocities, $d_x/d_t : d_y/d_t = d_x/d_y$, can all be determined graphically from the position of the tangent at points x and y. A similar result can also be obtained from the two other diagram systems. In the diagram shown in Fig. VIII-10, the actual pulping zone is outlined by the boundaries for $x = 0$, $x = 28$, and $y = 42$ and $y = 72$ (138,320).

4. COOKING PROCESSES AND EQUIPMENT

4.1 The Discontinuous Cook

The discontinuous or batch cooking process is the classical form of the alkaline pulping process, regardless of whether it is carried out with or

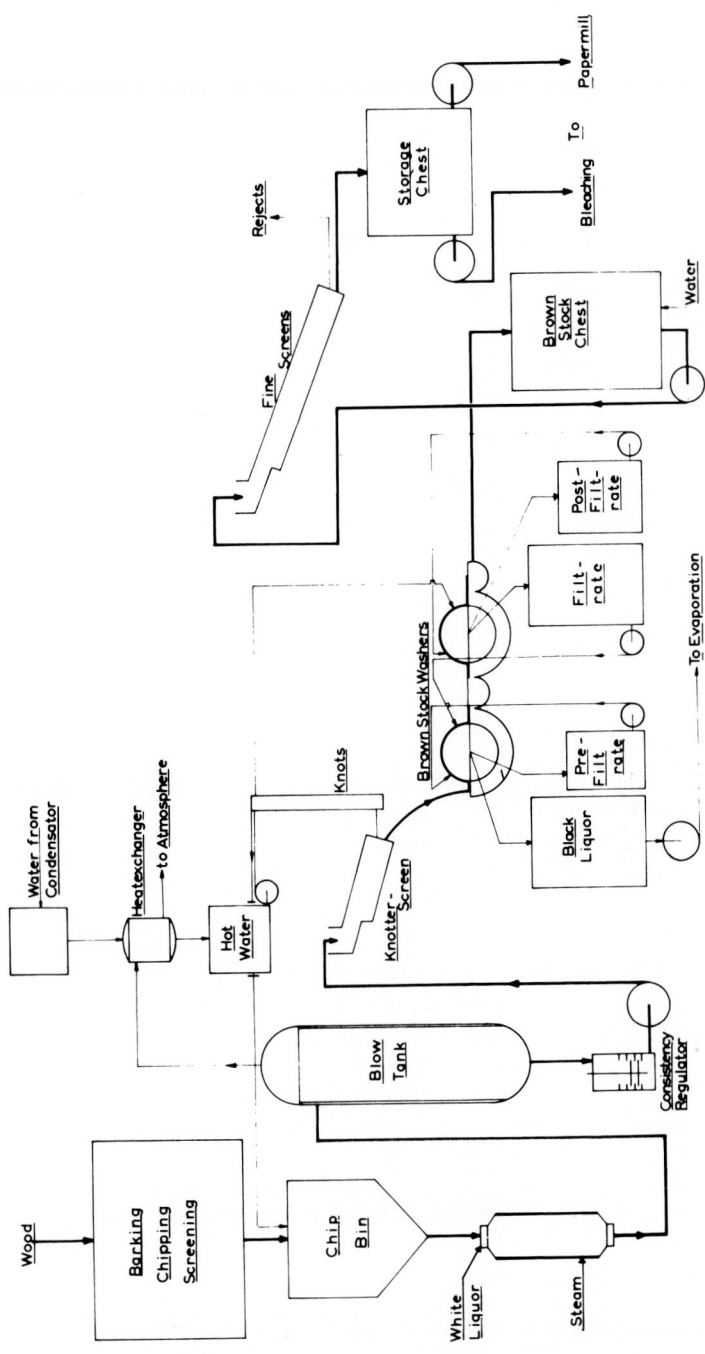

Figure VIII-11. General arrangement of a kraft pulp mill.

without a sulfur component. This process consists of the charging of the digester with wood chips, the introduction of the cooking liquor, heating to the maximum temperature, the cooking period at this temperature, the relief of the pressure, and the discharge of the digester. Figure VIII–11 shows a flow-sheet of the course of such a cook, including the subsequent processing of the pulp. The wood is received in the form of chips from the wood preparation plant, as has been described in Chapter VI. The chips are transferred from overhead chip bins into the digester by special-ly designed machines which assure that the chips are packed as uniformly and densely as possible. In some cases such feeders are operated by steam and thus a certain effect of steaming of the chips is achieved. After the cooking liquor has been added, steam heating is begun. Some mills heat the digester by the direct injection of steam, causing a certain dilution of the cooking liquor by condensate formation. Other mills accomplish the heating by the use of heat exchangers through which the liquor is pumped. With direct steam injection, the wood-to-liquor ratio of $1:2.1$ changes, toward the end of the cook, to $1:4$ or $1:5$. The moisture content of the wood chips, too, plays a role in the dilution. An alternate direct and indirect injection of steam is also practised. At the conclusion of the cook, the pulp is blown by the pressure of the digester into the blow tank where, through the release of the pressure, the chips are largely defibered. The liquor is separated from the pulp, as far as possible, without dilution. The relief steam is utilized for the preparation of hot water for the subsequent washing of the pulp. The spent black liquor contains about 1350 kg solids, from which about 600 kg chemicals per ton of pulp can be regenerated. With the blowing of the digester, the pulping process is practically com-plete. All subsequent operations deal with the processing of the pulp and the recovery of the chemicals from the black liquor.

All possible variations connected with the cooking time, temperature, and liquor composition and concentration can be utilized in the discon-tinuous cooking process. The objective is mainly to utilize as completely as possible the volume of the digester and to increase the throughput and the production capacity. As already mentioned, a rapid and complete im-pregnation of the wood chips with cooking liquor is of the utmost impor-tance. For this reason, an even distribution of the cooking liquor within the wood chips is necessary for a uniform delignification and a pulp free of screenings. Uniformly impregnated wood chips can be pulped in the vapor phase but, in this case, differences occur with the various wood species. The dissolution of the lignin begins within the temperature range of $140°$ to $150°C$. The impregnation period can be shortened by in-creasing the temperature, provided that it does not surpass $140°C$. Im-pregnated poplarwood chips could be cooked, within 10 minutes at $185°C$, to give a pulp with a low lignin content and good strength properties. High-

density hardwoods require an especially careful impregnation, continuing for as long as 30 minutes at temperatures of up to 140°C. Pulps, low in lignin content and screening residues, are thus obtained in a yield of 50% after a cooking time of 15 minutes at 182.5°C (204).

Whether a prior impregnation of the wood chips will involve an increased alkali consumption depends mainly on the temperature at which it is carried out. An impregnation with black liquor has been proposed, but in this case the temperature must not surpass 96° to 100°C (118). Other suggestions for shortening the cooking time include a preimpregnation outside the digester, with a deaeration by steaming and subsequent treatment with cooking liquor at 200 psi pressure. Such a process, however, is economical only in connection with a continuous cooking process (233). The cost alone should not be the deciding factor because the continuous process also affords the possibility of improving the quality of the pulp since the various cooking zones affect the individual wood components in different ways. Moreover, increased yields can be expected at the same degree of pulping, as well as more uniform steam consumption, reduced consumption of chemicals, less storage space, and, finally, considerably less space for the entire plant (97).

4.2 The Continuous Cooking Process

In 1938, the Swedish firm, Kamyr A.B., began the development of a continuous cooking process which, at that time, was designed only for alkaline pulping. It took six years for the first pilot plant, with a daily production of eight tons, to produce a satisfactory pulp. Although the operative prerequisites for a quick cook had not at that time been completely clarified, the main difficulties involved the construction of the equipment. Future developments will be concerned with the construction of a digester system with fewer moving parts, since these are exposed to especially great wear and tear (305).

The results obtained by American engineers are especially significant with regard to the technical development of the continuous kraft cooking process. First, a curtailment of the cooking time was striven for by increasing the reactive surface of the wood chips. The chips were broken down into smaller pieces, as already described in Section 3.1. In some cases, the chips were presteamed in a weakly alkaline soda solution containing 4.8 grams per liter sodium carbonate and were subsequently further reduced in size. This presteaming was not intended to be an impregnation of the chips with an alkaline cooking liquor but served only to neutralize any acids formed, since these would have a hydrolyzing effect (274, 278). The pulping process can therefore be considered to be an interface reaction at the boundary zone between the wood and the cooking

liquor. This reaction zone then moves slowly into the interior of the chips. A preimpregnation of the wood chips with cooking liquor containing active alkali accelerates the pulping process but causes a degradation of the cellulose and damage to the quality of the pulp. From this point of view, a continuous pulping is possible only when the reaction time can be reduced to 30 minutes or less. Such a short period excludes a complete impregnation of the chips before the reaction starts (276). In the light of the experimental results discussed in the previous section, the author is of the opinion that a preimpregnation with concentrated cooking liquor is not necessarily disadvantageous so long as the right conditions with regard to temperature and concentration are observed. The fact that, on presteaming with dilute soda solution for 10 minutes at 185°C, not only are the lignin-cellulose bonds loosened, but also the pentosan content is decreased by about 25%, shows that even a presteaming causes an attack on some of the wood components. Chips that have been presteamed and reduced in size can produce any type of pulp desired, with α-cellulose contents of up to 95%, in a quick cook by the selection of the correct alkali concentration, reaction time, and temperature. Thus, for example, scrub oak gave a pulp in 45.3% yield, containing 2.6% lignin, 92.2% α-cellulose, and 8.8% pentosans, within 15 minutes at 148 psi (about 181°C) with a cooking liquor containing 35 grams per liter active alkali (as Na_2O) and at 25% sulfidity. In another experiment, a pulp with 1.5% lignin, 96.6% α-cellulose, and 2.2% pentosans was obtained in 40% yield in only 10 minutes cooking time at 148 psi and with a liquor concentration of 80 grams per liter active alkali (275). Slash pine gave screened pulps in yields of up to 46% from presteamed shredded chips when cooked with cooking liquors containing 20 to 80 grams per liter active alkali and at pressures of between 115 and 175 psi (171° to 187°C). In most cases, the pentosan content was lower than in pulps which were obtained in a discontinuous process under identical experimental conditions, with the exception of the reaction time. The concentration of the cooking liquor affects chiefly the mechanical properties of the pulps; the strongest are obtained with the lowest liquor concentration. Cooking times of 30 minutes or less gave pulps which had 95% of the strength of the strongest commercial kraft pulps (269).

These data indicate that the kraft mill of the future will use two continuous digesters and uniform chips that have been reduced in size for a daily production of 300 tons. The cooking time should be 35 minutes and the temperature about 182°C, giving a pulp with about 10% residual lignin which can be removed in a multistage bleaching process (270, 271).

Returning now to the development of the apparatus for the continuous cooking process — so far as the sulfite process is concerned, this has been described in Chapter VII, Section 8.4. Digesters for continuous alkaline pulping do not differ greatly from those of the sulfite process although they

have been in use for a much longer time. For a trouble-free operation of a continuous cooking digester, the following essential conditions must be observed: (*a*) careful control of the flow of the chips and the liquor; (*b*) feeding of the chips into the digester without the use of mechanical force; (*c*) complete impregnation of the chips with cooking liquor; (*d*) uniform heating of the contents of the digester; (*e*) adequate cooking time; (*f*) dis-

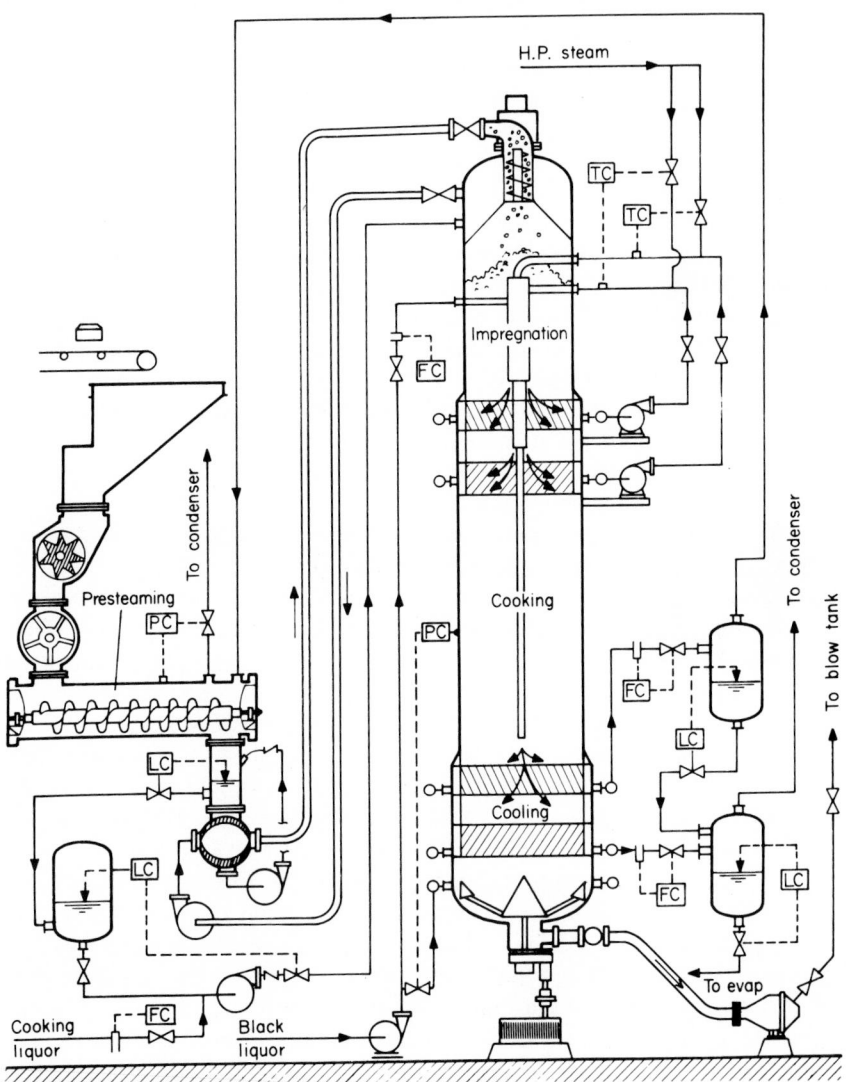

Figure VIII-12. Kamyr continuous kraft cooking system (20). (Reproduced from "Kraft Pulping Theory and Practice" by permission of Lockwood Publishing Co.)

charge of the digester without damaging the fibers and with recovery of heat and spent liquor.

Figure VIII–12 shows schematically a plant which largely meets these requirements. In brief, the operation is as follows: The chips enter the cooking system in a uniform, metered flow. They are then presteamed for 3 to 5 minutes in the steaming chamber to prepare them for the impregnation. A high-pressure feeding system then transfers the chips, without causing mechanical damage to them, into the digester. The feeder operates according to a two-cycle liquor-circulating system. A rotating plug opens the way to the high-pressure zone and the chips are carried by the top circulating cooking liquor to the top of the digester. The power consumption of the feeder is low, amounting to about 15 horsepower at a production of 300 tons per day.

The digester itself is divided into four zones. The actual reaction zone is in the middle, with a cooler zone above and below. The maintenance of the temperatures and pressures in the various zones is possible only if the digester is completely filled with liquor and is under hydraulic pressure of 40 to 50 psi in excess of the vapor pressure at the cooking temperature. The pressure in the middle zone is considerably higher because of the added vapor pressure. There is therefore no steam in the digester and consequently no steam bubbles can move upward. The transfer of heat by convection is prevented by the solidly packed column of chips which move continuously through the various zones.

During the first 45 minutes, the chips are impregnated at temperatures of 105° to 115°C. The temperature in the cooking zone is attained either by the introduction of hot cooking liquor or by the direct injection of steam. The cooking temperature depends upon the wood species and averages about the same as in normal kraft cooks. Because the delignified fibers are very susceptible to mechanical damage in the presence of the cooking liquor and at high temperatures, the cooking zone is followed by a diffusion-extraction zone. Here the hot, concentrated cooking liquor is displaced by cooler, or cold, weak liquor from the brown stock washer system and the pulp is strongly cooled. The discharge temperature usually lies at below 94°C. This diffusion-extraction system shortens the subsequent washing period of the pulp and increases the temperature and the solid content of the black liquor which then goes to the recovery plant.

The continuous cooking process requires the maintenance of a hydraulic excess pressure of the total cooking system. Before the installation of the diffusion-extraction zone, the digester contents were discharged at a relatively high temperature and the pulp thus produced sustained a considerable amount of damage. Further attempts were made to carry out the washing process as completely as possible inside the disgester. This required a rearrangement of the diffusion zone by the installation of two circular

strainers for the removal of the black liquor and its replacement by a cold, weak liquor from the first wash filter under its hydrostatic pressure of 175 psi (20,186).

Washing by diffusion and displacement is the classical method for the removal of the black liquor from pulp. The operation proceeds as follows: The pulp, impregnated and surrounded by black liquor, is washed from top to bottom with water of lower specific gravity, thus avoiding a mixing with the black liquor. The liquor in the fibers and nondefibered chips can be replaced only by diffusion or capillarity. The replacement of the surrounding liquor occurs comparatively rapidly, while the exchange in the fiber bundles by diffusion and capillary action is considerably slower. It is therefore important to know whether, in the washing, replacement or exchange of the liquor is the predominant factor. To remove all the liquor from the chips requires large amounts of water and a prolonged period of time because, after a certain time, an equilibrium is formed and the exchange comes to a stop. A diffusion-extraction washing system installed inside the digester therefore requires very careful control to achieve the highest efficiency. Figure VIII-13 shows schematically such an installation

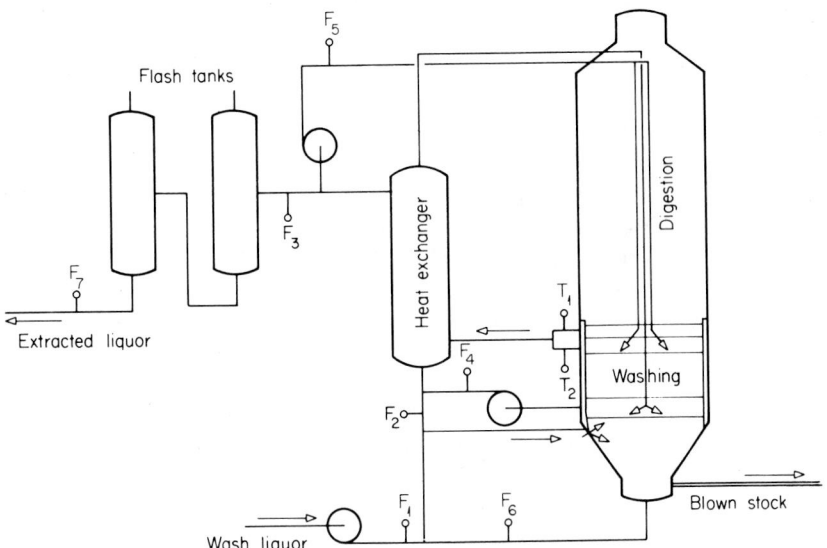

Figure VIII-13. Continuous digester showing temperature and flow measuring points (172).

and the points at which flow and temperature recordings are taken and at which samples can be removed. The concentration of the liquor can be measured at the following points: (*a*) at the pump where the wash water is added; (*b*) where the liquor comes from the pulp at the blow-line; (*c*) where the extracted liquor comes from the second flash tank.

The flow rates were recorded continuously at F_1, F_2, F_3, and F_7. The flow rates at F_4 and F_5 were maintained at 500 and 250 gallons per minute, respectively, while that at F_6 normally remained at 0. The temperature was recorded on either side of the washing zone, at T_1 and T_2. The rate of extraction could then be expressed in gallons per minute per ton of air-dried pulp per day and was termed the "extraction factor." This factor, multiplied by the daily production of brown stock, gives the amount of liquor to be extracted, in gallons per minute, after the flow rate has been measured and prior to flashing. To attain 100% efficiency, all the black liquor must be extracted. This means that 390 gallons per minute must be extracted when the extraction factor is 1.3 gallons per minute per air-dry ton per day with a production of 300 tons per day. With a figure above this amount, the black liquor will be diluted with water, and below this amount, a part of the liquor will be removed with the pulp. Since the washing stage cannot be prolonged at will, and as a result of the affinity of the dissolved material for the pulp fibers, the concentration of the black liquor at the optimum extraction flow of 1.3 times the daily production will be below 15.5°Bé and the black liquor removed with the pulp will have a correspondingly higher concentration than the black-liquor wash water. Flow and temperature measurements show that the loss of extracted liquor by flashing varies. When the extraction factor drops, the temperature at T_2 increases and, simultaneously, the concentration of the extracted liquor increases. The temperature of the descending black liquor, at T_1, is not greatly affected by variations in the extraction factor and remains almost constant at 167°C. The temperature at T_2 steadily decreases above an extraction factor of 1.27 because of the mixture of the hot black liquor with the rising, cooler (66°C) wash liquor. The temperature can therefore be used to establish the optimum washing conditions. It is obvious that the efficiency of the washing increases with decreasing concentration of the wash liquor and will be at a maximum at a concentration of zero, i.e., with water. It is therefore possible that the use of water as the extraction medium in the digester and the subsequent use of a vacuum filter would provide the same washing effect as that of a three-stage filter-wash outside the digester (123,167,172).

The discovery that black liquor, which is already enriched with the degradation products of wood, effects a mild dissolution of the noncellulosic components of the wood at the beginning of the cook had already led, at the end of the nineteenth century, to the development of a countercurrent cooking process in which, in a series of digesters, the wood was treated first with black liquor at low temperatures, then, with increasing temperature and concentration of the liquor, stepwise with more drastic reaction conditions and, finally, in a last step, with fresh white liquor. This process was not widely used because of its complicated equipment and circulatory system for the various liquors. The concept of a countercurrent

cook in a continuous digester has recently attracted new attention. This method differs basically from the countercurrent process mentioned above in that water is injected into the digester at the bottom in an amount sufficient not only to blow the pulp from the digester but also to move simultaneously upward in a continuous flow toward the descending chips. White liquor, heated indirectly by steam in the customary way, is added continuously to the circulating black liquor, which is removed through the middle ring strainer. The ascending water mixes with the cooking liquor and flows toward the descending wood chips. The black liquor is removed through a ring strainer at the upper part of the digester, passes through a heat exchanger, and is then pumped into the recovery system. The system described is operated with eucalyptus wood, which has a density of 1.4. It is questionable whether the movement of the wood in the digester would be the same with woods of lower density. This process has some disadvantages with regard to heat economy, and difficulties can arise when the wood chips are too highly defibered in the lower part of the digester; it has the advantage of giving pulps that are light in color and need little bleaching (335).

A newer development replaces the horizontal presteaming chamber by a vertical, combined presteaming and impregnation vessel from which the impregnated chips are passed directly into the top of the continuous digester and are exposed to the cooking temperature in the vapor phase without any temperature gradient. A secondary compensating circulation system in the liquid phase makes it possible to control the charge. This is accompanied by a reduction in the height of the digester which simplifies the installation, minimizes the incrustation, and increases the yield of turpentine. This new type of continuous digester permits many combinations of different cooking processes (9a, 306a).

The Kamyr system is widely used today and plants are in operation with an annual production of several million tons of pulp from various species of wood, annual plants, and agricultural wastes. In addition, modifications in the construction of continuous digesters have been proposed and tested. The production of semipulp has contributed extensively to the development of continuous and "high-yield" pulping. Even prior to the Second World War, suggestions for continuous cooking had been made. The equipment necessary for such a process consisted of a series of steaming and pressure vessels through which the plant material was passed in a continuous flow (1,316). Later, the digester system was modified and further developed in the United States. The Black and Clawson continuous digester illustrates the further development of such a semichemical pulper and is shown schematically in Fig. VIII-14. The wood chips are metered into the system by a feeder and are steamed in a presteamer at a low pressure, while being stirred around to ensure a uniform steaming. In tube 2,

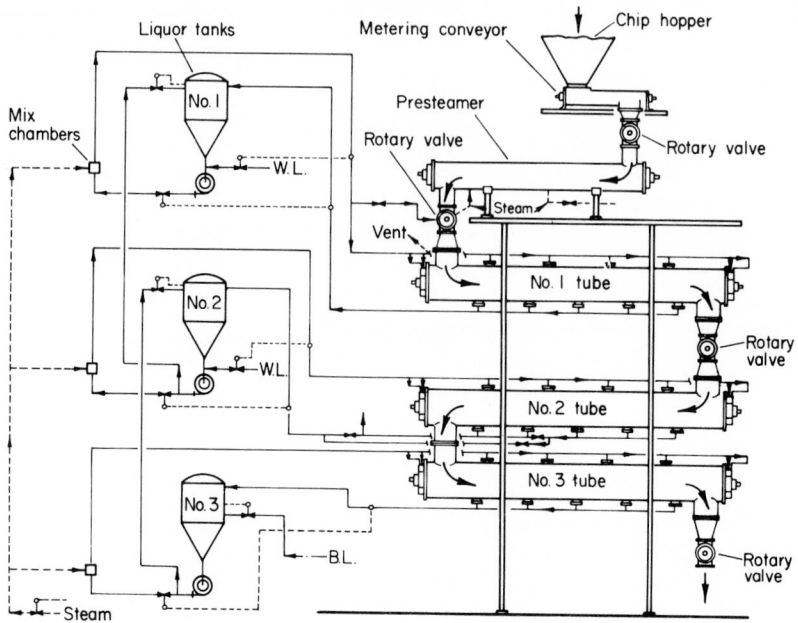

Figure VIII-14. Black Clawson kraft continuous digester system (353). (Reproduced from "Kraft Pulping Theory and Practice" by permission of Lockwood Publishing Co.)

which has its own liquor-circulating system, the wood is cooked under conventional cooking conditions. Additional chambers, each with its own circulating system, can be operated in such a way that the cooking conditions in each can be varied as to liquor concentration and temperature. The digesters are connected by rotary valves, thus permitting changes to be made in the throughput and, with it, the duration of the cook. A possible disadvantage may lie in the use of numerous rotating parts which, experience has shown, are subject to greater wear (353).

A somewhat different arrangement of the digesters is shown in Fig. VIII-15. Here the cylindrical reaction vessels (Bauer M and D tubes) are arranged at an angle of 45°. Each chamber is separated into two sections by a hollow midfeather along the length of the vessel. This midfeather is fitted with strainers which permit the withdrawal of the cooking liquor for the circulation system. The chips enter the reaction chamber at the upper end by a metering screw in the same way as in other systems. Before entering this chamber, the chips are moved by means of a screw conveyer through a horizontal presteaming vessel which is connected to the reaction vessel by a rotary valve. The level of the liquor in the digester is kept below the discharge valve so that the chips and the liquor are separated before they enter the next reaction vessel. This shows that here,

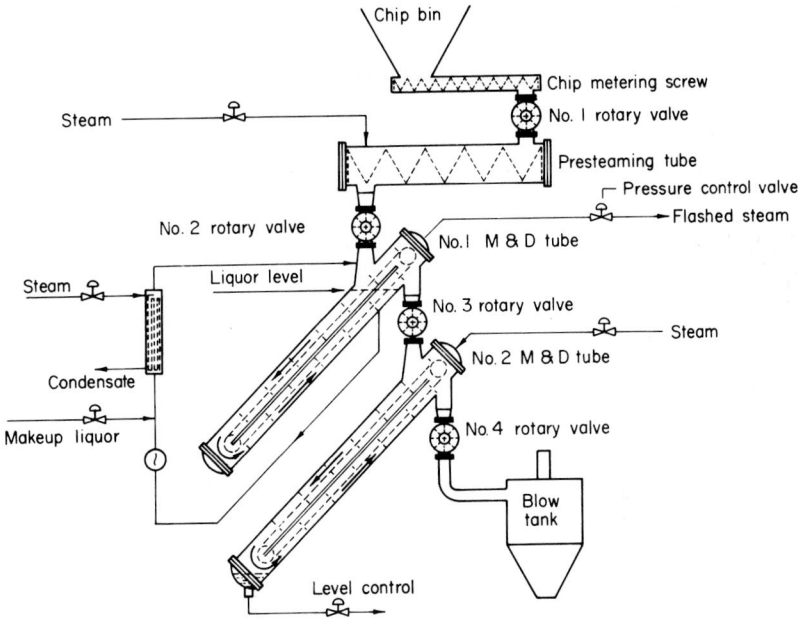

Figure VIII-15. Bauer M and D continuous digester system with cylindrical reaction vessels mounted at an angle of 45° (264). (Reproduced from "Kraft Pulping Theory and Practice" by permission of Lockwood Publishing Co.)

too, several reaction vessels are connected in a series and can be operated under the same or different reaction conditions (264).

As mentioned above, every continuous cooking system requires an accurate control of the reaction variables in each of the digesters, and the use of an automatic computer system has been suggested. More recent publications will provide information along these lines (174, 234, 384).

4.3 The Multistage Kraft Cook

Because of the heterogeneous nature of wood and the different types of reactions of the various wood components with the cooking liquor, every alkaline cook really takes place in several steps. Speaking here of multistage kraft cooks, it should be said that this stepwise reaction is attained by variations in the cooking liquor concentration or the temperature, or by other deliberate alterations in one or more of the reaction components. A distinction can be drawn between processes in which the same or related cooking agents are used in each step, and processes in which completely different liquors are used. For example, an acid-alkaline process, which uses a hydrolysis prior to the alkaline cook, belongs in the second group. In the process represented by the first group, the cook is

either interrupted or is separated into several steps by the removal of the cooking liquor and its substitution by the same or similar reagent. With modern kraft mills, using forced circulation, such exchanging operations can be carried out relatively easily. It is then necessary only to have the different cooking reagents, e.g., liquors of various compositions or concentrations, ready in storage tanks and to add them already heated to the right temperature. The problem is not so much one of exchanging the reaction agents as of maintaining the continuity of the cook and, later, of recovering the chemicals. The continuous cook with its different reaction zones within the digester already shows the basic characteristics of the multistage cooking process.

The first known proposal of a multistage cook was mentioned in a previous section. This process was modified in such a way that the wood was treated in a series of diffusers with black liquor at high pressures and temperatures, then with liquors of increasing concentration and decreasing temperatures until, in the last stage, fresh liquor was used at a low pressure and temperature. This is an example of a continuous countercurrent process (200). In contrast to this, other suggestions were to the effect that the cook should be started with a highly concentrated liquor, that this should then be drawn off or washed out, and the cook finished with the liquor that remained in the wood or with newly added, dilute liquor (281,301).

Still other suggestions provided for a steaming of the wood, which would serve primarily to facilitate the subsequent impregnation, then cooking the wood with a dilute cooking liquor, and finishing the cook with a normal liquor. For the countercurrent process (335), black liquor has been suggested for the first step and conventional kraft cooking liquor for the final cook (334,357). An initial cook with kraft cooking liquor and a final cook with lime have also been attempted (256).

On the basis of the concept that the cooking process can be divided into three integrated steps—in the first of which the chips are impregnated with the liquor, in the second the wood components react with the liquor, and in the third the dissolved wood components are extracted from the chips—a process was developed to accelerate and facilitate the necessary diffusion processes by the formation of suitable concentration gradients. Thus, the wood is impregnated with a relatively high-grade cooking liquor at a low temperature (about 70°C). The liquor is then drawn off and replaced either by water or by dilute liquor of such a concentration that it, together with the liquor remaining in the wood, forms a normal cooking liquor, suitable for the pulping of the wood species being used. The digester is then heated rapidly to the maximum temperature and the actual pulping process takes place in an extremely short period of time. Because of the marked concentration gradient between the liquor in the wood and that surrounding it, a reverse diffusion into the surrounding liquor and,

with it, an extraction of the dissolved wood components occur. By suitable adjustment of the composition and concentration of the liquor, of the wood-to-liquor ratio, and of the temperature and time, for each of the three stages, practically all types of pulp, ranging from high-strength pulps to pulps high in α-cellulose, can be prepared. There can be no doubt that this process can be carried out continuously as long as sufficient time is available for each of the reaction steps (388).

Such a two-stage continuous process consists in a preimpregnation with cooking liquor, a short high-temperature pulping step, and a low-temperature final step. In this case, therefore, the concentration and composition of the liquor remain unchanged and only the reaction temperature and duration are varied (240).

Crushing of the chips between two cooking stages should serve, first of all, to open up the reactive surfaces of the fibers and make them accessible to the cooking reagents. However, mechanical treatment of the micellar and intermicellar structure of the cell wall is less damaging in the alkaline pulping than in the acid cook. In spite of this, a crushing operation obviously has a detrimental effect on the strength properties of the pulp (35,191).

It has long been known that the alkali consumption is greatest at the beginning of the cook and this is caused, not by a reaction with the lignin, but by the dissolution of the hemicelluloses. It has therefore been suggested that the first part of the cook should be carried out with prehydrolyzed wood and a 2% sodium hydroxide solution, with a subsequent cook with 2.2% sodium hydroxide solution. Because the cooking liquor of the second stage is not exhausted as regards its alkali content, it can be reused for the first stage after the consumed alkali has been replaced. A considerable saving in alkali is made through the degradation, especially of the pentosans, during the prehydrolysis (91).

The migration of the reaction zone into the interior of the wood chips was studied by carrying out a preimpregnation of the chips with radioactive labeled polysulfide. These chips were then pulped with alkaline cooking liquors of various compositions. The result was an increase in the yield of pulp when the impregnation period, lasting from 20 to 140 minutes, was shortened. Sapwood from pine was impregnated more rapidly than heartwood. This was particularly apparent when the chips were presteamed. Chips with the greatest surface exposure in the tangential direction to the annual rings were impregnated more slowly than those that had been cut in the radial direction. Chips that were cut diagonally were most readily impregnated, especially when they had been presteamed. The impregnation temperature was increased by 6.5°C per minute from 20° to 98°C. The use of make-up black liquor can, under certain conditions, produce pulps with optimum yield and properties. The polysulfide pene-

trated only the very outer layers of the chips. In spite of this, the highest yield was obtained with an impregnation of short duration. During the subsequent alkaline cook, the polysulfide migrates ahead of the alkali, in this way forming a reaction zone between the sodium hydroxide and the polysulfide, and all reactions occur between these two reagents within a wide pH range which extends from pH 14 at the sodium hydroxide border to pH 9 at the polysulfide border. This reaction zone moves toward the center of the chips and, in this way, performs a specific delignification action and causes the dissolution of the decomposition products of the lignin in the cooking liquor (187).

4.4   The Acid-Alkaline Cook

The multistage acid-alkaline cook is characterized primarily by the fact that the pH value in each step lies far in the acid or in the alkaline region. Under these conditions, the acid cooking liquor of the first stage may be drawn off, or the pulp may even be washed, and an alkaline cook follow in the second step. However, it is also possible to neutralize the cooking liquor at the end of the first step and to carry out the second step without interruption or drainage of the digester. Such a process, for example, may use a sodium bisulfite solution with a high content of bound sulfur dioxide and a pH of about 4 in the first step. After the addition of a preheated sodium hydroxide solution, the alkali step then proceeds at pH 7.5 to 9.5. Such a procedure already closely resembles a combined sodium bisulfite-alkaline sodium monosulfite process. In the second step, a kraft cooking liquor may also be applied. This type of process is practical if a sulfite and a kraft pulp mill are combined or are located so that only one plant is necessary for the regeneration of the spent liquor (287).

The following reaction steps have been tested: sodium bisulfite-sodium monosulfite, sodium bisulfite-buffered sodium monosulfite, sodium bisulfite-sodium monosulfite + sodium hydroxide, and sodium bisulfite-kraft cooking liquor. Whereas the first three cooking sequences fall under the heading of multistage sulfite cooks and have been discussed there, the last is a two-stage acid-alkaline kraft cook. In this case, the bisulfite cook was interrupted at yields of 80%, 75%, 70%, and 65%. Thus the first stage gave a semipulp or, at 65% yield, "high-yield" pulp. After the first stage, the cooking liquor was removed and was replaced by a kraft cooking liquor. The necessary amount of active alkali ($NaOH + 1/2\ Na_2S$) was calculated from the alkali consumption curve for a normal kraft cook with the same yield. The alkaline cook was then carried out with excess and reduced amounts of alkali. It was found that no regular kraft cook could be successfully carried out with the reduced amount of alkali. With a 22% excess over the theoretical amount of alkali, 62% to 68% of the active

alkali was already consumed after 25 minutes cooking time. With the residual amount, the cook could be completed in a normal way. The high consumption of alkali at the beginning of the second stage parallels a marked decrease in yield, far greater than that of a normal kraft cook. Undoubtedly this loss in yield in the second step can be attributed to an attack on the carbohydrates, which consume most of the active alkali at the beginning of the cook. In this case, the carbohydrates had already largely been depolymerized during the first step, which, therefore, controls the degree of dissolution of the carbohydrates in the second step. The Ross diagram in Fig. VIII-16 shows these relationships. The delignification

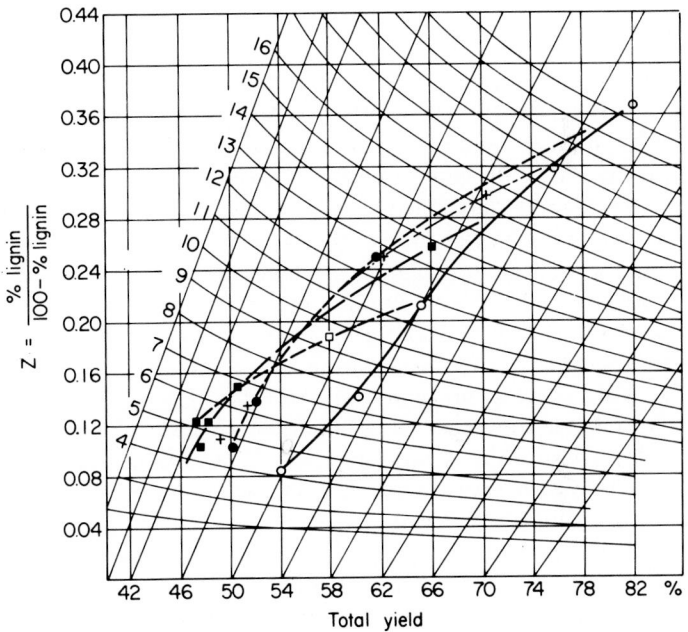

Figure VIII-16. Ross diagram of two-stage acid bisulfite-kraft cooking (82). ○=acid bisulfite cook; ●=first stage, 80% total yield; +=first stage, 75% total yield; ■=first stage, 65% total yield.

curves are shifted so close together that they show very little selectivity. The yields from the acid cooking stage affect the final result in such a way that the second stage, with regard to its selectivity, approaches that of a pure kraft cook in inverse proportion to the degree of cooking at the end of the first stage. By varying the degree of cooking in the first stage and the amount of active alkali in the second, and by varying the temperature and cooking time in either or both, pulps of the greatest variety in their physical properties are produced (82).

4.5   The Prehydrolysis

It has long been known that some components of wood are relatively easily hydrolyzed. The application of hydrolysis for the degradation of certain carbohydrates to low-molecular sugars is based on experience gained in the total hydrolysis of wood. This has been discussed in Chapter IV. On the basis of the known behavior of certain carbohydrates in an acid hydrolysis, a selective hydrolysis, directed chiefly toward the degradation of the hemicelluloses, was applied prior to the alkaline cook and, in this way, pulps with an especially low hemicellulose content could be produced even from woods containing a large proportion of this component. Such pulps, with a low content of pentosans and increased content of alkali-resistant α-cellulose were sought by the rayon industry. Unlike coniferous woods, which had become increasingly scarce in Europe, deciduous woods and annual plants, such as straw, could not be pulped satisfactorily by the conventional calcium bisulfite cooking process. The alkaline cooking process, on the other hand, gave pulps with high strength properties but unsuitable for the rayon industry because of their high pentosan content. Thus the alkaline pulping process, with a mild prehydrolysis, was developed (302).

During the Second World War, a number of pulp mills were built in Germany designed to use deciduous woods or straw in an alkaline cook with a prehydrolysis. The hydrolyzates were processed for the production of fodder yeast. By this process, deciduous woods and straw were treated at 120° to 130°C with 0.5% to 1% sulfuric acid and, in a second stage, with a conventional kraft cooking liquor (22% sulfidity) with a maximum cooking time and temperature of 3 hours and 165°C to produce pulps with a high α-cellulose content (110). In other experiments, concentrated acids and low temperatures were used in the prehydrolysis, e.g., 30% sulfuric acid at 70°C or 20% at 80°C. In this process, beechwood lost 25% of its weight and the lignin content of the prehydrolyzed wood remained almost constant. Obviously, the use of such high-grade acids would necessitate their recovery, thus making the processing of the hydrolyzates more difficult (188,194).

A comparison of the hydrolyzing action of acetic, sulfuric, and hydrochloric acids at concentrations of between 0.1% and 0.3% clearly showed that the hydrochloric acid is the most effective. Its optimal concentration was at 0.7%. After a prehydrolysis with hydrochloric acid and a subsequent kraft cook, beechwood gave pulps with 95% α-cellulose and 5.5% pentosans (90, 92). Many of the processes and plants which, during the war years, used mineral acids for a prehydrolysis, were later abandoned. Those that still used a prehydrolysis usually limited it to water or steam at elevated temperatures, about 180°C, for 15 minutes. A subsequent

alkaline cook at 160°C for 45 to 90 minutes gave pulps with 96% to 97% α-cellulose, 4.6% pentosans, and 1.1% to 1.6% lignin (333).

Eucalyptus wood, prehyrolyzed with water at 158°C, was subjected to a kraft cook with an alkali concentration of between 19% and 30%, at 135°C. Regardless of the alkali concentration, the xylan content of the pulps was practically the same, 2.8% to 3%, but the yield, bleachability, and viscosity of the pulp varied to a considerable degree. The pentosan content, 18% of the wood, dropped, after the hydrolysis, to 5%, but the lignin content was not affected. In spite of this, it must be assumed that the lignin-carbohydrate complex was changed to some extent in the hydrolysis and also that the xylan fraction in the prehydrolyzed wood was decreased (248).

Experiments were carried out with loblolly pine, Douglas fir, and sweet gum to study the combined action of the prehydrolysis, carried out

TABLE VIII-2

PREPARATION AND YIELD OF LOBLOLLY PINE SULFATE AND PREHYDROLYSIS SULFATE PULPS (329)

	Pulping process			
	Prehydrolysis sulfate		Sulfate	
	Digestion I	Digestion II	Digestion I	Digestion II
Prehydrolysis with 5 parts of water to 1 of wood				
Maximum temperature, °C	160	160	—	—
Time to maximum temperature, minutes	75	75	—	—
Time at maximum temperature, minutes	0	75	—	—
Sulfate pulping with about 5 parts of liquor to 1 of wood				
Total chemicals (NaOH+Na$_2$S), percent[a]	21.0	26.0	21.0	26.0
Active alkali (as Na$_2$O), percent	16.4	20.3	16.4	20.3
Sulfidity, based on active alkali, percent	25.5	25.5	25.5	25.5
Maximum temperature, °C	170	170	170	170
Time to maximum temperature, minutes	75	75	80	90
Time at maximum temperature, minutes	90	90	90	90
Unbleached pulp				
Yield of screened pulp, percent	41.5	35.0	44.8	42.3
Screenings, percent	0.2	0.1	0.9	0.3
Total yield, percent	42.0	35.1	45.7	42.6
Permanganate number	18.9	11.7	24.3	15.2

[a] Percentage values for chemicals and yields are based on weight of moisture-free wood.

with water, and kraft cooks under different cooking conditions. For comparison, two kraft cooks were made without a prehydrolysis. Table VIII–2 shows the reaction conditions of these cooks with loblolly pine and the yield and permanganate numbers of the unbleached pulps. As was to be expected, the yield and permanganate number of the cooks with a prehydrolysis are lower than those with a conventional kraft cook. This is especially apparent when the total consumption of active alkali was increased. The differences in the analytical data of these pulps are given in Table VIII–3. The prehydrolysis reduces the yield because a major part of the

TABLE VIII-3

ANALYSES OF LOBLOLLY PINE SULFATE AND PREHYDROLYSIS SULFATE PULPS (329)

	Kraft cooking		Kraft cooking with prehydrolysis	
	I	II	I	II
Yield, based on wood, %	44.8	42.3	41.8	35.0
Lignin, %	5.6	3.4	3.4	2.0
Permanganate number, %	24.3	15.2	18.9	11.7
α-Cellulose, %	87.1	88.3	89.0	92.6
β-Cellulose, %	0.1	3.1	2.2	3.3
γ-Cellulose, %	6.9	5.1	3.3	0.8
Pentosans				
TAPPI standard 223 m, %	9.3	7.8	6.4	3.0
Chromatographic method, %	9.8	8.6	7.2	3.8
Mannans (chromatographic), %	6.4	6.4	4.8	2.6
Solubility				
in alcohol-benzene, %	0.26	0.22	0.18	0.20
in hot 7.14% NaOH, %	8.2	7.6	6.7	3.6
in cold 18% NaOH, %	7.1	7.9	5.7	3.9
Fraction resistant to 18% NaOH, %[a]	86.5	87.9	90.3	93.3
Carboxyl content[b]	8.33	5.12	5.96	3.06
Ash, %	0.6	0.6	0.4	0.6
Brightness, %	26.1	32.1	27.9	34.5
Disperse viscosity, cps	38.1	20.8	33.7	16.1

  [a] Calculated by difference between 100 and the sum of cold 18% NaOH-soluble, alcohol-benzene-soluble, ash and lignin.
  [b] Milliequivalents per 100 grams.

hemicelluloses is dissolved. The lignin content, also, is somewhat lower, probably as a result of a weakening of a lignin-carbohydrate bond during the hydrolysis. The α-, alkali-resistant, and β-cellulose contents are increased, as is the brightness of the unbleached pulps. Table VIII–4 gives the yield of kraft pulps and their pentosan and mannan contents from four

woods, loblolly pine, Douglas fir, sweet gum, and Western hemlock. The cooks were carried out with and without a prehydrolysis. The results show that, with the first three woods, 47% to 59% of the pentosans and 76% to 94% of the mannans were removed when the prehydrolysis was omitted. With prehydrolysis, these data increase to 66% to 95% and 84% to 97%, respectively. The pentosans of the deciduous wood (sweet gum) were more strongly degraded in the prehydrolysis than were those of the coniferous woods. So far as the mannans are concerned, the results were more or less reversed (328,329,330,331).

TABLE VIII-4

REMOVAL OF PENTOSANS AND MANNANS FROM WOODS BY THE PULPING PROCESS (329)

Wood species	Pulp type	Pentosan[a]			Mannan[a]	
		Yield[b] (%)	Content[c] (%)	Removed[d] (%)	Content[c] (%)	Removed[d] (%)
Loblolly pine		—	8.9	—	11.8	—
	Sulfate	45	9.8	50	6.4	76
	Sulfate	42	8.6	59	6.4	76
	Prehydrolysis sulfate	42	7.2	66	4.8	91
	Prehydrolysis sulfate	35	3.7	85	2.5	93
Douglas fir		—	4.6	—	11.2	—
	Sulfate	45	5.5	47	5.2	79
	Sulfate	42	4.7	57	5.4	80
	Prehydrolysis sulfate	44	3.2	69	4.0	84
	Prehydrolysis sulfate	37	1.6	87	2.3	92
Sweet gum		—	21.0	—	3.2	—
	Sulfate	49	20.5	52	0.4	94
	Sulfate	47	22.1	50	0.4	94
	Prehydrolysis sulfate	46	12.4	73	0.4	94
	Prehydrolysis sulfate	36	2.9	95	0.3	97
Western hemlock		—	4.1	—	12.4	—
	Sulfite	50	2.6	68	7.1	71
	Sulfite	44	2.3	76	5.5	81

[a] Determined chromatographically.
[b] Screened pulp based on moisture-free wood.
[c] Based on weight of moisture-free material.
[d] The proportion removed by pulping is equal to:
$$100 - (\text{yield} \times \text{content in pulp})/\text{content in wood}.$$

Figure VIII–17 shows to what extent the conditions under which the prehydrolysis is carried out affect the properties and the yield of a kraft pulp. Western hemlock was subjected to a prehydrolysis with water, in a ratio of water to wood of 6:1, at different temperatures and for different periods of time. A kraft cook with 20% sodium hydroxide (as Na$_2$O,

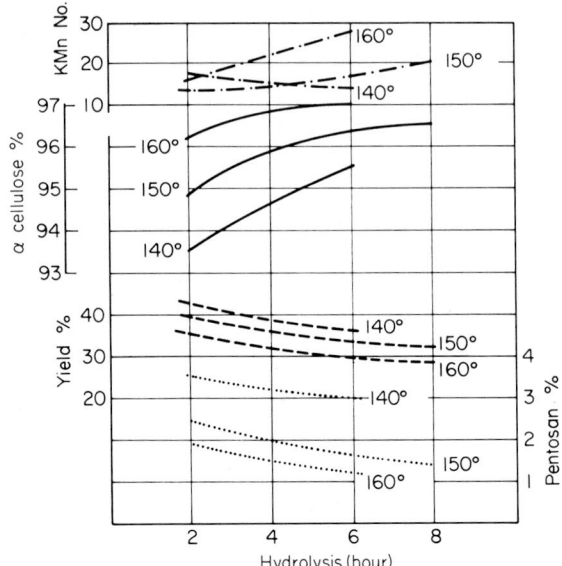

Figure VIII-17. Influence of prehydrolysis temperature and time on alpha cellulose content, pentosan and yield of Western hemlock (304).

based on the wood) and 25% sulfidity was then carried out for 1.5 hours to 166°C and for 2.5 hours at this temperature. The results show that, under suitable reaction conditions, pulps with a high α-cellulose content, a low pentosan content, good bleachability, desirable viscosity, high mechanical strengths, and good chemical reactivity can be produced. In comparing coniferous and deciduous woods, it was found that appreciable differences appear when the reaction conditions of the prehydrolysis are chosen so that the removal of the pentosans is as complete as possible.

Deciduous woods undergo a significantly smaller deactivation of the lignin, especially when the prehydrolysis is carried out at temperatures of over 170°C or is extended over a long period. The α-cellulose content of the kraft pulp is also considerably reduced with the increased temperature. Yield and pentosan content of the pulps decrease with increasing duration and temperature of the hydrolysis. High temperatures (185°C) give a pulp with a high lignin content, although the pentosan content is almost as low as that of cotton. A high proportion of screenings reduces the yield and indicates that the cellulose itself had undergone some degradation. A prehydrolysis with water for 2 hours at 160°C usually caused the pH to drop to 3.8, with that of deciduous woods being somewhat lower than that of coniferous woods. Acetic acid was the chief organic acid formed in the hydrolysis. The neutralization of such hydrolyzates requires 0.6% sodium hydroxide in the case of hemlock and 1.8% in the case of

white birch, both based on the original weight of the wood. When the water-to-wood ratio is reduced from 10:1 to 3:1, the pH drops from 3.6 to 3.3 at the conclusion of the hydrolysis (304).

The effects of temperature and time on the prehydrolysis with water have been confirmed unanimously by various investigators. For example, the prolongation of the hydrolysis to over 3 hours and increasing the temperature to over 160°C cause a noticeable loss in yield and damage to the cellulose. In general, the delignification is facilitated by the hydrolysis, but increased resistance on the part of a certain fraction of the lignin has also been observed. The addition of sodium borohydride in the kraft cook increases the yield and the β-cellulose and pentosan contents from prehydrolyzed wood but decreases the α-cellulose content. When the prehydrolysis is followed by a two-stage kraft cook, with cooking liquor of the same composition being used in each stage, the α-cellulose content increases by about 0.5% and the pentosan content decreases by about 0.3%, as compared with a one-stage cook. It is possible that this is caused by a stronger retention of the hemicelluloses by the pulp in the one-stage process. Temperature and time are the deciding factors in the prehydrolysis (219,355).

The monosaccharides obtained from the prehydrolysis of pinewood with water (2 hours, 160°C, pH 3.4) gave, on paper chromatographic analysis, 1.71% reducing substances of which 1.02% were identified as xylose, arabinose, galactose, glucose, mannose, and uronic acids. After inversion, the reducing substances increased to 2.5%. The amount of oligosaccharides was about 0.69% (327). For *Pinus radiata*, the optimum prehydrolysis conditions with water were found to be one hour at 170°C. A kraft cook, without or with such a prehydrolysis, at 170°C, 22% active alkali, and 18.7% sulfidity gave pulps with the analytical data as listed in Table VIII-5 (47).

In summary, it can be said that, by a suitable choice of the prehydrolysis conditions with regard to temperature and duration and of the con-

TABLE VIII-5

COMPARISON OF KRAFT AND PREHYDROLYSIS KRAFT PULPS FROM *Pinus radiata* (47)

	Normal kraft	Prehydrolysis kraft
α-Cellulose (%)	92.0	96.0
Pentosans (%)	7.0	1.8
Lignin (%)	4.6	2.7
TAPPI K	14.5	12.0
Kappa number	20.1	13.3
Intrinsic viscosity (dl/g)	6.9	5.1
Yield (%)	46.0	37.0
Ash (%)	0.35	0.16
Screenings (%)	0.29	0.03

ditions of the subsequent kraft cook—liquor concentration, sulfidity, liquor-to-wood ratio, temperature, and time—a cooking procedure can be developed for any wood species to give a pulp with the properties desired. For chemical conversion, the goal is to produce a pulp with high α-cellulose and low hemicellulose contents. For the manufacture of paper, with a few exceptions, the α-cellulose content is of minor importance, but the hemicellulose content has an important effect on the beatability of the pulp.

If the kraft cook in the continuous digester is preceded by a steaming phase, the prehydrolysis can also be applied during the continuous pulping process. In this case, as is shown in Fig. VIII–18, the upper, inner separator of the original Kamyr design (see Fig. VIII–12) has been moved out-

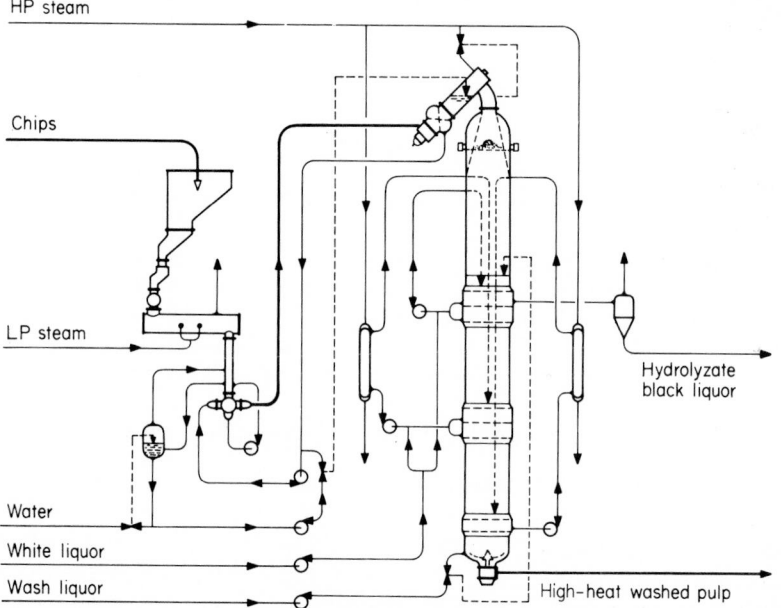

Figure VIII–18. Downflow, vapor-liquor-phase digester for prehydrolyzed kraft pulp (10). (Reproduced from "Kraft Pulping Theory and Practice" by permission of Lockwood Publishing Co.)

side and placed in an inverted slanted position with a strainer at its lower end. The wood chips are transported upward by a screw conveyer and enter the digester through an elbow. In this way they are free of water and carry only absorbed water into the digester. Steam is then injected directly and propels the chips through the digester. The amount of steam injected is regulated according to the temperature within the prehydrolysis zone. The level of the cooking liquor can be adjusted to any height in the prehydrolysis zone, but should be kept somewhat higher than the circula-

tion zone so that the two stages remain separate from each other. The cook itself can be carried out by a continuous current or by a counter-current process. Table VIII–6 gives the results of such kraft cooks with eucalyptus, pine, and birch woods (10).

TABLE VIII–6

<small>COOKING CONDITIONS AND PULP QUALITIES OF PREHYDROLYSIS KRAFT COOKING OF EUCALYPTUS, PINE AND BIRCH IN A CONTINUOUS KAMYR COOKING SYSTEM (10)</small>

	Eucalyptus		Pine	Birch
Daily output, air-dry tons	12	18	12	12
Temperature, °C	166	170	165	165
Prehydrolysis	166	170	165	165
Kraft cooking	155	158	162	157
High-temperature washing	135	135	—	—
Active alkali, % Na_2O on o.d. wood				
Transfer circulation	8.6	8.5	—	—
Cooking circulation	9.4	9.8	—	—
Total	18.0	18.3	19.0	20.6
Roe number	0.7	0.6	2.8	1.3
Screenings, %	0.9	0.9	3.0	1.0
Alkali resistance				
18% NaOH	97.8	97.5	96.4	96.5
10% NaOH	96.4	96.4	95.2	94.9
α-Cellulose	97.2	97.0	95.8	95.7
Viscosity (TAPPI)	65	56	35	60

4.6 The Chemical Processing of Kraft Pulps (Chemical Refining)

The chemical after-treatment of alkali pulps is, to some extent, a precursor of the prehydrolysis, which was introduced very much later. These two processes can be compared only in their final results and, even here, only partially. Certain portions of the hemicelluloses are resistant to alkaline solution, even at higher temperatures. During the alkaline cook, some hemicelluloses are reabsorbed by the cellulose under certain reaction conditions, and this phenomenon has recently been utilized advantageous-ly. The hemicelluloses are detrimental in the preparation of artificial fibers and cellulose derivatives. Attempts have therefore been made for many years to remove the hemicelluloses from the pulps. Such attempts include the use of oxidizing agents, followed by an extraction with dilute alkali or alkaline earth, or with reducing agents such as neutral sulfite and similar reagents. Numerous patents taken out from 1918 to 1921 testify to the great activity in this field. In the United States, too, this problem was investigated and numerous patents were taken out during the 1920s (390). Under the name "alpha fiber," a pulp, from which the residual

pentosans had largely been removed, was marketed. It was found, however, that, in contrast to those in the sulfite pulp, the pentosans remaining
in the kraft pulp were largely resistant toward alkalis at temperatures
below 150°C. This is not surprising because the pentosans were exposed,
during the cooking process, to strong alkaline solutions at high temperatures, and the hydrolysis, which is effective in the acid sulfite cook, is
lacking.

Investigations into the effect of mercerizing solutions upon the chemical
and mechanical properties of pulps showed that the changes in the x-ray
spectrum of the cellulose caused by the concentrated lyes paralleled the
basic changes in the properties of the pulps. x-Ray investigation (352)
showed that the mercerization of the cellulose with 18% sodium hydroxide
solution at 18°C is completed within one minute. A sodium hydroxide
solution of 12.5% to 19% gives a cellulose with an x-ray diagram of cellulose I, whereas a 21% to 45% solution gives the x-ray diagram of cellulose II. The latter also appears with dilute sodium hydroxide solutions at
0°C but disappears again on warming. A partial mercerization occurs
with 11% sodium hydroxide solution at 10°C and with 18% solution at
80°C. This explains why the best results are obtained with low sodium
hydroxide concentrations at a low temperature (169,389).

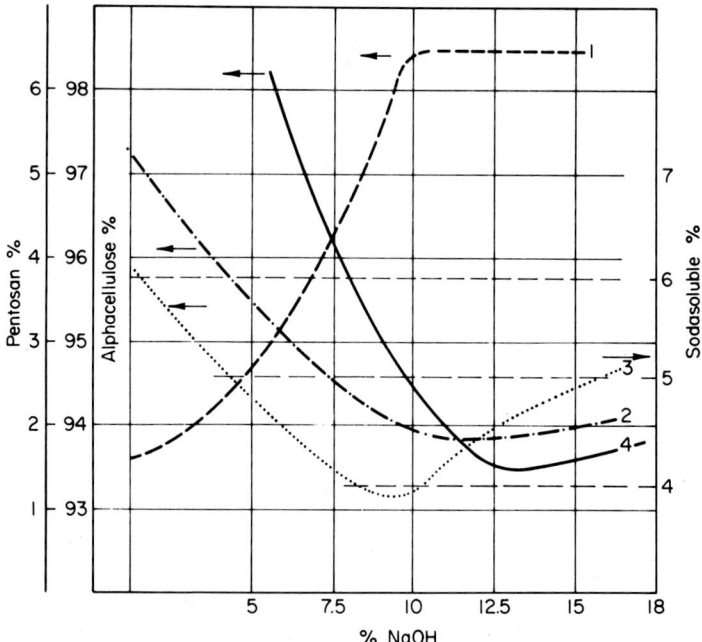

Figure VIII-19. Cold alkaline refining of unbleached kraft pulp at increasing NaOH
concentrations (303). 1 α-cellulose; 2 pentosan; 3 soda-soluble; 4 yield.

Figure VIII–19 shows the effect of the concentration of the sodium hydroxide solution, the temperature, and the reaction time on the properties of such "cold refined" pulps. It shows that, with the exception of the yield, all curves have a minimum at a sodium hydroxide concentration of 10%. That such minimums exist and are traceable to shifts in the crystal lattice of the cellulose had already been indicated (389). Figure VIII–20 clearly shows the detrimental effect of higher temperatures combined with high sodium hydroxide concentrations (303).

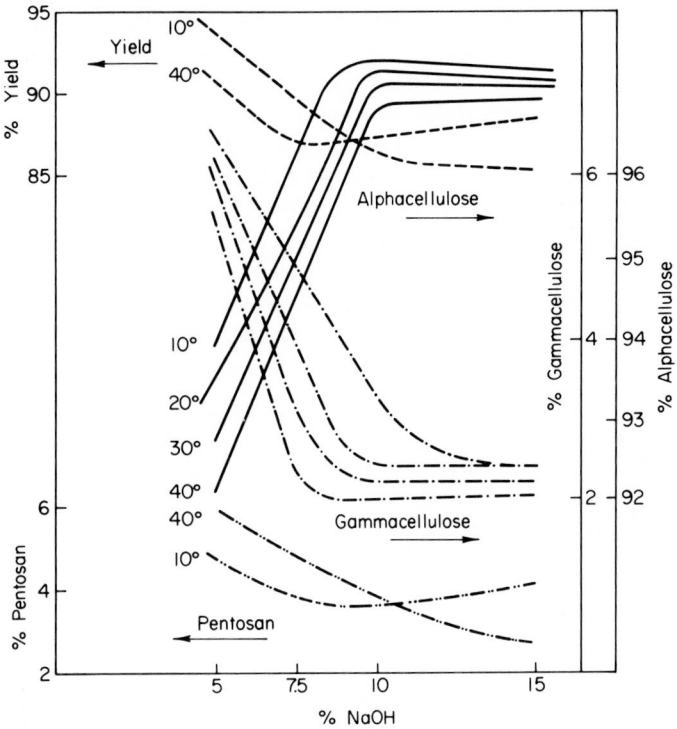

Figure VIII-20. Influence of temperature and NaOH concentration on treatment of unbleached kraft pulp (303).

It is known that the solubility of the hemicelluloses varies. The terms α-, β-, and γ-cellulose have no more than a technological meaning. It has also been proposed that the part that is dissolved by 5% potassium hydroxide solution should be termed fraction A, and that soluble in 24% potassium hydroxide solution, fraction B. The differentiation by determination of the total pentosan content in the wood and in the isolated Cross and Bevan cellulose, and by the uronic acid determination has been recommended. Some investigators consider the alkaline refining to be the

result of a series of reactions. Whereas cold refining produces only a few acid reaction products, hot refining causes an increase in such products and, at the same time, a degradation of the hemicelluloses; the latter then diffuse into the liquor. β-Cellulose and γ-cellulose are components of the cellulose structure and are rendered soluble in the degradation of the α-cellulose fraction. The insoluble residue consists chiefly of glucan. The soluble, but nondialyzable material is a mixture of degraded glucan and mannan with traces of xylan (88,279,398).

The pulps from loblolly pine, with and without prehydrolysis, listed in Tables VIII–2 and VIII–3, were treated with normal, cold white liquor with and without a prechlorination. In a third series of experiments, a mixture of sodium hydroxide and sodium chlorite was used as extraction medium. The pulps obtained from the prehydrolysis experiment were chlorinated and then treated with hot white liquor. Whereas an alkaline extraction with hot dilute sodium hydroxide solution proved to be ineffective with normal kraft pulps, it was effective with pulps that were obtained from prehydrolyzed chips. The same holds for a hot extraction in the presence of an oxidizing agent. This again confirms the fact that certain portions of the hemicelluloses can be extracted with alkaline solutions only after a prehydrolysis of the chips (329).

5. THE RECOVERY OF CHEMICALS FROM KRAFT SPENT LIQUORS

5.1 Brown Stock Washing

After the cook, the pulp is discharged into the blow tank (see Fig. VIII–11) and is diluted with black liquor to a consistency of about 2%. The mixture is then passed through a knotter where knots and incompletely cooked chips are removed and are then returned to the chip bin. The pulp is then passed over a series of rotary drum filters in order to recover the black liquor in a form as concentrated as possible, thus reducing the energy consumption in the later evaporation. Any unnecessary dilution of the liquor means an additional load for the evaporation plant. On the other hand, an incomplete recovery of the black liquor means loss in sulfate which must be made up in the regeneration of the alkali and the sulfur components. The washing procedure is complicated and consists of a series of diffusion processes; it is an extraction process which is dependent upon all the components that are capable of influencing the diffusion. Among such components is the degree of pulping. For a certain degree of pulping, the optimum conditions can be determined for the operation of the wash plant. Such a plant may consist of a series of three wash

filters, with repulping equipment installed between the first and second, and second and third filters. The amount of wash water used can be varied to accord with the degree of pulping. When one of these factors is changed, the operation of the plant must be adjusted to this change; for example, the speed of the rotary filter drums must be changed, the valves between the filters must be readjusted, or the position of the sprinkling system must be altered. With a pulp with a TAPPI kappa number of 34 and a daily production of 415 tons air-dry pulp, the total alkali content in the washed pulp increased from 43 pounds Na_2SO_4 per ton to 54 pounds per ton when the dilution factor was lowered from 2.5 to 0.9. A pulp with a kappa number of 29 to 31 could be washed to a lower alkali content even when the dilution factor in the daily production remained unchanged. It may therefore be assumed that the alkali content in the washed pulp at a given dilution factor increases with increasing production. Figure VIII–21 shows the relationships between the dilution factor and the alkali content in the washed pulp (201).

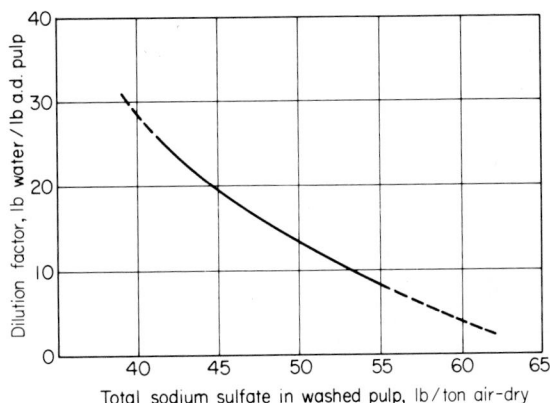

Figure VIII-21. Relation between dilution factor and total sodium sulfate in washed pulp. Pulp of TAPPI 34 hardness at 415 air-dry tons per 24 hr (201).

The efficiency in a multistage countercurrent washing plant can be expressed as the "displacement ratio" (DR). The DR is a function of the dilution factor and, under normal operating conditions, is independent of other variables, with the exception of the degree of pulping (kappa number) and the specific loading of the plant. With the use of the DR, the wash process can be expressed by the following equation:

$$S_m = \text{II} R (1 \rightarrow m) S_b$$

in which S_m is the solid content of the liquor (%) which leaves the washing zone with the washed pulp, S_b is the solid content of the original spent

liquor (%), and $IIR(1 \rightarrow m)$ is the product of the solids reduction ratios of each of the washing stages. The results of a series of experiments showed that the empirical relationship between displacement ratio and dilution factor can be expressed as a constant, K, which must be multiplied by the theoretical equivalent of the dilution ratio, as is shown in Fig. VIII–22 (291).

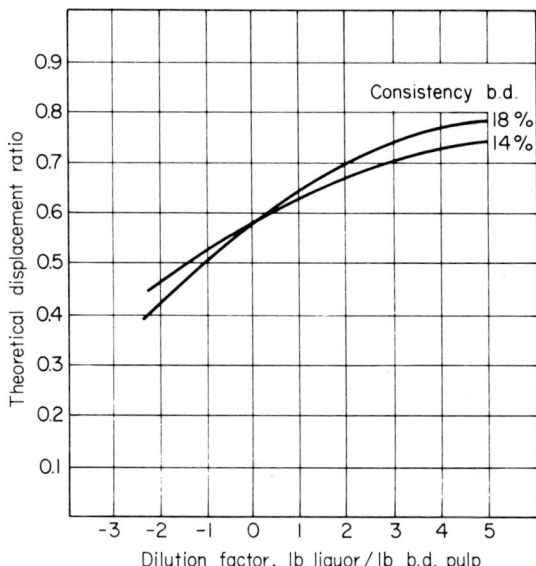

Figure VIII-22. Theoretical displacement ratio curve; effect of consistency (291).

$$\text{Theoretical DR} = 1 - \left[\frac{n\,W_p}{(n+1)\,W_p + D} \right]^n$$

W_p = Total weight of liquor with the pulp leaving the washers at solids $S\%$.
D = Dilution factor in lb of liquor entering a stage per lb of moisture-free pulp.
n = Number of showers (n) = 5.

The dilution of the spent liquor with wash water is of fundamental importance. With increasing amounts of wash water, the extraction of the dissolved material in the pulp increases, but, at the same time, the heat economy of the chemical recovery plant is affected unfavorably. Except for the small amount that remains on the pulp, the wash water goes through the filter and dilutes the black liquor. The displacement process is capillary in nature and proceeds rapidly. The removal of the black liquor from the fiber bundles occurs chiefly by diffusion and is, therefore, governed by the time element.

The degree of dilution (S) equals the amount of water added through the sprinklers minus the amount that remains in the pulp after enough has

been added to keep the concentration of the original liquor constant. The dilution factor can then be expressed by the following equation:

$$S = V - (L_f - \Delta L_0)$$

in which V is the amount of wash water, L_0 is the concentration of the original liquor, L_f is the liquor in the filter cake, and ΔL_0 is the original liquor in the filter cake. The purity of the pulp depends upon the volume of water used; the loss of alkali and the purity of the pulp therefore vary with the dilution.

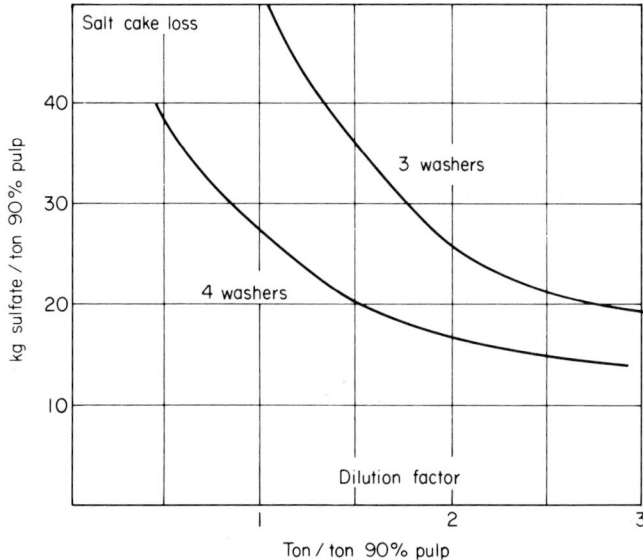

Figure VIII-23. Relation between total salt cake loss and dilution factor (199).

Figure VIII–23 shows the relationship between the total loss of alkali (in kg Na_2SO_4 per ton of pulp, 90% dry) and the degree of dilution (199).

When the brown stuff is washed in the digester, as is the case in the continuous cooking process, the conditions are somewhat different. The concept of displacement ratio is replaced by the term "filter entrainment" (E), which is defined as the weight amount of undisplaced liquor remaining in the filter discharge per unit of pulp on a dry basis. In the equation

$$E = W_p(1 - DR)$$

W denotes the ratio of 1 lb liquor per 1 lb air-dry pulp, and p is the pulp flow. Plant experience will give the range of E for each filter stage in a washing line. It has long been a rule of thumb that E will decrease logarithmically with the amount of wash water. If E is plotted logarithmically against W_s (the wash-water flow), a linear curve is obtained. Some authors

have concluded that the linearity should extend back to the E axis. In such a case, the curve would have to pass through the value $E = W_p$ when W_s becomes zero. The entire curve could then be expressed by a single "washer-efficiency parameter" leading to the equation

$$\frac{E}{W_p} = \left(\frac{1 - \text{washer efficiency}}{100}\right) \frac{W_s}{W_p}$$

in which W_p is lb liquid per lb air-dry pulp in the pulp cake and W_s is the same ratio within the wash-water flow (71,123).

On a wash filter, two processes occur successively: first, the black liquor is replaced by wash water, then the latter is replaced by air, although it is possible that the two processes may occur simultaneously. It can be shown mathematically that a linear semilogarithmic wash-displacement curve fulfills the necessary requirements only under certain conditions. The conditions are changed fundamentally when an air-displacement stage is added. The above equation can then be applied when the actual wash-displacement

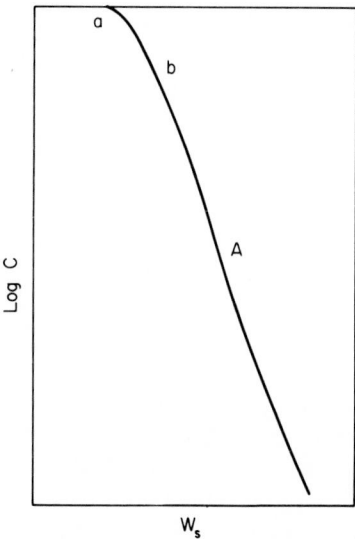

Figure VIII-24. Wash filtrate concentration curve (theoretical) (123).

curve conforms with the theoretical curve shown in Fig. VIII-24. In order to show the effect of air replacement, it is necessary to convert the filtrate concentration curve into an entrainment curve, which is shown in Fig. VIII-25. If one assumes that $E = 0$ and $W_s = \infty$, then a linear curve, identical with that in Fig. VIII-24, is obtained. The air front passing through the filter layer behaves quite differently from the wash-liquor front because air is not miscible with the liquid phase it is replacing, and it has

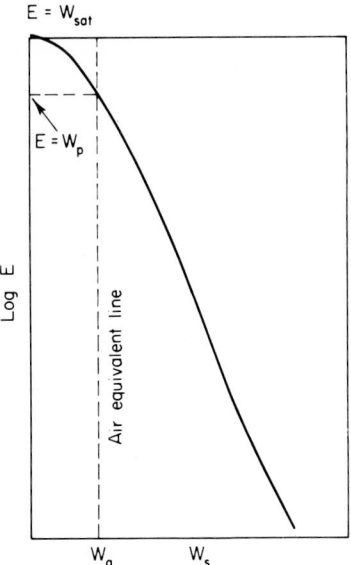

Figure VIII-25. Entrainment curve (theoretical) (123)

very different viscosity. When W_{sat} is the ratio of lb liquid per lb air-dry pulp in the filter cake before the air-replacement step, then the air reduces the filter sheet dilution from W_{sat} to W_p. To reduce the total entrainment of liquor and first wash water by a hypothetical second liquid from W_{sat} to W_p, a certain amount (W_a) of the second wash liquor would be required; W_a is a hypothetical amount of wash liquor, the displacement efficiency of which is equal to that of the air. From the wash entrainment curve in Fig. VIII-25 can be calculated how much of the second wash liquor would be required to reduce E from W_{sat} to W_p. This amount is termed the "air equivalent" (W_a). Assuming that the relative displacement of black liquor and first wash water is approximately the same, regardless of whether the second displacing medium is air or another liquid wash, the residual entrainment, E, of the black liquor in the filter cake would correspond to that amount which remains after a wash of W_s, plus a hypothetical or additional wash, equal to the air equivalent, W_a. This means that E can be obtained from the curve at the coordinate, $W_s = W_p + W_a$, in order to determine how much liquor has remained in the filter cake. The entrainment curves for washing plus air replacement then become identical in shape with that in Fig. VIII-25, but the starting point is shifted to the point $W_s = W_a$. If this shift in the starting point moves the E axis into the linear range, then the above equation is valid (123).

Changes in the production capacity of a mill can lead to major dif-

ficulties in the draining and washing of the pulp. As an example, a kraft pulp mill that operated with eight batch digesters increased its production from 600 to 1000 tons per day by the installation of a continuous digester before the filter wash plant had been set up for the increased production; this caused a sizeable loss in the recovery of the chemicals. Since the addition of wash water was limited by the capacity of the evaporation plant, each increase in production had to be compensated by a decrease in the dilution of the black liquor during the wash. Each reduction in the consistency of the filter layer meant that more liquor was removed from the pulp. The pro rata carry-over of alkali, adjusted to a certain production level and expressed as lb Na_2SO_4 per ton air-dry pulp, increased very rapidly with decreasing consistency of the filter layer and, at 14% consistency, amounted to 320 lb Na_2SO_4 per ton of pulp, while discharged liquor contained 3.26% Na_2SO_4, at a daily production rate of 650 tons. At a consistency of below 10%, there was a sharp increase in the amount of retained liquor, the drainage became insufficient, and strong liquor was carried over into the second wash filter, into the second seal tank below, and from there back into the sprinklers of the first wash filter; moreover, the wash zone in the continuous digester was adversely affected. Even after an increase in the capacity of the evaporation plant, the loss of alkali salts remained too great, although the dilution factor could be increased from 3.5 to 5.1. Regardless of whether the washing took place on the filter or in the continuous digester, the objective was still to obtain a pulp as pure as possible with a minimum amount of wash water. If the amount of extracted black liquor in the washing process equalled the amount formed in the cook, the dilution factor would be zero. Actual results indicated that the amount of black liquor formed in the cooking zone of a continuous digester was about 1.3 gallons per minute per ton, corresponding to a concentration equivalent of 15.4°Bé. For a continuous digester with a capacity of 300 tons per day, the optimal wash-liquor flow would then be 390 gallons per minute. With amounts above this figure, the black liquor would be diluted, while below this figure, the concentration would increase to above 15°Bé and the washing efficiency would be inadequate (171).

A recently developed continuous diffusion washer has a number of concentric strainer rings which oscillate in a vertical direction, with a stroke of about 4 inches, and move with about the same velocity as that of the pulp flow. The wash water enters the space between the rings through nozzles. Although this type of arrangement was designed more specifically for bleaching towers, it has been used successfully for the washing of brown stuff. A continuous kraft pulp digester moves the pulp through such a diffusion washer at 90°C, with a throughput of 60 tons per day and a consistency varying between 8% and 10%. Such a concentration range actually means a 20% change in the pulp throughput; with regard to the

flow of liquid, it means a change of only about 2%. During a prolonged investigation, the displacement ratio was between 0.94 and 0.95. The corresponding alkali losses (as Na_2SO_4) varied, in this one-stage washing operation, between 40 and 50 lb per ton of pulp. By interpolation, it is calculated that this is a loss of 70 to 90 lb with a batch digester without any wash, and 10 to 20 lb in a continuous digester with a high-temperature diffusion wash. If such a diffusion wash is followed by a multistage drum wash, the loss would be reduced to 15 lb per ton (see Fig. VIII-26) (306).

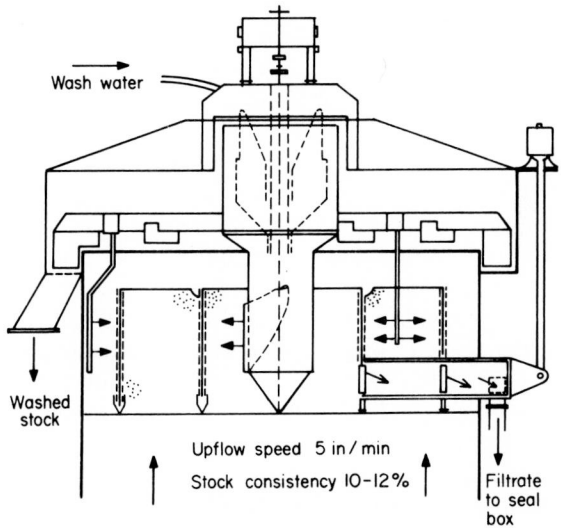

Figure VIII-26. Pulp washing by continuous diffusion (306).

The removal of the black liquor from the pulp can also be achieved by pressing. Here, the objective is to render the pulp as dry as possible, i.e., to leave a minimum amount of black liquor in it. In practice, one such operation is insufficient and several presses must be used successively, with the pulp being diluted with water after each pressing. Presses require little space and a moderate supply of power, and they form a completely closed system without causing the formation of any foam. Disadvantages include the danger of clogging, high initial installation costs when several presses are required, and, under certain conditions, higher fiber losses.

5.2 Evaporation of the Black Liquor

The black liquor from the washing is passed into a storage tank and, from there, it goes to the evaporation plant. The latter resembles a complicated chemical plant in which heat economy is the most important factor. Figure VIII-27 shows a flow-sheet of a multiple evaporator with six ag-

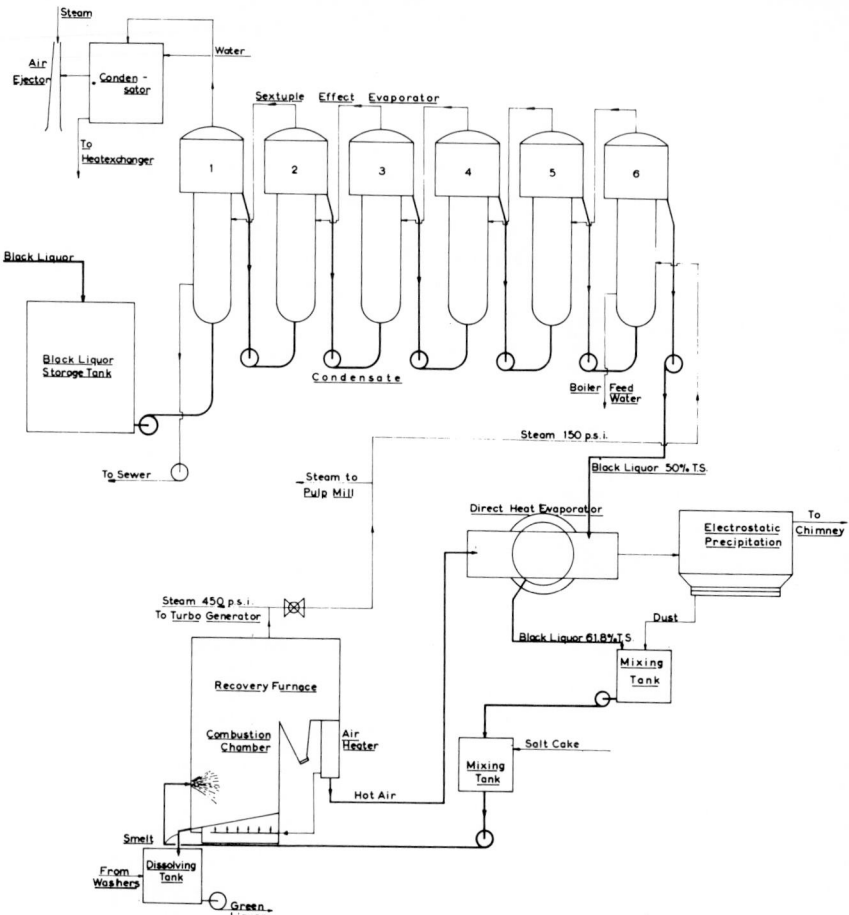

Figure VIII-27. General arrangement of kraft process chemicals recovery.

gregates, each of which is heated by steam from the preceding one. In this way one kilogram of steam can evaporate 4.7 kilograms or more of water. The liquor enters the first or second aggregate, which has the lowest temperature and is under a certain vacuum, produced by a steam ejector. The aggregates are connected by pumps, and pressure, or vacuum, and temperature increase from one aggregate to the next. The steam removed by the vacuum is passed into a condenser and the cooling water is passed through a heat exchanger. The black liquor, entering the evaporating system with a concentration of 15% to 18%, leaves it, normally, at a concentration of 48% to 54%.

Further concentration is limited by the increasing viscosity, and, in fact, the physical properties of the black liquor have an important effect

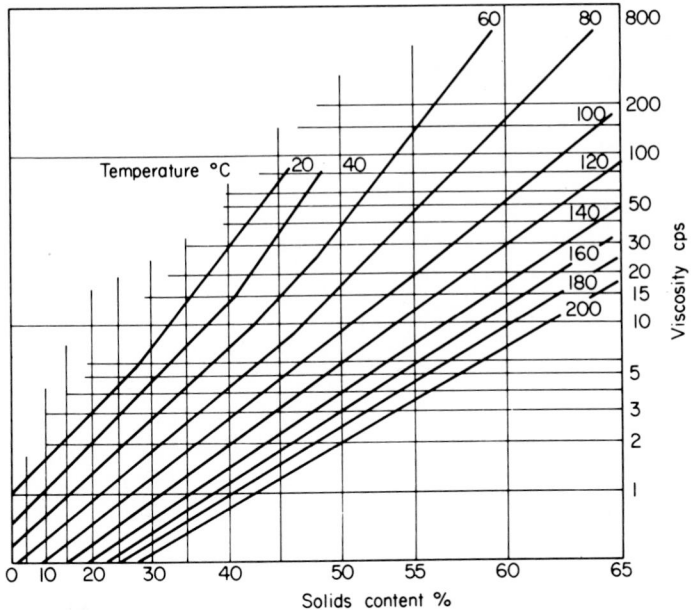

Figure VIII-28. Viscosity as function of black liquor solids content and temperature (166).

on the regeneration process in general. Figure VIII–28 shows the relationship between the solid content, the temperature, and the viscosity of the liquor. The composition of the liquor with regard to its inorganic components causes deviations in the viscosity, especially in the lower temperature region (166).

The following graphs show the variations in the physical data of the black liquor, depending upon the temperature, concentration, density, etc. Figure VIII–29 shows the relationship between the concentration of the black liquor, in °Bé, and the temperature. Figure VIII–30 shows the relationship between the specific gravity at 15°C and the concentration of the black liquor. Figure VIII–31 shows the solid content of the black liquor expressed in kilograms of water per 1000 kilograms of dry substance as a function of the concentration in °Bé at 90°C. Viscosity, specific gravity, and also the concentration prove to be widely dependent not only on the temperature but on each other. Of particular importance for the evaporation process and for the effective temperature gradient in each step of a multiple evaporation system is the increase in the boiling point as a function of the specific gravity. Figure VIII–32 shows the experimental results of three investigators. Even the conversion of Bé degrees into the corresponding specific densities at 15°C show marked divergencies in the different curves. In Fig. VIII–33, the relationships between steam pressure, temper-

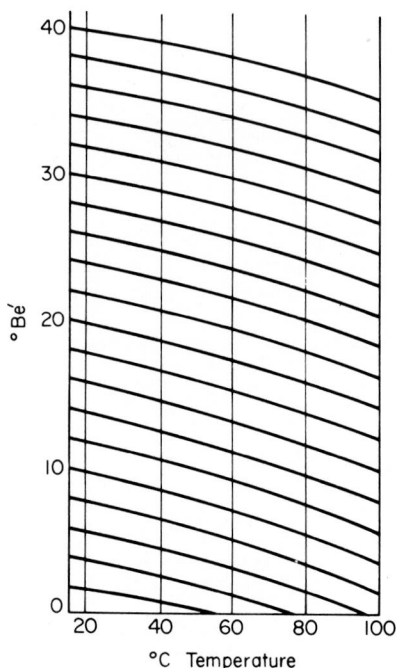

Figure VIII-29. Black liquor Bé degree as a function of temperature (25).

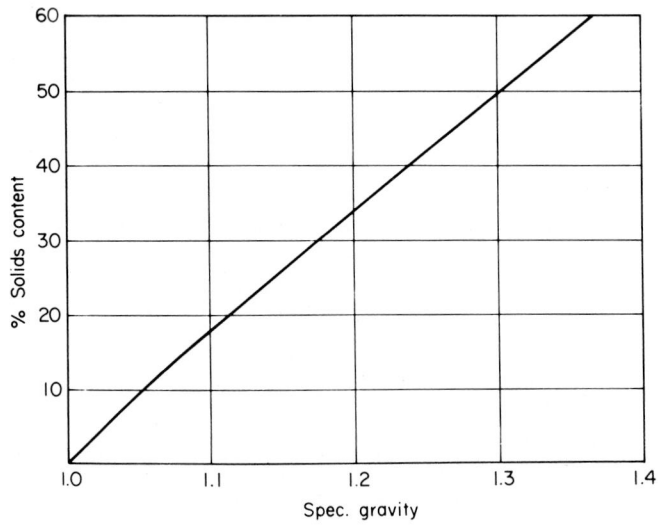

Figure VIII-30. Solids content of black liquor as function of specific gravity (25, 36).

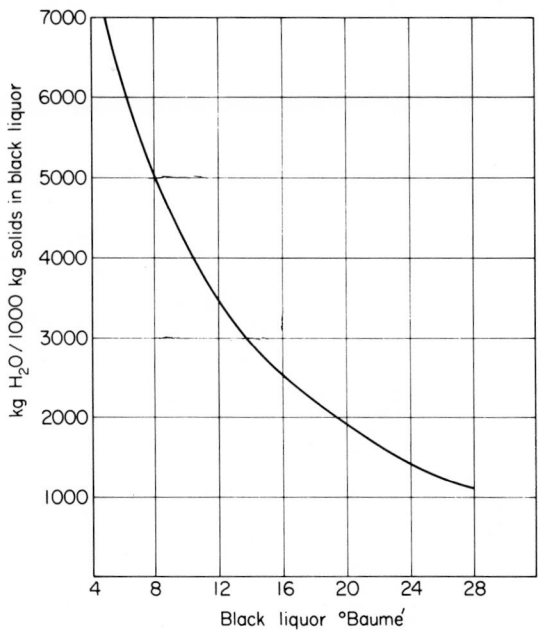

Figure VIII-31. Black liquor solids content in kilogram water to 1000 kilograms solids as function of Bé degree at 90°C (25).

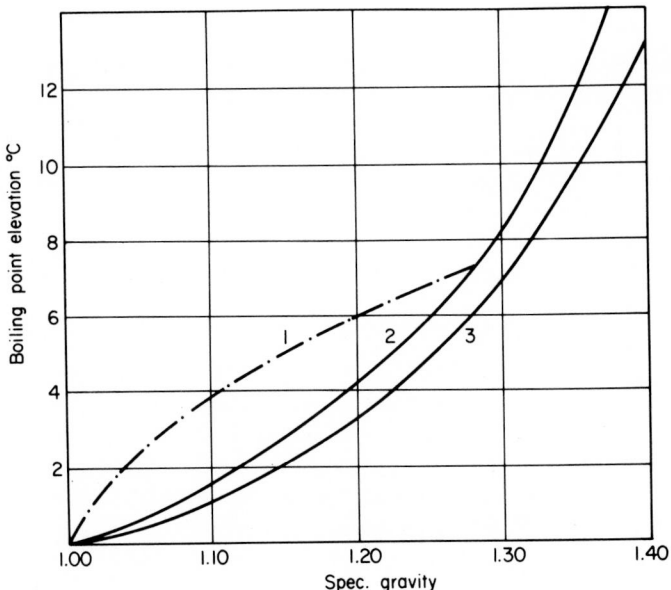

Figure VIII-32. Boiling point elevation as function of black liquors specific gravity. 1 (25), 2 (40), 3 (36).

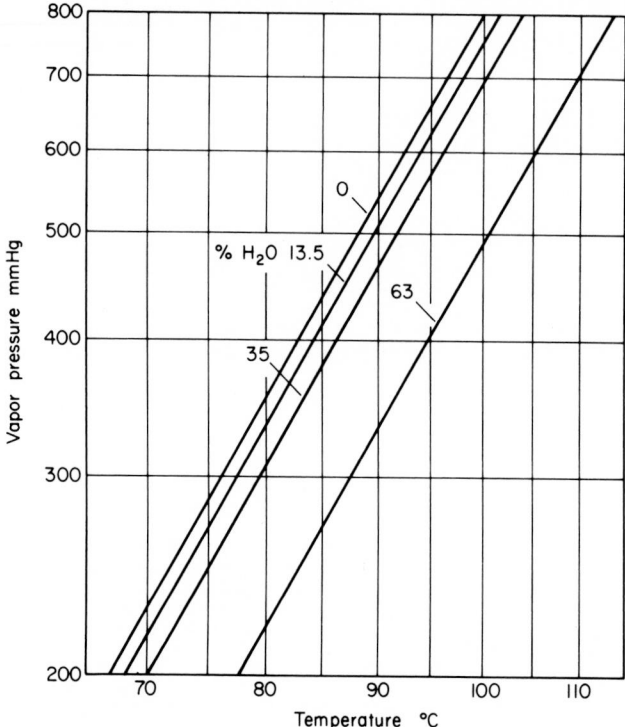

Figure VIII-33. Relationship between steam pressure and temperature at black liquor different solids content (208).

ature, and solid content of the black liquor are given. From this it is obvious to what extent the boiling point increases, especially at high concentrations (25,36,40,166,208,298).

Another important factor in the evaporation of the black liquor is its specific heat, which can be expressed by the following equation:

$$C = 0.98 - 0.0052\,S$$

in which C is the specific heat in cal per kg, and S is the solid content of the liquor in percent. Figure VIII-34 shows the specific heat as a function of the solid content, and Fig. VIII-35, the specific heat as a function of the specific gravity at 15°C. As expected, the specific heat decreases linearly with increasing solid content and specific gravity (25,36,208).

Many contradictory data have been reported on the heat value of the dry substance in black liquor. The average minimum value is given as 3300 kcal per kg dry substance, and the average maximum as 3930 kcal per kg. It seems that the type of preparation used for the sample for the calorific determination has a considerable effect on the results; when the evapo-

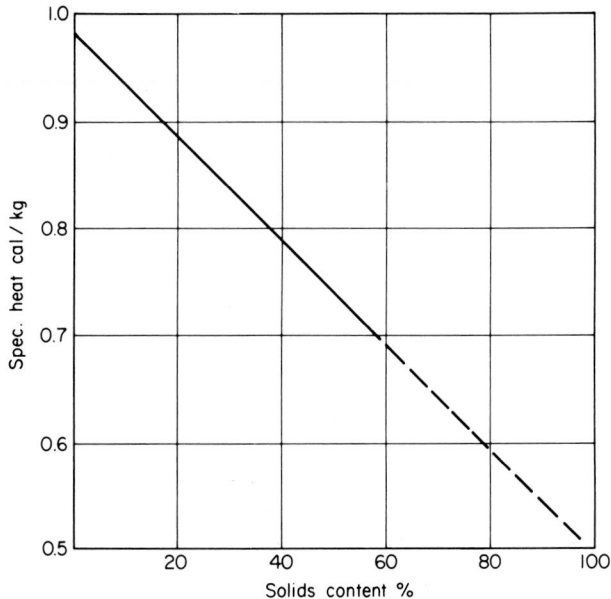

Figure VIII-34. Black liquor solids content as function of specific gravity (25, 34).

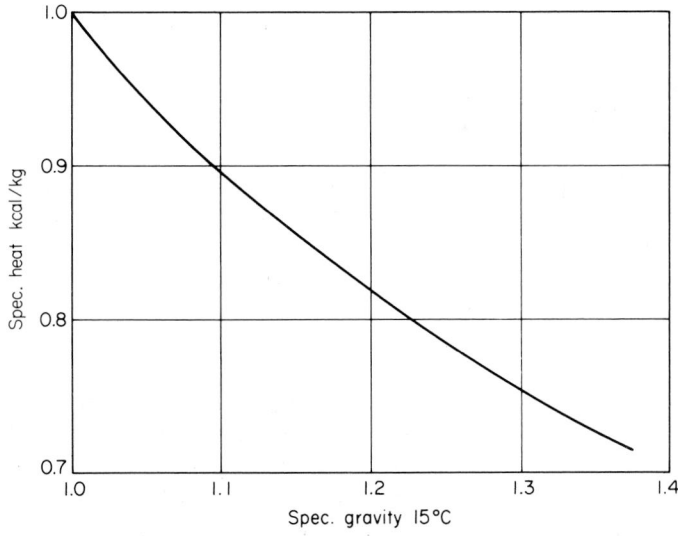

Figure VIII-35. Black liquor specific heat as function of specific weight at 15°C (25).

ration is carried out in a nitrogen atmosphere, the calorific value decreases from 4220 to 3970 kcal per kg, whereas on evaporation in a rotary drum it decreases to 3680 kcal per kg (25). The change in the heat-transfer co-

efficient with the increasing concentration of the solid content of the black liquor also has an appreciable effect on the evaporation. This is shown in Fig. VIII–36.

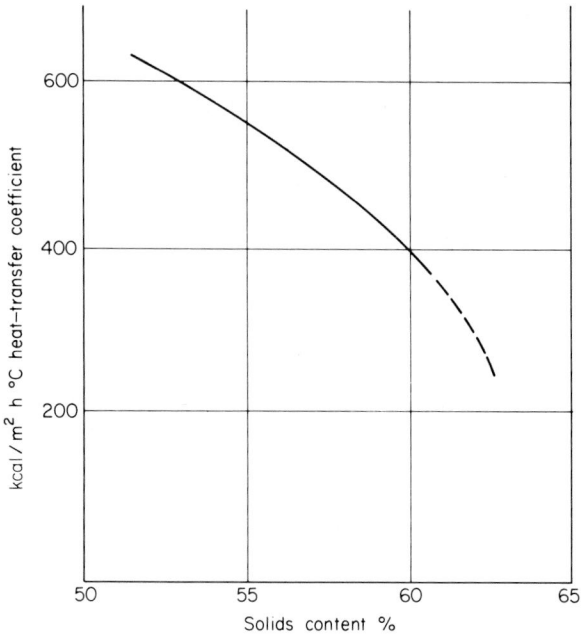

Figure VIII-36. Variation of the heat-transfer coefficient with increasing solids content of black liquor (see 298).

   In addition to the multiple effect vacuum evaporation, other processes, such as back-pressure evaporation and evaporation by thermal compression, are available. These methods have been discussed in Chapter VII, Section 9.2, for the evaporation of spent sulfite liquor.
   An investigation into the evaporation of black liquor by thermal compression has shown that the evaporation of such concentrated solutions, the boiling points of which are considerably above that of the solvent, is not practical. Deposits of solids and scaling hinder the heat transfer. On the other hand, a combination of the multiple evaporator plant and thermal compression seems to be advantageous. Such a combination can be operated in various ways. A thermal compression step for a preliminary concentration of the black liquor from 12% to 14% and from 17% to 18% is advantageous when the further concentration is carried out in a multiple evaporator. The preconcentration can also be carried out to a solid content of 50% when it is followed by a single effect evaporator. In this case the heat efficiency is not favorable because of the great increase in the boiling point of the concentrated liquor.

Another possibility consists in evaporating the liquor, first in a parallel three effect evaporator, then in a thermal compression unit, and returning the steam from the third unit, after compressing it, as superheated steam, to the first unit. In this method, although the compression ratio is substantially higher than in the other two processes, the amount of compressed steam is only about one-third the total amount evaporated (207,339).

A flash evaporator system is reported to be highly economical. The system utilizes three flash units, into the first of which hot liquor is sprayed. The steam from the flash evaporation can be used for the steaming of the wood chips. The residual liquor is released into the second and third flash units and the steam evolved is used in the five-stage evaporation. The black liquor from the third flash unit is passed into a storage tank or one of the

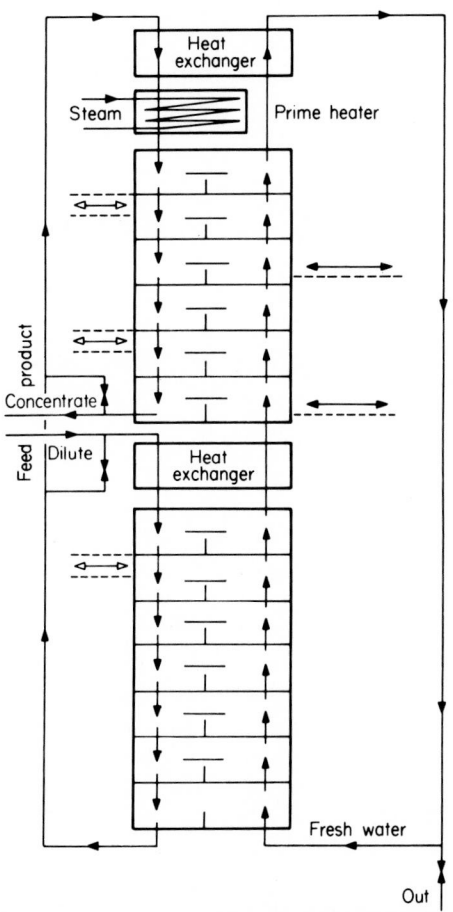

Figure VIII-37. Vapor reheat concentration of pulping black liquors (284).

last stages of the evaporation system. The use of a spray unit instead of an evaporating aggregate is said to effect an appreciable saving in investment capital and in steam consumption (12).

A related evaporation process uses a vapor reheat evaporation, shown schematically in Fig. VIII–37. This system consists of a ladderlike aggregate with 25 to 40 steps, although only a few are shown in the drawing. At the left side of each step, the liquor is flash evaporated, with the steam being condensed on the surface by a stream of fresh cool water pumped up at the right. The ladder is divided in the middle into two parts. At this point, the original dilute liquor, together with the reflux from the lower half of the system, enters the system through a heat exchanger, while the concentrated liquor is withdrawn from the bottom of the upper half. The black liquor and the dilute, less viscous reflux are recycled under reduced pressure in the lower half. The concentrated liquor is circulated in the upper half under superatmospheric pressure. The fresh water stream passes directly from the lower part through a heat exchanger into the upper part. The steam that originates in each step of the ladder is in an equilibrium with the liquor and the fresh water at the right half of the system, in spite of the fact that the evaporating liquor has a higher temperature because of its increased boiling point. The right side of the system, therefore, represents a reservoir for warm water and, from it, water of any desired temperature can be taken off at the appropriate step. A thermal equilibrium and control of the liquor flow can be achieved through the heat exchanger and by regulating the final withdrawal of the concentrated liquor. However, hot water from another part of the mill can be injected into the right side of the corresponding step and can release its heat. Similarly, steam can be injected at the left (or evaporation) side or at the right (or condensation) side of a certain step, or it can be withdrawn at any desired step. This type of operation can be recommended especially when relief steam from a continuous digester is available. One kilogram of steam can evaporate 20 kilograms or more of water. A particular advantage of this system is that no metallic heat-exchanger surfaces are needed and this keeps capital investment low. There are no heating surfaces that can become covered with scale. The elevation of the boiling point of the liquor plays a minor role, and the higher density may even be an advantage. In so far as volatile acids are present in the liquor they may be recovered during the course of the evaporation (284).

In the Bergström-Trobeck process, a concentrated liquor, with 50% to 55% solid content, from a multiple effect evaporator is treated in a concentrator with superheated steam of 425°C. The liquor and the steam are passed into a pressure tank in which a pressure of 10 atmospheres is maintained. The liquor is continuously withdrawn from the bottom of the tank and is concentrated to 85% to 90% solid content by very vigorous evapo-

ration. The almost dry material is then transported to the boiler, and the steam, which contains impurities, is, for example, used for the preheating of the boiler feed water. The calorific value of the organic substances in the liquor is increased to 76%, as compared with 66% in the usual injection-combustion processes, or with 51% in the older evaporation processes (368).

All methods of evaporation that use metallic contact surfaces for heat exchanging are subject to corrosion and incrustation problems. The black liquor contains large amounts of sodium ions which cause a marked decrease in the solubility of sodium sulfate and sodium carbonate salts and the double salt, $2Na_2SO_4 \cdot Na_2CO_3$, which are responsible for the scaling. It may be assumed that the ionic equilibrium favors the formation of the double salt and not that of the single salts. Indeed, the composition of the salt scale corresponds to 64% Na_2SO_4 and 27% Na_2CO_3, the same as the double salt. A temperature of around 100°C is important for the solubility ratio of the salts in the evaporation of the liquor. From calculations for the system $NaOH-Na_2SO_4-Na_2CO_3-H_2O$, it can be deduced that the organic and other anions present in the black liquor exert the same effect upon the solubility of the incrusting salts as the equivalent amount of hydroxyl ions. From this, a critical dry content of 52% of the black liquor is calculated. Although these figures represent only an approximate value, practical experience has shown that the incrustations occur at dry contents of 50% and higher of the black liquor. It has therefore been suggested that the concentrated liquor in the various evaporation stages should be replaced by less concentrated liquor, thus at least minimizing the incrustation (108, 143).

Once scales have been formed, they must be removed from time to time for reasons of heat economy. For this purpose, hydrochloric acid has been found to be effective, but it is very corrosive and therefore cannot be recommended. In the search for other agents, so-called "inhibitors" have been tested. Carbon steel can be treated with phosphoric or sulfamic acid in the presence of effective inhibitors—stainless steel can be treated even without inhibitors. Stainless chrome-nickel steel has a crystalline form known as "austenite"; it is not magnetic but shows a marked resistance to corrosion. On processing, this steel undergoes changes, the small carbon content being released from the austenitic structure, combining with the chromium, and precipitating as carbide as a separate phase along the grain boundaries. The metal then becomes susceptible to corrosion. Hydrochloric acid without inhibitors is especially harmful in places where the metal has been welded. There are numerous inhibitors on the market under special trade names. The presence of chlorides is particularly dangerous because they penetrate the oxide film which protects the stainless steel (79, 86).

The black liquor, with 48% to 54% solid content, coming from the

evaporation plant must be further concentrated before it can be burned under the boiler. If the evaporators used are unable to supply a black liquor of 60% or higher concentration, further evaporation to this concentration must be carried out in a suitable apparatus through direct contact of the liquor with the flue gases of the boiler. The disk evaporators used earlier have been modified by changing the vertically arranged disks to metal sheets which are placed in concentric circles between two vertical end-plates, thus greatly increasing the contact area between the liquor and the gases and assuring greater efficiency in the evaporation. These plate-cascade evaporators (see Fig. VIII–27) process from 100 to 350 tons per day. The temperature of the gases as they enter the evaporator is 320° to 400°C, as they leave, 140° to 165°C. In addition to cascade evaporators, cyclone evaporators are also used. These consist of cylindrical, vertical tanks into which the hot flue gases are blown tangentially and, simultaneously, concentrated liquor is sprayed. The sprayed liquor exposes a large surface to the hot gases; it collects at the bottom of the tank and is returned from there by a pump through spray nozzles to the top of the tank. The concentrated liquor runs down the walls of the tank, takes up the liquor spray, and is collected on a second bottom in the tank. From here it is transported to the mixing tank. Figure VIII–38 shows such a cyclone-type evaporator.

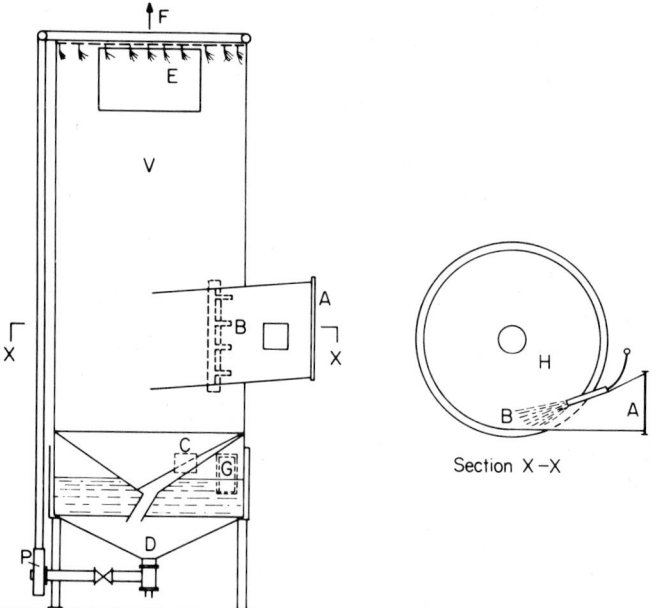

Figure VIII-38. Cyclone-type evaporator. V = vessel; B = gas inlet nozzles; C = conical bottom; D = liquor sump; E = wall washing nozzles; F = gas outlet; G = outlet to salt mixing tank; H = baffle; P = pump for recycling.

The black liquor coming from the cascade or cyclone evaporator has a solid content of up to 60%, a temperature of 80° to 110°C, a density of 1.2° to 1.35°Bé, and a pH of 9.5 to 13.0.

5.3 Combustion of the Evaporated Black Liquor and Addition of the Chemicals

The chemicals used up during the cook, or lost in the washing process, must be replenished. This is accomplished, in part, in the mixing tank where the solid content of the liquor is increased to 70%, largely through the addition of sodium sulfate or so-called "salt cake." It is the addition of the sodium sulfate that gives the kraft pulping process the name of "sulfate process," in spite of the fact that the sulfate does not take part in the cooking reaction. In the burning process, the sodium sulfate is reduced by the carbon regenerated from the organic substances to sodium sulfide, according to the equation

$$Na_2SO_4 + 2C = Na_2S + 2CO_2$$

Under normal operating conditions, 95% of the sulfur that is combined with the sodium is present as sodium sulfide and the remainder as sodium sulfate. The addition of sodium sulfate is not sufficient for the attainment of the requisite sulfidity in the white liquor. The more complete the recovery of alkali—expressed as Na_2O—the less sodium sulfate is needed to be added to the black liquor. If the sulfidity drops below 30%, the sulfide content in the black liquor decreases strongly, while the loss of hydrogen sulfide also becomes less during the evaporation. The formation of sulfur dioxide in the melt furnace decreases and the fumes consist chiefly of sodium carbonate and sodium sulfate. If the flue dust is recovered and returned to the regeneration cycle (mixing tank, in Fig. VIII-27), the soda-sulfate mixture, poorer in sulfur, when used instead of pure sulfate causes a further reduction in the sulfidity of the liquor (165).

Many mills with highly effective regeneration plants have therefore changed and now use elementary sulfur for the maintenance of the sulfidity. There are two possible ways of doing this: (*a*) the sulfur can be added to the white liquor before it is passed into the digester; (*b*) the sulfur can be added with the sulfate to the melt furnace. Neither method is entirely satisfactory. It has long been known that sulfur reacts, under certain conditions, with sodium hydroxide and sodium carbonate to form hydrogen sulfide and sodium thiosulfate. The former can be absorbed in the white liquor and used for the adjustment of the sulfidity. Thiosulfate can be added directly to the white liquor or to the black liquor before it enters the melt furnace. The reaction of the sulfur with alkali under pressure and at higher temperatures can be expressed by the following equations:

$$3S + 2H_2O \rightarrow 2H_2S + SO_2 \tag{1}$$

$$SO_2 + 2NaOH \rightarrow Na_2SO_3 + H_2O \tag{2}$$

$$Na_2SO_3 + S \rightarrow Na_2S_2O_3 \tag{3}$$

The overall reaction may be expressed by the equation

$$4S + 2NaOH + H_2O \rightarrow 2H_2S + Na_2S_2O_3 \tag{4}$$

With sodium carbonate, sodium sulfide, or lime, the following reactions take place:

$$4S + Na_2CO_3 + 2H_2O \rightarrow 2H_2S + Na_2S_2O_3 + CO_2 \tag{5}$$

$$4S + Na_2S + 3H_2O \rightarrow 3H_2S + Na_2S_2O_3 \tag{6}$$

$$3S + Ca(OH)_2 + H_2O \rightarrow 2H_2S + CaSO_3 \tag{7}$$

The choice of the most suitable alkali for the reaction with sulfur depends in every case upon the location of the mill. With sodium hydroxide or sodium carbonate as the alkaline medium, about half of the sulfur is converted into thiosulfate. If the presence of sodium thiosulfate does not interfere in the cooking process (see Section 1.3) it can be added directly to the white liquor; otherwise it is added to the black liquor and goes with the latter into the melt furnace (165, 237, 258). Opinions are divided as to the best place for the addition of the sulfur. Direct addition to the white liquor undoubtedly provides a simple method for the adjustment of the sulfidity. Addition to the furnace in a molten state has the advantage, besides the above-mentioned chemical considerations, that it can be carried out in a semicontinuous manner. In every case, the addition should be made as close as possible to the melt bed to assure a safe operation and an accurate metering (285, 289, 375).

In very big kraft mills with several furnaces, the transportation of large amounts of salt cake to the places destined for its use sometimes gives rise to difficulties. A hydraulic distribution system has been developed in which the sulfate is slurried in weak black liquor to give a concentrated suspension which is transported by pumps to the various places. This suspension can then be added by means of metering pumps in the required amounts (57). Different techniques present themselves when a sulfite pulp mill, using sodium or ammonium cooking liquor, and a kraft pulp mill are combined. In such a case the spent liquors of both mills can be processed together (103). Finally, the suggestion should be mentioned that the sodium sulfate obtained in the preparation of chlorine dioxide could be used for the makeup of the sulfidity. This presupposes that chlorine dioxide is used in at least one stage of the bleaching of the kraft pulp. It must then be decided on an individual basis whether the supply of chlorate or sulfate is more advantageous for the mill (296).

The heart of the entire chemical regeneration plant is the smelting hearth, which is more appropriately called the spent liquor steam boiler. Its purpose is not only to convert the salts present in the black liquor into a melt, but, more important, to burn the organic substances and, in this way, to provide the energy and heat requirements of the entire mill. Such boilers must assure a high degree of recovery of chemicals, a high production of steam, a high heat efficiency, and a high degree of safety in operation. The older boilers were spread-out units with a daily output corresponding to a production of 20 to 100 tons of pulp. The steam boiler was usually preceded by a rotary incinerator in which the burning of the organic substances took place, while the molten inorganic components were

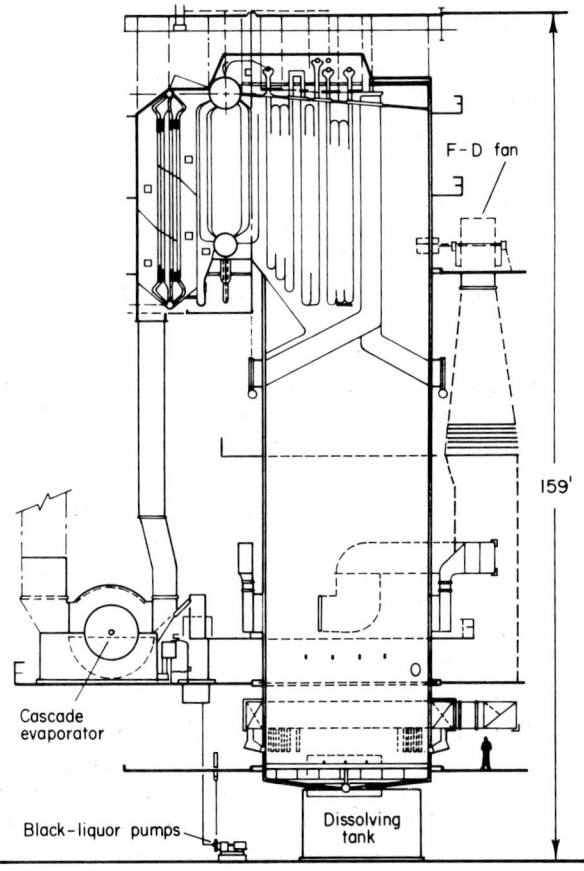

Figure VIII-39. Current 1000 ton basic recovery unit (338). (Reproduced from *Paper Trade Journal* by permission of Lockwood Publishing Co.)

collected in a smelter boiler and its heat was used for the regeneration of steam. The increase in daily production to several hundreds or even thousands of tons necessitated completely new boiler designs, in which the rotary oven, the melt hearth, and the steam regenerator were combined in one single unit. Up-to-date boiler units for a daily production of 1000 tons of pulp are in operation and it is impossible to say that this constitutes an upper limit to the capacity of such boilers. The advantages of such large units consist, first of all, in their high degree of economy, in the reduction of the labor force, and in a better control of operation. Figure VIII–39 shows such a recovery unit that permits the throughput of 1350 tons of dry substance per day from a liquor containing 65% to 68% solid content. The daily steam production reaches 225 tons at 42 atmospheres pressure and 380°C at the superheater steam outlet. To assure trouble-free operation, 32 soot blowers are used, so the yearly inspections with maintenance shutdowns are limited to two. A tangential arrangement of the boiler tubes assures a high heat economy and the flame direction creates maximum turbulence. Primary air is admitted at the bottom of the furnace and secondary air, in a tangential direction, above the fuel injection point. The melt is collected on a flat hearth. Difficulties in the take-off of the melt increase with the size and efficiency of the boiler, and the number of melt spouts must constantly be increased. In the place of one smelt dissolution tank in front of the furnace, two tanks, one on each side, are installed and the number of melt spouts is increased to such an extent that even the stoppage of one or more of them will not interrupt the operation. The number of injection nozzles for the black liquor also has to be greatly increased. For a unit of a mill producing 800 tons of pulp per day, twelve liquor injection guns are used (338).

The boiler is often the bottleneck of the entire regeneration plant and it is therefore necessary that the throughput of the solids should be carefully controlled. To reduce the temperatures in the carbonization zone and, with it, to lower the sublimation of sodium salts, the solid content of the injected liquor can be somewhat lowered, but it should not go below 64% since, if it does, the combustion will be incomplete. The supervision and control of the measurements of the air distribution in the combustion chamber are also important. The main difference between the American and Canadian and the Scandinavian designs is that, in the latter, more of the heat from the combustion of the black liquor is recovered as steam by the use of economizers, thus cooling the flue gases to about 260°F (127°C). The construction of new boiler units in the Scandinavian countries is carried out on a more conservative scale and is adjusted to the normal increase in pulp mill capacity by marginal investments. In Sweden, no direct-contact type evaporators are used and, to get satisfactory operation of the econ-

omizers, hot precipitators are placed between them and the boiler (94, 376).

One of the most dreaded occurrences in the operation of the boiler is the latent danger of an explosion. There are two main causes for such explosions: (a) the failure of the ignition of the injected liquor without its flow being immediately shut off; (b) contact between the melt and water. In the first case, error on the part of the operator, or malfunction of the auxiliary equipment is the likely cause. Older plants are usually equipped with so-called hearth torches which are used to ignite the liquor at the beginning of the operation of the boiler or in the event of occasional ignition failure. Such installations are inadequate for modern boilers because they do not operate automatically and a visual check is difficult. Newer boilers use a differential pressure system, but even this does not ensure a completely safe operation. In the United States, a black liquor advisory committee regularly undertakes a study of safety regulations and reports its findings. A principal requisite is that no water other than the black liquor should be added to the boiler and the concentration of the liquor should be adjusted in such a way that an immediate and complete evaporation of the water in the liquor is assured as the first step in the subsequent combustion. Devices must be installed to provide constant control of the concentration of the black liquor. Cooling with water of the auxiliary apparatus of the boiler should, whenever possible, be avoided and, when this is impossible, such cooling devices should operate under vacuum or very low pressure. Salt cake slurries and tall oil soaps are sources of additional water for the boiler and therefore should be eliminated, or they may be added to the black liquor and then be automatically checked through the control of the concentration (101,337,383).

Explosions that occasionally occur during the dissolution of the melt may be traced to the possible formation of elementary sodium during the combustion process. An examination of the chemical reaction equilibriums as a function of the combustion temperature has shown that the formation of elementary sodium is especially likely to occur when the boiler is operated at supernormal temperatures and the escape of the gaseous reaction products is permitted. Explosions can also originate from water leaking either inside the boiler or from cooling water pipe lines and coming into direct contact with the smelt hearth. There is no clear proof at this time as to whether these explosions are caused by gas explosions when water comes in contact with the smelt, or by explosive steam formation. However, it is certain that explosions do not always occur when water comes in contact with the smelt. It is possible that the degree of distribution in which the smelt is added to the solution tank is the cause of these explosions. Dispersion devices have therefore been suggested to ensure a fine distribution of the smelt before it comes in contact with the water (75, 84, 242).

5.4 Oxidation of the Black Liquor

Losses of sulfur during the evaporation of the black liquor and through the escape of sulfur-containing compounds with the steam are known to occur. When black liquor is exposed to the air for a prolonged time at 20°C, the hydrogen sulfide and mercaptans decrease as follows: from 1.180 grams H_2S and 1.108 grams CH_3SH per liter to 0.423 grams and 0.051 grams after 68 hours, and to 0.059 grams and 0.00 grams, respectively, after 260 hours. The oxidation occurs considerably more rapidly at higher temperatures (80°C) as follows (where CH_3SH calculated as H_2S is included):

Time (minute)	H_2S (g/liter)
0	1.99
5	1.02
10	0.38
15	0.14

That an oxidation occurs is shown by the fact that, when the experiments were carried out under a nitrogen atmosphere, practically no changes in the contents of these sulfur compounds occurred. The oxygen absorption at 100°C was as follows:

Time (minute)	Oxygen absorption (ml/ml black liquor)
1	1.00
5	2.05
30	2.05
60	5.20
90	9.60

The oxidation of the sulfide sulfur therefore occurs very rapidly. After a longer reaction period and at a higher temperature, oxygen is obviously consumed in other oxidative reactions. This observation had led to the idea of using the oxidation of the sulfur compounds in the black liquor before the evaporation to decrease the loss of sulfur and, simultaneously, to eliminate the pollution of air by hydrogen sulfide and mercaptans. The oxidation reaction of the sulfide sulfur is exothermic and takes place according to the following equation:

$$2Na_2S + 2O_2 + H_2O \rightarrow Na_2S_2O_3 + 2NaOH + 215.3 \text{ cal}$$

The heat formed during the oxidation of the black liquor amounts to 2% of its calorific heat value at a sodium sulfide content of 1 gram per liter. At this stage of oxidation, organic substances are oxidized only to a small extent or not at all. If the oxidation of the sulfide is carried out to com-

pletion, the heat formation increases to 3.6% of the calorific value. The chief oxidation product is thiosulfate, which is finally converted into sulfate. Polysulfide and polythionate are also formed under these reaction conditions and at temperatures of 70°C (29, 30, 230, 369).

The kinetics of black liquor oxidation has been investigated over a temperature range of 50° to 90°C. The rates of oxidation were found to vary in a complex manner, depending upon the partial pressure of oxygen, the concentration of sodium sulfide liquor, the rate of liquor agitation, and the chemical reaction taking place under the prevailing experimental conditions. The rate of oxidation of a weak black liquor at sodium sulfide concentrations of above 2.5 grams per liter reaches its maximum at a temperature of between 61° and 71°C. Although maximum rates occur at the indicated temperatures, the change in rate with a change in the temperature is very slight. Under some conditions the formation of elemental sulfur is favored by a low temperature, a low sulfide concentration, and a high oxygen pressure. The sulfur formed goes into solution during storage and regenerates some sulfide. This reversion of oxidized sulfide can be avoided by the application of the proper operating conditions (264a).

A comparison of the losses of sulfur in the evaporation of oxidized and nonoxidized black liquor is clearly in favor of the oxidized liquor. The loss of nonoxidized liquor amounted to 15.5 kg hydrogen sulfide, corresponding to 64.8 kg sodium sulfate, per ton of black liquor; with oxidized liquor of the same origin and composition, the loss was only 5.3 kg, corresponding to 22.2 kg sodium sulfate. The oxidation reaction occurs in two steps. As already mentioned, the sulfide sulfur is readily oxidized. It is important that the liquor to be oxidized should be distributed over a large surface and should be readily accessible to the oxygen. The effect of the temperature between 60° and 90°C is relatively small, because the partial pressure of the oxygen drops and that of the water vapor increases. One volume of black liquor requires 1 to 3 volumes of oxygen or 5 to 15 volumes of air. In the praxis, however, larger amounts of air are required. Sodium sulfide and methyl mercaptan are more readily oxidized in the black liquor than in aqueous alkali solution because the organic matter in the liquor obviously exerts a catalytic action (31, 400, 401).

The velocity of the oxidation is practically independent of the sulfide content of the black liquor. During the oxidation of 80% to 90% of the sulfide, 130% to 160% oxygen is consumed in the formation of thiosulfate, while the pH value of the liquor remains unchanged. Only after continued oxidation, involving a large amount of oxygen, does the pH value decrease. The use of oxidized liquor has the disadvantage that the evaporator tubes may become plugged and 3% to 6% of the heat available for the later production of steam is formed already in the oxidation unit (378).

Figure VIII–40 shows schematically a combined unit, consisting of an

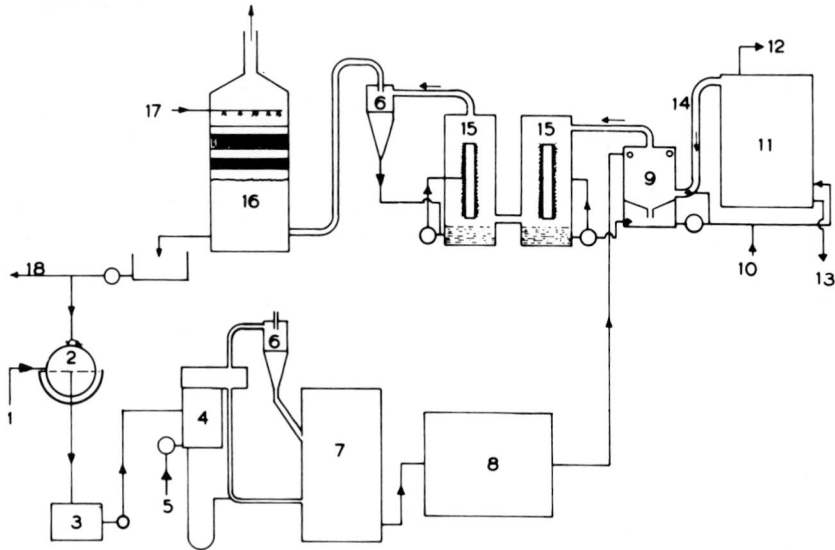

Figure VIII–40. Trobeck-Bergström-Tomlinson system for the oxidation of black liquor and secondary recovery of heat and chemicals (364). 1 black liquor from digesters and blow tank; 2 brown stock washer; 3 storage tank; 4 oxidation tower; 5 air supply; 6 cyclone; 7 foam tank; 8 multiple effect evaporator; 9 cyclonic evaporator; 10 salt cake addition; 11 recovery furnace; 12 steam to process; 13 smelt to causticizing; 14 flue gas; 15 deluge towers; 16 condensing tower; 17 cooling water.

oxidation tower, a multiple effect evaporation aggregate, a cyclonic evaporator, and a black liquor furnace. The black liquor that is not returned to the digester or the relief tank is passed into the oxidation tower. After oxidation, the black liquor is concentrated in the evaporation plant to a solid content of 50% to 55%. The concentrated liquor is then sprayed through nozzles into the cyclonic evaporator and exposed to the flue gases from the boiler. The black liquor is injected into the gas stream in such an amount that the temperature of the gas remains 20° to 25°C above its dew point. The gas stream is then passed into deluge towers where it is freed of suspended matter. The liquor is drawn off at such a rate that a concentration as uniform as possible is assured. The liquor from the deluge towers is added either to the original liquor before it reaches the evaporation plant, or directly to the cyclonic evaporator, depending upon its concentration. The concentrated liquor then goes from the cyclonic evaporator to the recovery furnace. The advantages of this combination process are a greater production of steam in the boiler, due to lower exit-gas temperature, and the additional production of considerable amounts of warm water. The advantages of the separation of the suspended matter in the flue gases and the far-reaching elimination of air pollution will be discussed in detail in the

following section (364). Finally, the advantages of the oxidation of the
black liquor are: the reduced corrosion within the evaporation plant, a
reduction in the sulfur losses, low hydrogen sulfide formation, improved
combustion in the liquor furnace, reduction of air pollution, simplification
of the caustification, increased sulfidity, and reduction in the lime consumption during the caustification.

Certain disadvantages counteract, in part, these advantages. The oxidation of dilute black liquor involves high investment costs, and the formation of foam causes difficulty. It has therefore been suggested that the
concentrated black liquor, with 44% solid content should be oxidized.
Figure VIII–41 shows schematically a unit for the oxidation of concentrated

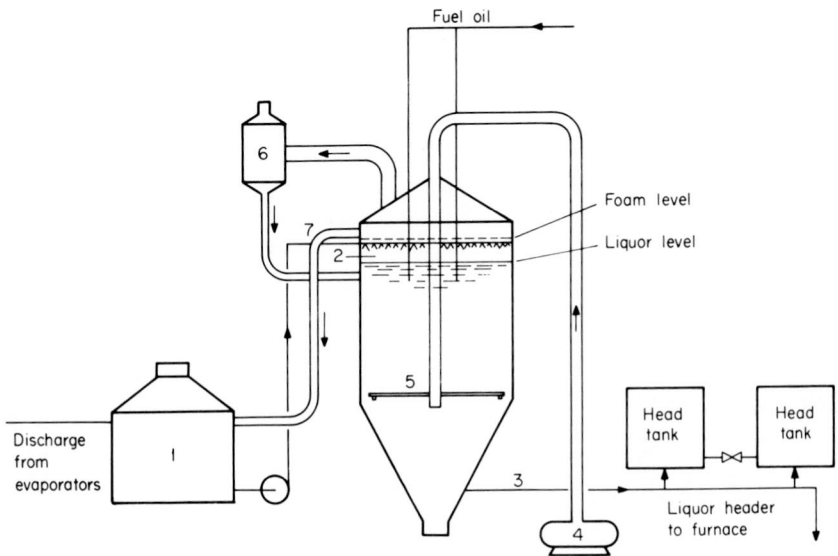

Figure VIII–41. Concentrated black liquor oxidation unit (222). 1 heavy liquor storage
tank; 2 liquor inlet nozzles; 3 oxidized liquor outlet; 4 blower; 5 air sparger; 6 cyclone
separator for air discharge; 7 overflow.

black liquor. The latter is sprayed from a storage tank by a pump on the
top of a pool of liquor about 12 feet high (3.6 meters) in a cylindrical oxidation tank. Air is blown in below the surface of the liquor through an
air sparger. The oxidation requires about 2.5 hours. The oxidized black
liquor is withdrawn from the conical bottom of the oxidation tank for
further concentration and final combustion. The foam problem is eased to
a considerable extent by the use of concentrated black liquor and can be
controlled by the addition of small amounts of fuel oil. The efficiency of
the unit is reported to be 98% to 100%. The density of the liquor increases
by about 2°Bé. The loss of sulfur in the form of hydrogen sulfide is de-

creased by 90%, but in the form of sulfur dioxide it is increased by 15% (77, 222).

Black or brown liquors which have been concentrated to about 50% solids can be fed into a fluidized bed-type roaster. The exhaust gas, containing steam, combustion products, sulfur dioxide, and some entrained solids, is passed into a cyclone and from there into a wet scrubbing system. Solid particles are removed and returned directly to the fluidized bed. The addition of other fuel is usually unnecessary since the combustion process is thermally self-supporting. If sodium is used as a base, a temperature of 650°C may be used (84a).

The advantages of the presence of polysulfide in the cooking liquor are partially nullified by the difficulties encountered in the recovery of polysulfides or sulfur and this has delayed the introduction of the polysulfide cook in the praxis. Sulfur or polysulfide can be recovered only in reduced form, as sulfide, as which it is accumulated in the system until the increased loss of chemicals is compensated by the addition of sulfur or polysulfide. Small additions of polysulfide can increase the corrosion by cathodic depolarization, but greater additions can prevent the corrosion by passivation of the digester walls. The formation of polysulfide during the oxidation of the black liquor can be greatly increased by the oxidation in a mixture with white liquor. In the black liquor the lignin is the first component to be oxidized. The oxidation products, possibly of partially quinoid character, may then affect the oxidation of the sulfide to polysulfide. At the start, thiosulfate and polysulfide are formed in about equal amounts from the sulfide, but with increasing oxidation the formation of thiosulfate increases and that of polysulfide decreases. The polysulfide concentration reaches its maximum when about 50% of the sulfide is oxidized. In this way, a polysulfide-containing cooking liquor should be obtained when a suitable mixture of white and black liquors is oxidized before it is added to the digester. The greater the sulfide content in the mixture, the greater will be the polysulfide content after the oxidation (221).

The differences in the efficiency of such black liquor oxidation plants that have been observed in various mills can be traced, in part, to differences in the composition of the black liquor. One of the main problems that should be solved by the oxidation of the black liquor is the prevention of odor formation and, connected with this, a decrease in the sulfur losses. There can be no doubt that further improvements in the design of such oxidation aggregates are to be expected, although the basic knowledge that has already been acquired from the plants now in operation is responsible for the fact that, today, black liquor oxidation is an integral part in the design of new mills and in the expansion of existing ones. So far as the odor problem is concerned—and this will be dealt with in greater detail later—it is known that an overloaded black liquor furnace can evolve several tons of

hydrogen sulfide daily and the relief gases can produce several hundred kilograms. With a well-controlled plant and loading of the black liquor furnace, the oxidation of the black liquor can largely eliminate air pollution (144).

5.5 The Recovery of Chemicals from the Flue Gases

The flue gases from the black liquor furnace carry with them many different kinds of particles. The larger ones are precipitated in the flue. Precipitations must be removed from the inside of the furnace, especially from the boiling tubes, because they would have an unfavorable effect upon the heat transfer. For this purpose, soot-blowers of various designs and arrangements are used. The addition of fuel oil has been suggested to facilitate the removal of the precipitations. The injection of a slurry of magnesium oxide through the soot-blowers should make the removal of the precipitate easier. The fume particles may be black or white, depending upon the degree of combustion. Their diameter lies between 0.2 and 0.3 μ and millions of them may be contained in one cubic centimeter. If these particles are not recovered from the flue gases, the loss of sodium in a kraft pulp mill would amount, on an average, to 22.5 kg sodium oxide, corresponding to a loss of 50 kg sodium sulfate per ton of pulp. To separate the suspended particles from the flue gases, electrostatic gas purification and a gas wash with scrubbers are used. The latter serve not only to remove the suspended particles but also to cool the gases, utilize the heat, and absorb the volatile sulfur compounds (67,350).

In the electrostatic gas purification, the aerosols are exposed to an electric field of a rapidly pulsating alternating current of 50,000 to 100,000 volts. The charging and discharging of the aerodispersoids are achieved by ionizing electrodes, whereas the precipitation occurs on collective electrodes with relatively large surfaces. The efficiency of a gas purification system depends upon the load that is passed through per time unit, and upon the composition and temperature of the gases. To allow the electrostatically charged particles to release their charge, a certain degree of humidity of the gas is necessary. If the gas leaves the boiler at about 230°C, the precipitate is almost dry. When the gas leaves the scrubber at 75° to 120°C, the powdery precipitate is rinsed off from the electrodes. When the precipitation electrodes consist of tubes, the gas coming from a scrubber at 110°C is taken up by a film of water that is formed on the inner surface of the tubes. The power consumption of such an electrostatic gas purification is relatively small and amounts to about 4 kwh per ton of pulp. The concentration of the dust at the entrance of the gases may amount to 4 to 15 grams per cubic meter, at the exit, to 0.13 to 0.80 grams per cubic meter. The velocity with which the charged particles move from the emission to

the collective electrode is governed by the voltage gradient (kV per cm). This voltage gradient depends upon the shaking device of the collecting electrode and the relative humidity content of the gas. If the velocity of the gas surpasses a certain optimum rate, the efficiency decreases. The electrodes must be made of corrosion-resistant material, and the isolators of the emission electrodes must be installed outside the gas stream to avoid flash-overs. As already mentioned, for a trouble-free and effective operation careful supervision of the load on the aggregate is necessary. This load is determined by the volume of the gas passing through the aggregate per time unit and by the amount of aerodispersoids in the gas. To determine the latter, a bolometer is used; it measures the amount of light that passes from a light source through the gas stream without being diverted or absorbed by the solid particles. The results are taken up by a receiver and graphically recorded. A linear relationship between the amount of dust in the gas and the light absorption is found and is expressed in weight units per gas volume (87,124,308,309).

To recover the heat and utilize the volatile sulfur compounds in the flue gases, a wet purification by scrubbers is employed. Such scrubbers consist, in principal, of upright cylindrical chambers into the lower part of which the gas is blown at a high velocity in a tangential direction while water is sprayed under pressure from a cylindrical shaft supplied with numerous spray nozzles. The water takes up the flue dust, partly by dissolving it and partly as a suspension, it is drained off at the lower part and returned by a circular route to the system until a certain degree of saturation is reached. Electron optical investigations have determined the distribution of the particle size of the aerodispersoids to be as follows:

smaller than 0.5 μ, 16.8%

0.5–1.0 μ, 53.2%

over 1 μ, 30%

Since 70% of the flue dust is below 1 μ in size, a certain reaction time [according to the diffusion theory (81,206)] is required to allow the liquid droplets, which are large as compared with the aerodispersoids, to come into contact with the small particles. It can be assumed that the droplets are surrounded by a streamline net of the gas which greatly minimizes the likelihood of collisions. Strong turbulence is created by a Venturi atomizer and the streamline net is disturbed. Simultaneously, the absorption liquor is injected at the narrowest part of the Venturi tube. The time required for the reaction is reached by the subsequent use of a retention tower. In this way, up to 99% of the suspended matter may be separated out. If the Venturi atomizer precedes the scrubber, the retention tower can be eliminated. The advantage of such a Venturi atomizer is that a recirculation of the absorption liquor, as with a normal scrubber, is avoided.

Thus the filtration of the solution, which otherwise is necessary to avoid the plugging of the nozzles in the scrubbers, can be omitted. Another advantage is that the investment costs are considerably lower than those of the electrostatic gas purification system. A disadvantage is the saturation of the flue gases with steam when the heat cannot be recovered from them (81, 235,348).

If a wet gas purification, as a second step, follows an electrostatic gas purification, devices with a relatively low pressure drop permit a 50% to 80% reduction in suspended matter in the gas stream coming from the electrostatic precipitation. This amount is somewhat less when the device is used after the Venturi atomizer. The separation capacity is found to be independent of the type of scrubbers used, but directly dependent upon the load of suspended materials in the gases before they are passed into the scrubber (37).

In recent years laws and regulations which require the reduction of the amount of air pollutants have been enacted. This has necessitated an improvement in the efficiency of the overall collection and removal of the odor of such pollutants and has led to the development of a two-stage evaporator-scrubber system. Such a system includes a more efficient and maintenance-free Venturi scrubber and provides a greater degree of thermal efficiency, greater dust collection efficiency, minimum maintenance, and highly desirable flexibility of operation (326a).

In another newer installation, a multiple Venturi gas-scrubber is combined with three cyclone evaporators and two electrostatic precipitators. This system removes about 99% of the dust emitted from the recovery furnace (399a). In plants where there is a demand for hot water, heat exchangers have been used to supply clean water from the heat recovered from the hot flue gases by gas scrubbers. The heat exchangers represent a large part of the investment and also cause trouble with scaling. In a more recent development, the flue gases are washed free of soot and dust before entering the heat-recovery stages (376a).

5.6 Clarification and Caustification of the Green Liquor

The melt from the smelting hearth is dissolved in water. The solution has a greenish color due to the presence of sodium-ferrous sulfide, ferrous sulfide, and ferrous cyanide—hence its name, "green liquor." The sulfur-iron component is present in a colloidal state. The carbonate present in the solution is converted into sodium hydroxide through the addition of lime, according to the following reaction:

$$Na_2CO_3 + CaO + H_2O \rightleftharpoons 2NaOH + CaCO_3$$

The reaction is reversible and occurs only to the extent of 85% to 90% under the most favorable conditions. The solubility of the calcium carbonate

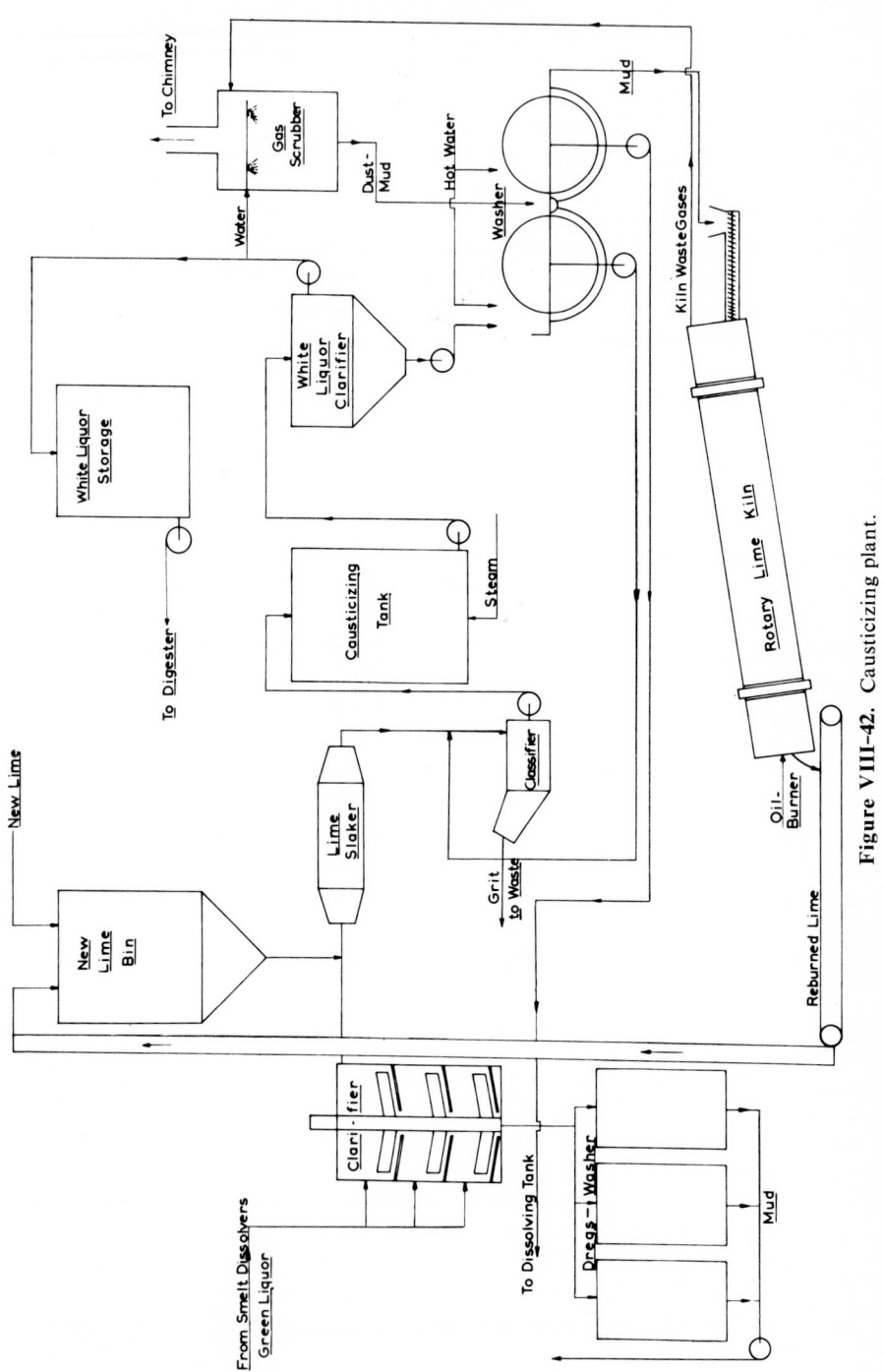

Figure VIII-42. Causticizing plant.

increases with increasing content of OH ions and shifts the reaction equilibrium toward the left side. The lime particles react first on their surface and are covered by a layer of calcium carbonate during the reaction, and this has an unfavorable influence on it.

Figure VIII–42 shows a flow-sheet of a causticizing plant. The green liquor is passed into a clarifying tank to separate the undissolved particles by sedimentation. Such a settling tank is often divided into several levels. The liquor enters at the top and is uniformly spread over the various levels. The sludge, after it has settled, is moved by paddles toward the central hollow axle to which they are attached, and runs through the shaft of the axle into the sludge-washing tank. The clarified liquor runs from the upper part of the clarifier through an overflow into the lime slaker. Similar installations are also used for the clarification of the white liquor.

The clarification of the green liquor can be improved by the addition of lime, but this increases the difficulties encountered in the washing of the sludge. The clarification, because of the kinetics of the reaction, is a time-consuming process and experiments have been carried out to facilitate and accelerate it. Green liquor cannot be clarified by centrifugation. The addition of coagulants, such as magnesium sulfate or magnesium chloride, does not provide any basic improvement. The sedimentation rate can be represented as a function of concentration and flow rate of the solid particles per unit of area. Solid flux is then a function of the concentration and can be determined from batch-settling experiments. Hindered sedimentation is then defined as sedimentation at concentrations greater than that at the point of inflection of the curve which shows the solid flux as a function of the concentration (197,209,297).

A system consisting of a combination of lime-slaker causticizer and a white liquor filter represents a substantial improvement. Besides the mud filter, only three operation units and three storage tanks are used; the latter are a green liquor, a weak liquor, and a white liquor tank. The weak liquor, coming from the wash zone of the white liquor filter and mud filter, flows to the melt-dissolving tank. Because a part of the lime mud from the caustification and calcination systems is taken away, the clarification of the green liquor can be omitted without running the danger of increasing the amount of nonreactive material in the system. Simultaneously, the washing of the dregs becomes unnecessary. Heated green liquor and hot recovered slaked lime are added in controlled amounts at the top of the slaker-causticizer. The latter has a diameter of 11 feet (3.35 meters) and a height of 31 feet (9.45 meters). A thorough mixing assures adequate circulation. The liquor, containing grit and other coarse particles, is withdrawn from the bottom and pumped over a vibrating screen, and the filtered liquor is returned to the causticizer. The latter, itself, is divided into several compartments through which the liquor, mixed with lime, flows from the top to the

bottom. From here, the completely causticized white liquor runs over a level control box and then to the white liquor filter. The reaction time in this combined causticizing section amounts to 130 minutes. The filtration of the white liquor is carried out on a special vacuum filter. A similar filter is used for the washing of the lime mud where the extraction is carried out to a sodium oxide content of 0.5% to 1% and a solid content of 55% to 75%. The lime mud is discharged into a pug mill and is then calcinized (217).

Several methods have been suggested for making the caustification plant simpler and more compact; for instance, the lime-mud wash can be replaced by a precoat filter. The operation of the white liquor clarifier and the lime-mud wash can be carried out on a belt filter, although such filters are complicated and difficult to handle. A new device is the pressure filter which can be used for the filtration of the white liquor, whereas the washing of the lime mud and its dewatering are carried out on a precoat filter (42).

If the slaking of the lime and the caustification are not carried out in one operation, as just described, the burned lime coming from the calcination kiln must be hydrated before it is added to the green liquor. According to one suggestion, differing from the process shown in Fig. VIII–42, the regenerated burned lime from the kiln is carried by a conveyer directly to the primary slaker, while the overflow from the latter goes to the secondary slaker into which the requisite additional new lime is also introduced. In this way, the vertical elevator to the lime storage bin becomes superfluous, the preheater for the green liquor can be dispensed with, and the addition of the new lime can be readily automated and controlled. A disadvantage is the lower storage capacity in the event of an interruption in the operation of the lime kiln. For the caustification of the green liquor, also, a similar two-step process can be used, in the first step of which a considerable excess of lime is used, and the lime mud obtained from the decantation is used for the caustification in the second step. The slaking of the lime is of special importance because the size and the shape of the lime particles exert a lasting effect on the reactivity of the lime (139,140).

Following further the course of the operation as shown in Fig. VIII–42, the liquor-lime suspension from the caustification tank is separated in the white liquor clarifier from the lime mud by decantation; the lime mud then goes to the mud washer. The mud formed from the dust in the gas washers of the kiln gases is also added to the mud washer. The washed lime mud, in a countercurrent, passes the hot flue gases in the kiln and is there converted into calcium oxide. This calcifying process has a considerable effect on the reactivity of the lime in the caustification. Here the temperature is of less importance than the rate of the decomposition of the carbonate. Prolonged treatment at a high temperature increases the reaction time but decreases the heating-up period. With short reaction periods and high temperatures,

the density of the lime increases but, simultaneously, its porosity decreases, and this adversely affects the reactivity of the lime in the caustification (265,377).

The dry content of the lime mud as it enters the kiln has a marked effect on the capacity of the kiln and the properties of the lime. A low dry content prolongs the reaction time and lowers the rate of decomposition. Predrying of the mud has been tested. Such a procedure, known as the "flash drying process," is shown schematically in Fig. VIII–43. The mud,

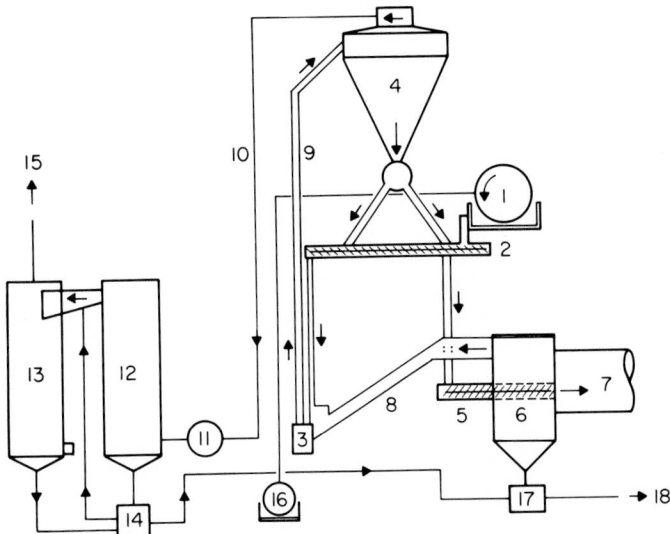

Figure VIII-43. Schematic diagram of lime flash dryer system. 1 filter; 2 paddle mixer; 3 cage mill; 4 cyclone separator; 5 paddle mixer; 6 kiln dust chamber; 7 kiln; 8 hot gas to cage mill; 9 gas and dust to cyclone; 10 cold gas; 11 fan; 12 peabody scrubber; 13 cyclone separator; 14 recirculating tank; 15 cyclone exhaust to atmosphere; 16 vacuum pump; 17 dust chamber pump; 18 to lime-mud washer.

coming from the wash filters, is led by means of a distribution system to a cage mill and then comes, in a finely divided state, into intimate contact with the hot flue gases from the kiln. The gases lose some heat through evaporation of the water from the mud; the latter is converted into almost dry dust and is separated in a cyclone. A part of the dry mud dust is mixed with lime mud from the filter, thus increasing the dry content of the latter, while another part, with 90% to 99% dry content, goes directly by means of a screw conveyer into the kiln. The cooled kiln gases pass a scrubber and a separator and the dust is returned as a thin suspension to the lime washer. To prevent a plugging of the screw conveyer, the feeding of the kiln is carried out with lime containing less than 1% humidity. This flash drying process increases the efficiency of the kiln to a considerable degree,

improves the quality of the lime, and reduces the losses of lime by 10% (152).

In contrast to these data on the optimum dry content of the predried lime mud before it enters the kiln, higher moisture contents, about 10%, are appropriate since, otherwise, the loss of lime, as dust, increases considerably. The predrying undoubtedly has the advantage of lower investment costs as compared with a theoretically possible extension of the length of the kiln to achieve equal efficiency (154). The hot-air predrying has stimulated a combination of the drying and calcinating, thus eliminating the kiln from the calcination. This procedure has been termed "fluidization," and is defined as consisting of a suspension of a solid—in this case, lime—in a gas so that the mixture assumes the properties of a liquid, with its own viscosity, density, and series of flow properties. When the flow of this liquid is slowed down, the suspended solid material begins to settle and the mixture becomes unhomogeneous. On the other hand, when the velocity of the gas stream is too great, the complete mixture is moved and the gas stream becomes a transportation medium. The flow rate and the settling out velocity of the suspended particles are kept in an equilibrium, allowing the gas to pass between the particles. Similar observations and apparative installations have been discussed in connection with the pyrite oven for flotation ores in Chapter VII, Section 1.3.

The mode of operation of such an installation, used as a calcinizing plant, is shown in Fig. VIII–44. In comparison with a kiln installation, its space requirements are small. The calcinated product is uniform and dust-

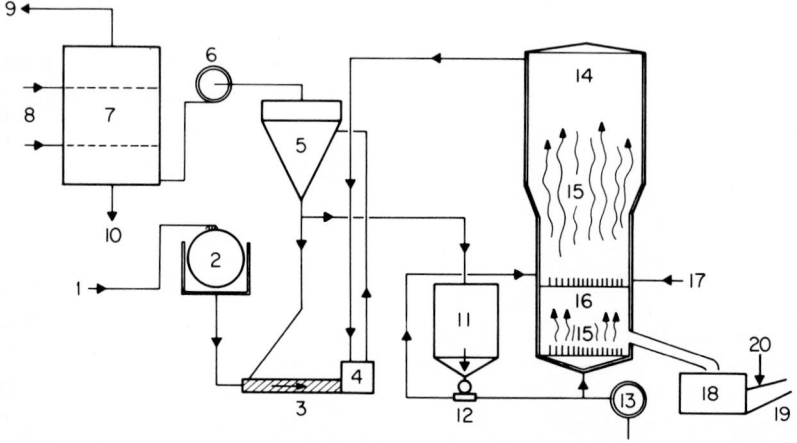

Figure VIII-44. Flow diagram of the Dorr-Oliver fluosolids lime reburning system. 1 lime mud; 2 filter; 3 paddle mixer; 4 cage mill; 5 cyclone; 6 exhaust fan; 7 scrubber; 8 water to scrubber; 9 carbon dioxide to carbonator; 10 effluent; 11 calciner feed bin; 12 feeder; 13 fluidizing air blower; 14 fluid solids calciner; 15 fluidized beds; 16 cooling compartment; 17 fuel (gas or oil); 18 pelletized lime; 19 slaker; 20 water inlet.

free and contains about 87% available active calcium oxide. The operation is simple and less subject to breakdowns and mishaps. The shut-down and resumption of operations are simpler and more rapid and the maintenance costs are considerably lower than those of a kiln plant. The heat requirement of 7.2 million BTU per ton (1.8 million kcal per ton) corresponds to that of a rotary kiln equipped with a good heat recovery installation (46, 154, 260). The efficiency of this, like all calcination plants, can easily be controlled by determining the content of noncalcinized calcium carbonate. For this purpose, a continuously or intermittently taken sample is decomposed with hydrochloric acid and the evolved carbon dioxide is measured. From the amount of the latter, the degree of calcination is determined. This method is quick and reliable and seems to be adequate for the control of the calcination, although the amount of inert material present is not taken into account (26).

The gases from any calcination plant contain not only considerable amounts of lime dust but also some alkali. The installation for the separation of the dust has been discussed in Section 5.5. Figure VIII–45 shows the

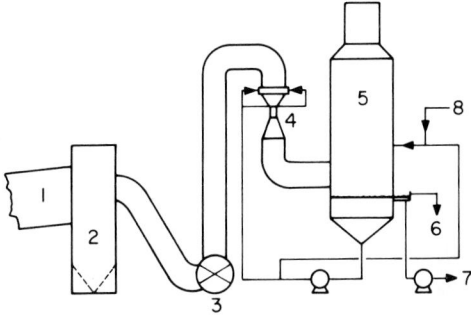

Figure VIII–45. Lime kiln Venturi scrubber (349). 1 kiln; 2 ash chamber; 3 L.D. fan; 4 S.F. Venturi; 5 cyclonic separator; 6 to sewer; 7 lime-sludge washers; 8 fresh water or kiln cooling water.

installation of a Venturi scrubber which has, in its upper part, a weir, from which water continuously flows, moistening the inner walls of the cone. The hot gases come into contact with the wet walls of the cone through a vertically inserted tube. In the nozzle-like construction, the water forms a kind of curtain through which the gas must pass. The gas disperses the water and, in this way, facilitates the collison with the dust particles. The dust and gas are separated in the cyclone separator and the liquor, containing the dust, is drained off at the bottom and returned to the cycle, while the gas escapes at the top. The operation of the Venturi requires a reduction in pressure. This drop can be attained by increasing either the gas velocity or the amount of water added. The recovery of the dust amounts to between

94% and 98%, and the drop in pressure is between 7 and 11 feet (2.10 to 3.35 meters) (water column) (78, 349).

5.7 Corrosion Problems

The installations in kraft pulp mills are made chiefly of iron and steel and are subject to various corrosive effects. During the decomposition of the wood, numerous organic compounds—in addition to the inorganic compounds used in the pulping process—accumulate in the cooking liquors; during the evaporation of the latter and the combustion of the concentrated black liquor, corrosive gases and vapors are formed. The corrosion occurs at various temperatures and concentrations of the solutions, the vapors, and the gases, and at various intervals of time. To this must be added the mechanical wear of the materials, especially those that are in motion or are in contact with the moving reaction mediums. All these factors increase the difficulty of detecting and controlling the occurrence of corrosion. Since these are special problems requiring special treatment, only more recent results which deal specifically with corrosion problems in kraft pulp mills will be considered here.

An electrochemical corrosion can occur not only in aqueous solutions but also in gases. It is caused by the formation of two electrode reactions on the metal surface which comes in contact with electrolytes. One reaction proceeds in cathodic direction, i.e., in the form of a reduction of the aggressive agents; the second, on the other hand, goes in an anodic direction, i.e., by dissolving the metals. Anodic and cathodic partial currents neutralize each other and the system appears to be dead.

If a metal is immersed in a neutral aqueous solution, it will dissolve, with the evolution of hydrogen, according to the electrochemical series of anodic dissolution. This will always be the case when the potential of the metal is less noble than that of hydrogen, which, in a neutral solution, is -0.41 V. Hydrogen exhibits overvoltage with various metals so that no visible hydrogen evolution takes place. If the overvoltage is discharged, e.g., by contact with a platinum wire, the hydrogen evolution sets in at the wire and the metal begins to dissolve. The covering of a not noble metal by a noble metal, therefore, under certain conditions, can lead to the appearance of further corrosion as soon as the corrosion-promoting electrolyte has access to both metals. This phenomenon is well-known with porous chromium layers. If the hydrogen overvoltage is great and the potential of the metal is not sufficiently negative to cause a hydrogen evolution, as is often the case with neutral or alkaline solutions, the oxygen is adsorbed on the metal or is dissolved in the electrolyte and acts in a depolarizing way. The rate of corrosion in this case is high at the start but decreases rapidly after the oxygen has been used up.

An investigation of the corrosion of carbon steel by kraft cooking liquors has shown that the maximum of the corrosion potential curves drops sharply with an increasing ratio of black liquor to white liquor. Simultaneously, the depolarization voltage increases with increasing black liquor content. This is particularly evident with oxidized black liquor. Dimethyl sulfide acts especially drastically when the mixture of black and white liquor is 1:1. The corrosive action of a white liquor increases sharply with the addition of black liquor, reaches a maximum at an addition of 50% to 60%, and then drops rapidly to zero. With increasing addition of black liquor, the depolarization voltage rises and the maximum of the corrosion voltage drops simultaneously. The rate of corrosion first follows the increasing depolarization voltage until it surpasses the maximum corrosion voltage; at that moment, the steel electrode becomes passive. In this state, the depolarization voltage equals the oxidation voltage and the corrosion voltage becomes practically zero. In this case, the corrosion is neither a "hydrogen evolution type," nor an "oxygen depolarization type," but can largely be attributed to the depolarizing action of sulfur substances, especially of the polysulfides, thiosulfates, dimethyl disulfide, and others (262).

The possibility of providing protection against an anodic corrosion has

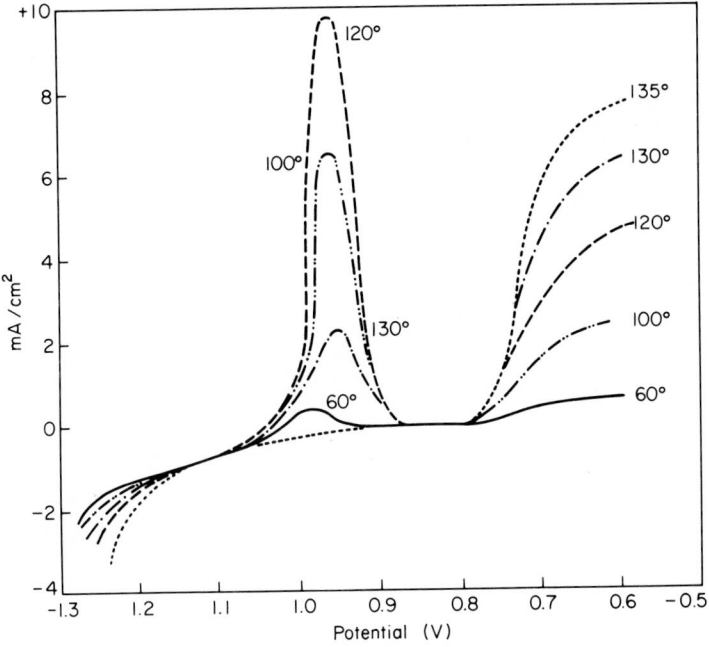

Figure VIII-46. Polarization curves measured during a laboratory cook in a mixture of 50% white and 50% nonoxidized black liquor (263).

been investigated with steel under the conditions of a kraft pulp cook. The required polarization curves had to be determined under rapidly changing reaction conditions because the sodium oxide concentration decreases quickly and the reaction temperature increases simultaneously. The results of such investigations are very instructive and are shown graphically in Fig. VIII–46. The curves were obtained from cooks with a liquor consisting of equal parts of white and nonoxidized black liquors. The negative current in the potential range from -1.2 to -1.05 V can be attributed to the reaction of oxidizing substances. The maxima of the current between -1.05 and -0.9 V are caused by corrosion, and the rapid increase between -0.8 and -0.7 V is due to reducing substances. Of special interest is the maximum reached at $120°C$ and the rapid decline commencing at $130°C$ and continuing with increasing temperature when the current reaches zero and below. This decline is interpreted as being caused by the decreasing alkalinity or the decrease in the concentration of nonhydrolyzed sodium sulfide. This is in agreement with the shift of the potential in the passive range during the later stages of the cook. Anodic corrosion protection is therefore possible; the current required for the passivation is markedly decreased (*a*) by increasing the addition of black liquor, (*b*) by a change from indirect to direct steam injection, (*c*) by a change from nonoxidized to oxidized black liquor, and (*d*) by a decrease in the concentration of nonhydrolyzed sodium sulfide (263).

In cooking liquors with very low polysulfide contents, the critical current density that is necessary for the passivation of carbon steel in white liquor is directly proportional to the third power of the hydroxyl ion concentration when the hydrolysis of the sodium sulfide is also taken into consideration. The addition of polysulfide causes a proportional decline in the critical current density. The current density required for the passivation is obviously lessened by the corrosion current. A sufficiently high proportion of polysulfide can bring about an adequate corrosion current and thus direct passivation without an auxiliary anodic current (220). The relationships between the corrosive action of the black liquor and the content of polysulfides, sodium hydroxide, and sodium sulfide have been confirmed repeatedly (231,300).

Another method for counteracting the corrosive action, especially of the cooking liquor, is the use of alloy steel containing chiefly chromium, nickel, and molybdenum. Because such steel as a construction material, e.g., for the inner lining of a digester, is very expensive, its use has been limited to sheets welded to conventional digester linings. This overlay welding is a very delicate operation and requires much experience, since, otherwise, the corrosion can even be intensified, for the reasons just discussed. Effective protection against anodic corrosion is probably safer and it requires a lower investment (45,61,115,127,307,385).

6. CONTROL OF AIR AND WATER POLLUTION

6.1 Analytical Control and Methods of Investigation

The problem of the elimination of the odor nuisance caused by kraft pulp mills has become increasingly important during the last few years. As already mention in Section 4, the odor is caused by sulfur compounds, primarily by hydrogen sulfide, methyl mercaptan, dimethyl sulfide, and, to a minor degree, by dimethyl disulfide. First, it must be decided how the source of these odors can be analytically detected and quantitatively determined. The physiological problem of odor detection will be discussed in the following section. The classical methods of precipitation with mercury and cadmium salts are used for distinguishing hydrogen sulfide, methyl sulfide, and dimethyl sulfide. The potentiometric titration with silver nitrate solution and a silver electrode can be used to identify inorganic sulfide and methyl sulfide, whereas the bromine oxidation method can be used for the determination of alkyl sulfide and dimethyl sulfide. The potentiometric titration proceeds according to the following equations:

$$S^{2-} + 2Ag^+ \rightarrow Ag_2S$$

$$CH_3S^- + Ag^+ \rightarrow CH_3SAg$$

The method requires the absorption of the sulfur compounds in an alkaline solution and this must be followed immediately by a titration to avoid oxidation. The ionization constant of methyl sulfide at pH 13 is 2.2×10^{-11}. The oxidation with bromine proceeds according to the following equations:

$$(CH_3)_2S + Br_2 + H_2O \rightarrow (CH_3)_2SO + 2HBr \tag{1}$$

$$CH_3SSCH_3 + 5Br_2 + 4H_2O \rightarrow 2CH_3SO_2Br + 8HBr \tag{2}$$

Instead of methyl, any other alkyl can be substituted, but this is of very minor importance so far as the kraft cooking process is concerned. The oxidation is carried out with a bromate-bromide solution because an aqueous bromine solution is too unstable. The oxidation reactions of hydrogen sulfide and of alkyl mercaptan proceed as follows:

$$H_2S + Br_2 \rightarrow S + 2HBr \tag{3}$$

$$2CH_3SH + Br_2 \rightarrow CH_3SSCH_3 + 2HBr \tag{4}$$

and then further, as in equation 2. These methods are accurate and are suitable for the investigation of the gases from kraft mills, but they require much time and this is incompatible with the rapidly changing composition of the gases. The total content of sulfidic sulfur in a gas can be determined by

catalytic oxidation. In this case, the gas mixture is passed through a combustion tube and all the sulfur compounds are oxidized to sulfur dioxide. The latter is absorbed in a hydrogen peroxide solution of known pH value. The sulfuric acid thus formed increases the acidity and, from this, the total sulfur content of the gas can be calculated. The combustion tube is heated to 950° to 1000°C (34).

A Beckman process gas chromatograph is used to control the performance of the recovery furnace and to provide data on the operation of the furnace. It has been found that the concentration of hydrogen sulfide in flue gases seems to be a more sensitive parameter of furnace operation than the concentration of oxygen and combustible materials (384a).

Mass spectroscopic investigations of the condensates of the relief gases from the digester have confirmed that the gas contains chiefly only hydrogen sulfide, methyl sulfide, dimethyl sulfide, and dimethyl disulfide, and that ethyl compounds and other sulfur-containing derivatives are not present. The acid components are absorbed in aqueous sodium hydroxide, the thio ether in benzene. After the separation of the solvents, the hydrogen sulfide and the methyl sulfide are determined potentiometrically with silver nitrate, dimethyl sulfide and dimethyl disulfide by conductivity titration after they have been separated by distillation. With these methods, the values shown in Table VIII-7 were obtained from technical relief gases.

TABLE VIII-7

ANALYSIS OF KRAFT MILL PROCESS STREAMS (302)

	H_2S	CH_3SH	CH_3SCH_3	CH_3SSCH_3
Stack flue gas, ppm/vol	493	42	—	2[a]
	759	68	—	5[a]
	534	64	—	2[a]
Digester relief gas, ppm/vol	131	5240	7350	4095
	138	4880	7000	3870
Digester blow gas condensate, mg/liter	71.7	232	171	125
	73.9	206	184	126
	71.7	227	187	128
	32.0	89	116	207
	33.6	85	120	212
Evaporator condensate, mg/liter	61	27	—	11[a]
	61	29	—	11[a]

[a] Calculated as CH_3SSCH_3 which may include some CH_3SCH_3.

For the identification of the gaseous and liquid samples, the method of partition chromatography between aqueous sodium hydroxide and carbon tetrachloride can be applied. Hydrogen sulfide and methyl sulfide, which are capable of forming salts, appear in the aqueous layer, and dimethyl sul-

fide in the carbon tetrachloride. Dimethyl sulfide and dimethyl disulfide can be separated by distillation. The unsaturated compounds are brominated and the sulfur components are determined by titration with bromate-bromide reagent. The determination of hydrogen sulfide and methyl sulfide is carried out by potentiometric titration. Dimethyl sulfide is often present in only small amounts and can be determined spectrographically after formation of an iodide addition compound (119).

The application of gas chromatography for the identification and the quantitative determination of the sulfur compounds in the relief gases suggests itself. For this purpose, columns charged with trimetacresyl phosphate and Carbowax 1540, with ethyl benzene as solvent, are recommended. Such columns permit the resolving of all known components in the gases of kraft pulp mills. The use of a two-column method makes possible the establishment of graphic illustrations for homolog series of mercaptans, ketones, esters, and normal alcohols. It should be emphasized that the accurate identification of an unknown by-product is only rarely possible by gas chromatography alone. The use of an ionization chamber, too, as an extremely sensitive detector has been suggested. A flame ionization detector can be used for the direct withdrawal of samples from the gas stream and their analysis. For this purpose it is not necessary to concentrate or dry the samples, as it is in the case of a thermal conductivity detector. The direct gas chromatographic analysis has its uncertainties and sources of error, but these can be minimized by taking into consideration the presence of water in the samples and by avoiding prolonged storage of the latter (3,33,65,359,397).

The technique for the quantitative gas chromatographic analysis of methyl mercaptan, methyl sulfide, and dimethyl disulfide in aqueous solutions requires a rigorous drying of the stream of nitrogen containing the stripped thiol and sulfides in quantities not exceeding five grams of each per liter. Because the passage of the gas stream through a drying tube containing one of the more commonly used desiccants results in total adsorption of the thiol and sulfides, magnesium perchlorate trihydrate has been specified as the preferred drying agent (296a,296b).

A new method for the isolation and analysis of the relief gases containing sulfur compounds uses the absorption or condensation of these gases on activated silica gel at $-195.8°C$ and subsequent transfer to a gas-liquid chromatographic column for the usual analytical separation by a stepwise increase in the temperature of the column. Because water vapors have a greater polarity than the odorous gases, the water must be removed first, so that no less polar compounds are eluted. The major part of the water is removed at $0°C$ by condensation and the remainder with the help of a drying agent (4). Hydrogen sulfide and sulfur dioxide may be adsorbed at room temperature on silica gel and desorbed at elevated temperatures (120°

to 150°C). The determination of these sulfur compounds is carried out by the gas chromatographic technique (10).

Another procedure that has been suggested uses coulometric methods for the continuous analysis of a stream containing oxidizable inorganic sulfur compounds. The titrant (bromine) required for the oxidation is regenerated in an electrolytic cell. The control consists of a cell for measuring the electromotive force, an amplifier, and a system of timers and relays. Its function consists in an automatic adjustment of the current level in the cell in order to maintain the titration at, or near, the equivalence point. The electrolysis current is related linearly to the concentration of the oxidizable compounds (392).

The quantitative measurement of sulfur dioxide, hydrogen sulfide, mercaptan, organic sulfide, and residual sulfur concentrations can be achieved with a recording electrolytic titrator. The equipment assures a rapid and reliable analysis of ambient air and of samples from kraft recovery furnace ducts, from oxidation tower vents, and from lime kiln stacks. Samples with concentrations ranging from 10 ppm to 800 ppm of hydrogen sulfide can be analyzed on selection of the proper range setting. The analysis requires 7 to 10 minutes and can be conducted in the laboratory or at the point of sample collection (358b).

For the continuous control of the suspended matter in the gases from the liquor boiler, an electrostatic precipitation has been suggested. In this way, the gas stream from which the sample is taken is measured by means of rotameters. The solids, which consist of alkali, are precipitated electrostatically and the gas is then washed with sodium hydroxide solution for the determination of hydrogen sulfide and methyl sulfide. By washing with 0.6% hydrogen peroxide solution, gaseous sulfur oxides can be determined. Finally, a gas sample for the chromatographic analysis can be withdrawn. The process permits a simultaneous determination of the content of suspended matter and of the acid gaseous sulfur compounds in the flue gases (228).

Returning, after having considered the analytical aspects, to the formation of these sulfur compounds: Comparative investigations of black liquors of the same composition at different temperatures show that deciduous woods produce more sulfur-containing organic compounds than coniferous woods do. This is obviously due to the sharp increase in the reactivity of the methoxyl groups of the lignin in the coniferous woods at the final stage of the cook. An increase in the temperature and a curtailment of the cooking time should lead to an appreciable reduction in the formation of such organic sulfur compounds. Figure VIII–47 shows the effect of the wood species on the formation of these compounds. Cooks that have been carried out at a lower temperature and for a shorter cooking time gave more methyl sulfide than dimethyl sulfide, whereas at higher tem-

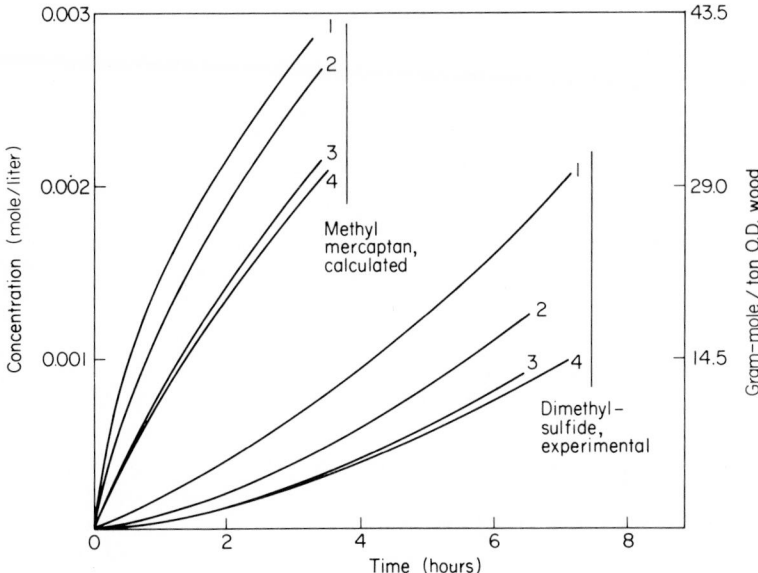

Figure VIII-47. Methyl mercaptan (calculated) and dimethyl sulfide produced from kraft cooking of several species (246). Liquor: 5% NaOH, 3.2% Na$_2$S; temperature: 175°C. 1 cottonwood; 2 aspen; 3 Douglas fir; 4 Western hemlock.

peratures and for longer cooking periods more of the latter and very little of the former were formed. From this it may be concluded that methyl sulfide is the primary reaction product and is consumed, with the formation of dimethyl sulfide, toward the end of the cook (246).

The following tables show the interesting relationships between the sulfur and gas concentrations at the various points at which the samples

TABLE VIII-8

Range of Sulfur Gas Concentrations Encountered in Kraft Mill Sampling (312)

Source	Gas concentration (ppm by volume)				
	SO$_2$	H$_2$S	RSH	RSR	RSSR
Digester vent	—	16–18,800	0–4370	3850–65,000	0–65,000
Blow gases	—	0–782	0–9840	522–46,900	0–10
Pulp washer	0.1–0.2	0–12	0–79	0	0.1–0.4
Evaporator, noncondensable	—	907–32,600	455–36,700	0–27,600	0–1278
Recovery furnace	4–798	14–1140	0–489	0–260	0–17
Smelt dissolving tank	0.5–70	10–44	0–212	0–91	0–4
Lime kiln	0–169	0–254	0–128	0–60	0–18
Tall oil cooking	2–822	5400–101,000	0–4660	0	103–769

TABLE VIII-9

EFFECT OF BLACK LIQUOR OXIDATION ON A FURNACE OPERATING AT DESIGN CAPACITY[a] (312)

	Without oxidation		With oxidation	
	SO$_2$	H$_2$S[b, c]	SO$_2$	H$_2$S[b, c]
	0.02	23.68	0.03	2.28
	0.01	24.87	0.11	3.41
	0.08	23.00	0.03	2.67
	0.11	23.46	—	1.08
			0.06	0.82
			0.38	0.49
			0.20	0.88
			0.02	0.77
Total	0.22	95.01	0.83	12.40
Mean	0.06	23.75	0.12	1.55
Average total emission		23.81		1.67

[a] lb S^{2-}/ton air-dry pulp.
[b] Determined by methylene blue method.
[c] Reduction with oxidation: 93%, 22.14 lb S/ton.

TABLE VIII-10

EFFECT OF BLACK LIQUOR OXIDATION ON A FURNACE OPERATING AT 2.2 × DESIGN CAPACITY[a] (312)

	Without oxidation					With oxidation				
	SO$_2$	H$_2$S[b]	RSH	RSR	RSSR	SO$_2$	H$_2$S[b]	RSH	RSR	RSSR
	0	10.10	0.80	0	0.71	0	5.35	0.60	0	0
	0	14.70	0.20	0	1.33	0	2.22	2.96	0	—
	0	15.75	—	0	3.44	0	7.41	—	0	—
	0	12.51	0.41	1.84	—	—	1.48	2.96	0	—
						—	5.03	1.70		
Total	0	53.06	1.41	1.84	5.48	0	21.49	8.72	0	0
Mean	0	13.26	0.47	0.46	1.83	0	4.30	2.18	0	0
Average total emission					16.02					6.48

[a] lb S^{2-}/ton air-dry pulp.
[b] Reduction with oxidation: 60%, 9.54 lb S/ton.

are taken (Table VIII-8), the effect of the oxidation of the black liquor on the sulfur content of the boiler gases of a liquor furnace operating with a planned capacity (Table VIII-9), and the same, taking into consideration the different components for a furnace operating with a 2.2–fold overload (Table VIII-10). When the sulfur in the SH ion in the black liquor is oxidized to a higher valency, sodium polysulfide, sodium thiosulfate, or

sodium sulfate can be formed. All these oxidation products are considerably more stable in solution than the SH ion is and they will decrease the amount of hydrogen sulfide which escapes from the liquor during the conversion into ash in the lower part of the furnace. Through the oxidation of the black liquor, the requirement of secondary air for the oxidation of the obnoxious sulfur compounds is decreased. Table VIII–10 clearly shows that the efficiency of the oxidation of the black liquor is appreciably reduced when the liquor boiler is overloaded (156).

6.2 The Odor Problem

The analytical detection of the amount of odorous substances in a given volume of air does not indicate whether these substances are perceptible by human sense of smell, and still less whether the odor is pleasant or unpleasant. Physiological problems enter here, and although certain precepts have been established, they are based on individual sensations and are therefore liable to subjective interpretation. Such physiological standards regarding the intensity of odors have been established for ethyl mercaptan. When the concentration of the ethyl mercaptan, in micrograms per liter, is plotted against the odor perceptibility coefficients, a linear function is obtained between the logarithm of the mercaptan concentration and the strength of the odor. Such a function follows, in general, the law of Weber-Fechner for many physiological sensations, such as light and sound intensity. Figure VIII–48 shows the relationship in which the intensity of the odor perceptibility has been evaluated by a more or less arbitrary scale by a number of observers. This scale is as follows:

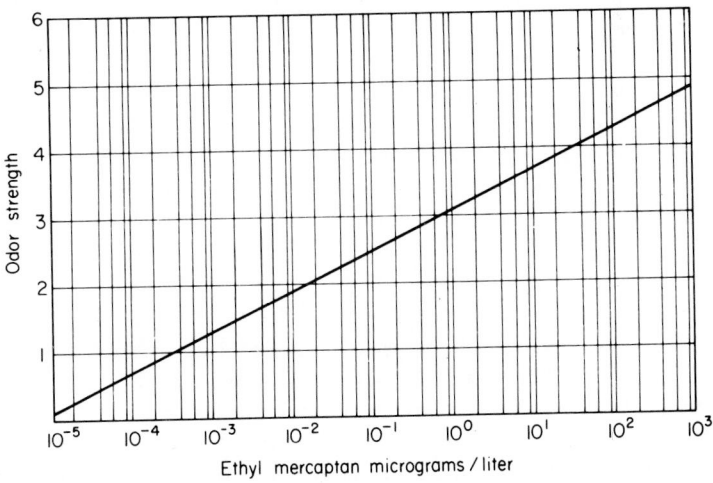

Figure VIII-48. Relationship between ethyl mercaptan content and odor strength (402).

Description of the odor	Intensity	Numerical evaluation
No detectable odor	No odor	0
Very weak but perceptible	Very faint	1
Weak but readily perceptible	Faint	2
Moderately intense	Easily noticeable	3
Cogent, forcible odor	Strong	4
Very strong, unbearable	Very strong	5

The problems of such an evaluation and scale are obvious. According to the scale, the odor threshold for ethyl mercaptan with an evaluation of 1 lies at a concentration of about 0.0001 mg per m^3 air, and for methyl mercaptan at 0.0005 mg per m^3 air. Other authors give for ethyl mercaptan 0.0002 mg per liter air, for hydrogen sulfide, 0.0011 mg per liter, and for dimethyl sulfide, 0.0011 mg per liter. These are enormous differences and can be explained only by the fact that the observers of the physiological evaluation have used different numerical scales (114, 121, 128, 402).

It must be mentioned, first of all, that the detection of weak odors is not possible in the presence of another strong odor. A weak, objectionable odor can be completely obscured by a strong but pleasant odor. This camouflage of the odor has been attempted on a technical basis. The intensity of an odor does not indicate what physiological effect it will have. Moreover, odor sensations of equal intensity quickly lead to habituation and, in this way, diminish the subjective power of perception. The impression of "pleasantness" or "unpleasantness" of any type of sensation depends, to a great extent, on its association with pleasant or unpleasant experiences of impressions of the past. Odor sensations of equal intensity may have antagonistic effects, so that the intensity of each single component appears to be minimized. This phenomenon, designated "counteraction," was discovered as early as 1895 by Zwaardemaker. When, therefore, mention is made of the threshold of perception, or of tolerance or intolerance of an odor sensation, all the details of the observer's sensibility or insensibility have to be taken into consideration as the primary factor. In the praxis, however, it is not the judgement of one or several observers that counts, but rather the reaction of a vast number, representing practically all degrees of subjective sensitivity, that must be taken into consideration.

As mentioned above, it is not sufficient to ascertain the perceptibility threshold of only one odorous component. Since the simultaneous presence of other odorous compounds can cause a strong shift to the negative, or, under certain conditions, to the positive side, the preceptibility threshold for the total odorous complex must be determined. The flue gases from kraft pulp mills contain not only hydrogen sulfide, methyl mercaptan, and dimethyl sulfide, but also sulfur dioxide, carbon dioxide, and other odorous

mixtures from the combustion of the organic substances and the inorganic suspended material. This is a typical example of an overlaying of odors and the quantitative amounts of the single components are subject to constant changes. It has also been suggested that so-called "odor profiles" should be set up, but this would not eliminate the objections to the analytical fixation of odor thresholds. That it is indeed impossible to predict odor threshold values on the basis of chemical analyses has been clearly shown by a comprehensive study recently published. Organoleptic determinations of the odor threshold must therefore be carried out on the flue gases as they are formed in the mill operation (27, 66, 96).

The limited valuation of the worth of the analytical flue gas control refers exclusively to the determination of the odor threshold. A constant and accurate control of the composition of the flue gases is necessary, if only for the recovery of the chemicals and heat. To estimate the odor nuisance from flue gases, a chemical and physioanalytical control has to be carried out over a long period of time and under various atmospheric conditions. Wind velocity and direction, temperature, and atmospheric moisture have a strong influence. Frequent changes in the direction of the wind offer an interesting criterion for turbulence in the atmosphere. For example, it has been found that the limit of the zone affected by the odor in the wind direction at midday is about 2500 meters, in the early morning before sunrise it is about 4000 meters, and in the evening after sunset it is about 8000 meters, from the source of the odor. These figures demonstrate the effect of the turbulence caused by solar heat and air currents. Concentration of the flue gases, wind velocity, and the height of the chimney have a lasting effect on the perceptibility limit. Thus, for example, an increase in the wind velocity from 1.6 to 16 km per hour causes an initial tenfold dilution of the malodorous gases from the stack, corresponding to about 90% decrease in odor perceptibility. On the other hand, at equal wind velocity, when the amount of odorous gases is reduced tenfold, the radius of the perceptibility border is reduced to one-quarter and that of the zone of annoyance, to one-sixteenth. The factor for the reduction of the malodorous stack gases by oxidation of the black liquor can be assumed to be about 100. This means that the radius of the perceptibility limit will be reduced by about one-sixteenth and that of the odor annoyance zone by about 250–fold. Increasing the height of the chimney without simultaneously decreasing the amount of stack gases does not cause an appreciable reduction in the odor annoyance, only a shift in the zones occurs; near the chimney there is an almost odor-free zone and the radius of the perceptibility zone is considerably enlarged and the odor-nuisance zone is widened (367, 403).

At ground level, the smoke concentration reaches its maximum at a distance of ten times the chimney height. The maximum concentration varies inversely with the square of the chimney height; at distances greater

than 50 times the chimney height, the concentration varies with the inverse square of the distance and is independent of the height of the chimney. This shows that, by doubling the chimney height, the maximum concentration at ground level can be substantially reduced. In direct opposition to this, and independent of the height of the chimney, the odor annoyance at greater distances and under certain local conditions can be intensified. To calculate the height of a chimney in order to reach a certain degree of dilution of the stack gases, the effects of local conditions on the flow of the gases must be taken into account. It is therefore of the utmost importance to take into consideration whether flat or hilly land or mountainous terrain is involved; the extent to which it is built up is also important (62, 64).

The importance of the oxidation of the black liquor for the elimination of malodorous gases has already been discussed in detail. But the vapors of the relief gases from the digesters also contain considerable amounts of mercaptans. Whereas the steam contains up to 400 mg methyl sulfide per liter in uncondensable gases, the condensate contains up to 1100 mg methyl sulfide per liter and up to 2200 mg dimethyl sulfide per liter. The direct addition of sulfur to the digester to increase the sulfidity, in particular, causes a sharp increase in mercaptan in the uncondensable gases. In this connection, the oxidation not only of the condensate but also of the uncondensable gases is important. These gases, therefore, are mostly passed into the black liquor oxidation towers after they have passed through a condenser. The addition of the gases has no effect on the efficiency of the oxidation tower (321, 322). Black liquor can absorb or release hydrogen sulfide in a direct-contact evaporator, depending upon the particular type of liquor and injected gas. When the concentration of sodium sulfide in the black liquor is high and the pH value and the hydrogen sulfide concentration of the gas stream are low, hydrogen sulfide is released from the liquor. At a high pH and low sodium sulfide content of the liquor, and a high content of hydrogen sulfide in the gas, hydrogen sulfide is absorbed. Regardless of the pH, absorption always takes place when the sodium sulfide content of the black liquor drops to zero (266).

The oxidation of the condensate of the steam from the digester is often carried out with chlorine or chlorine-containing bleaching agents. Whether this condensate is added to the chlorine tower in the bleaching plant as water for dilution or is treated in a washer with the uncondensable gases with such chlorine-containing bleaching agents is a matter of choice. Chlorine dioxide is preferred over chlorine because it leads directly to oxidation and not to substitution. The uncondensable gases are usually burned under the liquor furnace. For this, either an oil burner is used or, if possible, methane is mixed with the gases. With continuous digesters which release little relief steam, a special heat-recovery and odor-control tower is installed and the gases are passed into it. Cold water is injected into the tower to

condense the vapors and cool the gases. The cooled gases are treated in a scrubber with solutions of oxidizing agents, mostly bleaching agents (80, 104,131,311,365,399).

Finally, the question of odor masking must be briefly mentioned. Experiments in this direction were inspired by the above-mentioned possibility of overlaying odors by the addition of aromatic compounds, having strong but pleasant odors, to the digester. It need not be emphasized that such a procedure is only a makeshift and not a means of eliminating the odors. All the odorous substances present are dissolved in water. When such compounds are dissolved in water and the minimum concentration at which the faintest odor in the air above the water can be noticed, an odor recognition unit, or "scent unit," can be established. This scent unit is therefore physiologically connected with an odor identification, as opposed to an odor perceptibility unit. There is no simple relationship between these two units. The distribution of an odor between water and air does not follow a physical law, and the order of magnitude of the masking effect is not a direct function of the amount added or of the intensity of the odor. These investigations are of less technical than physiological interest in the attempts to eliminate air pollution (242,254).

A recently decribed process, known as the Sekor process (stripping effluents for kraft pulp mill odor reduction), has as its goal the separation of the volatile organic compounds from the relief vapors and condensates and, in this way, the reduction of the odor annoyance and of waste water pollution. The process uses steam-stripping with reflux to expel the volatile compounds and to separate them from the vapors by condensation. Hydrogen sulfide, methyl sulfide, dimethyl sulfide, and dimethyl disulfide can be removed from the vapors in this way with an efficiency of over 95% (178).

6.3 The Waste Water Problem

The type of processing used for the black liquor from a kraft pulp mill renders the loading of the drainage canal by waste water from the operations of the mill much less dangerous than is the case with the sulfite pulp mill. While in the latter, as long as calcium is used as the base, the processing of the spent sulfite liquor is uneconomical and therefore is not practised, in the kraft mill it is an economic necessity. But here, also, dilute wash water from the washing of the pulp and the lime sludge and from other sources in the mill goes into the drainage canal. To determine the influence on the oxygen demand of a drainage canal after the addition of waste water, the investigation must be carried out over a prolonged period of time, especially during all seasons, because the external conditions, such as flooding, air and water temperature, the addition of other waste waters, and other factors, all substantially affect the oxygen demand. The aut-

oxidation capacity of the drainage canal changes with flooding and with the temperature. The biological oxygen demand (BOD) is determined from the sum of the amount of oxygen in the water before the addition of the waste waters and the oxygen content of the water five days after the addition. During this time, the oxygen needed for the oxidation of the substances capable of being decomposed is consumed. From this figure, the minimum oxygen content necessary for the preservation of fish life must be deducted. The requirements that are enforced today for the cleanliness of the drainage canals, as regards oxygen-consuming or other dangerous waste waters, are extremely stringent, thus necessitating constant control of the waste waters.

Waste waters from kraft pulp mills are mostly dark-colored and this can be reduced by precipitation with lime, with the lime sludge being regenerated in the calcination plant, although it is difficult to dehydrate. Bacteria capable of reducing sulfate are widely distributed and are active within a pH range of 5.5 to 9.0 and at temperatures of up to 55°C. They cause a discoloration of the pulp through the formation of ferrous sulfide, they corrode iron and steel, and they cause the formation of hydrogen sulfide, especially in waste water basins. When the drainage canal contains large amounts of clay minerals, a complex compound is formed of these and organic substances present in the waste water of kraft pulp mills. In this way, the drainage canal can contribute to its own purification (28, 267,344).

The BOD of the waste water from the digester vapors and the evaporation condensate of a modern kraft pulp mill amounts to 5 to 10 kg per ton of pulp at 90% dry content, corresponding to 25% to 40% of the total amount of oxygen-consuming material. The BOD values of the digester condensate lie between 4 and 18 grams per liter, those of the evaporation condensates, between 0.5 and 2.0 grams per liter. Both condensates contain hydrogen sulfide, methyl sulfide, dimethyl sulfide, dimethyl disulfide, acetone, methanol, and ethanol. All these products are not only oxygen-consuming but also toxic to a high degree. An economically completely satisfactory method for the treatment of these condensates has not yet been found (312).

Various authors believe that, except for the condensates just mentioned, the toxicity of the waste water is not due to the sulfides and mercaptans and that their presence is not directly related to the BOD, as is generally assumed. They consider the waste waters from the bleaching plant to be especially toxic and to provide the greatest strain on the BOD. Clarification and oxidation tanks, as well as fermentation plants, can contribute to the lessening of the toxicity. Such fermentation plants differ from the active sludge treatment by lower nutrient requirements and diminished formation of secondary solid products. The BOD of waste water thus treated can be reduced by 90% (85,130,177).

7. BY-PRODUCTS OF THE KRAFT PULP INDUSTRY

7.1 Tall Oil

When coniferous woods are used in the kraft pulping process, the main by-product is tall oil. Its name is based on the Swedish word "tall," which means spruce, and, in fact, crude tall oil was first recovered in Sweden in 1901. The theoretical yield of tall oil soap amounts to 180 to 200 lb per ton of air-dry pulp from loblolly pine (*Pinus taeda*) and up to 300 lb from long-leaf pine (*Pinus palustris*). The production of crude tall oil in the years 1960 to 1961 was as follows:

Country	Production (1000 tons)	Country	Production (1000 tons)
U.S.	394	Japan	15
Sweden	50	Canada	8
Finland	37	France	6
U.S.S.R.	25	Norway	3

or a total of 538,000. It can be assumed that theses figures have increased in the meantime (183,185).

Depending upon the species of tree used, its location, the season when the tree was cut, and the duration of its storage, the yield and composition of the tall oil are subject to marked fluctuations. This can be seen in Table VIII–11. From these figures it is apparent that the yield of crude tall oil

TABLE VIII-11

Yield and Composition of Tall Oil from Various North American Regions (184)

	Yield (lb/ton a.d.p.)	Resin acids (%)	Fatty acids (%)	Neutral compound (%)
Southern states	90	46	40	14
Mid-Atlantic states	80	39	48	13
Canada	40	43	32	25
East of Cascades	50	30	41	29
West of Cascades	30	43	27	30

is considerably lower in Canada and in the northwestern states of the United States than in the eastern states, and, since the proportion of neutral components is high, the yield of acid products from such tall oil decreases (184). Scandinavian tall oil contains an average of 55% fatty acids, 35% resin acids, and about 10% unsaponifiable products. The fatty acid fraction can be separated by distillation into oleic, linoleic, and traces of palmitic acids

and an unsaturated acid with 18 carbon atoms and 3 double bonds. The principal component of the resin acid fraction is abietic acid, together with small amounts of dehydro- and dihydroabietic acids, dextro- and isodextro-pimaric acids, and palustric and neoabietic acids. These fatty and resin acids are present in the black liquor in a saponified state. Because of the high salt content, the acids are salted out from the liquor and float as a solid layer on top of it. Whereas formerly the soap was skimmed off and burned, today almost all pulp mills are equipped with plants to process the crude tall oil. In any case, the soap must be isolated before the evaporation of the black liquor to avoid a plugging of the boiler tubes (58).

The separation of the colloidally distributed soap in the black liquor occurs on cooling, or when the salt content of the liquor is increased. The soap still contains some volatile sulfur compounds and some lignin components. The separation of the soap can be facilitated by concentrating the liquor to 20% to 30% solid content. The addition of sodium sulfate (salt cake) also causes a better separation of the soap. The soap itself contains various amounts of tall oil, depending upon the wood species; that from Douglas fir, for example, consists of 53% tall oil, 7% black liquor, and about 40% water. For the separation of the free fatty or resin acids, the soap is acidified, usually with hot 30% to 50% sulfuric acid. Simultaneously, the colloidally dispersed lignin is precipitated (106,396).

The skimming off of the soap floating on the black liquor can be facilitated and its yield increased by adding about 0.5% to 1% (by volume) of an aromatic or terpene-like solvent with a medium boiling point, and simultaneously injecting air into the liquor. In this way the small particles are coagulated to larger aggregates and the formation of a stable foam is inhibited. The added solvents are continuously removed by distillation and returned to the cycle (41).

The acidification can be carried out either in batches or continuously, with both processes having their advantages and disadvantages. The process can also be carried out by automation. In addition, the soap can be isolated successfully by centrifugation or by the use of waste acid from a chlorine dioxide plant, but in this case there is a possibility of increased corrosion. The addition of solvents has already been mentioned, and one of these, naphtha, has been used in this "solvent process" and has led to the recovery of a crude tall oil with a low moisture, sulfuric acid, and lignin content and with a good color. In spite of every precaution and improvement, the losses of soap in the skimmer are still sizeable. A considerable portion of dissolved soap remains in the liquor. The chemical composition of the soap seems to have a greater effect on the loss than the type of operation of the skimmer (216,354,373).

The investigation of the tall oil soap is concerned primarily with the determination of its density, its water content, its free alkali, and its cal-

cium content. Gas chromatography provides information as to the chemical composition. For the determination of the density, standard methods are used. The same holds true for the determination of the refractive index. A description of these methods must be waived (2,317,318).

Gum and tall oil rosins are both products of the living tree. The differences in their composition are the results of the differences in the processes by which they are obtained. The chief changes are caused by the isomerization of the resin acids of the abietic acid type. The resin acids which are isolated from the tall oil rosin are subjected, during the isolation, to oxidative effects and high temperatures. Both factors contribute to an increased isomerization and lead to an increased yield of abietic acid. In the United States three types of resins are on the market: (a) the wood resin obtained on steam-distillation of the oleorosin extracted from pinewood stumps; (b) the tall oil rosin obtained by fractionated distillation of the tall oil; and (c) the gum rosin from oleorosin, an exudate of a demaged tree. Table VIII–12 shows the differences in the compositions of the various resins that are obtained in the recovery stages (317).

TABLE VIII–12

COMPOSITION OF ROSIN AND SOME OF ITS PRECURSORS (317)

	Levopimaric acid (%)	Abietic acid (%)	Palustric acid (%)	Neoabietic acid (%)	Dehydro-abietic acid (%)	Oxy acids (%)
Pine chips	14	14	12	12	2	21
Pinewood	19	11	13	11	2	11
Black liquor soap	0	27	7	9	3	25
Oleoresin	26	10	11	11	3	11
Gum rosin	0	20	18	17	5	7
Tall oil rosin	0	40	6	4	18	3

Levopimaric acid shows a marked decrease, even as the result of the action of only light and air on the wood chips. In the soap from the black liquor, levopimaric acid is no longer found and the portion of palustric and neoabietic acids also shows a strong decrease. Particularly noticeable is the high proportion of oxy acids in the black liquor soap. A further change in the composition occurs when the tall oil is distilled in order to obtain tall oil resin. During this distillation almost all the oxy acids disappear. Some are found in the rosin pitch, others are dehydrated and form dehydroabietic acid. In some cases, dehydrogenation can occur and the amount of dehydroabietic acid can increase to 20% (225).

Crude tall oil is used for the preparation of cleaning agents, in the production of which ionized and nonionized products can be obtained. Anionized products are formed mostly from sulfurated and sulfonated

groups, whereas nonionized products are polyoxyethylene polymers. The cationized products are chiefly ammonium derivatives. Plasticizers are obtained by the reaction of tall oil with fatty acids, whereas tall oil fatty acid esters can react with formaldehyde by acid catalysis to give condensation products that contain fewer unsaturated bonds, have a lower molecular weight, and a greater proportion of free hydroxyl groups, through the formation of intermediate ester groups and subsequent hydrolysis (11, 107, 225, 361).

7.2 Turpentine

Turpentine is a collective term for the volatile oils that can be obtained from coniferous woods. Gum turpentine results from the steam distillation of the resinous exudates of the tree; wood turpentine is obtained on extraction of tree stumps with suitable solvents; sulfite turpentine distills with the steam from the relief gases of the digester. Figure VIII–49 shows the

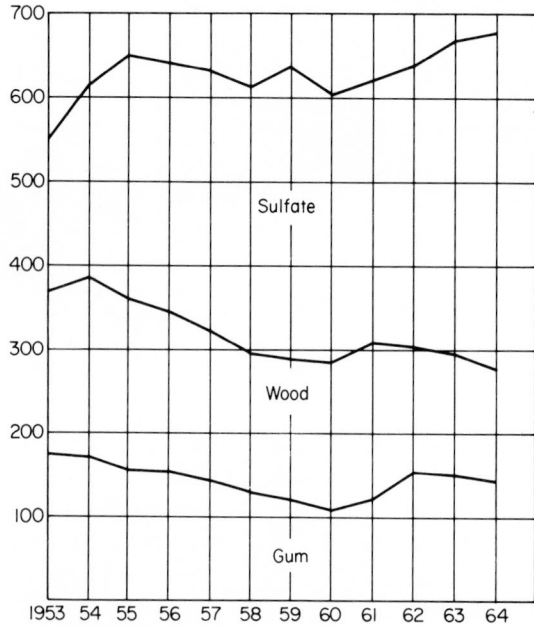

Figure VIII–49. United States turpentine distribution for 1953–1964 (crop year begins April 1) in thousands of 50-gal barrels (112).

distribution of turpentine production by the various methods and clearly indicates that the production from the kraft cooking process increases continuously while it decreases with the other methods. A comprehensive study of the yield of turpentine from over 30 different American and Can-

adian pulp woods shows not only that the yield varies with the wood species and its location, but also that the composition of the turpentine itself is subject to marked variations. Table VIII–13 gives the approximate composition of the turpentine from six selected woods, and Table VIII–14 gives a comparison of the composition of sulfate turpentine obtained from a kraft pulp mill with that from laboratory cooks of chips of the same origin. The experiments show that an increase in the yield of turpentine can be achieved by changing the cooking conditions. It has already been mentioned that the storage of the wood, especially when in the form of chips, has a lasting effect on the yield of tall oil and turpentine (112).

The yields of turpentine shown in Table VIII–13 are based on a ton of dry wood. Reports from 24 North American kraft pulp mills gave the average yields based on a ton of air-dried pulp, as listed in Table VIII–15.

The relief gases from the digester are passed into a cyclone to remove the carried-over particles of black liquor, the steam carrying the turpentine is condensed, and the turpentine floating on the condensate is separated in

TABLE VIII-13

COMPOSITION OF TURPENTINES OF VARIOUS NORTH AMERICAN
AND CANADIAN WOODS (112)

Wood species component (%)	Balsam fir (*Abies balsamea*)	Black spruce (*Picea mariana*)	Douglas fir (*Pseudotsuga menziesii*)	Eastern white pine (*Pinus strobus*)	Jack pine (*Pinus banksiana*)	Loblolly pine (*Pinus taeda*)
Light hydrocarbons	T[a]	T	T	—	—	—
α-Pinene	35.4	11.3	58.4	67.0	62.0	64.0
Camphene	1.0	T	2.4	2.9	2.0	1.3
β-Pinene	12.9	9.3	4.2	18.0	28.0	28.0
Myrcene	4.6	0.4	1.5	0.9	1.1	2.0
α-Turpinene	T	—	0.5	—	0.5	—
Limonene	1.3	0.4	6.3	0.9	2.8	1.5
β-Phellandrene	0.4	0.4	1.5	0.5	1.6	0.5
p-Cymene	0.8	—	0.5	—	—	—
α-Fenchol	3.4	0.4	0.6	T	T	—
Turpinen-4-ol	2.5	0.8	2.5	—	T	—
cis-β-Turpineol	0.4	—	T	2.6	—	—
β-Caryophyllene	2.1	—	T	T	T	—
trans-β-Turpineol	1.3	—	0.5	—	—	2.7
α-Turpineol	1.3	—	6.6	2.6	2.0	—
α-Caryophyllene	T	—	0.8	—	—	—
cis-Anethole	0.8	—	—	—	—	—
Unidentified compounds	31.8	77.0	11.7	4.1	T	—
Yield, gal/ton dry wood	0.93	0.076	0.6	1.05	0.81	0.97

[a] T = Trace

TABLE VIII-14

COMPARISON OF THE COMPOSITION OF TURPENTINE FROM PULP MILL
PRODUCTION AND LABORATORY COOK (112)

Component (%)	Pulp mill production			Laboratory cook		
	Oregon[a]	Washington[b]	Florida[c]	Oregon[a]	Washington[b]	Florida[c]
Light hydrocarbons	1.2	2.5	0.8	—	0.7	—
α-Pinene	52.0	47.0	58.8	53.9	51.6	59.3
Camphene	3.0	3.2	2.9	3.0	2.8	2.2
β-Pinene	3.6	3.8	20.6	3.0	2.1	21.2
Dipentenes	14.4	17.5	12.8	15.8	15.7	10.4
Pine oils	20.4	21.6	4.1	15.3	22.3	6.9
Unidentified components of higher boiling point than pine oil	5.4	4.4	—	—	4.8	—

[a] Wood furnish, 95% Douglas fir.
[b] 38% Douglas fir, 31% cedar, 28% hemlock, 3% alder.
[c] Mixed pines.

TABLE VIII-15

YIELD OF SULFATE TURPENTINE FROM DIFFERENT WOOD SPECIES (226)

Species	Yield (gal/ton air-dry pulp)
Slash and long-leaf pine	2.8–4.3
Loblolly, short-leaf and Virginia pine	1.5–2.7
Jack and white pine	1.5–3.1

a decantation tank. The crude turpentine, containing several types of impurities, is separated from mercaptans and thio ethers by fractionated distillation. The critical point in the recovery of turpentine is the separation of the black liquor particles and fibers from the steam. The design of the cyclone is therefore especially important (374).

The composition of the sulfate turpentine was shown in Table VIII–14. In general, it consists of about 50% to 60% α-pinene, 15% to 20% β-pinene, and 10% to 15% monocyclic terpenes. Their configuration is shown in Fig. VIII–50. The pinenes are reactive hydrocarbons which can readily be converted into a series of other compounds. By hydration, α-terpineol or terpinyl ethers can be obtained; isomerization leads to dipentene (limonene) or camphene, polymerization to dipolymers or higher polymers, and addition can cause the formation of a great number of organic compounds. On heating to higher temperatures, *allo*-cymene and myrcene are formed, in addition to dipentene. The latter is split at higher temperatures to form

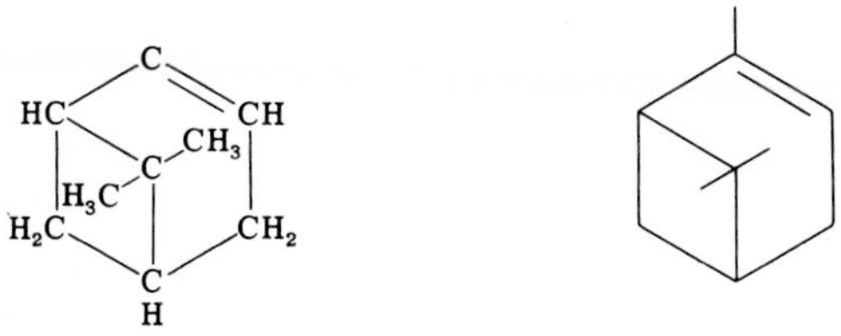

Alpha pinene
50 – 60%

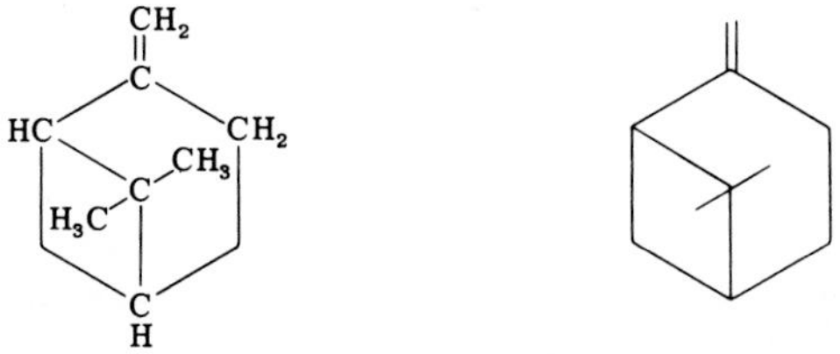

Beta pinene
15 – 20%

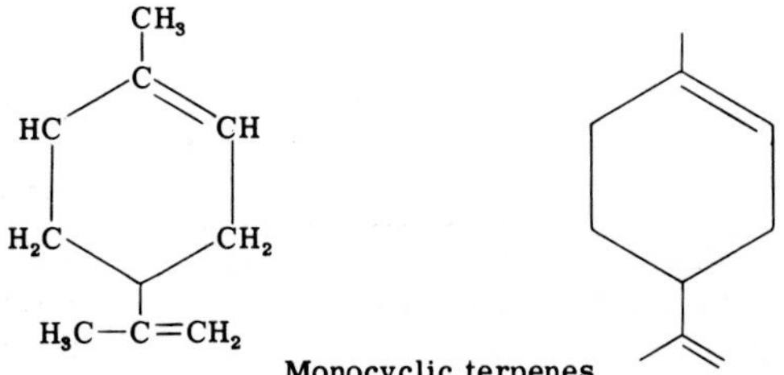

Monocyclic terpenes
10 – 15%

Figure VIII-50. Composition of sulfate turpentine (280).

isoprene, one of the starting materials for the production of butyl rubber. In the presence of dehydrating catalysts, sulfate turpentine is converted, in good yield, into *p*-cymene, which, in turn, can serve as the starting material for dimethylstyrene, *p*-methylacetophenone, and cuminic acid. Sulfate turpentine is especially important for the synthesis of flavoring and perfuming substances, such as menthol, geraniol, nerol, linaleol, citronellol, terpineol, camphor, isobornyl acetate, and other products. The natural peppermint oils are complex mixtures of mostly terpene-type compounds. Most of these compounds can be synthesized from α- and β-pinene. These pinenes are the major components of the sulfate turpentine. The preparation of industrially usable synthetic resins and polymers from pinenes shows promise and therefore the recovery and fractionation of turpentine is of steadily increasing importance (105, 280, 288, 294).

7.3 Lignin Precipitates

When the suspension of the lignin salt, precipitated from the black liquor, is heated, it melts and coagulates. This property is utilized for the recovery of the lignin. To separate the lignin by coagulation through acidification, it is necessary that the pH value should remain above 9.3. The acidification can be carried out with sulfuric acid or carbon dioxide. Both give the same product in equal yields. If the lignin is to be separated by filtration, the acidification must be carried out at a temperature a few degrees below the coagulation point of the lignin. The conversion of the lignin salt into the free lignin is best carried out at a temperature of 80°C and a concentration of about 20%, with the final pH value being kept at about 3 (252,253). The precipitation with carbon dioxide can also be carried out at a pressure of 4 to 5 atmospheres; this improves the yield and filterability of the lignin. The precipitation requires only 15 to 20 minutes and the temperature should not exceed 70° to 75°C. The lignin thus obtained can be utilized advantageously for the preparation of phenol resins. Numerous other possible uses for this lignin have also been suggested. The phenolic groups in alkali lignin react readily with formaldehyde and other aldehydes, with the formation of resins. Commercial alkali lignins are marketed under various trade names and are used as stabilizers for asphalt emulsions, as modifiers for lattices, as binders for printing colors, as dispersing agents, and many other purposes. Vanillin can be obtained in yields of up to 13% from black liquor lignin as long as the temperature during the cook has not greatly exceeded 140°C and the cooking time, 30 minutes (268).

7.4 Sulfur Compounds from Thiolignin

It has long been known that lignin is demethylated on heating in an alkaline medium in the presence of sulfide ions. Demethylation also occurs, to a small extent, under the conditions of the kraft pulping process. To increase the yield of dimethyl sulfide and of methyl mercaptan, the black liquor is heated for several hours, with the addition of sodium sulfide (164). A modification of this process provides for a two-stage reaction in which the primary reaction product is methyl sulfide if the reaction is carried out in such a way that the methyl sulfide formed is removed before it can react further. This can be accomplished by heating the mixture of lignin, sulfur, and alkali to about 220°C and distilling the volatile products. Under these conditions, a mixture of methyl sulfide and dimethyl sulfide is obtained in a yield of about 4%, based on the weight of the lignin, with 80% of the mixture being methyl sulfide. In the technical process, black liquor, with about 50% solid content, is heated at 200° to 230°C with a sufficient

amount of sulfur in a continuous reactor. The volatile organic products are distilled off from the discharged liquor by rapid distillation and, in this way, a crude product is obtained, containing about 85% dimethyl sulfide, 10% methyl sulfide, and small amounts of hydrogen sulfide and high-boiling products. This crude product is obtained in about 3% yield, based on the weight of the lignin. Not all methoxyl groups are split off and the percentage of reacting groups depends upon the reaction conditions. About 74% of the methoxyl groups split off form dimethyl sulfide. The latter boils at 37°C and has the characteristic mercaptan odor which contributes to the air pollution from kraft mills. It must be remembered that, during a normal kraft cook, only 0.09% of the lignin forms mercaptan. Dimethyl sulfide is miscible with most organic solvents, but immiscible with water (148,150,255).

The technical uses for dimethyl sulfide are limited. By oxidation, dimethyl sulfoxide [$(CH_3)_2SO$], usually called DMSO, is formed. For the oxidation, several processes are applicable: (*a*) oxidation in a vapor phase with a gas stream containing oxygen and in the presence of nitrogen oxides as catalysts (336); (*b*) treatment of the dimethyl sulfide with a solution of nitrogen dioxide in dimethyl sulfoxide, followed by separation from the solvent of the nitrogen oxides formed and reoxidation to nitrogen dioxide which is then returned to the reaction zone (179); (*c*) oxygen is passed into a solution of dimethyl sulfide in dimethyl sulfoxide in the presence of nitrogen oxides as catalysts (391). Dimethyl sulfoxide is an excellent solvent of high purity. Its boiling point is 190°C and its melting point, 18.5°C. It is miscible with water and most organic solvents except the benzene hydrocarbons. It dissolves cellulose esters and ethers, polyvinyl acetate, polyacrylic acid esters, chlorinated polyvinyl chloride, grafted vinyl chloride polymers, polyacrylnitrile, chlorinated caoutchouc, and many other resins. It can be used as a solvent for acetylene, nitrogen oxides, and ethylene oxide, and in the spinning of polyacrylnitrile, as an additive to paint removers, for lacquer layers that are difficult to remove, for the improvement of the cold-resistance and of the formation of films from dispersions of synthetic resins, and for many other purposes. The medicinal use of DMSO has not yet led to satisfactory clinical results (99, 168).

Besides its oxidation to dimethyl sulfone [$(CH_3)_2SO_2$] and its reduction to dimethyl sulfide, DMSO is readily reactive with anhydrides, with chlorinated reagents, toward alkylation, and with phenol. Such a reactive and widely applicable solvent naturally has attracted the attention of numerous chemists, with their investigations being directed primarily toward the improvement of the yield of DMSO from black liquor under varying reaction conditions (122).

7.5 Organic Acids and Furfural

Black liquor can also be used for the production of acetic and formic acids and thiolignin. For this purpose, the processing of the liquor includes its oxidation, acidification, filtration, drying of the filtration residue, extraction or distillation, and production of furfural. Table VIII–16 gives the composition of the solids and solutions from the precipitation of dilute black liquor (122, 210). From the black liquor filtrate from the acid-precipitation of the lignin, after neutralization with calcium carbonate, lactic, glycolic, and levulinic acids can be isolated, but the presence of oxalic,

TABLE VIII-16

CHARACTERISTICS OF SOLIDS AND LIQUORS FROM ACID PRECIPITATION
OF DILUTE BLACK LIQUORS (122)

	Dilute black liquor (% by weight)	
	Original	Oxidized
Total sulfur in precipitated solids	0.16	0.17
Sulfated ash in precipitated solids as NaOH	—	0.10
H_2S or SO_2 liberated on acidification as S	0.037	0.0075
Distillable acids from liquor acidified with 5% H_2SO_4 by weight as acetic acid	1.02	1.09
The same acidified with 4.1% H_2SO_4 as CH_3CO_2H	—	0.95
The same acidified with 3.2% H_2SO_4 as CH_3CO_2H	—	0.53
Ether-extractable acids from liquor acidified with 5% H_2SO_4 as CH_3CO_2H	0.94	—
Acid-soluble nonvolatile organic compounds in acidified liquor residue, as glucose	1.61	1.68
Furfural in acid distillate from liquor acidified with 5% H_2SO_4	—	0.11
Furfural obtainable from acidified liquor residue	—	0.088

tartaric, and trihydroxyglutaric acids is in doubt. The amount of free organic acids in the filtrate is widely dependent upon the pH of the solution; it increases from 2.9% (based on the dry content of the black liquor) at pH 3.95, to 9.3% at pH 1.75. The proportion of free acetic acid increases from 1.75% to 2.08%, that of formic acid, from 0.39% to 3.28%, and that of lactic acid from 0.65% to 3.15%; butyric, valeric, and propionic acids are present only in minor amounts. The total amount of these organic acids is 14.74% (based on the absolutely dry organic substance in the liquor) at pH 1.75. A modified method for the isolation of the hydroxy acids uses 72% sulfuric acid for the precipitation, after which the solution is purified with active carbon, concentrated to a syrup, and this is extracted

with a mixture of methyl alcohol and ethyl ether. The hydroxy acids consist chiefly of butyric and 2, 4–dihydroxybutyric acids and some C_5- and C_6-hydroxy acids in minute amounts.

In considering the feasibility of the recovery of such organic acids and the furfural from black liquor, it must be remembered that the heat economy of the kraft pulping process is always detrimentally affected when large amounts of black liquor are withdrawn in this way from the regeneration process. It is also questionable whether the products thus obtained can compete with similar products obtained by direct chemical synthesis (142, 210, 293, 295).

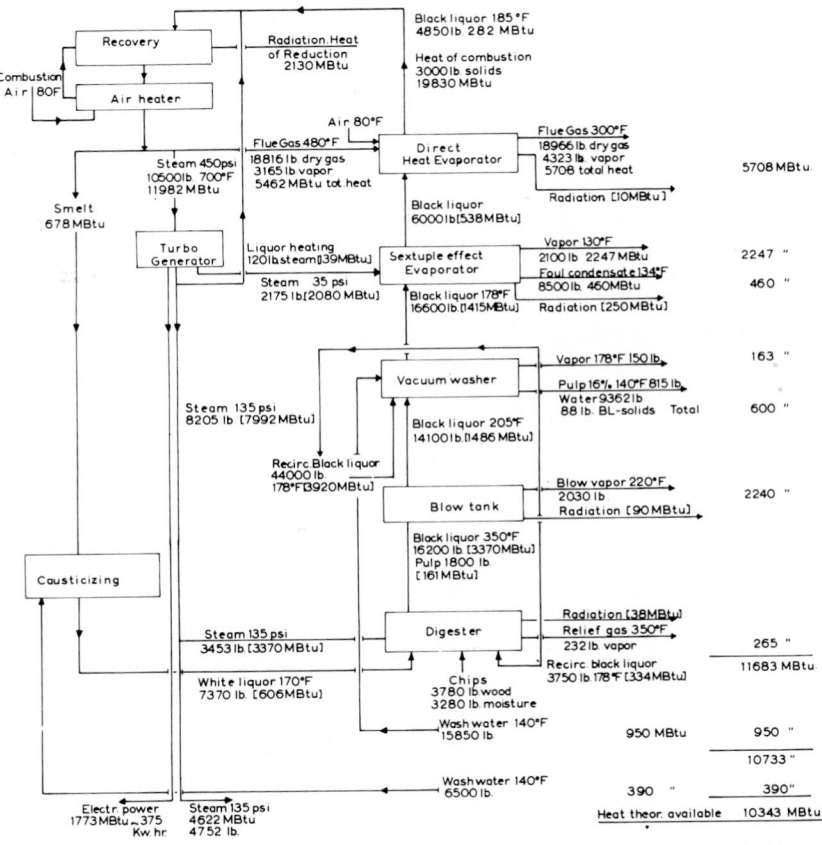

Figure VIII-51. Approximate heat balance in alkaline pulp mill (363). Based on 1 ton air-dry pulp (1800 lb M.F. pulp). Datum temperature 80°F. Assumed boiler feed water at 250°F, obtained from returned condensate.

8. HEAT ECONOMY—THE HEAT BALANCE OF THE KRAFT PULPING PROCESS

The entire heat and power requirements of a kraft pulp mill, with black liquor regeneration and controlled efficiency of operation, are covered by the combustion of the organic matter in the liquor. About 40% to 50% of the wood is dissolved in the liquor during the cook. As compared with other fuels, such as oil or coal, black liquor has a low calorific value. Whereas fuel oil produces about 18,000 BTU per lb (10,000 kcal per kg) and coal, depending upon its grade, 11,700 to 13,500 BTU per lb (6500 to 7500 kcal per kg), completely dry solids from black liquor produce 5900 to 7000 BTU per lb (3300 to 3900 kcal per kg) and a black liquor with 60% dry content, as it is injected into the furnace, only 3000 to 3400 BTU per lb (1700 to 1900 kcal per kg). It should be pointed out that black liquor consists of 40% water and only 30% combustible material, with the remaining 30% being inorganic components without any heat value. The entire regeneration plant must be adjusted, in accordance with heat engineering data, to the heterogeneous composition of the combustible material in the liquor in order to attain the high efficiency necessary for the plant. The heat balance of a kraft pulp mill, shown in Fig. VIII–51, is based on the production of one short ton of air-dry pulp with no oxidation of the black liquor before its combustion. As pointed out in Section 5.5, the oxidation of the liquor is an exothermic reaction and thus uses up a certain amount of heat (363,393).

As can be seen from Fig. VIII–51, the calorific units fed into the liquor furnace and taken from it are as follows:

	Heat value	
	MBTU	kcal
Black liquor (4850 lb or 2195 kg) with 61.8% dry content at 85°C	282	71,100
Combustion of organic material (3000 lb or 1360 kg)	19,830	4,977,200
Steam (120 lb or 54.4 kg) at 135psi (9.5 atm) from liquor preheating	139	35,000
Total heat produced	20,251	5,083,200
Radiation and reduction	2,130	516,700
Content of melt	678	170,860
Steam production (10,500 lb or 4755 kg) at 450 psi (32 atm) and 370°C	11,982	3,019,240
Flue gases	5,461	1,376,400
Total heat consumed	20,251	5,083,200

The amount of heat from the relief gases and vapors totals 11,683 MBTU (2,994,800 kcal) of which 1340 MBTU (33,700 kcal) has to be deducted for the preparation of warm water for the washing of the pulp and the caustification plant, so that 10,343 MBTU (2,607,100 kcal) remains and can be used for the warming of the water in the various departments if the necessary heat exchangers are available.

Figure VIII–52 shows a flow-sheet of the distribution of the amounts

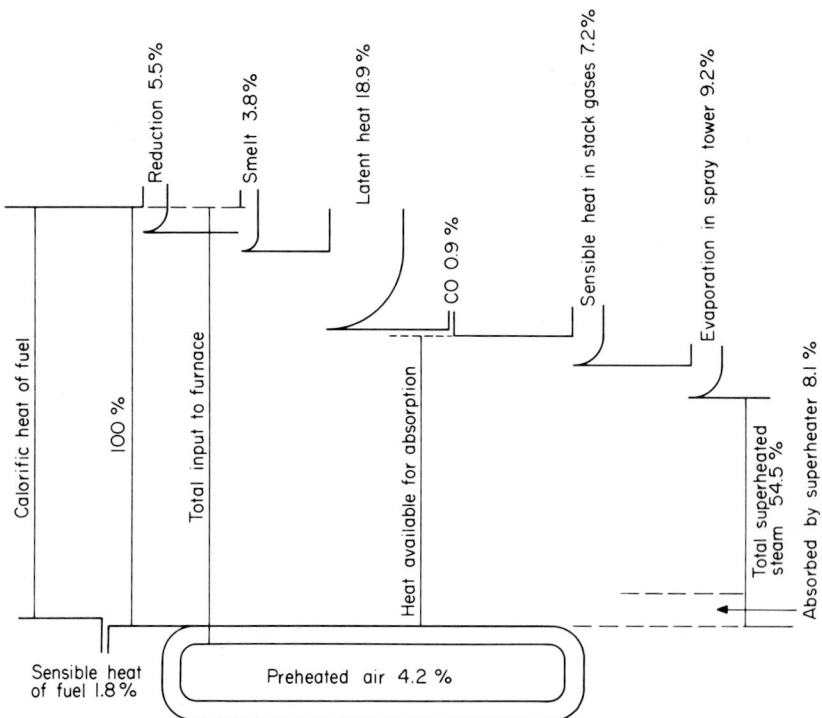

Figure VIII-52. Overall heat balance flow-sheet: distribution of the total heat supplied to the unit (393).

of heat units introduced into the furnace. The calorific heat and the sensible heat of the hot black liquor, upon which all other percentages are based, are taken as 100%. Small amounts of heat are supplied in the preheated air. Of the heat added, 54.5% leaves the furnace in the form of superheated steam. When the heat from the chemical reaction and from the evaporation of the water in the black liquor, which cannot be recovered, is deducted, then 70.9% of the heat remains and can be absorbed. About 9.2% of the total heat is consumed in the cyclone evaporator and 7.2% is lost in the stack gases.

The efficiency of the recovery boiler is lastingly affected by the temperature. As shown in Fig. VIII–53, the efficiency increases 1% with every

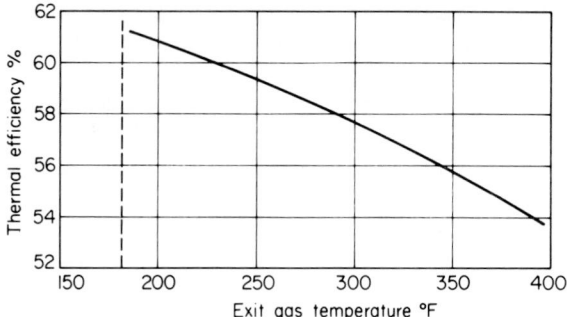

Figure VIII-53. Effect of exit gas temperature on recovery unit thermal performance (366).

20°C drop in the temperature of the exit gas. This increase in efficiency through the decrease in the exit gas temperature is more pronounced with a recovery boiler than with a normal steam boiler and is caused chiefly by the great loss of heat due to the high moisture content of the flue gases in the recovery system. By the use of a suitable evaporative Venturi scrubber,

TABLE VIII-17

COMPARISON OF HEAT LOSSES OF DIFFERENT EVAPORATIVE SYSTEMS (368)

Evaporative system losses	BT system (%)	Injection (%)	Rotary furnace (%)
Dry gases	5.5	5.5	5.5
Steam preheating	1.0	2.0	2.0
Steam evaporation	8.7	17.5	17.5
Total A	15.2	25.0	25.0
Losses by radiation	3.5	3.3	17.3
Melting and reaction heat	8.2	8.2	8.2
Incomplete combustion	0.7	0.7	0.7
Total B	12.4	12.2	26.2
Total A + B	27.6	37.2	51.2
Conversion in the boiler			
High-pressure steam	64.1	62.8	48.8
Low-pressure steam	5.4	—	—
Hot condensate from concentrator	2.9	—	—
Total C	72.4	62.8	48.8
Total A + B + C	100.0	100.0	100.0

exit gas temperatures of 85°C can be obtained (316,366). The lowering of the temperature of the exit gases is limited because of the danger of corrosion and the formation of deposits in the heat exchangers when they are operating at low temperatures. This danger is greatly increased by the formation of sulfuric acid, unless the heat exchangers are built of materials which resist corrosion (68).

The modified Trobeck-Bergström-Tomlinson process with black liquor oxidation, as discussed in Section 5.4 and schematically shown in Fig. VIII-40, is distinguished by a considerably lower heat loss. The losses in the heat introduced into the evaporative systems are shown in Table VIII-17.

REFERENCES

1. Abele, W., *Wochbl. Papierfabrik.* **40** (31/32), 125 (1942).
2. Abrams, E., *Tappi* **46** (2), 136A (1963).
3. Adams, D. F. and Koppe, R. K., *Tappi* **41** (7), 366 (1958).
4. Adams, D. F., Koppe, R. K. and Jungroth, D. M., *Tappi* **43** (6), 602 (1960).
4a. Ahlgren, P., Lemon, S. and Teder, A., *Acta Chem. Scand.* **21** (6), 1119 (1967).
4b. Ahlgren, P., Ishizu, A., Szabo, I. and Theander, O., *Svensk Papperstidn.* **71** (9), 355 (1968).
5. Ahlm, F., *Paper Trade J.* **113** (13), 115 (1941).
6. Ahlm, C. A. and Leopold, B., *Tappi* **46** (2), 102 (1963).
7. Alfredsson, B., Gedda, L. and Samuelson, O., *Svensk Papperstidn.* **64** (19), 694 (1961).
8. Alfredsson, B. and Samuelson, O., *Tappi* **46** (6), 379 (1963).
9. Alfredsson, B., Samuelson, O. and Sandstig, B., *Svensk Papperstidn.* **66** (18), 703 (1963).
9a. Annergren, G., Backlund, A. and Haglund, A., *Communications North Sulfate Colloquium* **8** (56), 66, Dec. 1967.
10. Annergren, G., Rydholm, S., Richter, J. and Backland, A., *Paper Trade J.* **149** (10), 42 (1965); *Tappi* **48** (7), 52A (1965).
11. Anon., *Tappi* **42** (6), 60A (1959).
12. Anon., *Chem. Eng. News* **44** (12), 46 (1966).
13. Aronowsky, S. J. and Gortner, R. A., *Ind. Eng. Chem.* **22**, 941 (1930).
14. Aurell, R., *Svensk Papperstidn.* **66** (23), 978 (1963).
15. Aurell, R., *Svensk Papperstidn.* **67** (2), 43 (1964); *Tappi* **48** (2), 80 (1965).
16. Aurell, R., *Svensk Papperstidn.* **67** (3), 89 (1964).
17. Aurell, R. and Hartler, N., *Tappi* **46** (4), 209 (1963).
18. Aurell, R. and Hartler, N., *Svensk Papperstidn.* **66** (3), 59, (4), 97 (1965).
19. Axelsson, S., Croon, I. and Enstrom, B., *Svensk Papperstidn.* **65** (18), 693 (1962).
20. Baldwin, N., *Tappi* **45** (7), 220A (1962).
21. Banfill, H. M., Wheeler, J. G. and Fergusson, J. M., *Pulp Paper Mag. Can.* **59** (C), 157 (1958).
22. Bard, J. W., *Tech. Assoc. Papers* **24**, 109 (1941).
23. Barefoot, A. C., Hitchings, R. G. and Ellwood, E. L., *Tappi* **47** (6), 343 (1964).
24. Barefoot, A. C., Hitchings, R. G. and Ellwood, E. L., *Tappi* **49** (4), 137 (1966).
25. Basberg, A., *Norsk Skogind.* **22**, 245 (1934).

26. Battershill, J. W., *Pulp Paper Mag. Can.* **67** (C), T-149 (1966).
27. Bergen, J. M. v., *Tappi* **39** (7), 195A (1956).
28. Berger, H. F. and Brown, R. I., *Tappi* **42** (3), 245 (1959).
29. Bergström, H., *Svensk Papperstidn.* **60**, 37 (1957).
30. Bergström, H. and Trobeck, K. G., *Svensk Papperstidn.* **42** (22), 554 (1939).
31. Bergström, H. and Trobeck, K. G., *Svensk Papperstidn.* **46** (15), 357 (1943).
32. Berthier, R., *Assoc. Tech. Ind. Papetiere, Bull.* **4**, 93 (1953).
33. Bethge, P. O. and Ehrenborg, L., *Svensk Papperstidn.* **70** (10), 347 (1967).
34. Bialkowsky, H. W. and DeHaas, G. G., *Pulp Paper Mag. Can.* **53** (10), 99 (1952); *Tappi* **36** (7), 330 (1953).
35. Bixler, A. L., *Tech. Assoc. Papers* **21**, 181 (1938).
36. Bjorkman, C. B. and Karlson, K. K., *Paperi Puu* **23** (7A), 82 (1941).
37. Blosser, R. O. and Cooper, H. B., *Paper Trade J.* **151** (11), 46 (1967).
38. Boesen, C. E., *Norsk Skogind.* **18** (2), 56 (1964).
39. Bogomolow, B. D. and Gelfand, E. D., *Bumazh. Prom.* (3), 3 (1966).
40. Boija, J. A., *Svensk Papperstidn.* **39**, 405 (1936).
41. Bolger, J. C., Tate, D. C. and Hopfenberg, H. B., *Tappi* **50** (5), 231, 247 (1967).
42. Boniface, A. and Mattison, R. J., *Paper Trade J.* **150** (22), 28 (1966).
43. Borlew, P. B., *Tappi* **39** (6), 184A (1956).
44. Borlew, P. B. and Pascoe. T. A., *Tech. Assoc. Papers* **29**, 166 (1946); **30**, 570 (1947).
45. Boro, P. C. and Blanchard, Z. S., *Tappi* **45** (2), 132A (1962).
46. Brandt, M. W., Krause, F. and Shafer, M., *Tappi* **47** (5), 137A (1964).
47. Brasch, D. J. and Free, K. W., *Tappi* **47** (4), 186 (1964); **48** (4), 245 (1965).
48. Brauns, F. E. and Grimes, W. S., *Tech. Assoc. Papers* **22**, 574 (1939).
49. Bray, M. W. and Curran, C., *Paper Trade J.* **97** (5), 30 (1933).
50. Bray, M. W. and Curran, C., *Tech. Assoc. Papers* **21**, 458 (1938).
51. Bray, M. W. and Martin, J. S., *Tech. Assoc. Papers* **14**, 214 (1931).
52. Bray, M. W. and Martin, J. S., *Tech. Assoc. Papers* **24**, 251 (1941).
53. Bray, M. W. and Martin, J. S., *Tech. Assoc. Papers* **30**, 388 (1947).
54. Bray, M. W., Martin, J. S. and Carpenter, J., *Tech. Assoc. Papers* **14**, 214 (1931).
55. Bray, M. W., Martin, J. S. and Schwartz, S. L., *Tech. Assoc. Papers* **22**, 382 (1939).
56. Bray, M. W., Schwartz, S. L. and Martin, J. S., *Tech. Assoc. Papers* **23**, 232 (1940).
57. Briston, O. J., Owen, J. J. and Yirak, J. J., *Tappi* **42** (5), 390 (1959).
58. Brunsvik, J. J., *Norsk Skogind.* **17** (4), 130 (1963).
59. Buxton, W. H., *Tappi* **46** (12), 183A (1963).
60. Byrd, L., Ellwood, E. L., Hitchings, R. G. and Barefoot, A. C., *Forest Prod. J.* **15** (8), 313 (1965).
61. Canavan, H. M., Blanchard, Z. S. and Rienhoff, H. Y., *Tappi* **45** (2), 129A (1962); **46** (9), 178A (1963); **47** (12), 159A (1964); **49** (2), 102A (1966).
62. Cancea, Ch. L., *Pulp Paper Mag. Can.* **65** (7), T-309 (1964).
63. Cann, E. D. and Roberson, W. B., *Tappi* **43** (2), 97 (1960).
64. Cate, F. L., *Tappi* **36** (5), 225 (1953).
65. Cave, G. C. B., *Tappi* **46** (1), 1, 5, 11, 15 (1963).
66. Cederlof, R., Edfors, M. L., Friberg, L. and Lindvall, T., *Tappi* **48** (7), 405 (1965).
67. Chappell, E. E. and Tester, J. W., *Tappi* **47** (4), 193 (1964).
68. Childs, G. D., *Tappi* **50** (6), 122A (1967).
69. Chirkin, C. S. and Tischenko, D. V., *Zh. Prikl. Khim.* **35** (1), 153 (1962).
70. Chirkin, C. S. and Tischenko, D. V., *Zh. Prikl. Khim.* **39** (2), 432 (1966).
71. Choudury, A. P. and Dahlstrom, D. A., *Am. J. Chem.* **3** (4), 433 (1957).
72. Christofferson, K. and Samuelson, O., *Svensk Papperstidn.* **63** (20), 729 (1960).
73. Clark, C. L., *Tappi* **33** (2), 108 (1950).

74. Clark, J. d'A., *Tech. Assoc. Papers* **26**, 462 (1943).
75. Clarke, W. W., *Tappi* **47** (2), 149A (1964).
75a. Clayton, D. W. and Sakai, A., *Pulp Paper Mag. Can.* **68** (12), T-619 (1967).
76. Clayton, D. W. and Stone, J. E., *Pulp Paper Mag. Can.* **64** (11), T-459 (1963).
77. Collins, T. T., *Tappi* **38** (8), 172A (1955).
78. Collins, T. T., *Tappi* **42** (1), 9 (1959).
79. Collins, T. T., *Tappi* **45** (9), 692 (1962).
80. Collins, T. T., *Paper Trade J.* **149** (22), 34 (1965).
81. Collins, T. T., Seaborne, C. R. and Anthony, A. W., *Tech. Assoc. Papers* **30**, 168 (1947).
82. Colombo, P., Corbetta, D., Pirotta, A. and Ruffini, G., *Assoc. Tech. Ind. Papetière, Bull.* **15** (5), 404 (1961); **16** (5), 370 (1962); **18** (6), 359 (1964).
83. Colombo, P., Corbetta, D., Pirotta, A. and Ruffini, G., *Svensk Papperstidn.* **67** (12), 505 (1964).
84. Combustion Engineering Co., U.S. Patent 3,122,421 (1964).
84a. Container Corp. of America, U.S. Patent 3,309,262, March 14, 1967.
85. Coogan, F. J. and Stovall, J. H., *Tappi* **48** (6), 94A (1965).
86. Cooke, W. P., *Tappi* **45** (9), 699 (1962).
87. Cooper, S. R. and Haskell, C. F., *Paper Trade J.* **151** (13), 58 (1967).
88. Corbett, W. M. and Kidd, J., *Tappi* **41** (3), 137 (1958).
89. Corbetta, D., Pirotta, A. and Ruffini, G., *Ind. Carta* **1**, 501 (1963).
90. Correns, E., *Cellulosechemie* **19** (5), 105 (1941).
91. Correns, E. and Jacopian, V., *Faserforsch. Textiltech.* **7**, 387 (1956).
92. Correns, E. and Schorning, P., *Cellulosechemie* **20**, 29 (1942).
93. Cowan, W. F., *Tappi* **42** (2), 152 (1959).
94. Craig, A. V., *Pulp Paper Mag. Can.* **66** (7), T-393 (1965).
95. Crandall, H. C. and Enderlein, G. F., *Tech. Assoc. Papers* **29**, 310 (1946).
96. Crocker, E. C., *Tappi* **45** (9), 169A (1962).
97. Crocker, E. C., *Tappi* **46** (9), 165A (1963).
98. Croon, I. and Enstrom, B. F., *Tappi* **44** (12), 870 (1961); *Svensk Papperstidn.* **65** (16), 595 (1962).
99. Crown Zellerbarch Co., *Chem. Eng. News* **43** (42), 39 (1965).
100. Daleski, E. J., *Tapp* **48** (6), 325 (1965).
101. Davis, A. S., *Paper Trade J.* **149** (34), 36 (1965).
102. Dean, J. C., Saul, C. M. and Turner, C. H., *Appita Proc.* **10**, 84 (1956).
103. DeHaas, G. G. and Amos, L. C., *Tappi* **50** (3), 75A (1967)
104. DeHaas, G. G. and Hansen, G. A., *Tappi* **38** (12), 732; (3), 134A (1955).
105. Derfer, J. M., *Tappi* **46** (9), 513 (1963); **49** (10), 117A (1966).
106. Devones, K. R., *Tappi* **46** (10), 167A (1963).
107. DiBella, E. P., Green, R., Kraft, W. M. and Gottesman, R. T., *Am. Oil Chem. Soc.* **43** (3), 196, 199 (1965).
108. Diedrichs, E. and Hedström, B., *Svensk Papperstidn.* **59** (15), 561 (1956).
109. Dillen, S. and Noreus, S., *Svensk Papperstidn.* **70** (4), 122 (1967).
109a. Dillen, S. and Noreus, S., *Svensk Papperstidn.* **71** (15), 509 (1968).
110. Dörr, R. E., *Wochbl. Papierfabik.* **37** (1), 1, (1939); 39, 267 (1941); *Cellulosechemie* **21**, 48 (1943).
111. Domansky, R., Kramar, A. and Ebringerova, A., *Assoc. Tech. Ind. Papetière, Bull.* **18** (3), 120 (1964).
112. Drew, J. and Pylant, G. D., *Tappi* **49** (10), 430 (1966).
113. Drewsen, E., U.S. Patent 996,225 (1911).
114. Dudley, H. C. and Dellavalle, J. M., *Tech. Assoc. Papers* **22**, 312 (1939).

115. Engineering Divison Corrosion Committee, *Tappi* **45** (2), 140A (1962).
116. Enkvist, T., *Svensk Papperstidn.* **51** (10), 225 (1948).
117. Enkvist, T., Ashorn, T. and Hastbacka, K., "Prom. Wood Chem. Symposium, Montreal 1961," p. 177, Butterworth, London, 1962.
118. Farkas, J., *Papir Cellulosa* **18**, 156 (1963).
119. Felicetta, V. F., Peniston, Q. P. and McCarthy, J. L., *Tappi* **36**, 425 (1953); **42** (4), 162A (1959).
120. Field, L., Drummont, P. E., Riggins, P. A. and Jones, E. A., *Tappi* **41** (12), 721, 727 (1958).
121. Fieldner, A. C., U.S. Bureau of Mines, Monograph No. 4 (1931).
122. Findlay, M. E., *Tappi* **43** (8), 183A (1960).
123. Fitch, B. Pitkin, W. H., *Tappi* **47** (10), 170A (1964).
124. Fogman, C. B., *Svensk Papperstidn.* **53**, 327 (1950).
125. Forman, L. V. and Niemeyer, D. D., *Tappi* Monograph No. 4, 167 (1947).
126. Franzon, O. and Samuelson, O., *Svensk Papperstidn.* **60** (23), 872 (1957).
127. Frombach, M. F., *Pulp Paper Mag. Can.* **64** (3), T-175 (1963).
128. Freudenberg, K. and Reichert, M., *Tappi* **38** (8), 165A (1955).
129. Frey-Wyssling, A., Mühlethaler, A. and Bosshard, K., *Holz Roh.-u Werkstoff* **13**, 245 (1955).
130. Gellman, I., *Tappi* **48** (6), 106A (1963).
131. Ghisoni, P., *Tappi* **37** (5), 201 (1954).
132. Gierer, J. and Lenz, B., *Svensk Papperstidn.* **68** (9), 334 (1965).
133. Gierer, J. Lenz, B., Noren, I. and Soderberg, S., *Tappi* **47** (4), 233 (1964).
134. Gierer, J. and Noren, I., *Acta Chem. Scand.* **16**, 1713 (1962).
135. Gierer, J. and Smedman, L. A., *Acta Chem. Scand.* **19** (5), 1103 (1965).
136. Gierer, J., Soderberg, S. and Thoren, S., *Svensk Papperstidn.* **66** (23), 990 (1963).
137. Gierer, J. and Wallin, N. H., *Acta Chem. Scand.* **19** (6), 1502 (1965).
138. Giertz, H. W., *Papier* **17** (5), 191 (1963).
139. Gillespie, D. C., *Tappi* **36** (4), 147 (1953).
140. Gillespie, D. C. and Johnston, I. W., *Tappi* **39** (7). 499 (1956).
141. Goldschmidt, Th. A.G., German Patent 1,144,242.
142. Green, J. W., *Tappi* **39** (7), 472 (1956).
143. Green, S. J. and Frattali, F. J., *J. Am. Chem. Soc.* **68**, 1789 (1946).
144. Guest, E. T., *Pulp Paper Mag. Can.* **66** (12), T-617 (1965).
145. Hägglund, E., *Svensk Papperstidn.* **49** (9), 204 (1946).
146. Hägglund, E., *Tappi* **32** (6), 241 (1949). *Holz Roh.-u Werkstoff* **11**, 251 (1953).
147. Hägglund, E., "Chemistry of wood." Academic Press, New York, 1951.
148. Hägglund, E. and Enkvist, T., U.S. Patent 2,711,430; reissue 24,293 (1956).
149. Hägglund, E. and Hedlund, W., *Wochbl. Papierfabrik.* **50** (49), 61 (1932).
150. Haller, A. K., *Tappi* **45** (8), 132A (1962).
151. Hamilton, J. K., Partlow, E. V. and Thompson, N. S., *Tappi* **41** (12), 803, 811 (1958).
152. Hammack, F. M., *Tappi* **44** (2), 168A (1961).
153. Hammond, R. N. and Billington, P. S., *Tappi* **32** (12), 563 (1949).
154. Hannan, Ph. I., *Tappi* **47** (2), 162A (1964).
155. Hanson, F. S., *Tech. Assoc. Papers* **23**, 163 (1940).
156. Harding, C. I. and Landry, J. E., *Tappi* **49** (8), 61A (1966).
157. Hartler, N., *Svensk Papperstidn.* **62** (13), 467 (1959); **65** (13), 513 (1962).
158. Hartler, N., *Svensk Papperstidn.* **68** (10), 369 (1965).
159. Hartler, N., *Svensk Papperstidn.* **69** (6), 191 (1966).
160. Hartler, N., *Tappi* **50** (3), 156 (1967).
160a. Hartler, N., Libert, J. and Teder, A., *Svensk Papperstidn.* **71** (3), 65 (1968).

161. Hartler, N. and Onisko, W., *Svensk Papperstidn.* **65** (22), 905 (1962).
162. Hartler, N. and Svensson, I. L., *Ind. Eng. Chem. Prod. Res. Devt.* **4** (2), 80 (1965).
163. Harvin, R. L., Hills, C. B., Rothrock, C. W. and Nolan, *Tappi* **33** (7), 338 (1950).
164. Hearon, W. H., McGregor, W. S. and Goheen, D. W., *Tappi* **45** (1), 28A (1962).
165. Heath, M. A., Bray, M. W. and Curran, C. E., *Tech. Assoc. Papers* **17**, 447 (1934).
166. Hedlund, J., *Svensk Papperstidn.* **54**, 408 (1951).
167. Hempsall, L. C., *Tappi* **46** (12), 160A (1963).
168. Herschler, R. J. and Jakob, S. W., *Tappi* **48** (6), 43A (1965).
169. Hess, K. and Trogus, K., *Z. Phys. Chem.* **B15** (2/3), 381 (1931).
170. Heuser, E., Shockley, W. and Kjellgren, R., *Tappi* **33** (2), 101 (1950).
171. Hill, D. A., *Paper Trade J.* **148** (51), 22 (1964).
172. Hill, D. A., *Paper Trade J.* **149** (6), 44 (1965); *Pulp Paper Mag. Can.* **66** (C), T-95 (1965).
173. Hinrichs, D. D., 20th Alk. Pulping Conference, Richmond, Sept. 1966.
174. Hoag, D. S., *Tappi* **47** (12), 734 (1964).
175. Hobden, J. F., *Appita Proc.* **10**, 99 (1956).
176. Holzer, W. F. and Lewis, H. F., *Tappi* **33** (2), 110 (150).
177. Howard, T. E. and Walden, C. C., *Tappi* **48** (3), 136 (1965).
178. Hruitford, B. F. and McCarthy, J. L., *Tappi* **50** (2), 82, 86; **50** (5), 270 (1967).
179. Hübnett, F. and Keim, K., U.S. Patent 2,935,533 (1960).
180. Hunter, H. O. and Wade, H. R., *Tappi* **39** (1), 179A (1956).
181. Jacopian, V., *Zellstoff Papier* **12** (9), 259 (1963).
182. Jacopian, V, and Casperson, G., *Faserforsch. Textiltech.* **17** (6), 267 (1966).
183. Jakobson, T., *Svensk Papperstidn.* **42**, 473, 513, 534, 600 (1939); **43**, 24, 44, 65, 109, 175 (1940); **46**, 128, 183, 283 (1943); **47**, 473, 492, 580, 605 (1944); **52**, 193, 290, 379, 451, 481 (1949); **53**, 369, 444, (1950); **55**, 432, 460, 483 (1952); **56**, 456, 491 (1953); **57**, 883, 926 (1954); **58**, 16 (1955).
184. Janes, W. J., *Tappi* **47** (5), 122A (1964).
195. Jansson, J., *Svensk Papperstidn.* **66** (20), 853 (1963).
186. Jansson, L. B., *Tappi* **46** (5), 296 (1963).
187. Jarvela, O. and Makkonen, H., *Tappi* **50** (3), 147 (1967).
188. Jayme, G., *Wochbl. Papierfabrik.* **38**, 277 (1940).
189. Jayme, G., *Tappi* **46** (7), 415 (1963).
189a. Jayme, G., *Holz Roh.-u Werkstoff* **24**, 449 (1966).
190. Jayme, G. and Gessler, H., *Papier* **17** (1), 5 (1963).
191. Jayme, G. and Grögaard, L., *Wochbl. Papierfabrik.* **38**, 149 (1940).
192. Jayme, G., Kohler, L. and Haas, W. L., *Papier* **10**, 495, 540 (1956).
193. Jayme, G. and Licht, W., *Papier* **6**, 450, 510 (1952); *Holzforschung* **9**, 33 (1955).
194. Jayme, G. and Schorning, P., *Holz Roh.-u Werkstoff* **3** (9), 273 (1940).
195. Jayme, G., Wagenbach, H. and Deloff, W., *Wochbl. Papierfabrik.* **37** 229, 240, 353, 361 (1939).
196. Jayme, G. and Wörner, G., *Holz Roh.-u Werkstoff* **10**, 244 (1952).
197. Jernqvist, A. S., *Svensk Papperstidn.* **68** (15), 506; (16), 545, (17), 578 (1965); **69** (11), 395 (1966).
198. Jones, E. D., Campbell, R. T. and Nelson, G. G., *Tappi* **49** (9), 410 (1966).
199. Jonsson, S. E., *Svensk Papperstidn.* **64** (15), 565 (1961).
200. Kienitz, G. A., *Holz Roh.-u Werkstoff* **1**, 30 (1937).
201. Klein, M., McCall, M. S. and Lientz, J. R., *Tech. Assoc. Papers* **30**, 562 (1947).
202. Kleinert, T. N., *Tappi* **49** (2), 53, (3), 126 (1966).
203. Kleinert, T. N. and Marraccini, L. M., *Tappi* **47** (10), 605 (1964); **48** (3), 165 (1965).

204. Kleinert, T. N. and Marraccini, L. M., *Tappi* **48** (3), 165, 170, (4), 214, 224, (5), 270 (1965).
205. Kleinert, T. N., Marraccini, L. M. and Dostal, E., *Tappi* **43** (3), 201 (1960).
206. Kleinschmidt, R. V. and Antony, A. W., see reference 82,
207. Kleinschmidt, R. V., White, F. P. and Bowen, E. C., *Tappi* **41** (2), 86 (1958).
208. Kobe, K. A. and Sorensen, A. J., *Pacific Pulp Paper Ind.* **13** (2), 12 (1939).
209. Kominek, E. G. and Kahn, J. M., *Tech. Assoc. Papers* **29**, 365 (1946).
210. Komshilow, N. F., *Tr. Karelsh. Filiala Akad. Nauk.* **38**, 45 (1963).
211. Kottek, J. F., *Appita* **12** (3), 88 (1958).
212. Kress, O. and McGregor, G. H., *Paper Trade J.* **96** (24), 40 (1933).
213. Kress, O. and Mosher, R. H., *Tech. Assoc. Papers* **27**, 649 (1944).
214. Kress, O. and Ratliff, F. T., *Tech. Assoc. Papers* **27**, 660 (1944).
215. Kringstad, K. and Kleppe, P. J., *Norsk Skogind.* **17** (11), 428 (1963); **18** (1), 13 (1964); **19** (9), 344 (1965).
216. Krumbein, J. P., *Tappi* **47** (5), 142A (1964).
217. Kuehl, J. F., Cornell, C. F. and Silverblatt, S. E., *Paper Trade J.* **150** (10), 52 (1966); *Tappi* **49** (11), 53A (1966).
218. Kulkarni, G. R. and Nolan, W. J., *Paper Ind.* **37** (2), 142 (1955).
219. Kurbegovic, M. and Bravar, M., *Papier* **21** (4), 170 (1967).
220. Landmark, P., *Norsk Skogind.* **15** (8), 342 (1961).
221. Landmark, P. A., Kleppe, P. J. and Johnsen, K., *Paper Trade J.* **148** (46), 48 (1964); *Tappi* **48** (5), 56A (1965).
222. Landry, J. E., *Tappi* **46** (12), 766 (1963).
223. Laroque, G. L. and Maass, O., *Can. J. Res.* **19B** (3), 1 (1941).
224. Laundrie, J. F., *Forest. Prod. Lab. Rept.* No. 2138, Dec. 1958.
225. Lawrence, R. V., *Tappi* **45** (8), 654 (1962).
226. Lawrence, W. P., *Tech. Assoc. Papers* **30**, 583 (1947).
227. Legg, G. W. and Hart, J. S., *Pulp Paper Mag. Can.* **61** (6), T-299 (1960).
228. Leonard, J. S., *Tappi* **49** (10), 84A (1966).
228a. Libert, J., *Communication North Sulfate Colloquium* **8** (56), 88 (Dec. 1967).
229. Lindberg, B., *Svensk Papperstidn.* **59** (15), 531 (1956).
230. Lindholm, I. and Stockman, L., *Svensk Papperstidn.* **65** (19), 755 (1962).
231. Lindholm, I. and Stockman, L., *Svensk Papperstidn.* **65** (17), 658 (1962).
232. Loras, V. and Loeschbrandt, F., *Norsk Skogind.* **10** (11), 402 (1956); Meddelese No. 118, Papir Forskningsinstutt; *Norsk Skogind.* **15**, 302 (1961).
233. Lumnus Co., U.S. Patent 3,165,436 (1965).
234. Mackay, W. B., *Pulp Paper Mag. Can.* **65** (C), T-107 (1964).
235. Maksimov, V. F. and Torf, A. I., *Bumazh. Prom.* **1964** (5), 6.
236. Martin, G. E., *Tappi* **33** (2), 84 (1950).
237. Martin, H. C., *Tappi* **42** (2), 108 (1959).
238. Martin. J. S., *Tappi* **32** (12), 534 (1949).
239. Martin, J. S. and Bray, M. W., *Tech. Assoc. Papers* **24**, 596 (1941).
240. Marvin, J. L., 20th Alkaline Pulping Conference. Richmond, Sept. 1966.
241. Mattson, V., *Tappi* **39** (2), 77 (1956).
242. May, B.T., *Tappi* **36**, (8), 374 (1953).
243. May, M. N., *Tappi* **35** (11) 511 (1952).
244. May, M. N. and Peckham, J. R., *Tappi* **41** (2), 90 (1958).
245. McIntosh, D. C., *Tappi* **46** (5), 273 (1963).
246. McKean, W. T., Hrutfiord, B. F. and Sarkanen, K. V., *Tappi* **48** (12), 699 (1965).
247. Meissner, H. P., Conway, E. R. and Muckley, H. S., *Ind. Eng. Chem.* **48**, 1347 (1956).

248. Meller, A., *Tappi* **33** (5), 248 (1950).

249. Meller, A., *Appita* **18** (2), 41 (1964).

250. Meller, A., *Holzforschung* **19** (4), 118 (1965).

250a. Meller, A., *Holzforschung* **22** (3), 88 (1968).

251. Meller, A. and Ritman, E. L., *Tappi* **47** (1), 55, (10), 634 (1964).

252. Merewether, J. W., *Holzforschung* **16** (1), 26 (1962).

253. Merewether, J. W., *Tappi* **45** (2), 159 (1962).

254. Meuly, W. C. and Tremaine, B. K., *Tappi* **36** (4), 154 (1953); **37** (8), 141A (1954).

255. Meyer, W. G. and Coma, J. G., *Chem. Eng. Progress* **54** (5), 178 (1958).

256. Miller, A., U.S. Patents 2,599,571, 2,599,572 (1952).

257. Miller, R. N., Swanson, W. H., Bray, M. W., Söderquist, R., Andrews, T. M. and Monsson, W. H., "Chemistry of the sulfite process." New York, 1928.

258. Mitchell, C. R. and Ross, J. H., *Forest Prod. Lab. Can. Quaterly Review* **15** 7, July-Sept. (1933).

259. Montano, J. and Hossfeld, R., *Tappi* **34** (10), 468 (1951).

260. Moran, J. and Wall, Ch., *Paper Trade J.* **149** (27), 31 (1965); *Tappi* **49** (3), 89A (1966).

261. Muckley, H. S., *Ind. Eng. Chem.* **48**, 1347 (1956).

262. Mueller, A., *Can. J. Res.* **34** (3), 162 (1956); *Tappi* **40** (3), 129 (1957).

263. Mueller, A., *Tappi* **42** (3), 179 (1959); *Pulp Paper Mag. Can.* **59** (1), T-3 (1959).

264. Munk, L., Todorski, Z. and Tomlinson, H. G., *Pulp Paper Mag. Can.* **65** (10), T-411 (1964).

264a. Murray, F. E., *Tappi* **42** (9), 761 (1959); *Pulp Pager Mag. Can.* **69** (1), T-26 (1968).

265. Murray, J. A., Fisher, H. and Sabean, D. W., *Paper Trade J.* **132** (22), 22 (1951).

266. Murray, F. E. and Rayner, H. B., *Tappi* **48** (10), 588 (1965).

267. Nelson, B. W., *Tappi* **46** (5), 277 (1963).

267a. Nilsson, E. R. and Oestberg, K., *Svensk Papperstidn.* **71** (3), 71 (1968).

268. Nikitin, W. M., Obolenskaya, A. W., Skatschkow, W. M. and Iwanenko, A. D., *Bumazh. Prom.* **38** (11), 14 (1963).

269. Nolan, W. J., *Tappi* **36** (9), 406 (1953).

270. Nolan, W. J., *Tappi* **40** (3), 170 (1957).

271. Nolan, W. J., *Tappi* **41** (4), 178A (1958).

272. Nolan, W. J., *Tappi* **41** (10), 567 (1958).

273. Nolan, W. J., *Tappi* **42** (4), 320 (1959).

274. Nolan, W. J. and Brown, W. F., *Tappi* **35** (9), 425 (1952).

275. Nolan, W. J. and Brown, W. F., *Tappi* **35** (11), 505 (1952).

276. Nolan, W. J. and Brown, W. F., *Pulp Paper Mag. Can.* **53** (8), 98 (1952).

277. Nolan, W. J. and McCready, D. W., *Tech. Assoc. Papers* **19**, 237 (1936).

278. Nolan, W. J., Harvin, R. L., Reeder, L. M. and Rothrock, C. W., *Tappi* **34** (12), 529 (1951).

279. Norman, A. G. *in* Ott and Spurlin, "Cellulose" Vol. I, pp. 463, 1954.

280. Norton, L., *Paper Trade J.* **120** (7), 36 (1945).

281. Olsen, F., U.S. Patent 2,061,205 (1931).

282. Olsson, J. E. and Samuelson, O., *Svensk Papperstidn.* **69** (20), 703 (1966).

283. O'Neil, F. W., Keller, E. L. and Martin, J. S., *Northeast Tech. Comm. Utilization of Beech, Beech Utilization Series* No. 17, 21 pp. 1958.

284. Othmer, D. F., *Tappi* **50** (3), 101 (1967).

285. Owen, H. M., *Tappi* **42** (5), 176A (1959).

286. Parks, L. R., *Tappi* **42** (4), 317 (1959).

287. Pascoe, T. A., Buchanan, J. S., Kennedy, E. H. and Sivola, G., *Tappi* **42** (4), 265 (1959).

288. Patureau, A. M., *Tappi* **48** (9), 68A (1965).
289. Peach, R. E. and Petersen, R. E., *Tappi* **47** (6), 167A (1964).
290. Peckham, J. R. and May, M. N., *Tappi* **43** (1), 45 (1960).
291. Perkins, J. K., Welsh, H. S. and Mappus, J. H., *Tappi* **37** (3), 83 (1954).
292. Pettersson, S. E. and Rydholm, S., *Svensk Papperstidn.* **64** (1), 4 (1961).
293. Pilyugina, L. G., Komshilow, N. F., Bachurian, G. V. and Djukiev, E. F., *Khim. Pererabotka Drevesiny Akad. Nauk Tekhn.* **14**, 9 (1964).
294. Powers, P. O., *Tech. Assoc. Papers* **31**, 632 (1948).
295. Radej, Z. and Kristofowa, Z., *Papir Cellulosa* **19** (6), 152 (1964).
296. Rapson, W. H., *Pulp Paper Mag. Can.* **66** (5), T–295 (1965).
296a. Rayner, H. B. and Murray, F. E., *Pulp Paper Mag. Can.* **69** (15), T–319 (1968).
296b. Rayner, H. B., Murray, F. E. and Williams, I. H., *Pulp Paper Mag. Can.* **68** (6), T–301 (1967).
297. Reed, A. E. and Gillespie, W. F., *Tappi* **32** (12), 529 (1949).
298. Regestad, S. O., *Svensk Papperstidn.* **54** (2), 35 (1951).
299. Reid, H. A., *Appita* **15** (5), 102 (1962).
300. Rennel, J. and Stockman, L., *Svensk Papperstidn.* **66** (21), 863 (1963).
301. Richter, G. A., U.S. Patent 1,088,044 (1930).
302. Richter, G. A., U.S. Patents 1,787,953, 1,816,343, 1,819,002 (1931); 1,880,043 (1932); 2,036,606 (1936).
303. Richter, G. A., *Tech. Assoc. Papers* **19**, 447 (1936); *Tappi* **38** (3), 129 (1955).
304. Richter, G. A., *Tappi* **39** (4), 193 (1956).
305. Richter, J., *Tappi* **32** (7), 330 (1949).
306. Richter, J., *Paper Trade J.* **150** (10), 63 (1966); *Tappi* **49** (6), 48A (1966).
306a. Richter, J., *Communications North Sulfate Colloquium* Ser. **8**, No. 56, 95 Dec. 1967.
307. Rienhoff, H. Y., *Tappi* **45** (2), 137A (1962).
308. Roberts, L. M., *Tech. Assoc. Papers* **24**, 119 (1941).
309. Roberts, L. M., Beaver, C. E. and Blessing, W. H., *Tech. Assoc. Papers* **31**, 651 (1948).
310. Rothrock, C. W. and Nolan, W. J., *Tappi* **35** (1), 29 (1952).
311. Ruus, L., *Svensk Papperstidn.* **66** (16), 554 (1963).
312. Ruus, L., *Svensk Papperstidn.* **67** (19), 751 (1964).
313. Saarnio, J. and Gustavsson, C., *Paperi Puu* **35**, 65 (1953).
314. Sakata, I., *J. Chem. Soc. Japan, Ind. Chem. Sect.* **66** (4), A 499, 504 (1963).
314a. Sanyer, N., *Tappi* **51** (8), 48A (1968).
315. Sanyer, N. and Laundrie, J. F., *Tappi* **47** (10), 640 (1964).
316. Sapp, J. E., *Tappi* **35** (6), 40A (1952).
317. Scan. Test 1:65, *Svensk Papperstidn.* **68** (22), 805 (1965).
318. Scan. Test 2:65, *Svensk Papperstidn.* **68** (23), 849 (1965).
319. Schegolew, W. P. and Nikitin, W. N., *Bumazh. Prom.* **39** (7), 3 (1964).
320. Schmied, J., *Chem. Zvesti* **15** (9), 677 (1961).
321. Schoening, M. A., Shemilt, L. W. and Wright, R. H., *Tappi* **36** (4), 176 (1953).
322. Schoening, M. A. and Wright, R. H., *Tappi* **35** (12), 564 (1952).
323. Schwartz, S. L. and Bray, M. W., *Paper Trade J.* **107** (12), 140 (1938).
324. Schwartz, S. L. and Bray, M. W., *Tech. Assoc. Papers* **22**, 600 (1939).
325. Schwartz, S. L. and Bray, M. W., *Tech. Assoc. Papers* **29**, 560 (1946).
326. Scopp, J. W., *Paper Trade J.* **144** (40), 39 (1960).
326a. Shah, I. S. and Mason, L., *Tappi* **50** (10), 27A (1967).
327. Sharkov, V. I., Kuibina, N. T. and Soloveva, Y. P., *Zh. Priklad. Khim.* **36** (7), 1579 (1963).
328. Simmonds, F. A., *Tappi* **46** (11), 145A (1963).

329. Simmonds, F. A. and Chidester, G. H., *Forest Prod. Lab. Rept.* **2189**, June 1960.

330. Simmonds, F. A., Kingsbury, R. M. and Martin, J. S., *Tappi* **38** (3), 179 (1955).

331. Simmonds, F. A., Kingsbury, R. M., Martin, J. S. and Mitchell, R. L., *Tappi* **39** (9), 641 (1956).

332. Simonson, R., *Svensk Papperstidn.* **66** (20), 839 (1963); **68** (8), 275, (15), 500 (1965).

333. Sirakoff, G., *Holz Roh.-u Werkstoff* **4** (6), 205 (1941).

334. Sloman, A. R., *Appita* **3**, 47 (1949); U.S. Patent 2,639,987 (1953).

335. Sloman, A. R., Dean, J. C., Howard, M. C. and Harbour, M. J., *Appita* **18** (2), 61 (1964).

336. Smedslund, T. H., *Paperi Puu* **33** (5), 185 (1951); U.S. Patent 2,581,050 (1952).

337. Smelt Water Research Group, *Paper Trade J.* **150** (42), 56 (1966).

338. Smith, E. L., *Paper Trade J.* **148** (44), 30 (1964).

339. Smith, H. L., *Tappi* **46** (12), 182A (1963).

340. Spence, G., *Paper Trade J.* **71** (11), 24 (1920).

341. Stamm, A. J., *Ind. Eng. Chem.* **27**, 1480 (1935).

342. Stamm, A. J., *Ind. Eng. Chem.* **27**, 401 (1935).

343. Stamm, A. J. and Loughborough, W. K., *J. Phys. Chem.* **39**, 121 (1935).

344. Starkey, R. L., *Tappi* **44** (7), 493 (1961).

345. Stone, J. E. and Nickerson, L., *Tappi* **42** (1), 51 (1959).

346. Stone, J. E. and Nickerson, L. F., *Pulp Paper Mag. Can.* **62** (9), T-429 (1961).

347. Strapp, R. K., *Pulp Paper Mag. Can.* **56** (C), 179 (1955).

348. Stuart, H. H. and Bailey, R. E., *Paper Trade J.* **148** (45), 50 (1964).

349. Stuart, H. H. and Bailey, R. E., *Tappi* **48** (5), 104A (1965).

350. Sultzer, N. W. and Beaver, C. E., *Tech. Assoc. Papers* **19**, 266 (1936).

351. Surewicz, W., *Tappi* **45** (7), 570 (1962).

352. Susich, G. V. and Wolf, W. W., *Z. Phys. Chem.* **8B** (3), 221 (1930).

353. Sutherland, D. G., *Tappi* **46** (8), 178A (1963).

354. Sylla, K. F., *Svensk Papperstidn.* **67** (3), 75 (1964).

355. Takahashi, S, Nishiyama, K. and Fukuda, Y., *Nippon Nogeikagaku Kaishi* **34**, 766, 772, 857 (1960).

356. Tappi Standard 0 400 p-54, June 1954.

357. Tarkomen, F., U.S. Patent 2,614, 923 (1952).

358. Taylor, L. B., *Tappi* **34** (1) 12 (1951).

358a. Teder, A., *Svensk Papperstidn.* **71** (5), 149 (1998).

358b. Thoen, G. N., DeHaas, G. G. and Austin, R. R., *Tappi* **51** (6), 246 (1968).

359. Thomas, E. W., *Tappi* **47** (9), 587 (1964).

360. Thompson, N. S., Peckham. J. R. and Thode, E. F., *Tappi* **45** (6), 433 (1962).

361. Thrush, R. E., *Am. Oil Chem. Soc.* **43** (3), 193 (1965).

362. Tischchenko, D. V. and Rozenberger, E. N., *Bumazh. Prom.* **32** (6), 7 (1957).

363. Tomlinson, G. H., "Pulp paper manufacture," p. 416. McGraw-Hill, New York, 1950.

364. Tomlinson, G. H. and Douglas, H. R., *Pulp Paper Mag. Can.* **53** (4), 96 (1952).

365. Tomlinson, G. H. and Fergusson, J. M., *Pulp Paper Mag. Can.* **57** (12) 119 (1956).

366. Trainor, J. W., *Tappi* **46** (6), 169A (1963).

367. Tremaine, B. K., *Pulp Paper Mag. Can.* **57** (8), 132 (1956).

368. Trobeck, K. G., *Svensk Papperstidn.* **53** (1), 8 (1950).

369. Trobeck, K. G., *Svensk Papperstidn.* **54** (18), 632 (1951).

370. Turner, C. H., *Appita Proc.* 6, 89 (1952).

371. Turunen, J., *Soc. Sci. Fennica, Commentations Phys.-Math.* **28** (9) 64 (1963).

372. Valeur, Ch., *Svensk Papperstidn.* **55** (20), 776 (1952).

373. Various, Chem. Prod. Sess. Papers, New York 1962, *Tappi* **46** (2), 126A (1963).

374. Various, Workshop on Turpentine Recovery, *Tappi* **50** (4), 106A (1967).
375. Vassie, J. E., *Tappi* **36** (8), 367 (1953).
376. Vegeby, A., *Tappi* **49** (7), 103A (1966).
376a. Vegeby, A., *Pulp Paper Mag. Can.* **69** (9). T-242 (1968).
377. Venemark, E., *Svensk Papperstidn.* **53** (1), 1 (1950).
378. Venemark, E., *Svensk Papperstidn.* **59** (18), 629 (1956).
379. Venemark, E., *Svensk Papperstidn.* **67** (5), 157 (1964).
380. Vroom, K. E., *Pulp Paper Mag. Can.* **58** (3), 228 (1957).
381. Wangaard, F. F., *Tappi* **45** (7), 548 (1962).
382. Wangaard, F. F., Kellogg, R. M. and Djerf, A., *Tappi* **50** (3), 109 (1967).
383. Wangerin, D. D., *Tappi* **45** (10), 162A (1962).
384. Walters, A. J., *Appita* **18** (1), 33 (1964).
384a. Walther, J. E. and Amberg, H. R., *Tappi* **50** (10), 108A (1967); **51** (11), 126A (1968).
385. Watson, T. R. B., *Pulp Paper Mag. Can.* **63** (4), T-247 (1962); **65** (10), T-415 (1964).
386. Wells, S. D. and Arnold, A. K., *Tech. Assoc. Papers* **24**, 156 (1941).
387. Wenzl, H. F. J., *Papier* **11** (19/20), 435 (1957).
388. Wenzl, H. F. J., U.S. Patent 2,874,044 (1959); Swedish Patent 156,071 (1954); Norwegian Patents 71,537 (1942); 89,629 (1954); Finnish Patent 29,715 (1955); Canadian Patent 579,306 (1959).
389. Wenzl, H. F. J. and Congehl, M., *Tech. Chem. Papier Zellstoff-Fabr.* **29** (2), 17 (1932).
390. Wenzl, H. F. J. and Ingruber, O. V., *Paper Trade J.* **150** (44), 48 (1966).
391. Wetterholm, G. A. and Fossan, K. R., U.S. Patent 2,702,824 (1955).
392. White, L. R., Kessler, R. B. and Hardecker, K. W., *Tappi* **47** (3), 129 (1964).
393. Wilcoxson, L. S. and Ely, F. G., *Tech. Assoc. Papers* **23**, 410 (1940).
394. Wilder, H. D. and Daleski, E. J., *Tappi* **47** (5), 270 (1964); **48** (5), 293 (1965).
395. Wilder, H. D. and Han, S. T., *Tappi* **45** (1), 1 (1962).
396. Wilhelmsen, L. A., *Tappi* **45** (12), 910 (1962).
397. Williams, I. H. and Murray, F. E., *Pulp Paper Mag. Can.* **67** (8), T-347 (1966).
398. Wise, L. E. and Jahn, E., "Wood chemistry" Vol. II, p. 1159. Reinhold, New York, 1952.
399. Woodward, E. R., *Tappi* **36** (5), 216 (1963).
399a. Wong, A., *Pulp Paper Mag. Can.* **69** (9), T-235 (1968).
400. Wright, R. H., *Tappi* **35** (6), 276 (1952).
401. Wright, R. H., *Tappi* **36** (2), 85 (1953).
402. Wright, R. H., *Pulp Paper Mag. Can.* **56** (4), 131 (1955).
403. Wright, R. H., Schoening, M. A. and Shemilt, L. W., *Tappi* **36** (4), 180 (1953).
404. Yllner, S. and Enstrom, B., *Svensk Papperstidn.* **59**, 229 (1956).
405. Zhigarow, Y. and Tischchenko, D. V., *Zh. Priklad Khim.* **33** (1), 14 (1962).

Author Index

Numbers in parentheses are reference numbers and indicate that an author's work is referred to although his name is not cited in the text. Numbers in italics show the page on which the complete reference is listed.

A

Abdelmoniem, S. A., 442(351), *505*
Abele, W., 564(1), *648*
Abrams, E., 636(2), *648*
Abranyants, S. V., 236(128), *249*
Adam, K., 264(40), 294(40), *297*
Adams, D. F., 624(3, 4), *648*
Adams, G. A., 123(1), 125(105), *149, 152*
Adams, M. F., 125(2), 127(69), *149, 151*
Adamski, Z., 163(174), *250*
Adkins, H., 132(107), 137(3), *149, 152*
Adler, E., 119(133), 133(4), 134(4, 5), *149, 150, 152,* 303(1), 312(1), *354,* 401(1), 403(2), 404(3), *496*
Agamalova, V. G., 167(149), *250*
Agarwal, A. K., *89*
Aggeryd, B., 433(421), *506*
Ahlborg, N. K., 483(61), *498*
Ahlen, L., 449(4), *496*
Ahlgren, P., 518(4A), 520(4B), *648*
Ahlm, C. A., 359(6), *648*
Ahlm, F., 515(5), *648*
Aho, O., 415(408), *506*
Akahori, T., 113(263), *156*
Akerlund, G., 147(12), *150*
Akhmina, 241(209, 210), *251*
Akhmina, E. I., 269(2), 272(1, 131) 294(1, 2), *296, 300*
Akim, L. E., 408(5), *496*
Aleksandrova, O. A., 161(35), 196(23), *246*
Alfredson, B., 137(66), *151,* 537(7, 8, 9), *648*
Algar, W. H., 81(1), *88*
Alhojarvi, J., 312(2), *354*

Allard, G. A., 421(182A), *501*
Allen, G. A., 393(303), *504*
Allgeier, R. J., 229(1), *245*
Allgulander, O., 451(6), *496*
Alm, A. A., 312(2), *354,* 433(195), 435(409), *501, 506*
Alsner, H., 161(186), *251*
Aman, A. K., (36), *246*
Amberg, H. R., 623(384A), *657*
Ambronn, H., 106(6), *150*
Amos, L. C., 601(103), *650*
Anderson, L., 139(99), *152*
Andreeva, Z. N., 269(2), 294(2), *296*
Andress, K. R., *150*
Andresson, T., 476(359), *505*
Andrews, T. M., 554(257), *654*
Andrievskaya, E. A., 260(43, 45), *297*
Annergreen, G., 304(4), 307(5), *354,* 400(8, 9), 455(9), 456(10), 460(10), *496,* 564(9A), 577(10), 578(10), *648*
Anthias, A., 123(8), *150*
Anthoni, B., 399(358), *505*
Anthony, A. W., 611(81), 612(81), *650*
Antony, A. W., 611(206), *653*
Ant-Wuorinen, O., 397(15), 463(14), *496*
Apel, A., 188(2), 229(2), *245*
Applegarth, D. A., 117(9), *150*
Appling, J. W., 95(272), 96(271), *156*
Arend, J. L., 339(9), *354*
Argue, G. H., 294(15), 295(15), *296*
Aries, R. S., 179(113), *249*
Arkhipov, M. I., 240(158), *250*
Arlt, H. G., 232(3), 242(3), *245*
Arnold, A. K., 516(386), *657*
Arnold, G. C., 455(16), *496*
Arnoud, J. E., 162(4), 232(4), *245*

659

W

Subject Index

A

Abiperm recovery process, 486
Acetic acid, formation in wood pyrolysis, 262, 274
   content in wood vinegar, 275
Acetone in pyroligneous acid, 276
Activation energy in the thermal decomposition of wood, 257
Agrifuran process, 236
Alkali fusion of sawdust, 290
Alkaline pulping, 507
   historical development, 510
   pulp production of principal countries, 511
Alpha cellulose, 95
   composition in different wood species, 99
   content of wood, 98
Aluminium-base sulfite pulping, 442
Ammonia, aqueous solutions, 384
   -base cooking, 385, 436
   -sulfur dioxide solutions, 384
Annual rings in wood, 34
Arabinogalactan, 124, 127
   average molecular weight, 126
   basic structure, 125, 126
   distribution within the tree, 126
   solubility, 125
Araboxylan, formation in kraft pulping, 531

B

Bagasse, hydrogenolysis of, 295
Bark, adhesion to wood, 323
   calorific value, 340
   chemical composition, 315
   compounds in the water extract of, 317
   content of various American woods, 319
   formation and growth factors, 34, 310
   presses, 343

   prospects for utilization, 344
   pyrolysis, 272, 275
   solubility of extractives of, 314
Barking, 322
   continuous devices, 325
   drums, 328
   hydraulic processes
     pocket devices, 327
     water jet devices, 334
   knife barkers, 324
   presteaming of wood for, 338
Bast layers in the tree, 34
Beating process mechanics of pulp, 86
   swelling of fibers in the, 87
Beechwood, cross-section, 43
Benzopyrene, formation in wood pyrolysis, 274
Biogenesis
   of cellulose, 114
   of hemicelluloses, 127
   of lignin, 139
Borohydride, in kraft pulping, 522, 534
Broad-leaf species of wood, average fiber length, 45
Butanol, from wood hydrolyzate, 216
Butylene glycol, 216
Butyric acid fermentation, 238

C

Cambium, 34
Campesterol, 318
Catechol from wood pyrolysis, 274
Cell formation, 33
Cell wall
   chain length distribution in, 116
   components, 531
   cytoplasmic membrane, 116
   delignification in sulfite pulping, 405, 407
   density of crystalline cellulose in, 75

684

L

M

O